MICROWAVE ENGINEERING
CONCEPTS AND FUNDAMENTALS

T0225531

MICROWAVE ENGINEERING
CONCEPTS AND FUNDAMENTALS

Ahmad Shahid Khan

CRC Press
Taylor & Francis Group
Boca Raton   London   New York

CRC Press is an imprint of the
Taylor & Francis Group, an **informa** business

CRC Press
Taylor & Francis Group
6000 Broken Sound Parkway NW, Suite 300
Boca Raton, FL 33487-2742

© 2014 by Taylor & Francis Group, LLC
CRC Press is an imprint of Taylor & Francis Group, an Informa business

First issued in paperback 2017

No claim to original U.S. Government works
Version Date: 20140114

ISBN 13: 978-1-138-07242-8 (pbk)
ISBN 13: 978-1-4665-9141-7 (hbk)

Library of Congress Cataloging-in-Publication Data

Khan, Ahmad Shahid.
    Microwave engineering : concepts and fundamentals / author, Ahmad Shahid Khan.
        pages cm
    Summary: "This book presents comprehensive material on microwave engineering. Though there are a number of books on this subject but none of them covers all the aspects in detail and with the equal emphasis. In the present text almost all topics including microwave generation, transmission, measurement and processing are given almost the equal weightage"-- Provided by publisher.
    Includes bibliographical references and index.
    ISBN 978-1-4665-9141-7 (hardback)
    1. Microwaves. 2. Microwave devices. I. Title.

TK7876.K467 2014
621.381'3--dc23   2013049468

Visit the Taylor & Francis Web site at
http://www.taylorandfrancis.com

and the CRC Press Web site at
http://www.crcpress.com

Contents

Preface...xxvii
Author... xxix

1 Introduction...1
   1.1   Introduction 1
   1.2   Microwave Frequency Bands 2
   1.3   Advantages 3
   1.4   Components of Microwave System 5
          1.4.1   Microwave Generation...5
          1.4.2   Microwave Processing 6
          1.4.3   Microwave Transmission..6
          1.4.4   Microwave Measurements 8
          1.4.5   Microwave Antennas 9
   1.5   Applications...9
          1.5.1   Communication..9
          1.5.2   Radars... 11
          1.5.3   Radio Astronomy.. 11
          1.5.4   Navigation... 12
          1.5.5   Home Appliances.. 12
          1.5.6   Industrial Heating 13
          1.5.7   Plasma Generation.. 13
          1.5.8   Weaponry System 13
          1.5.9   Spectroscopy... 13
   1.6   Health Hazards 13
   Descriptive Questions 14
   Further Reading... 14

2 Fundamentals of Wave Propagation... 17
   2.1   Introduction 17
   2.2   Basic Equations and Parameters... 17
   2.3   Nature of Media 19
   2.4   Wave in Lossless Media 19
   2.5   Wave in Lossy Media... 21
   2.6   Conductors and Dielectrics 22
   2.7   Polarisation 24
   2.8   Depth of Penetration... 26
   2.9   Surface Impedance... 27
   2.10   Poynting Theorem 27
   2.11   Reflection and Refraction.. 29
   2.12   Direction Cosines, Wavelength and Phase Velocity 31

2.13   Classification of the Cases of Reflection 32
2.14   Normal Incidence Cases 33
        2.14.1   Perfect Conductor 34
        2.14.2   Perfect Dielectric 35
2.15   Oblique Incidence 36
        2.15.1   Perfect Conductor 36
                 2.15.1.1   Perpendicular (or Horizontal) Polarisation 36
                 2.15.1.2   Parallel (or Vertical) Polarisation 36
        2.15.2   Perfect Dielectric 37
                 2.15.2.1   Perpendicular (Horizontal) Polarisation 39
                 2.15.2.2   Parallel (Vertical) Polarisation 39
                 2.15.2.3   Brewster's Angle 39
                 2.15.2.4   Total Internal Reflection 40
2.16   Parallel Plane Guide 40
2.17   Transverse Waves 42
        2.17.1   Transverse Electric Waves 42
        2.17.2   Transverse Magnetic Waves 44
2.18   Characteristics of TE and TM Waves 45
2.19   Transverse Electromagnetic Waves 47
2.20   Wave Impedances 48
2.21   Attenuation in the Walls of Parallel Plane Guide 51
2.22   Transmission Lines 53
2.23   Equations Governing Transmission Line Behaviour 54
        2.23.1   Transmission Line Equations 54
        2.23.2   Solution of Transmission Line Equations 56
                 2.23.2.1   Exponential Form 56
                 2.23.2.2   Hyperbolic Form 57
2.24   Lossless RF and UHF Lines with Different Terminations 58
2.25   Reflection Phenomena 61
2.26   Resonance Phenomena in Line Sections 62
2.27   Quality Factor of a Resonant Section 63
2.28   UHF Lines as Circuit Elements 63
2.29   Applications of Transmission Lines 66
        2.29.1   Quarter-Wave Section as Tuned Line 66
        2.29.2   Quarter-Wave Section as Impedance Transformer 67
        2.29.3   Quarter-Wave Section as Voltage Transformer 67
        2.29.4   Line Sections as Harmonic Suppressors 68
                 2.29.4.1   Suppression of Third Harmonics 68
                 2.29.4.2   Suppression of Even Harmonics 69
        2.29.5   Line Sections for Stub Matching 69
                 2.29.5.1   Single Stub Matching 69
                 2.29.5.2   Double Stub Matching 70
2.30   Types of Transmission Lines 70
2.31   Coaxial Cables 71
2.32   Limitations of Different Guiding Structures 73

Problems...84
Descriptive Questions 85
Further Reading 86

3 Waveguides 87
  3.1 Introduction 87
  3.2 Interrelation between Transmission Line and Waveguide...........87
      3.2.1 Impact of Frequency Change 89
  3.3 Rectangular Waveguide..90
      3.3.1 Transverse Magnetic Wave...93
      3.3.2 Transverse Electric Wave 96
      3.3.3 Behaviour of Waves with Frequency Variation.................96
      3.3.4 Possible and Impossible Modes 97
            3.3.4.1 Lowest Possible TM Mode 97
            3.3.4.2 Lowest Possible TE Mode.................................97
            3.3.4.3 Impossibility of TEM Wave 97
      3.3.5 Field Distribution for Different Modes............................98
            3.3.5.1 TE Wave..98
            3.3.5.2 TM Wave..99
      3.3.6 Excitation of Modes 100
            3.3.6.1 TE Modes..100
            3.3.6.2 TM Modes 100
      3.3.7 Wave Impedances 101
            3.3.7.1 TM Wave...101
            3.3.7.2 TE Wave 101
            3.3.7.3 Variation of Z_z, v and λ with Frequency..........102
      3.3.8 Circuit Equivalence of Waveguides 103
            3.3.8.1 TM Wave...103
            3.3.8.2 TE Wave 105
      3.3.9 Power Transmission in Rectangular Waveguide 106
      3.3.10 Attenuation in Waveguides.......................................107
      3.3.11 Quality Factor...108
  3.4 Circular Waveguide...109
      3.4.1 TM Waves...111
      3.4.2 TE Waves...112
      3.4.3 Solution of Wave Equation 112
            3.4.3.1 Field Expressions for TM Wave..........................114
            3.4.3.2 Field Expressions for TE Wave..........................115
      3.4.4 TEM Modes in Circular Waveguide 116
      3.4.5 Mode Designation in Circular Waveguide 117
      3.4.6 Field Distribution in Circular Waveguide........................118
      3.4.7 Mode Excitation..119
      3.4.8 Cutoff Frequencies 119
      3.4.9 Attenuation in Circular Waveguides 119
  3.5 Dielectric Waveguides..119

      3.5.1   Dielectric Slab Waveguide 120
      3.5.2   Dielectric Rod Waveguide 123
  3.6   Physical Interpretation of Wave Terminology 124
      3.6.1   Modes 124
      3.6.2   Mode Designations 124
      3.6.3   Dominant Mode 124
      3.6.4   Cutoff Frequency 125
      3.6.5   Usable Frequency Range 125
      3.6.6   Characteristic Impedance 125
      3.6.7   Guide Wavelength 125
      3.6.8   Phase Velocity 125
      3.6.9   Group Velocity 125
      3.6.10 Transformation of Modes 126
      3.6.11 Mode Disturbance 126
      3.6.12 Multimode Propagation 126
      3.6.13 Group Velocity Variations 127
      3.6.14 Wavefront Movement 128
      3.6.15 Spurious Modes 128
  3.7   Relative Merits of Waveguides 129
      3.7.1   Copper Losses 129
      3.7.2   Dielectric Loss 129
      3.7.3   Insulation Breakdown 129
      3.7.4   Power-Handling Capability 130
      3.7.5   Radiation Losses 130
  3.8   Limitations of Waveguides 130
  Problems 135
  Descriptive Questions 136
  Further Reading 137

4 Cavity Resonators 139
  4.1   Introduction 139
  4.2   Shapes and Types of Cavities 140
      4.2.1   Cavity Shapes 140
      4.2.2   Cavity Types 140
      4.2.3   Reentrant Cavities 142
            4.2.3.1   Resonant Frequency 143
  4.3   Cavity Formation 144
      4.3.1   Formation of a Rectangular Cavity 144
      4.3.2   Formation of a Cylindrical Cavity 145
  4.4   Fields in Cavity Resonators 145
      4.4.1   Rectangular Cavity Resonator 146
            4.4.1.1   Mode Degeneracy 148
      4.4.2   Circular Cavity Resonator 148
      4.4.3   Semi-Circular Cavity Resonator 150
  4.5   Quality Factor 150

4.6   Coupling Mechanism 154
4.7   Tuning Methods 154
4.8   Advantages and Applications 155
4.9   Dielectric Resonators 156
Problems 160
Descriptive Questions 160
Further Reading 161

5 **Microwave Ferrite Devices** 163
5.1   Introduction 163
5.2   Ferrites 165
5.3   Faraday's Rotation 167
5.4   Non-Reciprocal Ferrite Devices 168
        5.4.1   Gyrators 169
                5.4.1.1   Gyrator with Input–Output Ports Rotated
                          by 90° 169
                5.4.1.2   Gyrator with 90° Twist 170
        5.4.2   Isolators 170
                5.4.2.1   Faraday's Rotation Isolators 171
                5.4.2.2   Resonance Isolator 172
        5.4.3   Circulators 173
                5.4.3.1   Three-Port Circulators 173
                5.4.3.2   Four-Port Circulators 175
5.5   Ferrite Phase Shifter 178
5.6   Ferrite Attenuators 178
5.7   Ferrite Switches 179
5.8   YIG Filters 179
5.9   Figures of Merit of Ferrite Devices 180
Problems 182
Descriptive Questions 182
Further Reading 183

6 **Smith Chart** 185
6.1   Introduction 185
6.2   Characteristic Parameters of a Uniform Transmission Line 185
6.3   Polar Chart 189
6.4   Smith Chart for Impedance Mapping 191
6.5   Smith Chart for Admittance Mapping 195
6.6   Information Imparted by Smith Chart 197
        6.6.1   Mapping of Normalised Impedances 197
        6.6.2   Rotation by 180° 198
        6.6.3   Reflection and Transmission Coefficients 198
        6.6.4   Voltage Standing Wave Ratio 200
        6.6.5   Two Half-Wave Peripheral Scales 200
        6.6.6   Inversion of Impedance/Admittance 202

| | 6.6.7 | Equivalence of Movement on Transmission Line | 202 |

6.7   Advantages of Smith Chart 203
6.8   Smith Chart for Lossless Transmission Lines.............................. 204
      6.8.1   Evaluation of $\Gamma(d)$ for given $Z(d)$ and Z_0 204
      6.8.2   Evaluation of $\Gamma(d)$ and $Z(d)$ for Given Z_R, Z_0 and d.......... 205
      6.8.3   Evaluation of d_{max} and d_{min} for Given Z_R and Z_0.............. 206
      6.8.4   Evaluation of VSWR for Given Γ_R and Z_{R1}, Z_{R2} and Z_0.. 207
      6.8.5   Evaluation of $Y(d)$ for Given Z_R.................................... 208
      6.8.6   Mapping of a Multi-Element Circuit onto a Smith Chart 210
      6.8.7   Evaluation of Normalised Impedances 212
      6.8.8   Consideration of Attenuation Factor.............................. 213
6.9   Stub Matching 214
      6.9.1   Entry on Smith Chart 215
      6.9.2   Susceptance Variation of SC Stubs over a Frequency Band 215
      6.9.3   Stub for Changing the Phase of a Main-Line Signal 215
      6.9.4   Requirements for Proper Matching 216
      6.9.5   Selection of Appropriate Type and Length of Stub 217
      6.9.6   Impact of Rotation of Impedance 217
      6.9.7   Implementation of Single Stub Matching...................... 219
      6.9.8   Implementation of Double Stub Matching..................... 220
      6.9.9   Conjugate Matching Problems with Distributed Components... 222
Problems... 228
Descriptive Questions 229
Further Reading 229

7 Microwave Components... 231
7.1   Introduction 231
7.2   Waveguides and Its Accessories 232
      7.2.1   Waveguide Shapes 232
      7.2.2   Ridged Waveguides... 233
      7.2.3   Dielectric Loaded Waveguides 233
      7.2.4   Bifurcated and Trifurcated Waveguides 234
      7.2.5   Flexible Waveguide.. 234
      7.2.6   Impact of Dimensional Change 234
      7.2.7   Waveguide Joints 235
              7.2.7.1   Waveguide Flanges.. 235
              7.2.7.2   Waveguide Rotary Joints................................. 236
      7.2.8   Waveguide Stands 237
      7.2.9   Waveguide Bends.. 237
              7.2.9.1   *E*-Plane Bend 238
              7.2.9.2   *H*-Plane Bend 238

       7.2.10 Waveguide Corners 238
                  7.2.10.1 *E*-Plane Corner 239
                  7.2.10.2 *H*-Plane Corner 239
       7.2.11 Waveguide Twists 239
       7.2.12 Waveguide Transitions 240
  7.3 Input–Output Methods in Waveguides 240
       7.3.1 Probe Coupling 241
       7.3.2 Loop Coupling 242
       7.3.3 Slot/Aperture Coupling 243
       7.3.4 Input–Output Coupling in Circular Waveguides 243
  7.4 Coaxial to Waveguide Adapter 243
  7.5 Waveguide Junctions 245
       7.5.1 *E*-Plane Tee 245
       7.5.2 *H*-Plane Tee 246
       7.5.3 *EH* (or Magic) Tee 246
                  7.5.3.1 Impedance Matching in Magic Tee 247
       7.5.4 Hybrid Ring (Rat Race Tee) 248
  7.6 Directional Couplers 249
       7.6.1 Bathe-Hole Coupler 250
       7.6.2 Double-Hole Coupler 251
       7.6.3 Moreno Cross-Guide Coupler 251
       7.6.4 Schwinger Reversed-Phase Coupler 251
       7.6.5 Multi-Hole Directional Coupler 252
       7.6.6 Unidirectional and Bidirectional Couplers 252
       7.6.7 Short Slot, Top Wall and Branch Guide Couplers 252
       7.6.8 Figures of Merit 253
  7.7 Waveguide Terminations 256
       7.7.1 Fixed Matched Termination (or Matched Load) 257
       7.7.2 Adjustable Terminations or Moving Loads 258
       7.7.3 Water Loads 258
  7.8 Attenuators 259
       7.8.1 Fixed Attenuator 259
       7.8.2 Variable Attenuators 260
  7.9 Impedance Matching 261
       7.9.1 Inductive Irises 261
       7.9.2 Capacitive Irises 262
       7.9.3 Resonant Windows 262
       7.9.4 Posts and Screws 263
  7.10 Tuners 264
       7.10.1 *EH* Tuner 264
       7.10.2 Slide Screw Tuner 265
       7.10.3 Coaxial Line Stubs and Line Stretchers 265
       7.10.4 Waveguide Slug Tuner 266
  7.11 Phase Shifters 267
       7.11.1 Line Stretcher Phase Shifter 267

        7.11.2 Dielectric Vane Phase Shifter 268
        7.11.3 Linear Phase Shifter 268
        7.11.4 Circular Waveguide Phase Shifter 268
   7.12 Microwave Filters 269
        7.12.1 Band-Pass Filter 269
        7.12.2 Low-Pass Filter 269
        7.12.3 High-Pass Filter 270
        7.12.4 Parallel Resonant Filter 270
   7.13 Duplexers 270
   7.14 Diplexers 271
   7.15 Mode Suppressors 271
   Problems 275
   Descriptive Questions 276
   Further Reading 277

8 **Scattering Parameters** 279
   8.1   Introduction 279
   8.2   Properties of Scattering Matrices 281
        8.2.1 Order and Nature 281
        8.2.2 Symmetry Property 281
        8.2.3 Unity Property 281
        8.2.4 Zero Property 281
        8.2.5 Phase Shift Property 282
   8.3   Scattering Parameters for Networks with Different Ports 282
        8.3.1 1-Port Network 282
        8.3.2 2-Port Network 282
        8.3.3 3-Port Network 285
        8.3.4 *N*-Port Network 286
        8.3.5 Cascaded Networks 286
   8.4   Nature of Networks 287
        8.4.1 Lossless Network 287
        8.4.2 Lossy Network 287
   8.5   Types of *s*-Parameters 287
        8.5.1 Small-Signal *s*-Parameters 287
        8.5.2 Large-Signal *s*-Parameters 287
        8.5.3 Mixed-Mode *s*-Parameters 288
        8.5.4 Pulsed *s*-Parameters 288
   8.6   Scattering Matrices for Some Commonly Used
        Microwave Components 288
        8.6.1 H-Plane Tee 288
        8.6.2 E-Plane Tee 292
        8.6.3 *EH* Tee 295
        8.6.4 Directional Coupler 298
        8.6.5 Hybrid Ring 301

8.6.6   Circulators 301
      8.6.6.1   4-Port Circulator 302
      8.6.6.2   3-Port Circulator 302
8.6.7   Attenuator 303
8.6.8   Gyrator 304
8.6.9   Isolator 304
8.7   Electrical Properties of 2-Port Networks 305
    8.7.1   Complex Linear Gain 305
    8.7.2   Scalar Linear Gain 305
    8.7.3   Scalar Logarithmic Gain 305
    8.7.4   Insertion Loss 306
    8.7.5   Return Loss 306
      8.7.5.1   Input Return Loss 306
      8.7.5.2   Output Return Loss 306
    8.7.6   Reverse Gain and Reverse Isolation 306
    8.7.7   Voltage Reflection Coefficient 307
    8.7.8   Voltage Standing Wave Ratio 307
8.8   *s*-Parameters and Smith Chart 307
8.9   Scattering Transfer (or *T*) Parameters 311
    8.9.1   Conversion from *S*-Parameters to *T*-Parameters 312
    8.9.2   Conversion from *T*-Parameters to *S*-Parameters 312
Problems 314
Descriptive Questions 314
Further Reading 315

9   **Microwave Antennas** 317
9.1   Introduction 317
9.2   Antenna Theorems and Characteristic Parameters 317
    9.2.1   Antenna Theorems 317
    9.2.2   Antenna Characteristic Parameters 318
9.3   Types of Microwave Antennas 319
    9.3.1   Reflector Antennas 319
      9.3.1.1   Paraboloidal Reflector 321
      9.3.1.2   Truncated Paraboloid 321
      9.3.1.3   Orange-Peel Paraboloid 322
      9.3.1.4   Cylindrical Paraboloid 322
      9.3.1.5   Pillbox 322
      9.3.1.6   Cheese Antenna 322
      9.3.1.7   Spherical Reflector 323
      9.3.1.8   Cassegrain Antenna 323
      9.3.1.9   Corner Reflectors 324
    9.3.2   Horn Antennas 325
    9.3.3   Slot Antennas 326
    9.3.4   Lens Antennas 327

                    9.3.4.1   WG-Type Lens 331
                    9.3.4.2   Delay Lenses 331
                    9.3.4.3   Zoned Lens 332
                    9.3.4.4   Loaded Lens 333
          9.3.5   Frequency-Sensitive Antennas 333
    9.4   Antenna Arrays 333
          9.4.1   Types of Arrays 334
                    9.4.1.1   Broadside Array 335
                    9.4.1.2   End-Fire Array 336
                    9.4.1.3   Collinear Array 336
                    9.4.1.4   Parasitic Arrays 336
    9.5   Microstrip Antennas 337
          9.5.1   Feed Methods 338
          9.5.2   Characteristics of MSAs 339
          9.5.3   Merits of MSAs 340
          9.5.4   Demerits of MSAs 340
          9.5.5   Applications 340
    Descriptive Questions 340
    Further Reading 341

10 Microwave Measurements 343
    10.1   Introduction 343
    10.2   Klystron Power Supply 344
           10.2.1   Operating Procedure 345
           10.2.2   Reflex Klystron 346
           10.2.3   Klystron Mount 346
    10.3   VSWR Meter 347
    10.4   Travelling Wave Detection 348
           10.4.1   Slotted Section 348
           10.4.2   Tunable Probes 349
           10.4.3   Movable Shorts 350
           10.4.4   Frequency (or Wave) Meters 350
           10.4.5   Microwave Detectors 352
           10.4.6   Tunable Crystal Detector Mounts 353
           10.4.7   Bolometer 353
                    10.4.7.1   Barraters 354
                    10.4.7.2   Thermistor 355
                    10.4.7.3   Bolometer Mounts 355
    10.5   Qualities of Microwave Components and Devices 355
    10.6   Precautions 356
    10.7   Some Standard Norms 358
    10.8   Measurement of Basic Quantities 359
           10.8.1   Measurement of Wavelength 359
           10.8.2   Measurement of Frequency 360

10.8.2.1 Transmission Method..360
10.8.2.2 Reaction Method 360
10.8.3 Measurement of VSWR..361
10.8.3.1 Transmission Line Method..............................361
10.8.3.2 Twice Minimum Method 361
10.8.3.3 Reflectometre Method 362
10.8.4 Measurement of Phase Shift...362
10.8.4.1 Method I 363
10.8.4.2 Method II..363
10.8.5 Measurement of Quality Factor.......................................364
10.8.6 Measurement of Scattering Parameters............................365
10.8.7 Measurement of Insertion Loss/Attenuation...................365
10.8.7.1 Insertion or Power Ratio Method 366
10.8.7.2 Substitution Methods 367
10.8.7.3 Cavity Resonance Method 368
10.8.7.4 Scattering Coefficient Method...........................369
10.8.8 Measurement of Dielectric Constant 369
10.8.9 Measurement of Impedance...370
10.8.9.1 Bridge Method 370
10.8.9.2 Slotted Line Method 370
10.8.10 Measurement of Power 371
10.8.10.1 Bolometer Bridge Method 371
10.8.10.2 Calorimetric Method 372
10.8.11 Measurement of Electric Field Intensity...........................373
10.8.12 Measurement of Reflection Coefficient.............................374
10.8.12.1 Comparison Method......................................374
10.8.13 Measurement of Transmission Coefficient.......................375
10.8.13.1 Comparison Method......................................375
10.9 Some Practical Applications 375
10.9.1 Measurement of Gain of an Aerial Horn..........................375
10.9.2 *E* and *H* Plane Radiation Patterns 376
10.9.3 Measurement of Thickness of a Metallic Sheet.............376
10.9.4 Measurement of Thickness of a Dielectric Sheet 377
10.9.5 Measurement of Wire Diameter 377
10.9.6 Measurement of Moisture Content 377
10.9.7 Measurement/Monitoring Moisture Content in
Liquids..378
Descriptive Questions 379
Further Reading..379

11 Basics of Microwave Tubes..381
11.1 Introduction 381
11.2 Frequency Limitations of Conventional Tubes....................381
11.2.1 Inter-Electrode Capacitances...382
11.2.2 Lead Inductances 383

11.2.2.1 Impact of Lead Inductance and
Inter-Electrode Capacitance.................383
11.2.2.2 Gain: BW Product Limitation 385
11.2.3 Transit Time Effect..386
11.2.4 Skin Effect 387
11.2.5 Electrostatic Induction 387
11.2.6 Dielectric Loss 388
11.3 Influence of Fields on Motion of Charged Particles.....................388
11.3.1 Motion in Electric Field.......................................390
11.3.1.1 Cartesian Coordinate System 390
11.3.1.2 Cylindrical Coordinate System 391
11.3.2 Motion in Magnetic Field.....................................393
11.3.2.1 Cartesian Coordinate System 393
11.3.2.2 Cylindrical Coordinate System 395
11.3.3 Motion in Combined Electric and
Magnetic Field 397
11.3.4 Motion in Electric, Magnetic and an AC
Field...398
11.3.4.1 Cartesian Coordinates 398
11.3.4.2 Cylindrical Coordinates 402
11.4 Velocity Modulation 403
11.4.1 Influence of Electric Field on Electrons 403
11.4.2 Operation of Velocity-Modulated Tubes 404
11.4.2.1 Bunching Process 405
11.4.2.2 Current Modulation 405
11.4.2.3 Energy Extraction....................................406
11.5 Classification of Microwave Tubes 406
11.5.1 Linear Beam (or O-Type) Tubes 407
11.5.2 Crossed-Field (or M-Type) Tubes 407
11.5.2.1 Tubes Using Resonant Cavities 408
11.5.2.2 Tubes with Non-Resonant Cavities 409
Descriptive Questions 409
Further Reading 410

12 **Microwave Tubes**...411
12.1 Introduction 411
12.2 Klystron...411
12.2.1 Types of Klystron...411
12.2.1.1 Reflex Klystron 411
12.2.1.2 Two-Cavity Klystron................................412
12.2.1.3 Multi-Cavity Klystron 412
12.2.1.4 Extended Interaction Klystron 412
12.3 Two-Cavity Klystron 412
12.3.1 Constructional Features.......................................413
12.3.2 Operation 414

12.3.3 Mathematical Analysis 415
    12.3.3.1 Velocity Modulation 415
    12.3.3.2 Process of Bunching 417
12.3.4 Multi-Cavity Klystron Amplifiers 424
12.3.5 Tuning 425
    12.3.5.1 Synchronous Tuning 425
    12.3.5.2 Staggered Tuning 426
12.3.6 Two-Cavity Klystron Oscillator 426
12.3.7 Performance 426
12.3.8 Applications 426
12.4 Reflex Klystron 426
12.4.1 Constructional Features 427
12.4.2 Power Sources Required 428
12.4.3 Velocity Modulation 428
12.4.4 Process of Bunching 431
    12.4.4.1 Current Modulation 433
    12.4.4.2 Round-Trip Transit Angle 433
    12.4.4.3 Beam Current 433
    12.4.4.4 Power Output and Efficiency 434
    12.4.4.5 Electronic Admittance 435
    12.4.4.6 Equivalent Circuit 435
12.4.5 Modes of Operation 436
12.4.6 Debunching 438
12.4.7 Tuning 439
12.4.8 Applications 439
12.5 Travelling-Wave Tube 440
12.5.1 Simplified Model of TWT 441
12.5.2 Slow Wave Structures 441
    12.5.2.1 Helical Structure 441
    12.5.2.2 Coupled-Cavity Structure 444
    12.5.2.3 Ring-Loop Structure 444
    12.5.2.4 Ring-Bar Structure 445
12.5.3 Construction 445
12.5.4 Focussing 446
12.5.5 Input–Output Arrangements 448
12.5.6 Basic Operating Principle 448
12.5.7 Power Supply 449
12.5.8 Mathematical Analysis 450
    12.5.8.1 Brillouin Diagram 451
    12.5.8.2 Amplification Process 454
    12.5.8.3 Convection Current 456
    12.5.8.4 Axial Electric Field 457
    12.5.8.5 Wave Modes 457
    12.5.8.6 Output Power Gain 458
12.5.9 Characteristics 458

    12.5.10 Prevention of Oscillations..459
    12.5.11 Applications...459
 12.6 Magnetron..460
    12.6.1 Construction of Conventional Cylindrical Magnetron..460
    12.6.2 Working Mechanism...461
      12.6.2.1 Effect of Electric Field...461
      12.6.2.2 Effect of Magnetic Field.......................................461
      12.6.2.3 Effect of Combined Field......................................462
      12.6.2.4 RF Field..463
      12.6.2.5 Effect of Combined (E, B and RF) Fields...........464
    12.6.3 Modes of Oscillation..466
    12.6.4 Strapping...467
    12.6.5 Frequency Pushing and Pulling.......................................469
    12.6.6 Coupling Methods...469
    12.6.7 Magnetron Tuning...469
      12.6.7.1 Inductive Tuning..470
      12.6.7.2 Capacitive Tuning..470
    12.6.8 Mathematical Analysis..471
      12.6.8.1 Electron Motion..471
      12.6.8.2 Hull Cutoff Equations...472
      12.6.8.3 Cyclotron Angular Frequency............................473
      12.6.8.4 Equivalent Circuit...474
      12.6.8.5 Quality Factor..475
      12.6.8.6 Power and Efficiency..475
    12.6.9 Other Magnetrons..476
      12.6.9.1 Linear Magnetron..476
      12.6.9.2 Coaxial Magnetron..477
      12.6.9.3 Inverted Coaxial Magnetron................................478
      12.6.9.4 Voltage Tunable Magnetron.................................479
      12.6.9.5 Frequency-Agile Coaxial Magnetron.................480
      12.6.9.6 Negative-Resistance or Split-Anode
          Magnetron..482
      12.6.9.7 Electron-Resonance Magnetron..........................483
    12.6.10 Performance Parameters and Applications.....................484
    12.6.11 Arcing...485
      12.6.11.1 Baking-In Procedure..485
 12.7 Crossed-Field Amplifier..485
    12.7.1 Forward-Wave Crossed-Field Amplifier.........................485
    12.7.2 Backward-Wave Crossed-Field Amplifier.......................486
 12.8 Backward-Wave Oscillators...487
    12.8.1 *O*-Type Backward-Wave Oscillator.................................488
    12.8.2 Backward-Wave *M*-Type Cross-Field Oscillator............489
      12.8.2.1 Linear *M*-Carcinotron Oscillator.....................489
      12.8.2.2 Circular *M*-Carcinotron Oscillator...................489
    12.8.3 Applications and Performance Parameters.....................490

Problems..498
Descriptive Questions 499
Further Reading..500

13 Microwave Diodes...503
  13.1 Introduction..503
  13.2 Basics of Semiconductor Devices...503
       13.2.1 Properties of Semiconducting Materials.........................503
       13.2.2 Mechanism Involved...505
       13.2.3 Significance of Negative Resistance................................505
  13.3 Conventional Diodes 506
       13.3.1 Low-Frequency Conventional Diodes 506
              13.3.1.1 PN Junction Diodes...506
              13.3.1.2 Zener Diode..507
              13.3.1.3 Avalanche Diode 507
       13.3.2 Characteristic Parameters of Conventional Diodes........507
              13.3.2.1 Forward Voltage Drop (V_f)...............................507
              13.3.2.2 Peak Inverse Voltage 507
              13.3.2.3 Maximum Forward Current...................................507
              13.3.2.4 Leakage Current..508
              13.3.2.5 Junction Capacitance..508
       13.3.3 High-Frequency Limitations of Conventional Diodes.....508
       13.3.4 Optical Frequency Diodes 508
              13.3.4.1 Light-Emitting Diodes...508
              13.3.4.2 Laser Diodes...508
              13.3.4.3 Photodiodes..509
  13.4 Microwave Diodes 509
       13.4.1 Point-Contact Diode 509
              13.4.1.1 Construction 509
              13.4.1.2 Operation..510
       13.4.2 Step-Recovery Diode 511
              13.4.2.1 Construction 511
              13.4.2.2 Operation..512
              13.4.2.3 Applications 512
       13.4.3 PIN Diode..512
              13.4.3.1 Construction 512
              13.4.3.2 Operation..514
              13.4.3.3 Advantages..515
              13.4.3.4 Applications 515
       13.4.4 Tunnel Diode 515
              13.4.4.1 Construction 515
              13.4.4.2 Operation..516
              13.4.4.3 *V–I* Characteristics 517
              13.4.4.4 Advantages..518
              13.4.4.5 Applications 518

13.4.5   Schottky Diode...518
          13.4.5.1   Construction 519
          13.4.5.2   Fabrication..519
          13.4.5.3   Characteristics 520
          13.4.5.4   Advantages...520
          13.4.5.5   Limitations 520
          13.4.5.6   Applications 521
13.4.6   Varactor or Varicap Diode 521
          13.4.6.1   Operation..521
          13.4.6.2   Characteristics 522
          13.4.6.3   Applications 523
13.5   Transferred Electron Devices...523
          13.5.1   Valley Band Structure 524
                    13.5.1.1   Conductivity and Current Density....................526
                    13.5.1.2   Ridley–Watkins–Helsum Theory.......................528
                    13.5.1.3   Differential Negative Resistance.......................529
          13.5.2   High Field Domain...531
                    13.5.2.1   Properties of High Field Domain.....................533
          13.5.3   Modes of Operation..533
                    13.5.3.1   Gunn Diode Mode 534
                    13.5.3.2   Stable Amplification Mode 534
                    13.5.3.3   LSA Mode..534
                    13.5.3.4   Bias-Circuit Oscillation Mode 534
          13.5.4   Criterion for Mode's Classification....................535
          13.5.5   Gunn Diode 537
                    13.5.5.1   Gunn Effect 537
                    13.5.5.2   Current Fluctuations..................................538
          13.5.6   LSA Diode..538
          13.5.7   Indium Phosphide Diodes.................................540
          13.5.8   Cadmium Telluride Diodes................................541
          13.5.9   Advantages 541
          13.5.10   Applications..541
13.6   Avalanche Transit Time Devices...542
          13.6.1   Read Diodes...543
                    13.6.1.1   Operation..543
                    13.6.1.2   Output Power and Quality Factor 546
          13.6.2   IMPATT Diode...547
                    13.6.2.1   Principle of Operation 547
                    13.6.2.2   Construction 548
                    13.6.2.3   Working 548
                    13.6.2.4   Performance 550
                    13.6.2.5   Advantages and Limitations 551
                    13.6.2.6   Applications 552
          13.6.3   TRAPATT Diode 552
                    13.6.3.1   Operation..552

            13.6.3.2 Performance 553
            13.6.3.3 Advantages and Limitations 553
            13.6.3.4 Applications 554
       13.6.4 BARITT Diode 554
            13.6.4.1 Structures 555
            13.6.4.2 Operation...555
            13.6.4.3 Advantages and Limitations 556
            13.6.4.4 Applications 556
            13.6.4.5 Comparison of Power Levels............................556
  13.7 Parametric Devices 556
       13.7.1 Manley–Rowe Power Relations....................................558
       13.7.2 Classification of Parametric Devices............................561
       13.7.3 Parametric Amplifiers 562
            13.7.3.1 Parametric Up Converter 563
            13.7.3.2 Parametric Down Converter.............................564
            13.7.3.3 Negative Resistance Parametric Amplifiers.....564
            13.7.3.4 Degenerate Parametric Amplifiers or
                      Oscillator 565
       13.7.4 Relative Merits...566
  13.8 Masers..566
       13.8.1 Principle of Operation 567
       13.8.2 Types of Masers..568
            13.8.2.1 Ammonia Maser..568
            13.8.2.2 Ruby Maser 568
            13.8.2.3 Hydrogen Maser..569
       13.8.3 Maser Applications..570
  Problems...574
  Descriptive Questions 575
  Further Reading..576

14 Microwave Transistors 579
  14.1 Introduction 579
  14.2 Transistors and Vacuum Tubes..579
       14.2.1 Conventional Transistor..579
       14.2.2 Limitations of Conventional Transistors.....................580
  14.3 Microwave Transistors 581
       14.3.1 Bipolar Junction Transistors 581
            14.3.1.1 Transistor Structures 582
            14.3.1.2 Transistor Configurations 583
            14.3.1.3 Biasing and Modes of Operation 584
            14.3.1.4 Equivalent Model 585
            14.3.1.5 Transistor Characteristics.................................587
            14.3.1.6 Transistor Assessment Parameters...................588
            14.3.1.7 Limitations of Transistors 589
       14.3.2 Heterojunction Bipolar Transistors 591

            14.3.2.1 Materials and Energy Band Diagrams.............. 591
            14.3.2.2 Advantages... 594
            14.3.2.3 Applications 595
    14.4 Field Effect Transistors.. 595
        14.4.1 Junction Field Effect Transistor...................................... 595
            14.4.1.1 JFET Structure.. 595
            14.4.1.2 Types of Microwave FETs............................... 596
            14.4.1.3 Principles of Operation.................................. 597
            14.4.1.4 Comparison between FETs and BJTs................. 601
        14.4.2 Metal Semiconductor Field Effect Transistors................. 602
            14.4.2.1 Structure... 602
            14.4.2.2 Principles of Operation.................................. 603
            14.4.2.3 Small-Signal Equivalent Circuit....................... 605
            14.4.2.4 MESFET Characteristics................................. 606
            14.4.2.5 Advantages and Limitations of Using
                         Schottky Barrier Diode.................................. 610
            14.4.2.6 Applications 610
        14.4.3 Metal-Oxide-Semiconductor Field Effect Transistors 610
            14.4.3.1 Structure... 610
            14.4.3.2 Operation of n-Channel MOSFET 612
            14.4.3.3 Circuit Symbols 613
            14.4.3.4 Sub-Division of Enhancement Mode................. 617
            14.4.3.5 Materials... 620
            14.4.3.6 Advantages and Limitations 620
            14.4.3.7 Applications 620
            14.4.3.8 Comparison of MOSFETs with BJTs.................. 621
        14.4.4 High Electron Mobility Transistors................................. 621
            14.4.4.1 HEMT Structure.. 622
            14.4.4.2 Operation.. 623
            14.4.4.3 Current–Voltage Characteristics....................... 623
            14.4.4.4 HEMTs Fabrication Using IC Technology 624
            14.4.4.5 Materials Used... 625
            14.4.4.6 Advantages... 626
            14.4.4.7 Applications 626
    14.5 Metal-Oxide-Semiconductor Transistors.................................. 626
        14.5.1 NMOS Logic 627
            14.5.1.1 NMOS Structures.. 627
            14.5.1.2 NMOS Operation 628
        14.5.2 P-Type Metal-Oxide-Semiconductor Logic 629
        14.5.3 Complementary Metal-Oxide-Semiconductor 629
            14.5.3.1 CMOS Structures 630
            14.5.3.2 CMOS Operation... 631
            14.5.3.3 Power Dissipation... 632
            14.5.3.4 Analog CMOS... 633
            14.5.3.5 Advantages.. 633

14.5.3.6 Applications 633
14.6 Memory Devices 633
14.6.1 Memory Classification 633
14.6.2 Semiconductor Memories 634
14.6.2.1 Read-Only Memory 634
14.6.2.2 Programmable Read-Only Memory 635
14.6.2.3 Erasable Programmable Read-Only
Memory 635
14.6.2.4 Electrically Erasable Programmable
Read-Only Memory 635
14.6.2.5 Static Random Access Memory 636
14.6.2.6 Bipolar Static Random Access Memory 636
14.6.2.7 Dynamic Random Access Memories 636
14.6.2.8 Flash Memories 638
14.6.3 Some Other Memories 640
14.6.3.1 Ferroelectric RAM 640
14.6.3.2 Resistance RAM 640
14.6.3.3 Phase Change RAM 641
14.6.3.4 Ovonics Unified Memory 642
14.7 Charge-Coupled Devices 642
14.7.1 Structure and Working 642
14.7.2 Types of CCDs 646
14.7.3 Dynamic Characteristics 646
14.7.3.1 Charge Transfer Efficiency (η) 646
14.7.3.2 Frequency Response 647
14.7.3.3 Power Dissipation 647
14.7.4 Other Assessment Parameters 647
14.7.5 Advantages 648
14.7.6 Applications 648
Problems 655
Descriptive Questions 656
Further Reading 657

15 Planar Transmission Lines 659
15.1 Introduction 659
15.2 Striplines 661
15.2.1 Types of Striplines 663
15.2.2 Assessment Parameters 663
15.3 Microstrips 665
15.3.1 Types of Microstrips 666
15.3.2 Assessment Parameters 667
15.3.3 Approximate Electrostatic Solution 669
15.3.4 Microstrip Design and Analysis 672
15.3.5 Suspended Microstrips 674
15.4 Coplanar Waveguides 675

        15.4.1 Types of Coplanar Waveguides 675
15.5 Coplanar Strips.. 676
15.6 Slot Line... 677
15.7 Fin Lines... 678
15.8 Micromachined Lines... 678
15.9 Realisation of Lumped Elements 678
        15.9.1 Resistors 679
        15.9.2 Inductors 679
                15.9.2.1 Ribbon Inductors.. 679
                15.9.2.2 Spiral Inductors 680
        15.9.3 Capacitors... 680
                15.9.3.1 Inter-Digital Capacitor 680
                15.9.3.2 Woven Capacitor 681
        15.9.4 Monolithic Transformers 681
                15.9.4.1 Tapped Transformer 682
                15.9.4.2 Interleaved Transformer.................................. 682
                15.9.4.3 Stacked Transformer.. 682
15.10 Realisation of Microwave Components 682
        15.10.1 Realisation of Basic Stripline
                  Elements 682
        15.10.2 Realisation of Transmission Lines............................... 682
        15.10.3 Realisation of Stubs 683
        15.10.4 Realisation of Power Splitters 683
                15.10.4.1 T-Junction Power Splitter................................ 683
                15.10.4.2 Wilkinson Power Splitter 683
        15.10.5 Realisation of Couplers 685
                15.10.5.1 Proximity Couplers 686
                15.10.5.2 Lange Coupler... 688
                15.10.5.3 Branch-Line Coupler.. 689
                15.10.5.4 Rat-Race Coupler 690
        15.10.6 Resonators.. 690
        15.10.7 Filters 690
                15.10.7.1 Low-Pass Filters.. 691
                15.10.7.2 Band-Pass Filters.. 691
        15.10.8 Circuit Elements.. 691
Problems... 697
Descriptive Questions 698
Further Reading 698

16 **Microwave Integrated Circuits**.. 701
16.1 Introduction 701
16.2 Merits and Limitations of MICs... 702
        16.2.1 Merits.. 702
        16.2.2 Limitations... 703
16.3 Types of MICs 704

16.3.1 Hybrid ICs...704
16.3.2 Monolithic Microwave Integrated Circuit........................705
16.4 Materials Used...706
16.4.1 Substrates 706
16.4.2 Conducting Materials..706
16.4.3 Dielectric Materials...707
16.4.4 Resistive Materials..707
16.5 Fabrication Techniques..708
16.5.1 Copper Clad Board 708
16.5.2 Thin-Film Fabrication..708
16.5.3 Thick-Film Fabrication 708
        16.5.3.1 Photoimageable Thick-Film Process..................709
16.6 Fabrication Processes..710
16.6.1 Diffusion and Ion Implantation...710
16.6.2 Oxidation and Film Deposition 710
16.6.3 Epitaxial Growth Technique 710
        16.6.3.1 Vapour-Phase Epitaxy 710
        16.6.3.2 Molecular-Beam Epitaxy......................................710
        16.6.3.3 Liquid-Phase Epitaxy...710
16.6.4 Lithography 711
16.6.5 Etching and Photo-Resist..711
16.6.6 Deposition...711
        16.6.6.1 Vacuum Evaporation...711
        16.6.6.2 Electron Beam Evaporation 711
        16.6.6.3 Cathode (or dc) Sputtering..................................712
        16.6.6.4 Processes Used in Different
                 Technologies..712
16.7 Illustration of Fabrication by Photo-Resist Technique.................712
16.7.1 Planar Resistors..712
16.7.2 Planar Inductors...714
        16.7.2.1 Ribbon Inductor...714
        16.7.2.2 Round-Wire Inductor..714
        16.7.2.3 Single-Turn Flat Circular Loop Inductor 714
        16.7.2.4 Circular Spiral Inductor 714
        16.7.2.5 Square Spiral Inductor..715
16.7.3 Planar Film Capacitor...715
        16.7.3.1 Metal-Oxide-Metal Capacitor..............................715
        16.7.3.2 Interdigitated Capacitor 716
16.8 Fabrication of Devices 716
16.8.1 MOSFET 716
16.8.2 Micro-Fabrication of a CMOS
       Inverter 716
Problems..718
Descriptive Questions 718
Further Reading..719

Appendix A1: Maxwell's and Other Equations..721

Appendix A2: Solution to Equation 11.19..723

Appendix A3: Solution to Equation 11.25..727

Appendix A4: Solution to Equations 11.29 and 11.30...............................731

Appendix A5: Solution to Equation 11.42..735

Index...739

Preface

The subject of microwaves is normally offered at the undergraduate level to electronics and communication engineering students in most universities and technical institutions. At some places, an advanced course on microwaves is also taught at the postgraduate level. Some of its selected topics are also taught to students of diploma courses in the electronics and communication areas. Though a large number of books on microwave engineering are available, in my opinion none of these cover all aspects in detail and with equal emphasis. In this book, an effort has been made to present comprehensive material on microwave engineering. Also, almost all topics including microwave generation, transmission, measurements and processing are given equal weightage. The contents of the book are divided into 16 chapters, which are summarised as follows:

Chapter 1 is an introductory chapter which highlights the importance of microwaves.

Chapter 2 summarises the concepts of wave propagation, reflection and refraction, guided waves and transmission lines which form the basis for the understanding of microwaves.

Chapter 3 deals with the mathematical theory of waveguides.

Chapter 4 discusses different aspects of cavity resonators.

Chapter 5 deals with different ferrite devices.

Chapter 6 is devoted to the Smith chart and its applications.

Chapter 7 includes different types of microwave components.

Chapter 8 deals with scattering parameters and their properties.

Chapter 9 includes different types of antennas used in the microwave range.

Chapter 10 deals with the devices and techniques used in microwave measurements.

Chapter 11 discusses limitations of conventional tubes, behaviour of charged particles in different fields and the concept of velocity modulation.

Chapter 12 includes various types of microwave tubes.

Chapter 13 deals with different types of microwave diodes and parametric devices.

Chapter 14 describes different types and aspects of microwave transistors.

Chapter 15 deals with planar structures including striplines and microstrips.

Chapter 16 deals with different aspects of microwave integrated circuits.

Each of the above chapters contains a number of descriptive questions related to the contents therein. A number of solved and unsolved examples are added in those chapters that contain mathematical derivations or relations. MATLAB®-based solutions to unsolved examples are included in a solution manual.

I am thankful to all the reviewers whose comments on the contents and sample chapters were of immense value. In fact, their comments helped me to recast my book in the present form. I gratefully acknowledge the help rendered by Professor Wasim Ahmed, Professor M. Salim Beg, Professor Ekram Khan, Professor Shah Alam, Professor Omar Farooq and Professor M. Javed Siddiqui, all from the Department of Electronics Engineering, Z.H. College of Engineering and Technology, Aligarh Muslim University (AMU), Aligarh (India), in checking the manuscript. Thanks are also due to Mr. Ghufran, assistant professor, Jauhar College of Engineering, Rampur, for identifying the figures and page numbers from different books. I am also thankful to Professor Muzaffer A. Siddiqui who advised me to opt for CRC Press and to Dr. Gagandeep Singh of CRC Press for his repeated reminders to complete this task. I am also thankful to Dr. S. Suhail Murtaza (director) and Mr. Ashraf Ali of Mewat Engineering College (Wakf), Palla, Nuh, Haryana, for helping me in the preparation of the MATLAB-based solution manual. This manual is likely to be placed on the website of CRC Press. The cooperation of my family for allowing me to work for long hours deserves appreciation, too.

I am indeed indebted to my father (late) Abdur Rahman Khan, my mother (late) Mahmoodi Begum and my elder brother (late) Nawabuddin whose love, affection, moral and financial support helped in shaping my career. I am also indebted to my maternal uncle (late) Mumrez Khan who got me admitted to AMU, Aligarh, one of the most prestigious universities in India. I am also indebted to my teacher (late) Professor Jalaluddin, who not only guided, encouraged and inspired me to work hard, but also inducted me as a teacher in the department he once headed. I dedicate this book to all these (departed) noble souls.

Thanks are also due to Professor S.K. Mukerji, my teacher, PhD supervisor and former head of the Electrical Engineering Department, who inculcated in me a liking for the study of electromagnetic fields. I am also grateful to all my teachers who, ever, bit by bit enhanced my knowledge from the primary to the university level.

Ahmad Shahid Khan

Author

Ahmad Shahid Khan is a former professor and chairman of the Department of Electronics Engineering, Aligarh Muslim University (AMU), Aligarh, Uttar Pradesh, India. Professor Khan has nearly 42 years of teaching, research and administrative experience. He earned his BSc (engineering), MSc (engineering) and PhD degrees from AMU in 1968, 1971 and 1980, respectively. He joined the Department of Electrical Engineering of AMU in March 1972 as a lecturer. He was promoted to reader in 1985 and as professor in 1998. During a span of nearly 35 years, he served AMU as the chairman of the Department of Electronics Engineering, registrar of the university and estate officer (gazetted). After retirement in December 2006 from AMU, he served as the director of IIMT, Meerut, as the director of VCTM, Aligarh and as the director of JCET, Rampur. He also served as a professor in KIET (Ghaziabad) and IMS (Ghaziabad). All these engineering institutes are affiliated to U.P. Technical University, Lucknow. He is presently serving as a visiting professor at MEC (Wakf), Palla, Nuh, a college affiliated with MDU, Rohtak, Haryana.

Dr. Khan is a fellow of the Institution of Electronics and Telecommunication Engineers (India) and a life member of the Institution of Engineers (India), Indian Society for Technical Education and Systems Society of India. He has attended many international and national conferences and refresher/orientation courses in the emerging areas of electronics and telecommunication. His area of interest is mainly related to electromagnetics, antennas and wave propagation, microwaves and radar systems. He has published 23 papers mainly related to electromagnetics. He is a recipient of the Pandit Madan Mohan Malviya Memorial Gold Medal for one of his research papers published in the *Journal of the Institution of Engineers (India)* in 1978.

Dr. Khan edited a book titled *A Guide to Laboratory Practice in Electronics & Communication Engineering* in 1998, which was republished in 2002 by the Department of Electronics Engineering, AMU. He was a coauthor of the Indian version of the third edition of the book *Antennas for All Applications* published by Tata McGraw-Hill, New Delhi, 2006. The fourth edition of this book was published in April 2010 with a new title *Antennas and Wave Propagation* by Tata McGraw-Hill, New Delhi. Dr. Khan contributed six new chapters to this edition. Its 11th print appeared in November 2013.

1

Introduction

1.1 Introduction

Microwaves, as the word spells, are extremely short waves. Their region extends from approximately 1/3 meter to millimetre or even to sub-millimetre range of wavelengths. The term microwave is loosely coupled with frequencies of 1 GHz and above but is less than those of optical signals. This broad definition includes ultra-high frequency (UHF), super-high frequency (SHF) and extremely high-frequency (EHF) bands.

The existence of electromagnetic waves was predicted by J. C. Maxwell in 1864 and was first demonstrated by Heinrich Hertz in 1888 by building an apparatus to produce and detect signals in UHF region. In 1894, J. C. Bose publicly demonstrated radio control of a bell using millimetric waves. The first formal use of the term *microwave* was in 1931. The Hungarian engineer Zoltán Bay sent ultrashort radio waves to the moon in 1943. Perhaps, the word *microwave* in an astronomical context was first used in 1946 in an article 'Microwave Radiation from the Sun and Moon' by Robert Dicke and Robert Beringer.

When wavelengths of signals and dimensions of equipment are in close proximity, the apparatus and techniques may be described as belonging to microwaves. At this stage, the lumped-element circuit theory becomes inaccurate and as a consequence, the practical microwave technique tends to depart from the discrete resistors, capacitors and inductors used at lower frequencies. At microwaves, the distributed circuit elements and transmission line theory become more relevant and useful for the design and analysis of systems. Open-wire and coaxial transmission lines give way to waveguides and planar lines and lumped-element tuned circuits are replaced by cavity resonators or resonant cavities. The effects of reflection, refraction, diffraction, polarisation and absorption become more significant and are of practical importance in the study of wave propagation. Despite these deviations, the behaviour of microwaves is governed by the same equations of electromagnetic theory.

In view of their short wavelengths, microwaves offer some distinct advantages in many areas of applications. These include excellent directivity obtained with relatively small antennas. Thus, in many applications, even low-power transmitters effectively serve the purpose. These features make the microwave range suitable for both military and civilian applications. Besides, at microwaves, the size and weight of antennas and of other devices and components reduce. Since lesser weight and smaller size are the two key considerations in airborne and space vehicles, the equipment planning in all these applications is greatly simplified. The enhanced directivity allows the detection of smaller and farther targets with better clarity.

Apart from the above merits, microwaves do pose certain peculiar problems in generation, transmission and circuit design that are generally not encountered at lower frequencies. Conventional circuit theory requires an understanding of the behaviour of voltages and currents. At microwave range, the key role is played by field concepts. Since the electromagnetic field concepts entirely encompass the phenomenon of wave propagation and antenna theory, the proper understanding of field theory becomes absolutely essential at microwave range. As neither waves nor its field contents can be seen or conveniently measured, such an understanding poses problems to a common learner. In many such situations, one has to rely only on its effects and mathematical relations. Although there is no dearth of derived relations, in some of the cases these are based on simplifying assumptions. Besides, there are empirical relations that are based on experimental results. Another set of key tools, which greatly helps in the understanding of field theory and hence the microwaves, includes mental visualisation and imagination.

1.2 Microwave Frequency Bands

The entire frequency spectrum is generally divided into segments that are often referred to as frequency bands or simply bands. Such a division not only helps in identifying the applications but also the technologies used in different ranges of frequencies. Several such classifications are available on records. The numbers 1, 2 and so on assigned by international telecommunication union (ITU) to different bands correspond to extremely low frequency (ELF), super low frequency (SLF), ultralow frequency (ULF), very low frequency (VLF), low frequency (LF), medium frequency (MF), high frequency (HF), very high frequency (VHF), ultra high frequency (UHF), super high frequency (SHF), extremely high frequency (EHF) and terahertz frequency (THF). As a general rule, the frequency of a band lies between $3 \times 10^{n-1}$ and 3×10^n Hz, where n is the number of band. As an example for HF ($n = 7$), the frequency falls between 3×10^6 and 3×10^7 or 3 and 30 MHz.

As all the bands are not suitable for all applications, some of the bands are either earmarked or frequently used for a particular application. Thus, HF is used for telephones and telegraphs, VHF for television (TV) and frequency modulation (FM) broadcast, UHF for TV, satellite links, cellular communication telemetry and, microwave ovens SHF for microwave links and navigation and for military applications and space research. Radar in general covers UHF and SHF bands whereas ground wave radar operates in HF band.

The symbols assigned to different radio bands by North Atlantic Treaty Organization (NATO) are from A to M in continuity, whereas the symbols assigned to radar bands by Institute of Electrical and Electronics Engineers (IEEE) are HF, VHF, UHF, L, S, C, X, K_u, K, K_a, Q, V and W. The L to W IEEE bands spread over EHF and SHF bands. The UHF and upper frequency bands are further divided into segments. Tables 1.1 and 1.2 include two such divisions. It is to be noted that the waves over frequency ranges from 30 to 300 GHz and 300 to 3000 GHz are often referred to as millimetric and sub-millimetric waves.

In view of the above discussion, it can be noted that the band designations used by different organisations differ. Similarly, the band designations used by different countries also differ. In general, the IEEE band designation is internationally accepted.

It needs to be noted that when the radar was first developed at K band during the Second World War, the existence of nearby absorption band was not accounted due to the presence of water vapour and oxygen in the atmosphere. After realisation of the same, this problem was circumvented by further dividing the K band into lower K_u and upper K_a band. The mid K band has also been retained in some of the classifications.

1.3 Advantages

Microwaves have the following advantages in comparison to the lower-frequency range:

- In the field of communication, these provide very reliable means for transmission and reception of signals. In communication, they consume a proportionately smaller percentage of the total bandwidth of information channels where modulation rides on the main carrier. This inherently permits the use of more channels at lower carrier frequencies.

- In navigation and radar, microwave antennas provide extreme directivity because of short wavelengths that allow better focussing much the same way as lenses focus and direct light rays. At 10 GHz, a pencil beam of 1° beam width can be obtained by using 6.9-ft-diameter antenna, whereas at 10 MHz, this will require an antenna diameter of 6900 ft.

TABLE 1.1

Division of UHF, SHF, EHF and Upper Frequency Bands

Band Designation	L	S	C	X	K$_u$	K	K$_a$	Q	U	V	E	W	F	D
Frequency range (GHz)	1–2	2–4	4–8	8–12	12–18	18–26.5	26.5–40	33–50	40–60	50–75	60–90	75–110	90–140	110–170

TABLE 1.2

New U.S. Military Microwave Bands

Band	A	B	C	D	E	F	G	H	I	J	K	L	M
Frequency in GHz	0.1–0.25	0.25–0.5	0.5–1	1–2	2–3	3–4	4–6	6–8	8–10	10–20	20–40	40–60	60–100

- Unlike lower radio frequencies, microwaves are not reflected and practically not absorbed by the ionosphere. This property makes these suitable for space communication, including radio astronomy and satellite communication. Microwaves are widely used for telephone network, and TV systems, railway and military applications through line of sight (LOS) and tropospheric links and satellites.

- Microwaves provide an insight into the nature of molecules and their interaction. In particular, they provide valuable information about relaxation processes and electron spin resonance phenomena.

- Useful molecular resonances exist at microwave frequencies in diodes of certain crystal materials. In these crystals, energy can be generated or converted by means of atomic oscillations in and around the avalanche conditions. The Gunn, Read, Impatt diodes and so on are based on this property.

- These have extensive applications in the industry for testing, analysis and processing. For industrial heating, drying, cooking and physical diathermy, microwave energy is more easily directed, controlled and concentrated than LF energy.

- Microwave techniques may also be used in extremely fast computer applications. Pulses with very small widths are used in high-speed logic circuits.

1.4 Components of Microwave System

Any microwave system will require (i) a device that can generate a microwave signal of appropriate frequency, strength and of desired form, (ii) a transmission media through which this signal is guided from the point of origin to the location of destination, (iii) some components that can perform different desired tasks such as coupling, dividing, adding, directing, rotating, amplifying, attenuating, tuning, filtering and so on of the signal and (iv) some equipment and methods through which different parameters that spell the characteristics and behaviour of the signal can be accurately assessed. All these segments of the microwave system are briefly described below.

1.4.1 Microwave Generation

Radio frequency signals can be generated and amplified by using some conventional tubes (viz. triode, tetrode and pentode) and semiconductor devices (viz. junction transistors). At higher frequencies, the behaviour of these tubes and semiconductor devices becomes erratic and thus, these are replaced either by their modified versions or by those operating on altogether different principles. The limitations of conventional tubes at microwave frequencies include lead inductance, stray capacitance and transit time effect, which assumes significance as the frequency is raised. The most commonly used microwave tubes include klystron, reflex klystron, magnetron, travelling wave tube (TWT) and the backward wave oscillator (BWO). In view of the above limitations, a triode may be used up to 1 GHz, whereas the frequencies of operation for other tubes are of a much higher order; for example, a klystron can be used in the range of 0.4–70 GHz, a TWT can operate over 1–35 GHz and an amplitron can operate over a range of 0.4–9 GHz. These modified devices operate on the ballistic motion of electrons in vacuum under the influence of controlling electric or magnetic fields. These devices work in the density-modulated mode, instead of current-modulated mode. Thus, these work on the basis of clumps of electrons flying ballistically through them, rather than using a continuous stream.

The semiconductor devices are preferred wherever the systems can operate with relatively low power. These devices are of smaller size and lower weight. Besides, these can be easily integrated into the microwave-integrated circuits that generally use microstrip lines and similar other structures. These devices may include microwave diodes (viz. PIN, Tunnel and Varactor diodes), microwave transistors (viz. BJTs and HBTs), different types of field effect transistors (e.g. JFET, MOSFET, MESFET etc.), transferred electron devices (viz. Gunn, LSA, InP and CdTe diodes), avalanche transit time devices (viz. Read, Impatt, Trapatt and Baritt diodes) and some of the memory devices including the charged coupled devices. Maser is another device used to generate low-power and low-noise signals. It amplifies microwave signals in the same way as the laser amplifies light energy by stimulation process. Some commonly used microwave devices and components are shown in Figure 1.1.

The sun also emits microwave radiation, and most of it is blocked by the earth's atmosphere. The cosmic microwave background radiation (CMBR) is also a source of microwaves that supports the science of cosmology's Big Bang theory of the origin of the universe.

1.4.2 Microwave Processing

The word processor can be loosely assigned to the components that may couple, circulate, isolate, attenuate, match, tune and filter the microwave signals. Besides, the phase changers, duplexers, diplexers and mode suppressors can also be a part of this family.

1.4.3 Microwave Transmission

The transmission lines used at microwave range include coaxial cables, waveguides and planar lines. A coaxial line with low-loss dielectric materials is the commonly used transmission line in microwave range.

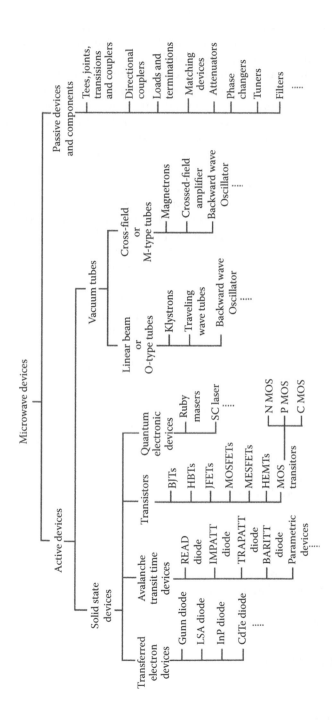

FIGURE 1.1
Microwave devices and components.

Semirigid coaxial lines with continuous cylindrical conductors outside perform well at microwave range. Its cross section, however, has to be much smaller than the signal wavelength for single-mode transmission. The reduction of cross section limits the power-handling capacity of these lines.

In high-power microwave circuits, coaxial lines are replaced by waveguides. These waveguides may be of different shapes and types. Rectangular waveguides are commonly employed for connecting high-power microwave devices as their manufacturing is relatively easier than that of circular waveguides. However, in certain components (e.g. rotary joints, isolators etc.), circular cross sections are required. A ridged waveguide provides broadband operation in comparison to a rectangular waveguide. Fin lines are commonly used in the millimetre-wave band. Planar lines are convenient in connecting the circuit components on a printed circuit board (PCB). The physical dimensions of these lines depend on the relative permittivity of the insulating material and the operating frequency band.

The transmission of microwaves requires the basic understanding of unguided and guided wave propagation, which in turn requires an insight of field theory. The unguided segment begins with the propagation of plane waves, which leads to many new concepts, for example, the wave equation obeyed by electric and magnetic fields, uniform and non-uniform plane waves, characteristic impedance of media, reflection and refraction of waves, travelling and standing waves, phase and group velocities, Brewster's angle, total internal reflection and so on. The concept of guided waves is equally important and needs to be gradually and systematically built up. The guiding structures at the initial stage include parallel planes and parallel wire lines. Although these structures, as such, are rarely used to guide the energy at microwave range, however, their study leads to some very useful concepts, for example, wave impedances, modes, cut-off frequency and so on. These concepts help in the understanding of transmission lines, stripline, microstrips and different forms of waveguides.

It needs to be noted that early microwave systems relied mostly on coaxial cables as transmission media. Despite their high bandwidth and convenience in test applications, the fabrication of complex microwave components with coaxial cables is a bit difficult. As frequency increases, losses in coaxial cables also increase. At this juncture, waveguides provide a better option. Transmission line techniques can only be applied to short conductors such as circuit board traces. Planar transmission lines can be used in the form of striplines, microstrips, slotted and coplanar lines. The waveguides and planar transmission lines are used for low-loss transmission of power.

1.4.4 Microwave Measurements

This is an important segment of a microwave system. The energy fed at the point of origination travels through the transmission media ultimately arrives at its destination. Its quality and quantity needs timely monitoring at each and every stage to ensure proper operation of the system. This monitoring includes measurement of various quantities and requires special equipment. The description of the devices and methods used for the measurement of different quantities at microwave ranges needs a thorough study and understanding as the conventional and simple measurement techniques employed at lower frequencies fail to yield the desired results.

1.4.5 Microwave Antennas

Antennas, as such, do not fall under any of the above four segments of the microwave system. These, however, are equally important as most of the services listed under the head of 'applications' require radiation of electromagnetic energy. Therefore, the proper selection and use of an antenna duly contribute towards the system performance.

In view of the above discussion, it is evident that each segment of the microwave system is equally important and contributes towards its proper functioning and required performance.

1.5 Applications

As illustrated in Figure 1.2, microwaves have a wide range of applications that encompass areas of communications, radars and navigation, home, industry, scientific research and the medical field. The frequencies allocated to industrial, scientific research and medical applications are 0.915, 2.45, 5.8, 10.525 and 20.125 GHz. The industrial control and measurement applications include (i) thickness measurement of metal sheets in rolling mills, (ii) continuous measurement of wire diameters, (iii) monitoring and measurement of moisture contents, (iv) motion sensors based on 'Doppler effect' and (v) applications based on thermal effects of microwaves. At microwave frequencies, skin depth in metals is very small, and a metal surface causes total reflection. Most of the measurements for industrial applications involve the determination of complex reflection and transmission coefficients. As the wavelength is small, the phase variations are quite rapid. Consequently, a small change in position or dimension gives rise to a significant phase change that can be detected and measured. Some of the applications noted above are further elaborated below.

1.5.1 Communication

Before the advent of optical systems, most of the long-distance communication systems were based on networks of microwave radio relay links

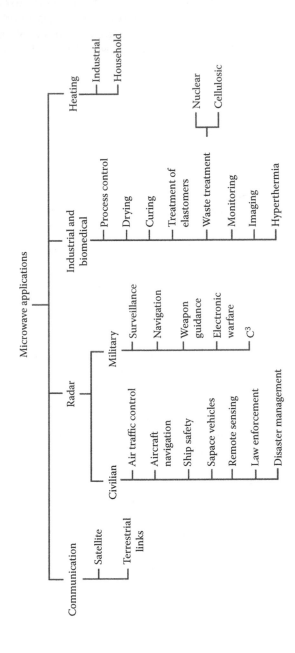

FIGURE 1.2
Applications of microwaves.

including those of LOS, tropospheres and satellites. Wireless local area network (LAN) protocols, such as Bluetooth, IEEE 802.11 and IEEE 802.11a specifications use 2.4 and 5 GHz Industrial, Scientific and Medical (ISM) bands. Long-range wireless Internet access services (up to about 25 km) use 3.5–4.0 GHz range. The worldwide interoperability for microwave access (WIMAX) services operate at the 3.65 GHz band. Metropolitan area network protocols, such as WIMAX are based on standards such as IEEE802.16, designed to operate between 2 and 11 GHz. Commercial implementations are in the 2.3, 2.5, 3.5 and 5.8 GHz ranges. Mobile broadband wireless access protocols operate between 1.6 and 2.3 GHz to provide mobility and in-building penetration characteristics similar to mobile phones but with vastly greater spectral efficiency. Some mobile phone networks, such as Global System for Mobile (GSM), use the low-microwave/high-UHF frequencies around 1.8 and 1.9 GHz. Digital Video Broadcasting–Satellite services to Handhelds (DVB-SH) and Satellite-Digital Multimedia Broadcasting (S-DMB) use 1.452–1.492 GHz, while proprietary/incompatible satellite radio uses around 2.3 GHz for Digital Audio Radio Service (DARS).

Microwaves are used in telecommunication transmissions because of their short wavelengths. This aspect results in smaller but highly directional antennas and in larger bandwidths. Most satellite systems operate in the C, X, K_a or K_u bands of the microwave spectrum. These frequencies not only allow larger bandwidths but also avoid crowding of UHF frequencies and are still staying below the atmospheric absorption of EHF frequencies. Satellite TV either operates in the C band for the traditional large-dish fixed satellite service or K_u band for direct broadcast satellite. Military communications primarily run over X or K_u-band links, with K_a band being used for Minstar.

1.5.2 Radars

Microwave radiations are used to detect the range, speed and other characteristics of remote objects. Radars are also used for applications such as air traffic control, weather forecasting, navigation of ships and speed limit enforcement. Microwave techniques involving Gunndiode oscillator and waveguides are used in motion detectors for automatic door openers.

Table 1.3 summarises microwave applications in the areas of communication and radars. The corresponding band and frequency ranges (in gigahertz) are also given for most of these services. In view of this table, one can easily appreciate the importance of microwaves.

1.5.3 Radio Astronomy

Microwaves are used in most of the radio astronomy systems. In general, these systems observe the naturally occurring microwave radiations. Some active radar experiments have also been conducted with objects in the solar system to determine distances or mapping invisible objects.

TABLE 1.3

Applications of Microwaves

Sl. No.	Application	Frequency Band(s)	Frequency Range(s) in GHz
1	Airborne weather avoidance radar	C	5.4
2	Antiaircraft weapon system	MMW	94 and 140
3	Astronomical radars	P,L, S and C	0.408, 0.43, 0.7, 1.295, 2.388 and 8
4	Automatic toll collection	P and C	0.905 and 5–6
5	Battlefield surveillance radar	MMW	70
6	Cellular phone	P	0.824–0.849 and 0.869–0.895
7	Cellular video	Ka	28
8	Collision avoidance radar	MMW	60, 77 and 94
9	Direct broadcast satellite	X	11.7–12.5
10	Global digital satellite	C	4–8
11	GPS	L	1.575 and 1.227
12	Ground-based radar and navigation	X	8–12
13	GSM	P	0.890–0.915 and0.935–0.960
14	Instrumentation radars	S and C	2.9, 5.6 and 1–4
15	Missile guidance seeker system with radiometric sensors	MMW	35 and 94
16	Paging	P	0.931–0.932
17	Personal communication system	L and S	1.85–1.99 and 2.18–2.20
18	Radio altimeters	C	4.2–4.4
19	Shipborne navigation radars	X	9.345–9.405
20	Special telecom services	Ka	27–40
21	Surveillance and acquisition radar	MMW	94
22	Synthetic aperture radar	P	0.23–1
23	Very Small Aperture Terminal (VAST) networks	Ku	12–18
24	Wide-area computer networks	MMW	60
25	Wireless LAN	S	2–4
26	Wireless LANs	S and C	2.4–2.48 and 5.4

1.5.4 Navigation

Microwaves are used in global navigation satellite systems including Chinese Beidou, American global positioning satellite (GPS) and Russian GLONASS in various frequency bands between 1.2 and 1.6 GHz.

1.5.5 Home Appliances

Microwave ovens using cavity magnetrons are the common kitchen appliances. A microwave oven passes (non-ionising) microwave radiations, at a frequency of about 2.45 GHz, through food. These result in dielectric heating by absorption of energy in water, fats and sugar contained in the food item. Water in the liquid state possesses many molecular interactions that broaden the absorption peak. In the vapour phase, isolated water molecules absorb at around 22 GHz, almost 10 times the frequency of the microwave oven.

1.5.6 Industrial Heating

Microwave heating is widely used in industrial processes for drying and curing the products.

1.5.7 Plasma Generation

Many semiconductor processing uses microwaves to generate plasma for purposes such as reactive ion etching and plasma-enhanced chemical vapour deposition. Microwave frequencies ranging from 110 to 140 GHz are used in stellarators and tokamak experimental fusion reactors to heat the fuel into a plasma state. The upcoming International Thermonuclear Experimental Reactor (ITER) is expected to range from 110 to 170 GHz and will employ electron cyclotron resonance heating.

1.5.8 Weaponry System

Less-than-lethal weaponry exists that uses millimetre waves to heat a thin layer of human skin to an intolerable temperature so as to make the targeted person move away. A 2-s burst of the 95 GHz focussed beam heats the skin to a temperature of 130°F (54°C) at a depth of 0.4 mm. The United States Air Force and Marines currently use this type of active denial system.

1.5.9 Spectroscopy

Microwave radiation is used in electron paramagnetic resonance spectroscopy, in X-band region (~9 GHz) in conjunction with magnetic fields of 0.3 T. This technique provides information on unpaired electrons in chemical systems, such as free radicals or transition metal ions. The microwave radiation can also be combined with electrochemistry as in microwave-enhanced electrochemistry.

1.6 Health Hazards

Microwaves can be considered as a form of non-ionising radiations since these do not contain sufficient energy to chemically change substances by ionisation. The term radiation is not to be confused with radioactivity. It has not yet been conclusively proved that microwaves cause significant adverse biological effects. Some studies, however, suggest that long-term exposure may result in a carcinogenic effect. This effect is different from the risks associated with very-high-intensity exposure, which can cause heating and burns like any heat source, and not a unique property of microwaves specifically. The injury from exposure to microwaves usually occurs from dielectric heating induced in the body. Such exposure can produce cataracts, because the microwave heating denatures proteins in the crystalline lens of the eye faster than the lens that can be cooled by the surrounding structures. The lens and cornea of the eye are especially vulnerable because they contain no blood vessels that can carry away heat. The intense exposure to radiations can produce heat damage in other tissues as well, up to and including serious burns that may not be immediately evident because of the tendency of microwaves to heat deeper tissues with higher moisture content.

Descriptive Questions

Q-1.1   Discuss the logic behind dividing the microwave frequencies into different bands.

Q-1.2   List the advantages offered by microwaves in comparison to lower-frequency range.

Q-1.3   List the various segments of a microwave system and discuss their significance in brief.

Q-1.4   Discuss the key applications that utilise the microwave frequency band.

Q-1.5   List the civilian and military applications of microwaves along with their frequency bands.

Further Reading

1. Adam, S. F., *Microwave Theory and Applications*. Prentice-Hall, Inc., New Jersey, 1969.

2. Chatterjee, R., *Microwave Engineering*. East West Press, New Delhi, 1990.

3. Das, A. and Das, S. K., *Microwave Engineering*. Tata McGraw-Hill, New Delhi, 2002.

4. Gandhi, O. P., *Microwave Engineering and Applications*. Pergamon Press, New York, 1981.

5. Gupta, K. C., *Microwaves*. Wiley Eastern, New Delhi, 1979.

6. Jakes, W. C., *Microwave Mobile Communication*. John Wiley & Sons Inc., New York, 1974.

7. Lio, S. Y., *Microwave Devices and Circuits*, 3rd ed. Prentice-Hall of India, New Delhi, 1995.

8. Liou, K. N., *An Introduction to Atmosphere Radiations*. Academic Press, California, 2002.

9. Pozar, D. M., *Microwave Engineering*. Addison-Wesley Publishing Co., New York, 1990.

10. Pozar, D. M., *Microwave Engineering*, 2nd ed. John Wiley & Sons, New Delhi, 1999.

11. Reich, H. J. et al. *Microwave Principles*, D. Van Nostrand, Reinhold Co. New York, 1957.

12. Skolnik, M. I., *Introduction to Radar Systems*, 3rd ed. Tata McGraw-Hill, New Delhi, 2001.

13. Terman, F. E., *Electronics and Radio Engineering*. McGraw-Hill, Tokyo, 1955.

14. Wheeler, G. J., *Introduction to Microwaves*. Prentice-Hall, Englewood Cliffs, NJ, 1963.

15. Young, L., *Advances in Microwaves*. Academic Press, New York, 1974.

2

Fundamentals of Wave Propagation

2.1 Introduction

In the introductory chapter it was noted that microwaves encompass a wide sphere of applications owing to their many inherent advantages. The study of microwave systems and devices requires the proper understanding of plane waves, transmission lines and related topics that are based on field theory. In general, these topics are included in the curriculum of electromagnetics given at a lower level. Since the concepts involved in these form the basis for microwaves, this chapter summarises some of the essential aspects of wave propagation in free space, between parallel planes and in transmission lines.

2.2 Basic Equations and Parameters

The behaviour of wave phenomena is governed by Maxwell's equations which are given below in their general form.

$$\nabla \times H = J + \frac{\partial D}{\partial t} = J + j\omega D = \sigma E + j\omega\varepsilon E = J + J_d \qquad (2.1a)$$

$$\nabla \times E = -\frac{\partial B}{\partial t} = -j\omega B = -j\omega\mu H \qquad (2.1b)$$

$$\nabla \cdot D = \rho_v \qquad (2.1c)$$

$$\nabla \cdot B = 0 \qquad (2.1d)$$

The quantities E (electric field intensity, V/m), H (magnetic field intensity, A/m), D (electric flux density, C/m²), B (magnetic flux density Web/m²), J (conduction current density, A/m²), J_d (displacement current density, A/m²) and ρ_v (volume charge density, C/m³) involved in Equation 2.1 are connected by the relations:

$$D = \varepsilon E \qquad (2.2a)$$

$$B = \mu H \qquad (2.2b)$$

$$J = \sigma E \qquad (2.2c)$$

where ε (permittivity, F/m), μ (permeability, H/m) and σ (conductivity, Mhos/m) are characterising parameters of the medium. For free space $\mu = \mu_0 = 4\pi \times 10^{-7}$ H/m and $\varepsilon = \varepsilon_0 = 8.854 \times 10^{-12}$ or $10^{-9}/36\pi$ F/m. In free space $\sigma = 0$, $J = 0$ and $\rho_v = 0$.

The manipulation of Maxwell's equation lead to another set of equations called wave equations. These wave equations for lossless and lossy media are

$$\nabla^2 E = \mu_0 \varepsilon_0 \frac{\partial^2 E}{\partial t^2} \quad \text{(Lossless media)} \qquad (2.3a)$$

$$\nabla^2 H = \mu_0 \varepsilon_0 \frac{\partial^2 H}{\partial t^2} \quad \text{(Lossless media)} \qquad (2.3b)$$

$$\nabla^2 E = \mu \sigma \frac{\partial E}{\partial t} + \mu \varepsilon \frac{\partial^2 E}{\partial t^2} \quad \text{(Lossy media)} \qquad (2.4a)$$

$$\nabla^2 H = \mu \sigma \frac{\partial H}{\partial t} + \mu \varepsilon \frac{\partial^2 H}{\partial t^2} \quad \text{(Lossy media)} \qquad (2.4b)$$

Wave equations spell that an electromagnetic wave can be conceived in terms of E and H. These further reveal that if a physical phenomenon occurs at one place at a given time and the same is reproduced at other places at later times such group of phenomenon constitutes a wave. The time delay is proportional to the space separation between the locations of two such occurrences. A wave need not be a repetitive phenomenon in time.

In view of inherent involvement of time (t) in the wave phenomenon normally *sinusoidal time variation* is assumed which is accounted by a factor $e^{j\omega t}$, where ω is the angular frequency measured in rad/s. Also in a wave travelling in an arbitrary direction r, *space variation* is accounted by multiplying all field quantities by a factor $e^{-\gamma r}$, where γ is called *propagation constant*. The parameter γ is composed of two quantities {i.e. $\gamma = (\alpha + j\beta)$}, where α is the attenuation constant or gain coefficient measured in Nep./m and β is the phase shift constant measured in rad/m. If α is positive, then the wave decays in amplitude with distance and it is called *attenuation constant*. If α is negative, then the wave grows, and α is called *gain coefficient*. *The phase constant β* is a measure of phase-shift per unit length. When $\alpha = 0$, $\beta \approx \omega\sqrt{(\mu\varepsilon)}$ is termed *wave number*. The value of β in free space {i.e. $\beta \approx \omega\sqrt{(\mu_0\varepsilon_0)}$} is referred to as *free space wave number*. It is also referred to as *spatial frequency*. In general, the study of plane waves and transmission lines involve Cartesian coordinate system.

2.3 Nature of Media

A wave on its course of journey may encounter media of varied nature and behave accordingly. The media encountered may be classified as linear or non-linear, homogeneous or non-homogeneous, isotropic or non-isotropic, source free, bilateral or unilateral and lossless or lossy. These are briefly described below.

A medium is referred to as *linear* if E and D (or B and H) bear a linear relation. If variation of E is non-linearly related with D (or that of H with B) the medium is termed as *non-linear*. As an example, B and H, in ferromagnetic materials are not linearly related. A *homogeneous* medium is one for which parameters ε, μ and σ are constant. If values of either of these parameters differ at different locations the media are called *heterogeneous* or *non-homogeneous*. A medium is *isotropic* if ε (or μ) is a scalar constant, so that D and E (or B and H) vectors have the same direction everywhere. If either of the media parameters is such that D and E (or B and H) have different orientations the media are termed as *anisotropic* or *non-isotropic*. The ionosphere is an example of anisotropic media. A *source free* media or region is one in which no impressed voltages or currents (i.e. no generators) are present. A *bilateral media* is one in which the wave on its forward and backward journey behaves similarly. In case of *unilateral media*, such as in magnetised ferrites the behaviour of forward and backward waves is not the same. A medium is said to be *lossless* if $\sigma = 0$. Free space or vacuum is an example of lossless media. If $\sigma \neq 0$ the media are considered to be *lossy*.

2.4 Wave in Lossless Media

With the assumption that E or H is the function of only one space coordinate (say x), and $E = E_y \, a_y$ only, wave equation (2.3) yields the following form of solution:

$$E_y = f_1(x - v_0 t) + f_2(x + v_0 t) \qquad (2.5a)$$

where,

$$v_0 = \frac{1}{\sqrt{\mu_0 \varepsilon_0}} \tag{2.5b}$$

The solution illustrated by Figure 2.1 involves forward and backward waves, where *forward wave* is the wave travelling in the positive x direction and is given by term f_1 and *backward wave* is the wave travelling in the negative x direction and is given by f_2.

The analysis of wave equations reveals that a wave progressing in an arbitrary direction will have no component of E or H in that direction. As both E and H fields are perpendicular to the direction of propagation and both lie in a plane that is transverse to the direction of propagation such a wave is referred to as *transverse electromagnetic wave*.

When a wave progresses in an assigned direction its amplitude and phase keeps changing. Joining of space locations of *equiphase or equiamplitude points* may result in planar configurations. Such configurations may be referred to as *equiphase or equiamplitude planes*. A wave having equiphase plane is termed a *plane wave*. If equiphase and equiamplitude planes are the same the wave is called a *uniform plane wave*. The uniform plane wave is a transverse wave. A perpendicular drawn to an equiphase plane is called a *ray*.

The manipulation of Maxwell's equations and that of the solution of wave equation yields a ratio $E/H = \sqrt{(\mu/\varepsilon)} = \eta$ (in Ω), which is called the *characteristic (or intrinsic) impedance* of the medium. For free space η is replaced by $\eta_0 = \sqrt{(\mu_0/\varepsilon_0)} = 120\,\pi \approx 377\,\Omega$.

In Figure 2.1 two waves referred to as forward and backward waves were shown. The backward wave is normally the reflected wave. When there are no reflections the wave is a progressing wave and is called a travelling *wave* wherein minimas and maximas keeps shifting in the forward direction.

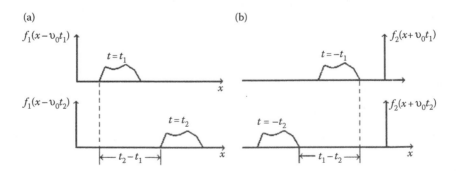

FIGURE 2.1
Two forms of wave. (a) Forward wave. (b) Backward wave.

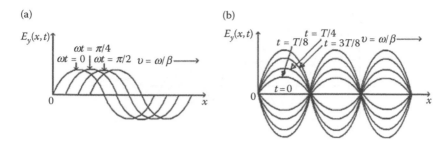

FIGURE 2.2
Wave in lossless media. (a) Travelling wave. (b) Standing wave.

When reflections are added the resulting wave is a *standing wave* wherein maxima and minima occur at the same space locations and the wave appears to be expanding and contracting in magnitude with time. The travelling and standing waves are shown in Figure 2.2.

On incorporation of time variation Equation 2.3a becomes

$$\nabla^2 E + \beta^2 E = 0 \tag{2.6}$$

The real part of the solution of Equation 2.6 representing only the forward wave with the assumption that E is a function of x & t and contains only E_y component, can be written as

$$\text{Re}[E_y(x,t)] = C_1 \cos(\omega t - \beta x) \tag{2.7}$$

By equating the argument of the cosine term of Equation 2.7 to a constant (conveniently to zero) an expression of velocity v (= $x/t = \omega/\beta$) is obtained. This velocity is called the *phase velocity*. The velocity with which the wave propagates in vacuum or free space given by v_0 (= c, where c is the velocity of light $\approx 3 \times 10^8$ m/s) is called the *wave velocity*. To differentiate between the two velocities the phase velocity is sometimes denoted by v_p.

2.5 Wave in Lossy Media

In view of the sinusoidal time variation, Equation 2.4a can be re-written as

$$\nabla^2 E - \gamma^2 E = 0 \tag{2.8a}$$

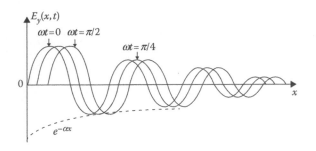

FIGURE 2.3
Travelling waves in lossy media.

where

$$\gamma = \sqrt{j\omega\mu(\sigma + j\omega\varepsilon)} = \alpha + j\beta \tag{2.8b}$$

As stated earlier the real part of solution to Equation 2.8a, representing only forward wave, is

$$\mathrm{Re}[E_y(x,t)] = C_1 e^{-\alpha x}\cos(\omega t - \beta x) \tag{2.9}$$

Plots of E_y from expressions of Equation 2.9 are shown in Figure 2.3, which represents a pattern of travelling wave with attenuation.

For lossy medium ($\sigma \neq 0$) the ratio $E/H = \eta = \sqrt{\{j\omega\mu/(\sigma + j\omega\varepsilon)\}}$.

2.6 Conductors and Dielectrics

In view of Equation 2.1a the magnitude of the ratio of conduction current density and displacement current density (for sinusoidal time variation) can be written as

$$\left|\frac{J}{J_d}\right| = \frac{|\sigma E|}{|\omega\varepsilon E|} = \frac{\sigma}{\omega\varepsilon} \tag{2.10}$$

This ratio $\{(\sigma/\omega\varepsilon) = 1\}$ is considered as the dividing line between conductors and dielectrics. For conductors $\sigma/\omega\varepsilon > 1$ while for dielectrics $\sigma/\omega\varepsilon < 1$. For good dielectric $\sigma/\omega\varepsilon \ll 1$ and for good conductor $\sigma/\omega\varepsilon \gg 1$. This is illustrated in Figure 2.4.

The ratio $\sigma/\omega\varepsilon$ is called *dissipation factor* of the dielectric and is denoted by D (not electric flux density). The parameter D is related with another

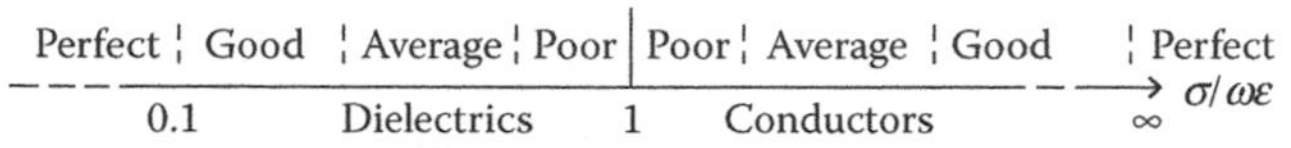

Perfect ┊ Good ┊ Average ┊ Poor	Poor ┊ Average ┊ Good ┊ Perfect

$$\xrightarrow{} \sigma/\omega\varepsilon$$

0.1   Dielectrics   1   Conductors   ∞

FIGURE 2.4
Dielectrics and conductors.

parameter φ {= $\tan^{-1} D \approx D$ (for small values of D)}, which is called the *power factor* or *loss tangent*.

Equation 2.8b can be appropriately manipulated to get the values of α and β. These are as follows:

$$\alpha = \omega\sqrt{\frac{\mu\varepsilon}{2}\left\{\sqrt{\left(1 + \frac{\sigma^2}{\omega^2\varepsilon^2}\right)} - 1\right\}} \tag{2.11}$$

$$\beta = \omega\sqrt{\frac{\mu\varepsilon}{2}\left\{\sqrt{\left(1 + \frac{\sigma^2}{\omega^2\varepsilon^2}\right)} + 1\right\}} \tag{2.12}$$

Since $\sigma/\omega\varepsilon \ll 1$ for good dielectrics, Equations 2.11 and 2.12 can be manipulated to yield

$$\alpha = \frac{\sigma}{2}\sqrt{\frac{\mu}{\varepsilon}} \tag{2.13a}$$

$$\beta = \omega\sqrt{\mu\varepsilon}\left(1 + \frac{\sigma^2}{8\omega^2\varepsilon^2}\right) \tag{2.13b}$$

In view of Equation 2.13 expressions for velocity and characteristic impedance get modified to

$$\upsilon = \frac{\omega}{\beta} \approx \upsilon_0\left(1 - \frac{\sigma^2}{8\omega^2\varepsilon^2}\right) \tag{2.14a}$$

$$\eta = \sqrt{\frac{j\omega\mu}{\sigma + j\omega\varepsilon}} \approx \sqrt{\frac{\mu}{\varepsilon}}\left(1 + \frac{j\sigma}{2\omega\varepsilon}\right) = \eta\left(1 + \frac{j\sigma}{2\omega\varepsilon}\right) \tag{2.14b}$$

When

$$\sigma = 0, \quad \eta \approx \sqrt{\left(\frac{\mu}{\varepsilon}\right)} = \eta_0 \tag{2.14c}$$

When

$$\sigma \neq 0, \quad \eta \approx \eta_0 \left(1 + \frac{j\sigma}{2\omega\varepsilon}\right) \tag{2.14d}$$

These equations reveal that the presence of small σ in good dielectrics results in slight reduction of velocity and in addition of a reactive component in characteristic impedance.

Similarly, when $\omega\varepsilon/\sigma \approx 0$, Equation 2.8b modifies to

$$\gamma = \alpha + j\beta \approx \sqrt{[j\omega\mu\sigma]} \tag{2.15a}$$

Thus

$$\alpha = \beta = \sqrt{[\omega\mu\sigma]}\angle 45° = \sqrt{\left[\frac{\omega\mu\sigma}{2}\right]} \tag{2.15b}$$

$$\upsilon = \frac{\omega}{\beta} = \sqrt{\left[\frac{2\omega}{\mu\sigma}\right]} \tag{2.16a}$$

$$\eta \approx \sqrt{\left[\frac{\omega\mu}{\sigma}\right]}\angle 45° \tag{2.16b}$$

Equations 2.16a,b reveal that as α is large for good conductors the wave greatly attenuates. As β is also large there is great reduction in wave velocity almost to the level of sound waves in air. The characteristic impedance is also very small and contains a reactive component always having an angle of 45°.

2.7 Polarisation

The plane of polarisation or simply polarisation of a wave is defined by the direction in which the E vector is aligned during the passage of at least one full cycle. In general, both magnitude and pointing of E vector vary during each cycle and map-out an ellipse in the plane normal to the direction of propagation, thus a wave is said to be elliptically polarised.

The ratio of minor-to-major axis of an ellipse is referred to as *ellipticity*, which is expressed in dBs and is less than 1 for any ellipse. The direction of major axis of an ellipse is called the *polarisation orientation*.

When ellipticity becomes ∞ dB (i.e. minor-to-major axis ratio is zero) a wave is said to be *linearly polarised*. Thus, the *linearly polarised wave* is a transverse wave whose electric field vector, at a point in a homogeneous isotropic medium, at all times lies along a fixed line.

When ellipticity is 0 dB (or minor-to-major axis ratio is unity) the wave is circularly polarised. Thus, a *circularly polarised wave* is a transverse wave for which the electric field vector at a point describes a circle. In case of circular polarisation the *sense of polarisation* is identified in reference to an observer. For an observer looking in the direction of propagation, the rotation of E vector in a transverse plane is *clockwise* for *right-hand polarisation* and *counterclockwise* for *left-hand polarisation*. Thus, a circularly polarised wave may have *left-hand or right-hand sense*. In case of elliptically polarised wave, the sense is taken to be the same as that of the predominant circular component.

As the polarisation of a uniform plane wave refers to the time behaviour of E at some fixed point in space, assume that a wave is travelling in the z direction and thus E may have E_x and E_y components only. If $E_y = 0$ and only E_x component is present, the wave is said to be *polarised in the x direction*. If $E_x = 0$ and only E_y is present, the wave is said to be *polarised in the y direction*. If both E_x and E_y are present and are in phase, the resultant E vector will change in magnitude but not in direction. Such a wave is called *linearly polarised*. If the phase difference between E_x and E_y is 180° the wave is again linearly polarised but vector direction will reverse. Figure 2.5 illustrates the linear polarisation.

If the phase difference between E_x and E_y is 90° the resultant E vector will not change in magnitude but will assume different directions. An envelope joining the tips of this vector at different time instants will appear to be a circle. Such a wave is called *circularly polarised*. If the phase difference between E_x and E_y is 270° the wave is again circularly polarised but the sense of polarisation will reverse, that is, the direction of the E vector will be opposite to that in case of the 90° phase difference. Figure 2.6 illustrates the circular polarisation.

If E_x and E_y have a phase difference other than 0°, 90°, 180° or 270° but lies between 0° and 180° the resultant E vector will change in magnitude as well in directions. If an envelope joins tips of this vector at different instants, it will form an ellipse and the wave is termed as *elliptically polarised*. The sense

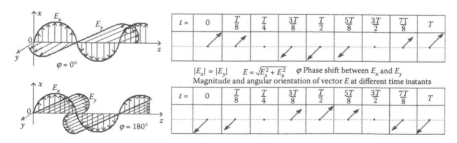

FIGURE 2.5
Linear polarisation.

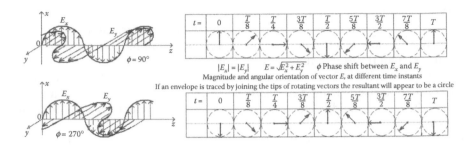

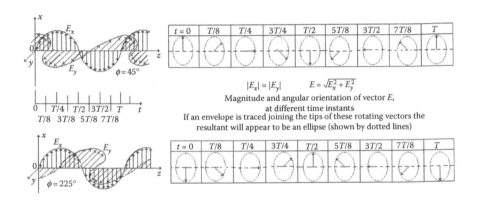

FIGURE 2.6
Circular polarisation.

FIGURE 2.7
Elliptical polarisation.

of polarisation will reverse for angles between 180° and 360° from that for angles between 0° and 180°. Figure 2.7 illustrates the elliptical polarisation.

Besides the above, a wave may be vertically or horizontally polarised. These polarisations are defined with reference to the surface of the earth which is assumed to be perfectly smooth and flat. Thus, a wave emerging from a vertical antenna is taken to be *vertically polarised,* whereas that from a horizontal antenna is *horizontally polarised.* In general, a horizontal antenna causes a perfect horizontally polarised wave, whereas a vertical antenna does not yield a purely vertically polarised wave.

2.8 Depth of Penetration

The depth of penetration is a measure of decrease of amplitude of the wave content to e^{-1} times its original value when it penetrates a surface. It is related to frequency and is as follows:

$$\delta = \frac{1}{\alpha} \approx \sqrt{\left(\frac{2}{\omega\mu\sigma}\right)}$$ (2.17)

2.9 Surface Impedance

At high frequency the flow of current is almost entirely confined to a very thin layer of the surface of the sheet of a conductor. The impedance (Z_s) offered by the surface to the flow of current is referred to as surface imped-ance Z_s and is given by

$$Z_s = \sqrt{\frac{(j\omega\mu\sigma)}{\sigma}} = \sqrt{\left(\frac{j\omega\mu}{\sigma}\right)} = \eta = R_s + jX_s$$ (2.18)

Where

$$R_s = X_s = \sqrt{\left(\frac{\omega\mu}{2\sigma}\right)}$$ (2.19a)

In view of Equation 2.17, Equation 2.19a can be written as

$$R_s = X_s = \frac{1}{(\sigma\delta)}$$ (2.19b)

2.10 Poynting Theorem

The first Maxwell's Equation 2.1a can be manipulated to yield the relation:

$$\int_v (E \cdot J)dv = -\frac{\partial}{\partial t}\int_v \left(\frac{\mu H^2}{2}\right)dv - \frac{\partial}{\partial t}\int_v \left(\frac{\varepsilon E^2}{2}\right)dv - \oint_s (E \times H) \cdot ds$$ (2.20)

Equation 2.20 involves four terms each of which has a specific meaning as noted below.

$$Term - 1: \quad \int_v (E \cdot J)dv$$ (2.21a)

Here, E has the unit of V/m and J of ampere/m^2, the unit of $E \cdot J$ is W/m^3. Thus, this term represents the power dissipated per unit volume provided E and J are in the same direction.

$$\text{Term-2:} \quad -\frac{\partial}{\partial t} \int_v \left(\frac{\mu H^2}{2} \right) dv \qquad (2.21b)$$

The term $\mu H^2/2$ represents the stored magnetic energy per unit volume, $\int_v (\mu H^2/2) dv$ is the total magnetic energy stored in a volume v. The term $(\partial/\partial t) \int_v (\mu H^2/2) dv$ gives the rate of change of magnetic energy stored in the volume. With the addition of negative sign the term $[-(\partial/\partial t) \int_v (\mu H^2/2) dv]$ indicates the rate of reduction of magnetic energy in the volume.

$$\text{Term-3:} \quad -\frac{\partial}{\partial t} \int_v \left(\frac{\varepsilon E^2}{2} \right) dv \qquad (2.21c)$$

The term $\varepsilon E^2/2$ represents the stored electric energy per unit volume, $\int_v (\varepsilon E^2/2) \, dv$ is the total electric energy stored in a volume v and $(\partial/\partial t) \int_v (\varepsilon E^2/2) \, dv$ is the rate of change of electric energy stored and $[-(\partial/\partial t) \int_v (\varepsilon E^2/2) \, dv]$ indicates the rate of reduction of electric energy in the volume.

$$\text{Term-4:} \quad -\oint_s (E \times H) \cdot ds \qquad (2.21d)$$

The term $E \times H$ bears the units $[(V/m) \times (A/m) = (W/m^2)]$ of power (P) per square metre. The term $\oint_s (E \times H) \cdot ds = \oint_s P \cdot ds$ represents the rate of flow of energy outward through the surface (s) enclosing the volume (v). Thus, Equation 2.20 can be written as

$$\oint_s P \cdot ds = -\int_v (E \cdot J) dv - \frac{\partial}{\partial t} \int_v \left(\frac{\mu H^2}{2} \right) dv - \frac{\partial}{\partial t} \int_v \left(\frac{\varepsilon E^2}{2} \right) dv \qquad (2.22)$$

The term $P(= E \times H)$ is called *Poynting Vector* and Equation 2.22 is referred to as *Poynting theorem*. As shown in Figure 2.8, power P_1, P_2 and P_3 leave the volume (v), thus $P(= P_1 + P_2 + P_3)$ is the total power that leaves the volume (v) through the surface (s).

If E and J are oriented in the same directions $E \cdot J$ will be positive. It is the case when the battery is drawing power from the main supply while charging. If E and J are oriented in the opposite directions $E \cdot J$ will be negative. It is the case of discharging of battery as it is now supplying power to the connecting equipment. Equation 2.22 follows the law of conservation,

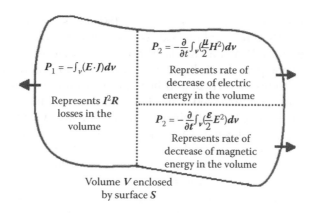

FIGURE 2.8
Components of Poynting vector.

that is, the rate of dissipation of energy in a volume must be equal to the rate at which the stored energy in volume v is decreasing and the rate at which energy is entering the volume from the outside. Thus, the relation $-\oint_s (E \times H) \cdot ds$ represents inward flow and $+\oint_s (E \times H) \cdot ds$ represents outward flow of the energy through the surface 's' of volume 'v'. The same can be summarised by a statement commonly known as *Poynting theorem*. The relation $P = E \times H$ represents an instantaneous power density measured in W/m^2. In view of cross product the direction of power flow will be perpendicular to the direction of orientation of E and that of H. If P is complex it can be written as

$$P = \frac{1}{2}(E \times H^*) \tag{2.23}$$

2.11 Reflection and Refraction

During the passage of its journey an electromagnetic wave may face the following situations. It may strike a surface with somewhat different media parameters or pass through it. If it strikes a perfect conductor it is totally reflected. In case of perfect dielectric the wave crosses the medium through refraction. In case of imperfect dielectric the wave is partially reflected and partially refracted. In the course of reflection or refraction the characteristics of striking wave may change in accordance with its polarisation. When a vertically polarised wave is reflected from a smooth surface, which is the interface between a normal propagation medium and with infinite impedance, there is no change in its characteristics. When a horizontally polarised wave is reflected from a smooth surface, there is 180° phase change (because of the coordinate system reversal in space when one looks in the reverse direction of propagation). When a circularly polarised wave is reflected from a smooth surface, the phase of its horizontal component is altered by 180° (and since the magnitudes of vertical and horizontal components are equal) the sense of the wave gets reversed. Circularly polarised antennas are unique in being entirely unable to see their own images in any symmetrical reflecting surface, since the reflected wave has its sense reversed and is therefore orthogonal to the polarisation of the antenna from which it originates. For an elliptically polarised wave, reflection is equivalent to altering the differential phase shift (e.g. the phase difference between the horizontally and vertically polarised linear components) by 180°. The new polarisation ellipse may be determined by combining the vertical and phase reversed (reflected) horizontal component vectorially.

The term striking wave means that E impinging on the boundary surface or on interface between the two media. This striking wave may cause induction of charges on the surface or a flow of current in the material to which this surface belongs. Whereas the charges so induced shall remain on the surface the current may remain confined to the skin or may penetrate a little deeper depending on characteristics of the material and the frequency. The striking process of a wave is normally represented by a straight line depicting a ray.

When a ray strikes an object the energy content of the wave can either get reflected back, get transmitted with or without refraction or get absorbed. Leaving aside the absorbing materials as they do not result in reflection or refraction the involvement of other materials lead to the terms 'reflection coefficient', 'refraction coefficient' and the 'transmission coefficient'. The *reflection coefficient* for E (or H) represents the ratio of the field contents of E (or H) in reflected and incident waves. The *refraction coefficient* represents the ratio of E (or H) in refracted and incident waves. Similarly, the *transmission coefficient* represents the ratio of E or H in transmitted and incident waves. Thus, the phenomenon of reflection and refraction is of utmost importance for the wave propagation and emanates from boundary conditions satisfied by electric and magnetic field quantities. A cursory look at boundary conditions particularly those for time-varying fields are quite helpful in understanding the wave behaviour. The boundary conditions to be satisfied depend on the type of field and the interface between the types of media. These for *time-varying field* are:

$$E_{t1} = E_{t2} \quad \text{and} \quad D_{n1} - D_{n2} = \rho_S \qquad (2.24a)$$

$$H_{t1} - H_{t2} = k \quad \text{or} \quad H_{t1} - H_{t2} = 0 \text{ (for } k = 0) \quad \text{and} \quad B_{n1} = B_{n2} \qquad (2.24b)$$

For highly conducting case when $\sigma = \infty$ the current becomes a surface current k:

$$E_{t1} = 0, \quad D_{n1} = \rho_s, \quad H_{t1} = k \quad \text{and} \quad B_{n1} = 0 \qquad (2.24c)$$

In Equation 2.24 D_t, E_t, B_t and H_t indicate tangential components D_n, E_n, B_n and H_n normal components of D, E, B and H, respectively. Suffixes 1 and 2 represent regions to which these quantities belong. It is to be mentioned that surface charge density (ρ_s) is considered a physical possibility for either dielectrics or perfect/imperfect conductors, but the surface current density (k) is assumed in connection with perfect conductors only.

2.12 Direction Cosines, Wavelength and Phase Velocity

In most of the cases a wave is normally assumed to be a plane wave. While dealing with such a wave travelling in an arbitrary direction its equation is to be written in terms of angles which the unit vector $ñ$ along the direction of travel (say S) makes with x, y and z axes, respectively. This unit vector is normal to the plane of constant phase F–F. If $E(x) = E_0 e^{-j\beta x}$ be the expression for a wave travelling in the positive x direction its equiphase planes can be given by $x =$ constant. Thus, for a wave travelling in an arbitrary direction it is necessary to replace x by an expression that when put equal to the constant gives an equiphase surface.

Let the equation of a plane be given by $ñ \cdot ř = a$. In view of Figure 2.9a $ñ \cdot ř = x \cos A + y \cos B + z \cos C$, where x, y, z are the components of a vector $ř$ and $\cos A$, $\cos B$, $\cos C$ are the components of unit vector $ñ$ along x, y, z axis. A, B and C are the angles shown. The equation of a plane wave travelling in the direction $ñ$ can now be written as

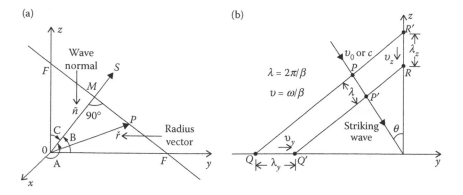

FIGURE 2.9
Direction cosines, wavelength and phase velocity. (a) Direction cosines. (b) Wavelength and velocity components.

$$E(r) = E_0 e^{-j\beta \hat{n} \cdot \vec{r}} = E_0 e^{-j\beta(x \cos A + y \cos B + z \cos C)} \qquad (2.25a)$$

In time varying form with the assumption of E_0 being complex ($E_0 = E_r + jE_i$) the above expression becomes

$$E(r,t) = \text{Re}[E_0 e^{-j(\beta \hat{n} \cdot \vec{r} - \omega t)}] = E_r \cos(\beta \hat{n} \cdot \vec{r} - \omega t) + E_i \sin(\beta \hat{n} \cdot \vec{r} - \omega t) \qquad (2.25b)$$

Since the wave propagation in general is a three-dimensional phenomenon, components of phase velocity have to be different along different space coordinates. Figure 2.9b illustrates oblique incidence of a ray on a surface along the y-axis. As can be seen, the wavelength (λ_y) and phase velocity (v_y) along this surface (or along perpendicular to the surface (λ_z and v_z)) are not the same as along striking ray (λ and v). Thus, if a striking ray makes an angle A with x-axis, B with y-axis and C with z-axis wavelengths and phase velocities can be obtained by using the following relations:

$$\lambda_u = \frac{2\pi}{[\beta \cos w]} = \frac{\lambda}{\cos w} \qquad (2.26a)$$

$$v_u = \frac{\omega}{[\beta \cos w]} = \frac{v}{\cos w} \qquad (2.26b)$$

where suffix u represents the coordinate x, y or z and angle w is the corresponding angle A, B or C. As long as angles A, B and C are not zero the wavelengths and phase velocities along the respective axes are greater than when measured along the wave normal (i.e. λ_y and $\lambda_z > \lambda$). Figure 2.9b shows the wavelengths (λ_y, λ_z) and phase velocities (v_y, v_z) along the y and z axes. These values depend on the striking angle, that is, v_y with which the crest moves along y axis becomes very large when θ (= angle C in Figure 2.9b) approaches zero.

2.13 Classification of the Cases of Reflection

Figure 2.10 illustrates the classification of the cases of reflection. These cases include reflection from conducting and dielectric interfaces. In both of these cases a striking ray may be perpendicular to the surface or may make an angle with the perpendicular drawn to the surface. The latter case of incidence is referred to as oblique incidence.

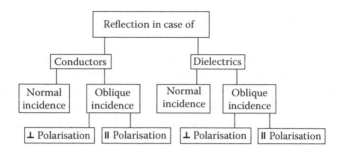

FIGURE 2.10
Classification of cases of reflection.

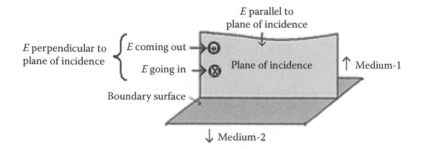

FIGURE 2.11
Illustration of *E*-field on the plane of incidence.

Figure 2.11 illustrates two orthogonal surfaces one of which is called a boundary surface, whereas the other is referred to as plane of incidence. It shows the perpendicular and parallel components of *E* vis-à-vis the plane of incidence.

As there is no restriction imposed on ray whether it belongs to a horizontally or vertically polarised wave; the *E*-vector may be perpendicular or parallel to the plane of incidence. Correspondingly *E*-vector may be parallel or perpendicular to the boundary surface.

2.14 Normal Incidence Cases

There can be two cases of normal incidence when a wave impinges on a perfect conductor and when it strikes a surface of perfect dielectric. It needs to be mentioned that in both of these cases the surfaces are assumed to be perfectly flat and smooth.

2.14.1 Perfect Conductor

As shown in Figure 2.12, the moment a wave strikes the surface of a *perfect conductor* it gets reflected back. Let E_i and E_r be the electric field intensities of incident and reflected waves at surface ($x = 0$). Since there is total reflection and the two (incident and reflected) waves travel in opposite directions the continuity of tangential E demands that $E_i = -E_r$. In view of the Poynting theorem ($P = E \times H$) the reversal of the direction of power flow requires phase reversal either of E or H vector. Thus, in case of H field there will be no phase reversal at $x = 0$ and $H_i = H_r$.

If a lossless medium is assumed and $E_i\, e^{-j\beta x}$ and $E_r\, e^{j\beta x}$ are taken to be the expressions of E for incident and reflected waves the total values of electric field $\{E_t(x, t)\}$ and magnetic field $\{H_t(x, t)\}$ at any point along x axis are obtained to be

$$E_t(x,t) = 2\,E_i \sin(\beta x)\sin(\omega t) \tag{2.27a}$$

$$H_t(x,t) = 2\,H_i \cos(\beta x)\cos(\omega t) \tag{2.27b}$$

Equations 2.27 are obtained by using Equations 2.24. As illustrated in Figure 2.13 both the Equations 2.26a and 2.26b represent standing waves for

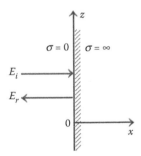

FIGURE 2.12
Normal incidence—perfect conductor.

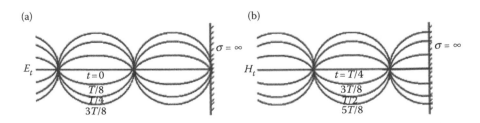

FIGURE 2.13
Standing waves for (a) E and (b) H fields.

E and H. The wave does not progress and its minima and maxima remain stagnated at the same space location of x. The magnitude of maxima of both E_t and H_t, however, keep changing at different time instants.

2.14.2 Perfect Dielectric

In this case (Figure 2.14), the relation $\eta = E/H$ and continuity of E_{tan} and H_{norm} at $x = 0$ leads to the following coefficients in terms of media parameters

Reflection coefficient for E:

$$\frac{E_r}{E_i} = \frac{\left(\sqrt{\varepsilon_1} - \sqrt{\varepsilon_2}\right)}{\left(\sqrt{\varepsilon_1} + \sqrt{\varepsilon_2}\right)} \tag{2.28a}$$

Transmission coefficient for E:

$$\frac{E_t}{E_i} = \frac{2\sqrt{\varepsilon_1}}{\left(\sqrt{\varepsilon_1} + \sqrt{\varepsilon_2}\right)} \tag{2.28b}$$

Reflection coefficient for H:

$$\frac{H_r}{H_i} = \frac{\left(\sqrt{\varepsilon_2} - \sqrt{\varepsilon_1}\right)}{\left(\sqrt{\varepsilon_1} + \sqrt{\varepsilon_2}\right)} \tag{2.28c}$$

Transmission coefficient for H:

$$\frac{H_t}{H_i} = \frac{2\sqrt{\varepsilon_2}}{\left(\sqrt{\varepsilon_1} + \sqrt{\varepsilon_2}\right)} \tag{2.28d}$$

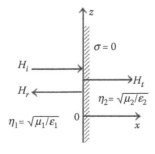

FIGURE 2.14
Normal incidence—perfect dielectric.

2.15 Oblique Incidence

In case of oblique incidence the polarisation of wave plays an important role and thus must be addressed first. In an obliquely striking ray if E is perpendicular to the plane of incidence and H is perpendicular to the boundary surface such a wave is termed as *perpendicularly (or horizontally) polarised wave*. Similarly, if E is parallel to the plane of incidence and H is parallel to the boundary surface it is called *parallelly (or vertically) polarised wave*. This terminology relates to the horizontal and vertical antennas as discussed earlier. These two cases are shown in Figure 2.15.

2.15.1 Perfect Conductor

2.15.1.1 Perpendicular (or Horizontal) Polarisation

In this case the wave has a standing wave distribution of E along z axis. The wavelength measured along the z-axis (λ_z) is greater than λ of the incident wave. The planes of zero E occur at even multiples of $\lambda_z/4$ from the reflecting surface, whereas planes of maximum E occur at odd multiples of $\lambda_z/4$ from the reflecting surface. The whole standing wave distribution of E is found to be travelling in the y direction with a velocity $v_y = v/\sin\theta$. This is the velocity with which the crest moves along y-axis. As shown in Figure 2.16 $\lambda_y = \lambda/\sin\theta$.

2.15.1.2 Parallel (or Vertical) Polarisation

In this case, H has a standing wave distribution in the z direction with plane of maximum H located at $z = 0$ and even multiple of $\lambda_z/4$ from the surface. The planes of zero H occur at odd multiple of $\lambda_z/4$ from the surface. Both

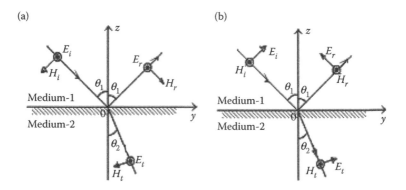

FIGURE 2.15
Illustration of two forms of polarisation. (a) Perpendicular or horizontal polarisation. (b) Parallel or vertical polarisation.

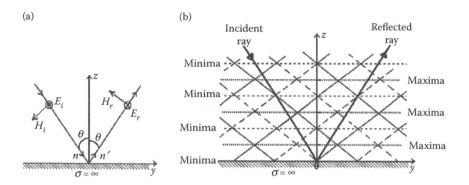

FIGURE 2.16
Perpendicular polarisation case. (a) Orientation of E and H field. (b) Resulting standing wave pattern along z-axis.

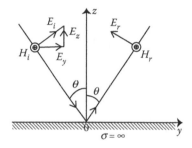

FIGURE 2.17
E in parallel polarisation.

E_z and E_y have standing wave distribution above the reflecting surface. The maxima for E_z occur at the plane and the odd multiples of $\lambda_z/4$ from the plane, whereas for E_y minima occur at the plane and even multiples of $\lambda_z/4$ from the plane. Figure 2.17 illustrates the E-field in parallel polarisation.

2.15.2 Perfect Dielectric

Figure 2.18 illustrates a boundary between two media which lies along y axis. Two incident rays are striking at points A and B. The energy contents of these rays are partly reflected and remain in medium 1. The remaining part passes through medium 2 through refraction. Since the velocity of wave depends on media parameters it is obviously different in the two media. By the time an incident ray travels from C to B the transmitted ray travels from A to D. Thus,

$$CB = AB \sin\theta_1, AD = AB \sin\theta_2, CB/AD = \sin\theta_1/\sin\theta_2 = \upsilon_1/\upsilon_2.$$

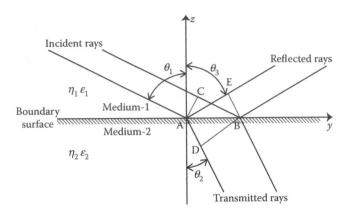

FIGURE 2.18
Incident, reflected and transmitted rays in case of oblique incident.

where

$$\upsilon_1 = \frac{1}{\sqrt{(\mu_1\varepsilon_1)}} = \frac{1}{\sqrt{(\mu_0\varepsilon_1)}} \quad \text{and} \quad \upsilon_2 = \frac{1}{\sqrt{(\mu_1\varepsilon_2)}} = \frac{1}{\sqrt{(\mu_0\varepsilon_2)}}.$$

Thus,

$$\frac{\sin\theta_1}{\sin\theta_2} = \sqrt{\left(\frac{\varepsilon_2}{\varepsilon_1}\right)} \quad \text{or} \quad \sin\theta_2 = \sqrt{\left(\frac{\varepsilon_1}{\varepsilon_2}\right)}\sin\theta_1 \tag{2.29}$$

Also, since AE = CB, it leads to the relation: $\sin\theta_1 = \sin\theta_3$ which is called 'Law of sines' or 'Snell's Law'.

According to the Poynting theorem $P = E \times H = E^2/\eta$, thus the power contents for

1. Incident wave striking AB is proportional to $E_i^2\cos\theta_1/\eta_1$
2. Reflected wave across AB is proportional to $E_r^2\cos\theta_1/\eta_1$ and
3. Transmitted wave is proportional to $E_t^2\cos\theta_2/\eta_2$.

These power contents can be equated to obtain the following relation:

$$\frac{\left(E_i^2\cos\theta_1\right)}{\eta_1} = \frac{\left(E_r^2\cos\theta_1\right)}{\eta_1} + \frac{\left(E_t^2\cos\theta_2\right)}{\eta_2} \tag{2.30a}$$

Equation 2.30a finally leads to

$$\left(\frac{E_r^2}{E_i^2}\right) = 1 - \left[\frac{\left\{\left(\sqrt{\varepsilon_2}\right)E_t^2\cos\theta_2\right\}}{\left\{\left(\pi\sqrt{\varepsilon_1}\right)E_i^2\cos\theta_1\right\}}\right] \tag{2.30b}$$

2.15.2.1 *Perpendicular (Horizontal) Polarisation*

In view of Figure 2.15a

$$E_t = E_i + E_r, \quad \text{or} \quad \left(\frac{E_r}{E_i}\right)^2 = \left(1 - \frac{E_t}{E_i}\right)^2$$

In view of Equation 2.30b

$$\frac{E_r}{E_i} = \frac{\left[\sqrt{\varepsilon_1}\cos\theta_1 - \sqrt{\varepsilon_2}\cos\theta_2\right]}{\left[\sqrt{\varepsilon_1}\cos\theta_1 + \sqrt{\varepsilon_2}\cos\theta_2\right]}$$

Also, since $\sin\theta_2 = \sqrt{(\varepsilon_1/\varepsilon_2)}\sin\theta_1$

$$\frac{E_r}{E_i} = \frac{[\cos\theta_1 - \sqrt{[(\varepsilon_2/\varepsilon_1 - \sin^2\theta_1]}}{[\cos\theta_1 + \sqrt{[(\varepsilon_2/\varepsilon_1) - \sin^2\theta_1]}} \tag{2.31}$$

2.15.2.2 *Parallel (Vertical) Polarisation*

For this case in view of Figure 2.15b

$$(E_i - E_r)\cos\theta_1 = E_t\cos\theta_2 \quad \text{or} \quad \frac{E_t}{E_i} = \{1 - (E_r - E_i)\}\left(\frac{\cos\theta_1}{\cos\theta_2}\right) \tag{2.32a}$$

In view of Equations 2.29, 2.30b and 2.32 we finally get,

$$\frac{E_r}{E_i} = \frac{\left[(\varepsilon_2/\varepsilon_1)\cos\theta_1 - \sqrt{\{(\varepsilon_2/\varepsilon_1) - \sin^2\theta_1\}}\right]}{\left[(\varepsilon_2/\varepsilon_1)\cos\theta_1 + \sqrt{\{(\varepsilon_2/\varepsilon_1) - \sin^2\theta_1\}}\right]} \tag{2.32b}$$

2.15.2.3 *Brewster's Angle*

Equation 2.32b leads to a no reflection condition provided the angle of incidence (θ_1) is

$$\theta_1 = \tan^{-1}\sqrt{\left(\frac{\varepsilon_2}{\varepsilon_1}\right)}. \tag{2.33}$$

Angle θ_1 is called Brewster's angle. No such angle can be obtained from Equation 2.31, that is, for perpendicular polarisation.

2.15.2.4 *Total Internal Reflection*

Equations 2.31 and 2.32 may be manipulated to take the form

$$\frac{E_r}{E_i} = \frac{(a - jb)}{(a + jb)} \quad \text{and} \quad \left|\frac{E_r}{E_i}\right| = 1 \qquad (2.34)$$

This condition is met only if $\varepsilon_1 > \varepsilon_2$ or $\sin^2 \theta_1 > \sqrt{(\varepsilon_2/\varepsilon_1)}$. Equation 2.34 spells that if θ_1 is large enough and medium 1 is denser than medium 2, there will be total internal reflection of wave when it strikes a dielectric obliquely. Unity magnitudes in the above equation signify that the reflection is total. The concept of total internal reflection is the basis of propagation of light waves in *optical fibres*.

2.16 Parallel Plane Guide

So far our study was confined to the waves travelling in space. These waves can be termed as unguided waves. During the course of this study many terms, namely plane waves, uniform plane waves among others were introduced. The uniform plane wave is the key player in almost all cases wherein communication is to be established through unguided waves. These unguided waves can further be classified as ground, space and sky waves depending on their media and mode of propagation. These may adopt straight line or curved paths under different conditions and can have polarisations of varied nature. In unguided waves the energy is free to adopt its own course in free space or the media it travels.

During the course of this study there was no consideration for the guidance of energy or for confining the same within some specified boundaries. In many practical applications (viz. power transmission and distribution lines, telephone lines comprising of wire pairs and coaxial cables) the waves are guided along or over the surfaces. However, in these cases energy is not confined within the structures except in case of coaxial cables. Besides, the waveguides, strip lines and microstrips are also some of the means through which the wave can be fully or partially confined within some permissible limits. Understanding of guiding principles followed by waves in any of the above physical systems is solely based on the basic Maxwell's equations. In the subsequent sections the guiding principles are explored by considering a parallel plane guide. The concepts to be developed here will later be applied to understand the working of transmission lines and waveguides.

Figure 2.19 shows two (perfectly conducting) parallel planes (separated by a distance 'a' along x-axis), wherein a wave is assumed to be propagating in the z direction. Thus, along x-axis the boundary conditions ($E_{tan} = 0$ and

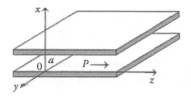

FIGURE 2.19
Configuration of two parallel planes.

$H_{norm} = 0$) are to be satisfied. As these planes are also assumed to be extended to infinity along y-axis no boundary conditions are to be met in this direction. The field in the y-direction can be assumed to be uniform and thus all the y derivatives will vanish. Along z-axis these planes may extend to any extent. Since the time and space variations are accounted by the same factors, that is, $e^{j\omega t}$ and $e^{-\gamma z} = e^{-(\alpha + j\beta)z}$ the Maxwell and wave equations can be written as

$$\nabla \times H = J + \frac{\partial D}{\partial t} = J + j\omega D = (\sigma + j\omega\varepsilon)E \qquad (2.35a)$$

$$= j\omega\varepsilon E \quad \text{(For } \sigma = 0 \text{ that is in free space)} \qquad (2.35b)$$

$$\nabla \times E = -\frac{\partial B}{\partial t} = -j\omega B = -j\omega\mu H \qquad (2.35c)$$

The wave equations for E and H are

$$\nabla^2 E = \gamma^2 E \qquad (2.36a)$$

$$\nabla^2 H = \gamma^2 H \qquad (2.36b)$$

Equations 2.35a and 2.35b can be written in differential form and the like terms can be equated. Since all y derivatives vanish the resulting equations get modified accordingly. In view of the assumed z-variation, all field quantities are to be multiplied by $e^{-\gamma z}$. These steps lead to

$$H_x = -\frac{\gamma}{h^2}\frac{\partial H_z}{\partial x} \qquad (2.37a)$$

$$H_y = -\frac{j\omega\varepsilon}{h^2}\frac{\partial E_z}{\partial x} \qquad (2.37b)$$

$$E_x = -\frac{\gamma}{h^2}\frac{\partial E_z}{\partial x} \qquad (2.38a)$$

$$E_y = -\frac{j\omega\mu}{h^2}\frac{\partial H_z}{\partial x} \qquad (2.38b)$$

In Equations 2.37 and 2.38 expressions of H_x, H_y, E_x and E_y are given in terms of E_z or H_z. Similar manipulation of Equations 2.36a and 2.36b gives

$$\frac{\partial^2 H}{\partial x^2} + h^2 H = 0 \qquad (2.39a)$$

$$\frac{\partial^2 E}{\partial x^2} + h^2 E = 0 \qquad (2.39b)$$

In Equations 2.38 and 2.39

$$h^2 = \gamma^2 + \omega^2\mu\varepsilon \qquad (2.40)$$

2.17 Transverse Waves

A wave may contain E_z, H_z or both E_z and H_z components at the same time. In case when only E_z is present, and $H_z = 0$ the H field is entirely transverse. Such a wave is called a *Transverse Magnetic (TM) or E wave*. Similarly, if H_z is present and E_z is zero, E field is entirely transverse, the wave in this case is referred to *Transverse Electric (TE) or H wave*. For treatment of these waves, solutions to Equations 2.39a and 2.39b are separately obtained in the following sections.

2.17.1 Transverse Electric Waves

When $E_z \equiv 0$ and $H_z \neq 0$, E_x and H_y are zero and only H_x and E_y components in Equations 2.37 and 2.38 survive. Since each of the field components obeys the wave equation, solution can be obtained for any of the surviving components. The wave equation for component E_y from Equation 2.39b can be written as

$$\frac{\partial^2 E_y}{\partial x^2} + h^2 E_y = 0 \qquad (2.41)$$

Solution to Equation 2.41 can be obtained by using method of separation of variable applying appropriate boundary conditions. Further in view of Equations 2.37 and 2.38 other field components can also be evaluated. The final expressions for different field components are

$$E_y = C_1\sin\left(\frac{m\pi x}{a}\right)e^{-\gamma z} \qquad (2.42a)$$

$$H_z = -\left[\frac{m\pi}{j\omega\mu a}\right] C_1 \cos\left(\frac{m\pi x}{a}\right) e^{-\gamma z} \tag{2.42b}$$

$$H_x = -\left[\frac{\gamma}{j\omega\mu}\right] C_1 \sin\left(\frac{m\pi x}{a}\right) e^{-\gamma z} \tag{2.42c}$$

where

$$h = m\pi/a \quad (m = 1, 2, 3, \ldots.) \tag{2.42d}$$

In Equation 2.42 each value of m specifies a particular field configuration or mode and the pattern so obtained is designated as $TE_{m,0}$ mode. The term mode spells the way in which fields distribute itself when an electromagnetic wave propagates.

It can be observed that all field components given by Equation 2.42 vanish on substitution of $m = 0$. Thus, the minimum value which can be assigned to m is 1. The second subscript in term $TE_{m,0}$ refers to the factor accounting the variation along y axis, which is zero in the present case. Since variation along the z-axis is governed by the factor $e^{-\gamma z}$ (which is independent of m) it will have the same (cos or sin) pattern irrespective of the components. If conductivity of the planes is infinite, the expressions of Equation 2.42 can be modified by replacing γ by $j\beta$ wherever it is involved. In view of these modified expressions, patterns for different field components can be obtained by assigning a numerical value to the separation 'a' and selecting 'm' in accordance with the mode. For $m = 1$, $\mu = \mu_0$ Equation 2.42 can be rewritten as

$$E_y = C_1 \sin\left(\frac{\pi x}{a}\right) e^{-j\beta z} \tag{2.43a}$$

$$H_x = -\left[\frac{\beta}{\omega\mu_0}\right] C_1 \sin\left(\frac{\pi x}{a}\right) e^{-j\beta z} \tag{2.43b}$$

$$H_z = -\left[\frac{\pi}{j\omega\mu_0 a}\right] C_1 \cos\left(\frac{\pi x}{a}\right) e^{-j\beta z} \tag{2.43c}$$

Equation 2.43 shows that E_y and H_x have sinusoidal variation, whereas H_z is cosinusoidal along x direction. The values of E_y and H_x are zero at $x = 0$ and $x = a$, and maximum in between 0 and a. The H_z in contrast is zero between $x = 0$ and a, and maximum at the surfaces of upper and lower conductors. The maximum magnitudes of all these components will be determined by the multiplying factors involved in their expressions. Since $H = \sqrt{(H_x^2 + H_z^2)}$ its

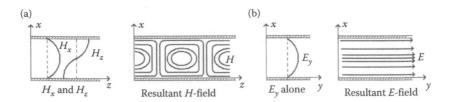

FIGURE 2.20
Field configuration for TE$_{1,0}$ mode. (a) Field distribution in x–z plane. (b) Field distribution in x–y plane.

resultant plot contains a loop-like structure which will keep repeating along the z-axis. By changing the value of m another set of plots can be similarly obtained. The field configurations for TE$_{1,0}$ mode is shown in Figure 2.20.

2.17.2 Transverse Magnetic Waves

In this case, $H_z \equiv 0$ and $E_z \neq 0$, E_y and H_x components are zero. The only surviving components in Equations 2.37 and 2.38 are E_x and H_y. In general, the solution can be obtained for any of the surviving field components. Here, the solution is obtained for H_y by adopting the same procedure as for TE wave. However, in this case the boundary conditions ($E_{tan} = 0$, $H_{norm} = 0$) cannot be directly applied to H_y. As an alternative first E_z is to be obtained in view of Equations 2.37 and 2.38 and then the boundary conditions are applied. Thus, the final expressions for surviving field components are

$$E_z = \left(\frac{m\pi}{j\omega\varepsilon a} \right) C_4 \sin\left(\frac{m\pi x}{a} \right) e^{-\gamma z} \qquad (2.44a)$$

$$H_y = C_4 \cos\left(\frac{m\pi x}{a} \right) e^{-\gamma z} \qquad (2.44b)$$

$$E_x = \left[\frac{\gamma}{j\omega\varepsilon} \right] C_4 \cos\left(\frac{m\pi x}{a} \right) e^{-\gamma z} \qquad (2.44c)$$

In Equation 2.44 'm' again specifies a particular field configuration or mode and the patterns so obtained are termed TM$_{m,0}$ modes. The minimum value, which can be assumed by m happens to be 1 in general case for TM waves. The second subscript '0' in TM$_{m,0}$ again refers to the factor accounting the variation along y axis, which is zero in the present case. Since variation along z-axis is governed by the factor $e^{-\gamma z}$ (which is independent of m) it will have the same (cos or sin) pattern irrespective of components. Equation 2.45 can again be modified by replacing γ by $j\beta$. Also a numerical value is to be assigned to 'a' and to integer m (say $m = 1$). Since $\varepsilon = \varepsilon_0$ Equation 2.44 reduces to

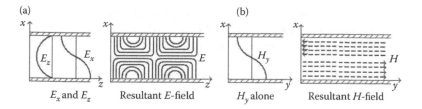

FIGURE 2.21
Field configuration for $TM_{1,0}$ mode. (a) Field distribution in x–z plane. (b) Field distribution in x–y plane.

$$E_z = \left(\frac{\pi}{j\omega\varepsilon_0 a}\right) C_4 \sin\left(\frac{\pi x}{a}\right) e^{-j\beta z} \tag{2.45a}$$

$$H_y = C_4 \cos\left(\frac{\pi x}{a}\right) e^{-j\beta z} \tag{2.45b}$$

$$E_x = \left[\frac{\beta}{\omega\varepsilon_0}\right] C_4 \cos\left(\frac{\pi x}{a}\right) e^{-j\beta z} \tag{2.45c}$$

As can be seen from Equation 2.45 E_z has sinusoidal variation, whereas H_y and E_x are cosinusoidal along x. The values of E_z are zero at $x = 0$ and $x = a$, and maximum at midway between these. H_y and E_x are zero between 0 and a, and maximum at the upper and lower conductor surfaces. As before their maximum magnitudes are determined by the multiplying factors involved in their expressions. In this case $E = \sqrt{(E_x^2 + E_z^2)}$ its resultant plot contains half-loop-like structure that keeps repeating along the z-axis. Again by assigning different values to m new sets of plots can be obtained. The patterns obtained from Equation 2.45 are illustrated in Figure 2.21.

2.18 Characteristics of TE and TM Waves

In view of Equations 2.40

$$\gamma = \sqrt{(h^2 - \omega^2 \mu\varepsilon)} = \sqrt{\left[\left(\frac{m\pi}{a}\right)^2 - \omega^2 \mu\varepsilon\right]} \tag{2.46}$$

Inspection of Equation 2.46 reveals the following three possibilities, which may result due to variation of frequency for specified values of a, μ and ε.

1. When $(m\pi/a)^2 > \omega^2\mu\varepsilon$, γ is purely real ($\gamma = \alpha$, $\beta = 0$), there is no wave propagation and it gets attenuated in accordance with the value of α. The expression for γ can be written as

$$\gamma = \sqrt{\left[\left(\frac{m\pi}{a}\right)^2 - \omega^2\mu\varepsilon\right]} = \alpha \qquad (2.47a)$$

2. When $(m\pi/a)^2 < \omega^2\mu\varepsilon$, γ is purely imaginary ($\gamma = j\beta$ and $\alpha = 0$) and thus the wave propagates without any attenuation. For this case the expression for γ becomes

$$\gamma = j\sqrt{\left[\omega^2\mu\varepsilon - \left(\frac{m\pi}{a}\right)^2\right]} = j\beta \quad \text{or} \quad \beta = \sqrt{\left[\omega^2\mu\varepsilon - \left(\frac{m\pi}{a}\right)^2\right]} \qquad (2.47b)$$

3. When $(m\pi/a)^2 = \omega^2\mu\varepsilon$, there is neither attenuation nor propagation. Thus, this frequency (ω of f) termed as critical or cutoff frequency and designated as ω_c or f_c is given as

$$\omega_c = \left(\frac{m\pi}{a}\right)\sqrt{\left(\frac{1}{\mu\varepsilon}\right)} \quad \text{or} \quad f_c = \left(\frac{m}{2a}\right)\sqrt{\left(\frac{1}{\mu\varepsilon}\right)} \qquad (2.47c)$$

4. The expression for wavelength (λ), which is the distance required for the phase to change by 2π radians, can be written as

$$\lambda = \left(\frac{2\pi}{\beta}\right) = \frac{2\pi}{\sqrt{\left[\omega^2\mu\varepsilon - (m\pi/a)^2\right]}} \qquad (2.47d)$$

5. Also the expression for velocity (v), can be written as

$$v = \lambda f = \frac{2\pi f}{\beta} = \frac{\omega}{\beta} = \frac{\omega}{\sqrt{\left[\omega^2\mu\varepsilon - (m\pi/a)^2\right]}} \qquad (2.47e)$$

Equation 2.47c reveals that for each value of m there is a corresponding cut-off frequency f_c. Further, velocity varies between $v = \infty$ at $f = f_c$ and $v = 1/\sqrt{\mu\varepsilon}$ at $f = \infty$. The velocity reduces to v_0 ($= c$) for $\mu = \mu_0$ and $\varepsilon = \varepsilon_0$ as frequency approaches infinity.

2.19 Transverse Electromagnetic Waves

In view of Equations 2.37 and 2.38 if both E_z and H_z are assumed to be absent all the field components must vanish. However, Equation 2.44 reveals that unlike TE wave in case of TM wave H_y and E_x do not vanish for $m = 0$ even if H_z becomes zero. Thus, the wave phenomena between parallel planes exist in spite of the absence of axial (E_z and H_z) components. Since the surviving field is entirely transverse such a wave is called 'Transverse Electro Magnetic (TEM)' wave. The equations for TEM wave are

$$H_y = C_4 e^{-j\beta z} \tag{2.48a}$$

$$E_x = \left[\frac{\beta}{\omega\varepsilon}\right] C_4 e^{-j\beta z} \tag{2.48b}$$

In view of Equation 2.48 E and H field patterns between two parallel planes can be obtained. Figure 2.22 illustrates the same.

Substitution of $m = 0$ into Equations 2.46 and 2.47 yields:

$$\beta = \omega\sqrt{\mu\varepsilon} \tag{2.49a}$$

$$\lambda = \frac{2\pi}{\omega}\frac{1}{\sqrt{\mu\varepsilon}} = \frac{c}{f} \tag{2.49b}$$

$$\upsilon = \frac{1}{\sqrt{\mu\varepsilon}} = c \quad \text{or} \quad \upsilon_0 \tag{2.49c}$$

$$f_c = 0 \tag{2.49d}$$

Unlike TE or TM wave the velocity of TEM wave is independent of frequency. Also, it has the lowest (zero) cut-off frequency. Thus, all the frequencies down to zero can propagate in TEM mode between parallel planes. Its velocity of propagation will remain of the order of free space velocity as long

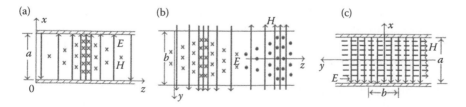

FIGURE 2.22
Field distribution for TEM wave. (a) In x–z plane. (b) In y–z plane. (c) In x–y plane.

as the space between the planes is a perfect dielectric. In case of imperfection the wave velocity will slightly reduce. All transmission lines used to carry power or signals at lower frequencies use TEM waves. Also, in view of Equation 2.48 the ratio between E and H for a travelling wave between parallel planes is obtained to be

$$\left| \frac{E_x}{E_y} \right| = \frac{\beta}{\omega \varepsilon} = \sqrt{\left(\frac{\mu_0}{\varepsilon_0} \right)} = \eta_0 \tag{2.50}$$

2.20 Wave Impedances

The concept of wave impedances emanates from complex Poynting vector given by Equation 2.23. In view of the components of E and H, P can be written as

$$P = P_x a_x + P_y a_y + P_z a_z \tag{2.51}$$

where the components of P are:

$$P_x = E_y H_z^* - E_z H_y^* \tag{2.52a}$$

$$P_y = E_x H_z^* - E_z H_x^* \tag{2.52b}$$

$$P_z = E_x H_y^* - E_y H_x^* \tag{2.52c}$$

When a wave travels it encounters the impedance offered by the environment across the cross-section which is perpendicular to the assigned direction of the wave. A wave emanating from a broadcast antenna may spread in all possible directions. It encounters various impedances in different directions. The impedances seen by waves travelling in different directions are referred to as *wave impedances*. Mathematically these impedances can be classified into the following two categories.

1. Impedances seen by the waves travelling along the positive axes or in the forward directions (across cross sections xy, yz and zx) are termed as $Z_{xy}^+ (= E_x / H_y), Z_{yz}^+ (= E_y / H_z), Z_{zx}^+ (= E_z / H_x)$ or $Z_{yx}^- (= -E_y / H_x), Z_{zy}^- (= -E_z / H_y), Z_{xz}^- (= -E_x / H_z)$.

2. Similarly, the expressions of impedances seen by the waves travelling along the negative axes or in the backward directions (across cross sections xy, yz and zx) can be written in terms of components of E and H.

In view of the above the components of Poynting vector of Equation 2.55 can be obtained as

$$P_x = \frac{1}{2}[Z_{yz}^+ H_z H_z^* - Z_{zy}^+ H_y H_y^*] = -\frac{1}{2}[Z_{yz}^- H_z H_z^* - Z_{zy}^- H_y H_y^*] \quad (2.53a)$$

$$P_y = \frac{1}{2}[Z_{zx}^+ H_x H_x^* - Z_{xz}^+ E_z H_z^*] = -\frac{1}{2}[Z_{zx}^- H_x H_x^* - Z_{xz}^- E_z H_z^*] \quad (2.53b)$$

$$P_z = \frac{1}{2}[Z_{xy}^+ H_y H_y^* - Z_{yx}^+ H_x H_x^*] = -\frac{1}{2}[Z_{xy}^- H_y H_y^* - Z_{yx}^- H_x H_x^*] \quad (2.53c)$$

In the beginning it was assumed that both TM and TE waves are travelling in the positive z-direction. Thus, in view of the above discussion the wave impedance for TE wave can be obtained by taking the ratio of field components E_y and H_x given by Equations 2.44a and 2.44b:

$$Z_{yx}^+ = -\frac{E_y}{H_x} = \frac{\omega\mu}{\beta} \quad (2.54)$$

Figure 2.23a shows the zig-zag path adopted by the TE wave as it travels down between the two parallel planes. In view of this figure $E = E_y \, a_y$ and $H = H_x \, a_x$ or $-H \cos \theta$
Thus,

$$Z_{yx}^+ = -\frac{E_y}{H_x} = \frac{E}{H \cos\theta} = \frac{\eta}{\cos\theta} = \frac{\eta}{\sqrt{(1 - \sin^2\theta)}} \quad (2.55)$$

In view of the relations:

$$\sin\theta = \frac{m\lambda}{2a} = \frac{\lambda}{\lambda_c} \quad \lambda_c = \frac{2a}{m} \quad \text{and} \quad f_c = \frac{1}{\lambda_c \sqrt{\mu\varepsilon}}$$

(a)   (b)

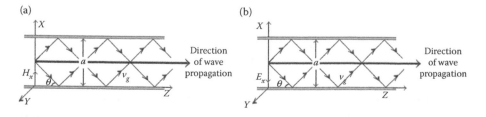

FIGURE 2.23
Zig-zag paths and field components of (a) TE and (b) TM waves.

$$Z_{yx}^+ = \frac{\eta}{\sqrt{(1-\sin^2\theta)}} = \eta \Bigg/ \sqrt{\left\{1-\left(\frac{m\lambda}{2a}\right)^2\right\}} = \eta \Bigg/ \sqrt{\left\{1\left(\frac{\lambda}{\lambda_c}\right)^2\right\}} = \eta \Bigg/ \sqrt{\left\{\left(1-\left(\frac{f_c}{f}\right)^2\right)\right\}}$$

(2.56)

In circuit theory the distributed networks can be represented by T or π networks and characteristic impedances for these networks are represented by Z_{0T} and $Z_{0\pi}$ respectively. As reproduced below the expression of TE wave (Equation 2.56) is similar to that of $Z_{0\pi}$

$$Z_{0\pi} = \frac{\sqrt{(L/C)}}{\sqrt{\{1-(f_c/f)^2\}}}$$

(2.57)

Similarly, the wave impedance for TM wave can be obtained by taking the ratio of field components H_y and E_x given by Equations 2.44b and 2.44c:

$$Z_{xy}^+ = \frac{E_x}{H_y} = \frac{\beta}{\omega\varepsilon}$$

(2.58)

Figure 2.23b shows the zig-zag path adopted by the TM wave between the two parallel planes. In view of this figure H is H_y and E is E_x or $E\cos\theta$
Thus,

$$Z_{xy}^+ = \frac{E_x}{H_y} = \frac{E\cos\theta}{H} = \eta\cos\theta = \sqrt{(1-\sin^2\theta)}$$

(2.59)

In view of the relations for, λ_c and f_c used earlier

$$Z_{xy}^+ = \eta\sqrt{\left\{1-\left(\frac{f_c}{f}\right)^2\right\}}$$

(2.60)

The expression for TM wave is similar to characteristic impedance of a T-network given as

$$Z_{0T} = \sqrt{\frac{L}{C}}\sqrt{\left\{1-\left(\frac{f_c}{f}\right)^2\right\}}$$

(2.61)

From the above it can be concluded that a TE wave sees the space as a π-network and the TM wave as T-network. In view of Equations 2.56 and 2.60 the variation of impedance with the frequency is shown in Figure 2.24. As illustrated, both TE and TM waves propagate only after the frequency

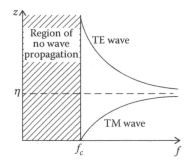

FIGURE 2.24
Variation of impedances with frequency.

attains a cutoff value f_c. Thus there exists a region of no wave propagation. In case of TE wave the impedance is infinity at f_c and gradually decreases to characteristic impedance η at $f = \infty$. For TM wave the impedance at f_c is zero and rises to η at very large (infinite) frequency.

It needs to be noted that an electromagnetic wave propagates at the speed of light provided its media of travel is air or vacuum. It however adopts a zig-zag path as shown in Figure 2.23. As a result, the time taken by the wave to arrive at the destination is more than what it would have taken along a straight line. This fact can alternatively be spelled in terms of the velocity of signal which is referred to as *group velocity* and is denoted by 'v_g'. This velocity is always less than the velocity of light in contrast to the phase velocity which is always greater than c.

2.21 Attenuation in the Walls of Parallel Plane Guide

If for the walls of parallel plane guide $\sigma \neq \infty$ the wave gets attenuated as it progresses along the walls. The attenuation (σ) can be evaluated by using the relation:

$$\alpha = \frac{\text{Power lost per unit length}}{(2 \times \text{Power transmitted})} \tag{2.62}$$

In view of Equation 2.62 the values of attenuation constant (α) for TE, TM and TEM waves can be obtained by using relevant field components. The values of attenuation constant for different modes are
For TEM wave:

$$\alpha = \frac{C_4^2 R_s b}{2 \times (1/2)\eta C_4^2 ba} = \frac{R_s}{\eta a} = \frac{1}{\eta a}\sqrt{\frac{\omega \mu_m}{2\sigma_m}} \tag{2.63}$$

For TE waves:

$$\alpha = \frac{2(m\pi)^2 R_s}{\beta \omega \mu a^3} \tag{2.64}$$

For TM waves:

$$\alpha = \frac{2\omega\varepsilon}{\beta a} R_s \tag{2.65}$$

where,

$$R_s = X_s = \sqrt{\frac{\omega\mu_m}{2\sigma_m}} \tag{2.66}$$

$$\beta = \sqrt{\omega^2\mu\varepsilon - \left(\frac{m\pi}{a}\right)^2} \tag{2.67}$$

In Equations 2.63 through 2.67 μ_m and σ_m are the values of μ and σ for metallic conductor of planes, a is the separation, b is the width of strip taken in y direction and η is the characteristic impedance. In deriving these relations it is assumed that losses have negligible effect on field distribution. Substitution of values of parameters, involved in expressions of α, leads to the curves of Figure 2.25. These curves lead to the following conclusions.

1. The attenuation for TEM wave increases in proportion to $\sqrt{\omega}$. As attenuation for TEM waves at lower frequencies is quite low these frequencies are used in power transmission.

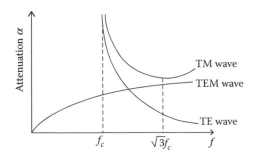

FIGURE 2.25
Variation of attenuation with frequency.

2. The losses for TE waves decrease as frequency is increased. It keeps decreasing until it approaches zero attenuation. Thus, TE wave is the best option at very high frequencies.

3. The curve for TM wave has a typical behaviour as it starts drooping from infinity, attains minima at $f = \sqrt{3}f_c$ and then rises again.

It is to be mentioned that the attenuation for TE and TM waves are functions of mode order m. Since f_c is the function of m it will keep shifting towards higher side for larger m but the shape of the curves will remain unaltered.

2.22 Transmission Lines

The theory of transmission line emanates from guided waves wherein relations for TE, TM and TEM waves were derived. Since TEM mode has the lowest cutoff frequency our study of transmission lines is confined to this mode. The propagation of TE and TM modes requires certain minimum separation between conductors (in terms of λ), whereas for TEM waves this distance may be much smaller than λ in all practical lines and for all powers for frequencies below 200 or 300 MHz. If a system of conductors guides this low-frequency TEM wave it is called 'Transmission Line'. Such a transmission line is illustrated in Figure 2.26 which is the modified form of Figure 2.19 with the difference that two narrow strips of width 'b' are shown (Figure 2.26a) in both the parallel planes running along the z-axis. In fact if these strips (Figure 2.26b) are carved out from the planes the configuration will appear as a transmission line comprising two narrow rectangular conductors

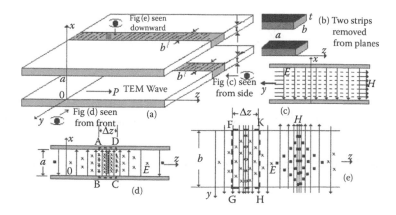

FIGURE 2.26
Configuration of transmission line and field distribution.

(or strips) of width '*b*', thickness '*t*', separated by a distance '*a*' and running all along the *z*-axis. The configuration also illustrates the field distribution between these strips (as these are) viewed from different sides.

A transmission line is in fact a material medium or structure that provides path for the energy to travel from one place to another. Its types include ladder lines, coaxial cables, dielectrics labs, striplines and waveguide or optical fibre. In general, a line is treated as transmission line if its length is greater than $\lambda/10$. The length of a transmission line may vary from millimetres or even less in processors to hundreds and thousands of kilometres in interconnection of long haul networks. Transmission lines always have at least two conductors between which a voltage can exist. In fact through the ordinary transmission line theory a distributed parameter configuration is represented by a lumped parameter configuration in order to facilitate application of circuit theory techniques. Thus, transmission line theory bridges the gulf between Maxwell's equations and the circuit theory to a great extent.

2.23 Equations Governing Transmission Line Behaviour

The transmission line equations and parameters involved therein can be obtained from the configuration shown in Figure 2.26c–e by applying e.m.f. and m.m.f. concepts.

2.23.1 Transmission Line Equations

For *lossless line* application of e.m.f. equation on loop ABCDA (Figure 2.26d) gives

$$\frac{dV}{dz} = -j\omega \left[\frac{\mu a}{b} \right] I = -j\omega L I \tag{2.68a}$$

where,

$$L = \frac{\mu a}{b} \text{ is the inductance per unit length.} \tag{2.68b}$$

For *lossy line* ($\sigma \neq \infty$) application of e.m.f. equation on the same loop gives

$$\frac{dV}{dz} = -[R + j\omega L] I = -ZI \tag{2.68c}$$

where, R is the resistance and Z ($= R + j\omega L$) is the impedance per unit length.

For *lossless line* application of m.m.f. equation on loop FGHKF (Figure 2.26e) gives

$$\frac{dI}{dz} = j\omega \left[\frac{\varepsilon b}{a}\right] V = -j\omega CV \qquad (2.69a)$$

where

$$C = \left(\frac{\varepsilon b}{a}\right) \text{ is the capacitance per unit length.} \qquad (2.69b)$$

For *lossy line* application of m m.f. equation on the same loop gives

$$\frac{dI}{dz} = -[G + j\omega C]V = -YV. \qquad (2.69c)$$

where

$$G\left(= \frac{b\sigma}{a}\right) \text{ is conductance and } Y(= G + jC) \text{ is admittance per unit length}$$

$$\qquad (2.69d)$$

Equations 2.68b and 2.69b verify the following conventional relations for velocity of propagation (v) and characteristic impedance (Z_0) in a network.

$$v = \frac{1}{\sqrt{(LC)}} = \frac{1}{\sqrt{(\mu\varepsilon)}} \qquad (2.70a)$$

$$Z_0 = \sqrt{\left(\frac{L}{C}\right)} = \sqrt{\left(\frac{\mu}{\varepsilon}\right)} \qquad (2.70b)$$

The differentiation of Equation 2.68c (w.r.t. z) and substitution of Equation 2.69c gives

$$\frac{d^2V}{dz^2} = -[R + j\omega L]\frac{dI}{dz} = [R + j\omega L][G + j\omega C]V = \gamma^2 V \qquad (2.71)$$

Similarly, differentiation of Equation 2.69c (w.r.t. z) and substitution of Equation 2.68c gives

$$\frac{d^2I}{dz^2} = -[G + j\omega C]\frac{dV}{dz} = [R + j\omega L][G + j\omega C]I = \gamma^2 I \qquad (2.72)$$

where

$$\gamma = \sqrt{(R + j\omega L)(G + j\omega C)} = \alpha + j\beta \qquad (2.73)$$

The parameter γ is a complex quantity called propagation constant. As before α and β represent attenuation and phase shift constants, respectively. Equations 2.71 and 2.72 are called transmission line equations. Their solutions can be written in exponential or hyperbolic forms. These two forms are mentioned in the next section.

2.23.2 Solution of Transmission Line Equations

2.23.2.1 Exponential Form

$$V = V'e^{\gamma z} + V''e^{-\gamma z}. \tag{2.74a}$$

$$I = I'e^{\gamma z} + I''e^{-\gamma z} \tag{2.74b}$$

Equations 2.74a and 2.74b are composed of two parts. The first part ($V'\,e^{\gamma z}$ and $I'\,e^{\gamma z}$) represents forward wave and the second part ($V''\,e^{-\gamma z}$ and $I''\,e^{-\gamma z}$) a backward wave. In these equations V' and V'' are the magnitudes of voltages of forward and backward waves and I' and I'' are the magnitudes of currents of forward and backward waves, respectively.

Equation 2.74 leads to a single parameter which describes the behaviour of a line provided it is uniform throughout its length. This parameter (denoted by Z_0) is referred to *characteristic impedance*. The transfer of maximum power to the load and minimum reflections towards the source can be ensured if the *load impedance* is equal to Z_0. In view of Equations 2.74a and 2.74b Z_0 can be written as

$$Z_0 = \frac{V'}{I'} = -\frac{V''}{I''} \tag{2.75a}$$

In view of the conventional circuit relation $Z_0 = \sqrt{(Z/Y)}$

$$Z_0 = \sqrt{\frac{R + j\omega L}{G + j\omega C}} \tag{2.75b}$$

The terminating impedance (Z_R) *at the receiving end of the line is given by*

$$Z_R = \frac{V}{I}\bigg|_{z=0} = \frac{V' + V''}{I' + I''} = Z_0 \frac{I' - I''}{I' + I''} = Z_0 \frac{V' + V''}{V' - V''} \tag{2.76}$$

The reflection coefficient (Γ) *due to mismatch or discontinuities is given as*

$$\Gamma = \frac{V'}{V'} = -\frac{I''}{I'} \tag{2.77}$$

Equations 2.76 and 2.77 can be manipulated to relate Γ with Z_0 and Z_R as

$$\Gamma = \frac{V''}{V'} = \frac{Z_R - Z_0}{Z_R + Z_0} = -\frac{I''}{I'} = \frac{Z_0 - Z_R}{Z_0 + Z_R} \qquad (2.78)$$

2.23.2.2 Hyperbolic Form

The alternative form of solutions of Equations 2.71 and 2.72 is

$$V = A_1 \cosh \gamma z + B_1 \sinh \gamma z \qquad (2.79a)$$

$$I = A_2 \cosh \gamma z + B_2 \sinh \gamma z \qquad (2.79b)$$

These equations can be manipulated to obtain general transmission line equations that relate the voltages and currents at the receiving and transmitting ends (Figure 2.27).

For lossy line

$$V_S = V_R \cosh \gamma l + I_R Z_0 \sinh \gamma l \qquad (2.80a)$$

$$I_S = I_R \cosh \gamma l + \frac{V_R}{Z_0} \sinh \gamma l \qquad (2.80b)$$

For lossless line

$$V_S = V_R \cos \beta l + j I_R Z_0 \sin \beta l \qquad (2.80c)$$

$$I_S = I_R \cos \beta l + \frac{j V_R}{Z_0} \sin \beta l \qquad (2.80d)$$

Input impedance of the transmission line can be obtained from Equation 2.80.

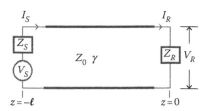

FIGURE 2.27
Circuit representation of transmission line.

$$Z_{in} = Z_s = \frac{V_s}{I_s} = \frac{V_R \cosh \gamma l + Z_0 I_R \sinh \gamma l}{I_R \cosh \gamma l + V_R \sinh \gamma l / Z_0} \tag{2.81a}$$

For $\gamma = j\beta$, Equation 2.81a can be manipulated to the form

$$Z_s = Z_0 \frac{Z_R \cos \beta l + j Z_0 \sin \beta l}{Z_0 \cos \beta l + j Z_R \sin \beta l} \tag{2.81b}$$

Equation 2.81 leads to the following cases of special interest:

a. When line is short circuited (SC) $Z_R = 0$, $V_R = 0$

$$Z_{in} = Z_{SC} = Z_0 \tanh \gamma l \tag{2.82}$$

b. When the line is open circuited (OC) $Z_R = \infty$, $I_R = 0$

$$Z_{in} = Z_{OC} = Z_0 \coth \gamma l \tag{2.83}$$

Product of Equations 2.82 and 2.83 gives

$$Z_{SC} Z_{OC} = Z_0^2 \tag{2.84}$$

2.24 Lossless RF and UHF Lines with Different Terminations

In terms of line parameters a line is considered to be lossless if $R \ll \omega L$ and $G \ll \omega C$. Equations 2.73 and 2.75b can be manipulated to obtain

$$\beta \approx \omega\sqrt{LC} \quad \text{and} \quad \alpha = 0 \tag{2.85a}$$

If $\alpha \neq 0$ but small, its approximate value from the expression of γ {Equations 2.73} is

$$\alpha = \frac{R}{2}\sqrt{\frac{C}{L}} + \frac{G}{2}\sqrt{\frac{L}{C}} \tag{2.85b}$$

In view of the above, Equations (2.80) and (2.81) get modified. These equations can be further modified on putting $\ell = x$, where x is any arbitrary distance. The resulting equations are

$$V_x = V_R \cos \beta x + j Z_0 I_R \sin \beta x \tag{2.86a}$$

$$I_x = I_R \cos \beta x + j \frac{V_R}{Z_0} \sin \beta x \tag{2.86b}$$

For $Z_R = R$ and $Z_0 = R_0$ Equation 2.86 can be written as

$$|V_x| = V_R \sqrt{\cos^2 \beta x + \left(\frac{R_0}{R} \right)^2 \sin^2 \beta x} \tag{2.87a}$$

$$|I_x| = I_R \sqrt{\cos^2 \beta x + \left(\frac{R}{R_0} \right)^2 \sin^2 \beta x} \tag{2.87b}$$

As illustrated in Figure 2.28, Equation 2.87 spells the nature of voltage and current distributions for different terminations. These cases include (a) line is SC, that is, $V_R = 0$, (b) line is OC, that is, $I_R = 0$, (c) line is terminate in characteristic impedance, that is, $Z_R = Z_0$ or $R = R_0$, (d) line is terminate in pure resistance, that is, $Z_R \neq Z_0$ or $Z_R = R \neq R_0$ and (e) line is terminated in a complex impedance $Z_R = R \pm jX$ or $Z_R \neq R_0$.

As shown in Figure 2.28, in case (a and b) there will be standing wave formation with either current or voltage minima or maxima at the terminating end. In (c) there is no standing wave formation. In (d) there will be minima or maxima of voltage or current at the termination but their magnitudes may

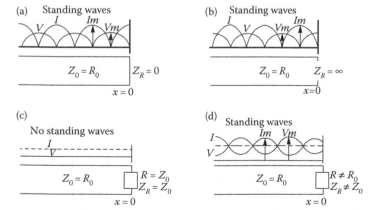

FIGURE 2.28
Voltage and current distribution for different terminations. (a) Line short circuited. (b) Line open circuited. (c) Line terminated in characteristic impedance. (d) Terminated impedance is not equal to characteristic impedance.

not be minimum or maximum. V and I will still have standing wave patterns. These patterns, however, depend on the relative values of R and R_0.

i. For $R < R_0$ it can be observed that

$$V_{max} = V_R\left(\frac{R_0}{R}\right), \quad I_{max} = I_R, \quad V_{min} = V_R \quad \text{and} \quad I_{min} = I_R\left(\frac{R}{R_0}\right)$$

Thus

$$\frac{V_{max}}{V_{min}} = \frac{I_{max}}{I_{min}} = \frac{R_0}{R} \tag{2.88a}$$

ii. For $R < R_0$,

$$V_{max} = V_R\,(R/R_0), \quad I_{max} = I_R, \quad V_{min} = V_R \quad \text{and} \quad I_{min} = I_R\,(R/R_0)$$

Thus,

$$\frac{V_{max}}{V_{min}} = \frac{I_{max}}{I_{min}} = \frac{R}{R_0} \tag{2.88b}$$

The ratio V_{max}/V_{min} is called the *voltage standing wave ratio* (VSWR) and I_{max}/I_{min} *current standing wave ratio* (CSWR). These ratios denoted by ρ or s are always greater than 1.

$$\rho = \frac{R}{R_0}\ (\text{for } R > R_0) \quad \text{and} \quad \rho = \frac{R_0}{R}\ (\text{for } R < R_0) \tag{2.89}$$

The standing wave ratio is related to the reflection coefficient as

$$\rho = \frac{[1+|\Gamma|]}{[1-|\Gamma|]} \tag{2.90}$$

In (e) there will be neither minima nor maxima of voltage or current at termination. The minima or maxima will always be displaced to either side of the terminating point. The direction and location of these displacements along with the standing wave measurements leads to the estimation of the values of reactance and their types.

Figure 2.29a illustrates a transmission line (with characteristic impedance $Z_0 = R_0$) terminated in an unknown impedance $R \pm jX$ at location c. Manipulation of the relevant mathematical relations yield expressions from which components (R and X) of unknown complex impedance can be evaluated.

$$R = \frac{\rho R_0}{\rho^2 \cos^2 \beta L_2 + \sin^2 \beta L_2} \tag{2.91a}$$

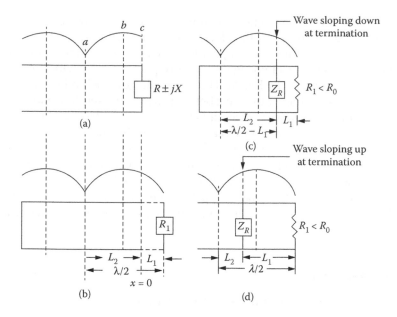

FIGURE 2.29
Voltage and current distributions for termination in complex impedance $Z = R \pm X$. (a) Z is located at $x = 0$, (b) Z is replaced by R at a distance L_1 from termination, (c) X is inductive and (d) X is capacitive.

$$X = \frac{-R_0(\rho^2 - 1)\sin \beta L_2 \cos \beta L_2}{\rho^2 \cos^2 \beta L_2 + \sin^2 \beta L_2} \tag{2.91b}$$

In view of Equation 2.91b for $R_1 < R_0$, when $L_2 < \lambda/4$, wave is sloping up, X is positive or inductive and when $\lambda/4 < L_2 < \lambda/2$, wave is sloping down, X is negative or capacitive.

2.25 Reflection Phenomena

The term reflection has frequently been used in earlier sections particularly in connection with OC and SC lines. In fact, reflection occurs even if there is slight mismatch on account of termination or imperfections in the transmission path itself. An equivalent circuit can be drawn in terms of its R, L, C and G for a line with loss and in terms of L and C for a lossless line. The burden of carrying energy contents of a propagating wave is shared by electric and magnetic fields alike. The portions of energy shared by each are accounted by the well-known relations $CV^2/2$ and $LI^2/2$.

When a wave propagates in a line and arrives at OC the end current (I) becomes zero and part of the energy shared by the magnetic field ($LI^2/2$)

becomes zero. But since the energy can neither be created nor destroyed, this burden is to be taken up by the electric field. In the expression $CV^2/2$, line parameter C ($= \varepsilon A/d$) cannot change unless area (A), separation (d) or ε in line gets altered. The change of voltage is the only possibility by which additional energy can be carried out by E-field. Thus, the voltage at OC end rises. The voltage at OC end is now higher than just before towards the sending end. The current starts flowing back, which amounts to travelling of the current wave. The voltage wave joins the race as it was earlier, but this time both (current and voltage) waves are the reflected ones and they will start sharing their own burdens as before. The current in fact remains zero at the OC point only momentarily. Similarly if the line is SC, the voltage (and hence E-field) becomes zero and part of energy shared by it is to be momentarily taken up by the magnetic field. This time the current rises at receiving end resulting in initiation of voltage wave.

2.26 Resonance Phenomena in Line Sections

Consider a quarter-wave section shown in Figure 2.30, having one of its end OC and the other as SC. Assume that the power is fed to this section at point 'a' (located near to the SC end). The wave (say of voltage) will travel down to end 'b' and will get reflected by OC without phase reversal. This reflected wave will travel back and after arriving at 'c' it will again get reflected by SC end, but this time with phase reversal. At this point the original signal fed at this corresponding instant and the reflected wave (with reversed phase) will be in the same phase and will get added increasing the net magnitude of the signal. The wave (with enhanced magnitude) will travel down to point 'b' and will get reflected from OC without phase reversal. It will again travel to point 'c', finds SC and gets reflected with phase reversal. Here, further addition will take place since the enhanced reflected wave and the original input have the same phase. This addition will further enhance the

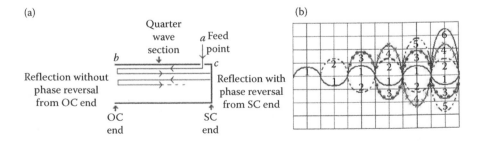

FIGURE 2.30
Resonance phenomenon. (a) Quarter-wave section and (b) voltage buildup.

magnitude. The process of reflection without and with phase reversal will continue and voltage buildup will also continue until the power supplied to the section met the I^2R loss requirement of the line. The process of voltage buildup (Figure 2.30b) is explained below.

1. The wave marked 1, 1, 1, .. is the original wave fed at point c which is almost adjacent to the SC end. It will remain so in case there is neither reflection nor phase reversal. Points a and b correspond to SC and OC locations in Figure 2.30a.

2. With first phase reversal, the two waves add resulting in a wave with higher magnitude of voltage. This wave is shown by 2, 2, 2. It will remain so if there are no further reflections and phase reversals.

3. With second phase reversal, wave magnitude further rises and is shown by 3, 3, 3. It will remain in its existing condition without further phase reversals and additions there from.

4. The process of voltage buildup may continue indefinitely if there is no restriction of power requirement.

2.27 Quality Factor of a Resonant Section

Equation 2.82 for input impedance of a shortened line can be written in an expanded form for $\alpha = 0$.

$$Z_S = Z_0 \tanh \gamma l = Z_0 \frac{\sinh \alpha l \cos \beta l + j \cosh \alpha l \sin \beta l}{\cosh \alpha l \cos \beta l + j \sinh \alpha l \sin \beta l} \tag{2.82}$$

Values of different terms involved in Equation 2.82 can be evaluated for $f = f_0$ and $f = f_0 + \delta f$. Substitution of $\delta f = \Delta f / 2$ and further manipulation and simplification of resulting equations leads to expression of quality factor.

$$Q = \frac{f_0}{\Delta f} = 2\pi f_0 \frac{L}{R} = \frac{\omega_0 L}{R} \text{ or simply } \frac{wL}{R} \tag{2.92}$$

2.28 UHF Lines as Circuit Elements

Above 150 MHz ($\lambda < 2$ m), the construction of ordinary lumped circuit elements becomes difficult. At the same time the required physical size of sections (in relation to wavelength) of transmission lines becomes sufficiently small. At this stage these line sections become good contenders to replace the lumped circuit elements in all practical applications. These line sections can conveniently be used up to frequencies of about 3000 MHz after which the size of sections become too small to be handled and waveguide technique begin to take over.

For lossless lines this takeover can be understood from Equation 2.81b by observing its different forms under different conditions. Equation 2.81b is reproduced below.

$$Z_s = Z_0 \frac{Z_R \cos \beta l + jZ_0 \sin \beta l}{Z_0 \cos \beta l + jZ_R \sin \beta l} \tag{2.81b}$$

When lossless line is SC or $Z_R = 0$, Equation 2.81b gives $Z_s = jZ_0 \tan\beta l$.

a. For $0 < l < \lambda/4$, $0 < \beta l < \pi/2$, Z_S is positive or inductive. Its equivalent circuit is shown as ① in Figure 2.31.

b. For $\lambda/4 < l < \lambda/2$, $\pi/2\,\pi < \beta l < \pi$, Z_S is negative or capacitive. Its equivalent circuit is shown as ② in Figure 2.31.

When lossless line is OC or $Z_R = \infty$ *Equation 2.81b gives* $Z_s = jZ_0 \cot\beta l$.

a. For $0 < l < \lambda/4$, $0 < \beta l < \pi/2$, Z_S is negative or capacitive. Its equivalent circuit is shown as; in ③ Figure 2.31.

b. For $\lambda/4 < l < \lambda/2$, $\pi/2 < \beta l < \pi$, Z_S is positive or inductive. Its equivalent circuit is shown as; in ④ Figure 2.31.

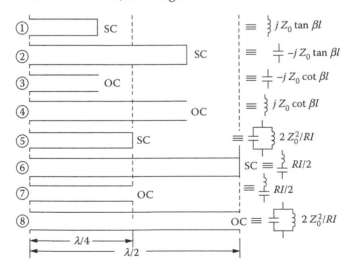

FIGURE 2.31
Line sections with equivalent elements for different lengths and terminations.

For lossy line different forms of Equation 2.81b emerging under different conditions are to be observed. The modified form of Equation 2.81b is as given below:

$$Z_s = Z_0 \frac{Z_R \cosh \gamma l + Z_0 \sinh \gamma l}{Z_0 \cosh \gamma l + Z_R \sinh \gamma l} \tag{2.81a}$$

When lossy line is SC (i.e. $Z_R = 0$) Equation 2.81b can be manipulated by letting $\gamma = \alpha + j\beta$ to yield:

a. For $l = \lambda/4$, $\beta l = \pi/2$, $\cos\beta l = 0$, $\sin\beta l = 1$, $\cosh\alpha l = 1$ and $\sinh\alpha l \approx \alpha l$ for small α where $\alpha \approx 1/2(R/Z_0 + GZ_0) \approx R/2Z_0$, $Z_S = 2Z_0^2/Rl$
This expression is also valid for SC line whose length is an odd multiple of $\lambda/4$.

b. For $l = \lambda/2$, $\beta l = \pi$, $\cos\beta l = -1$, $\sin\beta l = 0$ with α as above, $Z_S = Rl/2$
When lossy line is OC (i.e. $Z_R = \infty$) Equation 2.82a yields

c. For $l = \lambda/4$, $\beta l = \pi/2$, $\cos\beta l = 0$, $\sin\beta l = 1$, $\cosh\alpha l = 1$ and $\sinh\alpha l \approx \alpha l$ for small α where $\alpha \approx R/2Z_0$ as before, $Z_s = Rl/2$

d. Similar manipulation for $l = \lambda/2$ yields $Z_s = 2Z_0^2/Rl$

Cases (a), (b), (c) and (d) are shown by ⑤, ⑥, ⑦ and ⑧, respectively, of Figure 2.31. These equivalent elements can also be represented in terms of electrical length (βl) of the line as shown in Figure 2.32.

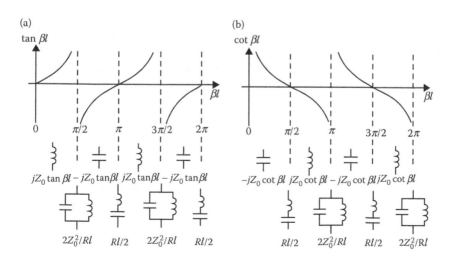

FIGURE 2.32
Circuit elements for different electrical lengths and terminations. (a) Short-circuited lines. (b) Open-circuited lines.

2.29 Applications of Transmission Lines

A transmission line can be used for high-frequency signal transfer over long or short distances with minimum power loss such as through down-lead from a TV or radio aerial to the receiver. It can also be used as circuit element, pulse generator, stub filter, stub for matching, harmonic suppressor, tuned circuit and as impedance and voltage transformer. Some of these applications are briefly discussed below.

2.29.1 Quarter-Wave Section as Tuned Line

In Figure 2.33 the line is divided into two parts. In left of the tapping, the line is SC and in right side it is OC. In view of Equation 2.81b the two impedances Z_{SC} and Z_{OC} as they appear from the location of tapping are

$$Z_{SC} = jZ_0 \tan \beta x = X_L \tag{2.93a}$$

$$Z_{OC} = -jZ_0 \tan \beta x = X_C \tag{2.93b}$$

At $x = 0$, $X_L = X_C = 0$ (or very low), and at $x = \lambda/4$, $X_L = X_C = \infty$ (or very high), thus Z_S can be given as

$$Z_S = \frac{2Z_0^2}{Rl} = R_s \tag{2.93c}$$

For given magnitude of voltage and current in quarter-wave section a certain fixed amount of input will be required to supply I^2R losses regardless of where the power is being fed. This input power $= V_S^2/R_S = V_S^2 Rl/2Z_0^2$, where V_S and R_S are voltage and input resistance, respectively, at the open end of the section. When power is fed at a distance x from the SC end and $V_X = V_S \sin \beta x$ is the voltage at this point

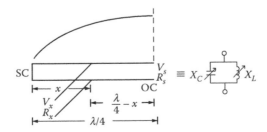

FIGURE 2.33
Transmission line as a tuned circuit.

$$\text{Power input} = \frac{V_X^2}{R_X} = V_S^2 \sin^2 \frac{\beta x}{R_X} = \frac{V_S^2}{R_S} = \frac{V_S^2 Rl}{2Z_0^2} \tag{2.94a}$$

Thus,

$$R_X = \left(\frac{2Z_0^2}{Rl}\right) \sin^2 \beta x \tag{2.94b}$$

Thus input resistance varies as square of sine of angular distance from the SC end.

2.29.2 Quarter-Wave Section as Impedance Transformer

Equation 2.81b for the particular case of a quarter-wave section, that is, for $\ell = \lambda/4$, thus $\beta\ell = \pi/2$, $\cos \beta\ell = 0$, and $\sin \beta\ell = 1$, yields the relation $Z_S Z_R = Z_0^2$, which spells matching or impedance inversion, that is, if $Z_R = R + jX_L$, $Z_S = R - jX_C$. If Z_S and Z_R are taken to be the normalised values ($z_s = Z_S/Z_0$ and $z_R = Z_R/Z_0$) the equation $Z_S Z_R = Z_0^2$ transforms to $Z_S Z_R = 1$ and $Z_S = 1/Z_R$ (i.e. input impedance is simply the reciprocal of output or terminating impedance). The meaning of matching is spelled by Figure 2.34. When the two lines with characteristic impedances Z_{01} and Z_{02}, are connected in series, they must conform to the equation $Z_S Z_R = Z_0^2$.

2.29.3 Quarter-Wave Section as Voltage Transformer

a. *In case of lossless line, Equations 2.80c and 2.80d for $l = \lambda/4$, $\beta l = \pi/2$*

$$V_S = I_R Z_0 \quad \text{and} \quad I_S = \frac{V_R}{Z_0}$$

$$\frac{V_S}{V_R} = \frac{I_R Z_0}{V_R} = \frac{Z_0}{Z_R} = \sqrt{\frac{Z_S}{Z_R}} \quad \text{or} \quad \left|\frac{V_S}{V_R}\right| = \sqrt{\frac{Z_S}{Z_R}} \tag{2.95}$$

For $Z_R = \infty$, $V_R = \infty$ and for $Z_R = 0$, $V_R = 0$

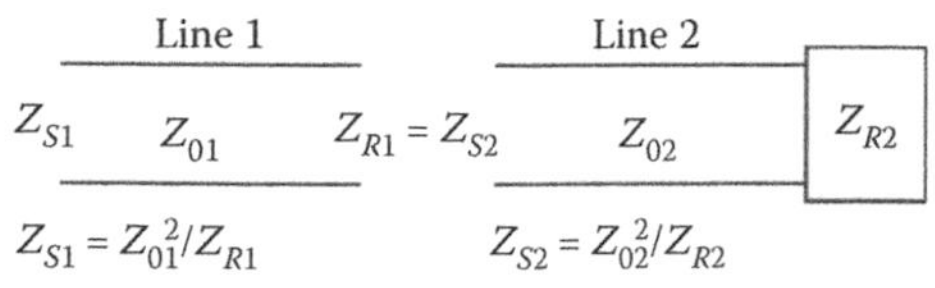

Line 1   Line 2

Z_{S1} Z_{01} $Z_{R1} = Z_{S2}$ Z_{02} Z_{R2}

$Z_{S1} = Z_{01}^2/Z_{R1}$ $Z_{S2} = Z_{02}^2/Z_{R2}$

FIGURE 2.34
Impedance transformation.

b. In case of *lossy line*, Equations 2.80a and 2.80b for $l = \lambda/4$, $\beta l = \pi/2$, $\cos \beta l = 0$, $\sin \beta l = 1$, $\cosh \alpha l \approx 1$ and $\sinh \alpha l \approx \alpha l$

$$V_S = jV_R \sinh \alpha l + jI_R Z_0 \cosh \alpha l$$

For the OC line $I_R = 0$ or $V_S = jV_R \sinh \alpha l$

$$\text{Thus,} \quad \left| \frac{V_R}{V_S} \right| = \frac{1}{\sinh \alpha l} \approx \frac{1}{\alpha l} = \frac{2Z_0}{Rl} \tag{2.96a}$$

$$\text{For } l = \lambda/4, \quad \left| \frac{V_R}{V_S} \right| = \frac{8Z_0}{R\lambda} = \frac{8Z_0 f}{R\lambda} \tag{2.96b}$$

$$\text{For } l = 3\lambda/4, \quad \left| \frac{V_R}{V_S} \right| = \frac{8Z_0}{3R\lambda} = \frac{8Z_0 f}{3R\lambda} \tag{2.96c}$$

2.29.4 Line Sections as Harmonic Suppressors

Transmission line sections can also be used for the suppression of harmonics. The elimination of third and all even harmonics is explained as below.

2.29.4.1 Suppression of Third Harmonics

This can be explained in view of Figure 2.35a and Equation 2.81b.

$$Z_S = Z_0 \frac{Z_R \cos \beta l + jZ_0 \sin \beta l}{Z_0 \cos \beta l + jZ_R \sin \beta l} = Z_0 \frac{Z_R + jZ_0 \tan \beta l}{Z_0 + jZ_R \tan \beta l} \tag{2.81b}$$

Let f_0 is the frequency to be transmitted by the line. At f_0, $l_1 = \lambda/12$, $l_2 = \lambda/6$, for OC put $Z_R = \infty$ and for SC $Z_R = 0$. Also $\beta l_1 = (2\pi/\lambda) \lambda/12 = 30°$ or $\cot \beta l_1 = \sqrt{3}$ and $\beta l_2 = (2\pi/\lambda) \lambda/6 = 60°$ or $\tan \beta l_2 = \sqrt{3}$. Thus,

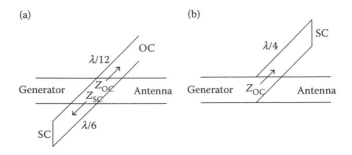

FIGURE 2.35
Suppression of (a) third harmonic and (b) even harmonics.

$$Z_{OC} = Z_0 \frac{1 + j(Z_0/Z_R)\tan \beta l_1}{(Z_0/Z_R) + j\tan \beta l_1} = -jZ_0 \cot \beta l_1 = -j1.7321Z_0 \text{ (capacitive)} \quad (2.97a)$$

$$Z_{SC} = jZ_0 \tan \beta l_2 = j1.7321Z_0 \text{ (inductive)} \quad (2.97b)$$

These two reactances are in the formation of a parallel resonance circuit, whose input impedance is infinite (or very high if resistance is present). It will appear to the wave as if nothing is connected across the transmission line.

At $f = 3 f_0$ the length of the line section $l_1 = \lambda/12$ becomes $\lambda/4$ and $l_2 = \lambda/6$ becomes $\lambda/2$.

The $\lambda/4$ section is OC, which is equivalent to series resonant circuit or SC at the point where it joins the transmission line. The $\lambda/2$ section is SC which will appear as OC at $\lambda/4$ and SC at $\lambda/2$ and thus is equivalent to a series resonant circuit as before at the connecting point. Thus, at the point of joining it is doubly SC and third harmonics are effectively suppressed.

2.29.4.2 Suppression of Even Harmonics

In case of even harmonics, at fundamental frequency $f = f_0$, $Z_S = \infty$ as if the stub is not connected. At frequencies $f = 2f_0$, $4f_0$, $6f_0$ ..., nf_0 (where n is an even number), lengths of the stub shown in Figure 2.35b will become $\lambda/2$, $2\lambda/2$, $3\lambda/2$ .... For all these lengths the line will appear as SC (i.e. $Z_S = 0$). Thus, all even harmonics will be effectively suppressed.

2.29.5 Line Sections for Stub Matching

2.29.5.1 Single Stub Matching

For single stub matching the related phasor diagram is shown in Figure 2.36. According to this diagram the load admittance Y_R is transformed by length d into an admittance $Y_A = G_A \pm jB_A$. Also, the Y_{SC} is transformed by the stub length s to Y_S. If $Y_S = +jB_A$ and $G_A = Y_C$, the sum of Y_A and Y_S is simply the

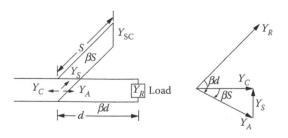

FIGURE 2.36
Single-stub matching.

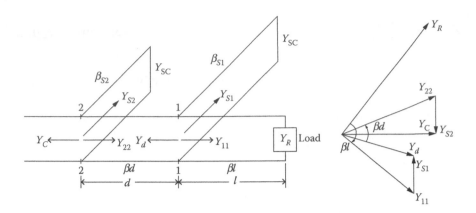

FIGURE 2.37
Double stub matching.

characteristic admittance Y_C and the line will be flat towards generator and there will be no reflection from the location of stub.

2.29.5.2 Double Stub Matching

The double stub matching phasor diagram is shown in Figure 2.37. According to this diagram the load admittance Y_R is transformed by length ℓ into an admittance $Y_{11} = G_{11} \pm jB_{11}$. Also, the Y_{SC} of stub 1 is translated by length s_1 to a pure susceptance $Y_{S1} = jb_1$ and Y_{SC} of stub 2 to $Y_{S2} = jb_2$. At location 1 Y_{S1} gets added to Y_{11} and partially neutralises the susceptance of Y_{11}, hence the net admittance becomes Y_d. This Y_d again gets transformed by line length d to a new value $Y_{22} = G_{22} \pm jB_{22}$. If addition of Y_{S2} to Y_{22} neutralises its susceptance part $\pm jB_{22}$ the matching is accomplished provided $G_{22} = Y_C$. Thus, $\pm jB_{22}$ and jb_2 must add to result in zero value of susceptance.

In general, the stub matching requires two variables. In single stub case the length of stub and distance d are the two variables, however, variation of d is not a convenient choice. In double-stub case two required variables are the stub lengths s_1 and s_2. Change in stub lengths is relatively more convenient with the help of shorting plungers. Thus, double-stub matching is normally preferred.

2.30 Types of Transmission Lines

The types of transmission lines include the following: (i) Single-wire lines (unbalanced lines, now not used); (ii) Balanced lines (consist of two conductors of the same type and equal impedance to ground and other circuits. Its formats include twisted pair, star quad and twin-lead); (iii) Lecher lines (form of parallel conductor used for creating resonant circuits at UHF range. These fill the gap between lumped components used at HF/VHF and resonant cavities used at UHF/SHF); (iv) Stepped lines (consist of multiple transmission line segments connected in series, used for broad impedance matching; (v) Coaxial cable (discussed in the next section); (vi) Waveguides (discussed in Chapter 3); (vii) Planar lines (discussed in Chapter 15); (viii) Optical fibres (solid transparent structure of diminishing dimensions, made of glass or polymer, used to transmit signal at optical, or near infrared, wavelengths).

2.31 Coaxial Cables

The field in coaxial lines is confined between inner and outer conductors. These can therefore be safely bent and twisted, within some limits, without any adverse effect. These can also be strapped to the conductive supports without inducing unwanted currents in them. Upto a few GHz these carry only the TEM waves and can carry TE and TM modes if the frequency is further raised. In case of more than one mode, bends and irregularities in the cable geometry can result in intermodal power transfer. Its most common applications include television and other signals with multiple megahertz bandwidth. The Z_0 of a coaxial cable may be 50 or 75 Ω. Coaxial cables are also used to make some microwave components.

Figure 2.38a and b shows the geometry and cross section of a coaxial cable. Its inner conductor has a radius 'a' and the outer conductor with radius 'b'. The gap between these conductors may be filled with air or some other dielectrics. Figure 2.38c illustrates the distribution of electric and magnetic fields for TEM mode which is the principal mode of propagation in coaxial cables. If V is the voltage between conductors and I is the current in each conductor E and H fields and other important parameters (for $a \leq \rho \leq b$) can be given as below.

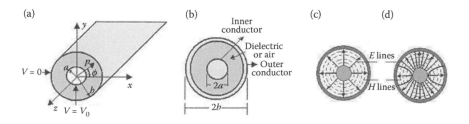

FIGURE 2.38
Coaxial cable. (a) Geometry, (b) cross section, (c) TEM mode and (d) TE_{11} mode.

Electric field intensity:

$$E = \frac{V}{\rho \ln(b/a)}$$ (2.98a)

Magnetic field intensity:

$$H = \frac{I}{2\pi\rho}$$ (2.98b)

Velocity of propagation for voltage and current waves:

$$v = \frac{v_0}{\sqrt{\mu_r \varepsilon_r}}$$ (2.99a)

Wavelength in coaxial cable:

$$\lambda = \frac{\lambda_0}{\sqrt{\mu_r \varepsilon_r}}$$ (2.99b)

High frequency inductance per unit length:

$$L = \frac{\mu_0 \mu_r}{2\pi} \ln(b/a)$$ (2.99c)

High frequency capacitance per unit length:

$$C = \frac{2\pi\varepsilon_0\varepsilon_r}{\ln(b/a)}$$ (2.99d)

Characteristic impedance:

$$Z_0 = \sqrt{\frac{L}{C}} = \frac{1}{2\pi}\sqrt{\frac{\mu}{\varepsilon}} \ln\frac{b}{a}$$ (2.99e)

Maximum allowable rms voltage:

$$V_{max} = \frac{aE_d}{\sqrt{2}} \frac{\ln(b/a)}{(b/a)}$$ (2.99f)

where E_d is the dielectric strength of the insulating material filled between the inner and outer conductors. All other parameters involved in the above equations were defined earlier.

The design of most of the coaxial components (viz. detectors, attenuators, filters, etc.) is based on TEM mode. When the frequency of operation exceeds a certain limit (referred to as cutoff frequency f_c) higher-order modes can also propagate. One such mode is designated as TE_{11} mode which has the lowest cutoff frequency. Its field pattern is shown in Figure 2.38d.

2.32 Limitations of Different Guiding Structures

Two-wire transmission lines used in conventional circuits become inefficient for transferring electromagnetic energy at microwave range. As illustrated in Figure 2.39a the field is not entirely confined between two conductors and therefore some of the energy escapes by radiation. In view of Figure 2.39b, the field remains confined between the conductors of a coaxial line. Thus, coaxial cable is relatively more efficient than two-wire line.

Waveguides are indisputably the key players at microwave range. If not all, the majority of the components required for different purposes in a microwave systems are carved out of waveguides. The mathematical theory of waveguides is discussed in Chapter 3 and the components referred to above are included in Chapter 7.

Waveguides are the most efficient way to transfer electromagnetic energy at higher frequencies. A circular waveguide is essentially a coaxial line without a central conductor. Figure 2.40 illustrates the cross section of a coaxial cable and that of a circular waveguide, which is obtained after removal of central conductor of the coaxial cable. In view of this illustration the relative spacing available, for the distribution of field, in case of waveguide is much more than that available in coaxial cable.

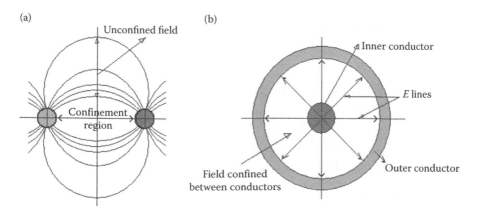

FIGURE 2.39
End views and field distributions. (a) Two-wire Tx line. (b) Coaxial cable.

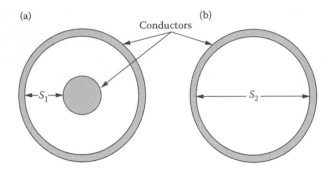

FIGURE 2.40
Relative spacing in coaxial cable and circular waveguide. (a) Coaxial cable. (b) Circular waveguide.

At microwave range the conventional circuit components lose their significance and almost all processing is done through components carved out of waveguides made of conducting materials. These waveguides may be of rectangular, circular, or elliptical shapes. Rectangular waveguides, however, have the lions share. Circular waveguides are made of conducting materials and dielectric waveguides are normally used for special applications. The optical fibres, which operate in tera-hertz (THz) range, are in essence circular dielectric waveguides.

EXAMPLE 2.1

A wave propagates in a non-magnetic media having relative dielectric constant ε_r. Find its value if (a) $\eta = 180\ \Omega$, (b) the wavelength at 10 GHz is 2 cm and (c) $\beta = 0.001$, $f = 25{,}000$ Hz.

Solution

The required relations are: $v = 1/\sqrt{(\mu\varepsilon)}$, $\eta = \sqrt{(\mu/\varepsilon)}$, $\beta = 2\pi/\lambda$ and $\lambda = v/f = v_0/[f\sqrt{(\mu_r\,\varepsilon_r)}]$.

For non-magnetic media $\mu_r = 1$ or $\mu = \mu_0 = 4\pi \times 10^{-7}$

a. $\eta = 180\ \Omega = \sqrt{(\mu/\varepsilon)} = \sqrt{(\mu_0/\varepsilon_0\varepsilon_r)} = \eta_0/\sqrt{\varepsilon_r}$
   or $\varepsilon_r = (\eta_0/\eta)^2 = (377/180)^2 = 4.38669$

b. $\lambda = 2 \times 10^{-2} = v/f = v/10^{10}$
   $v = 2 \times 10^8 = 1/\sqrt{(\mu\varepsilon)} = 1/\sqrt{(\mu_0\varepsilon_0\varepsilon_r)} = v_0/\sqrt{\varepsilon_r} = 3 \times 10^8/\sqrt{\varepsilon_r}$
   $\varepsilon_r = (3/2)^2 = 2.25$

c. $\beta = 0.001 = 10^{-3} = 2\pi/\lambda = 2\pi f/v = (2\pi \times 25{,}000)/(v_0/\sqrt{\varepsilon_r})$
   $= (5\pi \times 10^4)\sqrt{\varepsilon_r}/(3 \times 10^8) = (5\pi/3) \times 10^{-4}\sqrt{\varepsilon_r}$
   $\sqrt{\varepsilon_r} = 10^{-3}/(5\pi/3)10^{-4} = (30/5\pi) = 6/\pi$
   $\varepsilon_r = (6/\pi)^2 = 3.64756$

EXAMPLE 2.2

A wave propagates at 100 MHz in a dielectric media having some component of conductivity (σ) of the order of $10^{-5}\ \mho/m$ and $\mu = \mu_0$. Find the values of α, β, v and η.

Solution

In view of following of relations with $\mu = \mu_0 = 4\pi \times 10^{-7}$ and $\varepsilon = \varepsilon_0 = 10^{-9}/36\pi$

$$\eta = \sqrt{(\mu_0/\varepsilon_0)} = \eta_0 = 377\ \Omega \text{ and } v_0 = 1/\sqrt{(\mu_0\ \varepsilon_0)} = 1/(4\pi \times 10^{-7} \times 10^{-9}/36\pi)$$

$$\approx 3 \times 10^8 \text{m/s}$$

$$\alpha = (\sigma/2)\sqrt{(\mu/\varepsilon)} = \sigma\eta_0/2 = 10^{-5} \times 377/2 = 188.5 \times 10^{-5} \text{ Nep./m}$$

$$\beta = \omega \sqrt{[(\mu_0\varepsilon_0)} (1 + \sigma^2/8\omega^2\ \varepsilon^2)] = (2\pi \times 10^8/3 \times 10^8) \sqrt{[1 + 10^{-10}/\{8 \times (2\pi \times 10^8)^2} \times (10^{-9}/36\pi)^2\}]$$

$$= (2\pi/3) \sqrt{[1 + 10^{-10}/\{8 \times (4\pi^2 \times 10^{16})(10^{-18}/36 \times 36 \times \pi^2)]}$$

$$= (2\pi/3) \sqrt{[1 + 10^{-10}/\{8 \times (10^{16})(10^{-18}/9 \times 36)]}$$

$$= (2\pi/3) \sqrt{[1 + 0.02469 \times 10^{-8}]} \approx 2.0944 \text{ rad./m}$$

$$v \approx v_0 (1 - \sigma^2/8\omega^2\ \varepsilon^2) = 3 \times 10^8 [1 - 0.405 \times 10^{-6}] \text{ m/s}$$

$$\eta \approx \eta_0 (1 + j\sigma/2\omega\varepsilon) = 377 [1 + j10^{-5}/(2 \times 2\pi \times 10^8 \times 10^{-9}/36\pi)]$$

$$= 377 [1 + j10^{-6}/9)] = 377 [1 + j9 \times 10^{-4}]\ \Omega$$

EXAMPLE 2.3

Find the depth of penetration at 1000 Hz in (a) silver ($\sigma = 6.17 \times 10^7\ \mho/m$), (b) aluminium ($\sigma = 3.72 \times 10^7\ \mho/m$), (c) brass ($1.5 \times 10^7\ \mho/m$) and (d) fresh water ($\sigma = 10^{-3}\ \mho/m$).

Solution

In view of Equation 2.50 $\delta = \sqrt{[1/\pi f\mu\ \sigma]} = \sqrt{[1/\pi f\mu\}}\sqrt{[1/\sigma]}$
Given $f = 1000$ Hz, $\mu = \mu_0 = 4\pi \times 10^{-7}$
Thus, $\pi f\mu = 1000\ \pi \times 4\pi \times 10^{-7} = 4\pi^2 \times 10^{-4} = 39.478 \times 10^{-4}$

$$1/\pi f\mu = 253.3056 \quad \sqrt{[1/\pi f\mu\}} = 15.9155 \quad \text{and} \quad \delta = 15.9155/\sqrt{\sigma}$$

a. Silver: $\sqrt{\sigma} = \sqrt{(6.17 \times 10^7)} = 7855 \quad \delta = 15.9155/7855 = 0.00202$ m
b. Aluminium: $\sqrt{\sigma} = \sqrt{(3.72 \times 10^7)} = 6099 \quad \delta = 15.9155/6099 = 0.00261$ m
c. Brass: $\sqrt{\sigma} = \sqrt{(1.5 \times 10^7)} = 3872 \quad \delta = 15.9155/3872 = 0.00411$ m
d. Fresh water: $\sqrt{\sigma} = \sqrt{10^{-3}} = 0.03162277 \quad \delta = 15.9155/0.03162277 = 503.292$ m

EXAMPLE 2.4

Find the time average power density at $x = 1$ if for a uniform plane wave is given by $E = 30\ e^{-\alpha x} \cos (10^8 t - \beta x)\ a_z$ V/m and the dielectric constant of the propagating media is (a) $\varepsilon = \varepsilon_0$, $\mu = \mu_0$ (b) $\varepsilon_r = 2.26$, $\sigma = 0$ and (c) $\varepsilon_r = 3.4$, $\sigma/\omega\varepsilon = 0.2$.

Solution

In view of the given expression of E, $E_{0z} = 30$ and $\omega = 10^8$

a. $\eta = \eta_0 = \sqrt{(\mu_0/\varepsilon_0)} = \sqrt{(4\pi \times 10^{-7}/8.854 \times 10^{-12})} \approx 377\ \Omega$
$P_{z,av} = E_{0z}^2/2\eta = (30)^2/(2 \times 377) = 1.1936 a_x$
b. $\eta = \sqrt{(\mu_0/\varepsilon_0\varepsilon_r)} = \eta_0\sqrt{(1/\varepsilon_r)} = 377\sqrt{(1/2.26)} = 250\ \Omega$
$P_{z,av} = E_{0z}^2/2\eta = (30)^2/(2 \times 250) = 1.7944 a_x$
c. $\eta = \eta_0\sqrt{(1/\varepsilon_r)} = 377\sqrt{(1/3.4)} = 204\ \Omega$

Given the loss tangent $\tan\theta = \sigma/\omega\varepsilon = 0.2$ or $\theta = 11.31°$ and $\cos\theta = 0.98056$

$$\sigma = 0.2\ \omega\varepsilon = 0.2 \times 10^8 \times 3.4 \times 8.854 \times 10^{-12} = 6.02072 \times 10^{-4}\ \mho/m$$

$$\alpha = \sigma\eta/2 = 6.02072 \times 10^{-4} \times 204/2 = 0.06141\ \text{nepers/m}$$

$$P_{z,av} = (E_{0z}^2/2\eta)\ e^{-2\alpha x}\cos\theta = \{(30 \times 30)/2 \times 204\}\ e^{-2\times0.061411} \times 0.98056$$

$$= 2.163\ e^{-0.12282} = 2.163 \times 0.886 = 1.9164 a_x$$

EXAMPLE 2.5

The conductivity (σ) and depth of penetration (δ) at 1000 Hz for silver and aluminium slabs are as follows: (a) silver $\sigma = 6.17 \times 10^7\ \mho/m$, $\delta = 0.00202$ m and (b) aluminium $\sigma = 3.72 \times 10^7\ \mho/m$, $\delta = 0.00261$ m. Find the surface resistance of these slabs at 1000 Hz. Also, find the change if this frequency is raised to 10 MHz.

Solution

In view of Equation 2.55c $R_s = 1/(\sigma\delta)$, thus at 1000 Hz for:

a. Silver: $R_s = 1/(6.17 \times 10^7 \times 0.00202) = 80.235 \times 10^{-5} = 0.80235\ m\Omega$
b. Aluminium: $R_s = 1/(3.72 \times 10^7 \times 0.00261) = 102.995 \times 10^{-5}$
   $= 1.02995\ m\Omega$

As $\delta = \sqrt{[1/\pi f\mu\sigma]} = [1/\sqrt{(\pi\mu\sigma)}]\ (1/\sqrt{f})$ and $R_s = (1/\sigma)\ (1/\delta) = (1/\sigma)\ \sqrt{(\pi\mu\sigma)}\ \sqrt{f}$
At $f = 1000$ Hz, $R_s = (1/\sigma)\ \sqrt{(\pi\mu\sigma)}\ \sqrt{1000}$
and at $f = 10$ MHz, $R_s = (1/\sigma)\ \sqrt{(\pi\mu\sigma)}\ \sqrt{10^7}$

Thus, the change due to increase in frequency from 1000 Hz to 10 MHz is $R_s(10\ \text{MHz})/R_s(1000\ \text{Hz}) = \sqrt{10^7}/\sqrt{10^3} = \sqrt{10^4} = 100$ times.
The values at 10 MHz are:

a. Silver: $R_s = 80.235\ m\Omega$
b. Aluminium: $R_s = 102.995\ m\Omega$

EXAMPLE 2.6

Calculate the percentage of reflected and transmitted powers for a uniform plane wave for which the transmission and reflection coefficients for E and H are given as below:

	Reflection Coefficient for:		Transmission Coefficient for:	
	$E = \Gamma_E$	$H = \Gamma_H$	$E = T_E$	$H = T_H$
a.	0.227977	−0.227977	1.228	0.772
b.	−0.227977	0.227977	0.772	1.228
c.	0.02817	−0.02817	1.0282	0.97179

Solution

Assuming the magnitudes of both the field (E & H) contents of incident wave to be unity, that is, $E_i = 1$ and $H_i = 1$, thus the incident power $= E_i$ $H_i = 1$.

Let the reflection coefficient for E is $\Gamma_E = E_r/E_i$, thus $E_r = \Gamma_E$
Also, the reflection coefficient for H is $\Gamma_H = H_r/H_i$, thus $H_r = \Gamma_H$
The reflected power $= E_r H_r = \Gamma_E \Gamma_H$
Similarly, the transmission coefficient for E is $T_E = E_t/E_i$, thus $E_t = T_E$
And the transmission coefficient for H is $T_H = H_t/H_i$, thus $H_t = T_H$
Thus, the transmitted power $= E_t H_t = T_E T_H$

In view of the values given in table:

a. Reflected power $= \Gamma_E \Gamma_H = 0.227977 \times -0.227977 = -0.05197$ or 5.197%
   Transmitted power $= T_E T_H = 1.228 \times 0.772 = 0.948016$ or 94.8016%
b. Reflected power $= \Gamma_E \Gamma_H = -0.227977 \times 0.227977 = -0.05197$ or 5.197%
   Transmitted power $= T_E T_H = 0.772 \times 1.228 = 0.948016$ or 94.8016%
c. Reflected power $= \Gamma_E \Gamma_H = 0.02817 \times -0.02817 = 0.0007935$ or 0.07935%
   Transmitted power $= T_E T_H = 1.0282 \times 0.97179 = 0.999194478$ or 99.92%

EXAMPLE 2.7

Find the reflection coefficients for a parallel polarised wave obliquely incident on the interface between two regions making an angle of 30° with the perpendicular drawn at the boundary surface. The relative dielectric constants of the two regions are (a) $\varepsilon_1 = 2.53$ and $\varepsilon_2 = 1$ (b) $\varepsilon_1 = 1$ and $\varepsilon_2 = 2.53$ (c) $\varepsilon_1 = 2.53$ and $\varepsilon_2 = 2.26$.

$E_r/E_i = [(\varepsilon_2/\varepsilon_1) \cos\theta_1 - \sqrt{\{(\varepsilon_2/\varepsilon_1) - \sin^2\theta_1\}}]/[(\varepsilon_2/\varepsilon_1) \cos\theta_1 - \sqrt{\{(\varepsilon_2/\varepsilon_1) - \sin^2\theta_1\}}]$

Given $\theta = 30°$ $\cos\theta = 0.866$ $\sin\theta = 0.5$ $\sin^2\theta = 0.25$

a. $\varepsilon_1 = 2.53$ and $\varepsilon_2 = 1$, $\varepsilon_2/\varepsilon_1 = 1/2.53 = 0.395257$

$E_r/E_i = [0.345257 \times 0.866 - \sqrt{\{0.395257 - 0.25\}}]/[0.395257 \times 0.866 + \sqrt{\{0.395257 - 0.25\}}]$
$= [0.3423 - \sqrt{0.14525}]/[0.3423 + \sqrt{0.14525}]$

$$= [0.3423 - 0.381126]/[0.3423 + 0.381126]$$

$$= -0.038826/0.7234 = -0.05367$$

b. $\varepsilon_1 = 1$ and $\varepsilon_2 = 2.53$, $\varepsilon_2/\varepsilon_1 = 2.53$

$$E_r/E_i = [2.53 \times 0.866 - \surd\{2.53 - 0.25\}]/[2.53 \times 0.866 + \surd\{2.53 - 0.25\}]$$

$$= [2.53 \times 0.866 - \surd2.28]/[2.53 \times 0.866 + \surd2.28]$$

$$= [2.191 - 1.51]/[2.191 + 1.51]$$

$$= 0.681/3.701 = 0.184$$

c. $\varepsilon_1 = 2.53$ and $\varepsilon_2 = 2.26$, $\varepsilon_2/\varepsilon_1 = 2.26/2.53 = 0.89328$

$$E_r/E_i = [0.89328 \times 0.866 - \surd\{0.89328 - 0.25\}]/\times [0.89328 \times 0.866$$
$$+ \surd\{0.89328 - 0.25\}]$$

$$= [0.77358 - \surd0.64328]/[0.77358 + \surd0.64328]$$

$$= [0.77358 - 0.802]/[0.77358 + 0.802]$$

$$= -0.02842/1.57158 = -0.0180837$$

EXAMPLE 2.8

A wave propagates between two parallel planes separated by 3 cm. The space between planes is filled by a non-magnetic material having relative dielectric constant ε_r, where (a) $\varepsilon_r = 1$ (b) $\varepsilon_r = 2.26$ and (c) $\varepsilon_r = 2.53$. Find the cutoff frequencies for $m = 1, 2$ and 3.

Solution

At cutoff frequency ω_c (or f_c) there is no propagation of wave, that is,
$\gamma = \surd(h^2 - \omega^2\mu\varepsilon) = \surd[(m\pi/a)^2 - \omega^2\mu\varepsilon] = 0$
    Thus, $\omega_c^2\mu\varepsilon = (m\pi/a)^2$ or $\omega_c = 2\pi f_c = [1/\surd\mu\varepsilon]\,(m\pi/a)$
Since $\mu = \mu_0$ and $\varepsilon = \varepsilon_0\varepsilon_r$ $1/\surd\mu\varepsilon = 1/\surd(\mu_0\varepsilon_0\varepsilon_r) = 1/\surd(\mu_0\varepsilon_0)$ $1/\surd\varepsilon_r = v_0/\surd\varepsilon_r$

$$f_c = v_0\,(m/2a)/\surd\varepsilon_r$$

$$v_0/2a = 3 \times 10^8/2 \times 3 \times 10^{-2} = 0.5 \times 10^{10}\quad f_c = m\{(v_0/2a)/\surd\varepsilon_r\}$$

a. $\varepsilon_r = 1$ $\surd\varepsilon_r = 1$ $1/\surd\varepsilon_r = 1$ $(v_0/2a)/\surd\varepsilon_r = 0.5 \times 10^{10}/1 = 5 \times 10^9$
    The cutoff frequency $f_c =: 5$ GHz $(m = 1)$ 10 GHz $(m = 2)$
    15 GHz $(m = 3)$
b. $\varepsilon_r = 2.26$ $\surd\varepsilon_r = 1.50333$ $1/\surd\varepsilon_r = 0.6652$

$$(\upsilon_0/2a)/\sqrt{\varepsilon_r} = 0.5 \times 10^{10}/0.6652 = 7.516 \times 10^9$$

The cutoff frequency $f_c =$: 7.516 GHz $(m = 1)$ 15.032 GHz $(m = 2)$ 22.548 GHz $(m = 3)$

c. $\varepsilon_r = 2.53\sqrt{\varepsilon_r} = 1.5906$ $1/\sqrt{\varepsilon_r} = 0.6287$

$$(\upsilon_0/2a)/\sqrt{\varepsilon_r} = 0.5 \times 10^{10}/0.6287 = 7.953 \times 10^9$$

The cutoff frequency $f_c =$: 7.953 GHz $(m = 1)$ 15.906 GHz $(m = 2)$, 23.859 GHz $(m = 3)$

EXAMPLE 2.9

The cutoff frequency of a wave is 80% of its operating frequency. Find the wave impedances for TE and TM waves if its characteristic impedance is 120π.

Solution

The required relations are:
For TE wave: $Z_{yx}^+ = \eta/\sqrt{\{1 - (f_c/f)^2\}}$

For TM wave: $Z_{xy}^+ = \eta/\sqrt{\{1 - (f_c/f)^2\}}$

Given $f_c = 0.8f$ $f_c/f = 0.8$ $(f_c/f)^2 = 0.64$

$1 - (f_c/f)^2 = 0.36$ $\sqrt{\{1 - (f_c/f)^2\}} = \sqrt{0.36} = 0.6$

For TE wave: $Z_{yx}^+ = \eta/\sqrt{\{1 - (f_c/f)^2\}} = 120\pi/0.6 = 200\pi = 628.318\ \Omega$
For TM wave: $Z_{xy}^+ = \eta/\sqrt{\{1 - (f_c/f)^2\}} = 120\pi \times 0.6 = 72\pi = 226.1946\ \Omega$

EXAMPLE 2.10

A 10 GHz wave propagates between two parallel planes separated by 5 cm. Find the attenuation constants (α) for TE_{10} wave if the space between planes is filled with a material having $\mu_r = 1.2$ and $\sigma_m = 5.8 \times 10^7$ $\mho/m$ and $\varepsilon_r = 2$.

Solution

The required relations are given below:

$$\alpha = \frac{2(m\pi)^2 R_s}{\beta\omega\mu a^3}$$

$$R_s = \sqrt{\frac{\omega\mu_m}{2\sigma_m}} \quad \beta = \sqrt{\omega^2\mu\varepsilon - (m\pi/a)^2}$$

Given: $m = 1$ $a = 5 \times 10^{-2}$ m $f = 10^{10}$ Hz

$$\mu_r = 1.2 \quad \varepsilon_r = 2 \qquad \sigma_m = 5.8 \times 10^7\ \mho/m$$

$$a^3 = 125 \times 10^{-6}\ (m\pi)^2 = \pi^2 = 9.6996$$

$$m\pi/a = \pi/5 \times 10^{-2} \ (m\pi/a)^2 = \pi^2 \times 10^4/25 = 0.39478 \times 10^4$$

$$\omega = 2\pi f = 2\pi \times 10^{10} \qquad \omega^2 = 4\pi^2 \times 10^{20} = 39.4784 \times 10^{20}$$

$$\mu = \mu_m = 1.2 \times 4\pi \times 10^{-7} \qquad \varepsilon = 2 \times 8.854 \times 10^{-12}$$

$$\mu_m\varepsilon = 1.2 \times 4\pi \times 10^{-7} \times 2 \times 8.854 \times 10^{-12} = 267.0303 \times 10^{-19}$$

$$\mu_m/2\sigma_m = 1.2 \times 4\pi \times 10^{-7}/2 \times 5.8 \times 10^7 = 1.3 \times 10^{-14}$$

$$\omega^2\mu\varepsilon = 39.4784 \times 10^{20} \times 267.0303 \times 10^{-19} = 105419$$

$$\omega^2\mu\varepsilon - (m\pi/a)^2 = 10.5419 \times 10^4 - 0.39478 \times 10^4 = 10.1471 \times 10^4$$

$$\beta = \sqrt{\{\omega^2\mu\varepsilon - (m\pi/a)^2\}} = \sqrt{(10.1471 \times 10^4)} = 3.1855 \times 10^2 \text{ rad/m}$$

$$R_s = \sqrt{(\omega \, \mu_m/2\sigma_m)} = \sqrt{2\pi \times 10^{10} \times 1.3 \times 10^{-14}} = 2.858 \times 10^{-2} \ \Omega$$

$$\alpha = 2 \times (m\pi)^2 R_s/\beta\omega\mu a^3 = \{2 \times 9.6996 \times 2.858 \times 10^{-2}\}$$
$$/\{3.1855 \times 10^2 \times 2\pi \times 10^{10} \times 4.8\pi \times 10^{-7} \times 125 \times 10^{-6}\}$$
$$= 0.554429136/3772.75 = 1.46956 \times 10^{-4} \text{ nepers/m}$$

EXAMPLE 2.11

Two strip-shaped conductors of width b (= 5 cm) and thickness t (= 1 cm) in a two parallel wire transmission line are separated by a distance a (= 10 cm). Calculate the inductance, capacitance and conductance, velocity of propagation and the characteristic impedance if $\mu = \mu_0$, $\varepsilon = \varepsilon_0$ and $\sigma = 3.72 \times 10^7$ ℧/m.

Solution

Given $a = 10$ cm, $b = 5$ cm, $t = 1$ cm

$$\mu = \mu_0, \quad \varepsilon = \varepsilon_0 \quad \sigma = 3.72 \times 10^7 \ \Omega/\text{m}$$

The required relations are as follows: $L = \mu \, a/b \quad C = (\varepsilon \, b/a) \quad G = b\sigma/a$

$$v = 1/\sqrt{LC} \quad Z_0 = \sqrt{(L/C)}$$

$$L = \mu \, a/b = 4\pi \times 10^{-7} \times 10 \times 10^{-2}/5 \times 10^{-2} = 8\pi \times 10^{-7} = 2.51327 \ \mu\text{H/m}$$

$$C = (\varepsilon \, b/a) = 8.854 \times 10^{-12} \times 5 \times 10^{-2}/10 \times 10^{-2} = 4.427 \ \mu\mu\text{F/m}$$

$$G = b\sigma/a = 3.72 \times 10^7 \times 5 \times 10^{-2}/10 \times 10^{-2} = 18.6 \ \text{M℧/m}$$

$$v = 1/\sqrt{LC} = 1/\sqrt{(8\pi \times 10^{-7} \times 4.427 \times 10^{-12})} = 1/\sqrt{11.12626 \times 10^{-18}}$$

$$= 1/3.3356 \times 10^{-9} = 2.9979 \times 10^{9} \text{ m/s}$$

$$Z_0 = \sqrt{(L/C)} = \sqrt{(8\pi \times 10^{-7}/4.427 \times 10^{-12})} = \sqrt{56.7715 \times 10^{4}} = 753.4686 \ \Omega$$

EXAMPLE 2.12

A 10 cm long lossless transmission line has a characteristic impedance of 50 Ω. Find its input impedance at 50 MHz if it is terminated in (a) OC, (b) SC and (c) 10 pF capacitor.

Solution:

Given $f = 50$ MHz $Z_0 = 50 \ \Omega$ $l = 10$ cm $= 0.1$ m

$$Z_{in} = Z_0 \frac{Z_R \cos \beta l + j Z_0 \sin \beta l}{Z_0 \cos \beta l + j Z_R \sin \beta l}$$

$$\omega = 2\pi f = 2\pi \times 50 \times 10^{6} = 3.1415927 \times 10^{8} \text{ rad/s} \qquad v = 3 \times 10^{8} \text{ m/s}$$

$$\beta = \omega/v = 3.1415927 \times 10^{8}/3 \times 10^{8} = 1.04719 \text{ rad/m}$$

$$\beta l = 1.04719 \times 0.1 = 0.104719 = 6°$$

$$\sin \beta l = \sin 6 = 0.1045277 \qquad \cos \beta l = \cos 6 = 0.994522$$

$$\tan \beta l = \tan 6 = 0.1051 \qquad \cot \beta l = \cot 6 = 9.51436$$

a. Line is OC or $Z_R = \infty$ $Z_{in} = -jZ_0 \cot\beta l = 50 \times 9.51436 = -j475.718 \ \Omega$
b. Line is SC or $Z_R = 0$ $Z_{in} = jZ_0 \tan\beta l = j50 \times 0.1051 = j5.255 \ \Omega$
c. Line is terminated in 10 pF capacitor or $Z_R = 1/j\omega C = -j/2\pi f C$

$$= -j/(3.1415927 \times 10^{8} \times 10 \times 10^{-12}) = -j/(3.1415927 \times 10^{-3}) = -j318.31 \ \Omega$$

$$Z_{in} = 50\frac{-j318.31 \times 0.994522 + j50 \times 0.1045277}{50 \times 0.994522 + 318.31 \times 0.1045277} = 50\frac{-j316.566 + j5.226385}{49.7221 + 33.2727}$$

$$= j50\frac{-316.566 + 5.226385}{49.7221 + 33.2727} = j50\frac{-311.34}{82.9948} = -j187.566$$

EXAMPLE 2.13

A 50 MHz transmission line with characteristic impedance of 50 Ω is terminated in 200 Ω. Find its reflection coefficient (Γ) and VSWR (s). Also find its quality factor (Q) if this line is a resonant section and its inductance is L (= 0.4 μH), capacitance C (= 40 pF/m), resistance is R (= 0.1 Ω) and conductance G (= 10^{-5} $\mho$/m).

Solution

Given $Z_R = 200\ \Omega$ $Z_0 = 50\ \Omega$ $L = 0.4\ \mu H/m$

$$R = 0.1\ \Omega/m \quad C = 40\ pF/m \quad G = 10^{-5}\ \mho/m$$

$$\Gamma = (Z_R - Z_0)/(Z_R + Z_0) = (200 - 50)/(200 + 50) = 150/250 = 0.6$$

$$S = 1 + |\Gamma|/1 - |\Gamma| = (1 + 0.6)/(1 - 0.6) = 1.6/0.4 = 4$$

$$\omega = 2\pi f = 2\pi \times 50 \times 10^6 = 3.1415927 \times 10^8\ rad/s$$

$$Q = \omega L/R = 3.1415927 \times 10^8 \times 0.4 \times 10^{-6}/0.1 = 3.1415927 \times 400 = 1256.637$$

$$\upsilon = 1/\sqrt{LC} = 1/\sqrt{(0.4 \times 10^{-6} \times 40 \times 10^{-12})}$$
$$= 1/\sqrt{16 \times 10^{-18}} = 1/4 \times 10^{-9} = 0.25 \times 10^9\ m/s$$

Q can also be obtained by using the relation $Q = \beta/2\alpha$

$$\beta = \omega/\upsilon = 3.1415927 \times 10^8/0.25 \times 10^9 = 0.31415927/0.25 = 1.256637\ rad/m$$

$$\alpha \approx 0.0005\ nepers/m$$

from Equation 2.75b and the relation $\alpha \approx R/2Z_0$ for small R and G.

$$Q = \beta/2\alpha = 1.256637/2 \times 0.0005 = 1256.635$$

EXAMPLE 2.14

Calculate the inductance (L) and capacitance (C), velocity of propagation (υ), wavelength (λ) and the characteristic impedance (Z_0) for a lossless coaxial line operating at 3 GHz and having

a. $a = 1$ mm, $b = 3$ mm, $\mu_r = \varepsilon_r = 1$
b. $a = 1$ mm, $b = 3$ mm, $\mu_r = 1$ and $\varepsilon_r = 3$
c. $a = 1$ mm, $b = 5$ mm, $\mu_r = 1$ and $\varepsilon_r = 3$

Solution

Given $f = 3$ GHz $\upsilon_0 = 3 \times 10^8\ m/s$

$$\lambda_0 = \upsilon_0/f = 3 \times 10^8/3 \times 10^9 = 10\ cm$$

The required relations are $L = (\mu_0\mu_r/2\pi)\ln(b/a)\ C = 2\pi\varepsilon_0\varepsilon_r/\ln(b/a)$

$$\upsilon = \upsilon_0/\sqrt{\mu_r\varepsilon_r} \quad \lambda = \lambda_0/\sqrt{\mu_r\varepsilon_r} \quad Z_0 = 60\sqrt{\mu_r\varepsilon_r}\ \ln b/a$$

a. $a = 1$ mm, $b = 3$ mm, $\log (b/a) = \log (3) = 1.09861$

$$\mu_r = 1\ \mu_0\ \mu_r/2\pi = 4\pi \times 10^{-7}/2\pi = 2 \times 10^{-7}$$

$$\varepsilon_r = 1 \qquad 2\pi \, \varepsilon_0 \varepsilon_r = 2\pi \times 10^{-9}/36\pi = 10^{-9}/18 = 55.5555 \times 10^{-12}$$

$$L = (\mu_0 \, \mu_r/2\pi) \log (b/a) = 2 \times 10^{-7} \times 1.09861 = 0.219722 \times 10^{-6}$$
$$= 0.219722 \; \mu H/m$$

$$C = 2\pi \, \varepsilon_0 \, \varepsilon_r/\log (b/a) = 55.5555 \times 10^{-12}/1.09861 = 50.5689 \, pF/m$$

$$v = v_0/\sqrt{(\mu_r \varepsilon_r)} = 3 \times 10^8 \; m/s$$

$$\lambda = \lambda_0/\sqrt{(\mu_r \, \varepsilon_r)} = 10 \; cm$$

$$Z_0 = 60 \, \sqrt{(\mu_r \, \varepsilon_r)} \log (b/a) = 60 \times 1.09861 = 65.9166 \; \Omega$$

b. $a = 1$ mm, $b = 3$ mm, $\log (b/a) = \log (3) = 1.09861$

$$\mu_r = 1 \quad \mu_0 \, \mu_r/2\pi = 4\pi \times 10^{-7}/2\pi = 2 \times 10^{-7}$$

$$L = (\mu_0 \, \mu_r/2\pi) \log (b/a) = 2 \times 10^{-7} \times 1.09861 = 0.219722 \times 10^{-6}$$
$$= 0.219722 \; \mu H/m$$

$$\varepsilon_r = 3 \quad 2\pi \, \varepsilon_0 \, \varepsilon_r = 2\pi \times 3 \times 10^{-9}/36\pi = 10^{-9}/18 = 166.66666 \times 10^{-12}$$

$$C = 2\pi \, \varepsilon_0 \, \varepsilon_r/\log (b/a) = 3 \times 166.66666 \times 10^{-12}/1.09861 = 151.7067 \; pF/m$$

$$v = v_0/\sqrt{(\mu_r \, \varepsilon_r)} = v_0/\sqrt{3} = 3 \times 10^8/\sqrt{3} = 1.732 \times 10^8 \; m/s$$

$$\lambda = \lambda_0/\sqrt{(\mu_r \, \varepsilon_r)} = 10/\sqrt{3} = 5.7735 \; cm$$

$$Z_0 = 60 \, \sqrt{(\mu_r \, \varepsilon_r)} \log (b/a) = 60 \times \sqrt{3} \times 1.09861 = 65.9166 \times \sqrt{3} = 114.171 \; \Omega$$

c. $a = 1$ mm, $b = 5$ mm, $\log (b/a) = \log (5) = 1.6034$

$$\mu_r = 1 \quad \mu_0 \, \mu_r/2\pi = 4\pi \times 10^{-7}/2\pi = 2 \times 10^{-7}$$

$$L = (\mu_0 \, \mu_r/2\pi) \log (b/a) = 2 \times 10^{-7} \times 1.6034 = 0.32068 \; \mu H/m$$

$$\varepsilon_r = 3 \quad 2\pi \, \varepsilon_0 \, \varepsilon_r = 2\pi \times 3 \times 10^{-9}/36\pi = 10^{-9}/18 = 166.66666 \times 10^{-12}$$

$$C = 2\pi \, \varepsilon_0 \, \varepsilon_r/\log (b/a) = 166.66666 \times 10^{-12}/1.6034 = 103.946 \; pF/m$$

$$v = v_0/\sqrt{(\mu_r \, \varepsilon_r)} = v_0/\sqrt{3} \; 3 \times 10^8/\sqrt{3} = 1.732 \times 10^8 \; m/s$$

$$\lambda = \lambda_0/\sqrt{(\mu_r \, \varepsilon_r)} = 10/\sqrt{3} = 5.7735 \; cm$$

$$Z_0 = 60 \, \sqrt{(\mu_r \, \varepsilon_r)} \log (b/a) = 60 \times \sqrt{3} \times 1.6034 = 166.63 \; \Omega$$

PROBLEMS

P-2.1 Two lossless dielectric materials are characterised by μ_{r1}, ε_{r1} and μ_{r2}, ε_{r2}, respectively. Find μ_{r2}, ε_{r2} if $\mu_{r1} = 3$, $\varepsilon_{r1} = 2$ and (a) $v_1 = v_2$ and $\eta_2 = \eta_1/2$ (b) $\eta_2 = \eta_1$ and $\lambda_2 = \lambda_1/2$ (c) $\eta_2 = 3\eta_1$ and $\beta_2 = 2\beta_1$.

P-2.2 Find α, β, v and η for a wave propagating at 50 MHz in a conducting medium having $\sigma = 5.8 \times 10^7$ $\mho/m$ and $\mu = \mu_0$.

P-2.3 Find the depth of penetration at 500 Hz in (a) copper ($\sigma = 5.8 \times 10^7$ $\mho/m$), (b) sea water ($\sigma = 4$ $\mho/m$) and (c) sandy soil ($\sigma = 10^{-5}$ $\mho/m$).

P-2.4 The conductivity (σ) and depth of penetration (δ) at 1000 Hz for slabs made of different materials are as follows: (a) copper $\sigma = 5.8 \times 10^7$ $\mho/m$, $\delta = 0.00209$ m and (b) brass $\sigma = 1.5 \times 10^7$ $\mho/m$, $\delta = 0.00411$ m. Find the surface resistance of these slabs at 1000 Hz. Also find the change if this frequency is raised to 10 MHz.

P-2.5 Find the reflection and transmission coefficients for E and H of a uniform plane wave travelling in region-1 normally incident of the surface of region-2. The relative dielectric constants are (a) $\varepsilon_1 = 2.53$ and $\varepsilon_2 = 1$, (b) $\varepsilon_1 = 1$ and $\varepsilon_2 = 2.53$, (c) $\varepsilon_1 = 2.53$ and $\varepsilon_2 = 2.26$.

P-2.6 Find the reflection coefficients for a parallel polarised wave obliquely incident on the interface between two regions making an angle of 45° with the perpendicular drawn at the boundary surface. The relative dielectric constants of the two regions are (a) $\varepsilon_1 = 2.53$ and $\varepsilon_2 = 1$ (b) $\varepsilon_1 = 1$ and $\varepsilon_2 = 2.53$ (c) $\varepsilon_1 = 2.53$ and $\varepsilon_2 = 2.26$.

P-2.7 Find the Brewster's angle if the relative dielectric constants for the two regions are (a) $\varepsilon_1 = 2.53$ and $\varepsilon_2 = 1$ (b) $\varepsilon_1 = 1$ and $\varepsilon_2 = 2.53$ (c) $\varepsilon_1 = 2.53$ and $\varepsilon_2 = 2.26$.

P-2.8 A 10 GHz, TE_{10} wave propagates between two parallel planes separated by 5 cm. The space between planes is filled by a non-magnetic material with $\varepsilon_r = 2.26$. Find β, λ and v.

P-2.9 A 20 GHz, TE_{10} wave propagates between two parallel planes separated by 3 cm. Find wave impedance if the space between planes is occupied by a material having $\mu_r = 2$ and $\varepsilon_r = 2.5$.

P-2.10 A TM_{10} wave propagates in air dielectric between two parallel planes 5 cm apart. Find the variation of attenuation constants (α) if $\sigma = 5 \times 10^7$ $\mho/m$ and cutoff frequency is 10 GHz.

P-2.11 A transmission line carries a sinusoidal signal of 10^9 rad/s. Find the velocity of propagation if its parameters are as follows: inductance L ($= 0.4$ $\mu H/m$) and capacitance C ($= 40$ pF/m) and (a) $R = 0$, $G = 0$ (b) $R = 0.1$ Ω/m, $G = 10^{-5}$ $\mho/m$ and (c) $R = 300$ Ω/m, $G = 0$.

P-2.12 A 20 cm long lossless transmission line has a characteristic impedance of 50 Ω. Find its input impedance at 50 MHz if it is terminated in a: (a) OC (b) SC and (c) 10 pF capacitor.

P-2.13 A 100 MHz transmission line has inductance L ($= 0.4\,\mu H/m$), capacitance C ($= 40\,pF/m$), resistance R ($= 0.1\,\Omega/m$) and conductance G ($= 10^{-5}\,\mho/m$). Find propagation velocity, phase shift constant (β), characteristic impedance (Z_0) and attenuation constant (α) if it is terminated in 200 Ω.

P-2.14 A 300 MHz transmission line of characteristic impedance 50 Ω is terminated in unknown impedance $R + jX$. The voltage standing wave ratio on the line is 3. Find R and X if the distance (L_2) between the first minima and the terminating end is (a) 70 cm (b) 30 cm.

Descriptive Questions

Q-2.1 Discuss the types of media which may be encountered by an electromagnetic wave.

Q-2.2 What is the logic of assuming a sinusoidal time variation?

Q-2.3 Explain the terms plane wave, uniform plane wave, travelling wave and standing wave.

Q-2.4 Discuss the criteria through which the conductors and dielectrics can be distinguished.

Q-2.5 Discuss the impact of small conductivity in dielectrics on α, β, υ and η.

Q-2.6 What do you understand by linearly, circularly and elliptically polarised waves?

Q-2.7 Explain the meaning of depth of penetration and surface impedance.

Q-2.8 State and explain Poynting theorem. How it is related to wave propagation?

Q-2.9 Explain the terms wave velocity and phase velocity.

Q-2.10 What do you understand by perpendicular and parallel polarisations?

Q-2.11 Explain Brewster's angle and conditions for total internal reflection.

Q-2.12 Explain the meaning of TE, TM and TEM waves.

Q-2.13 What is the concept of wave impedances?

Q-2.14 How are voltage and current distributions affected by different terminating impedances?

Q-2.15 Explain reflection and resonance phenomena in transmission line sections.

Q-2.16 Discuss the use of transmission line sections as circuit elements at higher frequencies?

Q-2.17 How are line sections used as tuned line, voltage transformer and impedance transformer?

Q-2.18 Can line section be used as harmonic suppressors? If yes, how?

Q-2.19 Discuss single and double-stub matching. Add necessary figures.

Q-2.20 Discuss different types of transmission lines along with their applications.

Q-2.21 Discuss the limitations of different guiding structures.

Further Reading

1. Alder, R. B. et al., *Electromagnetic Energy Transmission and Radiation*. MIT Press, Cambridge, MA, 1969.
2. Brown, R. G. et al., *Lines, Waves and Antennas*, 2nd ed. John Wiley & Sons, New York, 1970.
3. Dworsky, L. N., *Modern Transmission Line Theory and Applications*. John Wiley & Sons, New York, 1979.
4. Everitt, W. L., *Communication Engineering*, McGraw-Hill, New York, 1937.
5. Hayt, W. Jr. and Buck, J. A., *Engineering Electromagnetics*, 6th ed., McGraw-Hill Publishing Co. Ltd., New Delhi, 2001.
6. Jesic, H. (ed.), *Antenna Engineering Handbook*, Tata McGraw-Hill, New York, 1961.
7. Johnson, W. C., *Transmission Lines and Networks*, McGraw-Hill, New York, 1950.
8. Jordon, E. C. and Balmain, K. G., *Electromagnetic Waves and Radiating Systems*, Prentice-Hall Ltd., New Delhi, 1987.
9. Moore, R. K., *Travelling-Wave Engineering*, McGraw-Hill, New York, 1960.
10. Ramo, S., Whinnery, J. R. and Van Duzer, T., *Fields and Waves in Communication Electronics*, John Wiley & Sons, Inc., New York, 1965.
11. Sinnema, W., *Electronic Transmission Technology: Lines and Applications*. Prentice-Hall, Inc., Englewood Cliffs, NJ, 1979.
12. Seely, S. and A. D. Poularikas, *Electromagnetics, Classical and Modern Theory and Applications*. Marcel Dekker, Inc., New York, 1979.
13. Skilling, H. H., *Fundamentals of Electric Waves*, John Wiley & Sons, Inc., 2nd ed., New York, 1948.
14. South worth, G. E., *Principles and Applications of Waveguide Transmission*. Van Nostrand Company, Princeton, NJ, 1950.
15. Assadourian, F. and Rimai, E. Simplified theory of microstrip transmission systems. *Proc. IRE*, 40, 1651–1657, 1952.

3

Waveguides

3.1 Introduction

In Chapter 2, various aspects of waves and transmission lines were summarised. In connection with transmission lines, it was noted that the propagation of TE and TM modes requires certain minimum separation between conductors. It was also noted that waveguides are the key components for guiding electromagnetic energy at microwave range as these can support TE and TM modes. The propagation of these modes becomes a necessity as the TEM mode is not suitable owing to increased attenuation at higher frequencies. Many of the features of waveguides resemble those of transmission lines and the concept of reflection, transmission, current flow in conductor skin, properties of quarter- and half-wave sections, effects of discontinuities and so on hold equally good for both. This is because of the fact that there is an interrelation between transmission line and waveguide. The same is being explored in the subsequent section. Besides the interrelation, this chapter also deals with the mathematical theory of waveguide. In waveguides, the energy is carried out by both electric and magnetic fields. The existence and behaviour of these fields cause energy to travel in accordance with the Poynting theorem. At the end, physical interpretation of various terms, the relative merits and limitations of waveguides are included.

3.2 Interrelation between Transmission Line and Waveguide

In Chapter 2, it was stated that most of the concepts developed, in connection with the parallel wire transmission line, are equally valid for coaxial cables and waveguides. The transmission line in itself is an offshoot of parallel plane guide. The coaxial line too is composed of two parallel (cylindrical) planes but of unequal surface areas. Removal of its inner conductor results in yet another configuration referred to as cylindrical waveguide. Similarly, there exists a possibility that a rectangular waveguide may be related either to the two-wire line or to the parallel plane guide. The relation with parallel plane guide is no mystery as will be evident from the mathematical theory of waveguide in the following sections. The relation with two-wire line, however, requires some stretch of imagination.

Consider an ordinary two-wire transmission line shown in Figure 3.1a. It contains two parallel wires with ordinary insulators. The proper operation demands that these insulators must offer very high impedance to the ground. At very high frequencies, a low-impedance insulator may act as a short circuit between line and ground. As per the definition of capacitance, an ordinary insulator may behave like the dielectric of a capacitor formed by the wire and the ground. As per the relation of capacitive reactance ($Xc = 1/\omega C$), the overall impedance decreases with the increase in frequency. In Chapter 2, it was noted that a quarter-wave line acts as an impedance transformer. If it is shorted at one end, it will transform the zero impedance to infinity at its other (open) end. Thus, a quarter-wave section shorted at one end is a better high-frequency insulator in comparison to the conventional insulators.

Figure 3.1b shows a quarter-wave auxiliary line that is connected at some location to the two-wire ordinary transmission line (Figure 3.1a). This auxiliary line will have no effect on the transmission line behaviour in any sense. This is mainly because the shorted auxiliary line will act as an open circuit at the location of its connection. This amounts to as if nothing is connected across the two wires at that location. The impedance of this shorted quarter-wave section is very high at the open-end junction with the two-wire line. Such an insulator may be referred to as a metallic insulator and may be placed anywhere along a two-wire line. It is, however, to be noted that such a quarter-wave section acts as an insulator only at the frequency that corresponds to quarter wavelength. This aspect severely limits the bandwidth, efficiency and application of the two-wire line.

Figure 3.2 illustrates a gradual increase of metallic insulators on each side of a two-wire line. As more and more quarter-wave insulators are added, they start coming closer to each other. When these sections are quite densely

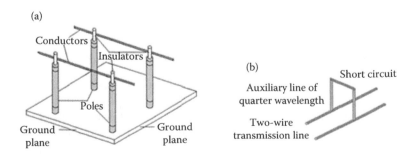

FIGURE 3.1
Formation of a metallic insulator. (a) Two wire line structure and (b) auxilliary and two lines.

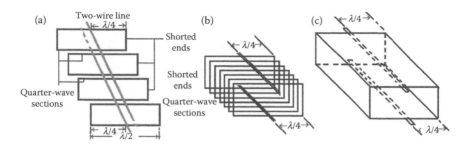

FIGURE 3.2
Number of metallic insulators: (a) 4, (b) 7 and (c) ∞.

located, the configuration takes the shape of a hollow rectangular pipe called waveguide and the lines become part of its walls. The energy now travels within the hollow waveguide instead of travelling along the two-wire line.

Despite this gradual transformation, the operation of fields on a transmission line and in a waveguide is not exact. During this transformation, the electromagnetic field configurations also undergo many changes. The unconfined field of a two-wire line is now confined within the waveguide walls. As a result, the waveguide does not operate like a two-wire line that is completely shunted by quarter-wave sections. Had it been so, the use of a waveguide would have been limited to a single frequency corresponding to the wavelength equal to four times the length of the quarter-wave sections. At these wavelengths, the energy cannot efficiently pass through waveguides. It would be only a small range of frequencies of somewhat shorter wavelengths that would be efficiently transmitted.

Once the waveguide is formed, its dimensions need to be labelled. Its wider dimension that determines the range of operating frequency is labelled as '*a*' and the narrower dimension that determines the power-handling capability is labelled as '*b*'.

3.2.1 Impact of Frequency Change

Figure 3.3 illustrates the impact of frequency change. This demonstrates the ability of a waveguide of given dimensions to transport more than one frequency. A waveguide may be considered to be composed of upper and lower quarter-wave sections and a central solid conductor called a bus bar. In this figure, the distance between the given points m and n is equal to the distance between p and q. Both these distances are equal to one quarter-wavelength. The distance between n and p is the width of the bus bar.

Let '*a*' and '*b*' dimensions of the waveguide be constant. As the frequency is increased, the required length of quarter-wave sections decreases. This causes the bus-bar width to increase. Theoretically, a waveguide can operate at infinite number of frequencies higher than the designed frequency, provided the length of each quarter-wave section approaches zero and the width

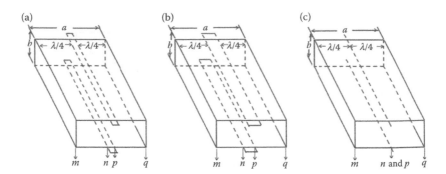

FIGURE 3.3
Impact of change of frequency. (a) Normal operating frequency, (b) more than operating frequency and (c) less than operating frequency.

of the bus bar continues to increase to fill the available space. However, in practice, an upper frequency limit is determined by modes of operation. The concept of modes has already been discussed in connection with the parallel plane guide. In case the frequency of the signal is decreased and the two quarter-wavelengths exceed the dimension 'a' of a waveguide, energy will no longer pass through it. This sets the lower frequency limit often referred to as cutoff frequency of waveguide. In practical applications, dimension 'a' is usually of the order of 0.7λ at the operating frequency. This allows a waveguide to handle only a small range of frequencies both above and below the frequency of operation. Dimension 'b' commonly varies between 0.2λ and 0.5λ and is governed by the breakdown voltage of the dielectric, which is usually air. In centimetric and millimetric ranges, waveguide is the key player in a majority of applications wherein the wave is guided by the four walls.

3.3 Rectangular Waveguide

To explore the guiding principles of wave propagation through a waveguide, Figure 3.4 shows a rectangular waveguide along with its dimensions and coordinate system. The wave is assumed to be propagating along its z-axis. As before, the time variation is accounted by $e^{j\omega t}$ and the space variation along the z-axis by $e^{-\gamma z} = e^{-(\alpha+j\beta)z}$, where parameters α, β and γ have the same meaning as discussed in Chapter 2. Since the waveguide configuration contains four conducting walls, boundary conditions along the x- and y-axes include the continuity of E_{tan} and H_{norm} at $x = 0$ and a, for all values of y and at $y = 0$ and b for all values of x. In case of waveguides the behaviour of waves is also governed by Maxwell's equations, which are written as below.

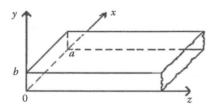

FIGURE 3.4
Geometry of a rectangular waveguide.

$$\nabla \times H = \frac{\partial D}{\partial t} = j\omega D = j\omega\varepsilon E \quad \text{for } J = 0 \tag{3.1a}$$

$$\nabla \times E = -\frac{\partial B}{\partial t} = -j\omega B = -j\omega\mu H \tag{3.1b}$$

Also, the wave equations for E and H are

$$\nabla^2 E = \gamma^2 E \tag{3.2a}$$

$$\nabla^2 H = \gamma^2 H \tag{3.2b}$$

where

$$\gamma = \sqrt{[(\sigma + j\omega\varepsilon)j\omega\mu]} \tag{3.3a}$$

$$\gamma = j\omega\sqrt{\mu\varepsilon} = j\beta \quad (\text{for } \sigma = 0) \tag{3.3b}$$

$$\beta = \omega\sqrt{\mu\varepsilon} \text{ (for any medium)} \quad \text{and} \quad \beta = \beta_0 = \omega\sqrt{\mu_0\varepsilon_0} \text{ (for free space)} \tag{3.3c}$$

After incorporating time and space variations, Equations 3.1a and 3.1b can be written as

$$\frac{\partial H_z}{\partial y} - \frac{\partial H_y}{\partial z} = \frac{\partial H_z}{\partial y} + \gamma H_y = +j\omega\varepsilon E_x \tag{3.4a}$$

$$\frac{\partial H_x}{\partial z} - \frac{\partial H_z}{\partial x} = \frac{\partial H_z}{\partial x} + \gamma H_x = -j\omega\varepsilon E_y \tag{3.4b}$$

$$\frac{\partial H_y}{\partial x} - \frac{\partial H_x}{\partial y} = j\omega\varepsilon E_z \tag{3.4c}$$

$$\frac{\partial E_z}{\partial y} - \frac{\partial E_y}{\partial z} = \frac{\partial E_z}{\partial y} + \gamma E_y = -j\omega\mu H_x \tag{3.5a}$$

$$\frac{\partial E_x}{\partial z} - \frac{\partial E_z}{\partial x} = \frac{\partial E_z}{\partial x} + \gamma E_x = +j\omega\mu H_y \tag{3.5b}$$

$$\frac{\partial E_y}{\partial x} - \frac{\partial E_x}{\partial y} = -j\omega\mu H_z \tag{3.5c}$$

Similarly, the wave Equations 7.2a and 7.2b for E_z and H_z in differential form (for $\sigma = 0$) are

$$\frac{\partial^2 E_z}{\partial x^2} + \frac{\partial^2 E_z}{\partial y^2} + \frac{\partial^2 E_z}{\partial z^2} = \frac{\partial^2 E_z}{\partial x^2} + \frac{\partial^2 E_z}{\partial y^2} + \gamma^2 E_z = -\omega^2 \mu\varepsilon E_z \tag{3.6a}$$

$$\frac{\partial^2 H_z}{\partial x^2} + \frac{\partial^2 H_z}{\partial y^2} + \frac{\partial^2 H_z}{\partial z^2} = \frac{\partial^2 H_z}{\partial x^2} + \frac{\partial^2 H_z}{\partial y^2} + \gamma^2 H_z = -\omega^2 \mu\varepsilon H_z \tag{3.6b}$$

Substitution of (i) H_y from Equation 3.5b in Equation 3.4a, (ii) H_x from Equation 3.5a in Equation 3.4b, (iii) Equation 3.4b in Equation 3.5a and (iv) Equation 3.4a in Equation 3.5b and some mathematical manipulations lead to the following:

$$E_x = -\frac{\gamma}{h^2}\frac{\partial E_z}{\partial x} - \frac{j\omega\mu}{h^2}\frac{\partial H_z}{\partial y} \tag{3.7a}$$

$$E_y = -\frac{\gamma}{h^2}\frac{\partial E_z}{\partial y} + \frac{j\omega\mu}{h^2}\frac{\partial H_z}{\partial x} \tag{3.7b}$$

$$H_x = -\frac{\gamma}{h^2}\frac{\partial H_z}{\partial x} + \frac{j\omega\varepsilon}{h^2}\frac{\partial E_z}{\partial y} \tag{3.7c}$$

$$H_y = -\frac{\gamma}{h^2}\frac{\partial H_z}{\partial y} - \frac{j\omega\varepsilon}{h^2}\frac{\partial E_z}{\partial x} \tag{3.7d}$$

In Equations 3.7

$$h^2 = \omega^2 \mu\varepsilon + \gamma^2 \tag{3.8}$$

The components of E have to satisfy the following boundary conditions:

$$E_x = E_z = 0 \quad \text{at } y = 0 \text{ and } y = b \tag{3.9a}$$

$$E_y = E_z = 0 \quad \text{at } x = 0 \text{ and } x = a \tag{3.9b}$$

In Section 2.17, the concept of transverse wave was introduced. In waveguides also a similar situation emerges and a wave may have either $E_z = 0$ or $H_z = 0$. These two resulting waves may be assigned different nomenclatures. Thus, if $H_z \equiv 0$ the wave is called *transverse magnetic (TM) wave* and if $E_z \equiv 0$ it is termed as *transverse electric (TE) wave*. This terminology is derived from the condition that either H or E field is entirely transverse. These two cases of TE and TM waves are separately discussed below.

3.3.1 Transverse Magnetic Wave

Since $H_z \equiv 0$ and $E_z \neq 0$, only Equation 3.2a, that is, $\nabla^2 E = \gamma^2 E$ needs to be solved. This solution can be obtained by employing the method of separation of variables. As E_z may be a function of all the three coordinates (z, y and z), we can write

$$E_z(x, y, z) = E_z(x, y)\, e^{-\gamma z} \tag{3.10a}$$

where

$$E_z(x, y) = X(x)\, Y(y) \quad \text{or} \quad \text{simply } XY \tag{3.10b}$$

$X(x)$ means that X is the function of x alone and $Y(y)$ shows that Y is a function of y alone.

The substitution of Equation 3.10b in Equation 3.2a (for $\sigma = 0$) gives

$$Y\left(\frac{d^2X}{dx^2}\right) + X\left(\frac{d^2Y}{dy^2}\right) + \gamma^2 XY = -\omega^2 \mu\varepsilon XY \tag{3.11}$$

Division of Equation 3.11 by XY and manipulation of terms gives

$$\frac{1}{X}\left(\frac{d^2X}{dx^2}\right) + h^2 = A^2 \tag{3.12a}$$

and

$$\frac{1}{Y}\left(\frac{d^2Y}{dy^2}\right) = -A^2 \tag{3.12b}$$

Equation 3.12 leads to the following set of equations:

$$\left(\frac{d^2X}{dx^2}\right) + (h^2 - A^2)X = \left(\frac{d^2X}{dx^2}\right) + B^2X = 0 \tag{3.13a}$$

and

$$\left(\frac{d^2Y}{dy^2}\right) + A^2Y = 0 \tag{3.13b}$$

where

$$B^2 = h^2 - A^2 \quad \text{or} \quad h^2 = A^2 + B^2 \tag{3.13c}$$

Solutions of Equations 3.13a and 3.13b can be written as

$$X = C_1\cos(Bx) + C_2\sin(Bx) \tag{3.14a}$$

$$Y = C_3\cos(Ay) + C_4\sin(Ay) \tag{3.14b}$$

Equation 3.10b can now be written as

$$E_z = XY = \{C_1\cos(Bx) + C_2\sin(Bx)\}\{C_3\cos(Ay) + C_4\sin(Ay)\} \tag{3.15a}$$

The application of boundary conditions separately on $X(x)$ and $Y(y)$ leads to $E_z = 0$ at $x = 0$; thus, $C_1 = 0$. Also, $E_z = 0$ at $y = 0$; thus, $C_3 = 0$. Equation 3.15a thus reduces to

$$E_z = \{C_2\sin(Bx)C_4\sin(Ay)\} = C\sin(Bx)\sin(Ay) \tag{3.15b}$$

In Equation 3.15b, $C = C_2 C_4$.
Since $E_z = 0$ at $x = a$, this condition will be met only if $Ba = m\pi$ or when

$$B = \frac{m\pi}{a} \tag{3.16a}$$

Also, $E_z = 0$ at $y = b$; this condition will be met only if $Ab = n\pi$ or when

$$A = \frac{n\pi}{b} \tag{3.16b}$$

In Equations 7.16a and 7.16b, both m and n are positive integers and may take values 0, 1, 2 and so on.

In view of relations (3.16a and 3.16b), Equation 3.15b takes the form

$$E_z = C \sin\left(\frac{m\pi x}{a}\right) \sin\left(\frac{n\pi y}{b}\right) \tag{3.17a}$$

On replacing C by E_{0z} (which represents the magnitude of E_z), Equation 3.17a can be written as

$$E_z = E_{0z} \sin\left(\frac{m\pi x}{a}\right) \sin\left(\frac{n\pi y}{b}\right) \tag{3.17b}$$

In view of Equations 3.13c, 3.16a and 3.16b

$$h^2 = A^2 + B^2 = \left(\frac{n\pi}{b}\right)^2 + \left(\frac{m\pi}{a}\right)^2 = h_{nm}^2 \tag{3.18a}$$

Thus,

$$h = h_{nm} = \sqrt{\left(\frac{n\pi}{b}\right)^2 + \left(\frac{m\pi}{a}\right)^2} \tag{3.18b}$$

The parameter h is sometimes referred to as cutoff wave number. In view of Equations 3.7 and 3.17b, different field components after the incorporation of field variation along the z-axis (for $\alpha = 0$, $\gamma = j\beta$) and on merging the involved terms are

$$E_z = E_{0z} \sin(Bx)\sin(Ay)e^{-j\beta \cdot z} \tag{3.19a}$$

$$E_x = E_{0x} \cos(Bx)\sin(Ay)e^{-j\beta \cdot z} \tag{3.19b}$$

$$E_y = E_{0y} \sin(Bx)\cos(Ay)e^{-j\beta \cdot z} \tag{3.19c}$$

$$H_x = H_{0x} \sin(Bx)\cos(Ay)e^{-j\beta \cdot z} \tag{3.19d}$$

$$H_y = H_{0y} \cos(Bx)\sin(Ay)e^{-j\beta \cdot z} \tag{3.19e}$$

where $E_{0x} = CB(-j\beta/h^2)$, $E_{0y} = CA(-j\beta/h^2)$, $H_{0x} = CA(j\omega\varepsilon/h^2)$ and $H_{0y} = CB(-j\omega\varepsilon/h^2)$.

Also in view of Equations 3.8, 3.16 and 3.18

$$\gamma = \sqrt{\{h^2 - \omega^2\mu\varepsilon\}} = \sqrt{\left[\left\{\left(\frac{m\pi}{a}\right)^2 + \left(\frac{n\pi}{b}\right)^2\right\} - \omega^2\mu\varepsilon\right]} \tag{3.20}$$

3.3.2 Transverse Electric Wave

In this case, $E_z \equiv 0$ and $H_z \neq 0$, and only Equation 3.2b, that is, $\nabla^2 H = \gamma^2 H$, needs to be solved. Its solution can be obtained by following the steps of Equations 3.10 through 3.19. Since the boundary conditions of continuity of $E_{\tan}$ and H_{norm} cannot be directly applied to H_z, Equations 3.7a and 3.7b are to be used to obtain E_x and E_y. Then, the boundary conditions (Equations 3.9a and 3.9b) are to be applied to these components. The finally obtained expressions after incorporation of variation along the z-axis are

$$H_z = C\cos(Bx)\cos(Ay)e^{-j\beta \cdot z}$$

On replacing C by H_{0z} (which represents the magnitude of H_z), H_z can be rewritten as

$$H_z = H_{0z}\cos(Bx)\cos(Ay)e^{-j\beta \cdot z} \tag{3.21a}$$

In view of Equations 3.7 and 3.21a, different field components after the incorporation of field variation along the z-axis ($\alpha = 0$, $\gamma = j\beta$) and on merging the involved terms are

$$E_x = E_{0x}\cos(Bx)\sin(Ay)e^{-j\beta \cdot z} \tag{3.21b}$$

$$E_y = E_{0y}\sin(Bx)\cos(Ay)e^{-j\beta \cdot z} \tag{3.21c}$$

$$H_x = H_{0x}\sin(Bx)\cos(Ay)e^{-j\beta \cdot z} \tag{3.21d}$$

$$H_y = H_{0y}\cos(Bx)\sin(Ay)e^{-j\beta \cdot z} \tag{3.21e}$$

where $E_{0x} = CA(j\omega\mu/h^2)$, $E_{0y} = CB(-j\omega\mu/h^2)$, $H_{0x} = CB(j\beta/h^2)$, $H_{0y} = CA(j\beta/h^2)$.

Parameters A and B in Equation 3.21 have the same meaning as given by Equation 3.16. Thus, Equations 3.18 and 3.20 are equally valid for TM and TE waves.

3.3.3 Behaviour of Waves with Frequency Variation

A critical study of Equation 3.20 leads to the following outcome:

1. If $\{(m\pi/a)^2 + (n\pi/b)^2\} > \omega^2\mu\varepsilon$, γ is purely real (i.e., $\gamma = \alpha$), the wave only attenuates and does not progress. The attenuation constant is obtained to be

$$\alpha = \gamma = \sqrt{\left[\left\{\left(\frac{m\pi}{a}\right)^2 + \left(\frac{n\pi}{b}\right)^2\right\} - \omega^2\mu\varepsilon\right]} \tag{3.22a}$$

2. If $\{(m\pi/a)^2 + (n\pi/b)^2\} > \omega^2\mu\varepsilon$, γ is purely imaginary (i.e., $\gamma = j\beta$), the wave progresses (theoretically) without attenuation. The phase shift constant of the wave is

$$\beta = \gamma = \sqrt{\omega^2\mu\varepsilon - \left\{\left(\frac{m\pi}{a}\right)^2 + \left(\frac{n\pi}{b}\right)^2\right\}} \quad \text{or} \quad \beta^2 = \omega^2\mu\varepsilon - h_{nm}^2 \quad (3.22b)$$

3. Lastly, if $\{(m\pi/a)^2 + (n\pi/b)^2\} > \omega^2\mu\varepsilon$, the angular frequency $\omega = \omega_c$ ($= 2\pi f_c$), where f_c is referred to as the cutoff or critical frequency and is given by

$$f_c = \left\{\frac{1}{2\pi\sqrt{\mu\varepsilon}}\right\}\sqrt{\left\{\left(\frac{m\pi}{a}\right)^2 + \left(\frac{n\pi}{b}\right)^2\right\}} \quad (3.22c)$$

3.3.4 Possible and Impossible Modes

In view of Equations 3.19 and 3.21, the following conclusions can be drawn:

3.3.4.1 Lowest Possible TM Mode

In case of the TM wave (Equation 3.19), it can be observed that for m or $n = 0$, all field quantities vanish. Thus, $TM_{0,0}$, $TM_{0,1}$ or $TM_{1,0}$ modes cannot propagate and the lowest possible mode for TM wave is $TM_{1,1}$.

3.3.4.2 Lowest Possible TE Mode

In case of TE wave (Equation 3.21) for $m = 0$ and $n \neq 0$, $B = 0$, $H_x = E_y = 0$, H_z, H_y and E_x survive, whereas for $n = 0$ and $m \neq 0$, $A = 0$, $H_y = E_x = 0$, H_z, H_x and E_y survive. When $m = n = 0$, only H_z exists and there is no surviving component of E. In view of the above, it can be noted that the lowest possible TE modes are $TE_{0,1}$ and $TE_{1,0}$.

3.3.4.3 Impossibility of TEM Wave

The substitution of $m = n = 0$ reveals that unlike parallel planes there is no possibility of TEM wave in a waveguide. The reason for this is obvious because for the existence of TEM waves in a hollow waveguide H lines must lie in a transverse plane. Also, in a non-magnetic material, the relation $\nabla \cdot B$ (or $\nabla \cdot H$) $= 0$ requires H lines to be in the form of closed loops in the plane perpendicular to the direction of propagation. In view of the first Maxwell's equation ($\nabla \times H = J + \partial D/\partial t$), the magneto-motive-force (m.m.f.) around each of the close loops must be equal to the axial current through the loop. It is possible either in the presence of a conductor with current density J or the displacement current, which in turn requires the presence of E_z. But neither conductor nor E_z is present and thus TEM wave cannot exist inside a hollow waveguide of any shape.

3.3.5 Field Distribution for Different Modes

In view of Equations 3.19 and 3.21, field distribution for TM and TE modes can be obtained by substituting values of m and n in the following manner.

3.3.5.1 TE Wave

Substitution of $m = 1$ and $n = 0$ in the expressions of Equation 3.21 results in $E_x = H_y = 0$. The real parts of the surviving components E_y, H_x and H_z are

$$H_z = H_{0z}\cos(\pi x/a)\cos(\beta z) \tag{3.23a}$$

$$E_y = E_{0y}\sin(\pi x/a)\sin(\beta z) \tag{3.23b}$$

$$H_x = H_{0x}\sin(\pi x/a)\sin(\beta z) \tag{3.23c}$$

where H_{0x}, E_{0y} and H_{0z} are the magnitudes of H_z, E_y and H_x, respectively.

In view of Equation 3.23, E has only E_y component, whereas H has H_x and H_z components. Thus, E and H can be written as

$$E = E_y = E_{0y}\sin(\pi x/a)\sin(\beta z) \tag{3.24a}$$

$$H = \sqrt{\left[H_x^2 + H_z^2\right]} \tag{3.24b}$$

Field distribution obtained from Equation 3.23 in different planes are shown in Figure 3.5.

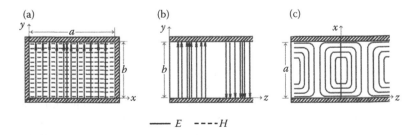

$$\underline{\quad\quad} E \quad \text{----} H$$

FIGURE 3.5
TE_{10} mode in different planes (2D view). (a) x–y plane, (b) y–z plane and (c) x–z plane.

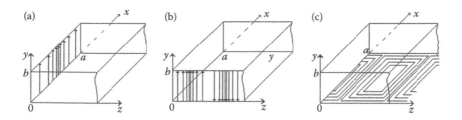

FIGURE 3.6
TE$_{10}$ mode in different planes (3D view). (a) x–y plane, (b) y–z plane and (c) x–z plane.

In view of Equation 3.23c, E_y is sinusoidal along the x-axis and becomes zero at $x = 0$ and 'a' and maximum at $x = a/2$. E is also sinusoidal along the z-axis. These two cases obtained by keeping either z or x constant are shown in Figure 3.5a and b. When both x and z are considered simultaneously, the resulting field appears as illustrated in Figure 3.5c. The plots of a, b and c are again illustrated in Figure 3.6a, b and c, respectively, for clarity. This figure illustrates a 3D view of the field distributions shown in Figure 3.5.

Figure 3.7 illustrates various field distributions for different TE modes in the x–y plane (i.e., across the waveguide cross section). These modes can be obtained by changing values of m and n in Equation 3.19 and obtaining appropriate relations.

3.3.5.2 TM Wave

Substitution of values of m and n in Equation 3.19 and appropriate manipulations may lead to different TM modes. Figure 3.8 illustrates some such

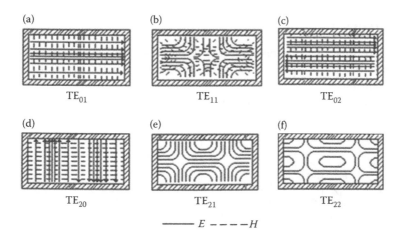

FIGURE 3.7
TE modes in the x–y plane: (a) TE$_{01}$, (b) TE$_{11}$, (c) TE$_{02}$, (d) TE$_{20}$, (e) TE$_{21}$ and (f) TE$_{22}$.

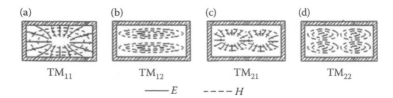

FIGURE 3.8
TM modes in the x–y plane: (a) TM_{11}, (b) TM_{12}, (c) TM_{21} and (d) TM_{22}.

field patterns. An appropriate computer programme for the selected values of involved parameters may provide a very clear picture of the field distribution for TM and TE waves in different planes. In Figure 3.8, the illustration of the TM mode field distribution is confined to the waveguide cross section.

3.3.6 Excitation of Modes

The excitation of modes for TM and TE modes are described as given below.

3.3.6.1 TE Modes

Figure 3.9 illustrates the excitation of various TE modes. The location of a probe or inner conductor of the coaxial cable in a broader or narrower wall will result in the excitation of a particular mode. As the wave starts progressing in the direction of propagation, the field orientation of patterns changes appropriately.

3.3.6.2 TM Modes

Figure 3.10 illustrates the excitation of four different TM modes. As can be seen from the figure, probes or inner conductors of the coaxial cable are inserted into the cross section of a waveguide in different numbers and at different locations for exciting different modes.

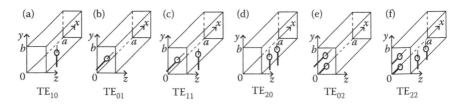

FIGURE 3.9
Excitation of TE modes: (a) TE_{10}, (b) TE_{01}, (c) TE_{11}, (d) TE_{20}, (e) TE_{02} and (f) TE_{22}.

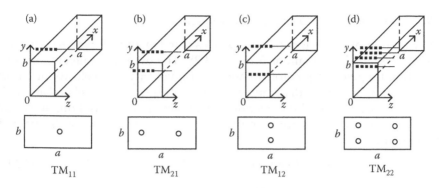

FIGURE 3.10
Excitation of TM modes: (a) TM_{11}, (b) TM_{21}, (c) TM_{12} and (d) TM_{22}.

3.3.7 Wave Impedances

The concept of wave impedance was introduced in Section 2.20. As it is the impedance that is seen by the wave when it travels in the direction of propagation, the values of impedances for TE and TM waves are to be evaluated in view of the relevant field expressions. In waveguide, the wave is travelling in the z direction and hence it will encounter impedance which is the ratio of the orthogonal E and H components across the x–y plane.

3.3.7.1 TM Wave

The wave impedance for TM wave is given as follows:

$$E_x/H_y = -E_y/H_x = \beta/\omega\varepsilon \tag{3.25a}$$

In view of Equations 3.22b, 3.22c and 3.18

$$\beta = \omega\sqrt{(\mu\varepsilon)}\sqrt{\{1-(f_c/f)^2\}} \quad \text{and} \quad \beta/\omega\varepsilon = \sqrt{(\mu/\varepsilon)}\sqrt{\{1-(f_c/f)^2\}} \tag{3.25b}$$

$$Z_z(TM) = \beta/\omega\varepsilon = \eta\sqrt{\{1-(\omega_c/\omega)^2\}} = \eta\sqrt{\{1-(f_c/f)^2\}} \tag{3.26}$$

3.3.7.2 TE Wave

The wave impedance for the TE wave is given as follows:

$$E_x/H_y = -E_y/H_x = \omega\mu/\beta \tag{3.27a}$$

In view of Equations 3.22b, 3.22c and 3.18

$$\omega\mu/\beta = \sqrt{(\mu/\varepsilon)}[1/\sqrt{\{1-(\omega_c/\omega)^2\}}] \tag{3.27b}$$

$$Z_z(TE) = \omega\mu/\beta = \eta/\sqrt{\{1-(\omega_c/\omega)^2\}} = \eta/\sqrt{\{1-(f_c/f)^2\}} \tag{3.28}$$

3.3.7.3 Variation of Z_z, v and λ with Frequency

The variation of characteristic impedances (for TM and TE waves) with frequency is given by Equations 3.26 and 3.28. The other parameters, namely, operating frequency (f), cutoff frequency (f_c), velocity (v), velocity at cutoff (v_c), operating wavelength (λ) and cutoff wavelength (λ_c) are related and will be as discussed below. These relations are obtained in view of Equations 3.22 and the general relations, for example, $v = f\lambda$, $v = \omega/\beta$ and $\lambda = 2\pi/\beta$ obtained in Chapter 2.

$$v = f\lambda = \omega/\beta = v_0/\sqrt{\{1 - (\omega_c/\omega)^2\}} = v_0/\sqrt{\{1 - (f_c/f)^2\}} \tag{3.29a}$$

$$\lambda = \lambda_g = 2\pi/\beta = 1/[f\sqrt{(\mu/\varepsilon)}\sqrt{\{1 - (\omega_c/\omega)^2\}}] = \lambda_0/[\sqrt{\{1 - (\omega_c/\omega)^2\}}]$$
$$= \lambda_0/\sqrt{\{1 - (f_c/f)^2\}} \tag{3.29b}$$

$$\lambda_c = v_c/f_c = 2\pi v_c/2\pi f_c = 2\pi v_c/\omega_c = 2\pi/\sqrt{\{(m\pi/a)^2 + (n\pi/b)^2\}} \tag{3.29c}$$

Since $(\omega_c/\omega)^2 = (\lambda_0/\lambda_c)^2$, the value of λ parallel to waveguide walls can be written as

$$\lambda = (\lambda_0\lambda_c)/\sqrt{(\lambda_c^2 - \lambda_0^2)} \quad \text{or} \quad \lambda_0 = (\lambda\lambda_c)/\sqrt{(\lambda_c^2 + \lambda^2)} \tag{3.29d}$$

In view of Equations 3.26, 3.28 and 3.29, variations of different parameters are shown in Figure 3.11.

Figure 3.12 illustrates the relative values of cutoff frequencies for some modes in a rectangular waveguide for $a = 1.07$ cm, $b = 0.43$ cm, $f = 15$ GHz and $\varepsilon_r = 2.08$. The actual values of these cutoff frequencies (in GHz) for different modes are TE_{11} (9.72), TE_{20} (19.44), TE_{01} (24.19), TE_{11} and TM_{11} (26.07) and TE_{21} and TM_{21} (31.03). The characteristic parameters for TE and TM waves in rectangular waveguide are summarized in Table 3.1. The mathematical relation corresponding to each parameter is also indicated. Similarly Table 3.2 gives comparison between TE and TM waves in rectangular waveguide.

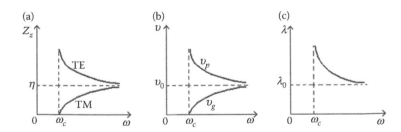

FIGURE 3.11
Variation of Z_Z(TE), Z_Z(TM), v_p, v_g and λ with ω. (a) Variation of impedance, (b) variation of velocity and (c) variation of wavelength.

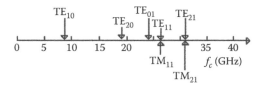

FIGURE 3.12
Cutoff frequencies of some modes.

TABLE 3.1

Characteristic Parameters for TE and TM Waves in Rectangular Waveguide

Parameter	Mathematical Relation (Equation Number)	
Cutoff wave number	$h = \omega_c\sqrt{(\mu\varepsilon)} = \sqrt{\{(m\pi/a)^2 + (n\pi/b)^2\}}$	(3.18b)
Attenuation constant	$\alpha = \sqrt{[\{(m\pi/a)^2 + (n\pi/b)^2\} - \omega^2\mu\varepsilon]}$	(3.22a)
Phase shift constant	$\beta = \omega\sqrt{(\mu\varepsilon)}\sqrt{\{1 - (f_c/f)^2\}}$	(3.22b)
Cutoff frequency	$f_c = \{1/2\sqrt{(\mu\varepsilon)}\}\sqrt{\{(m/a)^2 + (n/b)^2\}}$	(3.22c)
Phase velocity	$v_p = v_0/\sqrt{\{1 - (f_c/f)^2\}}$	(3.29a)
Guide wavelength	$\lambda_g = \lambda/\sqrt{\{1 - (f_c/f)^2\}}$	(3.29b)
Cutoff wavelength	$\lambda_c = 2/\sqrt{\{(m/a)^2 + (n/b)^2\}}$	(3.29c)
Wave impedance (TM wave)	$Z_z = \eta\sqrt{\{1 - (f_c/f)^2\}}$	(3.26)
Wave impedance (TE wave)	$Z_z = \eta/\sqrt{\{1 - (f_c/f)^2\}}$	(3.28)

TABLE 3.2

Comparison between TE and TM Waves in Rectangular Waveguide

	TE Mode	TM Mode
Entirely transverse field	Electric	Magnetic
Mean energy transmission through	H_z (since $E_z \equiv 0$)	E_z (since $H_z \equiv 0$)
Dominant mode	TE_{10}	TM_{11}
Minimum value of m and n	0, 1 or 1, 0	1, 1
Cutoff wavelength for dominant mode	$2a$	$2ab/\sqrt{(a^2 + b^2)}$

3.3.8 Circuit Equivalence of Waveguides

The basic Maxwell's equations $\nabla \times H = \partial D/\partial t$ (for $J = 0$) and $\nabla \times E = -\partial B/\partial t$ after incorporating time variation ($e^{j\omega t}$) and space variation ($e^{-\gamma z}$) leads to the following relations.

3.3.8.1 TM Wave

Substitution of $H_z = 0$ in Equations 3.4a, 3.5b and 3.7d leads to the following expressions:

$$-\frac{\partial H_y}{\partial z} = j\omega\varepsilon E_x \tag{3.30}$$

$$-\frac{\partial E_z}{\partial x} + \frac{\partial E_x}{\partial z} = j\omega\mu H_y \tag{3.31}$$

$$H_y = -\frac{j\omega\varepsilon}{h^2}\frac{\partial E_z}{\partial x} \tag{3.32}$$

Since $\nabla \cdot B$ or $\nabla \cdot H = 0$, $\{(\partial H_x/\partial x) + (\partial H_y/\partial y)\} = 0$. Also, since $H_z \equiv 0$, $(\nabla \times E)_z = 0$. This indicates that in the x–y plane, E has no curl (i.e. the voltage around a closed path is zero). Thus, in the x–y plane, E may be written as the gradient of scalar potential V (as the curl of a gradient is identically zero). The relevant expression is

$$E_x = -\frac{\partial V}{\partial x} \tag{3.33}$$

From Equations 3.30, 3.32 and 3.33

$$\frac{\partial}{\partial z}\left(-\frac{j\omega\varepsilon}{h^2}\frac{\partial E_z}{\partial x}\right) = j\omega\varepsilon\frac{\partial V}{\partial x} \quad \text{or} \quad \frac{\partial}{\partial z}(-\frac{j\omega\varepsilon}{h^2}E_z) = j\omega\varepsilon V \tag{3.34}$$

Also, from Equations 3.31 and 3.32

$$\frac{\partial E_x}{\partial z} - \frac{\partial E_z}{\partial x} = -\omega^2\frac{\mu\varepsilon}{h^2}\frac{\partial E_z}{\partial x}$$

In view of Equation 3.33

$$\frac{\partial V}{\partial z} = \left(\omega^2\frac{\mu\varepsilon}{h^2} - 1\right)E_z = -\left(j\omega\mu + \frac{h^2}{j\omega\varepsilon}\right)\left(\frac{j\omega\varepsilon}{h^2}E_z\right) \tag{3.35}$$

In Equation 3.35, the term $j\omega\varepsilon E_z$ is the longitudinal displacement current density and $1/h^2$ has the dimension of area; thus the term $(j\omega\varepsilon E_z/h^2)$ represents a current I_z flowing in the z direction. Equations 3.34 and 3.35 can be rewritten as

$$\frac{\partial I_z}{\partial z} = -j\omega\varepsilon V = -YV \tag{3.36a}$$

$$\frac{\partial V}{\partial z} = -\left(j\omega\mu + \frac{h^2}{j\omega\varepsilon}\right)I_z = -ZI_z \tag{3.36b}$$

where $Y = j\omega\varepsilon$ and $Z = j\omega\mu + (h^2/j\omega\varepsilon)$.

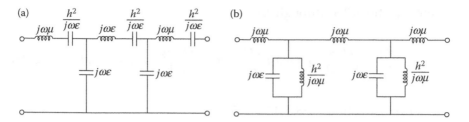

FIGURE 3.13
Equivalent circuits of TE and TM waves. (a) TM wave and (b) TE wave.

Equation 3.36 leads to a form of circuit that is termed as an equivalent circuit for TM wave and is shown in Figure 3.13a along with the values of involved components. The characteristic impedance for TM wave obtained in view of Equation 3.36 is given below:

$$Z_0(\text{TM}) = \sqrt{\frac{Z}{Y}} = \sqrt{\left[\frac{j\omega\mu + (h^2/j\omega\varepsilon)}{j\omega\varepsilon}\right]} = \sqrt{\frac{\mu}{\varepsilon}}\sqrt{\{1 - (\omega_c/\omega)^2\}}$$
$$= \eta\sqrt{\{1 - (\omega_c/\omega)^2\}} = Z_z(\text{TM}) \tag{3.37}$$

3.3.8.2 TE Wave

Substitution of $E_z = 0$ in Equations 3.4b, 3.5b and 3.7a leads to the following expressions:

$$\frac{\partial H_z}{\partial y} - \frac{\partial H_y}{\partial z} = j\omega\varepsilon E_x \tag{3.38}$$

$$\frac{\partial E_x}{\partial z} = j\omega\mu H_y \tag{3.39}$$

$$E_x = -\frac{j\omega\mu}{h^2}\frac{\partial H_z}{\partial y} \tag{3.40}$$

In this case, $E_z \equiv 0$, thus $(\nabla \times H)_z = 0$. It is now possible to define a scalar (magnetic) potential U in the x–y plane and to write H as a gradient of U. Its relevant expression is given as

$$H_y = -\frac{\partial U}{\partial y} \tag{3.41}$$

From Equations 3.39 through 3.41

$$\frac{\partial}{\partial z}\left(-\frac{j\omega\mu}{h^2}\frac{\partial H_z}{\partial y}\right) = -j\omega\mu\frac{\partial U}{\partial y} \quad \text{or} \quad \frac{\partial}{\partial z}\left(\frac{j\omega\mu}{h^2}H_z\right) = -j\omega\mu U \quad (3.42a)$$

Also, from Equations 3.38, 3.40 and 3.41

$$\frac{\partial H_z}{\partial y} - \frac{\partial H_y}{\partial z} = \omega^2\frac{\mu\varepsilon}{h^2}\frac{\partial H_z}{\partial y}$$

$$\frac{\partial U}{\partial z} = -\left(\frac{h^2}{j\omega\mu} - j\omega\varepsilon\right)\left(\frac{j\omega\mu}{h^2}H_z\right) \quad (3.42b)$$

In Equation 3.42b, the term $\{(j\omega\mu/h^2)H_z\}$ has the dimension of voltage (say V_1) and U has the dimension of current (say I_1). Equations 3.42a and 3.42b can be rewritten as below:

$$\frac{\partial V_1}{\partial z} = -j\omega\mu U = -ZI_1 \quad (3.43a)$$

$$\frac{\partial I_1}{\partial z} = -\left(\frac{h^2}{j\omega\mu} - j\omega\varepsilon\right)V_1 = -YV_1 \quad (3.43b)$$

where $Z = j\omega\mu$ and $Y = j\omega\varepsilon + (h^2/j\omega\mu)$.

Equation 3.43 leads to an equivalent circuit for the TE wave shown in Figure 3.13b.

The expression for characteristic impedance for TE wave is given below:

$$Z_0(\text{TE}) = \sqrt{\frac{Z}{Y}} = \sqrt{\left[\frac{j\omega\mu}{j\omega\varepsilon + (h^2/j\omega\mu)}\right]} = \sqrt{\frac{\mu}{\varepsilon}}\Big/\sqrt{\{1 - (\omega_c/\omega)^2\}}$$

$$= \eta\Big/\sqrt{\{1 - (\omega_c/\omega)^2\}} = Z_z(\text{TE}) \quad (3.44)$$

3.3.9 Power Transmission in Rectangular Waveguide

Power transmission through a rectangular waveguide can be calculated under the assumptions that there are no reflections in the waveguide and the waveguide length is much larger than the wavelength. For the dominant TE_{10} mode, the power carried by the waveguide with dimensions 'a' and 'b' in the assumed (z) direction of propagation is given by,

$$P = \text{Re}\left[(1/2)\int_0^a\int_0^b E \times H^* \, dxdy\right] \tag{3.45}$$

In view of Equations 3.23b and 3.23c written in the modified form as $E_y = E_{y0}\sin(x\pi/a)\,e^{-j\beta z}$ and $H_x = H_{x0}\sin(\pi x/a)\,e^{-j\beta z}$,

$$P = \text{Re}\left[(1/2)\int_0^a\int_0^b E_{y0}\sin(\pi x/a)e^{-j\beta z} \times H_{x0}\sin(\pi x/a)e^{j\beta z} \, dxdy\right] \tag{3.46}$$

Since $H_{x0} = E_{y0}/Z_z(\text{TE}) = -E_{y0}/Z_z(\text{TE})$.
In view of Equation 3.28, $H_{x0} = -(\beta/\omega\mu)\,E_{y0}$.
Thus, Equation 3.46 becomes

$$P = \text{Re}\left[(1/2)\int_0^a\int_0^b E_{y0}^2 \times (-\beta/\omega\varepsilon)\sin^2(\pi x/a) \, dxdy\right] = \frac{1}{4}E_{y0}^2\frac{\beta}{\omega\mu}ab \tag{3.47}$$

3.3.10 Attenuation in Waveguides

The attenuation in waveguides is mainly due to its finite conductivity. It results in some (small) surface resistance in the conductor walls, which in turn results in losses. As a consequence, the wave decays as it progresses in the direction of propagation. This decay is accounted by attenuation constant (α in Nep./m) for different modes and can be given by the relation:

$$\alpha(\text{TE}_{10}) = \frac{R_s\{1 + (2b/a)(f_c/f)^2\}}{\eta b\sqrt{1 - (f_c/f)^2}} = \frac{1}{\eta b}\left(\frac{\mu\pi f}{\sigma\{1 - (f_c/f)^2\}}\right)^{1/2}\left[1 + \frac{2b}{a}\left(\frac{f_c}{f}\right)^2\right] \tag{3.48}$$

The general expressions for $\alpha(\text{TE}_{mn})$ and $\alpha(\text{TM}_{mn})$ modes in terms of f_{cmn} cutoff frequencies are

$$\alpha(\text{TE}_{mn}) = \frac{2R_s/b\eta}{\sqrt{1 - (f_{cmn}/f)^2}}$$

$$\left[(1 + b/a)(f_{cmn}/f)^2 + \{(\delta_{0n}/2) - (f_{cmn}/f)^2\}\frac{(b/a)\{(b/a)m^2 + n^2\}}{(b/a)^2m^2 + n^2}\right] \tag{3.49}$$

where

$$\delta_{0n} = \begin{cases}1 \\ 2\end{cases} \text{ for } \begin{array}{l}n = 0 \\ n \neq 0\end{array} \quad \alpha(\text{TM}_{mn}) = \frac{2R_s/b\eta}{\sqrt{1 - (f_{cmn}/f)^2}}\left[\frac{(b/a)^3m^2 + n^2}{(b/a)^2m^2 + n^2}\right] \tag{3.50}$$

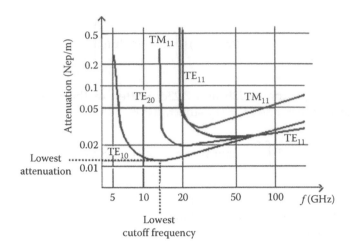

FIGURE 3.14
Variation of attenuation with frequency.

Figure 3.14 illustrates the variation of attenuation constant (α) with the frequency for a few TE_{mn} and TM_{mn} modes in a rectangular waveguide. From this figure, it can be observed that TE_{10} has the lowest cutoff frequency and minimum attenuation as compared to other modes. This is the reason that this mode is referred to as dominant mode and is the most commonly used mode. The numerical figure of the cutoff frequency can be worked out by substituting specified values of 'a' and 'b'. As per these characteristics, the attenuation tends to infinity when f is close to the cutoff, decreases towards an optimum frequency and then increases almost linearly with f.

3.3.11 Quality Factor

The quality factor (Q) is closely related to the attenuation constant, and for an ordinary transmission line carrying a TEM wave, it can be given as

$$Q = \omega \text{ (energy stored in circuit/energy lost per second)} \qquad (3.51a)$$

For TEM wave in lossless lines, the velocity v represents both (phase and group) velocities. In cases of TE and TM modes, these two velocities are connected by the relation $v_p\, v_g = c^2$.

In a waveguide, energy transmitted per second = v_g (energy stored per unit length).

Thus, energy stored per unit length = power transmitted/v_g, or

$$Q = (\omega/v_g) \text{ (power transmitted/power lost per unit length)}$$

$$= (\omega/2\,\alpha v_g) = (\omega v_p/2\alpha c^2) = \omega/2\alpha c\ \sqrt{\{1 - (\omega_c^2/\omega^2)\}} \qquad (3.51b)$$

In comparison to the transmission lines, waveguides have much lower attenuation and thus can have high Q. In filters, waveguides sections are used as resonators. The low attenuation aspect greatly improves the filter performance.

3.4 Circular Waveguide

The study of circular waveguides is required mainly in view of their involvement in many microwave components, namely, circulators, isolators and so on. Figure 3.15 illustrates such a waveguide with a circular cross section of radius 'a' wherein the wave is assumed to propagate in the z direction. As before, the time variation is accounted by the factor $e^{j\omega t}$ and the space variation along the z-axis by $e^{-\gamma z}$ (where $\gamma = \alpha + j\beta$). The parameters α, β and γ are again the attenuation, phase shift and propagation constants, respectively. It is further assumed that the waveguide is made of conducting material with $\sigma = \infty$. Thus, continuity conditions of E_{tan} and H_{norm} holds at the surface of the conductor, that is, for $0 \le \phi \le 2\pi$ and $0 \le z \le \infty$. The waveguide is assumed to be filled with air or a dielectric for which $\sigma = 0$ and thus $\gamma = j\beta$. In view of the shape of the waveguide cylindrical coordinate system (ρ, ϕ, Z) is to be employed. In this case too, the basics governing Maxwell's equations for perfect dielectric media are

$$\nabla \times H = j\omega\varepsilon E \tag{3.52a}$$

$$\nabla \times E = -j\omega\mu H \tag{3.52b}$$

The wave equations for H and E are

$$\nabla^2 H = -\gamma^2 H \tag{3.53a}$$

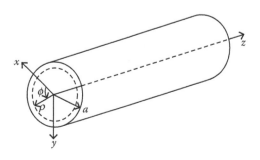

FIGURE 3.15
Geometry of circular waveguide.

$$\nabla^2 E = -\gamma^2 E \qquad (3.53b)$$

From Equation 3.52a, the governing equations in differential form are

$$\frac{1}{\rho}\frac{\partial H_z}{\partial \phi} - \frac{\partial H_\phi}{\partial z} = j\omega\varepsilon E_\rho \qquad (3.54a)$$

$$\frac{\partial H_\rho}{\partial z} - \frac{\partial H_z}{\partial \rho} = j\omega\varepsilon E_\phi \qquad (3.54b)$$

$$-\frac{1}{\rho}\frac{\partial}{\partial \rho}(\rho H_\phi) - \frac{1}{\rho}\frac{\partial H_\rho}{\partial \phi} = j\omega\varepsilon E_z \qquad (3.54c)$$

Also from Equation 3.52b

$$\frac{1}{\rho}\frac{\partial E_z}{\partial \phi} - \frac{\partial E_\phi}{\partial z} = -j\omega\mu H_\rho \qquad (3.55a)$$

$$\frac{\partial E_\rho}{\partial z} - \frac{\partial E_z}{\partial} = j\omega\mu H \qquad (3.55b)$$

$$\frac{1}{\rho}\frac{\partial}{\partial \rho}(\rho E_\phi) - \frac{1}{\rho}\frac{\partial E_\rho}{\partial \phi} = j\omega\mu E_z \qquad (3.55c)$$

After incorporating z derivatives, Equations 3.54 and 3.55 become

$$\frac{1}{\rho}\frac{\partial H_z}{\partial \phi} + \gamma H_\phi = -j\omega\varepsilon E_\rho \qquad (3.56a)$$

$$-\gamma H_\rho + \frac{\partial H_z}{\partial \rho} = j\omega\varepsilon E_\phi \qquad (3.56b)$$

$$\frac{1}{\rho}\frac{\partial}{\partial \rho}(\rho H_\phi) - \frac{1}{\rho}\frac{\partial H_\rho}{\partial \phi} = j\omega\varepsilon E_z \qquad (3.56c)$$

$$\frac{1}{\rho}\frac{\partial E_z}{\partial \phi} + \gamma E_\phi \frac{\partial E_\phi}{\partial z} = -j\omega\mu H_\rho \qquad (3.57a)$$

$$-\gamma E_\rho + \frac{\partial E_z}{\partial r} = -j\omega\mu H_\phi \qquad (3.57b)$$

$$\frac{1}{\rho}\frac{\partial}{\partial\rho}(\rho E_\phi) - \frac{1}{\rho}\frac{\partial E_\rho}{\partial\phi} = -j\omega\mu E_z \qquad (3.57c)$$

After manipulating Equations 3.56 and 3.57, the following relations can be obtained:

$$H_\rho = \frac{j\omega\varepsilon}{h^2}\frac{1}{\rho}\frac{\partial E_z}{\partial\phi} - \frac{\gamma}{h^2}\frac{\partial H_z}{\partial\rho} \qquad (3.58a)$$

$$H_\phi = -\frac{j\omega\varepsilon}{h^2}\frac{\partial E_z}{\partial\rho} - \frac{\gamma}{h^2}\frac{1}{\rho}\frac{\partial H_z}{\partial\phi} \qquad (3.58b)$$

$$E_\rho = -\frac{\gamma}{h^2}\frac{\partial E_z}{\partial\rho} - \frac{j\omega\mu}{h^2}\frac{1}{\rho}\frac{\partial H_z}{\partial\phi} \qquad (3.58c)$$

$$E_\phi = -\frac{\gamma}{h^2}\frac{1}{\rho}\frac{\partial E_z}{\partial\phi} + \frac{j\omega\mu}{h^2}\frac{\partial H_z}{\partial\rho} \qquad (3.58d)$$

In the above equations, γ and h^2 have the same meaning as indicated by Equations 3.3 and 3.8. Here also, we can have two different modes of propagation wherein either $H_z \equiv 0$ or $E_z \equiv 0$. These are again referred to as TM and TE waves. The resulting expressions for these waves are as follows.

3.4.1 TM Waves

For this case, $H_z \equiv 0$; thus, the expressions of field components given by Equation 3.58 become

$$E_\rho = -\frac{\gamma}{h^2}\frac{\partial E_z}{\partial\rho} \qquad (3.59a)$$

$$E_\phi = -\frac{\gamma}{h^2}\frac{1}{\rho}\frac{\partial E_z}{\partial\phi} \qquad (3.59b)$$

$$H_\rho = \frac{j\omega\varepsilon}{h^2}\frac{1}{\rho}\frac{\partial E_z}{\partial\phi} \qquad (3.59c)$$

$$H_\phi = -\frac{j\omega\varepsilon}{h^2}\frac{\partial E_z}{\partial\rho} \qquad (3.59d)$$

3.4.2 TE Waves

For this case, $E_z \equiv 0$; thus, the field components given by Equation 3.58 become

$$E_\rho = -\frac{j\omega\mu}{h^2}\frac{1}{\rho}\frac{\partial H_z}{\partial \phi} \tag{3.60a}$$

$$E_\phi = \frac{j\omega\mu}{h^2}\frac{\partial H_z}{\partial \rho} \tag{3.60b}$$

$$H_\rho = -\frac{\gamma}{h^2}\frac{\partial H_z}{\partial \rho} \tag{3.60c}$$

$$H_\phi = -\frac{\gamma}{h^2}\frac{1}{\rho}\frac{\partial H_z}{\partial \phi} \tag{3.60d}$$

3.4.3 Solution of Wave Equation

In view of Equations 3.59 and 3.60, all the field components are related to E_z or H_z. E_z and H_z can be obtained by solving Equations 3.53a and 3.53b for E or H. These are obtained as below.

Equations 3.53a and 3.53b can be written in differential forms in cylindrical coordinates as

$$\frac{1}{\rho}\frac{\partial}{\partial \rho}\left(\rho\frac{\partial E_z}{\partial \rho}\right) + \frac{1}{\rho^2}\frac{\partial^2 E_z}{\partial \phi^2} + h^2 E_z = 0 \tag{3.61a}$$

$$\frac{1}{\rho}\frac{\partial}{\partial \rho}\left(\rho\frac{\partial H_z}{\partial \rho}\right) + \frac{1}{\rho^2}\frac{\partial^2 H_z}{\partial \phi^2} + h^2 H_z = 0 \tag{3.61b}$$

To obtain the solution of Equations 3.61a and 3.61b, we can proceed as follows:

$$\text{Let } E_z \quad \text{or} \quad H_z = R(\rho)\, S(\phi) \tag{3.62}$$

In Equation 3.61, R is a function of ρ alone and S is a function of ϕ alone. Substitution of E_z in Equation 3.61a and of H_z in Equation 3.61b leads to the following:

$$S\frac{d^2 R}{d\rho^2} + \frac{1}{\rho}S\frac{dR}{d\rho} + \frac{R}{\rho^2}\frac{d^2 S}{d\phi^2} + RSh^2 = 0 \tag{3.63}$$

Division of Equation 3.63 by RS, addition and subtraction of n^2/ρ^2 in the resulting equation and rearrangement of terms results in

$$\left[\frac{1}{R}\frac{d^2R}{d\rho^2} + \frac{1}{R\rho}\frac{dR}{d\rho} + h^2 - \frac{n^2}{\rho^2}\right] + \left[\frac{1}{S\rho^2}\frac{d^2S}{d\phi^2} + \frac{n^2}{\rho^2}\right] = 0 \qquad (3.64)$$

Separation of R and S terms gives

$$\left[\frac{1}{R}\frac{d^2R}{d\rho^2} + \frac{1}{R\rho}\frac{dR}{d\rho} + h^2 - \frac{n^2}{\rho^2}\right] = 0 \qquad (3.65a)$$

$$\left[\frac{1}{S\rho^2}\frac{d^2S}{d\phi^2} + \frac{n^2}{\rho^2}\right] = 0 \qquad (3.65b)$$

In Equation 3.65a, replace ρ by ρh and rewrite the equation:

$$\left[\frac{d^2R}{d(\rho h)^2} + \frac{1}{(\rho h)}\frac{dR}{d(\rho h)} + \left(1 - \frac{n^2}{(\rho h)^2}\right)R\right] = 0 \qquad (3.66)$$

Equation 3.66 is of the form of Bessel's equation and its solution can be written as

$$R(\rho h) = A_{n1}J_n(\rho h) + B_{n1}Y_n(\rho h) \qquad (3.67)$$

In Equation 3.67, J and Y represent Bessel's functions of first and second kinds, respectively. The suffix n (with J and Y) indicates their order and the term (ρh) indicates argument of J and Y.

Equation 3.65b can be rewritten as

$$\frac{d^2S}{d\phi^2} + n^2S = 0 \qquad (3.68)$$

Its solution can be written as

$$S = C_{n1}\cos(n\phi) + D_{n1}\sin(n\phi) \qquad (3.69)$$

Substitute Equations 3.67 and 3.69 in Equations 3.62 to get

$$E_z \text{ or } H_z = [\{A_{n1}J_n(\rho h) + B_{n1}Y_n(\rho h)\}\{C_{n1}\cos(n\phi) + D_{n1}\sin(n\phi)\}] \qquad (3.70)$$

In Equation 3.70, if $\rho = 0$, $\rho h = 0$, $Y_n(\rho h) = \infty$, Y_n does not represent a physical field and B_{n1} may be set to zero. A_{n1} can now be merged with arbitrary constants C_{n1} and D_{n1}. This leads to new arbitrary constants as noted below:

$$A_n \ (\text{or } C_n)(= A_{n1}C_{n1}) \qquad\qquad (3.71a)$$

$$B_n \ (\text{or } D_n)(= A_{n1}D_{n1}) \qquad\qquad (3.71b)$$

Though E_z and H_z have the same form of solution, in order to differentiate between the two, the symbols used for their arbitrary constant are changed. Thus, for TM wave

$$E_z = J_n(\rho h)\{A_n \cos(n\phi) + B_n \sin(n\phi)\} \qquad\qquad (3.72)$$

For TE wave the expression takes the form

$$H_z = J_n(\rho h)\{C_n \cos(n\phi) + D_n \sin(n\phi)\} \qquad\qquad (3.73)$$

In expressions (3.72 and 3.73), the relative values of A_n and B_n determine the orientation of the field in a waveguide. In circular waveguide, ϕ axis (for any value of 'n') can be oriented to make either A_n or B_n equal to zero. For $\phi = 0$, $2\pi A_n = 0$ and for $\phi = \pi/2, 3\pi/2$ $B_n = 0$. Select $\phi = \pi/2$ to get $B_n = 0$.

3.4.3.1 Field Expressions for TM Wave

The final form of expression for E_z is

$$E_z = A_n J_n(\rho h) \cos(n\phi) \qquad\qquad (3.74a)$$

In view of Equations 3.74a and 3.59, the other field components for TM wave are

$$E_\rho = -\frac{\gamma}{h^2} A_n J_n(\rho h) \cos(n\phi) \qquad\qquad (3.74b)$$

$$E_\phi = -\frac{\gamma}{h^2} \frac{n}{\rho} A_n J_n'(\rho h) \sin(n\phi) \qquad\qquad (3.74c)$$

$$H_\rho = \frac{-j\omega\varepsilon}{h^2} \frac{n}{\rho} A_n J_n'(\rho h) \sin(n\phi) \qquad\qquad (3.74d)$$

$$H_\phi = -\frac{j\omega\varepsilon}{h^2} A_n J_n(\rho h) \cos(n\phi) \qquad\qquad (3.74e)$$

3.4.3.2 Field Expressions for TE Wave

Following the procedure adopted for TM wave, the final form of expression for H_z is

$$H_z = C_n J'_n(\rho h) \cos(n\phi) \tag{3.75a}$$

In view of Equations 3.75a and 3.60, the other field components for TE wave are

$$E_\rho = -\frac{j\omega\mu}{h^2}\frac{n}{\rho} C_n J_n(\rho h) \sin(n\phi) \tag{3.75b}$$

$$E_\phi = \frac{j\omega\mu}{h^2} C_n J'_n(\rho h) \cos(n\phi) \tag{3.75c}$$

$$H_\rho = -\frac{\gamma}{h^2} C_n J'_n(\rho h) \cos(n\phi) \tag{3.75d}$$

$$H_\phi = -\frac{\gamma}{h^2}\frac{n}{\rho} C_n J_n(\rho h) \sin(n\phi) \tag{3.75e}$$

In Equations 3.74c, 3.74d, 3.75a, 3.75c and 3.75d, J'_n presents the derivative of $J_n(\rho h)$ with respect to ρ or h. Also all expressions of Equations 3.74 and 3.75 are presumed to be multiplied by $e^{-j\beta z}$. All these expressions involve $J_n(x)$ term, thus the variation of only $J_n(x)$ is shown in Figure 3.16.

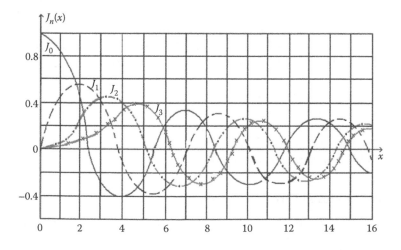

FIGURE 3.16
Bessel's function of first kind.

TABLE 3.3

Zeroes of $J_n(x)$

N	pth Zero of $J_n(x)$			
	P = 1	P = 2	P = 3	P = 4
0	2.405	5.520	8.645	11.792
1	3.832	7.106	10.173	13.324
2	5.136	8.417	11.620	14.796
3	6.380	9.761	13.015	—
4	7.588	11.065	14.372	—
5	8.771	12.339	—	—

TABLE 3.4

Zeroes of $J'_n(x)$

N	pth zero of $J'_n(x)$			
	P = 1	P = 2	P = 3	P = 4
0	3.832	7.016	10.173	13.324
1	1.841	5.831	8.536	11.706
2	3.054	6.706	9.969	13.170
3	4.201	8.015	11.346	—
4	5.317	9.282	12.682	—
5	6.416	10.520	13.989	—

In case of TM wave, $E_z = 0$ at the conducting wall of the waveguide (i.e., at $\rho = a$). This requires $J_n(ha) = 0$ where 'a' is the radius of the waveguide. Figure 3.16 illustrates that $J_n(ha)$ is zero infinite number of times for any integer value of n. The value of arguments at which Bessel's function becomes zero are referred to as zeroes of Bessel's function. Some of the zeroes of Bessel's function of different orders are given in Table 3.3, where 'p' is the number of zeroes.

In case of TE, wave boundary condition cannot be directly applied to H_z. Therefore, first $E\phi$ is to be obtained, which is tangential to the waveguide wall. The application of boundary conditions leads to $J'_n(ha) = 0$. Again $J'_n(ha)$ will be zero for all zeroes of $J'_n(ha)$. The values of some of the zeroes of J'_n are given in Table 3.4. In J'_n, n is the order of the derivative of Bessel's function.

3.4.4 TEM Modes in Circular Waveguide

In this case, $E_z = H_z = 0$. This means that electric and magnetic fields are completely transverse to the direction of propagation. This mode, therefore, cannot exist in a hollow waveguide irrespective of its shape. It can propagate only in a two-conductor system such as coaxial cable and two-wire lines.

3.4.5 Mode Designation in Circular Waveguide

In the TM case, the wave equation is solved for E_z. At the conductor surface, $E_z = 0$ and thus $J_n(ha) = 0$. In view of Equation 3.8, for a propagating wave $h^2 < \omega^2 \mu \varepsilon$. This requires that either h is small or the frequency is high. To meet this requirement, only a few roots of $J_n(ha) = 0$ will be of practical importance. Such first few roots are $(ha)_{01} = 2.405$, $(ha)_{11} = 3.85$, $(ha)_{02} = 5.52$ and $(ha)_{12} = 7.02$. These roots can thus be designated as $(ha)_{nm}$. In $(ha)_{nm}$, the first suffix (n) is the order of Bessel's function and the second suffix is the number of root. Similarly, the first few roots of $J'_n(ha) = 0$ are $(ha)'_{01} = 3.83$, $(ha)'_{11} = 1.84$, $(ha)'_{02} = 7.02$ and $(ha)'_{12} = 5.33$. These roots are designated as $(ha)'_{nm}$. In view of the above, the TM modes can be referred as TM_{01}, TM_{11}, TM_{02} and TM_{12}.

In TE wave, $E_z \equiv 0$ and the field components are given by Equation 3.75. In view of the zeroes of $J_n(ha) = 0$ and $J'_n(ha) = 0$, TE modes can be referred to as TE_{01}, TE_{11}, TE_{02} and TE_{12}.

In these modes

$$h_{nm} = \frac{(ha)_{nm}}{a} \tag{3.76a}$$

$$h'_{nm} = \frac{(ha)'_{nm}}{a} \tag{3.76b}$$

In view of Equation 3.76, various components of TM and TE waves given by Equations 3.74 and 3.75 can be written in the modified form by replacing h (Equation 3.18b) by h_{nm} or h'_{nm}. For circular waveguide, the expressions for h'_{nm}, h_{nm}, β_{nm}, f_c, v_p, λ_g, Z_z (for both TM and TE waves) obtained in view of Equations 3.76a, 3.76b, 3.22b, 3.22c, 3.29a, 3.29b, 3.26 and 3.28 are given in Table 3.5. It can be noted that the relations for f_c, v_p, λ_g, Z_z(TM) and Z_z(TE) are the same as obtained in case of the rectangular waveguide.

TABLE 3.5

Characteristic Parameters for TE and TM Waves in Circular Waveguide

Characteristic Parameter	TE Wave		TM Wave	
Cutoff wave number	$h'_{nm} = \dfrac{(ha)'_{nm}}{a} = \omega_c\sqrt{\mu\varepsilon}$	(3.76a)	$h_{nm} = \dfrac{(ha)_{nm}}{a} = \omega_c\sqrt{\mu\varepsilon}$	(3.76b)
Phase shift constant	$\beta_{nm} = \sqrt{(\omega^2\mu\varepsilon - h'^2_{nm})}$	(3.76c)	$\beta_{nm} = \sqrt{(\omega^2\mu\varepsilon - h^2_{nm})}$	(3.76d)
Cutoff frequency	$f_c = \dfrac{h'_{nm}}{2\pi\sqrt{\mu\varepsilon}}$	(3.76e)	$f_c = \dfrac{h_{nm}}{2\pi\sqrt{\mu\varepsilon}}$	(3.76f)
Phase velocity	$v_p = c/\sqrt{1 - (f_c/f)^2}$ (for both TE and TM modes)			(3.76g)
Guide wavelength	$\lambda_g = \lambda/\sqrt{1 - (f_c/f)^2}$ (for both TE and TM modes)			(3.76h)
Wave impedance	$Z_{TE} = \eta/\sqrt{1 - (f_c/f)^2}$	(3.76i)	$Z_{TM} = \eta\sqrt{1 - (f_c/f)^2}$	(3.76j)

3.4.6 Field Distribution in Circular Waveguide

Field distributions for some TM and TE modes across the cross section of a circular waveguide are shown in Figure 3.17. Figure 3.18 illustrates field distributions for the same modes across section a–a. Figure 3.19 shows H field distributions across section b–b and Figure 3.20 gives E field distributions across section a–a for a few modes.

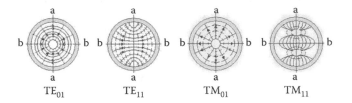

FIGURE 3.17
TE and TM modes in circular waveguide.

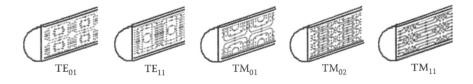

FIGURE 3.18
Field distributions across a–a for TE and TM modes.

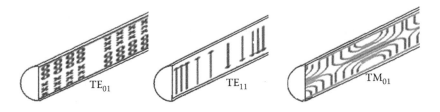

FIGURE 3.19
E field across section b–b for TE_{01}, TE_{11} and TM_{01} modes.

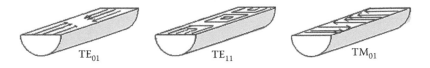

FIGURE 3.20
H field across section b–b for TE_{01}, TE_{11} and TM_{01} modes.

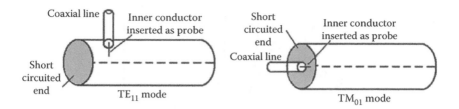

FIGURE 3.21
Excitation of TE_{11} and TM_{01} modes in a circular waveguide.

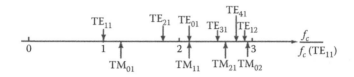

FIGURE 3.22
Cutoff frequencies of TE and TM modes.

3.4.7 Mode Excitation

The method of excitation of TE_{11} and TM_{01} modes is shown in Figure 3.21.

3.4.8 Cutoff Frequencies

Figure 3.22 illustrates some of the cutoff frequencies for TE and TM modes in circular waveguide, relative to the cutoff frequency of the dominant TE_{11} mode.

3.4.9 Attenuation in Circular Waveguides

Attenuation constant (α) for TE and TM modes can be obtained by adopting the method used for a rectangular waveguide. Variation of α (in dBs/m) in a (copper) waveguide with $a = 2.54$ cm for TE_{11}, TE_{01} and TM_{01} are shown in Figure 3.23, wherein the location of arrows indicates cutoff frequencies.

3.5 Dielectric Waveguides

A dielectric waveguide can have rectangular- or circular-shaped geometry. Some of these may be aided with ground planes. Figure 3.24 illustrates some of the dielectric waveguides.

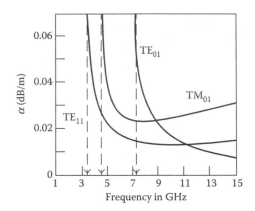

FIGURE 3.23
Attenuation in a circular waveguide.

3.5.1 Dielectric Slab Waveguide

A dielectric slab (or slab waveguide) can support a wave in a similar manner as it is supported by the metallic waveguides. Under the conditions of total internal reflection (Section 2.15.2.4) with a reflection coefficient of suitable phase shift, a wave can propagate unattenuated by bouncing back and forth between two surfaces of a dielectric slab. The problem is similar to that of the electric field and current flow within the conductor and may be similarly tackled.

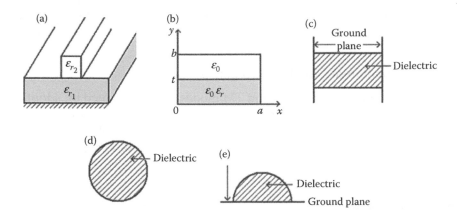

FIGURE 3.24
Different types of dielectric waveguides. (a) Geometry of rectangular dielectric waveguide, (b) partially filled rectangular dielectric waveguide, (c) *H*-plane type dielectric waveguide, (d) circular rod dielectric waveguide and (e) dielectric waveguide with ground plane.

The geometry of a dielectric slab is shown in Figure 3.25a, whereas its regions and involved parameters are shown in Figure 3.25b. The slab, with thickness d is assumed to be large along the x-axis. It extends from $-a/2$ to $+a/2$ along the y-axis and from $-\infty$ to $+\infty$ along z-axis. The configuration is divided into two regions. One is the air region that extends from 0 to $+\infty$ along the x-axis. This region is referred to as outside (or O) region. The other region confined to the thick dielectric slab is called the internal (or I) region. Solutions separately obtained in these regions must match at the common boundary in view of the continuity of tangential E and normal H. The solutions require certain simplifying assumptions which are as follows:

1. The wave is propagating in the z direction and the field variation along z is characterised by $e^{-\gamma z}$. The values of propagation constant (γ) in regions O and I are γ_0 and γ_1, respectively.
2. There is no variation of the field along the x direction; thus, all x derivatives vanish (i.e., $\partial/\partial x \equiv 0$).
3. The variation along y is unknown and is to be evaluated.

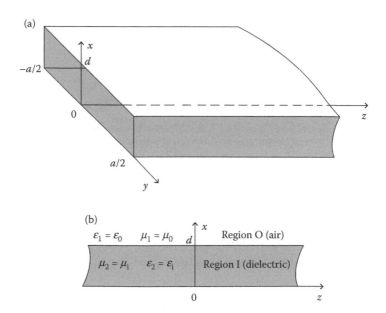

FIGURE 3.25
Dielectric slab waveguide. (a) Geometry and (b) regions and parameters involved.

In view of these assumptions, the Maxwell's equations in region O and I can be written as follows:

Region O (Air Region)		Region I (Dielectric Slab Region)	
$\dfrac{\partial E_z}{\partial y} + \gamma_o E_y = -j\omega\mu_o H_x$	(3.77a)	$\dfrac{\partial E_z}{\partial y} + \gamma_i E_y = -j\omega\mu_i H_x$	(3.78a)
$-\gamma_o H_x = j\omega\varepsilon_o E_y$	(3.77b)	$-\gamma_i H_x = j\omega\varepsilon_i E_y$	(3.78b)
$-\dfrac{\partial H_x}{\partial y} = j\omega\varepsilon_o E_z$	(3.77c)	$-\dfrac{\partial H_x}{\partial y} = j\omega\varepsilon_i E_z$	(3.78c)

Equations 3.77 and 3.78 can be manipulated to yield

$\dfrac{\partial^2 H_x}{\partial y^2} + \gamma_o^2 H_x + \omega^2\mu_o\varepsilon_o H_x = 0$	(3.79a)	$\dfrac{\partial^2 H_x}{\partial y^2} + \gamma_i^2 H_x + \omega^2\mu_i\varepsilon_i H_x = 0$	(3.80a)
$\dfrac{\partial^2 H_x}{\partial y^2} + h_o^2 H_x = 0$	(3.79b)	$\dfrac{\partial^2 H_x}{\partial y^2} + h_i^2 H_x = 0$	(3.80b)
where $h_o^2 = \gamma_o^2 + \omega^2\mu_o\varepsilon_o$	(3.79c)	$h_i^2 = \gamma_i^2 + \omega^2\mu_i\varepsilon_i$	(3.80c)

Solution to Equation 3.79b can be written as:

$$H_x = C_1\, e^{h_o y} + C_2\, e^{-h_o y} \qquad (3.81a)$$

The first term of Equation 3.81a becomes infinite at $y = \infty$ and does not represent a physical field. It may be eliminated by putting $C_1 = 0$ to get

$$H_x = C_2\, e^{-h_o y} \qquad (3.81b)$$

Incorporation of z-variation in Equation 3.81b gives

$$H_x = C_2\, e^{-h_o y}\, e^{-\gamma_o z} = H_{xo}\, e^{-\gamma_o z} \qquad (3.83a)$$

$$\text{where } H_{xo} = C_2\, e^{-h_o y} \qquad (3.83b)$$

In view of the finite thickness of slab, Equation 3.80b yields the following two forms of the solutions:

$$H_x = C_3 \sin h_i y \qquad (3.82a)$$

$$H_x = C_4 \cos h_i y \qquad (3.82b)$$

Inclusion of z-variation in Equations 3.82a and 3.82b gives

$$H_x = C_3 \sin h_i y\, e^{-\gamma_i z} = H_{xi}\, e^{-\gamma_i z} \qquad (3.84a)$$

$$\text{where } H_{xi} = C_3 \sin h_i y \qquad (3.84b)$$

$$H_x = C_4 \cos h_i y\, e^{-\gamma_i z} = H_{xi}\, e^{-\gamma_i z} \qquad (3.84c)$$

$$\text{where } H_{xi} = C_4 \cos h_i y \qquad (3.84d)$$

Since the values of H_x given by Equations 3.83b, 3.84b and 3.84d are to be matched at the surface of slab, the two propagation constants have to be equal, that is, $\gamma_o = \gamma_i$. Also, $\mu_i = \mu_0$ for air and the dielectric. Thus, Equations 3.79c and 3.80c become

$h_o^2 = \gamma_o^2 + \omega^2\mu_o\varepsilon_o$	(3.85)	$h_i^2 = \gamma_o^2 + \omega^2\mu_o\varepsilon_i$	(3.86)

In view of Equations 7.79b and 7.7c

In view of Equation 7.80b and 7.80c

$E_{yo} = [j\gamma_o/\omega\varepsilon_o]H_{xo}$	(3.87a)	$E_{yi} = [j\gamma_o/\omega\varepsilon_i]H_{xi}$	(3.88a)
$E_{zo} = -[jh_o/\omega\varepsilon_o]H_{xo}$	(3.87b)	$E_{zi} = -[jh_i/\omega\varepsilon_i]H_{xi}$	(3.88b)

The surface impedance looking down on the surface of slab is

$$Z_s = -E_z/H_x = jh_o/\omega\varepsilon_o \qquad (3.89)$$

If h_o is positive real (necessary for the existence of a wave), Z_s is a positive reactance. Thus, a surface wave can exist over a reactive surface. In case of the TM wave, this reactance has to be inductive in nature. For dielectric slab, tangential field components are to be matched at the dielectric surface. The two cases noted in Equations 3.84b and 3.84d are to be considered separately.

Case I

$$H_{xi} = C_3 \sin h_i y \qquad (3.84b)$$

Equating Equations 3.83b and 3.84b for the continuity of tangential H at $y = a/2$ to get

$$C = C_2/C_3 = e^{h_0 a/2} \sin (h_i a/2) \qquad (3.90a)$$

$$e^{h_0 a/2} = C/\sin (h_i a/2) \qquad (3.90b)$$

For continuity of tangential E equate Equations 3.87b and 3.88b after putting values of H_{xo} from Equation 3.83b and H_{xi} from Equation 3.84b. It gives after manipulation:

$$-(\varepsilon_o h_i/\varepsilon_1) \cot (h_i a/2) = h_0 \qquad (3.90c)$$

Case II

$$Hx_i = C_4 \cos h_i y \qquad (3.84d)$$

Equating Equations 3.83b and 3.84d for the continuity of tangential H at $y = a/2$ to get

$$C' = C_2/C_4 = e^{h_0 a/2} \cos (h_i a/2) \qquad (3.91a)$$

$$e^{h_0 a/2} = C'/\cos (h_i a/2) \qquad (3.91b)$$

For continuity of tangential E equate Equations 3.87b and 3.88b after putting values of H_{xo} from Equation 3.83b and H_{xi} from Equation 3.84d. It gives after manipulation:

$$(\varepsilon_o h_i/\varepsilon_1) \tan (h_i a/2) = h_0 \qquad (3.91c)$$

Equations 3.85 and 3.86 for h_0 and h_i may be combined to give

$$h_0^2 + h_i^2 = \omega^2 \mu(\varepsilon_1 - \varepsilon_0) \qquad (3.92)$$

Finally, h_0 may be eliminated (from Equations 3.84 and 3.85) to give a transcendental equation. The resulting transcendental equations for the two cases may be solved graphically to give h_i. Parameters γ_o and h_0 then may be obtained for both the cases.

Figure 3.26 shows a dielectric waveguide with ground plane. In this case, the z component of the E field is zero at $y = 0$ so that the solution applies as well to a dielectric slab of thickness $a/2$ over a perfectly conducting plane. Besides, dielectric slab waves can also be guided by dielectric rods (optical fibre) or by corrugated periodic surfaces.

3.5.2 Dielectric Rod Waveguide

At millimetric wave range, the precise fabrication of ordinary waveguide becomes difficult. At these frequencies, the dielectric rod waveguide can be safely used as a low-loss transmission line. These exhibit good propagation characteristics for frequencies greater than 100 GHz. In rod waveguides, energy follows the principle of total internal reflection as in the case of optical fibres. As shown in Figure 3.27, the wave in rod waveguides travels in a zig-zag fashion, and remains confined within the (rod) waveguide provided its angle of launch or angle of incidence (θ_i) is greater than the critical angle (θ_c), where $\theta_c = \tan^{-1} \sqrt{(1/\mu_r \, \varepsilon_r)}$.

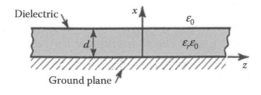

FIGURE 3.26
Dielectric slab waveguide with a ground plane.

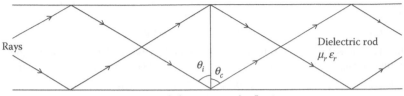

Zigzag path due to internal reflection

FIGURE 3.27
Zigzag path of a wave in a rod waveguide.

3.6 Physical Interpretation of Wave Terminology

This section includes the physical interpretation of some important terms introduced in connection with the waveguides.

3.6.1 Modes

An electromagnetic wave can propagate in various (ways or) modes. The time taken by the wave to move down the waveguide varies with the mode. Each mode has a cutoff frequency below which it cannot propagate.

3.6.2 Mode Designations

In a wave, if the electric field is at right angles to the direction of travel, the wave is said to be following 'transverse electric (TE) mode' and if the magnetic field is at right angles to the direction of travel, it is said to be following the 'transverse magnetic (TM) mode'. The above nomenclature is true for parallel plane guide, waveguides and for planar lines. In the TEM mode, both E and H fields are perpendicular to the direction of propagation. Waves in power lines are of the TEM mode since its cutoff frequency is zero. Waves of the TEM mode cannot propagate in a hollow waveguide irrespective of the shape of its cross section.

3.6.3 Dominant Mode

The mode with the lowest cutoff frequency is termed as dominant mode. The TE_{10} mode has the lowest cutoff frequency for rectangular waveguide wherein it has one half-cycle along the broader dimension of the waveguide. The wavelength (λ) is related to 'a' as '$a = \lambda c/2$', where c is the velocity of light. Modes with next higher cutoff frequency are TE_{01} and TE_{20}. In a circular waveguide, TE_{11} is the mode with the lowest cutoff frequency followed by TM_{01}.

3.6.4 Cutoff Frequency

It is the frequency below which there is no propagation of wave. For the TE_{10} mode in the rectangular waveguide with $a = 2b$, the cutoff frequency is given by $f_c (= c/2a)$. For circular waveguide of radius 'a', the cutoff or critical frequency is given by

$$f_c = \frac{h'_{nm}}{2\pi\sqrt{\mu\varepsilon}} \quad \text{for TM wave} \quad \text{and} \quad f_c = \frac{h_{nm}}{2\pi\sqrt{\mu\varepsilon}} \quad \text{for TE wave}$$

3.6.5 Usable Frequency Range

When a wave progresses along its path, it gets dispersed. Dispersion mainly occurs between cutoff frequency (for the TE_{10} mode in rectangular waveguide) and twice that of the cutoff frequency; thus, it is not desirable to use a waveguide at the extremes of this range. Single-mode propagation has lesser dispersion and thus is highly desirable.

3.6.6 Characteristic Impedance

Characteristic impedances $Z_z(TE)$ and $Z_z(TM)$ are the functions of frequency for both rectangular and circular waveguides and are given by the relations $Z_Z(TE) = \eta/\sqrt{\{1 - (f_c/f)^2\}}$ and $Z_Z(TM) = \eta\sqrt{\{1 - (f_c/f)^2\}}$.

3.6.7 Guide Wavelength

The wavelength in rectangular or circular waveguide (λ_g) is longer than that in free space at the same frequency. For both TE and TM wave, it is given by $\lambda_g = \lambda/\sqrt{\{1 - (f_c/f)^2\}}$.

3.6.8 Phase Velocity

The phase velocity (v_p) of the wave along the walls of a waveguide is not a real velocity. This in fact gives the rate of change of phase and is always greater than c. For both TE and TM wave in rectangular waveguide, it is given by $v_p = c/\sqrt{\{1 - (f_c/f)^2\}}$.

3.6.9 Group Velocity

In a hollow waveguide having air or vacuum inside, the electromagnetic wave propagates at the speed of light (c) but follows a zig-zag path. Thus, the overall speed at which the signal travels the guide is always less than c. This velocity of the signal is called group velocity and is denoted by 'v_g'. In view of the relation $v_p\, v_g = c^2$ and the expression of v_p the expression v_g is obtained to be $v_g = c\sqrt{\{1 - (f_c/f)^2\}}$.

3.6.10 Transformation of Modes

In view of the mathematical theory developed for rectangular and circular waveguides, several modes of wave propagation were illustrated. The dominant modes for both are reillustrated in Figure 3.28 wherein the firm lines represent the E field and dotted lines the H field. Although in most of the cases rectangular waveguides are used, in some of the components, involvement of the circular waveguide is unavoidable. Very often, waveguide components of rectangular and circular shapes follow each other. It therefore needs to be noted that a wave with the TE_{10} mode of rectangular guide transforms into the TE_{11} mode when it enters a circular guide. The opposite is also true, that is, a TE_{11} mode in the circular guide transforms into a TE_{10} mode in the rectangular guide.

The above transformation can be justified if a rectangular waveguide (Figure 3.4) is hammered till it takes the shape of a circular waveguide (Figure 3.15). Its TE_{10} mode (Figure 7.5) will transform into the TE_{11} mode (Figure 3.17) of circular waveguide. Similarly, if a circular waveguide is hammered to take the shape of a rectangular waveguide, the TE_{11} mode of a circular waveguide will transform to the TE_{10} mode in a rectangular waveguide.

3.6.11 Mode Disturbance

If in a waveguide the propagation is assumed to be along the z-axis, an analogy can be drawn between the field corresponding to a particular mode and the movement of rice-shaped items (of $\lambda/2$ lengths) in a hollow pipe. As shown in Figure 3.29a corresponding to the TE_{10} mode these rice-shaped items move in a single line one after the other. Figure 3.29b corresponds to the TE_{20} mode. In this case, two (upper and lower) adjacent objects move in the opposite directions in order to match the alternating nature of the field. In case of an increase of frequency, the cross section of the waveguide will not correspond to $\lambda/2$ or its integer multiple. This will correspond to the movement of items in a disturbed fashion as illustrated in Figure 3.29c.

3.6.12 Multimode Propagation

Figure 3.30 illustrates two rays one of which represents a lower-order mode and the other a higher-order mode. The nature of propagation of these two

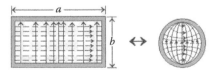

TE$_{10}$ mode in rectangular   TE$_{11}$ mode in circular
waveguide   waveguide

FIGURE 3.28
Transformation of modes.

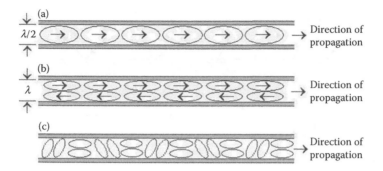

FIGURE 3.29

Propagation of different modes. (a) Dominant TE_{10} mode, (b) TE_{20} mode and (c) combination of modes with the increase of frequency.

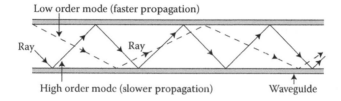

FIGURE 3.30

Multimode propagation.

modes can be explained in view of Equation 3.30a. According to this velocity relation, there will be faster propagation of the lower-order mode and slower propagation of the higher-order mode. Correspondingly, the wave will strike the two walls more frequently in case of higher-order mode than in case of the lower-order mode.

3.6.13 Group Velocity Variations

In view of Equations 3.18, 3.22 and 3.29, it is evident that both velocity 'v' and wavelength 'λ' are inversely proportional to m and n. If the frequency 'ω' remains the same and m and n are increased (i.e., order of mode is higher), the wave will progress at slower velocity. Furthermore, the figure shows the locations of wave incidence on the waveguide walls. These locations are much nearer for higher-order mode in comparison to the lower and thus the wavelength along the wall is bound to be less. Figure 3.31 shows the paths followed by rays with the variation of group velocity which varies with frequency.

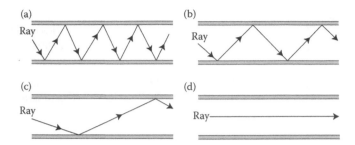

FIGURE 3.31
Variation of group velocity with frequency. (a) f just above f_c, (b) $f > f_c$, (c) $f \gg f_c$ and (d) f approaches infinity.

3.6.14 Wavefront Movement

Similarly, Figure 3.32 illustrates the progressing wavefronts with alternating polarity in a waveguide. The *wavefront* is nothing but an equiphase plane. In view of the alternating nature of the signal, the polarity of phase and hence the direction of the wavefront alternates.

3.6.15 Spurious Modes

For the TE_{10} mode, the separation between two narrower walls has to be $\lambda/2$ at the frequency of operation. Waveguides are generally designed at the central frequency of the band of operation. In case the frequency is lowered or raised within the band itself, the condition of $\lambda/2$ separation cannot be met. It will amount to the addition of other modes than that has been excited. This phenomenon is analogous to the presence of harmonics in a distorted signal.

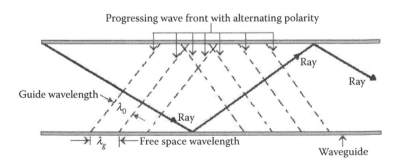

FIGURE 3.32
Movement of wavefronts along a waveguide.

3.7 Relative Merits of Waveguides

The assessment of waveguides in terms of different performance parameters in comparison to two-wire and coaxial lines is as discussed in the following section.

3.7.1 Copper Losses

The well-known relation '$R = \rho L/A$' indicates that the larger the area (A) of the current-carrying conductor, the lesser will be the resistance (R); the larger the length (L) of the conductor, the larger will be the resistance; and the larger the resistivity (ρ) of the conducting material, the larger will be the resistance. Also, if the resistance is large, more will be the copper losses. In view of this relation, two-wire line, coaxial cable and the waveguides are assessed in the following paragraphs.

The relatively small surface area of a two-wire line results in large copper losses. In case of the coaxial cable, the surface area of the outer conductor is large and that of the inner conductor is relatively small. With the increase in frequency, the current tries to confine to the skin of the conductor. In view of this phenomenon referred to as the 'skin effect', at microwave frequencies, the current-carrying area of the inner conductor reduces only to a small layer of the conductor surface. Thus, the skin effect tends to increase the effective resistance of the conductor. The energy transfer in the coaxial cable is caused by electromagnetic field motion. The magnitude of the field is limited by the size of the inner conductor. As the size of the inner conductor is further reduced by the skin effect, the energy transmission by the coaxial cable becomes less efficient. In comparison to the two-wire line and the coaxial cable, the copper losses (I^2R) in the waveguide are greatly reduced in view of its large surface area.

3.7.2 Dielectric Loss

The conductors of two-wire lines and coaxial cables are separated by (air or other) dielectric. A potential difference across the two wires results in heating of the dielectric, which in turn results in power loss. This loss is often referred to as dielectric loss. The dielectric in a waveguide is air, which has much lower dielectric loss than in conventional insulating materials. Thus, dielectric loss in waveguides is lower than in two-wire and coaxial lines.

3.7.3 Insulation Breakdown

Earlier it was noted that if a line is not terminated in its characteristic impedance, there will be reflections and hence standing waves. Standing waves are stationary by nature and the voltage of standing waves at the points of greatest magnitude may become large enough to break down the insulation between line conductors. As standing waves contain high-voltage spikes or 'nodes', the breakdown usually occurs at these stationary nodes. In practical applications, the breakdown of insulation in lines is more frequent and severe. Waveguides too may be subject to dielectric breakdown. In a waveguide, the standing wave may cause arcing, which may damage its surface and may decrease its efficiency of energy transfer.

3.7.4 Power-Handling Capability

The power-handling capability is related to the distance between conductors. In view of Figure 3.3, conductors in a waveguide have greater separation than in coaxial line. Thus, waveguides can handle more power than coaxial lines of the same size.

3.7.5 Radiation Losses

In view of Figure 2.39, the field is not confined in case of two-wire line, whereas it is confined in case of coaxial cable and waveguides. Thus, in comparison to the two-wire line, the radiation losses are much less in coaxial cables and waveguides.

3.8 Limitations of Waveguides

As the waveguide width has to be about a half wavelength for the dominant mode at the frequency of operation, the use of waveguides below 1 GHz becomes almost impractical. The lower frequency limits are primarily determined by the physical size of waveguides. Thus, waveguides are used only at frequencies that fall under UHF and higher bands. Also, waveguides are rigid hollow pipes and their proper installation and operation require special components. The inner surfaces of waveguides are often coated with highly conducting materials such as silver, gold or their alloys for reducing the losses due to the skin effect. The last two measures increase the cost of waveguide systems and decrease its practicality, and at frequencies other than microwaves, it becomes almost impractical to use waveguide systems.

EXAMPLE 3.1

A TE_{10} mode is propagated in a rectangular waveguide with dimensions $a = 4$ cm and $b = 2$ cm. Find the signal frequency if the distance between the adjacent minima and maxima of the propagating wave is 5 cm.

Solution

Given the TE_{10} mode, $m = 1$ and $n = 0$, $a = 4$ cm, $b = 2$ cm and the distance between the adjacent minima and maxima = $\lambda_g/4 = 5$ cm or $\lambda_g = 20$ cm.

The cutoff wavelength of the wave is given by

$$\lambda_c = 2/\sqrt{\{(m/a)^2 - (n/b)^2\}} = 2/\sqrt{\{(1/a)^2\}} = 2a = 2 \times 4 = 8 \text{ cm}$$
$$\lambda_g = \lambda_0/\sqrt{\{1 - (\lambda_0/\lambda_c)^2\}}$$

On squaring both sides and on simplification

$$\lambda_g^2 = \lambda_0^2/\{1 - \left(\lambda_0/\lambda_c\right)^2\} \quad \lambda_0^2 = \lambda_g^2\{1 - \left(\lambda_0/\lambda_c\right)^2\}$$

$$\lambda_0^2\lambda_c^2 = \lambda_g^2\lambda_c^2 - \lambda_g^2\lambda_0^2 \quad \lambda_0^2\lambda_c^2 + \lambda_g^2\lambda_0^2 = \lambda_g^2\lambda_c^2$$

$$\lambda_0^2 = \lambda_g^2\lambda_c^2/(\lambda_c^2 + \lambda_g^2) \quad \lambda_0^2 = 400 \times 64/464 \quad \text{or} \quad \lambda_0 = 7.4278 \text{ cm}$$

$$\lambda_0 = v_0/f_0 \quad \text{or} \quad f_0 = v_0/\lambda_0 \quad f_0 = 3 \times 10^8/7.4278 \times 10^{-2} = 4.03887 \text{ GHz}$$

EXAMPLE 3.2

Determine (a) the cutoff frequency f_c, (b) the phase velocity v_p and (c) the guide wavelength λ_g for a TE_{11} wave of 10 GHz travelling in an air-filled rectangular waveguide having $\beta_g = 1.1$ rad/m in the z direction. Also write the expressions of different surviving field components of the electric field intensity if the magnetic field intensity expression is given by $H_z = H_0 \cos (\pi x/\sqrt{6}) \cos (\pi y/\sqrt{6})$ A/m.

Solution

Given $m = 1$, $n = 1$, $f_0 = 10$ GHz, $\beta_g = 1.1$ rad/m $\mu = \mu_0$, $\varepsilon = \varepsilon_0$ and $H_z = H_0 \cos (\pi x/\sqrt{6}) \cos (\pi y/\sqrt{6})$

In general, the expression of H_z is written as $H_z = H_{0z} \cos (m\pi x/a) \cos (n\pi y/b)$ A/m.

Comparison of two H_z expressions for the TE_{11} mode gives $a = b = \sqrt{6}$
For an air dielectric: $v_0 = 1/\sqrt{(\mu\varepsilon)} = 3 \times 10^8$ m/s
For $f_0 = 10$ GHz, $\omega = 2\pi f_0 = 2 \times 3.14157 \times 10 \times 10^9 = 62.83185 \times 10^9$

a. $f_c = \{1/2\sqrt{(\mu\varepsilon)}\}\sqrt{\{(m/a)^2 - (n/b)^2\}} = \{v_0/2\}\sqrt{(1/a)^2 + (1/b)^2}$

   $= \{(3 \times 10^8)/2\}\sqrt{\{(1/\sqrt{6})^2 - (1/\sqrt{6})^2\}} = (1.5 \times 10^8) \times 5.77\} = 8.66$ GHz

   $f_c/f_0 = 8.66/10 = 0.866$, $(f_c/f_0)^2 = 0.75$

   $\{1 - (f_c/f_0)^2\} = 0.25$ and $\sqrt{\{1 - (f_c/f_0)^2\}} = 0.5$

   $v_0 = f_0 \lambda_0$ $\lambda_0 = v_0/f_0 = 3 \times 10^8/10 \times 10^9 = 0.03$ m

b. $v_p = \omega/\beta_g = 62.83185 \times 10^9/1.1 = 57.12 \times 10^9$ m/s

   $v_g = c^2/v_p = (3 \times 10^8)^2/57.12 \times 10^9 = 0.15756 \times 10^7$ m/s

c. $\lambda_g = \lambda_0/\sqrt{\{1 - (f_c/f_0)^2\}} = 0.03$ m$/0.5 = 0.06$ m

In view of Equation 3.21, E_x and E_y are the two surviving components of E for the TE wave. These are as given below:

$$E_x = C A (j\omega\mu/h^2) \cos(m\pi x/a) \sin(n\pi y/b)e^{-j\beta z}$$
$$= E_{x0} \cos(m\pi x/a) \sin(n\pi y/b) \, e^{-j\beta z}$$

$$E_y = C\,B\,(-j\omega\mu/h^2)\,\sin(m\pi x/a)\cos(n\pi y/b)\,e^{-j\beta z}$$
$$= E_{y0}\sin(m\pi x/a)\cos(n\pi y/b)\,e^{-j\beta z}$$

Since $m = n = 1$, $a = b = \sqrt{6}$ and $\beta_g = 1.1$ rad/m

$$E_x = E_{x0}\cos(\pi x/\sqrt{6})\sin(\pi y/\sqrt{6})\,e^{-j1.1z}\ \text{V/m}$$
$$E_y = E_{y0}\sin(\pi x/\sqrt{6})\cos(\pi y/\sqrt{6})\,e^{-j1.1z}\ \text{V/m}$$

EXAMPLE 3.3

Calculate the values of inductive and capacitive reactances of equivalent circuits for (a) TE_{11} and (b) TM_{11} of 10 GHz wave propagating in an air-filled hollow rectangular waveguide with $a = 5$ cm and $b = 2.5$ cm. Illustrate these values in equivalent circuits and calculate corresponding $Z_0(TM)$ and $Z_0(TE)$.

Solution

Given $m = n = 1$ for both TE and TM waves, $a = 5$ cm and $b = 2.5$ cm.
For air, $\mu = \mu_0$, $\varepsilon = \varepsilon_0$, $f = 10$ GHz and $\omega = 2\pi \times 10^{10}$
$\omega\mu_0 = 2\pi \times 10^{10} \times 4\pi \times 10^{-7} = 7.895 \times 10^4$　$\omega\varepsilon_0 = 2\pi \times 10^{10} \times 10^{-9}/36\pi = 0.55555$
$h^2 = (m\pi/a)^2 + (n\pi/b)^2 = (\pi/5 \times 10^{-2})^2 + (\pi/2.5 \times 10^{-2})^2 = 1.974 \times 10^4$
$h^2/\omega\varepsilon_0 = 1.974 \times 10^4/0.556 = 3.55 \times 10^4$
$h^2/\omega\mu_0 = 1.974 \times 10^4/7.895 \times 10^4 = 0.25$

$$Z_0(TM) = \sqrt{\left[\frac{j\omega\mu + (h^2/j\omega\varepsilon)}{j\omega\varepsilon}\right]}$$

$$= \sqrt{\{(7.895 \times 10^4 - 3.55 \times 10^4)/0.556\}}$$

$$= 279.55\ \Omega$$

$$Z_0(TE) = \sqrt{\frac{j\omega\mu}{j\omega\varepsilon + (h^2/j\omega\mu)}}$$

$$= \sqrt{\{(7.895 \times 10^4)/(0.556 - 0.25)\}}$$

$$= 508\ \Omega$$

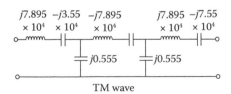

TM wave

FIGURE 3.33
Equivalent circuit for a TM wave.

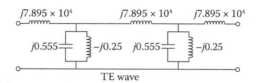

FIGURE 3.34
Equivalent circuit for a TE wave.

EXAMPLE 3.4

An air-filled rectangular waveguide having 4 cm × 2 cm cross section and made of copper operates at 10 GHz in the TE_{10} mode. Calculate its attenuation constant.

Solution

Given $m = 1$, $n = 0$, $a = 4$ cm and $b = 2$ cm
$\sigma = 5.8 \times 10^7$ ℧/m for copper $\mu = \mu_0$ $\varepsilon = \varepsilon_0$
$\eta = \sqrt{(\mu_0/\varepsilon_0)} = 377\ \Omega$ $1/\eta b = 0.132626$ $2b/a = 1$
$v_0 = 1/\sqrt{(\mu_0\varepsilon_0)} = 3 \times 10^8$ m/s $v_0/2 = 1.5 \times 10^8$ m/s
$f = 10$ GHz $\omega = 2\pi \times 10^{10}$
$f_c = \{1/2\sqrt{(\mu_0\varepsilon_0)}\}(1/a) = v_0/2a = 1.5 \times 10^8/(0.04) = 3.75$ GHz
$f_c/f = 3.75/10 = 0.375$ $(f_c/f)^2 = 0.140625$
$1 - (f_c/f)^2 = 0.86$ $\sqrt{\{1 - (f_c/f)^2\}} = 0.92736$
$(2b/a)\ (f_c/f)^2 = 1 \times 0.140625 = 0.140625 + (2b/a)\ (f_c/f)^2 = 1.140625$
$\mu\pi f = 4\pi \times 10^{-7} \times \pi \times 10^{10} = 39.478 \times 10^3$
$\sigma(1 - (f_c/f)^2) = (5.8 \times 10^7 \times 0.86) = 4.988 \times 10^7$
$\sqrt{[\mu\pi f/\{\sigma(1 - (f_c/f)^2)\}]} = \sqrt{\{(39.478 \times 10^3)/(4.988 \times 10^7)\}} = 0.0281$
$(1/\eta b)\ \sqrt{[\mu\pi f/\{\sigma(1 - (f_c/f)^2)\}]}\ \{1 + (2b/a)\ (f_c/f)^2\}$

Thus, in view of Equation 3.50a, the attenuation constant for the TE_{10} mode is

$$\alpha\ (TE_{10}) = 0.132626 \times 0.0281 \times 1.140625 = 0.004253\ \text{Nep./m}$$

EXAMPLE 3.5

A rectangular waveguide with dimensions 6 cm × 4 cm operates at 5 MHz in the TM_{11} mode. Compute (i) cutoff frequency f_c, (ii) guide wavelength λ_g, (iii) phase constant β_g, (iv) phase velocity v_p, (v) group velocity v_g and (iv) wave impedance Z for the case when the waveguide is filled with (a) air and (b) a dielectric having $\mu_r = 1$ and $\varepsilon_r = 2.1$.

Solution

Given the TM_{11} mode, $m = 1$, $n = 1$, $a = 6$ cm, $b = 4$ cm and $\mu = \mu_0$
$f = 5 \times 10^9$ Hz $\omega = 2\pi \times 5 \times 10^9 = \pi \times 10^{10}$
$\sqrt{\{(m/a)^2 + (n/b)^2\}} = \sqrt{\{(10^2/6)^2 + (10^2/4)^2\}} = 30.046$

Case (a) $\varepsilon_r = 1$ or $\varepsilon = \varepsilon_0$

$\sqrt{(\mu\varepsilon)} = \sqrt{(\mu_0\varepsilon_0)} = \sqrt{(4\pi \times 10^{-7} \times 10^{-9}/36\pi)}$
$\qquad = 0.333 \times 10^{-8}$

$\upsilon_0 = 1/\sqrt{(\mu_0\varepsilon_0)} = 3 \times 10^8$

$1/2\sqrt{(\mu\varepsilon)} = \upsilon_0/2 = 1.5 \times 10^8$

i. Cutoff frequency f_c
$f_c = \{1/2\sqrt{(\mu\varepsilon)}\}\sqrt{\{(m/a)^2 + (n/b)^2\}}$
$\qquad = 1.5 \times 10^8 \times 30.046$
$\qquad = 4.5069 \times 10^9$ Hz

ii. Guide wavelength λ_g
$\lambda = \upsilon_0/f = 3 \times 10^8/5 \times 10^9 = 0.06$ m

$f_c/f = 4.5069 \times 10^9/5 \times 10^9 = 0.90138$
$(f_c/f)^2 = 0.8125$
$1 - (f_c/f)^2 = 0.1875$
$\sqrt{\{1 - (f_c/f)^2\}} = 0.433$
$\lambda_g = \lambda/\sqrt{\{1 - (f_c/f)^2\}} = 0.06/0.433$
$\qquad = 0.13856$ m

iii. Phase constant β_g
$\beta_g = \omega\sqrt{(\mu\varepsilon)}\sqrt{\{1 - (f_c/f)^2\}}$
$\qquad = \pi \times 10^{10} \times 0.333 \times 10^{-8} \times 0.433$
$\qquad = 44.89$ rad/m

iv. Phase velocity υ_p
$\upsilon_p = \upsilon_0/\sqrt{\{1 - (f_c/f)^2\}}$
$\qquad = 3 \times 10^8/0.433$
$\qquad = 6.9284 \times 10^8$ m/s

v. Group velocity υ_g
$\upsilon_g = \upsilon_0 \sqrt{\{1 - (f_c/f)^2\}}$
$\qquad = 3 \times 10^8 \times 0.433 = 6.9284 \times 10^8$
$\qquad = 1.299 \times 10^8$ m/s

vi. Wave impedance Z_g
$\eta = \sqrt{(\mu_0/\varepsilon_0)} = 377$

$Z_g = \eta \sqrt{\{1 - (f_c/f)^2\}} = 377 \times 0.433$
$\qquad = 163.241\ \Omega$

Case (b) $\varepsilon_r = 2.1$ or $\varepsilon = 2.1\ \varepsilon_0$

$\sqrt{\mu\varepsilon} = \sqrt{(\mu_0\varepsilon_0)}\sqrt{\varepsilon_r} = 0.333 \times 10^{-8} \times \sqrt{2.1}$
$\qquad = 0.333 \times 10^{-8} \times 1.4491$
$\qquad = 0.483 \times 10^{-8}$

$\upsilon = 1/\sqrt{(\mu_0\varepsilon_0\varepsilon_r)} = 1/0.483 \times 10^{-8}$
$\qquad = 2.0702 \times 10^8$

$1/2\sqrt{(\mu\varepsilon)} = \upsilon/2 = 1.0351 \times 10^8$

i. Cutoff frequency f_c
$f_c = \{1/2\sqrt{(\mu\varepsilon)}\}\sqrt{\{(m/a)^2 + (n/b)^2\}}$
$\qquad = 1.0351 \times 10^8 \times 30.046$
$\qquad = 3.11 \times 10^9$ Hz

ii. Guide wavelength λ_g
$\lambda = \upsilon/f = 2.0702 \times 10^8/5 \times 10^9$
$\qquad = 0.0414$ m

$f_c/f = 3.11 \times 10^9/5 \times 10^9 = 0.622$
$(f_c/f)^2 = 0.386884$
$1 - (f_c/f)^2 = 0.613116$
$\sqrt{\{1 - (f_c/f)^2\}} = 0.783$
$\lambda_g = \lambda/\sqrt{\{1 - (f_c/f)^2\}} = 0.06/0.783$
$\qquad = 0.0766$ m

iii. Phase constant β_g
$\beta_g = \omega\sqrt{(\mu\varepsilon)}\sqrt{\{1 - (f_c/f)^2\}}$
$\qquad = \pi \times 10^{10} \times 0.483 \times 10^{-8} \times 0.783$
$\qquad = 118.81$ rad/m

iv. Phase velocity υ_p
$\upsilon_p = \upsilon/\sqrt{\{1 - (f_c/f)^2\}}$
$\qquad = 2.07 \times 10^8/0.783$
$\qquad = 2.6436 \times 10^8$ m/s

v. Group velocity υ_g
$\upsilon_g = \upsilon \sqrt{\{1 - (f_c/f)^2\}}$
$\qquad = 2.07 \times 10^8 \times 0.783$
$\qquad = 1.62081 \times 10^8$ m/s

vi. Wave impedance Z_g
$\eta = \sqrt{(\mu_0/\varepsilon_0\varepsilon_r)} = 377/\sqrt{\varepsilon_r} = 377/\sqrt{2.1}$
$\qquad = 260.155$

$Z_g = \eta \sqrt{\{1 - (f_c/f)^2\}} = 260.155 \times 0.783$
$\qquad = 203.7\ \Omega$

EXAMPLE 3.6

An air-filled circular waveguide carries a TE_{11} mode at 5 GHz. Its cutoff frequency is 80% of the operating frequency. Calculate its (i) diameter and (ii) phase velocity.

Solution

Given $m = 1$, $n = 1$, $\mu = \mu_0$ and $\varepsilon = \varepsilon_0$

$f = 5 \times 10^9$ Hz $f_c = 0.8f = 4 \times 10^9$ Hz

$v_0 = 3 \times 10^8$ m/s $v_0/2 = 1.5 \times 10^8$ m/s $\lambda = v_0 = 3 \times 10^8/2.5 \times 10^9 = 0.12$ m

From Equation 3.79b, $h'_{nm} = \{(ha)'_{nm}/a\} = \omega_c \sqrt{\mu\varepsilon}$ and

From Table 7.4 $(ha)'_{11} = 1.841$; thus, $h'_{11} = 1.841/6 \times 10^{-2} = 30.6833$

$$f_c = (ha)'_{nm}/2\pi a \sqrt{\mu\varepsilon} = \{(ha)'_{11}/2\pi a\} v_0$$

$$a = (ha)'_{nm}/2\pi f_c \sqrt{\mu\varepsilon} = \{(ha)'_{11}/2\pi f_c\} v_0$$

$$= 1.841 \times 3 \times 10^8/(2\pi \times 4 \times 10^9) = 2.2 \text{ cm}$$

Thus, the diameter of circular waveguide = $2a$ = 4.4 cm

$$f_c/f = 0.8 \ (f_c/f)^2 = 0.64 \quad 1 - (f_c/f)^2 = 0.36 \quad \sqrt{1 - (f_c/f)^2} = 0.6$$

$$v_p = v_0/\sqrt{1 - (f_c/f)^2} = 3 \times 10^8/0.6 = 5 \times 10^8 \text{ m/s}$$

EXAMPLE 3.7

A circular waveguide carries the dominant mode. Its cutoff frequency is 10 GHz. Calculate the inner diameter of the guide if it is filled with (a) air and (b) a dielectric having $\mu_r = 1$ and $\varepsilon_r = 2.5$.

Solution

Given dominant (TE_{11}) mode, $m = 1$, $n = 1$ and $f_c = 10 \times 10^9$ Hz

From Equation 3.79b, $h'_{nm} = \{(ha)'_{nm}/a\} = \omega_c \sqrt{\mu\varepsilon}$

From Table 7.4, $(ha)'_{11} = 1.841$ $h'_{11} = 1.841/a$

a. $\mu_r = 1$ or $\mu = \mu_0$ $\varepsilon_r = 1$ or $\varepsilon = \varepsilon_0$	b. $\mu_r = 1$ or $\mu = \mu_0$ $\varepsilon_r = 2.5$ or $\varepsilon = 2.5\varepsilon_0$
$v_0 = 1/\sqrt{(\mu_0\varepsilon_0)} = 3 \times 10^8$ m/s	$v = 1/\sqrt{(\mu_0\varepsilon_0\varepsilon_r)} = 1.897 \times 10^8$ m/s
$f_c = (ha)'_{nm}/2\pi a \sqrt{\mu_0\varepsilon_0}$	$f_c = (ha)'_{nm}/2\pi a \sqrt{\mu\varepsilon}$
$\quad = \{(ha)'_{11}/2\pi a\} v_0$	$\quad = \{(ha)'_{11}/2\pi a\} v$
$a = (ha)'_{nm}/2\pi f_c \sqrt{\mu\varepsilon}$	$a = (ha)'_{nm}/2\pi f_c \sqrt{\mu\varepsilon}$
$\quad = \{(ha)'_{11}/2\pi f_c\} v_0$	$\quad = \{(ha)'_{11}/2\pi f_c\} v$
$= 1.841 \times 3 \times 10^8/(2\pi \times 10 \times 10^9)$	$= 1.841 \times 1.897 \times 10^8/(2\pi \times 10 \times 10^9)$
$= 0.897 \times 10^{-2}$ m	$= 0.5558$ cm
Diameter of circular waveguide	Diameter of circular waveguide
$= 2a = 1.758$ cm	$= 2a = 1.111$ cm

PROBLEMS

P-3.1   A rectangular waveguide with air dielectric and dimensions $a = 4$ cm and $b = 2$ cm carries a signal of 5 GHz. Compute the following for the TE_{10} mode: (a) cutoff frequency f_c, (b) guide wavelength λ_g, (c) phase shift constant in guide β_g, (d) phase velocity v_p, (e) group velocity v_g and (f) waveguide impedance Z_g.

P-3.2 An air-filled hollow rectangular waveguide carrying a 20 GHz signal has $a = 5$ cm and $b = 2.5$ cm. Calculate the impedance encountered by the signal if the propagating mode is (a) TE_{11} and (b) TM_{11}.

P-3.3 An air-filled waveguide with dimensions 2 cm × 1 cm operating at 40 GHz delivers 300 W power through TE_{10} mode in the direction of propagation. Calculate the values of the electric field intensity in the guide.

P-3.4 An air-filled rectangular waveguide of 3 cm × 2 cm cross section has an attenuation factor of 5 Nep./m and operates at 8 GHz in the TE_{10} mode. Calculate its quality factor.

P-3.5 An air-filled circular waveguide of 12 cm diameter propagates a TE_{11} wave and operates at 2.5 GHz. Calculate its (i) cutoff frequency and (ii) guide wavelength and wave impedance.

P-3.6 Identify the possible modes that can propagate in an air-filled circular waveguide of 5 cm diameter if its cutoff frequency is 6 GHz.

P-3.7 An air-filled circular waveguide (CWG) of 8 cm diameter carries (a) the TE_{21} mode and (b) the TM_{01} mode. Compute the cutoff frequency for both the modes. If the guide is filled with a dielectric having $\varepsilon_r = 4$, by what factor will the cutoff frequency change? If the cutoff frequency is to be maintained at the same level, by what factor will the diameter be modified?

Descriptive Questions

Q-3.1 Using the method of separation of variables, obtain the expression of E_z for TM wave and explain the terms involved in it.

Q-3.2 Discuss the conclusions drawn from the expression $\gamma = \sqrt{[\{(m\pi/a)^2 + (n\pi/b)^2\} - \omega^2 \mu \varepsilon]}$.

Q-3.3 What are the lowest possible TE and TM modes in rectangular waveguides and why?

Q-3.4 Discuss the impossibility of the TEM mode in rectangular waveguide.

Q-3.5 Illustrate E and H distributions for at least three TE and three TM modes in the x–y plane.

Q-3.6 Discuss excitation of TM and TE modes in rectangular waveguides.

Q-3.7 Obtain expressions of $Z_z(\text{TE})$ and $Z_z(\text{TM})$ in terms of cutoff frequency.

Q-3.8 Illustrate the variation of $Z_Z(\text{TE})$, $Z_Z(\text{TM})$, v_p, v_g and λ with ω.

Q-3.9 Derive required relations and draw an equivalent circuit for a TM wave. Show the values of components involved therein.

Q-3.10 Illustrate the field distribution across the cross section of a circular waveguide for at least two TM and two TE modes.

Q-3.11 Discuss the mode designation in a circular waveguide.

Q-3.12 Explain the terms (i) mode, (ii) mode designation, (iii) dominant mode, (iv) mode transformation, (v) mode disturbance, (vi) spurious modes and (vii) multimode propagation.

Q-3.13 Discuss cutoff frequency, usable frequency range and frequency of operation.

Q-3.14 How does the guide wavelength differ from the free space wavelength?

Q-3.15 Discuss the relations between wave, phase and group velocities and explain their significance.

Further Reading

1. Adam, S. F., *Microwave Theory and Applications*. Prentice-Hall, Inc., New Jersey, 1969.
2. Brown, R. G. et al., *Lines, Waves and Antennas*, 2nd ed. John Wiley, New York, 1970.
3. Chatterjee, R., *Microwave Engineering*. East West Press, New Delhi, 1990.
4. Collins, R. E., *Foundation for Microwave Engineering*. McGraw-Hill, New York, 1966.
5. Das, A. and Das, S. K., *Microwave Engineering*. Tata McGraw-Hill, New Delhi, 2002.
6. Gandhi, O. P., *Microwave Engineering and Applications*. Pergamon Press, New York, 1981.
7. Gupta, K. C., *Microwaves*. Wiley Eastern, New Delhi, 1979.
8. Jordan, E. C. and Balmain, K. G., *Electromagnetic Waves and Radiating Systems*, 2nd ed. Prentice-Hall, New Delhi, 1987.
9. Lio, S. Y., *Microwave Devices & Circuits*, 3rd ed. Prentice-Hall of India, New Delhi, 1995.
10. Pozar, D. M. *Microwave Engineering*, 2nd ed. John Wiley & Sons, New Delhi, 1999.
11. Reich, H. J. et al., *Microwave Principles*, D. Van Nostrand, Reinhold Co. New York, 1957 (an East West edition).
12. Rizzi, P. A., *Microwave Engineering—Passive Circuits*. Prentice-Hall, New Delhi, 2001.

13. Saad, T. and Hansen, R. C., *Microwave Engineer's Handbook*, Vol. 1. Artech House, Dedham, MA, 1971.
14. Wheeler, G. J., *Introduction to Microwaves*. Prentice-Hall, Englewood Cliffs, NJ, 1963.

4

Cavity Resonators

4.1 Introduction

In an ordinary electronic equipment, a resonant (or tuned) circuit consists of an inductive coil and a capacitor, which are connected either in series or in parallel. These circuits comprising lumped parameters L and C are not without some resistance in the lower frequency range. The presence of resistance in circuits results in losses and lowers their Q. Besides, in view of the resonance frequency relation $\{f_0 = 1/2\pi\sqrt{(LC)}\}$, the requirement of L and C becomes too small at high frequencies. This may include requirement of lumped or distributed parameters. However, at higher frequencies, stray inductance and capacitance (Figure 4.1) attain greater significance in case of lumped parameters. Thus, at microwave range, conventional tank circuit is generally replaced by a cavity resonator.

The resonant frequency of a circuit can be increased by reducing the capacitance, or inductance, or both. Eventually, a point is reached beyond which further reduction, particularly of inductance, becomes impractical. At this point, the conventional circuit oscillates at the highest frequency. The upper limit for a conventional resonant circuit is between 2 and 3 GHz. At these frequencies, the inductance may comprise only one-half-turn coil. Also, the capacitance may simply be its stray capacitance. Tuning a half-turn coil is difficult and tuning its stray capacitance is even more difficult. Also, such a circuit can handle only a small current.

While discussing the properties of transmission lines (Chapter 2), it was noted that a quarter-wave section can act as a resonant circuit. Since the basic principles of transmission lines are equally valid to the waveguides, a quarter-wave-long waveguide can act as a resonant circuit. Thus, the quarter-wave section of a hollow waveguide can be considered as a resonant cavity with similar conditions as applied to the quarter-wave section. In general, any space that is completely enclosed by conducting walls is capable of containing oscillating electromagnetic fields. Such an enclosure possessing resonant properties can be termed as resonant cavity or cavity resonator. It is to be mentioned that a low-frequency circuit has only one resonance frequency, whereas a resonant cavity can have many resonant frequencies.

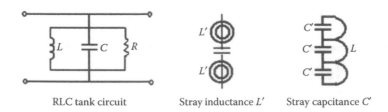

RLC tank circuit   Stray inductance L' Stray capcitance C'

FIGURE 4.1
RLC resonance circuit and its parasitic elements.

4.2 Shapes and Types of Cavities

The primary frequency of any resonant cavity is mainly determined by its physical size and shape. In general, the smaller the cavity, the higher its resonant frequency. A cavity can be carved out of a waveguide, a coaxial cable and from a spherical structure. Despite the variation in shapes, the basic principles of operation for all cavities remain the same.

4.2.1 Cavity Shapes

Figure 4.2 illustrates some of the commonly used shapes of cavities. Some of these involve more than one geometrical structure. In all these figures, the lengths of cavities are $p\lambda/2$, where $p = 1, 2, 3,....$ Also, their extreme ends are short circuited.

4.2.2 Cavity Types

The cavities can be classified in various ways. The first classification is based on shapes of cavities. In the second classification, a cavity may be termed

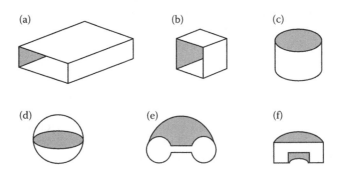

FIGURE 4.2
Different cavity shapes. (a) Rectangular, (b) cubical, (c) cylindrical, (d) spherical, (e) doughnut shaped and (f) cylindrical ring.

as absorption type or transmission type, which specifies the capability of a cavity to transmit or absorb electromagnetic energy. In an absorption-type cavity, there is maximum absorption, whereas in transmission-type cavity, there is maximum transmission of energy at resonant frequencies. Figure 4.3 illustrates four types of cavities, which include a half-wave cavity, a quarter-wave cavity, a quarter-wave coaxial cavity and a spherical cavity with reentrant cones.

The phenomenon of resonance in a line section was explained in Section 2.26. Therein, the buildup of signal through the resonance phenomenon in a quarter-wave section was shown in Figure 2.30. The cavities or waveguides not terminated by a short circuit (Figure 4.3b) will be terminated by free space having finite (characteristic) impedance ($\eta_0 = 377\ \Omega$) and thus perfect reflection will not be possible. Also, a half-wave cavity will be resonant at frequency f and its even multiples, that is, at $2f$, $4f$ and so on, whereas the quarter-wave cavity will exhibit resonance at f and its odd multiples, that is, $3f$, $5f$ and so on. In case of spherical cavity (Figure 4.3d), the first resonance frequency will correspond to $\lambda_r = 2.28a$, and the second for $\lambda_r = 1.4a$.

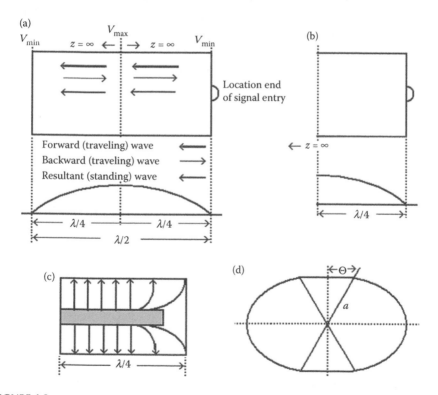

FIGURE 4.3
Types of cavities. (a) Half-wave cavity, (b) quarter-wave cavity (right half of (a)), (c) $\lambda/4$ coaxial cavity and (d) spherical cavity with reentrant cones.

4.2.3 Reentrant Cavities

At lower frequencies, a cavity resonator can be represented by a lumped component resonant circuit. Such a representation becomes increasingly inaccurate as operating frequencies increase, particularly beyond the VHF range.
To maintain resonance at these frequencies, the inductance and capacitance
must be reduced to a minimum level. The inductance can be reduced by
using a short wire. A reentrant cavity helps in the reduction of capacitance.
A reentrant cavity, one of the most useful forms of cavities, is used in klystrons and microwave triodes. In reentrant cavity, the metallic boundaries
extend into the interior of the cavity. The shapes of some reentrant cavities
are shown in Figure 4.4. Details of coaxial cavity are shown in Figure 4.4a
and that of radial cavity in Figure 4.4b.

In view of the structure of coaxial cavity shown in Figure 4.4a, there is
a considerable decrease in inductance and resistance losses. The radiation
losses also reduce due to its self-shielding enclosure. The exact value of the
resonant frequency of the given cavity structure cannot be obtained. Its
approximate value can, however, be evaluated by using transmission line
theory discussed in Chapter 2 and by the following steps.

The characteristic impedance of a coaxial cable given by Equation 2.99e is
reproduced as follows:

$$Z_0 = \frac{1}{2\pi} \sqrt{\frac{\mu}{\varepsilon}} \ln \frac{b}{a} \text{ ohms} \tag{4.1}$$

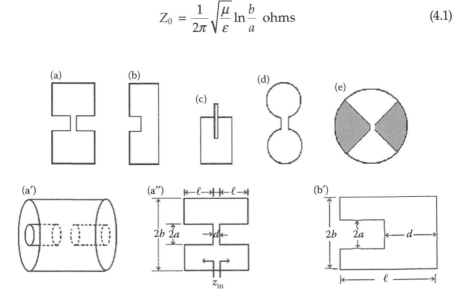

FIGURE 4.4
Different types of reentrant cavities with details of coaxial cavity. (a) Structural and dimensional details of coaxial cavity, (b) structural and dimensional details of radial cavity, (c) tunable, (d) toroidal, (e) butterfly.

The structure of a coaxial cavity can be obtained by shorting a coaxial line at its both ends and joining these at the centre by a capacitor. Input impedance to each shorted coaxial line (of length ℓ) is given by

$$Z_{in} = jZ_0 \tan \beta l \ \text{ohms} \tag{4.2}$$

In view of Equation 4.1, Equation 4.2 becomes

$$Z_{in} = j\frac{1}{2\pi}\sqrt{\frac{\mu}{\varepsilon}} \ln \frac{b}{a} \tan \beta l = jX_{in} \ \text{ohms} \tag{4.3}$$

The inductance (L) of the cavity is given by

$$L = \frac{2X_{in}}{\omega} = \frac{1}{\pi\omega}\sqrt{\frac{\mu}{\varepsilon}} \ln \frac{b}{a} \tan \beta l \tag{4.4a}$$

The capacitance (C_g) of the gap is

$$C_g = \frac{\varepsilon \pi a^3}{d} \tag{4.4b}$$

As per the condition of resonance ($\omega L = 1/\omega C_g$)

$$\omega L = \frac{1}{\pi}\sqrt{\frac{\mu}{\varepsilon}} \ln \frac{b}{a} \tan \beta l = \frac{1}{\omega C_g} = \frac{d}{\omega \varepsilon \pi a^3} \tag{4.5a}$$

This gives

$$\tan \beta l = \frac{dv}{\omega a^3 \ln(b/a)} \tag{4.5b}$$

where $v = 1/\sqrt{\mu\varepsilon}$ is the velocity of a wave in any medium.

4.2.3.1 Resonant Frequency

The resonant frequency of this cavity is obtained by solving Equation 4.5b. As values of tan βl infinitely repeat after each wavelength, Equation 4.5b will give an infinite number of solutions with higher values of frequencies. Thus, this cavity can support an infinite number of resonant frequencies or modes of oscillation. In general, the shorted coaxial line cavity stores more magnetic energy than electrical energy; the balance of electrical energy appears in the gap. It is at the resonance that these two energies are equal.

The inductance and capacitance of another commonly used (radial) cavity shown in Figure 4.4b with details in b′) are as follows:

$$L = \frac{\mu l}{2\pi} \ln \frac{b}{a} \tag{4.6a}$$

$$C = \varepsilon_0 \left[\frac{\pi a^2}{d} - 4a \ln \frac{0.765}{\sqrt{l^2 + (b-a)^2}} \right] \tag{4.6b}$$

The resonance frequency for this cavity is given by

$$f_r = \frac{c}{2\pi\sqrt{\varepsilon_r}} \left[al \left\{ \frac{a}{2d} - \frac{2}{l} \ln \frac{0.765}{\sqrt{l^2 + (b-a)^2}} \right\} \ln \frac{b}{a} \right]^{-1/2} \tag{4.7}$$

where c is the velocity of light in vacuum or free space.

4.3 Cavity Formation

4.3.1 Formation of a Rectangular Cavity

A cavity resonator of rectangular shape is shown in Figure 4.5. It is simply a section of rectangular waveguide, which is closed at both the ends by conducting plates. The distance between the two end plates is $\lambda/2$, which correspond to the frequency at which the resonance occurs. The magnetic and electric field patterns in the cavity are also shown.

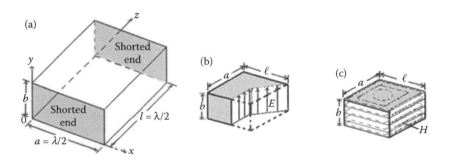

FIGURE 4.5
Rectangular cavity resonator. (a) Shape of resonator, (b) E-field patterns for a simple mode and (c) H-field patterns for a simple mode.

4.3.2 Formation of a Cylindrical Cavity

Figure 4.6a through e illustrates the step-wise formation of a cylindrical resonant cavity from an infinite number of quarter-wave transmission line sections. Figure 4.6a shows a quarter-wave section, whereas Figure 4.6b illustrates its equivalent resonant circuit containing small values of inductance and capacitance. The open-circuited ends of two or more line sections (shown in Figure 4.6a) can be joined together in parallel without affecting the resonance frequency. Figure 4.6c shows the equivalent circuit obtained by joining three quarter-wave sections. In Figure 4.6d, eight such sections are joined in parallel. In relation to the connections, it may be mentioned that although the current-carrying ability of several $\lambda/4$ sections is more than that of any one section, there will still be no change in the resonant frequency. This can be justified as the addition of inductances in parallel lowers the total inductance, whereas the addition of capacitance in parallel increases the total capacitance by the same proportion. Thus, the resonant frequency for a single section or more parallel sections shall remain the same. Besides, the increase in the number of current paths also decreases the total resistance and thus Q of the resonant circuit increases. Figure 4.6c and d shows the intermediate steps of the development of the cavity. If the number of these parallel line sections is raised to infinity, such joining will lead to the formation of a cylindrical cavity shown in Figure 4.6e. This cavity will have a diameter of $\lambda/2$ at the resonant frequency. It may be noted that as there are infinite number of parallel paths, the resistance must tend to zero and Q must tend to infinity.

4.4 Fields in Cavity Resonators

In Chapter 3, it was seen that the field behaviour in both rectangular and circular waveguides is governed by the basic Maxwell's equations. Also, this

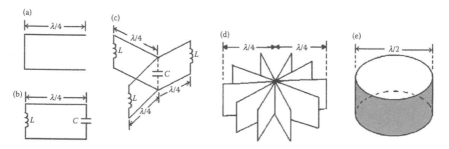

FIGURE 4.6
Formation of a cylindrical cavity resonator. (a) Quarter-wave section, (b) equivalent LC circuit L and C are small, (c) three quarter-wave sections joined, (d) eight quarter-wave sections joined to form a cylindrical cavity and (e) resulting cylindrical resonant cavity.

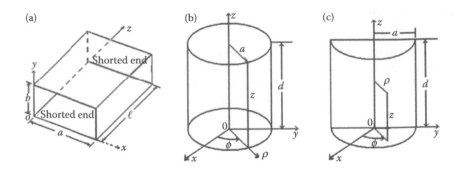

FIGURE 4.7
Geometry of different cavity resonators. (a) Rectangular, (b) circular and (c) semi-circular.

field has to satisfy the continuity conditions of tangential electric and normal magnetic fields. As no restriction was imposed on the waveguide length, the variation along the z-axis was accounted by a factor $e^{-\gamma z}$ (or by $e^{-\beta z}$ in free space). The situation in case of cavities is different as these have a definite length along the z-axis. Figure 4.7 illustrates the geometries of three cavity resonator wherein their lengths (l or d) are taken to be finite.

4.4.1 Rectangular Cavity Resonator

The geometry of a rectangular cavity and the coordinate system to be used are illustrated in Figure 4.7a. The solution to the wave equation for this cavity can be obtained by using the procedure followed in case of rectangular waveguide. The field so obtained involves almost similar terms along the x- and y-axes as were involved in case of the waveguide. Since the length of cavity is finite along the z-axis, some forms of harmonics in z are to be assumed. These harmonics have to meet the boundary conditions at the two shorted ends. The mathematical relations obtained for the relevant terms are

$$E_z = E_{0z} \sin\left(\frac{m\pi x}{a}\right) \sin\left(\frac{n\pi y}{b}\right) \cos\left(\frac{p\pi z}{l}\right) \quad \text{(for TM}_{mnp} \text{ mode)} \qquad (4.8)$$

$$H_z = H_{0z} \cos\left(\frac{m\pi x}{a}\right) \cos\left(\frac{n\pi y}{b}\right) \sin\left(\frac{p\pi z}{l}\right) \quad \text{(for TE}_{mnp} \text{ mode)} \qquad (4.9)$$

$$h_{nmp} = \sqrt{\{(m\pi/a)^2 + (n\pi/b)^2 + (p\pi/l)^2\}} \qquad (4.10)$$

$$\beta^2 = \omega^2 \mu\varepsilon - h_{nmp}^2 \quad \text{(for both TE}_{mnp} \text{ and TM}_{mnp} \text{ modes)} \qquad (4.11)$$

$$f_r = \frac{1}{2\sqrt{\mu\varepsilon}} \sqrt{\left(\frac{m}{a}\right)^2 + \left(\frac{n}{b}\right)^2 + \left(\frac{p}{l}\right)^2} \quad \text{(for both TE}_\text{mnp} \text{ and TM}_\text{mnp} \text{ modes)} \quad (4.12)$$

where f_r is the resonant frequency of cavity.

In Equations 4.8 through 4.12, m and $n = 1, 2, 3, \ldots$ and $p = 0, 1, 2, 3, \ldots$.

For a rectangular waveguide, the corresponding expressions are given by Equations 3.19a, 3.21a, 3.18b, 3.22b and 3.22c for cutoff frequency f_c. In these equations, m and $n = 0, 1, 2, 3,\ldots$. In view of the above relations, the modes in a cavity are described in terms of fields along the x-, y- and z-axes. As can be noticed, the mode representation in waveguides requires two subscripts, whereas in cavities, three subscripts are used. The modes in rectangular cavity are designated as TE$_{m,n,p}$ and TM$_{m,n,p}$. The first subscript (m) indicates the number of half wavelengths along the x-axis, the second subscript (n) indicates the number of half wavelengths along the y-axis and the third subscript (p) indicates the number of half wavelengths along the z-axis. The parameter p is related by $p = 2l/\lambda_g$, where l is the length of the cavity and λ_g is the guide wavelength. The physical length of the cavity is given by $l = p\lambda_g/2$, where

$$\frac{1}{\lambda_g^2} = \frac{1}{\lambda_0^2} - \frac{1}{\lambda_c^2} \quad \text{or} \quad \lambda_g = \frac{\lambda_0}{\sqrt{1 - (\lambda_0/\lambda_c)^2}} \quad (4.13)$$

The guide-length in waveguide filled with air is $\lambda_g = \lambda_0$ and that of waveguide filled with other dielectric is $\lambda_g = \lambda_0/\sqrt{\varepsilon}$. The propagation in waveguides is possible for any cross section, provided λ_c corresponds to the mode cross section. Thus, from Equation 4.13

For $\lambda_0 \ll \lambda_c$, denominator ≈ 1 and $\lambda_g = \lambda_0$

For $\lambda_0 \gg \lambda_c$, denominator is imaginary and there is no wave propagation

For $\lambda_0 = \lambda_c$, denominator $= \infty$, $\lambda_g = 0$

$$\text{Normally, } \lambda_g = \lambda_0, \; v_p = f\lambda_g, \; c = f\lambda_0; \text{ thus, } v_p = c \quad (4.14)$$

In Equation 4.14, v_p is the phase velocity. No intelligence or modulation travels at the frequency if $v_p = c$. The velocity of the modulation envelope is called the group velocity (λ_g). When a modulation signal travels in waveguide, it keeps on slipping backward with respect to the carrier.

$$v_p = \frac{c\lambda_g}{\lambda_0} \quad \text{and} \quad v_g = \frac{c\lambda_0}{\lambda_g} \quad \text{and thus} \quad v_p v_g = c^2 \quad (4.15)$$

For cubic cavity (i.e. $a = b = l$), the resonant length is

$$\lambda_r = 2/\sqrt{\left[\left(\frac{m}{a}\right)^2 + \left(\frac{n}{b}\right)^2 + \left(\frac{p}{l}\right)^2\right]} = 2a/\sqrt{\left[(m)^2 + (n)^2 + (p)^2\right]} \quad (4.16a)$$

From Equation 4.16a, the expression of resonant length for TE_{101} mode (i.e. $m = p = 1, n = 0$) reduces to

$$\lambda_r = \sqrt{2}a \quad (4.16b)$$

The approximate number of finite resonances (N) from practicable limit can be given in terms of the volume of cavity (V) and the minimum wavelength (λ_m) by the following relation:

$$N = \frac{(8\pi V)}{(3\lambda_m^3)} \quad (4.17)$$

The dominant mode for a cavity with unequal dimensions, that is, for $a > b > l$ is TE_{101}.

4.4.1.1 Mode Degeneracy

If there are many modes having the same resonant frequency, the situation is called mode degeneracy.

If $a \neq b \neq \ell$, there will be twofold degeneracy.

If $a = b, \neq \ell$ or $a = \ell, \neq b$ or $b = \ell, \neq a$, there will be fourfold degeneracy.

If $a = b = \ell$, there will be 12-fold degeneracy (six each for TE and TM modes).

The slightest manufacturing defect can cause a separation and degeneration of modes.

4.4.2 Circular Cavity Resonator

A circular cavity is a circular waveguide in which its two ends are shorted by metal plates. Figure 4.7b illustrates such a formation. Here also, the solutions of Maxwell's equations have to satisfy the same boundary conditions as in case of circular waveguide with the exception along the z-axis. Along this axis, a harmonic function is to be added to meet the required conditions. The resulting field expressions for the circular cavity of length l are

$$H_z = H_{0z}J_n'\left(h_{mn}'\rho\right)\cos(n\varphi)\sin\left(\frac{q\pi z}{l}\right) \quad \text{(for } TE_{mnq} \text{ mode)} \quad (4.18)$$

$$E_z = E_{0z}J_n(h_{mn}r)\cos(n\varphi)\sin\left(\frac{q\pi z}{l}\right) \quad \text{(for TM}_{nmp} \text{ mode)} \tag{4.19}$$

The separation equations for TE and TM modes are

$$h^2 = (h'_{nm})^2 + \left(\frac{p\pi}{l}\right)^2 \quad \text{(for TE mode)} \tag{4.20}$$

$$h^2 = (h_{nm})^2 + \left(\frac{p\pi}{l}\right)^2 \quad \text{(for TM mode)} \tag{4.21}$$

The resonant frequency of the cavity is given by

$$fr = \frac{1}{2\sqrt{\mu\varepsilon}}\sqrt{(h'_{nm})^2 + \left(\frac{p\pi}{l}\right)^2} \quad \text{(for TE mode)} \tag{4.22}$$

$$fr = \frac{1}{2\sqrt{\mu\varepsilon}}\sqrt{(h_{nm})^2 + \left(\frac{p\pi}{l}\right)^2} \quad \text{(for TM mode)} \tag{4.23}$$

In the above equations, $h'_{nm} = (ha)'_{nm}/a$ and $h_{nm} = (ha)_{nm}/a$. Also, E_{0z} and H_{0z} are the amplitudes of electric and magnetic field intensities, respectively, of the wave travelling in the z-direction.

In Equations 4.18 through 4.23, $n = 0, 1, 2,..., m = 1, 2, ..., q = 1, 2, 3,...$ and $p = 0, 1, 2, 3, ...$

In view of these equations, it can be noted that TM_{110} is the dominant mode for $2a > d$ and TE_{111} mode is dominant when $d \geq 2a$.

For circular waveguide, the expressions of H_z and E_z are given by Equations 3.75a and 3.74a, respectively. Expressions of h_{nm} and h'_{nm} are given by Equations 3.76a and 3.76b and the cutoff frequencies for TE and TM waves are given by Equations 3.76e and 3.76f, respectively. In these equations, $m = 0, 1, 2, 3,...$ and $n = 1, 2, 3,....$

The resonant lengths for cylindrical cavity with different modes are given as follows.

$$\lambda_r = 2/\sqrt{[(1/0.853d)^2 + (p/l)^2]} \text{ for TE}_{11p} \text{ mode} \tag{4.24a}$$

$$\lambda_r = 2/\sqrt{[(1/0.653d)^2 + (p/l)^2]} \text{ for TM}_{01p} \text{ mode} \tag{4.24b}$$

$$\lambda_r = 2/\sqrt{[(1/0.420d)^2 + (p/l)^2]} \tag{4.24c}$$

for TE_{01p} mode (with low attenuation and high Q)

TABLE 4.1

Summary of Mode Designations in Different Configurations

Coaxial Cables	Parallel Planes	Waveguides	Rectangular Cavities	Circular Cavities
TEM mode if length is of the order of half, one or two wavelengths TE$_{11}$ and higher modes at higher frequencies	TE$_{10}$, TE$_{20}$, ..., TE$_{m0}$ and TM$_{10}$, TM$_{20}$, ..., TM$_{m0}$ modes	TE$_{mn}$ and TM$_{mn}$ modes	TE$_{mnp}$ and TM$_{mnp}$ modes	TE$_{mnp}$ and TM$_{mn0}$ modes

The summary of mode designations in different configurations is given in Table 4.1.

4.4.3 Semi-Circular Cavity Resonator

A semi-circular resonator with radius 'a' and length 'ℓ' is shown in Figure 4.7c. The field expressions for TE and TM mode are as follows:

$$H_z = H_{0z} J_n\left(h'_{nm}\rho\right)\cos(n\phi)\sin\left(\frac{p\pi z}{l}\right)$$

(4.25)

(for TE$_{nmp}$ mode in circular cavity)

where $n = 0, 1, 2, ..., m = 1, 2,...$ and $p = 1, 2, ...$

$$E_z = E_{0z} J_n(h_{nm}\rho)\sin(n\phi)\cos\left(\frac{p\pi z}{l}\right)$$

(4.26)

(for TM$_{nmp}$ mode in circular cavity)

where $n = 1, 2,.... m = 1, 2,....$ and $p = 0, 1, 2, ...$

The expressions for h and f_r for TE and TM modes for circular and semi-circular cavities are the same. The values of n, m and p, however, differ from those for circular cavity. Also, TE$_{111}$ mode is dominant if $\ell > a$, and TM$_{110}$ mode is dominant for $\ell < a$.

4.5 Quality Factor

The quality factor 'Q' of a resonant circuit is defined as the ratio of maximum energy stored to the average power loss and can be written as

$$Q = 2\pi \text{ (Maximum energy stored '}W\text{'/energy dissipated per cycle '}P\text{')} \approx V/(A\delta)$$

(4.27)

where V is the volume of the cavity, A is its surface area and $\delta \{=\sqrt{2}/(\omega\mu\sigma)\}$ is the depth of penetration. In view of Equation 4.27, values of Q can be obtained by varying the dimensional parameters. The conditions for maximum Q for different cavities are as follows:

1. *For rectangular cavity:* When $a = b = \ell$

   where a and b are the dimensions of broader and narrower walls and ℓ is its length.

2. *For cylindrical cavity:* When $d = \ell$

   where d is the diameter of the cavity and l is its length.

3. *For coaxial cavity:* When $b/a = 3.6$

   where a and b are the inner and outer radii of coaxial cable.

4. *For spherical cavity:* When $\theta = 33.5°$

Where θ is the angle of reentrant cone

The equivalent circuit of an unloaded resonator can be represented either as a series or a parallel combination of R, L and C. The resonant frequency and quality factor Q_0 of such a circuit can be written as

$$f_0 = \frac{1}{2\pi\sqrt{LC}} \tag{4.28a}$$

and

$$Q_0 = \frac{\omega_0 L}{R} \tag{4.28b}$$

Another definition of Q may be related to the variation of amplitude of the output emerging out from a circuit (or cavity) with frequency f. Thus, in view of Figure 4.8, the quality factor can be written as

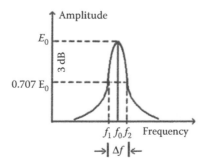

FIGURE 4.8
Variation of amplitude of E with frequency.

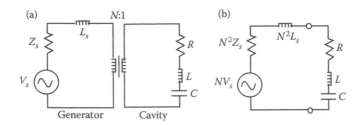

FIGURE 4.9
Coupled cavity and its equivalent. (a) Coupled circuit and (b) equivalent circuit.

$$Q_0 = \frac{f_0}{\Delta f} = \frac{f_0}{f_2 - f_1} \tag{4.29}$$

As shown in Figure 4.9a, a cavity with circuit parameters R, L and C is coupled to a signal generator having series inductance L_s and internal impedance Z_s, through an ideal transformer of ratio N:1. Its equivalent circuit is given in Figure 4.9b. The quality factor Q_0 of the system is given by

$$Q_0 = \frac{\omega_0 L_S}{R + N^2 Z_s} \quad \text{for} \left| N^2 L_s \right| \ll \left| R + N^2 Z_s \right| \tag{4.30}$$

The Q given by Equation 4.30 can be referred to as unloaded Q or Q_{UL}. As cavity resonators are used to couple power generally through small holes made in their walls, these holes result in loss of power. Since such power loss reduces the Q of cavity, the new value of Q is referred to as loaded Q_L. It is given by

$$Q_L = \frac{\omega_0 L_S}{R(1 + k)} = \frac{Q_0}{1 + k} \tag{4.31a}$$

where

$$k = \frac{N^2 Z_s}{R} \tag{4.31b}$$

Parameter k is the coupling coefficient of the system. Different Qs can be related as

$$\frac{1}{Q_L} = \frac{1}{Q_{UL}} + \frac{1}{Q_{Ext}} \tag{4.32a}$$

where Q_{Ext} is the Q of the external circuit and is given as

$$Q_{EXT} = \frac{Q_0}{k} = \frac{\omega_0 L}{kR} \tag{4.32b}$$

Here, the coupling mechanism referred to above can be classified as follows.

1. When $k = 1$, the resonator and signal generator are perfectly matched

$$Q_L = \frac{Q_{EXT}}{2} = \frac{Q_0}{2} \tag{4.33}$$

   In Equation 4.33, Q_{Ext} includes the Q of couplers or exciters.

2. When $k > 1$, the resonator terminals are at a voltage maximum in the input line at resonance. At this point, the normalised impedance at voltage maximum is simply the standing wave ratio (ρ), that is, $k = \rho$ and

$$Q_L = \frac{Q_0}{1 + \rho} \tag{4.34}$$

3. When $k < 1$, the resonator terminals are at a voltage minimum and line at resonance. At this point, the input terminal impedance is equal to the reciprocal of the standing wave ratio, that is, $k = 1/\rho$ and

$$Q_L = \frac{\rho Q_0}{1 + \rho} \tag{4.35}$$

Figure 4.10 illustrates the variation of standing wave ratio (ρ) with the coupling coefficient.

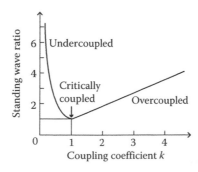

FIGURE 4.10
Variation of standing wave ratio with coupling coefficient.

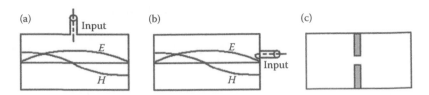

FIGURE 4.11
Coupling mechanism for cavities. (a) Probe coupling, (b) loop coupling and (c) aperture coupling.

4.6 Coupling Mechanism

Cavity resonator is a metallic enclosure whose physical dimensions are directly related to the operating frequency. Since it is a complete enclosure, it requires a small opening for feeding energy into it or for taking it out. The mechanism of feeding or taking out is referred to as coupling. This coupling mechanism has to be so located so as to excite a desired mode into the cavity. Also, the larger the opening, the tighter the coupling and the greater the effect on the resonant frequency. As in waveguides, the methods of coupling the energy to a cavity include (a) probe coupling, (b) loop coupling and (c) aperture, iris, hole or slot coupling. These are illustrated in Figure 4.11.

As the cavity resonators are energised in a manner similar to that used for waveguides, these will have the similar field distributions for a particular excited mode. In probe coupling, a straight wire is inserted at the location of maximum electric field intensity of the desired mode. In loop coupling, the loop is to be placed at the location of maximum magnetic field intensity. The cavities shown in Figure 4.11a and b are energised in TE mode. The wave will reflect back and forth along the z-axis and will result in the standing wave formation. These resulting field configurations within the cavity are in accordance with the boundary conditions to be satisfied for the excited mode. The slot coupling may be used when the cavity is to be inserted in between the components of a running passage of the system. The operating principles of probes, loops and slots are the same, irrespective of their application to a cavity or a waveguide. Also, depending on the location, a probe, loop, or hole can excite more than one mode.

4.7 Tuning Methods

The resonant frequency of a cavity can be varied either by changing its volume, capacitance, or inductance. This variation of frequency is referred to as

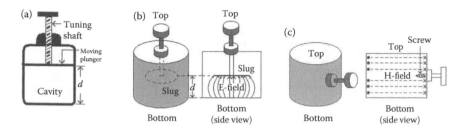

FIGURE 4.12
Tuning methods for cavities. (a) By changing the volume, (b) by changing the capacitance and (c) by changing the inductance.

tuning. All these methods are of mechanical nature and may vary with the application. The tuning by changing the volume, referred to as *volume tuning*, is illustrated in Figure 4.12a. As per the illustration, change of distance d will result in change of inductance and capacitance and hence change in frequency. The resonant frequency will be higher for lesser volume and lower for larger volume. The change in *d* can be obtained by moving a plunger in the cavity.

Capacitive tuning of a cavity is shown in Figure 4.12b. It uses an adjustable metallic slug or screw that is located in the area of maximum E field. The distance d represents the distance between two capacitor plates. The movement of the slug changes the distance between the two plates and hence the capacitance. This change of capacitance results in changed resonant frequency.

To accomplish *inductive tuning*, a non-magnetic slug is placed in the area of maximum H lines. This is shown in Figure 4.12c. The changing H induces a current in the slug that sets up an opposing H field. The opposing field reduces the total H field in the cavity, and thus the total inductance reduces. This reduction in inductance, by inward movement of the slug, increases the resonant frequency. Similarly, to decrease the resonant frequency, the inductance needs to be increased by moving the slug outward.

4.8 Advantages and Applications

Resonant cavities have many advantages and applications at microwave frequencies. These have a very high-quality factor and can be built to handle relatively large amount of power. Cavities with Qs exceeding 30,000 are not uncommon. In view of their high Q, the pass band of cavities is quite narrow and thus these can be tuned very accurately. Their rugged and simple construction is an added advantage.

The cavity resonators are used over a wide range of frequencies and for diverse applications. At microwave range, these replace the conventional

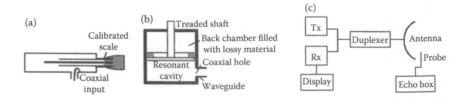

FIGURE 4.13
Cavity applications. (a) Coaxial wave meter, (b) cylindrical wave meter and (c) echo-box set-up.

tuned circuits. Most microwave tubes and transmitting devices use cavities in some form to generate microwave energy. *Fixed cavities* are used as circuit elements in filters and in frequency-controlled microwave oscillators. *Tunable cavities* are used as measuring instrument in the form of wave-meters and receiver pre-selectors. These wave-meters determine the frequency of the energy travelling in a waveguide, since conventional measuring devices do not work well at microwave range. Both fixed and tunable cavities can be used as echo boxes. Cavities can also be used to measure ε if frequency f_0 and quality factor Q are known. For all the above applications, cavities are to be tested to assess their f_0, Q, VSWR, insertion loss and effect of temperature before use. Some of the applications of cavities are shown in Figure 4.13.

4.9 Dielectric Resonators

A dielectric resonator (also referred to as dielectric resonator oscillator (DRO)) is an electronic component that exhibits resonant property. It functions by trapping energy in an extremely narrow band of frequencies within the confines of resonator volume. The resonance in the dielectric resonator closely resembles that of a circular hollow metallic waveguide. Energy is reflected back into the resonator, resulting in negligible radiation losses by presenting a large change in permittivity at the boundary of the resonator.

The basic difference between the resonance in dielectric resonator and that in metallic resonator lies in terms of their boundary conditions. In dielectric resonator, the boundary is defined by large change in permittivity, whereas in circular waveguide it is defined by a conductor. The dielectric resonators generally consist of a 'puck' of ceramic that has a large permittivity and a low dissipation factor. Their resonance frequencies are mainly governed by the physical dimensions of the puck and the dielectric constant of the material. They can be loaded by using probe (for E-field) or loop (for H-field) coupling methods.

The dielectric resonators and resonant metal cavities differ in their operations, mainly in terms of field distributions. In the case of metal cavity, the electric and magnetic fields are zero outside the walls. These fields are not zero outside the dielectric walls of the resonator. Alternatively, it can be stated that in case of metal cavities, open circuit boundary conditions are fully satisfied, whereas in dielectric resonators these are *approximately* satisfied. However, in dielectric resonators, electric and magnetic fields away from the resonator walls considerably reduce from their maximum values. At a given resonant frequency, most of the energy is stored in the resonator for a sufficiently high dielectric constant (ε_r). Dielectric resonators exhibit extremely high Q, comparable to a metal cavity.

The resonant modes that can be excited in dielectric resonators include (i) TE, (ii) TM and (iii) hybrid electromagnetic (HEM). Theoretically, there can be an infinite number of modes in each of these three groups. The desired mode is usually selected in view of the application requirements. TE_{01n} mode is generally used in most non-radiating applications, but other modes can have certain advantages for specific applications.

The frequency, Q and overall temperature coefficient of a dielectric resonator depends on its surroundings. The effect of cavity size on the overall resonant system can be described in terms of the ratio of the diameter of the cavity (D) and that of the puck (d). The effect of the cavity becomes negligible when the value of this ratio approaches 3. The frequency of dielectric resonators can be tuned by using a tuning plate, a dielectric plug or a dielectric disc. The tuning plate provides a coarse adjustment relative to the distance with reduced unloaded Q. The plug or disc provides finer frequency adjustment without affecting the unloaded Q much.

The resonant frequency (in GHz) of TE_{01n} mode for an isolated cylindrical dielectric resonator with radius a and length L (both in mm) can be approximately given by the relation

$$f_{GHz} = \frac{34}{a\sqrt{\varepsilon_r}} \left(\frac{a}{L} + 3.45 \right) \tag{4.36}$$

The accuracy of the above relation is approximately 2%, provided the values of a, L and ε_r satisfy the following limits:

$$0.5 < (a/L) < 2 \quad \text{and} \quad 30 < \varepsilon_r < 50 \tag{4.37}$$

Equations 4.36 and 4.37 are true only for isolated resonators. Since in most applications the dielectric resonators are enclosed in conducting cavities, the actual values of resonant frequencies are bound to differ. As the conducting walls of the enclosing cavity approach the dielectric resonator, change in boundary conditions and field containment start to affect resonant frequencies. The size and type of the material enclosing the cavity can drastically

affect the performance of the resonant circuit. The impact on resonant frequency for TE_{01n} mode due to metallic enclosure can be described in terms of
type of energy storing field. Thus, if the stored energy of the displaced field
is mostly electric, its resonant frequency will decrease, and if it is magnetic,
its resonant frequency will increase.

The dielectric resonators are commonly sensitive to temperature variation
and mechanical vibrations and may require compensating techniques to sta-
bilise the circuit performance over temperature and frequency.

The dielectric resonators are reduced-size alternatives to bulky waveguide
filters and low-cost alternatives to electronic oscillators, frequency-selective
limiter and slow-wave circuits. In addition to the above advantages of dielectric resonators over conventional metal cavity resonators, they have lower
weight, vast material availability and ease of manufacturing. These may
have an unloaded Q on the order of tens of thousand.

Dielectric resonators find applications as filters (e.g. cellular base station
filters, band-pass and band-stop), combiners, frequency-selective limiters,
DROs (diode, feedback, reflection, transmission and reaction-type), duplexers and dielectric resonator antenna (DRA) elements. These are commonly
used in satellite communications, GPS antennas and wireless internets.

EXAMPLE 4.1

An air-filled rectangular cavity with width $a = 5$ cm, height $b = 3$ cm and
length $l = 10$ cm operates in dominant TE_{101} mode. Calculate its (a) cutoff
wave number, (b) resonant frequency and (c) phase constant.

Solution

Given $a = 5$ cm, $b = 3$ cm, $l = 10$ cm, $\mu = \mu_0$ $\varepsilon = \varepsilon_0$

For TE_{101} $m = 1$, $n = 0$ $p = 1$ $v_0/2 = 1/2\sqrt{(\mu\varepsilon)} = 1.5 \times 10^8$

a. *Cutoff wave number (h)*

$$h = \sqrt{\{(m\pi/a)^2 + (n\pi/b)^2 + (p\pi/l)^2\}} = \sqrt{\{(\pi/5 \times 10^{-2})^2 + (\pi/10 \times 10^{-2})^2\}} = 70.248$$

b. *Resonant frequency (f_r)*

$$h/\sqrt{\pi} = \sqrt{\{(m\pi/a)^2 + (n\pi/b)^2 + (p\pi/l)^2\}}/\sqrt{\pi} = \sqrt{\{(m/a)^2 + }$$
$$(n/b)^2 + (p/l)^2\} = 70.248/\sqrt{\pi} = 39.633$$
$$f_r = \{1/2\sqrt{(\mu\varepsilon)}\} \, h/\sqrt{\pi} = 1.5 \times 10^8 \times 39.633 = 59.4495 \times 10^8$$

c. *Phase constant (β_g)*

$$\beta_g = p\pi/l = \pi/10 \times 10^{-2} = 31.415$$

EXAMPLE 4.2

An air-filled rectangular cavity with width $a = 4$ cm and height $b = 4$ cm
operates in dominant TE_{101} mode at 8 GHz resonant frequency. Calculate
the length of the cavity.

Solution

Given $a = 4 \times 10^{-2}$ $b = 4 \times 10^{-2}$

For TE_{101} $m = 1$, $n = 0$ $p = d$

$\mu = \mu_0 = 4\pi \times 10^{-7}$ $\varepsilon = \varepsilon_0 = 8.854 \times 10^{-12} = 10^{-9}/36\pi$

$v_0/2 = 1/2\sqrt{(\mu\varepsilon)} = 1.5 \times 10^8$ $f_r = 8 \times 10^9$

$f_r = \{1/2\sqrt{(\mu\varepsilon)}\}\sqrt{\{(m/a)^2 + (n/b)^2 + (p/d)^2\}}$

$8 \times 10^9 = 1.5 \times 10^8\sqrt{\{(1/a)^2 + (1/d)^2\}}$

$53.33 = \sqrt{\{(1/4 \times 10^{-2})^2 + (1/d)^2\}}$

$2844.44 = 2500 + (d/l)^2$ $(d/l)^2 = 2844.44 - 2500 = 344.44$

$(d/l) = \sqrt{344.44} = 18.559$ $d = 1/18.559 = 5.388 \times 10^{-2} = 5.388$ cm

EXAMPLE 4.3

Calculate the resonant frequency (f_r) of an air-filled circular cavity having a length of 10 cm and radius of 4 cm operates in (a) TE_{101} mode and (b) TM_{011} mode.

Solution

Given length $l = 10 \times 10^{-2}$ radius $a = 4 \times 10^{-2}$

$\mu = \mu_0 = 4\pi \times 10^{-7}$ $\varepsilon = \varepsilon_0 = 8.854 \times 10^{-12} = 10^{-9}/36\pi$ $1/2\sqrt{(\mu\varepsilon)} = v_0/2 = 1.5 \times 10^8$

$$f_r = \{1/2\pi\sqrt{(\mu\varepsilon)}\}\sqrt{\{(\{(ha)'_{nmp}/a)^2 + (p\pi/l)^2\}}$$

$$= 0.4775 \times 10^8\sqrt{\{(\{(ha)'_{nmp}/a)^2 + (p\pi/l)^2\}}} \quad \text{for TE modes}$$

$f_r = \{1/2\pi\sqrt{(\mu\varepsilon)}\}\sqrt{\{(\{(ha)_{nmp}/a)^2 + (p\pi/l)^2\}}$

$= 0.4775 \times 10^8\sqrt{\{(\{(ha)_{nmp}/a)^2 + (p\pi/l)^2\}}$ for TM modes

For TE_{101} $m = 1$, $n = 0$, $p = 1$

$f_r = 0.4775 \times 10^8\sqrt{\{(\{(ha)'_{nmp}/a)^2}}$
   $+ (p\pi/l)^2\}$

From Table 3.4 $(ha)'_{101} = 1.841$

$f_r = 0.4775 \times 10^8\sqrt{\{(1.841/4 \times 10^{-2})^2}}$
   $+ (\pi/10 \times 10^{-2})^2\}$

$= 0.4775 \times 10^8\sqrt{\{2118.3 + 986.96\}}$

$f_r = 2.66$ GHz

For TM_{011} $m = 0$, $n = 1$, $p = 1$

$f_r = 0.4775 \times 10^8\sqrt{\{(ha)_{nmp}/a)^2}}$
   $+ (p\pi/l)^2\}$

From Table 3.3 $(ha)_{011} = 2.045$

$f_r = 0.4775 \times 10^8\sqrt{\{(2.045/4 \times 10^{-2})^2}}$
   $+ (\pi/10 \times 10^{-2})^2\}$

$= 0.4775 \times 10^8\sqrt{\{2613.765 + 986.96\}}$

$f_r = 2.865$ GHz

EXAMPLE 4.4

An air-filled circular cavity of 6 cm diameter operates at resonant frequency of 10 GHz. Calculate its length if it operates in (a) TE_{211} mode and (b) TM_{111} mode.

Solution

Given radius $a = 3 \times 10^{-2}$ $f_r = 10^{10}$ Hz

$\mu = \mu_0 = 4\pi \times 10^{-7}$ $\varepsilon = \varepsilon_0 = 8.854 \times 10^{-12} = 10^{-9}/36\pi$ $1/2\sqrt{(\mu\varepsilon)} = v_0/2 = 1.5 \times 10^8$

$$f_r = \{1/2\pi \sqrt{(\mu\varepsilon)}\} \sqrt{\{((ha)'_{nmp}/a)^2 + (p\pi/d)^2\}}$$
$$= 0.4775 \times 10^8 \sqrt{\{((ha)'_{nmp}/a)^2 + (p\pi/d)^2\}} \quad \text{for TE modes}$$
$$f_r = \{1/2\pi\sqrt{(\mu\varepsilon)}\}\sqrt{\{((ha)_{nmp}/a)^2 + (p\pi/d)^2\}}$$
$$= 0.4775 \times 10^8 \sqrt{\{((ha)_{nmp}/a)^2 + (p\pi/d)^2\}} \quad \text{for TM modes}$$

For TE$_{101}$ $m = 2, n = 1, p = d$	**For TM$_{011}$** $m = 1, n = 1, p = d$
$f_r = 0.4775 \times 10^8 \sqrt{\{((ha)'_{nmp}/a)^2}$	$f_r = 0.4775 \times 10^8 \sqrt{\{((ha)_{nmp}/a)^2}$
$+ (p\pi/d)^2\}$	$+ (p\pi/d)^2\}$
From Table 7.4 $(ha)'_{211} = 3.054$	From Table 7.3 $(ha)_{111} = 3.832$
$10^{10} = 0.4775 \times 10^8 \sqrt{\{(3.054/3 \times 10^{-2})^2}$	$10^{10} = 0.4775 \times 10^8 \sqrt{\{(3.832/3 \times}$
$+ (\pi/dl)^2\}$	$10^{-2})^2 + (\pi/d)^2\}$
$209.4 = \sqrt{\{10363.24 + (\pi/d)^2\}}$	$209.4 = \sqrt{\{16315.8 + (\pi/d)^2\}}$
$43858.4 = 10363.24 + (\pi/d)^2$	$43858.4 = 16315.8 + (\pi/d)^2$
$(\pi/d) = 183$	$(\pi/d) = 165.96$
$d = \pi/183 = 1.72 \text{ cm}$	$d = \pi/165.96 = 1.89 \text{ cm}$

PROBLEMS

P-4.1 Calculate the (a) cutoff wave number, (b) resonant frequency and (c) phase constant for an air-filled rectangular cavity operating in dominant TE$_{101}$ mode. It has width $a = 6$ cm, height $b = 4$ cm and length $l = 8$ cm.

P-4.2 An air-filled rectangular cavity with width $a = 6$ cm and height $b = 4$ cm operates in dominant TE$_{101}$ mode at 12 GHz resonant frequency. Calculate its length.

P-4.3 Calculate the resonant frequency of an air-filled circular cavity operating in (a) TE$_{101}$ mode and (b) TM$_{011}$ mode. It has 8 cm length and 5 cm radius.

P-4.4 An air-filled circular cavity of 5 cm diameter operates at $f_r = 8$ GHz in (a) TE$_{211}$ mode and (b) TM$_{111}$ mode. Calculate its length.

Descriptive Questions

Q-4.1 Discuss the various shapes and types of resonant cavities.

Q-4.2 Illustrate half-wave, quarter-wave and quarter-wave coaxial cavities.

Q-4.3 Illustrate a spherical cavity with reentrant cones.

Q-4.4 Draw the shapes of different forms of reentrant cavities.

Q-4.5 With the aid of necessary figures, discuss the formation of a cylindrical cavity.

Q-4.6 Write the expressions for quality factor of loaded and unloaded cavities.

Q-4.7 Discuss the coupling and tuning methods used in resonant cavities.

Q-4.8 List the application of resonant cavities.

Further Reading

1. Adam, S. F., *Microwave Theory and Applications*. Prentice-Hall, Inc., New Jersey, 1969.
2. Alder, R. B. et al., *Electromagnetic Energy Transmission and Radiation*. MIT Press, Cambridge, MA, 1969.
3. Chatterjee, R., *Microwave Engineering*. East West Press, New Delhi, 1990.
4. Collin, R. E., *Foundations for Microwave Engineering*. McGraw-Hill, New York, 1966.
5. Das, A. and Das, S. K., *Microwave Engineering*. Tata McGraw-Hill, New Delhi, 2002.
6. Gandhi, O. P., *Microwave Engineering and Applications*. Pergamon Press, New York, 1981.
7. Gupta, K. C., *Microwaves*. Wiley Eastern, New Delhi, 1979.
8. Lance, A. L., *Introduction to Microwave Theory and Measurements*. McGraw-Hill, New York, 1964.
9. Liao, S. Y., *Microwave Devices and Circuits*, 3rd ed. Pearson Education Inc, Delhi, 2003.
10. Pozar, D. M., *Microwave Engineering*, 2nd ed. John Wiley & Sons, New Delhi, 1999.
11. Rizzi, P. A., *Microwave Engineering: Passive Circuits*. Prentice-Hall, Inc., Englewood Cliffs, NJ, 1988.
12. Saad, T. and Hansen, R. C., *Microwave Engineer's Handbook*, Vol. 1. Artech House, Dedham, MA, 1971.

5

Microwave Ferrite Devices

5.1 Introduction

Matter is composed of atoms and molecules. Atoms in themselves are composed of tiny particles, including electrons. As illustrated in Figure 5.1, these electrons not only revolve around a nucleus but also spin about their own axis. The orbiting and spinning electrons are considered to be tiny current loops. The two loops with I_{orb} and I_{spin} currents encompassing infinitesimal surface areas S_{orb} and S_{spin} result in two magnetic dipole moments m_{orb} ($= I_{orb} S_{orb}$) and m_{spin} ($= I_{spin} S_{spin}$). As is evident from Table 5.1, in all major magnetic materials, the contribution of spinning electrons is more prominent than that of revolving electrons; thus, its influence on the magnetic property is more significant.

These dipole moments (hereafter referred to as dipoles) can be conceived as two vectors oriented in different arbitrary directions. The magnetic property of a material is solely governed by the behaviour of these dipoles. Figure 5.2 shows that when these (m_{orb} and m_{spin}) dipoles come under the influence of an external (applied) magnetic field (B_{app}), they try to align themselves with the applied magnetic field. As a result of this alignment, the internal magnetic field (B_{int}) is also affected. The impact of this effect, which is different for different magnetic materials, can be observed in Table 5.1.

In *paramagnetic* materials, atoms and hence spinning electrons are sufficiently separated and do not influence each other. Almost similar is the case of *diamagnetic* materials wherein the influence of the external magnetic field remains almost insignificant. In *ferromagnetic* substances, the electrons are close enough to reinforce the effect of each other and the resulting magnetic field is much stronger. Thus, ferromagnetic materials have fairly large magnetic permeabilities. Unfortunately, conductivities of these materials are also fairly high. The application of high-frequency magnetic fields to such materials results in large eddy currents and hence, in significant power loss.

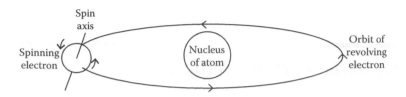

FIGURE 5.1
Revolving and spinning electrons.

TABLE 5.1

Magnetic Materials, Magnetic Moments and Relative Values of Flux Densities

No.	Material	Magnetic Moments	Value of B	Remarks
1	Diamagnetic	$m_{orb} + m_{spin} = 0$	$B_{int} < B_{app}$	B_{int} and B_{app} are almost equal
2	Paramagnetic	$m_{orb} + m_{spin} \Rightarrow 0$	$B_{int} > B_{app}$	B_{int} and B_{app} are almost equal
3	Ferromagnetic	$\|m_{spin}\| \gg \|m_{orb}\|$	$B_{int} \gg B_{app}$	Contains the domain, adjacent domains align and add to the result in large internal flux density
4	Anti-ferromagnetic	$\|m_{spin}\| \gg \|m_{orb}\|$	$B_{int} \approx B_{app}$	Contains the domain, equal adjacent domains oppose, result in total cancellation and no enhancement in internal flux density
5	Ferrimagnetic	$\|m_{spin}\| > \|m_{orb}\|$	$B_{int} > B_{app}$	Contains domains, unequal adjacent domains oppose, no total cancellation and enhancement in internal flux density
6	Super paramagnetic	$\|m_{spin}\| \gg \|m_{orb}\|$	$B_{int} > B_{app}$	Contains the matrix of magnetic and non-magnetic materials

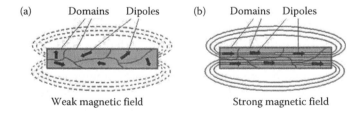

FIGURE 5.2
(a) Unmagnetised and (b) magnetised material.

5.2 Ferrites

The nature of alignments in different materials is shown in Figure 5.3. In the case of ferromagnetic materials, these dipoles align in the same direction in different domains and thus aid in the creation of a strong magnetic field. In *anti-ferromagnetic* materials, the alignment is equal and anti-parallel and thus leaves no scope for the enhancement of internal magnetic field. Yet, in another material, the alignment is unequal and anti-parallel. This material is referred to as a *ferrimagnetic material or ferrite*.

Ferrites have two separate crystal lattices. Electrons that spin in each lattice influence their neighbours and result in a sufficiently strong magnetic field. From Figure 5.3, it can be noted that in ferrimagnetic materials, the alignments of spins of two lattices are in opposite directions with respect to the applied field; thus, the two fields may get totally neutralised. But, since the resulting alignments in individual lattices are not the same, there is only a partial cancellation and thus, a net magnetic effect remains. The main advantage of ferrites is that they have fairly high resistivity (as high as 10^{14} times that of metals) and sufficiently high permeability (μ of the order of 5×10^3). Thus, at high frequencies, the eddy currents are greatly reduced.

Ferrites are made by sintering a mixture of metallic oxides. Their general chemical compositions can be written as $M{-}O{-}Fe_2O_3$, where M stands for a divalent metal, namely, cobalt, iron, magnesium, manganese, zinc and so on. Ferrites are transparent to electromagnetic waves. Thus, the requirement of magnetic materials for high-frequency applications is fully met by ferrites. The basic principle of operation of ferrite devices is based on the fact that the spin axes of electrons wobble at a natural resonant frequency under the influence of an external magnetic field. This aspect is shown in Figure 5.4.

As seen in Figure 5.2, in the absence of an external magnetic field, dipoles are randomly oriented and the combined effect is almost negligible. With the application of a static magnetic field, these dipoles try to get aligned along the direction of the field. If the applied static magnetic field ($\mathbf{H}_{app} = \mathbf{H}_0$)

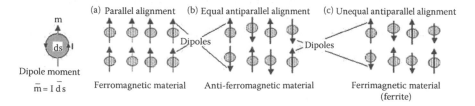

FIGURE 5.3
Nature of alignments in different magnetic materials.

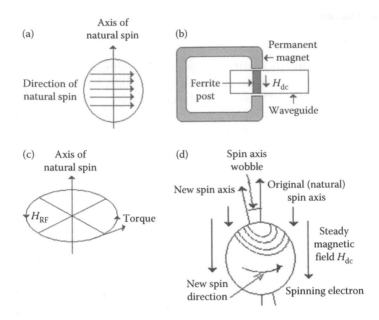

FIGURE 5.4
Information related with the electron spin. (a) Natural spin, (b) location of ferrite post, (c) torque due to RF field and (d) spin axis wobble.

is sufficiently large to cause saturation in ferrites (i.e., all the dipoles are perfectly aligned in the direction of the field), it results in the total dipole moment per unit volume (or magnetisation) **M**. Thus, the total internal magnetic flux density ($\mathbf{B}_{int} = \mathbf{B}_0$) can be written as

$$\mathbf{B}_0 = \mu \, (\mathbf{H}_0 + \mathbf{M}) \tag{5.1}$$

When an electromagnetic wave propagates, an alternating current (AC) field is superimposed over direct current (DC) magnetic field; the net magnetisation gets modified. The static magnetic field $\mathbf{H}_0$ contains only one component in the direction of its application (say **z**) (i.e., $\mathbf{H}_0 = H_0 \mathbf{a}_z$), whereas the time-varying (AC) field **H** may have all the three components (i.e., $\mathbf{H} = H_x \mathbf{a}_x + H_y \mathbf{a}_y + H_z \mathbf{a}_z$).

The permeability of ferrites assumes the form of a tensor that can be written as

$$[\mu] = \begin{bmatrix} \mu_1 & j\mu_2 & 0 \\ -j\mu_2 & \mu_1 & 0 \\ 0 & 0 & 1 \end{bmatrix} \tag{5.2}$$

In view of the permeability tensor, the *B–H* relation can be written as

$$
\begin{bmatrix} B_x \\ B_y \\ B_z \end{bmatrix} = \begin{bmatrix} \mu_1 & j\mu_2 & 0 \\ -j\mu_2 & \mu_1 & 0 \\ 0 & 0 & 1 \end{bmatrix} \begin{bmatrix} H_x \\ H_y \\ H_z + H_0 \end{bmatrix}
\tag{5.3}
$$

A linearly polarised wave propagating in a ferrite medium in the direction of the static magnetic field may be conceived to be composed of two oppositely rotating circularly polarised waves. Both of these will face different environments vis-à-vis values of permeabilities and thus will have different propagation constants.

As stated earlier, ferrite is a composite material with useful magnetic properties and with high resistance to the flow of current. Ferrites, in fact, have sufficient resistance and may be classified as semiconductors. The compounds used in its composition can be compared to those used in transistors wherein the magnetic and electrical properties can be manipulated by the proper selection of atoms in the right proportions and that too over a wide range.

5.3 Faraday's Rotation

Figure 5.5 shows a circular waveguide that is totally or partially filled with ferrite material. This waveguide supports a TE_{11} mode of a plane wave and is magnetised along its axis. Figure 5.5a illustrates the rotation of polarisation of a wave travelling in the forward direction while Figure 5.5b shows the rotation of a wave travelling in the backward direction.

Assume that the plane polarised wave has a magnetic field vector in a given direction. This vector may be considered to have two equal counter-rotating vectors, that is, the two circularly polarised waves. If the ferrite is magnetised somewhere in the spin resonance region, one of these two counter-rotating vectors will travel through a medium that has a relative permeability of about unity. The other component will experience altogether a different permeability as it will rotate in the opposite direction coinciding with the direction of natural precession. While the wave travels in the opposite direction, both these components will experience the same permeability and rotate with distance, an equal and opposite amount as they travel down the waveguide and thus preserve the plane of polarisation. However, if one component rotates more than the other, as will be the case with a magnetised ferrite, then the resultant vector and the plane of polarisation will also rotate. This effect is known as *Faraday's rotation*. This phenomenon is more prominent at optical frequencies and at radio frequencies (RFs) in the ionospheric region possessing the tensor permittivity.

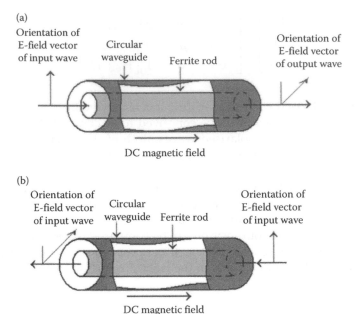

FIGURE 5.5
Rotation of polarisation of waves. (a) Wave travelling in forward direction and (b) wave travelling in backward direction.

In view of this unique effect, many useful devices are made to operate at microwave frequencies. These include gyrators, isolators, circulators, phase changers, attenuators, modulators and microwave switches. Some of these devices are described in the following sections.

5.4 Non-Reciprocal Ferrite Devices

Figure 5.6 shows the devices that allow propagation of the wave only in one direction. These, referred to as non-reciprocal devices, mainly include gyrators, circulators and isolators. The working of these devices is solely based on the principle of Faraday's rotation. All non-reciprocal devices are constructed by placing a piece of ferrite in a waveguide.

In Figure 5.5, a ferrite rod in a circular waveguide was subjected to a magnetic field. This field can be adjusted to make the electron wobble frequency equal to the frequency of the travelling wave. The energy travelling down the waveguide from left to right will set up a rotating magnetic field that rotates through the ferrite material in the same direction as the natural wobble of electrons. The aiding magnetic field increases the wobble of the ferrite

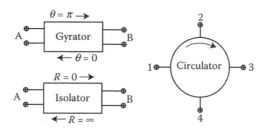

FIGURE 5.6
Non-reciprocal devices.

electrons so much so that almost all the energy in the waveguide is absorbed and dissipated as heat. The magnetic fields caused by energy travelling from right to left rotate in the opposite direction through the ferrite and have very little effect on the amount of electron wobble. In this case, the fields attempt to push the electrons in the direction opposite the natural wobble but no large movements occur. Since no overall energy exchange takes place, the effect on energy travelling from right to left is almost insignificant. This rotation of polarisation of the wave while travelling in the forward and backward directions was illustrated earlier in Figure 5.5.

5.4.1 Gyrators

A gyrator is a two-port device in which the plane of polarisation of the E vector of the propagating wave rotates by 180° in one direction and remains unaffected in the other. It contains rectangular waveguides at its two ends and a circular waveguide segment in between. A thin ferrite rod is placed well inside the circular waveguide that is properly tapered on both the ends to avoid reflections due to abrupt dimensional change. The length of the ferrite rod along with its bias is designed to rotate the plane of polarisation by 90° in the (assumed) clockwise direction. The gyrator along with its ports and ferrite rod is oriented in the direction of wave propagation (assumed to be along z-axis). As illustrated in Figure 3.28, the signal travelling from a rectangular waveguide to a circular waveguide, the dominant TE_{10} mode in the rectangular waveguide transforms into the TE_{11} mode in the circular waveguide. The reverse is also true, that is, the TE_{11} mode of the circular waveguide will automatically transform into the TE_{10} mode in the rectangular waveguide when the wave progresses that way. In the following section, two types of gyrators are given.

5.4.1.1 Gyrator with Input–Output Ports Rotated by 90°

As illustrated in Figure 5.7, port-2 of the device is rotated by 90° with respect to port-1. When the signal enters port-1, it travels down the circular

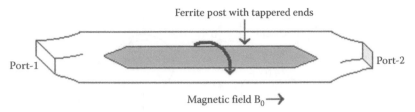

FIGURE 5.7
Gyrator with waveguide ports rotated by 90°.

waveguide wherein the ferrite post rotates it by 90° in the (assumed) clockwise direction. Since port-2 is rotated by 90°, the wave further rotates by 90° before it emerges from port-2. Thus, the output appearing at port-2 is 180° out of phase with respect to the input at port-1. When the signal enters port-2, it will first rotate by −90° due to the orientation of the port and then by +90° by the ferrite post. Thus, the signal at port-1 will appear without any rotation.

5.4.1.2 Gyrator with 90° Twist

In this case, ports 1 and 2 are in the same plane but port-1 is followed by a 90° twist. This twist rotates the plane of polarisation by 90° clockwise for the forward wave and by 90° anti-clockwise for the backward wave. Thus, the plane of polarisation of the wave will rotate by 180° on the forward path and by 0° on the backward journey. The configuration is shown in Figure 5.8.

5.4.2 Isolators

An isolator is a device that passes signals only in one direction and totally attenuates in the other. Such one-way transmission of electromagnetic waves

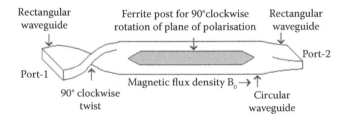

FIGURE 5.8
Gyrator with a twist at the input end of the waveguide.

is achieved by using gyromagnetic resonance phenomenon excited in ferrite materials when biased to resonate by an external steady magnetic field. Thus, a ferrite isolator is a non-reciprocal two-port component. Since it has to pass the microwave signal with low loss in the forward direction and absorbs the entire energy in the reverse direction, its insertion loss must be low in one direction and high in the other. This property of an isolator must be maintained over the frequency band of interest just as for any other component. An additional figure of merit used to describe an isolator is the ratio (in dBs) of the reverse attenuation to the forward insertion loss. In view of the above, the effect of load mismatch on the signal sources is minimised. These are useful for maintaining signal source stability, eliminating long-line and frequency-pulling effects in microwave signal sources. Since this device uses a permanent magnet, all magnetic materials should be kept at a safe distance.

5.4.2.1 Faraday's Rotation Isolators

Construction wise, Faraday's rotation isolator is similar to that of a gyrator except that it employs either a 45° twist at the input port or 45° rotation of the output port, instead of the 90° twist or rotation used in gyrators. Also, the ferrite rod rotates the signal only by 45° instead of 90° rotation introduced in the case of the gyrator.

Figure 5.9 illustrates an isolator, which contains two thin properly tapered resistive cards (vanes). These cards are placed in both the input and output ports of the device and are parallel to the broader walls of the input and output waveguides. A card allows unattenuated propagation if the E-vector is perpendicular to it and absorbs the entire energy if it is parallel to the card. In an isolator, the wave is again assumed to be travelling in the z direction and the rotation caused by the ferrite post is taken to be clockwise.

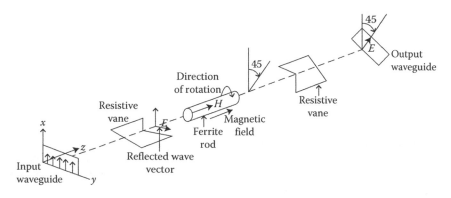

FIGURE 5.9
Faraday's rotation isolator.

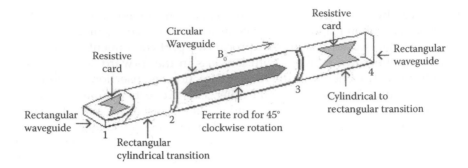

FIGURE 5.10
Faraday's rotation isolator with input and output ports at 45°.

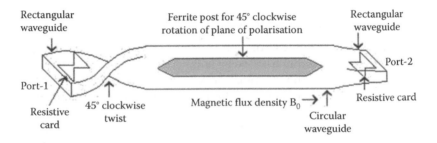

FIGURE 5.11
Faraday's rotation isolator with a 45° twist and 45° Faraday's rotation.

Figure 5.10 illustrates an isolator with input and output ports rotated by 45°, whereas Figure 5.11 illustrates an isolator with a 45° twist at the input port.

The orientation and availability of the signal at different locations in these two isolators can be worked out in view of Figure 5.9 that spells the basic operating principle.

5.4.2.2 Resonance Isolator

Faraday's rotation isolators lose their utility when the power level to be handled exceeds about 2 kW. In such applications, another class of isolators, commonly called resonance isolators, gains the ground. In these, a ferrite core is asymmetrically placed in a rectangular waveguide. Its location has to be such that when the wave travels in one direction, it gets totally absorbed, while in the reverse direction it remains unaffected. In principle, it may be constructed by using circular or rectangular waveguides. A rectangular waveguide, however, has an edge over circular waveguides since it gives

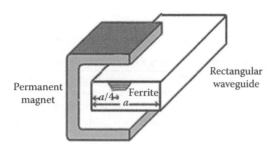

FIGURE 5.12
Resonance isolator.

a more compact structure, makes construction simpler and possesses more peak power-handling capacity.

As shown in Figure 5.12, the tapered ferrite slab is located at $a/4$ distance from one of the broader walls of width 'a' of the rectangular waveguide. The dominant TE_{10} mode in the rectangular waveguide is under the influence of the magnetic field and in this unique location of ferrite, the wave gets circularly polarised. This polarisation is in one direction for its forward travel and in the other direction for backward travel. This results in propagation in one direction and attenuation in the other. By proper design, the forward attenuation of the wave can be minimised.

The demerits of this resonator include (i) requirement of the higher static field to produce resonance, (ii) relatively low backward-to-forward attenuation ratio and (iii) due to temperature rise, the ferrite core may reach the Curie point. The summary of the merits and limitations of both Faraday's rotation isolator and that of resonance isolators are further listed in Table 5.2.

5.4.3 Circulators

Circulators are non-reciprocal devices that allow an input at one port to appear only at the next adjacent port around a circle. Thus, in a way, it circulates the signal from one port to the next. This circulation may be clockwise or anti-clockwise and is solely governed by the direction of magnetisation. These devices may contain three or four ports. Under the category of three-port devices, the ports may be located so as to form a T or a Y junction at the location of the ferrite post. Figure 5.13 illustrates two such circulators.

5.4.3.1 Three-Port Circulators

Three-port circulators allow microwave energy to flow in clockwise (or anti-clockwise) direction with negligible loss and provide high isolation in the opposite direction. The microwave energy entering port-1 will go

TABLE 5.2

Comparison of Faraday's Rotation and Resonance Isolators

Sl. No.	Aspect of Comparison	Faraday's Rotation Isolator	Resonance Isolator
1	Constituents	One circular waveguide with two rectangular waveguides at its each end	One rectangular waveguide
2	Location of ferrite core	In circular waveguide	In the circularly polarised magnetic field
3	Power handling capability	Low	High
4	Effect of temperature on power handling capacity	No	Yes
5	Static field requirement for resonance to occur	Low	High
6	Ratio of backward-to-forward attenuation	More	Less
7	Nature of construction	Complicated	Simple

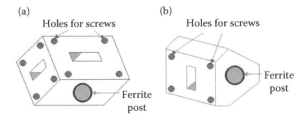

FIGURE 5.13
Three-port circulators. (a) T-type and (b) Y-type.

to port-2 with very little attenuation, whereas there would be practically no transmission to port no. 3 from port-1. The small attenuation suffered in transmission from port-1 to port-2, port-2 to port-3 or port-3 to port-1 is known as *insertion loss*, which is generally very small. The attenuation from port-1 to port-3, port-3 to port-2 or port-2 to port-1 is known as *isolation*, which is quite high. The insertion loss and isolation are referred to as figures of merit of circulators. For properly matched ports, these can be given in dBs as follows.

$$\text{Insertion loss (dB)} = -10 \log_{10}\left\{ \frac{(P_{out})_{port(n)}}{(P_{in})_{port(n-1)}} \right\} \tag{5.4}$$

$$\text{Isolation (dB)} = 10 \log_{10}\{(P_{out})_{port(n+1)}/(P_{in})_{port(n-1)}\} \tag{5.5}$$

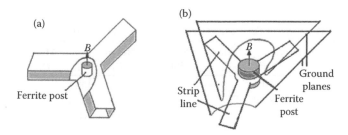

FIGURE 5.14
Interior of the three-port circulators. (a) Made of waveguide and (b) made of stripline.

Beside the above, VSWR, frequency of operation, number of ports, carrier and pulse power handling capability, linearity of phase and so on are the other important parameters of circulators.

If one of the ports (say port-2) is terminated by a matched load, this combination can be used as an isolator. It will allow energy to flow from port-3 to port-1, but any reflected energy entering port-1 will be absorbed by port-2 and will not reach port-3. To such a typical setup, klystron is attached to port-3 and the remaining test bench is attached to port-1 so that any reflection from port-1 may not disturb its frequency or power output. A microwave ferrite prism is mounted inside a precisely milled three-port junction and is biased by a permanent magnet fixed on the top and bottom of the junction. Both T and Y circulators are available. Figure 5.14 illustrates the interior of two three-port circulators made of waveguides and striplines.

5.4.3.2 Four-Port Circulators

A four-port circulator is shown in Figure 5.15. Its construction and features are similar to that of an isolator of Figure 5.10 with resistive cards but with two additional ports. The working of the circulator is explained in Table 5.3. This table includes the angular positions of different ports, and the signal

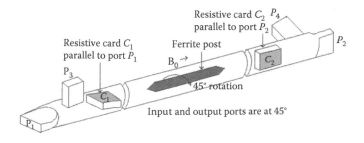

FIGURE 5.15
Four-port circulator with two additional ports.

TABLE 5.3

Illustration of Working of Four-Port Circulator Shown in Figure 5.16

P_1	C_1	Before P_3	After P_3	P_3	Ferrite Post	Before P_4	P_4	After P_3	C_2	P_2
Angle 0°	0°		90°			135°			45°	45°
$E\bot a$	$E\bot C_1$	E	$E\bot P_3$	E		E	$E\bot P_4$	E	$E\bot C_2$	$E\angle45°$
▶	NL		NC			$\angle45°$	NC	$\angle45°$	NL	$E\bot a$ ▲
$E=0$	$E\|C_1$	E	$E\|P_3$	E	+45°	E	$E\bot P_4$	E	$E\bot C_2$	$E\angle45°$
	TA	$\angle90°$	▲	$\angle90°$	Clockwise	$\angle45°$	NC	$\angle45°$	NL	$E\bot a$
$E=0$	$E\|C_1$	E	◀▶	E	rotation	E	$E\|P_4$	E	$E\|C_2$	$E=0$
	TA	$\angle90°$		$\angle90°$		$\angle135°$	▲	$\angle135°$	TA	
$-E\bot a$	$E\bot C_1$	E	$E\bot P_3$	E		E	◀▶	E	$E\|C_2$	$E=0$
▲	NL	$\angle180°$	NC	$\angle180°$		$\angle135°$		$\angle135°$	TA	

Meaning of symbol: All angles are with reference to port-1, field intensity (E), broader wall (a), no loss (NL), total absorption (TA), no coupling (NC), perpendicular ($\bot$), parallel ($\|$), signal in, forward flow (▶), signal in, backward flow (◀), signal in, flow in both directions (◀▶) and signal out (▲).

orientations at different locations. The ports in the table are designated as P_1, P_2, P_3 and P_4 respectively. The resistive cards are named as C_1 and C_2. Other symbols used are given as a footnote to the table. The angle of port P_1 is taken as the reference for all the angles marked. This table also indicates the location of the entering and exit ports of the signal by different symbols.

Another four-port circulator shown in Figure 5.16 is made by using two magic tees and a gyrator. The working of this circulator is explained in Table 5.4. The table is self-explanatory since it includes information about magnitudes of signals at different ports marked 1, 2, 3, 4, a, b, c and d and indicates the location of entering and exit ports by two different symbols.

Figure 5.17 shows a four-port circulator having two waveguide sections and phase shifters.

Figure 5.18 illustrates four-port circulators with ports located 90° apart from their adjacent ports. The figure also shows the use of these ports for isolating different devices.

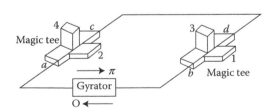

FIGURE 5.16
Four-port circulator comprising a gyrator and two magic tees.

TABLE 5.4

Illustration of Working of Four-Port Circulator Shown in Figure 5.17

	Signal at Port	1	b	d	3	2	a	c	4
Port	1	E▶	E/2	E/2	0	E▲	E/2	E/2	0
	2	0	−E/2	E/2	E▲	E▶	E/2	E/2	0
	3	0	E/2	−E/2	E▶	0	−E/2	E/2	E▲
	4	E▲	E/2	E/2		0	−E/2	E/2	E▶

Symbols: Input (▶) and output (▲).

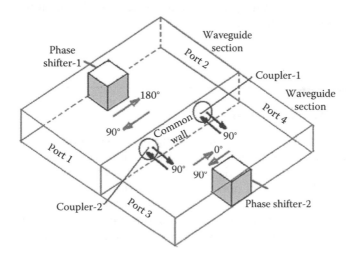

FIGURE 5.17
Four-port circulator comprising two waveguide sections and phase shifters.

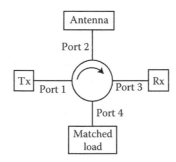

FIGURE 5.18
Decoupling or isolating different devices by using a four-port circulator.

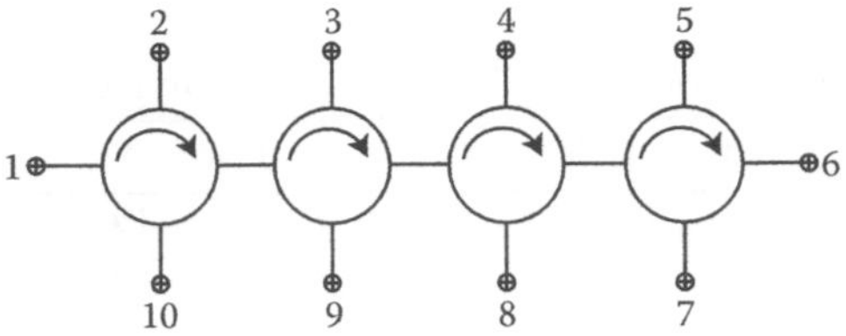

FIGURE 5.19
Cascaded circulators for increasing the number of ports.

In the case of requirement of more ports, the circulator of Figure 5.18 can be cascaded. Such cascading is shown in Figure 5.19 wherein four circulators are cascaded to yield 10 ports.

5.5 Ferrite Phase Shifter

In Figure 5.12, if the ferrite is located along the z-axis of the waveguide at $a/2$ distance instead of $a/4$, the resulting configuration will act as a phase shifter. The location of the magnetic field is kept as such. The phase shift will attain maxima near the region of maximum attenuation. Since phase shifts in forward and backward directions are the same, it becomes a reciprocal device. If the position of ferrite is shifted away from the centre, the phase shift in the forward and reverse directions is not the same and the device becomes non-reciprocal.

5.6 Ferrite Attenuators

A ferrite attenuator may be constructed to attenuate a particular microwave frequency and to pass others unaffected. In the case of ferrite phase shifter, a ferrite post is located in the centre of the waveguide. If this ferrite is so located that the magnetic fields caused by its electrons are perpendicular to the direction of propagation in the waveguide, a steady external magnetic field will cause the electrons to wobble at the same frequency as that of the energy to be attenuated. In view of this coincidence, the energy in the waveguide always adds to the wobble of electrons. The direction of spin axis of electrons changes due to the addition of energy during the wobble motion. The force causing the increase in wobble is the energy in the waveguide.

Thus, the energy in the waveguide is attenuated by the ferrite and is given off as heat. The energy in the waveguide at a frequency other than that of wobble frequency of the ferrite remains largely unaffected because it does not increase the amount of electron wobble. The resonant frequency of electron wobble can be varied over a limited range by changing the strength of the applied magnetic field.

5.7 Ferrite Switches

Ferrite switches are switchable isolators with internal or external magnetising circuits. An internal circuit is placed within the ferrite junction whereas the external magnetising circuit comprises an external magnet. The design of switches with internal magnetising circuits is a bit difficult. Also, these are slower but consume lesser power. Switches with external magnets are faster with switching speed between 10 and 50 μs. A high-power switch can be built by using a differential phase shifter. However, it has a limitation as its performance deteriorates due to eddy currents, temperature rise and hysteresis loss.

5.8 YIG Filters

Yttrium iron garnets (YIGs) are one of the ferrites commonly used at microwave frequencies. Resonance in YIG takes place when microwave frequency coincides with the electron precession and is determined by the product of the gyrometric ratio and the DC magnetic field. The unloaded Q of a highly polished sphere or a disk of YIG is comparable or even higher than those of most of the microwave circuits. Since this resonance is highly dependent on interception of the field, and the orientation of spheres does not affect the same YIG, spherical resonators are more popular. Besides, the tuning characteristic of spherical resonators is linear. The typical diameter of a YIG sphere is about 0.05–0.06 mm and its unloaded Q typically ranges between 2000 and 10,000.

Figure 5.20 shows an idealised band-pass filter containing two orthogonal coils (loops) and a small ferrite sphere. An analytical relation reveals that the unloaded quality factor reaches zero at RF frequency given by $f_0 = \omega_m/6\pi$, which is about 1670 MHz for a spherical YIG resonator. To make a multistage filter, more than one YIG filters can be cascaded.

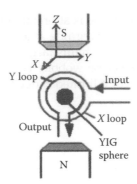

FIGURE 5.20
Idealised loop coupled to a YIG resonator as a band-pass filter.

5.9 Figures of Merit of Ferrite Devices

When a device is inserted as an element in a circuit, it may result in a loss in accordance with its nature. It may absorb power or reflect a part of the energy of the propagating wave. Some of such effects are common to all the components whereas some are specific to particular devices. The parameters to be accounted in ferrite devices are given below.

Insertion loss: If P_1 is the power at port-1 and P_2 is the power at port-2, the loss between ports 1 and 2 is referred to as insertion loss (denoted by L_{12}) and is given by the relation

$$L_{12} = 10 \log_{10} P_1/P_2 \tag{5.6}$$

Isolation: If all the ports are identical, the losses between different ports (L_{12}, L_{23}, L_{34} and L_{41}) are the same. The isolation (I) between ports 1 and 4 is given by the relation:

$$I_{14} = 10 \log_{10} P_1/P_4 \tag{5.7a}$$

For a four-port circulator with identical ports

$$I_{14} = I_{21} = I_{32} = I_{41} \tag{5.7b}$$

Cross-coupling: This term is valid only for those devices in which there is a possibility of coupling between wrong ports. As an example, in a four-port forward circulator, port-1 should mainly couple to port-2 and partly to port-4 but not to port-3. If it couples to port-3, the coupling coefficient can be given as

$$C_{13} = 10 \log_{10} P_1/P_3 \tag{5.8a}$$

In the case of a cross-directional coupler (a non-ferrite device) where all the four ports are identical

$$C_{13} = C_{24} = C_{31} = C_{42} \tag{5.8b}$$

It is to be noted that all the above parameters are negative by nature and will be again discussed in Chapter 7 in connection with the directional couplers.

EXAMPLE 5.1

An isolator has an insertion loss of 0.5 dB and isolation of 20 dB. Find the power at the output ports n and $n + 1$ if the input power at port $(n - 1)$ is 25 mW.

Solution

In view of Equation 5.4,

$$\text{Insertion loss (dB)} = -10 \log_{10}\{(P_{out})_{port(n)}/(P_{in})_{port(n-1)}\}$$

$$0.5 = -10 \log_{10}(P_{out})_{port(n)} + 10 \log_{10}(25 \times 10^{-3})$$

$$0.05 = -\log_{10}(P_{out})_{port(n)} + \log_{10}(25 \times 10^{-3}) = -\log_{10}(P_{out})_{port(n)} - 1.602$$

$$\log_{10}(P_{out})_{port(n)} = 1.62 - 0.05 = -1.67$$

$$P_{out}|_{port(n)} = 21.38 \text{ mW}$$

In view of Equation 5.5,

$$\text{Isolation (dB)} = -10 \log_{10}\{(P_{out})_{port(n+1)}/(P_{in})_{port(n-1)}\}$$

$$20 = -10 \log_{10}\{(P_{out})_{port(n+1)} + \log_{10}(25 \times 10^{-3})\}$$

$$2 = -\log_{10}(P_{out})_{port(n+1)} - 1.62$$

$$\log_{10}(P_{out})_{port(n+1)} = 3.62$$

$$P_{out}|_{port(n+1)} = 0.24 \text{ mW}$$

EXAMPLE 5.2

A microwave device is fed at the input with 20 mW while its output is 15 mW. Find its insertion loss.

Solution

In view of Equation 5.6, the insertion loss is

$$L_{12} = 10 \log_{10} P_1/P_2 = 10 \log_{10} (20 \times 10^{-3}/15 \times 10^{-3}) = 10 \log_{10} (4/3) = 1.249 \text{ dB}$$

EXAMPLE 5.3

The input to a microwave device is 50 mW, whereas its output at an isolated port is 0.5 mW. Find its isolation loss.

Solution

In view of Equation 5.7, the isolation loss is

$$I_{14} = 10 \log_{10} P_1/P_4 = 10 \log_{10} (50 \times 10^{-3}/0.5 \times 10^{-3}) = 10 \log_{10} (100) = 20 \text{ dB}$$

EXAMPLE 5.4

The input to a four-port circulator at port-1 is 40 mW, whereas the output at its port is 2 mW. Find the cross-coupling loss in dBs.

Solution

In view of Equation 5.8a, the cross-coupling loss is

$$C_{13} = 10 \log_{10} P_1/P_3 = 10 \log_{10} (40 \times 10^{-3}/2 \times 10^{-3}) = 10 \log_{10} (20) = 13.01 \text{ dB}$$

PROBLEMS

P-5.1 An isolator has an insertion loss of 0.25 dB and isolation of 15 dB. Find the power at its output ports n and $n + 1$ if the input power at port $(n - 1)$ is 30 mW.

P-5.2 A microwave device is fed at the input with 30 mW, whereas its output is 5 mW. Find the insertion loss of the device.

P-5.3 Find the isolation loss if the input to a microwave device is 60 mW and the output at an isolated port is 1.5 mW.

P-5.4 A four-port circulator is fed at port-1 with 50 mW power, whereas its output at port-3 is 4 mW. Find the cross- coupling loss in the circulator in dBs.

Descriptive Questions

Q-5.1 What are ferrites and how do they differ from other magnetic materials?

Q-5.2 Explain Faraday's rotation and its role in non-reciprocal ferrite devices.

Q-5.3 Discuss the working of gyrators.

Q-5.4 What is the purpose served by an isolator and how?

Q-5.5 How does the working of Faraday's isolator differ from that of resonance isolator?

Q-5.6 Discuss the working of different types of circulators.

Q-5.7 Explain the working of ferrite phase shifter and ferrite attenuator.

Q-5.8 Discuss the required relations of the figure of merits of ferrite devices.

Further Reading

1. Bowness, C., Microwave ferrites and their applications. *Microwave. J.* 13–21, July–August 1958.
2. Chatterjee, R., *Microwave Engineering*. East West Press, New Delhi, 1990.
3. Collin, R. E., *Foundations for Microwave Engineering*. McGraw-Hill, New York, 1966.
4. Das, A. and Das, S. K., *Microwave Engineering*. Tata McGraw-Hill, New Delhi, 2002.
5. David, M. P., *Microwave Engineering*, 2nd ed. John Wiley, New Delhi, 1999.
6. Gandhi, O. P., *Microwave Engineering and Applications*. Pergamon Press, New York, 1981.
7. Gupta, K. C., *Microwaves*. Wiley Eastern, New Delhi, 1979.
8. Lance, A. L., *Introduction to Microwave Theory and Measurements*. McGraw-Hill, New York, 1964.
9. Lax, B. and Button, K. J. *Microwave Ferrites and Ferrimagnetics*. McGraw-Hill, New York, 1962.
10. Liao, S. Y., *Microwave Devices and Circuits*. 3rd ed. Pearson Education Inc, Delhi, 2003.
11. Lio, S. Y., *Microwave Devices and Circuits*, 3rd ed. Prentice-Hall, New Delhi, 1995.
12. Rizzi, P. A., *Microwave Engineering: Passive Circuits*. Prentice-Hall, Englewood Cliffs, NJ, 1988.
13. Saad, T. and Hansen, R. C., *Microwave Engineer's Handbook*, Vol. 1. Artech House, Dedham, MA, 1971.
14. Soohoo, R. F., *Theory and Applications of Ferrites*. Prentice-Hall, Englewood Cliffs, NJ, 1960.

Q-5.6 Discuss the working of three-phase circulator.
Q-5.7 Explain the working of gyrator, isolator and terminators.
Q-5.8 Discuss the required relations of the figure of merits of varactor.

Further Reading

6

Smith Chart

6.1 Introduction

Almost all practical high-frequency circuits are composed of two or more transmission lines. These lines are interspersed with series and shunt elements. The problems involving such elements require a large number of numerical and algebraic manipulations for their solutions. Smith chart is an effective tool that significantly reduces these manipulations.

Smith chart is a circular plot containing a large number of interlaced circles. With its proper use, matching impedances with apparent complicated structures can be made without any computation. It requires correct reading of the values along appropriate circles. It can be visualised as a polar plot of the complex reflection coefficient (Γ) or can be defined mathematically as one-port scattering parameter s_{11}.

Although this chart primarily emanates from the relations belonging to two-wire transmission line, it is equally applicable to waveguides, coaxial lines, planar lines and other structures, which may be used to transmit or guide electromagnetic energy. This chart with a pair of coordinates for impedance and admittance possess the following useful characteristics:

- It represents all possible (real, imaginary or complex) values of impedances and admittances.
- It provides an easy method for conversion of impedance to admittance and vice versa.
- It provides a graphical method for determining the impedance and admittance transformation due to the length of the transmission line.

6.2 Characteristic Parameters of a Uniform Transmission Line

Figure 6.1 illustrates a two-wire transmission line that is assumed to be uniform vis-à-vis its conductor's cross sections, spacing between wires and field

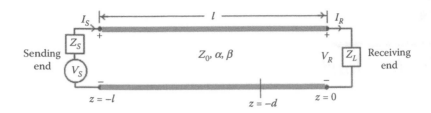

FIGURE 6.1
Uniform transmission line with an AC source and terminating impedance Z_L.

distribution therein. The line is fed an AC source at the sending end and terminated at the receiving end by an impedance of an arbitrary value. This also illustrates parameters used in subsequent sections.

An equivalent circuit of this line is shown in Figure 6.2. According to the maximum power transfer theorem, maximum power from the source to the load is transferred only when source impedance equals the complex conjugate of the load impedance. In mathematical terms

$$R_S \pm jX_S = R_L \pm jX_L \tag{6.1}$$

This condition maximises the energy transferred from the source to the load. Further, this condition is required to avoid the reflection of energy from the load back to the source, particularly in a high-frequency environment.

In view of Equation 2.74 (reproduced as Equation 6.2), the voltage (V) and the current (I) at any location along the direction (z) of the wave propagation on a line with $Z_L \neq Z_0$ are

$$V = V'e^{\gamma z} + V''e^{-\gamma z} \quad \text{and} \quad I = I'e^{\gamma z} + I''e^{-\gamma z} \tag{6.2}$$

The validity of Equation 6.2 relates to Figure 6.1, wherein $z = 0$ point is taken at the receiving end, $z = -l$ at the sending end and $z = -d$ somewhere in between. Both these equations are composed of two terms, one representing a forward wave and the other a backward (or reflected) wave. The parameter $\gamma (= \alpha + j\beta)$ is the propagation constant, V' and V'' are the magnitudes of voltages of forward and backward waves and I' and I'' are the magnitudes of

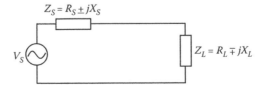

FIGURE 6.2
Illustration of the condition for maximum power transfer.

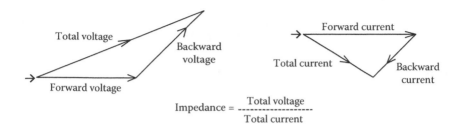

FIGURE 6.3
Wave voltage and current phasors.

currents of forward and backward waves, respectively. Figure 6.3 illustrates these voltages and currents in the form of phasors.

Equations 2.75 through 2.77 are reproduced with new numbers as shown below:

$$\frac{V'}{I'} = Z_0 = -\frac{V'}{V''} \tag{6.3}$$

$$Z_L = \left.\frac{V}{I}\right|_{z=0} = \frac{V' + V''}{I' + I''} = Z_0 \frac{I' - I''}{I' + I''} = Z_0 \frac{V' + V''}{V' - V''} \tag{6.4}$$

$$\frac{V''}{V'} = -\frac{I''}{I'} = \Gamma e^{2\gamma z} = |\Gamma|\angle(2\gamma z) \tag{6.5}$$

where Z_0 is the characteristic impedance for the forward wave, $-Z_0$ is the characteristic impedance for the backward wave, Z_L (or Z_R) is the terminating (or load) impedance and Γ is the reflection coefficient at the load end.

In view of Equation 6.5, the reflection coefficient (Γ) is the ratio of reflected to forward voltage (or current) at any arbitrary point on the line. Sometimes, at the load end, Γ is represented by Γ_L and at $z = -d$, by Γ_d or $\Gamma(d)$. Similarly, at an arbitrary location $-d$, Z_L is replaced simply by Z or $Z(d)$. Equations 6.3 through 6.5 can now be manipulated to obtain the relations

$$\Gamma = \frac{V''}{V'} = \frac{Z_L - Z_0}{Z_L + Z_0} \tag{6.6a}$$

$$\Gamma_L = -\frac{I''}{I'} = \frac{Z_0 - Z_L}{Z_0 + Z_L} \tag{6.6b}$$

$$\Gamma = \frac{Z - Z_0}{Z + Z_0} \tag{6.7a}$$

$$\Gamma = \frac{Y_0 - Y}{Y_0 + Y} \tag{6.7b}$$

where

$$Y = 1/Z \tag{6.7c}$$

A Smith chart is developed by examining the load where the impedance must be matched. Instead of considering its impedance directly, its reflection coefficient Γ_L can also be used to characterise a load (such as admittance, gain and *transconductance*). The reflection coefficient is more useful when dealing with RF frequencies. The reflection coefficient is defined in terms of reflected and incident voltages. These quantities are illustrated in Figure 6.4.

The quantum of reflected signal depends on the degree of mismatch between the source and the load impedances. From Equations 6.7a and 6.7b

$$Z = Z_0 \frac{1 + \Gamma}{1 - \Gamma} \quad \text{and} \quad Y = Y_0 \frac{1 - \Gamma}{1 + \Gamma} \tag{6.8}$$

If Z and Y are complex, the Γ is also complex. Besides, as and where possible, it is always desirable to reduce the number of unknown parameters. The characteristic impedance Z_0 is often constant and real, such as 50 Ω, 75 Ω, 100 Ω and so on. In view of the industry-supplied data for Z_0 with the specific manufactured guiding structure, the impedance Z (or Z_L) can be divided by Z_0 to obtain a dimensionless value, which is referred to as normalised impedance. Equations 6.8a and 6.8b can now be rewritten in the transformed form of normalised impedance (z) and normalised admittance (y) given by Equation 6.9:

$$z = \frac{Z}{Z_0} = \frac{1 + \Gamma}{1 - \Gamma} \quad \text{and} \quad y = \frac{Y}{Y_0} = \frac{1 - \Gamma}{1 + \Gamma} \tag{6.9}$$

Since both z and y are complex quantities, these can be written as

$$z = r + jx \quad \text{and} \quad y = g + jb \tag{6.10}$$

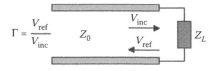

FIGURE 6.4
Illustration of reflection coefficient.

where r is the normalised resistance ($r = R/Z_0$), x is the normalised reactance ($x = X/Z_0$), g is the normalised conductance ($g = G/Y_0$) and b is the normalised susceptance ($b = B/Y_0$).

The voltage standing wave ratio (VSWR) and current standing wave ratio (CSWR) were defined by Equations 2.88 and 2.89. In case the line is terminated in pure resistance, Z is to be replaced by R and (for lossless line) Z_0 by R_0, the expressions for SWR can be rewritten as

$$\text{VSWR} = \frac{V_{max}}{V_{min}} = \frac{R}{R_0} \quad \text{and} \quad \text{CSWR} = \frac{I_{max}}{I_{min}} = \frac{R}{R_0} \qquad (6.11)$$

The value ρ that denotes both the VSWR and CSWR is always greater than unity. Thus, in terms of normalised resistance

$$\rho = R/R_0 = r \ \ (\text{for } R > R_0) \quad \text{and} \quad \rho = R_0/R = 1/r (\text{for } R < R_0) \qquad (6.12)$$

In view of Equations 6.9a and 6.12, SWR (ρ) and reflection coefficient (Γ) can be written as

$$\rho = \frac{1+|\Gamma|}{1-|\Gamma|} \quad \text{or} \quad \Gamma = \frac{\rho-1}{\rho+1} \qquad (6.13)$$

6.3 Polar Chart

For a lossless line, the domain of the reflection coefficient is a circle of unitary radius in the complex plane. The same is shown in Figure 6.5. This is also the domain of the Smith chart.

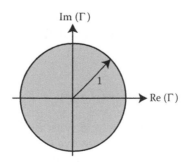

FIGURE 6.5
Domain of reflection coefficient.

The substitution of Equation 6.2 in Equation 6.4 results in

$$Z = \frac{V}{I} = \frac{V'e^{j\gamma l} + V''e^{-j\gamma l}}{I'e^{j\gamma l} + I''e^{-j\gamma l}} = R + jX \tag{6.14}$$

In view of Equations 6.3, 6.5 and 6.14

$$Z = \frac{V'(e^{j\gamma l} + \Gamma e^{-j\gamma l})}{I'(e^{j\gamma l} - \Gamma e^{-j\gamma l})} = Z_0 \frac{e^{j\gamma l} + \Gamma e^{-j\gamma l}}{e^{j\gamma l} - \Gamma e^{-j\gamma l}} = Z_0 \frac{(e^{j\gamma l} + \Gamma e^{-j\gamma l})(e^{-j\gamma l} - \Gamma e^{j\gamma l})}{(e^{j\gamma l} - \Gamma e^{-j\gamma l})(e^{-j\gamma l} - \Gamma e^{j\gamma l})}$$

$$= Z_0 \frac{(1 - \Gamma^2) - \Gamma(e^{2j\gamma l} - e^{-2j\gamma l})}{(1 + \Gamma^2) - \Gamma(e^{2j\gamma l} + e^{-2j\gamma l})} = Z_0 \frac{(1 - \Gamma^2) - 2j\Gamma \sin 2\gamma l}{(1 + \Gamma^2) - 2\Gamma \cos 2\gamma l} \tag{6.15}$$

In view of the normalisation of impedance (z), Equation 6.15 leads representing normalised resistance (r) and reactance (x) which can be written as

$$r = \frac{1 - \Gamma^2}{1 + \Gamma^2 - 2\Gamma \cos 2\gamma l} \tag{6.16a}$$

and

$$x = \frac{-2\Gamma \sin 2\gamma l}{1 + \Gamma^2 - 2\Gamma \cos 2\gamma l} \tag{6.16b}$$

Since Y and Z may be complex, in view of Equation 6.7 the Γ may also be complex, that is

$$\Gamma = (z - 1)/(z + 1)\} = p + jq \tag{6.17}$$

where the real and imaginary parts of Γ are denoted by

$$p \ (= \mathrm{Re}\ [\Gamma] = \Gamma \cos 2\gamma l) \tag{6.18a}$$

$$q \ (= \mathrm{Im}\ [\Gamma] = \Gamma \sin 2\gamma l) \tag{6.18b}$$

and

$$\Gamma^2 = (p^2 + q^2) \tag{6.18c}$$

Equation 6.18c is an equation of circle with radius Γ and centre at $p = 0$, $q = 0$. The two extreme conditions for a wave on a transmission line are that there is either no reflection ($\Gamma = 0$) or total reflection ($\Gamma = 1$). Thus, the circles representing Γ may have minimum and maximum values of radii of 0 and 1, respectively.

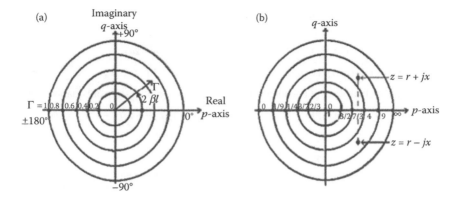

FIGURE 6.6
Polar chart.
(a) Complex reflection coefficient and (b) equivalent normalised impedance.

The collection of these concentric circles shown in Figure 6.6a is referred to as a *polar chart*. In this chart, p (= Re [Γ]) and q (= Im [Γ]) are the real and imaginary axes. In the expressions of p and q, the parameter γ is the propagation constant and ℓ is the length of the line in terms of wavelength. Thus, $2\gamma\ell$ represents the angle of Γ, which it makes with the real axis corresponding to $z = 0$. This angle rotates the phasor if the point at which Γ is to be evaluated moves along the line length between $z = 0$ and $z = -\ell$. Such a rotation was noted in connection with the stub matching in Figures 2.36 and 2.37.

In polar chart, the positive angles are taken to be in the counter-clockwise direction. For any passive impedance (i.e. Re $Z \geq 0$) and real Z_0, the magnitude of the reflection coefficient (Γ) is less than or equal to one. Thus, all real values of Γ can be plotted on polar chart having a maximum radius of unity.

Equation 6.9 relates the normalised impedance z with the reflection coefficient (Γ). This relation leads to Table 6.1 wherein the equivalent values of z corresponding to the values of Γ of Figure 6.6a are given. These corresponding values are plotted in Figure 6.6b.

6.4 Smith Chart for Impedance Mapping

The substitution of p and q from Equation 6.18 in Equation 6.16 leads to the following expressions:

$$r = \frac{1 - p^2 - q^2}{1 + p^2 + q^2 - 2p} \qquad (6.19a) \qquad x = \frac{-2q}{1 + p^2 + q^2 - 2p} \qquad (6.19b)$$

Equation 6.19 can be manipulated as below:

$r + r p^2 + r q^2 - 2pr = 1 - p^2 - q^2$

$p^2(1+r) + q^2(1+r) + p(-2r) = 1 - r$

$p^2 + q^2 - p\,2\,r/(1+r) = (1-r)/(1+r)$ (6.20a)

Add $\{r/(1+r)\}^2$ on both sides of Equation 6.20a

$[p^2 - \{2p\,r/(1+r)\} + \{r/(1+r)\}^2] + q^2$

$= (1-r)/(1+r) + \{r/(1+r)\}^2$

$[p - \{r/(1+r)\}]^2 + q^2 = \{1/(1+r)\}^2$ (6.20b)

Equation 6.20b is an equation of a circle with p as the horizontal coordinate and q as the vertical coordinate

The coordinates of the centre of the circle are

$p = r/(1+r)$ and $q = 0$, (6.20c)

The radius of the circle $= 1/(1+r)$

The radius is maximum for $r = 0$ with value 1

Since $q = 0$ at the centre

$[p - \{r/(1+r)\}]^2 = \{1/(1+r)\}^2$ or $p - r/(1+r) = 1/(1+r)$ (6.20e)

Consider the positive sign only.

Equation 6.20 leads to Table 6.2 and Figure 6.7.

$x + xp^2 + xq^2 - 2px = -2q$

$p^2 x + q^2 x - 2px + 2q = -x$

$p^2 + q^2 - 2p + 2q/x = -1$ (6.21a)

Add $(1/x^2)$ on both sides of Equation 6.21a

$(p^2 - 2p + 1) + (q^2 + 2q/x + 1/x^2) = 1/x^2$

$(p-1)^2 + (q + 1/x)^2 = (1/x)^2$ (6.21b)

Equation 6.21b is also an equation of a circle with p as the horizontal coordinate and q as the vertical coordinate

The coordinates of the centre of this circle are

$p = 1, q = -1/x$ (6.21c)

Also the length of the radius $= |1/x|$ (6.20d)

Since x is reactance

When x is inductive (or positive)

$q = -1/x$ (6.21e)

When x is capacitive (or negative)

$q = 1/x$ (6.21f)

Equation 6.21 leads to Table 6.3 and Figure 6.8.

TABLE 6.1

Equivalent Values of Reflection Coefficient and Normalised Impedance

$\Gamma\angle 0°$	$0\angle 0°$	$0.2\angle 0°$	$0.4\angle 0°$	$0.6\angle 0°$	$0.8\angle 0°$	$1.0\angle 0°$
z	1	2/3	3/7	1/4	1/9	0
$\Gamma\angle 180°$	$0\angle 180°$	$0.2\angle 180°$	$0.4\angle 180°$	$0.6\angle 180°$	$0.8\angle 180°$	$1.0\angle 180°$
z	1	3/2	7/3	4	9	∞

The following conclusions can be drawn from Figure 6.7:

- The locus of centres of constant r circles lies on $q = 0$ plane or p-axis.
- These centres gradually move from $p = 0$ to $p = 1$ as r changes from 0 to ∞.
- The centre of the largest constant r circle lies at $p = 0, q = 0$ with unity radius and corresponds to $r = 0$.

TABLE 6.2

Constant 'r' Circles from Equation 6.20

Normalised Resistance	Coordinates of the Centre		Radius
r	q	$p = r/(1+r)$	$R = 1/(1+r)$
0.0	0	0	1
0.25	0	0.2	0.8
0.5	0	0.333	0.667
1.0	0	0.5	0.5
4.0	0	0.8	0.2
9.0	0	0.9	0.1
∞	0	1.0	0.0

TABLE 6.3

Constant x Circles from Equation 6.21

Normalised Reactance	Coordinates of the Centre		Radius
X	p	$q = -1/x$	$R = 1/|x|$
0	1	$-\infty$	∞
0.1	1	−10	10
0.5	1	−2	2
1	1	−1	1
5	1	−0.2	0.2
10	1	−0.1	0.1
∞	1	0.0	0
−10	1	0.1	0.1
−5	1	0.2	0.2
−1	1	1	1
−0.5	1	2	2
−0.1	1	10	10
−0	1	∞	∞

- The centre of the smallest constant r circle lies at $p = 1, q = 0$ with zero radius and corresponds to $r = \infty$.
- All constant r circles pass through infinite impedance point located at $p = 1, q = 0$.
- The horizontal line is the $x = 0$ line. It is the resistance axis of the impedance coordinates with values ranging from $r = 0$ to $r = \infty$.

The following conclusions can be drawn from Figure 6.8:

- The locus of centres of constant x circles lies on $p = 1$ axis.

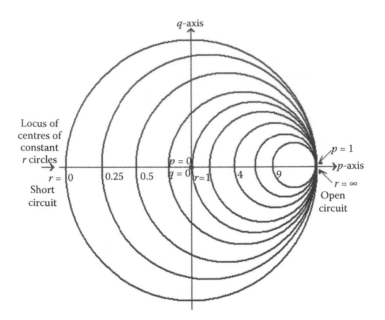

FIGURE 6.7
Constant r circles obtained from Table 6.2.

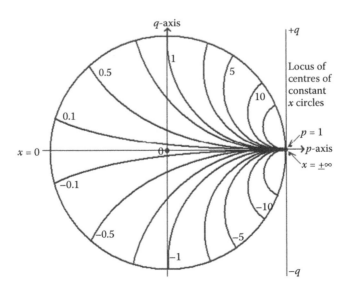

FIGURE 6.8
Constant x circles* obtained from Table 6.3.

- As the normalised x may assume positive or negative values, the centres of constant x circles will gradually move away from $p = 1$, $q = 0$.
- The movement of centres along the q axis will be in the direction opposite to the sign of x.
- The centre of the largest constant x circle will lie at $p = 1$, $q = \pm \infty$ with infinite radius and will correspond to $x = 0$.
- The centres of smallest constant x circles will lie at $p = 1$, $q = 0$ with zero radius and will correspond to $x = \infty$.
- All impedances on the outermost ($r = 0$) circle are purely reactive ranging from 0 to $\pm\infty$.
- This agrees that for pure reactance (i.e. at the outer most circle) $|\Gamma| = 1$.

The superposition of the two sets of circles shown in Figures 6.7 and 6.8 leads to the graphical configuration of Figure 6.9a, which is referred to as the Smith chart. Since this chart is composed of constant r and constant x circles, it is also called an impedance chart.

6.5 Smith Chart for Admittance Mapping

The Smith chart built by considering impedance $Z (= R + jX)$ can be used to analyse resistance and reactance parameters in both series and parallel. In series case, the new elements can be added and their effects can be determined simply by moving along the circle to their respective values. For the addition of elements in parallel, the Smith chart built by considering admittance $Y (= G + jB)$ provides an easier option. The Smith chart can now be used for line parameters, by shifting the space reference to the admittance location. The numerical values representing conductance and susceptance can be read on the chart thereafter. On the impedance chart, the correct reflection coefficient is always represented by the vector corresponding to the normalised impedance $z (= r + jx)$. Charts specifically prepared for admittances are modified to give the correct reflection coefficient in correspondence with normalised admittance $y (= g + jb)$. Analytically, the normalised admittance (y) is the reciprocal of normalised impedance (z) and thus

$$y = g + jb = 1/(r + jx) = (r - jx)/\{(r + jx)\,(r - jx)\} = (r - jx)/(r^2 + x^2) \quad (6.22)$$

$$\text{Thus, } g = r/(r^2 + x^2) \quad \text{and} \quad b = -x/(r^2 + x^2) \quad (6.23)$$

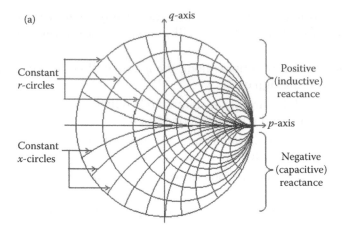

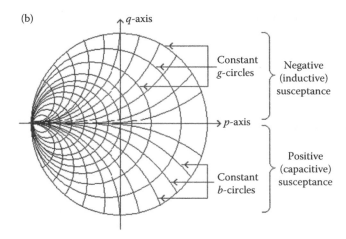

FIGURE 6.9
Formation of a Smith chart. (a) Impedance chart and (b) admittance chart.

In view of Equation 6.9

$$\Gamma = \frac{z-1}{z+1} = \frac{1-y}{1+y} = -\frac{y-1}{y+1} \qquad (6.24)$$

Equation 6.24 shows that the expression of Γ involving y is of opposite sign of that involving z $[\Gamma = \{(z-1)/(z+1)\}]$ from Equation 6.9a, that is, $\Gamma(y) = -\Gamma(z)$. Thus, if z is known, the signs of $\Gamma(z)$ can be inverted to obtain $\Gamma(y)$. This will represent a point situated at the same distance from $(0, 0)$, but in the opposite direction. This result can also be obtained by rotating $\Gamma(z)$ by 180° around the centre point.

An admittance Smith chart (Figure 6.9b) can be obtained by rotating the whole impedance Smith chart by 180° and every point on the impedance Smith chart can be converted into its admittance counterpart by 180° rotation around the origin of the complex Γ plane. This process eliminates the necessity of building another chart. The intersecting point of all the circles (for constant conductance (g) and constant susceptance (b)) is at the point (–1, 0) automatically. Such a plot makes the addition of the parallel elements easier.

In the Smith chart, the short circuit (SC) is located at the far left of the horizontal (resistance) axis. At this location, as the reflected voltage cancels the incident voltage, the potential is zero. It requires the voltage reflection coefficient to be –1 or a magnitude of 1 at an angle of 180°. Also, the open circuit (OC) is located at the far right of the same axis. At this location, the reflected and the incident voltages are equal and in phase, thus the reflection coefficient is +1.

The Smith chart represents both impedance and admittance plots. To use it as an admittance chart, turn it through 180° about the centre point. The directions 'towards the generator' and 'towards the load' remain the same. The contours of (constant normalised) resistance (r) and reactance (x) now become contours of (constant normalised) conductance (g) and susceptance (b), respectively.

The plot of the Smith chart is usually confined to inside of the region bounded by $|\Gamma| = 1$ circle. Outside this region, there is reflection gain, that is, in this outside region, the reflected signal is larger than the incident signal. This situation can arise only for *r less than 0*, that is, for the negative values of real part of the load impedance. Thus, the perimeter of the Smith chart as usually plotted is the $r = 0$ circle, which is coincident with the $|\Gamma| = 1$ circle. The $r = 1$ circle passes through the centre of the Smith chart. The point $\Gamma = 1$, angle 0 is a singular point at which r and x are multi-valued.

6.6 Information Imparted by Smith Chart

The Smith chart represents both, the impedance and the admittance, and is accordingly referred to as impedance chart (Figure 6.9a) or admittance chart (Figure 6.9b). Its utility can be understood by exploring the information it imparts. The following features have the special reference to the impedance chart.

6.6.1 Mapping of Normalised Impedances

All (real, imaginary and complex) values of normalised impedances can be mapped onto the Smith chart. Figure 6.10 illustrates the mapping of such values on the Smith chart. It is to be noted that the impedances (i) on the real

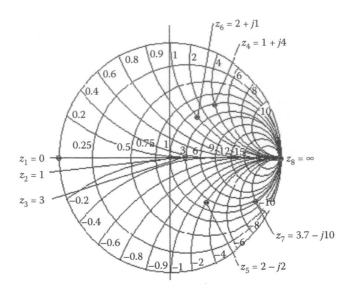

FIGURE 6.10
Normalised impedances on a Smith chart.

axis are always real, (ii) on the outermost circle are always imaginary and (iii) on intersections of constant r and constant x circles are always imaginary.

6.6.2 Rotation by 180°

If a point representing impedance (or admittance) plotted on the Smith chart is rotated by 180° in either direction, it will result in new values of admittance (or impedance). Consider the illustration of Figure 6.11 containing two points A and B. As the upper half of the Smith chart contains positive x and the lower part negative x, the points A and B will correspond to $z_1 = r_1 + jx_1$ and $z_2 = r_2 - jx_2$, respectively. If point A is rotated by 180°, the resulting point A′ will represent a normalised admittance $y_1 = g_1 - jb_1$, which is the reciprocal of z_1. Similarly, the rotation of B by 180° will result in B′ whereas $y_2 = g_2 + jb_2$ is the reciprocal of z_2. In case, A′ and B′ are the original points representing y_1 and y_2 and if both of these are rotated by 180°, the resulting values will represent normalised impedances z_1 and z_2, respectively.

6.6.3 Reflection and Transmission Coefficients

Equation 6.9 relates the normalised z (or y) to the reflection coefficient. Also in view of Equation 6.18, p and q axes are the real Γ and imaginary Γ axis. Thus, if z is located on the chart (Figure 6.12), the length oz will represent the magnitude of the reflection coefficient $|\Gamma|$ and length o′z will represent the magnitude of the transmission coefficient $|T|$. In case the complex values of

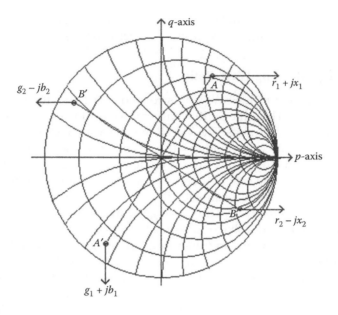

FIGURE 6.11
Normalised impedance and admittance on a Smith chart.

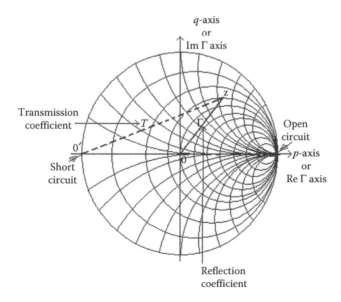

FIGURE 6.12
Reflection and transmission coefficients on a Smith chart.

Γ and T are required, the angles made by the phasor oz and o'z with the real axis can be measured and the respective values can be evaluated.

6.6.4 Voltage Standing Wave Ratio

In view of Equation 6.11, VSWR (ρ) is related to r ($\rho = r$ for $R > R_0$) or $1/r$ ($\rho = 1/r$ for $R < R_0$). Thus, the Smith chart can be used to evaluate voltage (or current) standing wave ratios. The VSWR may have an integer or non-integer real value. To find its value, with $p = 0$, $q = 0$ point as the centre and $p = k$ (k corresponds to r) as the radius draw a circle on the Smith chart. The crossing of this circle on one side of the real axis will correspond to $p = r$ and on other side to $p = 1/r$. These locations will correspond to voltage maxima and voltage minima, respectively. Thus, voltage minima *correspond* to the imped-ance minima ($1/r$) and voltage maxima to impedance maxima (r) on the p-axis. Since ρ is always greater than unity, ρ always corresponds to r with the assumption of $R > R_0$. As such, there is no circle for ρ since it has only one value and that too corresponds to r. Also, ρ is related to Γ by Equation 6.13b; $[\rho = \{(1 + |\Gamma|)/(1 - |\Gamma|)\}$ and $\Gamma = \{(\rho - 1)/(\rho + 1)\}]$ the value of one of these can be evaluated if the other is known. Thus, ρ corresponds to the value of r at the location wherever a constant Γ circle (Figure 6.13) crosses the real axis.

6.6.5 Two Half-Wave Peripheral Scales

In Section 2.29.2, it was noted that a quarter-wave section of transmission line acts as an impedance inverter. In case two quarter-wave sections are connected in series, the resulting half-wave section will simply repeat the original impedance (admittance). This aspect was illustrated in Figure 2.32 wherein the elements/circuits repeat after every half-wave. Thus, on the Smith chart, the total movement around its periphery must correspond to

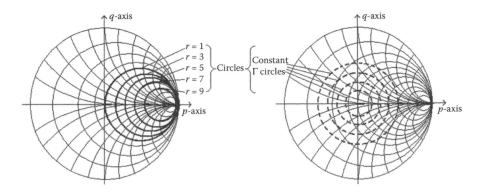

FIGURE 6.13
VSWR and Γ circles on a Smith chart.

the half-wavelength movement along the transmission line to arrive at the same impedance (admittance) value.

The Smith chart being a polar plot of complex reflection coefficient represents the ratio of the complex amplitudes of the backward and forward waves. Irrespective of the direction of propagation, the phase of a wave changes during its travel. The amount of change corresponds to the travelled distance. Thus, the total phase shift of a wave during its to and fro journey from a reference point is twice that of the phase shift introduced in one-way travel. There is 360° or 2π rad phase shift for $\lambda/2$ travelled distance. A further $\lambda/2$ travel from the reference point will add another 360° phase shift. Thus, for a reference point a whole wavelength away, there is a phase shift of 720°, as the round trip is twice that of the whole wavelength. It, therefore, represents a further complete cycle of the complex reflection (Smith) chart. Thus, in moving back one whole wavelength from the reference point, the round trip distance is actually increasing by two whole wavelengths, so the Smith chart is circumnavigated twice.

As illustrated in Figure 6.14, the periphery of the Smith chart is provided with two scales, which represent the wavelength. Both scales start at zero on the short-circuit end. In one of these scales, the value of the wavelength increases in the clockwise direction whereas in the second, it increases in the anti-clockwise direction. Both these scales coincide at the real axis with a value of a quarter wavelength. The entire periphery in either direction is covered by half wavelength.

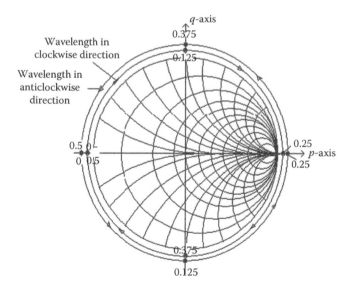

FIGURE 6.14
Wavelength scales on a Smith chart. (a) Smith chart for impedance and (b) Smith chart for admittance.

6.6.6 Inversion of Impedance/Admittance

The movement towards the generator on the transmission line corresponds to the movement of the locus on the Smith chart in the clockwise direction and that towards the load to the anti-clockwise direction. Also, 'θ' degree movement along the line in either direction moves a point on the locus by '2θ' degrees on the chart because the reflected wave must transverse the round-trip distance moved. As the impedance repeats itself every half wavelength along a uniform transmission line, the one-time movement around the chart gives the same impedance, and the quarter-wavelength movement results in the inversion of the impedance as shown in Figure 6.15.

6.6.7 Equivalence of Movement on Transmission Line

Figure 6.16 illustrates the new values of impedances in view of the movement on the transmission line from the assigned original location a–a. As evident from the figure, the impedance $r_a + jx_a$ seen across a–a (to be read a point 'a' on the Smith chart) will change to $r_b - jx_b$ at b–b, that is, at ℓ_2 distance towards the load (to be read a point 'b' on the Smith chart) and will change to $r_c + jx_c$ at c–c, that is, at ℓ_1 distance towards the generator (to be read a point 'c' on the Smith chart).

It is to be mentioned that any physical length of a transmission becomes the variable electrical length over a frequency band. Thus, any fixed impedance will spread out to a locus when viewed through a connected transmission

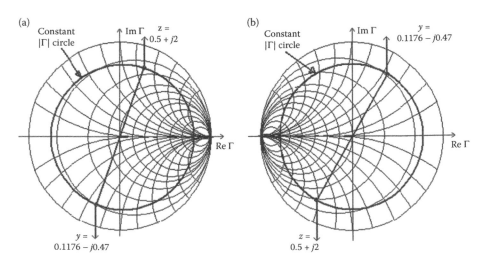

FIGURE 6.15
Inversion of (a) impedance and (b) admittance values.

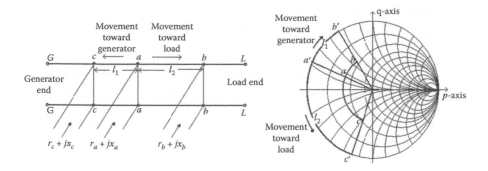

FIGURE 6.16
Equivalent of movement on transmission line on a Smith chart.

line. This aspect facilitates wideband matching close to the device or at discontinuity.

In general, the same Smith chart is used as impedance or admittance chart. It contains the same axis labelled as 'inductive reactance or capacitive susceptance'. It is, however, to be noted that in the Smith chart short circuit, open circuit, inductive reactance and capacitive reactance are the moving targets.

6.7 Advantages of Smith Chart

- The Smith chart is the direct graphical representation of the complex reflection coefficient in the complex plane.
- It is a Riemann surface, in that it is cyclical in numbers of half wavelengths along the line as the standing wave pattern repeats every half wavelength.
- It may be used as impedance or admittance calculator just by turning it through 180°.
- The region of interest in most of the cases lies inside $\Gamma = 1$ circle.
- Transformation along the (lossless) line results in a change of the angle, and not in the modulus or radius of Γ. Thus, plots may be made quickly and simply.
- Many of the more advanced properties of microwave circuits, such as noise figure and stability regions, can be mapped onto the Smith chart as circles.
- As the 'point at infinity' represents the limit of very large reflection gain, it never needs to be considered for practical circuits.

- The real axis maps to the SWR variable. A simple transfer of the plot locus to the real axis at constant radius gives a direct reading of the SWR.

- In commercially available Smith chart, the angular scale at the edge is divided into 1/500 of a wavelength an equivalent of 0.72°. Also, the reflection coefficient scale can be read to a precision of 0.02. These limits are quite sufficient in most of the practical applications. If more precision is required, its segments, can be enhanced by using a photocopier.

6.8 Smith Chart for Lossless Transmission Lines

If an impedance at an arbitrary distance d from the load end is identified as $Z(d)$ and the characteristic impedance (Z_0) of the line is known, the first step $Z(d)$ is to be divided by Z_0 to get the normalised impedance $z(d) = r + jx$, where r and x are the normalised resistance and reactance, respectively. The Smith chart is obtained by superimposing constant r circles and the segments of constant x circles. The intersection of these two curves gives the value of complex reflection coefficient in the complex plane, that is, it provides directly the magnitude and the phase angle of $\Gamma(d)$.

6.8.1 Evaluation of $\Gamma(d)$ for given $Z(d)$ and Z_0

For given $Z(d)$ $(= 25 + j100\ \Omega)$ and Z_0 $(= 50\ \Omega)$, the normalised $z(d)$ $(= 0.5 + j2)$.

- Locate the intersection of $r = 0.5$ and $x = 2$ circles on the Smith chart.
- Draw a line from the origin of the Smith chart to the point of this intersection.
- This line represents a phasor, which gives the value of the complex reflection coefficient.
- Read its magnitude and phase angle as $|\Gamma(d)| = 0.8246$ and $\angle\Gamma(d) = 50.9°$.
- Obtain $\Gamma(d)$ $(= 0.52 + j0.64)$ in complex form from values of $|\Gamma(d)|$ and $\angle\Gamma(d)$.

These steps are illustrated in Figure 6.17.

In case $\Gamma(d)$ is given and $Z(d)$ is to be obtained, a complex point is to be located that can represent the given reflection coefficient $\Gamma(d)$ on the chart. This point can be obtained with the knowledge of magnitude and phase of Γ $(=|\Gamma|e^{-j\theta})$. Corresponding to this point the values of r and x can be read. These are the values of constant r and constant x circles touching the tip of

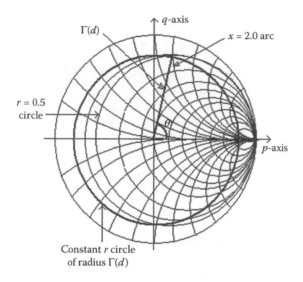

FIGURE 6.17
Evaluation of $\Gamma(d)$ for given $Z(d)$ and Z_0.

the phasor $\Gamma(d)$. As r and x are the normalised values giving $z(d)$ ($= r + jx$), the actual $Z(d)$ can be obtained on multiplying $z(d)$ by Z_0.

6.8.2 Evaluation of $\Gamma(d)$ and $Z(d)$ for Given Z_R, Z_0 and d

Let $Z_R = 25 + j100$ Ω, $Z_0 = 50$ Ω and $d = 0.18\lambda$. In view of the relation $\Gamma(d) = \Gamma_R$ $\exp(-j2\beta d) = \Gamma_R$, the magnitude of the reflection coefficient is constant along a lossless transmission line terminated by a specified load. Thus, on a complex plane, a circle with the centre at the origin and radius $|\Gamma_R|$ can represent all possible reflection coefficients found along the transmission line. Using this constant $|\Gamma_R|$ circle on the Smith chart, the values of the line impedance at any location can be determined as below:

- As Z_R and Z_0 are known, z_R is evaluated to be $0.5 + j2$, map this z_R on the Smith chart.
- Identify the load reflection coefficient Γ_R or the normalised impedance z_R on the Smith chart.
- Draw a circle that has its origin at the centre of the chart and passes through point z_R.
- Draw a circle of constant reflection coefficient amplitude $|\Gamma(d)| = |\Gamma_R|$.
- Evaluate the value of angle θ { $= 2\beta d = 2(2\pi/\lambda) \times 0.18\lambda = 2.262$ radians $= 129.6°$}.

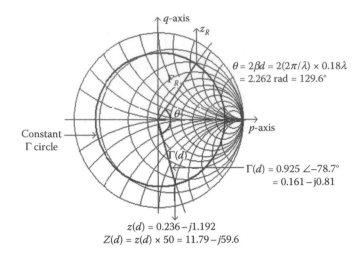

FIGURE 6.18
Evaluation of $\Gamma(d)$ and $Z(d)$ for given Z_R, Z_0 and d. (a) $Z_R = 25 + j100\ \Omega$ and $Z_R = 25 - j100\ \Omega$.

- Start from the point representing the load z_R $(=0.5 + j2)$ and move along the circle in the clockwise direction, by an angle θ $(= 129.6°)$ or a distance 0.18λ.

- This new location on the chart corresponds to the location d on the transmission line.

- Values of $\Gamma(d)$ $(= 0.925\angle{-78.7°})$ and $z(d)$ $(= 0.236 - j1.192)$ can be read from the chart.

These steps are illustrated in Figure 6.18.

6.8.3 Evaluation of d_{max} and d_{min} for Given Z_R and Z_0

- To evaluate the distance of the voltage maxima (d_{max}) and minima (d_{min}) from the load for $Z_R = 25 \pm j100\ \Omega$ and $Z_0 = 50\ \Omega$. In view of given Z_R and Z_0, obtain z_R $(= 0.5 \pm j2)$.

- Locate the (given) reflection coefficient Γ_R or normalised impedance z_R on the Smith chart.

- Draw a circle having a radius equal to the constant reflection coefficient amplitude $|\Gamma(d)| = |\Gamma_R|$. This circle intersects at two points on the negative and positive sides of the real Γ axis.

- Identify these intersecting points as the location of d_{max} (when $\Gamma(d)$ = real positive) and d_{min} (when $\Gamma(d)$ = real negative).

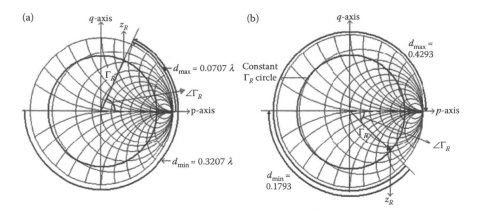

FIGURE 6.19
Evaluation of d_{max} and d_{min} for given Z_R and Z_0. (a) $Z_R = 25 + j100\ \Omega$ and (b) $Z_R = 25 - j100\ \Omega$.

- Read the normalised distances (in terms of wavelength) on the periphery of chart for the two cases by moving clockwise. These distances are
  - For $Z_R = 25 + j100\ \Omega$, $d_{max} = 0.0707\lambda$ and $d_{min} = 0.3207\lambda$
  - For $Z_R = 25 - j100\ \Omega$, $d_{max} = 0.4293\lambda$ and $d_{min} = 0.1793\lambda$
- These values can also be computed from the angles between vector Γ_R and the real axis. The angles measured for the two cases are
  - $\theta_{max} = 50.9°$ and $\theta_{min} = 230.9°$ and (b) $\theta_{max} = 3019.1°$ and $\theta_{min} = 129.1°$
- Since $\theta_{min} = 2\beta d_{min}$ and $\theta_{max} = 2\beta d_{max}$, the values of distances obtained are
  - For $Z_R = 25 + j100\ \Omega$, $d_{max} = 0.0707\lambda$ and $d_{min} = 0.3207\lambda$
  - For $Z_R = 25 - j100\ \Omega$, $d_{max} = 0.4293\lambda$ and $d_{min} = 0.1793\lambda$

These values are equal to the measured values. All the above steps are shown in Figure 6.19.

6.8.4 Evaluation of VSWR for Given Γ_R and Z_{R1}, Z_{R2} and Z_0

Let $Z_{R1}\ (= 25 + j100\ \Omega)$, $Z_{R2}\ (= 25 - j100\ \Omega)$ and $Z_0\ (= 50\ \Omega)$. In view of Equations 6.9 and 6.13, the VSWR can be re-written in terms of $z(d_{max})$ and $|\Gamma_R|$ at a maximum location of the standing wave pattern:

$$z(d_{max}) = \frac{1 + \Gamma(d_{max})}{1 - \Gamma(max)} = \frac{1 + |\Gamma_R|}{1 - |\Gamma_R|} = \text{VSWR} \qquad (6.25)$$

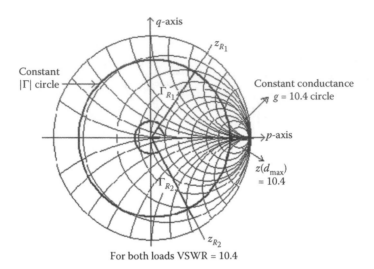

For both loads VSWR = 10.4

FIGURE 6.20
Evaluation of VSWR for known Γ_R, Z_{R1}, Z_{R2} and Z_0.

VSWR is always real and ≥ 1. Thus, VSWR is obtained on the Smith chart, simply by reading the value of the (real) normalised impedance, at the location d_{max}, where Γ is real and positive.

- The normalised values of the given impedances are $z_{R1} = 0.5 + j2\,\Omega$, $Z_{R2} = 0.5 - j2\,\Omega$
- Locate Γ_R or z_{R1} and Z_{R2} on the Smith chart.
- Draw the circle of constant reflection coefficient amplitude $|\Gamma(d)| = |\Gamma_R|$.
- Find the intersection of this circle with the real positive axis for the reflection coefficient (corresponding to the transmission line location d_{max}).
- A circle of constant normalised resistance will also intersect this point.
- Read the value of normalised resistance $r\ (= 10.4)$.
- In view of Equation 6.12, VSWR = 10.4 for both the loads.

All the above steps are shown in Figure 6.20.

6.8.5 Evaluation of Y(d) for Given Z_R

Let $Z_R = 25 + j100\,\Omega$. Equation 6.9 relates the normalised impedance and admittance as

$z(d) = \{1 + \Gamma(d)\}/\{1 - \Gamma(d)\}$ $\qquad$ $y(d) = \{1 - \Gamma(d)\}/\{1 + \Gamma(d)\}$

Also, $\Gamma\{d + (\lambda/4)\} = -\Gamma(d)$

Thus,

$$z\left(d + \frac{\lambda}{4}\right) = \frac{1 + \Gamma(d + \lambda/4)}{1 - \Gamma(d + \lambda/4)} = \frac{1 - \Gamma(d)}{1 + \Gamma(d)} = y(d) \qquad (6.26)$$

This equality is only valid for normalised impedance and admittance. The actual values are

$$Z\{d + (\lambda/4)\} = Z_0 \cdot z\{d + (\lambda/4)\} \quad \text{and} \quad Y(d) = Y_0 \cdot y(d) = y(d)/Z_0$$

where $Y_0 = 1/Z_0$ is the characteristic admittance of the transmission line.

- Plot Γ_R or the normalised load impedance $Z_r (= 0.5 + j2)$ on the Smith chart.
- Draw the circle of constant reflection coefficient amplitude $|\Gamma(d)| = |\Gamma_R|$.
- The normalised admittance $y(d)$ is located at a point on the circle of constant $|\Gamma|$, which is diametrically opposite to the normalised impedance.
- Read the value of $y(d)$ (= $0.11765 - j0.4706$). This value is the same as $z\{d + (\lambda/4)\}$, that is, $z(d)$ rotated by 180° in either direction.
- Find $Y(d) = y(d)/Z_0$ (= $0.002353 - j0.009412$).

All the above steps are shown in Figure 6.21. If the space reference is shifted to the admittance location in the Smith chart (see Section 6.5), it can be used

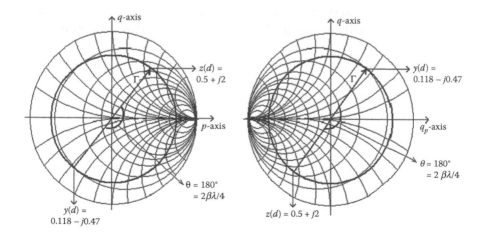

FIGURE 6.21
Evaluation of $Y(d)$ for given Z_R and Z_0.

for line admittances. With such shifting, the numerical values representing admittances can be directly read.

Inspection of Figures 6.9a and 6.9b reveal that the related impedance and admittance are on opposite sides of the same Smith chart. The imaginary parts always have different signs in impedance and admittance charts. Therefore, a positive (inductive) reactance corresponds to a negative (inductive) susceptance, while a negative (capacitive) reactance corresponds to a positive (capacitive) susceptance.

6.8.6 Mapping of a Multi-Element Circuit onto a Smith Chart

The Smith chart can be used to solve the problems containing elements in series and parallel mixed together. Such solutions, however, require switching/conversion of chart from z to y or y to z mode. To understand the process, consider a multi-element circuit of Figure 6.22a wherein the elements shown are normalised with $Z_0 = 50\ \Omega$. The series reactance (x) is positive for inductance and negative for capacitance and the susceptance (b) is positive for capacitance and negative for inductance. This multi-element circuit is simplified by breaking the same into its elements and is shown in Figure 6.22b. This process of mapping is described in the following steps:

- Start from the right side of the circuit and move towards its left side in accordance with the following steps.
- Its right most elements include a resistor and an inductive reactance both with values of 1. Thus, at the starting point, $z = 1 + j1$. Locate this as a series point on the chart at $r = 1$ circle and $x = 1$ circle. Mark this point as A on the chart.
- As the next element is a shunt element switch over to the admittance mode of the Smith chart by rotating the whole plane by 180°. This conversion will lead to $y = 1/(1 + j1) = 0.5 - j0.5$. This new point is marked as A'.
- Add the shunt element by moving along the constant conductance ($g = 0.5$) circle by a distance corresponding to $b = -0.3$. As the value of b to be added is negative and the new value of the admittance has

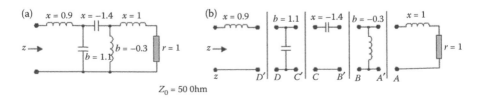

FIGURE 6.22
Multi-element circuit and its individual elements.

to be $y = 0.5 - j0.8$, the movement has to be in the counter-clockwise direction. Mark this point as B.

- As the next element is a series element, switch back again to the impedance mode of the Smith chart. Since B represents admittance, it is to be converted into impedance, which requires 180° rotation of the plane. This conversion leads to point B′. The impedance value at this point is $z = 1/(0.5 - j0.8) = 0.562 + j0.9$.

- Add the series element by moving along the resistance circle. Again, as the value to be added is negative, this movement has to be in the counter-clockwise direction by a distance, which corresponds to $x = -1.4$. Mark this point as C at which $z = 0.562 - j0.5$.

- As the next element is a shunt element, switch back again to the admittance mode. This conversion will lead to $y = 1/(0.562 - j0.5) = 1 + j0.8837$. Mark this point as C′.

- Add the shunt element by moving along the conductance circle. As the value to be added is positive ($b = 1.1$), the movement has to in the clockwise direction. Mark this point as D at which $y = 1 + j1.9837$.

- As the next element is in series, reconvert to the impedance mode and mark this point as D. This conversion gives $z \approx 1/(1 + j2) = 0.2 - 0.4$. Add the last element (the series inductor $x = 0.9$) by moving along the constant r circle and mark this point as D′.

- Determine the value of z shown at the left most point. As z is located at the intersection of resistor circle 0.2 and reactance circle 0.5, the value of z is determined to be $0.2 + j0.5$. In view of the assumed characteristic impedance of 50 Ω, the value of $Z = 10 + j25\ \Omega$.

All the above steps are shown in Figure 6.23.

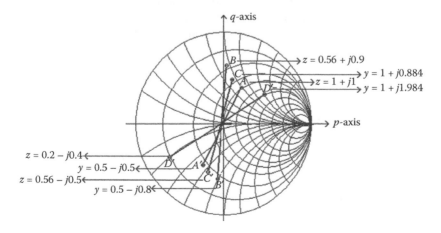

FIGURE 6.23
Circuit parameters mapped on a Smith chart.

6.8.7 Evaluation of Normalised Impedances

Consider a lossless 52 cm long transmission line (Figure 6.24) wherein a voltage minima occurs 18 cm from the load end and the adjacent minima are 20 cm apart. If VSWR = 2.5, the normalised impedances at the load end (z_L) and input end (z_{in}) can be evaluated as under.

- In view of the distance between two minima (20 cm), the wavelength $\lambda = 2 \times 20 = 40$ cm.
- The distance of the load from the minima (18 cm) in terms of wavelength = $(18/40)\lambda = 0.45\lambda$.
- The distance of the input terminals from the minima is (52 – 18) 34 cm = $(34/40)\lambda = 0.85\lambda$.
- One complete round of the chart requires a movement of 0.5λ. Thus, the 0.85λ movement around the periphery of the chart is the same as 0.35λ.
- Draw $\rho = 2.5$ circle on the chart.
- The impedance maxima is at the intersection of $\rho = 2.5$ circle with $r = 2.5$ circle and the impedance minima is at the intersection of $\rho = 2.5$ circle with $r = 0.4$ circle.
- Enter the chart at the location of voltage minima.
- Rotate towards the load on $\rho = 2.5$ circle by 0.45λ. Read the impedance z_R (= $0.43 + j0.27$).
- Rotate towards the generator on $\rho = 2.5$ circle by 0.35λ. Read the impedance z_{in} (= $0.89 - j0.89$).

In view of Figure 6.25, the distance covered from a to b by moving along the $\rho = 2.5$ circle in the anti-clockwise direction (towards the load) is 0.45λ. Also, the distance covered from a to c by moving along the $\rho = 2.5$ circle in the clockwise direction (towards the generator) is 0.35λ.

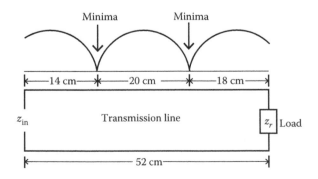

FIGURE 6.24
Details of transmission line and associated data.

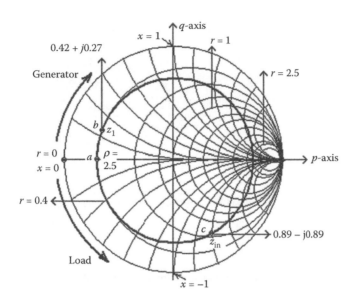

FIGURE 6.25
Evaluation of z_r and z_{in}.

6.8.8 Consideration of Attenuation Factor

In all earlier discussion and examples, the line was considered to be lossless. In lossy lines, a factor called attenuation constant (α) needs to be accounted along with the impact of the velocity of propagation (v). It is to be noted that commercially available Smith chart also contains some auxiliary scales, including those for (i) reflection coefficient, (ii) transmission loss and (iii) standing waves. These radially scaled parameters are externally located and are parallel to the real axis of the Smith chart with centres and ends of chart and scales aligned.

The above can be understood by taking the example of an antenna (Figure 6.26), which is connected to a load (in the form of TV receiver having 75 ohm input impedance) through a 50 ft, 300 ohm transmission line. The relative velocity (v_p) of the propagating wave on the line is 82% of the velocity of light (i.e. $v_p = 0.82 \times c = 2.46 \times 10^8$ m/s). This line has an attenuation factor α (=2 dB per 100 ft) at 100 MHz, thus causing a total loss of ($\alpha l = 2 \times 50/100$)1 dB. The impedance into which this antenna works at 100 MHz is to be evaluated. Corresponding to the velocity of propagation (v_p) and given frequency, the wavelength can be calculated to be $\lambda = v_p/f = (2.46 \times 10^8/100 \times 10^6) = 2.46$ m. Also, the line length ℓ in terms of wavelength is $\ell \times \lambda/2.46 = (50/3.281) \times (\lambda/2.46)\} = 6.19\lambda$. And lastly, the normalised terminating impedance is worked out to be ($z_r = Z_r/R_0 = 75/300$) 0.25, which corresponds to the standing wave ratio ρ (=4). With the above information, the problem can be worked out as under.

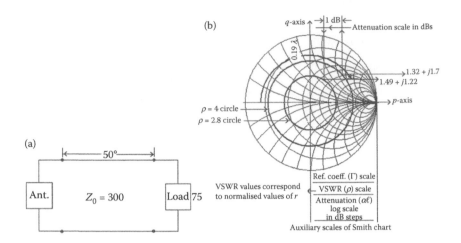

FIGURE 6.26
Consideration of attenuation factor. (a) Antenna circuit and (b) Smith chart solution.

- Enter the chart at z_r (=0.25).
- Rotate on the circle of constant ρ (=4) by 6.19λ (or by 0.19λ) towards the generator.
- Read the value of z_{in} (=1.32 + j1.7) on the $\rho = 4$ circle.
- Transfer from the $\rho = 4$ circle to αl scale.
- Move 1 dB left on the αl scale.
- Return from the αl scale to the $r =$ constant circle, which now corresponds to $\rho = 2.8$.
- Read the new value of z_{in}' (=1.49 + j1.22) on the $\rho = 2.8$ circle.
- Calculate $Z_{in}' = R_0 \cdot z_{in}'[\{ = 300\ (1.49 + j1.22)\} = (447 + j366)$ ohms].

6.9 Stub Matching

In Section 2.29.5, it was noted that stubs are the auxiliary lines connected across the main transmission lines. These introduce small phase delays to cancel reflections and are thus used for impedance matching. Figures 2.36 and 2.37 illustrate that a transmission line of specific physical length may behave as a short circuit, an open circuit, an inductive reactance or a capacitive reactance. Since the electrical length depends on the frequency of operation, this aspect of transmission line behaviour is a function of frequency. The auxiliary lines connected across the main transmission lines may be short circuited or open circuited at their farther or unconnected end. These are referred to as short-circuited or open-circuited stubs.

6.9.1 Entry on Smith Chart

While dealing with the problem of short-circuited stub, entry on the Smith chart is to be made at the short circuit point. Thereafter, clockwise movement towards the generator is to be made on the rim of the chart. The angular distance to be moved has to correspond to the length of the stub so as to arrive at its other end. Further, the angular distance to be moved has to be in accordance with the requirement of the reactive element to be introduced. To remain confined to the inductive region, this movement has to be less than 180° For more than 180° movement, stub input becomes capacitive. At exactly one-quarter wavelength, the impedance is infinite (an open circuit) and at one-half wavelength, the impedance becomes zero (i.e. again a short circuit). In case of open-circuit stub, entry on the Smith chart is made at the open circuit point.

6.9.2 Susceptance Variation of SC Stubs over a Frequency Band

Consider a short-circuited stub that is quarter-wavelength long at the mid-band of (say) 3:1 frequency band. It will be of $\lambda/8$ length at the lower end and $3\lambda/8$ length at the higher end of the band. Its normalised susceptance varies from –1.0 (inductive) at f_{low} to zero (open circuit) at midband to +1.0 (capacitive) at f_{high}. If Y_0 is the characteristic admittance of the stub, the unnormalised susceptance of this stub varies between $\pm 1.0 \cdot Y_0$ provided Y_0 of the stub and that of the main line are the same. At each frequency, the normalised susceptance of the stub may be added to the normalised admittance of the load to yield the normalised admittance of the parallel combination. If Y_0 of the stub and that of the main line differ, the stub's susceptance is to be renormalised by Y_0 of the main line before adding to the normalised admittance of the load.

Assume that both the stub and the main line have $Z_0 = 50$ ohm. In this case, the unnormalised susceptance variation of the shorted stub is ± 0.02 ($\pm 1.0Y_0$). If Z_0 of the main line is 50 ohm and instead of ± 1.0, a normalised susceptance variation of only ± 0.4 is required, the characteristic admittance of the stub has to be 0.4 times that of the main line or $Y_0 = 0.4 \times 0.02 = 0.008$. The Z_0 for the stub is now 125 ohm (i.e. 50/0.4). As its susceptance varies less over the band, it will have lower Q. It is to be mentioned that the unnormalised values are rarely needed. The normalised values may be renormalised by the ratio of the characteristic impedances involved. Figure 6.27 illustrates the above-described procedure.

6.9.3 Stub for Changing the Phase of a Main-Line Signal

An open-circuited stub less than a quarter-wavelength long retards the phase (i.e. adds phase delay). Similarly, a short-circuited stub less than a quarter-wavelength long will advance the phase. An open-circuited stub will have a transmission coefficient $(1 + \Gamma)$ with a negative phase angle if it is moved clockwise from the OC position. Figure 6.28 illustrates the phase delays (ϕ_1 and ϕ_2)

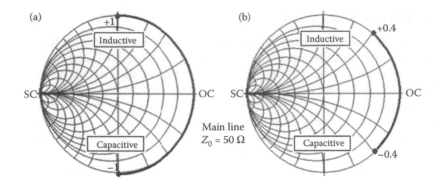

FIGURE 6.27
Variation of susceptance for stubs of different Z_0 with the same main line. (a) SC stub $Z_0 = 50$ Ω and (b) SC stub $Z_0 = 125\ \Omega$.

of 50-ohm and 25-ohm open-circuited stubs in shunt with a 50-ohm main line. The figure shows that the amount of phase delay increases as the Z_0 of the stub decreases. In principle, a larger Y_0 produces a larger unnormalised susceptance since a wider stub looks like a larger capacitor. This also illustrates mismatched results, which justify the addition of stubs in pairs to cancel the reflections.

6.9.4 Requirements for Proper Matching

Proper matching of a system over a frequency band depends on the magnitude of mismatch, desired bandwidth and the complexity of matching circuit. At a particular frequency, any impedance mismatch can be perfectly matched to the characteristic impedance of the transmission line, as long as it is not on the rim of the chart (or $|\Gamma| = 1$). Such matching can be achieved with single stub with length less than $\lambda/4$. The theoretical aspects

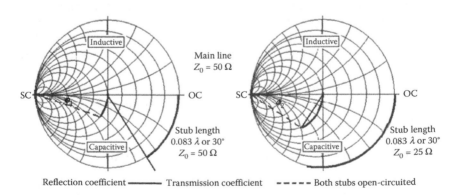

FIGURE 6.28
Phase delays (ϕ_1 and ϕ_2) introduced by 50 Ω and 25 Ω open-circuited stubs.

of single stub matching were discussed in Section 2.29.5 and illustrated by Figure 2.36.

The matching of an unmatched line generally requires two variables. These variables are provided by the movement of the stub along the line and the variation of the stub length. Since both these parameters belong to the same stub, such matching is referred to as single stub matching. In single stub matching the location of the stub on the line is to be precisely adjusted for proper results. This adjustment requires frequent mechanical movements along the line. With the passage of time, adjustment becomes erroneous and makes results inaccurate. These errors can be avoided by using double stub matching in which the locations of both the stubs are fixed (Figure 2.36). The two needed variables are provided by the lengths of the two stubs.

6.9.5 Selection of Appropriate Type and Length of Stub

In case of SC stub, enter at the SC point and move along the periphery until the $r = 1$ (or $g = 1$) circle is reached. To find the appropriate type and length of stub to be connected in series (or shunt) with the main line, move along this circle to the origin. If the far end of the stub is either short or open (or any pure reactance), its input end is also a pure reactance (or susceptance) so that it does not affect the resistance (or conductance) component of the mainline impedance (or admittance).

The addition of a stub in parallel with a transmission line is relatively easier. If at the joining point (of stub and main line) the resulting admittance is the sum of the stub's input susceptance and the main line admittance, the use of an admittance chart is preferred. In such a case, first the mismatched point is rotated around the origin until it reaches $g = 1$ circle. In the next step, the characteristic admittance and length of the stub is chosen such that its input susceptance is equal and opposite to the main line susceptance indicated on the $g = 1$ circle. Figure 6.29 illustrates two such cases. In the first case, the point is moved towards the generator by 39° of line and a short-circuited stub is added (Figure 6.29a) to provide normalised inductive susceptance (=j0.8). In the second case, the point is moved towards the generator by 107° of line and an open-circuited stub is added (Figure 6.29b) to provide normalised capacitive susceptance (= −j0.8).

6.9.6 Impact of Rotation of Impedance

At one frequency, the required normalised susceptance can be obtained from a stub of any characteristic impedance simply by adjusting its length. Thus, there can be an infinite number of possible solutions as the length of the stub can be adjusted to any value. If at a particular frequency the length of the stub is increased by an integer multiple of $\lambda/2$, its input susceptance will

(a)   (b)

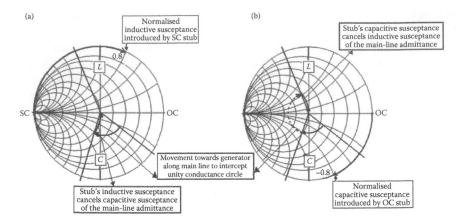

FIGURE 6.29
Length of (a) SC stub and (b) OC stub.

remain unchanged. But over a frequency band, the susceptance will vary considerably and noticeable differences in values will appear.

Figure 6.30 illustrates that a normalised load impedance z_L (=0.3 + j0.5) transforms into another impedance z_d (=1.6 + j1.7) by a distance of 0.12 λ towards the generator. Consider a coaxial cable with a characteristic impedance Z_0 (= 75 ohms) at a frequency (of say 146 MHz). If we further assume that the velocity factor of the cable is 0.67, then the velocity of waves on the cable is 0.67c (c = velocity of light in free space = 3 × 10^8 m/s) = 2 × 10^8 m/s and the wavelength λ (=v/f) = on the cable is 1.37 m. Therefore, a load impedance Z_L [{ = (0.3 + j0.5) Z_0} = (22.5 + j37.5) ohms] produces an impedance Z_d [=(1.6 + j1.7) Z_0 = 120 + j127.5 ohms] on the cable, at a distance of 0.12 λ = 1.37 × 0.12 = 16.4 cm from the load. Thus, there is a dramatic effect of even short lengths (in terms of a wavelength) of the cable.

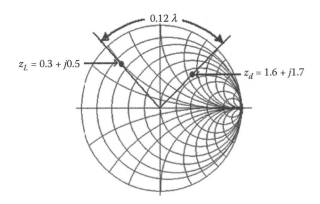

FIGURE 6.30
Effect of rotation of impedance over a short distance (in λ) on a Smith chart.

6.9.7 Implementation of Single Stub Matching

Consider a parallel wire transmission line (Figure 6.31a), which is terminated in an unknown impedance resulting in VSWR = 3. The first voltage minima and the first voltage maxima on this line occur, respectively, at 54 cm and 204 cm from the terminating end. A single stub matching section is to be so designed that the line remains flat (i.e. VSWR becomes unity) to the left of the line.

- The distance between first voltage minima and first voltage maxima = 204 − 54 = 150 = $\lambda/4$, thus λ = 600 cm.
- Draw the $\rho = 3$ circle on the Smith chart.
- Y_{min} coincides with r_{min} (or z_{min}).
- Enter the chart at r_{min} on the real axis of the chart at $\rho = 3$ circle.
- From r_{min}, rotate in the clockwise or anti-clockwise direction. These two rotations lead to the following results.

Anti-clockwise rotation

- Rotate in an anti-clockwise direction until the real part of $y_{in} = 1$ (at a) is reached.
- The distance to move from r_{min} to a is ℓ_1 [=(0.333−0.25)λ] = 0.083 λ = 49.8 cm.
- The location of stub at 'a' is at (54 − 49.8) = 4.2 cm from z_r.
- At 'a', the normalised susceptance is +j1.16 and hence the stub to be connected must introduced a susceptance of −j1.16 to neutralise the reactive component.

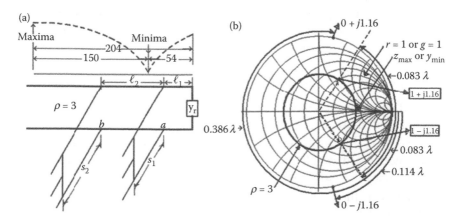

FIGURE 6.31
Single stub matching. (a) Stub locations and (b) Smith chart.

- Locate $x = -j1.16$ on the $r = 0$ circle.
- The stub length (s_1) required to introduce this suscep- tance $= (0.364 - 0.25)\lambda = 68.4$ cm.
- Thus, the stub connected at 'a' is 4.2 cm away from the load and its length is 68.6 cm.

Clockwise rotation

- Rotate in the clockwise direction until the real part of $y_{in} = 1$ (at b) is reached.
- The distance to move from r_{min} to b is also 49.8 cm.
- The location of the stub at 'b' is at $(54 + 49.8) = 103.8$ cm from z_r.
- At 'b', normalised susceptance $= -j1.16$, hence stub must have a sus- ceptance of $+j1.16$.
- Locate $x = +j1.16$ on the $r = 0$ circle.
- Stub length (s_2) required to introduce this susceptance $= (0.25 + 0.136)$ $\lambda = 231.6$ cm.
- Thus, the stub connected at 'a' is 103.8 cm away from the load and its length is 231.6 cm.

The implementation of single stub matching is illustrated in Figure 6.31b. In view of the outcome of the two rotations, it can be noted that location 'a' is to be preferred as in this case the stub has a shorter length and makes longer segment of the line flat.

6.9.8 Implementation of Double Stub Matching

Consider a parallel wire transmission line that is terminated in a known admittance y_r $(=1.23-j0.51)$ and is connected with two stubs (Figure 6.32a), whose lengths are to be calculated. The first stub is connected at a distance $l (= 0.1\lambda)$ from the terminating end, whereas the second stub is d $(= 0.4\lambda)$ dis- tance apart from the first stub.

The implementation requires the following steps:

- Find the centre of the $g = 1$ circle by bisecting its diameter on the chart and mark it as 'c_1'.
- Take 'o' as the centre and distance between 'o' and 'c_1' as the radius. Rotate towards the load to point 'c_2' by encompassing a segment of the circle of circumference 0.4 λ.
- Take 'c_2' as the centre and distance between 'o' and 'c_2' as the radius and draw a complete circle. By this process, the $g = 1$ circle gets relo- cated on the Smith chart.
- Locate $y_r = 1.23 - j0.51$ on the Smith chart.

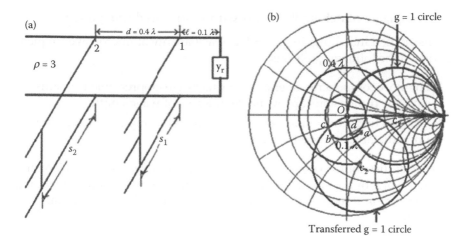

FIGURE 6.32
Double stub matching. (a) Location of stubs and (b) Smith chart.

- Take 'o' as the centre and the distance between 'o' and 'y_r' as the radius. Move from the y_r location towards the generator by 0.1λ and arrive at a point b. This rotation changes y_r (=$1.23 - j0.51$) to y_{11} ($0.7 - j0.3$).

- Move along the $g = 0.7$ circle towards the generator until it touches the transferred $g = 1$ circle at point c. This movement changes only the susceptance as it is carried out along a constant g circle. The value of admittance at c is y_d (=$0.7 - j0.14$). Thus, stub-1 introduces a susceptance that equals the difference between y_d and y_{11} (i.e. $b_1 = +j0.16$).

- Take o as the centre and the distance between o and c as the radius. Rotate towards the generator by 0.4λ and arrive at a point d. This rotation changes y_d (= $0.7 - j0.14$) to y_{22} ($1.0 - j0.4$).

- Move along the $g = 1$ circle towards the generator until it touches the point o. This movement changes only the susceptance as it is carried out along the constant g circle. The value of admittance at this point is y_{in} (= $1 + j0$). Thus, stub-2 introduces a susceptance that equals the difference between y_{in} and y_{22} (i.e. $b_2 = +j0.4$).

- The matching is thus finally achieved.

- As the susceptance to be introduced by stub-1 is b_1, enter the chart at $y_r = \infty$ and rotate clockwise along the outer periphery (i.e. along $g = 0$ circle) to find $b_1 = +j0.16$. Read this moved distance as $s_1\{ = (0.25 + 0.025)\lambda = 0.275\lambda\}$.

- As the susceptance to be introduced by stub-2 is b_2, enter the chart again at $y_r = \infty$ and rotate clockwise along the $g = 0$ circle to find $b_2 = +j0.4$. Read this moved distance as $s_2\{= (0.25 + 0.06)\lambda = 0.31\lambda\}$.

The implementation of double stub matching is illustrated in Figure 6.32b.

6.9.9 Conjugate Matching Problems with Distributed Components

The distributed matching, in general, is feasible only at microwave frequencies due to the following reasons:

1. For most components operating at this range, appreciable transmission line dimensions are available in terms of wavelengths.
2. The electrical behaviour of many lumped components becomes unpredictable at these frequencies.

In case of distributed components, the effects on reflection coefficient and impedance of moving along the transmission line must be allowed for using the outer circumferential scale of the Smith chart, which is calibrated in wavelengths.

EXAMPLE 6.1

A line with characteristic impedance of 50 Ω is terminated in the impedances of $Z_1 = 100 + j50\ \Omega$, $Z_2 = 75 - j100\ \Omega$, $Z_3 = j200\ \Omega$, $Z_4 = 150\ \Omega$, $Z_5 = \infty$ (open circuit), $Z_6 = 0$ (short circuit), $Z_7 = 50\ \Omega$, $Z_8 = 184 - j90\ \Omega$. (a) Calculate the normalised values of these impedances, (b) locate these on the Smith chart and (c) find the corresponding values of reflection coefficients.

Solution

The normalised values of impedances can be obtained simply by dividing the terminated impedances by the characteristic impedance. Thus, the corresponding normalised impedances are $z_1 = 2 + j0.5$, $z_2 = 1.5 - j2$, $z_3 = j4$, $z_4 = 3$, $z_5 = \infty$, $z_6 = 0$, $z_7 = 1$, $z_8 = 3.68 - j1.8$. These values are marked on the Smith chart in Figure 6.33a.

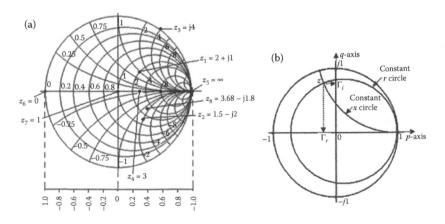

FIGURE 6.33
Parameters represented on a Smith chart. (a) Normalised impedances and (b) real and imaginary reflection coefficients

Every impedance point marked on the Smith chart corresponds to the intersection point of a constant resistance circle and of a constant reactance circle. Draw the rectangular coordinates projection on the horizontal and vertical axes. The values of these projections (Figure 6.33) correspond to the real part of the reflection coefficient (Γ_r) and the imaginary part of the reflection coefficient (Γ_j). The values of these projections are scaled to fall between –1 and 0 or 0 and 1 along both the real or imaginary axis of the chart. The scaling along the p-axis is shown in Figure 6.33a. The corresponding values of reflection coefficients (Γ) obtained from the Smith chart are $\Gamma_1 = 0.4 + j0.2$, $\Gamma_2 = 0.51 - j0.4$, $\Gamma_3 = 0.875 + j0.48$, $\Gamma_4 = 0.5$, $\Gamma_5 = 1$, $\Gamma_6 = -1$, $\Gamma_7 = 0$, $\Gamma_8 = 0.96 - j0.1$.

EXAMPLE 6.2

Find the values of normalised impedances (in rectangular form) corresponding to the following values of the reflection coefficients (in polar form): (a) 0.63 $\angle 60°$, (b) 0.73 $\angle 125°$ and (c) 0.44 $\angle{-}116°$.

Solution

The value of the reflection coefficient in the polar form represents a magnitude and an angle. This magnitude and angle can be marked on the Smith chart. Correspondingly, the three given values can be marked as P_1, P_2 and P_3.

The first value ((a) 0.63 $\angle 60°$) represented in the polar form has an angle of 60°. This angle can be plotted by using the circumferential (reflection coefficient) angle scale to find the $\angle 60°$ graduation and a ruler to draw a line that passes through this (circumferential point) and the centre of the Smith chart. The length of the line OP_1 (=0.63) is scaled to P_1 assuming the Smith chart radius to be unity. For example, if the actual radius of the chart is 100 mm, the length OP_1 will be 63 mm. The value of impedance at the tip of point P_1 represents the normalised impedance. This value read in view of constant r and constant x circles is 0.8 + j1.4. Similarly, the other two values of the reflection coefficients (b) 0.73$\angle 125°$ and (c) 0.44 $\angle{-}116°$ can be marked as P_2 and P_3. The impedances corresponding to point P_2 and P_3 are read as (b) 0.2 + j0.5 and (c) 0.5 – j0.5, respectively. The points P_1, P_2 and P_3 along with their distances from origin o and angles formed by OP_1, OP_2 and OP_3 with the p-axis are shown in Figure 6.34.

EXAMPLE 6.3

Find the values of normalised admittances (in rectangular form) corresponding to the following values of the normalised impedances: (a) 0.8 + j1.4 and (b) 0.1 + j0.22.

Solution

The value of normalised impedances 0.8 + j1.4 and 0.1 + j0.22 are marked on the Smith chart as P_1 and P_2. To change the normalised impedance to the normalised admittance or vice versa, the point representing the value of normalised impedance is to be moved through exactly 180° at

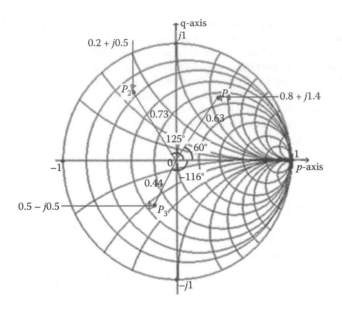

FIGURE 6.34
Illustration of reflection coefficients of Example 6.2.

the same radius. Thus, the movement of point P_1 representing a normalised impedance $(0.8 + j1.4)$ changes this to the equivalent normalised admittance marked as point P_1'. This point is obtained by drawing a line that passes through P_1, centre of the Smith chart (0) to P_1' having an equal radius in the opposite direction (i.e. $0P_1 = 0P_1'$). This is equivalent to moving the point through a circular path of exactly 180°. The value at P_1' read as $y = 0.3 - j0.54$ is the corresponding normalised admittance. This value can be verified by using the relation $y = 1/z$. Similarly, the point P_2 (representing $z = 0.1 + j1.4$) can be moved by 180° to P_2' and the value of admittance can be obtained as $y = 1.8 - j3.9$. The steps for solution are shown in Figure 6.35.

EXAMPLE 6.4

A 300 Ω, 100 MHz lossless line of one wavelength long is terminated in $60 - j50$ Ω. Using the Smith chart, find the impedance of the line at a distance of (a) 30 cm, (b) 90 cm and (c) 120 cm from the termination.

Solution

Given $Z_0 = 300$ Ω, $Z_R = 60 - j50$ Ω.
   The normalised impedance $z_r = Z_R/Z_0 = (60 - j50)/300 = 0.2 - j0.166$.
   $f = 100$ MHz, $\lambda = v/f = 3 \times 10^8/100 \times 10^6 = 3$ m.
   The length of the line = $\lambda = 3$ m.
   Locate $z_r = 0.2 - j0.166$ on the Smith chart and mark this point as z_r. Since the centre of the Smith chart lies at $p = q = 0$, mark this point as 0. Draw a circle with centre at 0 and distance between 0 and z_r as the radius.

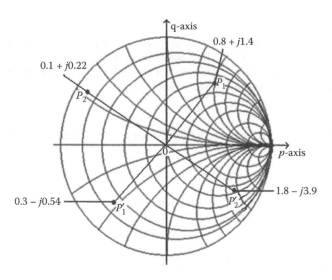

FIGURE 6.35
Normalised impedances and their equivalent normalised admittances.

1. As 30 cm = 0.1 λ, move clockwise (i.e. towards the generator) along the circle from z_r by 0.1 λ and mark this end point as A. The value at A is the input impedance z_i (=0.24 + j0.481).
2. As 90 cm = 0.3λ, move clockwise (i.e. towards the generator) along the circle from z_r by 0.3 λ and mark this end point as B. The value at B is the input impedance z_i (=0.83 + j1.63).
3. As 120 cm = 0.4 λ, move clockwise (i.e. towards the generator) along the circle from z_r by 0.4 λ and mark this end point as C. The value at C is the input impedance z_i (=0.39 − j0.93).

These steps are shown in Figure 6.36.

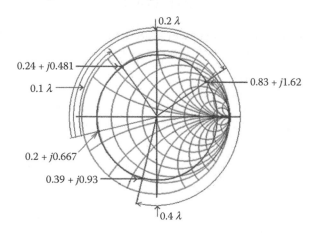

FIGURE 6.36
Normalised values of input impedances at different locations of the line.

EXAMPLE 6.5

A lossless air-spaced transmission line of characteristic impedance $Z_0 = 50\ \Omega$, operating at 800 MHz, is terminated with a series combination of resistor ($R = 17.5\ \Omega$) and an inductor ($L = 6.5$ nH). Evaluate the matching elements of this line using the Smith chart.

Solution

The reactance (X_L) of the inductor forming part of the termination at 800 MHz is

$$X_L = j\omega L = j2\pi f L = j2\pi \times 800 \times 10^6 \times 6.5 \times 10^{-9} = j32.67\ \Omega$$

The impedance (Z_T) of the combination is given by

$$Z_T = 17.5 + j32.67\ \Omega$$

The normalised impedance (z_T) is

$$z_t = Z_T/Z_O = (17.5 + j32.67)/50 = 0.35 + j0.653$$

This normalised impedance is plotted on the Smith chart as point *P*. The line OP is extended to the wavelength scale where it intersects at the wavelength $L_1 = 0.098\ \lambda$. As the transmission line is loss free, a circle with the centre of the Smith chart is drawn through the point P to represent the path of the constant magnitude reflection coefficient due to the termination. At point *Q*, the circle intersects with the unity circle of constant normalised impedance at $z_Q = 1 + j1.52$.

The extension of the line $0Q$ intersects the wavelength scale at $L_2 = 0.177\lambda$; therefore, the distance from the termination to this point on the line is given by

$$L_2 - L_1 = 0.177\lambda - 0.098\ \lambda = 0.079\lambda$$

Since the transmission line is air-spaced, the wavelength at 800 MHz in the line is the same as that in free space and is given by $\lambda = c/f (= 3 \times 10^8/ 800 \times 10^6 = 375$ mm), where c is the velocity of light in free space and f is the frequency in hertz. It gives the position of the matching component ($375 \times 0.079 =$) 29.6 mm from the load.

The conjugate match for the impedance at Q ($z_Q = 1 + j1.52$) is $Z' = -j1.52$.

As the Smith chart is still in the normalised impedance mode, the required series capacitor (C_m) is

$$Z' = -j1.52 = -j/(2\pi f\, C_m\, Z_0) \quad \text{or} \quad C_m = 1/\{(1.52)\,(2\pi f)\, Z_0\}$$
$$= 1/(1.52 \times 2\pi \times 800 \times 10^6 \times 50) \approx 2.6\ \text{pf}$$

Thus, to match the termination at 800 MHz, a capacitor of 2.6 pF must be placed in series with the transmission line at a distance of 29.6 mm from the termination. The steps involved in matching are shown in Figure 6.37.

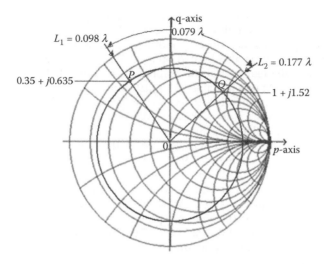

FIGURE 6.37
Illustration of steps involved in distributed transmission line matching.

EXAMPLE 6.6

A lossless transmission line having VSWR = 2.5 is terminated in an impedance $z_L = 2.1 + j0.8$. Find the location (d) and length (d') of the stub to make the line flat.

Solution

Given VSWR = s = 2.5 and $z_L = 2.1 + j0.8$.

Figure 6.38a shows the stub-matching arrangement, whereas Figure 6.38b illustrates the following steps for the solution through the Smith chart.

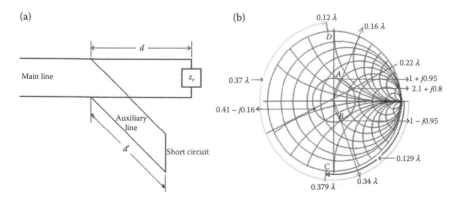

FIGURE 6.38
Single stub matching. (a) Line diagram and (b) Smith chart.

A circle for s = 2.5 is drawn. The load impedance $z_L = 2.1 + j0.8$ is located on the Smith chart. It falls on the s = 2.5 circle and corresponds to 0.22λ on the peripheral scale. Move by 180° (or 0.25λ) to find $y_L = 0.41 - j0.16$. The admittance y_L is also located on s = 2.5. Move clockwise to find a point where the real part of the admittance becomes 1. There are two such locations. At one its value is $1 + j0.95$ and at another it is $1 - j0.95$. The first value corresponds to 0.16λ and the second to 0.34λ, both towards the generator. Thus, a stub can be located at either location but the length of the line is more flat in the first case.

To compensate the susceptance $\pm j0.95 \ \Omega$, the stub should be able to introduce a susceptance of $\mp j0.95 \ \Omega$. To find the lengths of two stubs, one to introduce $-j0.95 \ \Omega$ and the other $+j0.95 \ \Omega$, the chart is to be entered at the location of infinite admittance, that is, at p = 1. Mark C for susceptance $-j0.95 \ \Omega$ and D for $+j0.95 \ \Omega$. Points C and D correspond to 0.379λ and 0.12λ on the outer periphery of the chart. Thus, stub at C will require a length of 0.129λ and 0.37λ. This shows that stub corresponding to point A will not only provide a larger flat length but will also be shorter.

PROBLEMS

P-6.1 A line with characteristic impedance of 75 Ω is terminated in the impedances of $Z_1 = 100 + j50 \ \Omega$, $Z_2 = 75 - j100 \ \Omega$, $Z_3 = j200 \ \Omega$, $Z_4 = 150 \ \Omega$, $Z_5 = \infty$ (open circuit), $Z_6 = 0$ (short circuit), $Z_7 = 50 \ \Omega$, $Z_8 = 184 - j90 \ \Omega$. (a) Calculate the normalised values of these impedances, (b) locate these on the Smith chart and (c) find the corresponding values of reflection coefficients.

P-6.2 Find the values of normalised impedances (in rectangular form) corresponding to following values of reflection coefficients (in polar form): (a) 0.73 ∠60°, (b) 0.63 ∠125° and (c) 0.44 ∠−126°.

P-6.3 Find the values of normalised admittances (in rectangular form) corresponding to the following values of the normalised impedances: (a) $0.6 + j1.4$ and (b) $0.22 + j0.22$.

P-6.4 A 300 Ω, 100 MHz lossless line of one wavelength long is terminated in $60 - j50 \ \Omega$. Using the Smith chart, find the impedance of the line at a distance of (a) 40 cm, (b) 80 cm and (c) 100 cm from the termination.

P-6.5 A lossless air-spaced transmission line of characteristic impedance $Z_0 = 75 \ \Omega$, operating at 800 MHz, is terminated with a series combination of resistor ($R = 17.5 \ \Omega$) and inductor ($L = 6.5$ nH). Evaluate the matching elements of this line using the Smith chart.

P-6.6 A lossless transmission line having VSWR = 3 is terminated in an impedance $z_L = 2.1 + j0.8$. Find the location (d) and length (d′) of the stub to make the line flat.

Descriptive Questions

Q-6.1 Obtain relevant relations that lead to a polar chart.

Q-6.2 Tabulate equivalent values of reflection coefficient and normalised impedance.

Q-6.3 Derive appropriate relations to draw constant r and constant x circles.

Q-6.4 How do the radii and centres of these circles vary with the values of *r* and *x*?

Q-6.5 Illustrate the difference between admittance and impedance charts.

Q-6.6 List the information imparted by a Smith chart.

Q-6.7 List the type of problems that can be solved by using a Smith chart.

Q-6.8 Map suitable multi-element circuits onto a Smith chart.

Q-6.9 Explain how a Smith chart can be used to solve stub-matching problems.

Q-6.10 How is the attenuation factor accounted for by the Smith chart in a lossy line?

Further Reading

1. Alder, R. B. et al., *Electromagnetic Energy Transmission and Radiation*. MIT Press, Cambridge, MA, 1969.
2. Brown, R. G. et al., *Lines, Waves and Antennas*, 2nd ed. John Wiley, New York, 1970.
3. Collins, R. E., *Foundation for Microwave Engineering*. McGraw-Hill, New York, 1966.
4. Everitt, W. L., *Communication Engineering*. McGraw-Hill, New York, 1937.
5. Gandhi, O. P., *Microwave Engineering and Applications*. Pergamon Press, New York, 1981.
6. Gonzalez, G., *Microwave Transistor Amplifiers Analysis and Design*, 2nd ed. Prentice-Hall, New Jersey, 1997.
7. Hayt, W. Jr. and Buck, J. A., *Engineering Electromagnetics*, 4th ed. McGraw-Hill Publishing Co. Ltd., New Delhi, 2001.
8. Johnson, W. C., *Transmission Lines and Networks*. McGraw-Hill, New York, 1950.
9. Jordan, E. C. and Balmain, K. G., *Electromagnetic Waves and Radiating Systems*, 2nd ed. Prentice-Hall Ltd, New Delhi, 1987.
10. Liao, S. Y., *Engineering Applications of Electromagnetic Theory*. West Publishing Company, St. Paul, MN, 1988.
11. Moore, R. K., *Travelling-Wave Engineering*. McGraw-Hill, New York, 1960.

12. Pozar, D. M., *Microwave Engineering*, 3rd ed. John Wiley & Sons, Inc., New Delhi, 2009.
13. Ramo, S., Whinnery, J. R. and Van Duzer, T., *Fields and Waves in Communications Electronics*. John Wiley & Sons, New York, 1965.
14. Reich, H. J. et al., *Microwave Principles*, CBS Publishers and Distributors, New Delhi, 2004.
15. Rizzi, P. A., *Microwave Engineering: Passive Circuits*. Prentice-Hall, Inc., Englewood Cliffs, NJ, 1988.
16. Saad, T. and Hansen, R. C. *Microwave Engineer's Handbook*, Vol. 1. Artech House, Dedham, MA, 1971.
17. Sinnema, W., *Electronic Transmission Technology: Lines and Applications*. Prentice-Hall, Inc., Englewood Cliffs, NJ, 1979.
18. Smith, P. H., An improved transmission line calculator. *Electronics*, 17(1), 130, January 1944.

7

Microwave Components

7.1 Introduction

In general, all communication systems require effective guiding and efficient processing elements. At UHF and above the conventional transmission, generation, reception and measuring equipment become either grossly inefficient or altogether defunct. At microwave range, these elements have to be more sophisticated and perfect. The requirement becomes more and more stringent with the increase of frequency. This is due to the fact that with the increase of frequency, the power-handling capability of components declines and the wave propagation becomes susceptible to the physical dimensions of components. The shapes and roughness of surfaces of components also significantly affect the propagating waves.

Figure 7.1 shows a waveguide system for simple radar. It involves various components, each of which plays its own role. These roles may include guiding, coupling, circulating, attenuating, matching, tuning, filtering and measuring of field contents of the wave. Since waveguides are only hollow metal pipes, their installation and the physical handling have many similarities to the ordinary plumbing. As evident from Figure 7.1, a system made of waveguides too may require bending, twisting, joining and coupling. The fabrication and installation of all components is to be done in such a way that no power is lost due to leakage, reflection and scattering. The design of waveguides, in many ways, differs from that of pipes to carry liquids or gases as it is mainly governed by the frequency and power level of electromagnetic energy to be handled. In this chapter, some of these aspects mainly related with the transmission segment of microwave systems are discussed. The subsequent sections encompass different types of microwave components required for proper transmission and processing.

Furthermore, it needs to be mentioned that the waveguide system may not only comprise rectangular waveguides but may also have components made of circular waveguides. Besides, the involvement of coaxial cables in the formation of some of the components is also unavoidable. With the advent of planar transmission lines and advances in microwave integrated circuit technology, the formation of hybrid systems is now a reality.

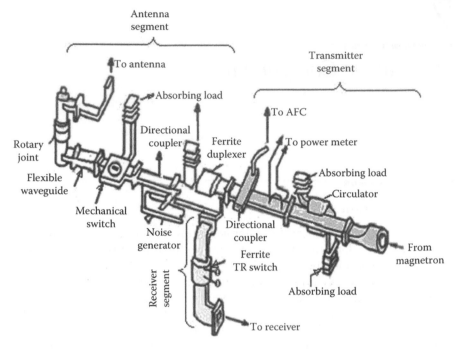

FIGURE 7.1
A simple radar system containing different microwave components.

7.2 Waveguides and Its Accessories

Waveguides are hollow pipes made of copper and their inner surfaces are generally coated with highly conducting material such as silver or its alloys for reducing the attenuation. These are compact and can be easily integrated with active devices. Waveguides may have rigid or flexible structures. They generally have high power-handling capability, but are bulky and expensive. Seamless waveguides as per the international specifications are available up to 2 m length. These are used in conjunction with the flanges.

7.2.1 Waveguide Shapes

The waveguides can have cross sections of rectangular, circular or elliptical shape. As far as the applications are concerned, rectangular waveguides have the lion's share. Circular waveguides are mostly used for making components such as isolators, circulators and so on. Elliptical waveguides may be used only in very special applications.

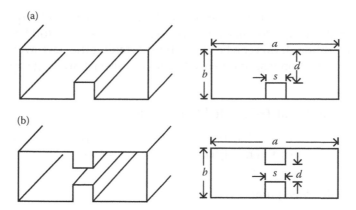

FIGURE 7.2
Ridged waveguides. (a) Cross section and dimensions of singly ridged waveguide and (b) cross section and dimensions of doubly ridged waveguide.

7.2.2 Ridged Waveguides

The useful bandwidth of a rectangular waveguide is about 40%. To obtain a larger bandwidth, ridged waveguides (Figure 7.2) are used, wherein the cross section (generally) of broader wall is shortened from one or both sides. Such a change amounts to the insertion of a capacitor (or an inductor in case of shortening of narrower walls) in the passage of wave. Besides providing larger bandwidth of an octave or more, for a given TE_{10} cutoff frequency, its cross section is smaller than that of an ordinary waveguide. Thus, the over-all waveguide structure reduces in size. Ridged waveguides, however, have higher losses and lower power-handling capacity.

7.2.3 Dielectric Loaded Waveguides

If a metallic waveguide is loaded with dielectric (Figure 7.3), it is equivalent to the capacitive loading. Its usable bandwidth is substantially greater than that

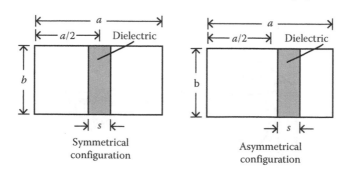

FIGURE 7.3
Dielectric-loaded waveguides. (a) Symmetrical configuration; (b) asymmetrical configuration

of an ordinary rectangular waveguide and thus its behaviour is equivalent to that of ridged waveguide. Dielectric loading decreases the cutoff frequency of TE_{10} mode. Although this can be raised by shortening the narrower wall, this enhances conduction loss and hence lowers its power-handling capability.

7.2.4 Bifurcated and Trifurcated Waveguides

Such waveguides (Figure 7.4) are used when the signal is to be bifurcated or trifurcated to feed different adjacent/auxiliary circuits.

7.2.5 Flexible Waveguide

Flexible waveguides (Figure 7.5) allow special bends, which may be required in some special applications. Thus, these are used only where no other reasonable solution is available. It consists of a specially wound ribbon of conducting material, most commonly brass, with the inner surface plated with chromium. Since flexible waveguides have more power losses mainly due to their imperfect inner surface, these are made only of short lengths.

7.2.6 Impact of Dimensional Change

The impact of change in waveguide cross section is illustrated in Figure 7.6. As evident from this figure, the change in broader wall (Figure 7.6b) is equivalent to the insertion of an inductance whereas the change in narrower wall

FIGURE 7.4
(a) Bifurcated and (b) trifurcated waveguides.

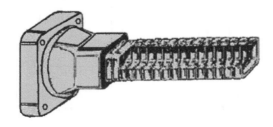

FIGURE 7.5
Flexible waveguide.

(a)   (b)

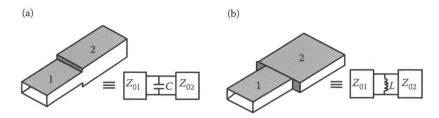

FIGURE 7.6
Dimensional change in waveguide and its equivalent. (a) Along narrower wall and (b) along broader wall.

(Figure 7.6a) amounts to the insertion of a capacitance. In view of the field distribution across the cross section of rectangular waveguide, the dimensional change results in squeezing of either the electric field or the magnetic field. The disturbance of the electric field can be viewed as insertion of a capacitance, whereas that of the magnetic field as insertion of inductance. In Figure 7.6, Z_{01} and Z_{02} are the characteristic impedances of the two waveguide sections marked as 1 and 2.

7.2.7 Waveguide Joints

In view of Figure 7.1, it can be observed that an entire waveguide system cannot be molded into one single piece and is composed of different components. The waveguides and its components are in the form of small sections, which are to be joined. There are three basic types of waveguide joints: (i) permanent, (ii) semi-permanent and (iii) rotating joints. The permanent joint is a factory-welded joint (Figure 7.7) and requires no maintenance. The other two categories are discussed below.

7.2.7.1 Waveguide Flanges

Sections of waveguide are often taken apart for maintenance and repair. Semi-permanent joints may be either in the form of simple *cover flange* (Figure

FIGURE 7.7
Rectangular and circular waveguides with permanent flanges. (a) Rectangular waveguide and (b) Circular waveguide

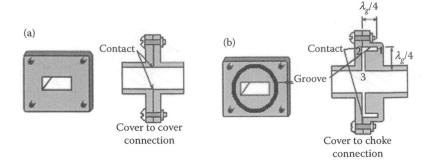

FIGURE 7.8
Simple cover and choke flanges. (a) Simple cover flange and (b) choke flange.

7.8a) or choke joint (or *choke flange*) (Figure 7.8b). Both of these are made of brass, precisely machined and fitted to the outer dimensions of waveguides. Cover flanges are simple, cheap and thus most commonly used in laboratories. Choke flanges, however, reduce the leakage of energy resulting in improved VSWR of the system and thus are used in actual setups.

The choke joints provide good electromagnetic continuity between sections of waveguide with very little power loss. A cross-sectional view of a choke joint is shown in Figure 7.8b. *A pressure gasket is to be provided between the two metal surfaces to form an airtight seal.* As can be seen, a passage for leakage of wave is apparently allowed between points marked as 1, 2 and 3. Both the slot lengths between 1 and 2 and 2 and 3 are exactly $\lambda/4$ from the broader wall of waveguide. The wave however sees it differently. At point 1, there is a short circuit, which transforms into high impedance at point 2 and this high impedance in turn results at point 3 as a perfect short circuit. This effect creates a good electrical connection between the two sections that permits energy to pass with very little or no leakage.

7.2.7.2 Waveguide Rotary Joints

Whenever a stationary rectangular waveguide is to be connected to a rotating antenna, a rotating joint is to be used. A circular waveguide is normally used in a rotating joint since the rotation of the rectangular waveguide will result in distortion of field pattern. The rotating section of the joint, illustrated in Figure 7.9a, uses a choke joint to complete the electrical connection with the stationary section. The circular waveguide is designed so that it operates in the TM_{11} mode. The attached rectangular sections prevent the circular waveguide to operate in the wrong mode. Distance $3/4\lambda$ shown is to provide a high impedance to avoid the waveguide to operate in any unwanted mode. This is the most common design used for rotating joints, but other types may also be used in specific applications. Figure 7.9b shows a coaxial line rotary joint.

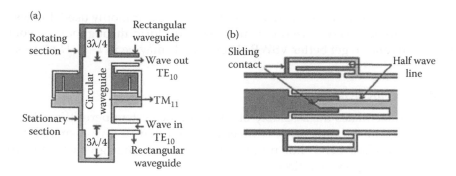

FIGURE 7.9
Rotary joints. (a) Waveguide rotary joint and (b) coaxial rotary joint.

7.2.8 Waveguide Stands

A waveguide stand (Figure 7.10) keeps a test bench perfectly levelled to avoid reflections due to bending. The stand shown in Figure 7.10a may be used for levelling one particular band whereas the stand of Figure 7.10b can accept components of many bands for example, C, S, X and K. Both of these stands are adjustable in height and can be locked in any position by means of a screw.

The size, shape and dielectric material of a waveguide must be constant throughout its length for energy to move from one end to the other without hindrances and reflections. Any abrupt change in its size or shape can cause reflections and loss of energy. As evident from Figure 7.1, in a practical system, a waveguide may have to be bent, twisted, bifurcated, trifurcated or tapered to change its shape from rectangular to circular or vice versa. To meet these requirements, certain components called bends, corners, twists and transitions are needed.

7.2.9 Waveguide Bends

Waveguide bends are used wherever change in the direction of transmission is desired. As such there is no bar and a bend may be designed for

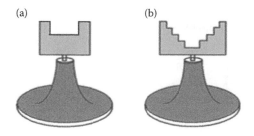

FIGURE 7.10
Waveguide stands. (a) Single band waveguide stand and (b) universal waveguide stand.

any angle. Angles with 90° and 45° bends are more commonly used. In these bends, waveguide cross section is to be kept uniform throughout to avoid reflections and to get better VSWR. These bends may be in *E*- or *H*-planes.

7.2.9.1 E-Plane Bend

This bend facilitates the assembly of a waveguide system permitting direction change in the *E*-plane. It is precisely fabricated to give a smooth 90° bend. The *E*-plane bend shown in Figure 7.11 has a radius greater than two wavelengths to prevent reflections.

7.2.9.2 H-Plane Bend

H-plane bend (Figure 7.12) is used in the assembly of a waveguide system wherein a smooth directional change in *H*-plane is required. In this case also, the radius of bend must be greater than two wavelengths to prevent reflections.

7.2.10 Waveguide Corners

Neither the *E* bend along dimension '*a*' nor the *H* bend along '*b*' changes the normal mode of operation. A sharp bend in either dimension may also be used if it meets certain requirements. Such bends are often referred to as corners. In some of the applications, waveguide corners are used, in lieu of bends, whenever change is required in the direction of transmission. Since in case of corners, maintaining the uniformity of the cross section is a bit

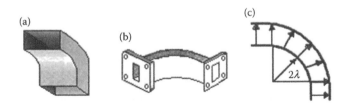

FIGURE 7.11
E-plane bend. (a) Without flange, (b) with flange and (c) Bend's curvature.

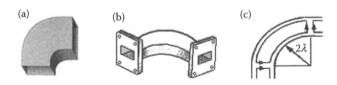

FIGURE 7.12
H-plane bend. (a) Without flange, (b) with flange and (c) Bend's curvature.

difficult in comparison to bends, the likelihood of reflections in corners is relatively more than in bends. These too can be classified in accordance with the planes of bending. Corners too may be designed for any required angle. In case of 45° bends, ports in the bend are $\lambda/4$ apart. The reflections that occur at the 45° bends cancel each other, leaving the fields as though no reflections have occurred.

7.2.10.1 E-Plane Corner

E-plane corner facilitates direction change in the *E*-plane. In order to have better cross-section uniformity, it is mitered (or flattened) on one side or on both the sides of the junction of two orthogonal arms. These are shown in Figure 7.13.

7.2.10.2 H-Plane Corner

H-plane corner is used where a smooth directional change is required in the *H*-plane. As shown in Figure 7.14, these too can be mitered on one or on both the sides.

7.2.11 Waveguide Twists

Sometimes the field of an electromagnetic wave is to be rotated so that it appropriately matches the load in terms of its phase. This may be accomplished by mechanically twisting the waveguide (Figure 7.15). To make the

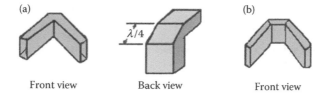

FIGURE 7.13
E-plane corners. (a) Single mitered and (b) double mitered.

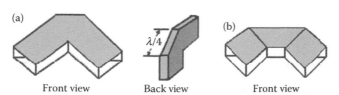

FIGURE 7.14
H-plane corners. (a) Single mitered and (b) double mitered.

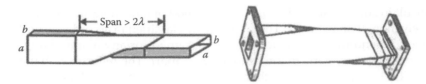

FIGURE 7.15
Waveguide twists.

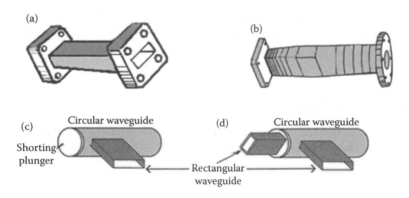

FIGURE 7.16
Waveguide transitions. (a) Tappered transition, (b) rectangular to circular transition, (c) rectangular to circular (single mode) transition and (d) double mode transition.

twist smooth, gradual and without disturbing the uniformity of its cross section the length of twisted part has to be sufficiently long.

7.2.12 Waveguide Transitions

These are used when a setup is formed of components of different shapes and sizes. Figure 7.16a shows a tapered transition, which involves two rectangular waveguides of different cross sections. Figure 7.16b shows rectangular to circular waveguide transition. In both the cases, uniformity and smoothness of cross sections is desired throughout the transition. Figures 7.16c and 7.16d also illustrate involvement of waveguides of rectangular and circular cross sections but in these cases smooth transition is no more feasible.

7.3 Input–Output Methods in Waveguides

The energy from a source is to be fed into a waveguide and to be taken out at some other location for measurement or other purposes. The three commonly used means to inject or remove energy from waveguides are probes, loops and slots. Slots are also termed as apertures or windows.

7.3.1 Probe Coupling

Small probe carrying microwave energy inserted into a waveguide acts as quarter-wave antenna. As illustrated in Figure 7.17a, a current-carrying conductor is inserted into the waveguide. The most efficient place to locate the probe is in the centre of a broader wall and parallel to the narrower wall. Further, it has to be at one quarter-wavelength distance from the end of the waveguide, which is to be shorted permanently or through a movable (plunger) short (Figure 7.17b). Thus, when a probe is located at the point of highest efficiency, the E field is of considerable intensity. Figure 7.17c shows the probe and its equivalence, which is normally referred to as electric or electrostatic coupling. The point illustrated in Figure 7.17b is at the maximum E field corresponding to the dominant mode. There is maximum energy transfer (coupling) at this point. The quarter-wavelength spacing is required to propagate the energy in the forward direction.

Figure 7.17d shows that the current carried by the probe sets up an E field in accordance with its direction. These E lines later detach themselves from the probe. Figure 7.17e illustrates the location of the probe along with the alternating nature of E field. In case a lesser degree of energy transfer or loose coupling is desired, it can be reduced by reducing the probe insertion or by moving it out of the centre of the E field, or by shielding it. If the degree of coupling is to be frequently varied the probe is to be made retractable so that the depth of insertion can be changed easily. The frequency, bandwidth and power handling capability are determined by the size and shape of the probe. As the diameter of a probe increases, the bandwidth

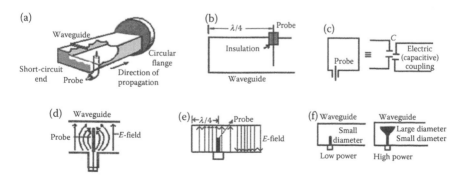

FIGURE 7.17
Probe coupling in a rectangular waveguide. (a) Location of probe, (b) probe at $\lambda/4$ distance from shorted end, (c) probe and its equivalence, (d) detached E-field, (e) E-field and probe and (f) broadband probes.

increases. A probe similar in shape to a door knob is capable of handling much higher power and a larger bandwidth than a conventional probe. The power-handling capability of a probe is directly related to its surface area. Two broadband probes are illustrated in Figure 7.17f. The energy removal from a waveguide is just the reversal of the injection process.

7.3.2 Loop Coupling

In this case, the energy is injected by setting up a magnetic field in the waveguide. This can be done by inserting a small current-carrying loop into the waveguide as shown in Figure 7.18a. Figure 7.18b shows the loop along with necessary insulation. Figure 7.18c illustrates the loop and its equivalence, which is termed as magnetic coupling. As shown in Figure 7.18d, the magnetic field builds up around the loop and expands to fill the waveguide. If the frequency of the loop current is within the waveguide bandwidth, energy will get transferred to the waveguide. For effective coupling, the loop is to be so inserted so that the magnetic field is maximum at the point of insertion. Figure 7.18e shows four loop locations for obtaining maximum magnetic field strength.

In some applications, only a small fraction of power is to be coupled or removed. In such cases, the coupling is reduced by moving or rotating the loop. The order of coupling will depend on the number of H lines encircled or intercepted. The power-handling capability of a loop is proportional to its diameter. Thus, when the loop diameter is increased, its power-handling capability also increases. Its bandwidth can be increased by increasing the size of the wire used for the loop formation. The removal of energy through the loop follows the same principle as that by a probe. As the loop is inserted into a waveguide having an H field, a current is induced in the loop, which can be fed to the desired device.

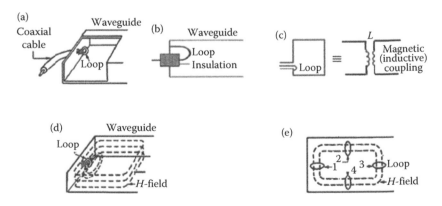

FIGURE 7.18
Loop coupling in a rectangular waveguide. (a) Location of loop, (b) loop and insulation, (c) loop and its equivalence, (d) loop at maximum H-field and (e) possible locations for loop.

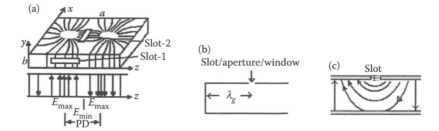

FIGURE 7.19
Slot coupling in a waveguide. (a) Current flow in waveguide walls and locations of slots for dominant mode, (b) slot at maximum E-field and (c) expansion of field in waveguide.

7.3.3 Slot/Aperture Coupling

Slot or aperture coupling (Figure 7.19) is used when loose coupling is desired. In this method, energy enters through a small slot and E field expands into the waveguide. These E lines expand first across the slot and later across the waveguide interior. When energy is injected or removed through slots from the waveguide, there is a likelihood of reflection of energy. For minimising reflections, the size of the slot is to be proportioned in accordance with the frequency of operation. When energy leaves, a field is formed around the end of the waveguide. This field results in impedance mismatch and hence in the formation of standing waves. This formation leads to energy loss or efficiency reduction. This can be avoided by using an impedance matching method.

7.3.4 Input–Output Coupling in Circular Waveguides

All the coupling methods (discussed above) for rectangular waveguides are equally applicable to circular waveguides. The location and method of insertion of probe into a circular guide was illustrated earlier in Figure 3.21.

7.4 Coaxial to Waveguide Adapter

As microwave systems may contain components made of a waveguide and a coaxial cable, transition from waveguide to coaxial or vice versa is needed. There are many ways to join a coaxial line to a waveguide, which minimise the reflection losses. In all these systems, the outer conductor of the coaxial line is connected to the wall of the waveguide. The inner conductor of the

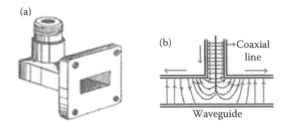

FIGURE 7.20
Coaxial to waveguide adopter. (a) Physical configuration and (b) field distribution.

coaxial line extends (like a short antenna) into the waveguide and some portion of it is parallel to the *E* field in the guide. One end of the waveguide is short circuited. The configuration of a coaxial to waveguide adopter and the transition of field are shown in Figure 7.20.

The transition can be matched by varying four parameters, the diameter and depth of penetration of the probe, the distance to the end plate or short circuit and (occasionally) the distance from the centre of the waveguide. A match of 1:2 over a band is easily obtained. These transitions are made with a very short section of coaxial line attached to the waveguide. The other end of the coaxial line has a connector, which can be joined with a standard cable connector. The methods of coupling are illustrated in Figure 7.21.

The discussions so far have been concentrated only to the components, which transfer the energy from one point to another. These components do nothing more than feeding, coupling, guiding or changing the direction of travel. There are components that modify the energy in some fashion during its transit and change the basic characteristics or power level of the electromagnetic energy. The more common among these include hybrid junctions and directional couplers, which are discussed in the subsequent sections.

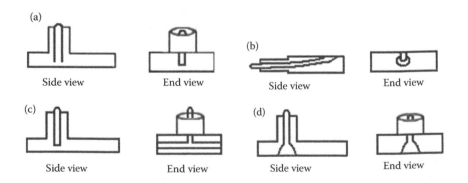

FIGURE 7.21
Feeding techniques in coaxial to waveguide adopter. (a) Probe feeding, (b) end-on feeding, (c) cross bar feeding and (d) door knob feeding.

7.5 Waveguide Junctions

Microwave systems may require the formation of junctions in order to add or divide the signals. While energy travelling through a waveguide reaches a junction, it is affected in different ways. It is therefore necessary to understand the basic operating principles of junctions. The tee junctions are the most simple and most commonly used. The two basic types of junctions are referred to as the *E*-type and the *H*-type. Tees are used to join three waveguide lines in order to combine, divide or compare power levels. The T-junctions can also be termed power dividers or combiners, depending on their way of utilisation in a microwave system. These may be used in conjunction with frequency meter, VSWR meter or other detection devices. Another class of junctions called hybrid junctions is a bit more complicated. Magic tee and hybrid ring belong to this category.

7.5.1 *E*-Plane Tee

E-plane tee (or *E*-type T-junction) consists of a section of flanged waveguide mounted at right angle to the broader wall of main waveguide. The power fed into the input arm splits equally in the other two arms but these two outputs are 180° out of phase. This tee is also called *series tee*. If the two signals of equal amplitude and the same phase are fed into port 2 and port 3 of the series tee, the output at arm one will be zero. This tee along with the addition and division of fields at its three ports in three different cases of inputs is shown in Figure 7.22. For clarity, the magnetic field lines have been omitted in all the cases shown.

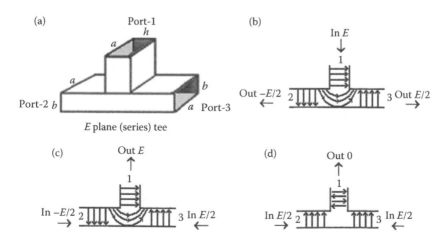

FIGURE 7.22
E-plane tee. (a) Physical configuration, (b) input at port 1, (c) out of phase inputs at port 2 and 3 and (d) in-phase inputs at ports 2 and 3.

7.5.2 *H*-Plane Tee

When a short piece of waveguide is fastened perpendicular to the narrower wall of a waveguide, the resulting three-port junction is a Calledean *H*-plane tee. The perpendicular arm is usually the input arm and the other two arms are in shunt to the input. For this reason, this tee is called a *shunt tee*. Power fed into the input arm splits equally and in phase in the other two arms. If two signals of equal amplitude and the same phase are fed into the two arms of this tee, these will combine and add in the input arm. This tee along with the division of fields for different locations of inputs and appearance of corresponding outputs is illustrated in Figure 7.23.

7.5.3 *EH* (or Magic) Tee

It is a combination of *E*-plane and *H*-plane tees. The combination is a four-port four-arms device and is referred to as *EH* tee or *hybrid tee*. Its collinear arms are called *side arms*. The arm, which makes an *H*-plane tee with the side arm is called the *H*-arm or *shunt arm*. The fourth arm which makes an *E*-plane tee with the side arms is called the E-arm or *series arm*. The *EH* tee and input–output at its different ports looking from different sides are shown in Figure 7.24.

If a signal is fed into *H*-arm (port 4), the power divides equally and in phase in the two side arms with no coupling to the *E*-arm. When a signal enters *E*-arm (port 3), it also divides equally into two side arms but this time the two halves are 180° out of phase and there is no coupling to the *H*-arm. If power is fed into a side arm, it divides equally into the shunt and series

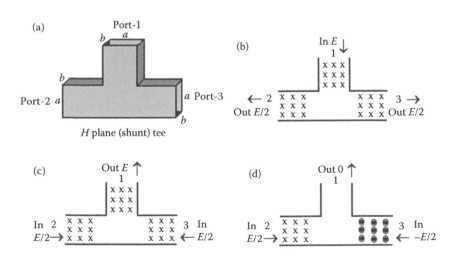

FIGURE 7.23
H-plane tee. (a) Physical configuration, (b) input at port 1, (c) in-phase inputs at port 2 and 3 and (d) out of phase inputs at ports 2 and 3.

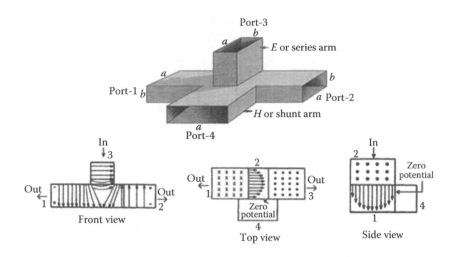

FIGURE 7.24
EH tee and input–output at its different ports. (a) *EH* tee, (b) front view (field distribution), (c) top view (field distribution) and (d) side view (field distribution)

arms and with no coupling to the collinear side arm. Owing to this unusual behaviour, this tee is also called *magic tee*.

If two signals are fed into the two side arms, they combine in the other two ports. The signal combines in phase in the *H*-arm and 180° out of phase in the *E*-arm. Thus, *H*-arm is a *sum arm* and the *E*-arm is the *difference arm*. The original signals do not have to be in phase, or equal in magnitude. The output at *H*-arm will be the vectorial sum of the two inputs at the side arms and the output at the *E*-arm will be the vectorial difference between the two signals.

EH tees are used in bridge circuits to compare impedances in transmission systems, for VSWR measurements and as balanced mixer in automatic frequency control circuits. The most common application of this type of junction is as the mixer section for microwave radar receivers.

7.5.3.1 Impedance Matching in Magic Tee

When an input signal is fed into port 1 as shown in Figure 7.25, a portion of the energy is coupled to port 4 as it would be in an *E*-type T-junction. An equal portion of the signal is coupled through port 3 because of the action of the *H*-type junction. Port 2 has two fields across it that are out of phase with each other. Therefore, the fields cancel, resulting in no output at port 2. The reverse of this action takes place if a signal is fed into port 2, resulting in outputs at ports 3 and 4 and no output at port 1. Unfortunately, when a signal is applied to any port of a magic tee, the energy at the output ports is affected by reflections that are the result of impedance mismatch at the junctions. In view of these reflections, there is power loss since all the energy fed into the

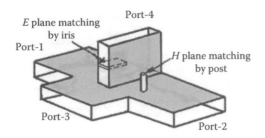

FIGURE 7.25
Impedance matching in magic tee.

junction does not reach the load and standing waves are produced, which may result in internal arching. As a result, the maximum power handling capability of magic T is greatly reduced. To reduce these reflections, impedance matching is to be employed provided the shape of junctions remains unaltered. The matching may be accomplished by using a post to match the *H*-plane, or an iris to match the *E*-plane. Although these methods reduce reflections, the power handling capability of *EH* tee further reduces.

7.5.4 Hybrid Ring (Rat Race Tee)

It is a type of hybrid junction that overcomes the power limitation of the magic T. The hybrid ring is also called a rat race tee. It is actually a modified magic T. The hybrid ring is constructed of rectangular waveguides molded into a circular pattern. The arms are joined to this circular ring to form *E*-type T-junctions. Figure 7.26 shows the required dimensions (in terms of wavelengths) for a hybrid ring to operate properly.

Hybrid ring finds application in high-powered radars and communications systems. It is used to isolate a transmitter from a receiver. During

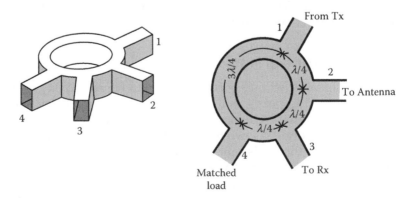

FIGURE 7.26
Rat race tee (hybrid ring).

transmit period, the hybrid ring couples energy from the transmitter to the antenna and allows no energy to reach the receiver. During receive cycle, it couples energy from the antenna to the receiver and allows no energy to reach the transmitter.

Its operation can be understood from the input–output appearance of signal at different ports. When the wave with content E enters port 1 it gets divided into two equal halves, that is, $E/2$ each. At port 2, the two halves add to E. At port 3, the two halves add to zero. At port 4, the two halves add to become $-E$. If no output is required, this port may be matched terminated. When the wave enters port 2, it again divides into two equal halves. At port 3, the two halves add to appear as E. At port 4, the two halves add to zero, and at port 1, the output becomes $-E$ so this port is to be matched terminated. Similarly, when wave enters through port 3, it appears at port 4 as E, at port 1 as zero and at port 2 as $-E$, which is again to be matched terminated. Since only one of the four ports can be terminated in a matched load, let the transmitter input be fed to port 1. It will appear at port 2 where an antenna may be connected. On reception, the signal enters port 2. This signal will appear at port 3, which can be connected to the receiver. In case there are some reflections at the receiver terminal, this signal will travel to port 4, which can be terminated in a matched load.

7.6 Directional Couplers

Directional coupler is a *four-port* device that allows energy sampling from within a waveguide. Such sampled energy is often required for qualitative and quantitative monitoring of signals. It has two transmission lines, which are coupled in such a way that the output at a port of one transmission line depends on the direction of propagation in the other. Figure 7.27 illustrates

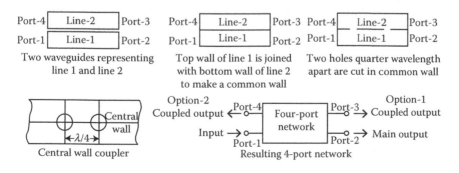

FIGURE 7.27
Formation of a directional coupler. (a) Two waveguides representing line-1 and line-2, (b) top wall of line-1 is joined with bottom wall of line-2, (c) two are cut in common wall quarter wavelength apart, (d) central wall coupler and (e) resulting 4-port network.

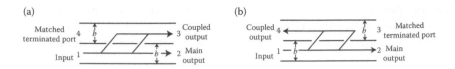

FIGURE 7.28
(a) Forward and (b) backward directional couplers.

the formation of a four-port network representing a directional coupler. Accordingly, line 1 containing port 1 and port 2 is coupled to line 2 containing port 3 and port 4. Line 1 is called the *main line* whereas line 2 is referred to as *auxiliary line*.

In an ideal coupler, a signal entering port 1 will travel to port 2 and a predetermined portion of this signal will appear at one of the other two ports (3 or 4). There will be zero output at the remaining port. If the main signal travels in the reverse direction from port 2 to port 1, the small-coupled signal will appear at the port at which the output was zero in the first case.

When a signal travels from port 1 to port 2, the coupled signal can appear either at port 3 or port 4 depending on the coupling mechanism used. If the signal in the side transmission line travels in the same direction as the main signal, the coupler is called *forward directional coupler*. In this, the coupled signal output appears at port 3 when the input is fed to port 1. If the coupler output is at port 4, it is called *backward directional coupler*. These are shown in Figure 7.28.

7.6.1 Bathe-Hole Coupler

It has a single centre offset hole in the common broader wall between the two waveguides. The coupling between port 1 and port 3 is controlled by the offset location of the hole, which is assessed by angle (θ) between the axis of the two guides. The coupling also depends on the frequency of operation and the size of the hole. This coupler is shown in Figure 7.29a.

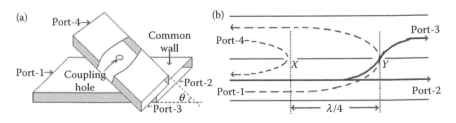

FIGURE 7.29
Bathe-hole coupler. (a) Physical configuration and (b) signal flow.

7.6.2 Double-Hole Coupler

In this coupler, two waveguides are coupled through two holes in the common broader wall. These holes are separated by $\lambda_g/4$. This coupler may also be classified as Bathe-hole coupler. Its working is illustrated in Figure 7.29b. Accordingly, when the wave travels from port 1 to port 3, it travels the same distance due to holes X and Y and adds constructively at port 3. The fraction of the wave travelling in the reverse direction, that is, the wave coming through hole-Y, travels $\lambda_g/4$ more distance than that coming through hole-X. Thus, these two waves arriving at port 4 destructively add due to 180° phase difference between the two waves. As a result, port 4 is not coupled to the input port. In view of $\lambda_g/4$ separation, its directivity is sensitive to the frequency variations.

7.6.3 Moreno Cross-Guide Coupler

It consists of two waveguide sections joined perpendicularly with broader walls adjoining the area of contact. One waveguide wall is removed to improve the efficiency of the device. Coupling between two guides is accomplished by the crossed slots separated by a distance of $\lambda_g/4$, which are cut in the common wall. Each pair of crossed slots may act like a Bathe-hole coupler if lengths of the two slots are properly chosen. Such proper selection may reduce the back scattering and enhances the forward scattering to double than that obtained from a single slot.

These couplers (Figure 7.30) are designed to operate as one-way power monitors in waveguide transmission lines. These are also useful in the measurement of reflections, reduction of attenuation, decoupling microwave instruments such as frequency meters, mixing microwave signals and for the measurement of receiver sensitivity. The unterminated cross-guide coupler may be used to sample incident and reflected power simultaneously. In terminated couplers, termination absorbs all the incident power impinging on it.

7.6.4 Schwinger Reversed-Phase Coupler

In this coupler, two waveguides are so placed that the field radiated by the two $\lambda_g/4$ distant slots are 180° out of phase. As a result, the directivity becomes less

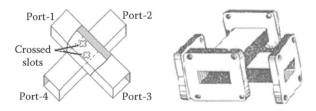

FIGURE 7.30
Cross-guide directional coupler.

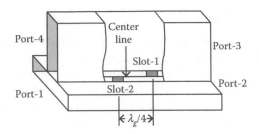

FIGURE 7.31
Schwinger reversed-phase coupler.

susceptible to frequency variations. Its construction is shown in Figure 7.31. In this, the coupling slots are oppositely located to the centre line of the broader wall of one waveguide and that of the narrower wall of the other.

7.6.5 Multi-Hole Directional Coupler

Figure 7.32 shows a multi-hole directional coupler. It is a four-port network with the fourth port terminated into a matched load. The main waveguide and an auxiliary (secondary) waveguide have a common broader wall containing the coupling holes. These holes are separated by a distance of $\lambda_g/4$. Each coupler has a high degree of uniformity of coupling, minimum frequency sensitivity (+0.5 dB) and high directivity.

7.6.6 Unidirectional and Bidirectional Couplers

The couplers that sample energy only in one direction of travel are called unidirectional couplers, whereas if the sampling is done in both the directions, these are termed as bidirectional. The bidirectional couplers are widely used in radar and communications systems.

7.6.7 Short Slot, Top Wall and Branch Guide Couplers

Figure 7.33 illustrates three different types of couplers, which include (a) short slot coupler, (b) top wall coupler and (c) branch guide coupler.

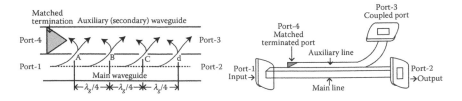

FIGURE 7.32
Multi-hole directional coupler.

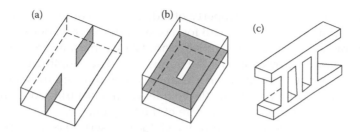

FIGURE 7.33
Three other couplers. (a) Short slot coupler, (b) top wall coupler and (c) branch guide coupler.

In *short slot coupler*, a small slot is cut in the common wall between two waveguides, which are joined along with their narrower walls. In *top wall coupler*, the broader wall is common between two waveguides. An aperture/slot is provided for coupling in this common wall. In branch guide couplers, two main waveguides are joined by a number of auxiliary guides. These auxiliary waveguides may be of rectangular or circular shapes. Some of these couplers find their applications in the formation of microwave filters and other devices.

7.6.8 Figures of Merit

The characterising parameters of a directional coupler are referred to as figures of merit. If powers at ports 1, 2, 3 and 4 shown in Figure 7.34 are P_1, P_2, P_3 and P_4, respectively, these parameters can be defined in terms of these powers using the symbols of the forms $K_{a,b}$. This indicates that parameter K at port a is due to an input at port b. These characterising parameters expressed in dBs are as given below:

Coupling coefficient (C): The coupling coefficient or coupling factor (in dBs) is the ratio of input power (P_1) to the coupled power output in the auxiliary arm (P_3).

$$C_{3,1} = -10\log_{10}\left(\frac{P_3}{P_1}\right) \qquad (7.1)$$

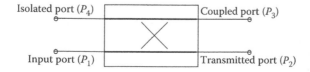

FIGURE 7.34
Powers at different ports.

It represents the primary property of a directional coupler and is a negative quantity. It cannot exceed 0 dB for a passive device, and does not exceed –3 dB in practice. Although a negative quantity, by some authors, it is defined as a *positive* quantity by dropping the minus sign. The coupling factor varies with the frequency. Although this variance can be reduced by some design methods theoretically, a perfectly flat coupler cannot be built. Coupling accuracy is generally specified at the centre of the frequency band.

Insertion loss (IL): It is the loss introduced due to insertion of a component in a line. It was earlier given in relation to three-port circulator by Equation 5.4 and in relation to general ferrite devices by Equation 5.6. In directional coupler, it specifies the ratio of total output power from all the ports to the input power. If the device inserts loss in the line, the output power has to be less than the input power. It is mainly because some of the power may be absorbed by the device and some get reflected due to mismatch. The insertion loss (in dBs) is given by the relation

$$IL = 10 \log_{10} \left[\frac{(P_2 + P_3 + P_4)}{P_1} \right]. \tag{7.2}$$

Isolation (I): It was earlier given by Equations 5.5 and 5.7. It can be defined as the difference in signal levels between the input port (1) and the isolated port (4) when two other ports (2 and 3) are match terminated.

$$I_{41} = -10 \log_{10} \left(\frac{P_4}{P_1} \right) dB \quad \text{or } 10 \log_{10} \left(\frac{P_1}{P_4} \right) dB \tag{7.3a}$$

In case input, (port 1) and isolated (port 4) are terminated by matched loads isolation can be defined between the two output ports (2 and 3). In this case, one of the output ports is used as the input and the other is considered as the output port.

$$I_{32} = -10 \log_{10} \left(\frac{P_3}{P_2} \right) dB \tag{7.3b}$$

Thus, the isolation between input and isolated ports may be different from the isolation between the two output ports. Isolation can be obtained from coupling plus return loss. The isolation should be as high as possible. In practice, isolated port is never completely isolated and some RF power remains. Waveguide directional couplers provide the best isolation.

In an ideal directional coupler, insertion loss is only the coupling loss whereas in a real directional coupler, it is a combination of coupling loss, dielectric loss, conductor loss and VSWR loss. Coupling loss becomes less significant above 15 dB coupling as compared to other types of losses. It however depends on the frequency range. Insertion loss due to coupling can be appreciated in view of the following data:

Coupling in dBs	3	6	10	20	30
Insertion/coupling loss in dBs	3	1.25	0.458	0.0436	0.00435

Directivity (D): It is a measure of performance of a directional coupler (in dBs) and is defined as the ratio of power output at the coupled port to the unwanted signal at the uncoupled port. Both these ports belong to the auxiliary arm. In mathematical terms, it is given as

$$D_{3,4} = -10\log_{10}\left(\frac{P_4}{P_3}\right) = -10\log_{10}\left(\frac{P_4}{P_1}\right) + 10\log_{10}\left(\frac{P_3}{P_1}\right). \tag{7.4a}$$

In view of Equations 7.1 and 7.3a, Equation 7.4 can be written as

$$D_{34} = I_{41} - C_{31} \tag{7.4b}$$

Equation 7.4b indicates that the directivity is directly related to isolation. As it depends on the cancellation of two wave components, it is a frequency-sensitive parameter. It is very high at the design frequency and should be as high as possible. In practice, it is not possible to build a perfect coupler to cover even a narrow frequency band. Waveguide directional couplers have the best directivity. Directivity is not directly measurable, and is calculated from the difference of the isolation and coupling measurements as given by Equation 7.4b.

Thus, if the signal in the fourth arm is 1% of the coupled signal, the directivity is 10 dBs less than the incident power, and output of the isolated arm is 40 dBs less than the incident power or 30 dBs below the coupled signal. Normally, only three of the four ports are used in microwave circuits and the unused port is terminated in a matched load. The component, thus, looks like a three-port network though it is still a four-port network. Ideally, no power is coupled to port 4; therefore, the directivity is infinite.

Directional couples are extensively used in test setups and for various other purposes in microwave systems. They are ideal for monitoring power levels, frequency, reflection coefficient, noise levels and pulse shapes. They are also used for the connecting equipment in the main and subsidiary lines. Broadband applications include measuring low-level VSWR and tuning waveguide systems. These are used as reflectometer, directional power dividers, variable impedance devices, balanced duplexers and fixed attenuators.

7.7 Waveguide Terminations

Electromagnetic wave is often passed through a waveguide in order to transfer the energy from a source into space. The impedance of a waveguide does not match the impedance of space. An abrupt change in impedance results in standing waves. If there is no proper impedance matching, the presence of standing waves may decrease the efficiency of the system. To avoid the formation of standing waves the change in the impedance at the end of a waveguide has to be gradual. Such gradual change can be obtained by terminating the waveguide with a funnel-shaped horn. The type of horn to be used depends on the frequency and the desired radiation pattern. Horns are simple antennas and have several advantages over other impedance-matching devices, such as their large bandwidth and simple construction. Chapter 9 includes some details about horns.

A waveguide may also be terminated in a resistive load that is matched to the characteristic impedance of the waveguide. The resistive load is often referred to as dummy load. Its only purpose is to absorb all the incident energy without causing standing waves. To terminate a waveguide, several special arrangements are made. One of such methods is to fill the end of the waveguide with a graphite and sand mixture. This arrangement is illustrated in Figure 7.35a. When the wave enters the mixture, it induces a current in the mixture, causing dissipation of energy in the form of heat. Another method (Figure 7.35b) is to use a high-resistance rod located at the centre of the electric field. This field induces current that flows in the rod. The power loss in the high-resistance rod again dissipates in the form of heat.

A wedge of a highly resistive material can also be used for terminating a waveguide. As shown in Figure 7.35c, the plane of the wedge is to be placed

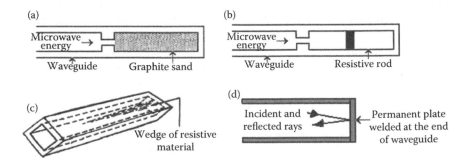

FIGURE 7.35
Terminations of different forms. (a) Waveguide with graphite and sand mixture, (b) waveguide with a high resistance rod, (c) waveguide with a resistive wedge and (d) waveguide with a welded end plate.

perpendicular to the magnetic lines of force. Wherever these *H* lines cut through the wedge, a current will flow in the wedge, which will result in power loss in the form of heat. As this wedge is tapered, very little energy will reach the end of the waveguide and thus there will be minimum reflections. All the above three terminations either radiate or absorb the energy without reflections. In some cases, it is required that all of the energy gets reflected from the end of the waveguide. As shown in Figure 7.35d, this can be easily accomplished by permanently welding a metal plate at the end of the waveguide.

Barring the case of total reflection, some of the microwave circuits require loads of varied nature. As discussed above, some of the loads may completely absorb the incident power. Yet, another class of loads may partially absorb the power incident on it. In case of complete absorption, the load is called *matched load* or *matched termination* and in case of partial absorption, it is termed as *variable load* or *variable termination*. A typical example of termination requirement is a directional coupler in which only three of the four ports are used.

A load and an attenuator are similar but the load is to be located at the end of waveguide, whereas an attenuator may be connected anywhere in a microwave system. Attenuator works by putting carbon vane or flap into the waveguide. The induced current in carbon results in attenuation.

7.7.1 Fixed Matched Termination (or Matched Load)

Matched loads are necessary in the laboratory when other microwave components are being developed and tested. To check the match at a discontinuity, it is necessary that no other source of reflection is present in the circuit. This can be accomplished by placing a matched load beyond the discontinuity to absorb all the power that impinges on it.

A simple form of matched load in a waveguide is a piece of resistive card placed in the guide parallel to the electric field. The front of the card is tapered, so that it presents no sudden discontinuity to the signal. If the card is long enough to absorb almost all the power, the reflection from the far end will be sufficiently small so that the net reflection is negligible.

Any lossy material can be used as a matched load as long as provision is made to avoid reflection from the front end. Materials commonly used for solid loads are lossy dielectrics, loaded with carbon or powdered metal, wood or sand or anything, which is not a good conductor. These are used in numerous test bench set-ups such as in standing wave ratio measurement, as non-reflective termination in a waveguide network and as dummy antennas in low power tests. These may also be used as accurate reproducible termination for directional couplers, magic tee and other similar microwave devices. Commercially available terminations have VSWR better than 1.05. Some commonly used low-power loads are shown in Figure 7.36.

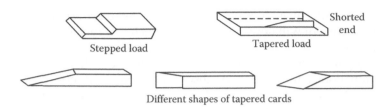

FIGURE 7.36
Low power loads.

7.7.2 Adjustable Terminations or Moving Loads

These consist of a section of a waveguide that is mounted with a low reflection load (Figure 7.37). This load can be moved by an external plunger. The tapered termination block may be moved longitudinally by at least half wavelength at the lowest waveguide frequency. At a particular frequency, the moving load provides a lower VSWR than that obtained by a fixed termination.

7.7.3 Water Loads

In order to measure some properties of liquids such as their dielectric constants, special types of structures are required that not only contain these liquids but can also be connected to the microwave system as a load. These are often referred to as water loads. These can be of non-circulating (Figure 7.38a) and circulating (Figure 7.38b) types. In the non-circulating type, a glass tube attached to the waveguide is filled with the liquid. In the circulating type, provision is made for inlet and outlet of water (or liquid) at appropriate locations. These tubes may be inclined to improve the impedance matching. Besides, these glass tubes may also be tapered for uniform heat dissipation.

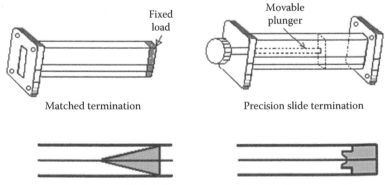

FIGURE 7.37
Waveguide and coaxial matched terminations. (a) Matched termination, (b) precision slide termination and (c) cards in low power coaxial termination.

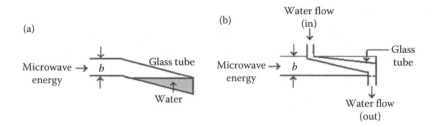

FIGURE 7.38
Water loads. (a) Non circulating and (b) circulating.

7.8 Attenuators

In microwave terminology, any dissipative element inserted in the energy field is called an attenuator. An attenuator may be of fixed or variable type.

7.8.1 Fixed Attenuator

It consists of a resistive or lossy surface called 'card' placed in a waveguide section and adjustable to any position between two narrow walls. This card absorbs a fixed amount of power and transmits the rest. The amount of attenuation is the reduction in power (expressed in dBs). If the power output of an attenuator is one-fourth of the input, it is called the 6 dB attenuator or pad.

The attenuation of a fixed pad varies with the frequency. At higher frequencies, there are more wavelengths in the lossy material and the loss in the material itself is dependent on the frequency. Thus, it is necessary to specify frequency as well as attenuation of a fixed attenuator.

Attenuators are used in the isolation of waveguide circuits, padding of microwave oscillators, power level adjustments and so on. If there is no isolation, the load may produce frequency and power changes in the signal. It may also cause reflections and other undesirable effects and can even damage the oscillator tuning. Two of the fixed attenuators are shown in Figure 7.39.

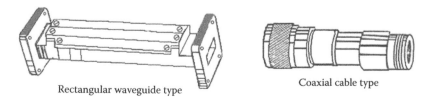

Rectangular waveguide type

Coaxial cable type

FIGURE 7.39
Fixed attenuators. (a) Rectangular waveguide and (b) Coaxial cable type

7.8.2 Variable Attenuators

It is frequently necessary to vary or control the amount of power flowing at a given position in a circuit. Since a piece of resistive card in a waveguide intercepts maximum power when it is at the centre of the guide and practically no power is intercepted when it is nearer to the side wall, it is possible to build a simple variable attenuator making use of this property. The resistive card is supported by two thin horizontal rods, which protrude through small holes in a narrow wall. A knob and gears control the movement of the card from the wall to the centre. The horizontal rods introduce negligible reflection since they are perpendicular to the electric field. To avoid any mismatch, the card is to be properly tapered at both ends. A variable attenuator is shown in Figure 7.40.

For more accurate controlled attenuation, the resistive card is replaced by a piece of glass, which has metalised film on one side. The method of operation is the same but the metalised glass is more stable with time and less sensitive to environmental changes than the resistive card. A calibrated dial may be used to read the required attenuation level.

Figure 7.41 illustrates another type of variable attenuator which is commonly referred to as flap-type attenuator. It contains a movable carbon flap, which is inserted into a slit cut in the top broader wall of waveguide. Its controlled insertion results in the introduction of variable attenuation.

The third type of variable attenuator is called the rotary vane attenuator and is shown in Figure 7.42. The attenuation to be introduced is controlled by the angle of rotation in the manner shown.

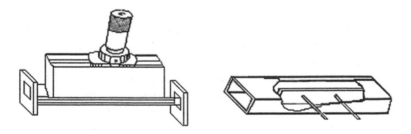

FIGURE 7.40
Waveguide-type variable attenuator with card movement mechanism.

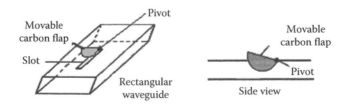

FIGURE 7.41
Flap-type variable attenuator.

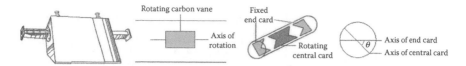

FIGURE 7.42
Rotary vane attenuator.

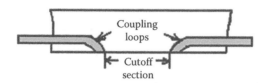

FIGURE 7.43
Coaxial cutoff attenuator.

If E is the amplitude of the input signal, A is the amplitude of output signal, θ is the angle between end cards and the central card and E is parallel to the central card then the output can be given by $A = E \cos^2\theta$. The attenuation (in dBs) = $20 \log_{10}\{E/(E \cos^2\theta)\} = 20 \log_{10}(s^2\theta)$. This attenuation is a function of frequency.

Another class of attenuator called coaxial cutoff attenuator is shown in Figure 7.43. It contains two coupling loops separated by a cutoff section.

The variable attenuators present two problems. First, just as fixed pads, they are frequency sensitive; thus, it is necessary to calibrate them at each operating frequency. In the second place, they produce phase shift with the attenuation variation.

7.9 Impedance Matching

As stated earlier, the impedance of a waveguide is not always matched to their load devices. The mismatch results in standing waves, loss of power and reduction in power-handling capability, and an increase in frequency sensitivity. To avoid the above, some impedance-changing devices are to be placed near the source of standing waves. Such devices include irises or windows, conducting posts and conducting screws. These are described in the following sections.

7.9.1 Inductive Irises

An iris is a metal plate that contains an opening to allow passage to the waves. It is located in the transverse plane. An inductive iris, its equivalent

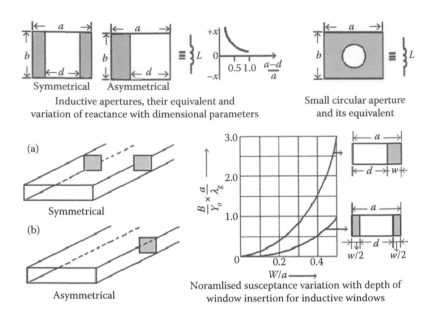

Inductive apertures, their equivalent and variation of reactance with dimensional parameters

Small circular aperture and its equivalent

Noramlised susceptance variation with depth of window insertion for inductive windows

FIGURE 7.44
Inductive irises. (a) Symmetrical and (b) asymmetrical.

circuit and the variation of normalised susceptance with the depth of window insertion are illustrated in Figure 7.44. These windows may be symmetrical or asymmetrical. An iris introduces a shunt inductive reactance across the waveguide that is directly proportional to the size of the opening. The edges of the inductive iris are perpendicular to the magnetic plane.

7.9.2 Capacitive Irises

Figure 7.45 illustrates an iris that introduces a shunt capacitive reactance in the waveguide. These act in the same way as the inductive irises. Again, the reactance is directly proportional to the size of the opening, but in these their edges are perpendicular to the electric plane.

7.9.3 Resonant Windows

The iris illustrated in Figure 7.46 has a portion across the magnetic plane and another segment across the electric plane. This forms an equivalent parallel *LC* circuit across the waveguide. At the resonant frequency, the iris acts as a high shunt resistance. Above or below the resonance, such an iris acts either as a capacitive reactance or as an inductive reactance.

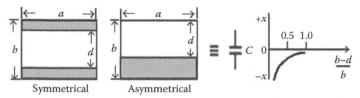

Capacitive aperture, its equivalence and variation of reactance
with dimensional parameters

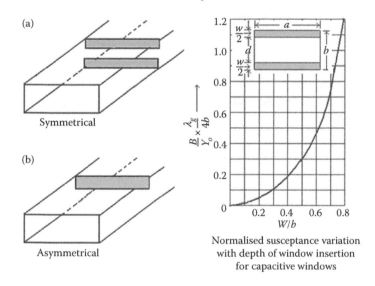

Normalised susceptance variation
with depth of window insertion
for capacitive windows

FIGURE 7.45
Capacitive irises. (a) Symmetrical and (b) asymmetrical.

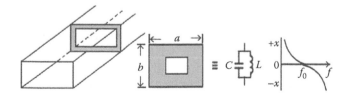

FIGURE 7.46
Parallel resonant diaphragm, its equivalent and reactance variation.

7.9.4 Posts and Screws

These are made of conducting material and can also be used for imped-ance matching. Figure 7.47 illustrates two basic methods of using posts and screws. A post or a screw that only partially penetrates into the waveguide

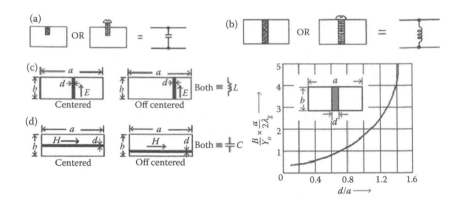

FIGURE 7.47
(a) Penetrating conducting posts and screws, (b) extended conducting posts and screws, (c) vertical thin wire or post and (d) horizontal thin wire or post.

acts as a shunt capacitive reactance. When the same post or screw extends further and touches the bottom wall, it acts as a shunt inductive reactance. Since the variation of insertion in case of screws is more convenient, its reactance can be easily changed.

7.10 Tuners

A microwave system can be tuned by using the following methods.

7.10.1 *EH* Tuner

If the shunt and series arms of the magic tee are both terminated by movable short circuiting plungers, the *EH* tee is referred to as *EH* tuner. A signal entering a side arm is divided equally into two shorted arms but the two halves of the signal are completely reflected. The relative phases of the reflected signals are determined by the position of the plungers. The signals can thus be made to combine so that they produce any desired reflection coefficient in the input side arm. By the same technique, *EH* tuner in front of a mismatch load can be made to produce a reflection that will exactly cancel the one from the load.

EH tuners (Figure 7.48) are used for the adjustment of waveguide systems to tune out residual VSWR of various microwave components such as crystal and bolometer mounts, low-power loads, antennas and other devices, which cause discontinuities in microwave transmission lines. Waveguide tuner is exceptionally useful for tuning out waveguide mismatch. It provides low

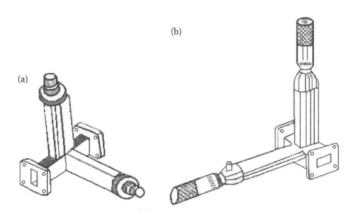

FIGURE 7.48
EH tuners. (a) Conventional type and (b) precision type.

RF leakage and possesses an inherent high-power capacity. In this tuner, VSWR value up to 20 can be reduced to 1.02 with proper adjustment of movable shorts.

7.10.2 Slide Screw Tuner

Slide screw tuners (Figures 7.49 and 7.50) are used for tuning out discontinuities or to match the impedance of components in the waveguide transmission lines. It consists of a slotted section in which a screw may be inserted whose penetration and position can be adjusted and locked.

7.10.3 Coaxial Line Stubs and Line Stretchers

Stub or tuning stub is the most commonly used coaxial line susceptance device. It consists of a section of coaxial line, which is connected in parallel with the main line and short circuited at the far end. In order to vary the susceptance, the stub length can be varied by means of a movable short circuited plunger. The value of susceptance can be varied to become positive

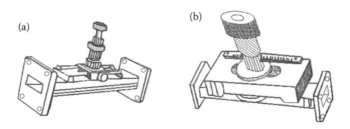

FIGURE 7.49
Slide screw tuners. (a) Conventional type and (b) precision type.

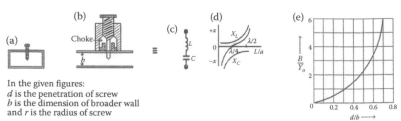

In the given figures:
d is the penetration of screw
b is the dimension of broader wall
and *r* is the radius of screw

Note: Sharpness of the resonance is inversely proportional to diameter of screw.

FIGURE 7.50
Tuning screw details. (a) Tuning screw without choke, (b) tuning screw with half wave transmission line choke, (c) equivalent circuit, (d) variation of susceptance with the ratio of insertion and dia of screw (e) normalised susceptance as a function of screw insertion.

or negative by changing the stub length over a stretch of half wavelength. Since due to physical limitations the stub length cannot be made zero, it is often allowed to exceed the half wavelength limit. The plungers used may be of contacting and non-contacting type. Contacting plunger (Figure 7.51a) remains in contact with inner and outer conductors whereas in a non-contacting plunger, there are gaps between plungers and the two conductors. When matching is to be accomplished by means of single stub only, the distance between the load and the stub needs to be varied. The structure of a variable-length section or line stretcher is shown in Figure 7.51b.

7.10.4 Waveguide Slug Tuner

Matching over a small range of VSWR can be accomplished by using a single quarter-wave dielectric slug (Figure 7.52a), which is adjustable in position and depth into the waveguide. The addition of a second slug (Figure 7.52b) will square the maximum VSWR for which matching is accomplished. Two such slugs have to be half wave distant from centre to centre. Change of slug position without a change in their separation will only vary the phase of the standing wave produced by the tuner. The magnitude of VSWR may

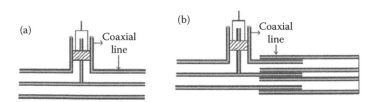

FIGURE 7.51
Coaxial line tuners. (a) Contacting plunger and (b) line stretcher.

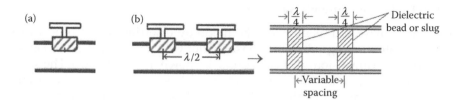

FIGURE 7.52
Waveguide slug tuners. (a) Single slug tuner and (b) double slug tuner.

be adjusted by changing the relative spacing of slugs or its insertion into the waveguide.

7.11 Phase Shifters

Phase shifters are used for matching and for introducing the required phase shift in the passage of waves. It serves the same purpose in waveguides as that of the variable-length section of coaxial line called line stretcher.

7.11.1 Line Stretcher Phase Shifter

One such stretcher is shown in Figure 7.53a. It consists of a waveguide section with longitudinal slots in its broad faces. With the application of pressure on the other faces, the cross section 'a' is reduced. Such reduction results in an increase of phase velocity. This increase in velocity within the adjustable section has the same effect beyond the adjustable section as a decrease of length of the section. Thus, the adjustment of the section produces a change of phase at all points beyond the section. For gradual dimensional change on application of pressure, the slot is made sufficiently long and thus no serious reflections occur.

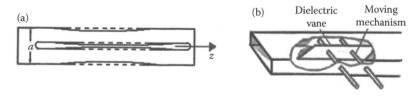

FIGURE 7.53
Phase shifters. (a) Line stretcher type and (b) dielectric vane type.

7.11.2 Dielectric Vane Phase Shifter

A variable phase shifter may also be made by inserting a slab of low-loss dielectric into a waveguide. This phase shifter with a moving mechanism of slab along the dimension '*a*' of waveguide is shown in Figure 7.53b. The spacing of supporting rods and the taper of the slab is chosen to give maximum reflection. The maximum effect of the slab is obtained when it is in the centre of the guide.

7.11.3 Linear Phase Shifter

A linear phase shifter (Figure 7.54) consists of several thin stacked slabs. These stacks, arranged mainly to avoid sudden discontinuities, are divided into moving and stationary segments. The four line lengths (shown) introduce phase shifts β_1, β_2, β_3 and β_4. The net phase shift $\Delta\phi$ due to displacement '*x*' of the central slab can be evaluated by Equation 7.5 provided $(\beta_1 + \beta_3) > (\beta_2 + \beta_4)$, which is true for 0.3*a* separation shown.

$$\Delta\phi = [(\beta_1 + \beta_3) - (\beta_2 + \beta_{4)})]x \tag{7.5}$$

7.11.4 Circular Waveguide Phase Shifter

As shown in Figure 7.55, a circular waveguide phase shifter with a dielectric slab contains a rotating section in the centre and two non-rotating sections at

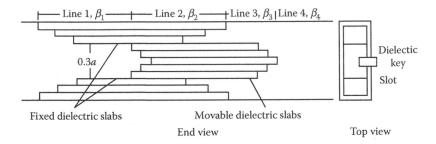

FIGURE 7.54
Linear phase shifter.

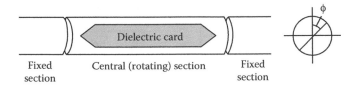

FIGURE 7.55
Phase shifter made of circular waveguide.

the ends. The end sections may be made of rectangular waveguides in conjunction with the rectangular to circular transitions. Adjustment of phase shift is made by rotating the central part. The phase shift in electrical degrees is twice the number of degrees of rotation of the central section. Thus, this type of phase shifter can be calibrated.

7.12 Microwave Filters

At microwave frequencies, the filtering action is to be achieved through waveguide components. A waveguide with dimensions to support the dominant mode in itself is a high-pass filter. For pass-band and other type of filtering, more complex arrangements are needed.

7.12.1 Band-Pass Filter

As shown in Figure 7.56a it contains two irises located at a short distance d and a tuning screw located between these. This distance d determines the centre frequency. Adjustment of tuning screw also provides proper selection of central frequency. The magnitude of susceptance introduced determines the bandwidth of the filter. In case a flat pass-band with steep skirts is not obtained from a single-cavity filter, a number of sections may be cascaded. Spacing between such filter sections may be an odd multiple of quarter wavelength. This odd number in most cases is taken to be three since it minimises the undesirable interaction between cavities and has the overall minimum length. As the number of sections are increased, the pass-band become flatter but at the cost of increased insertion loss.

7.12.2 Low-Pass Filter

As shown in Figure 7.56b, a low-pass coaxial filter may be constructed by inserting a series of capacitance disks into the line. At low frequency, these disks may be considered as shunt capacitances, but at higher frequencies, these act as short lengths of low impedance line.

FIGURE 7.56
Filters. (a) Band pass and (b) low pass.

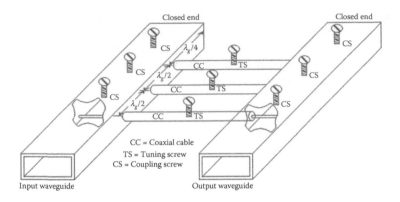

FIGURE 7.57
Parallel resonant filter.

7.12.3 High-Pass Filter

High pass coaxial filters can be formed by using short ($<\lambda/8$) stub lines, which are short circuited and placed at intervals along the line. The configuration of Figure 7.56b can also be made to act as band-pass filter if the short-circuited stub lines are $\lambda/4$ long at the centre of the pass-band.

7.12.4 Parallel Resonant Filter

Figure 7.57 shows a parallel resonant filter in which the input–output waveguides are coupled together by several resonators made of coaxial transmission lines with centre tuning screw for alignment purposes. These coaxial resonators are coupled to input and output guides by means of screws, which distort electric field in the main guides. The coupling depends on whether the screw is inserted into the bottom or top of the guide. The coaxial resonators are placed half wavelength apart to keep them effectively in parallel. The ends of guides are short circuited at quarter wavelength from the last resonator. It keeps the resonators at a high voltage position along the main guide.

7.13 Duplexers

Duplexer finds application mainly in radars and is generally included in the curriculum of radar systems. The following paragraph only briefly describes its purpose.

There is vast difference between the powers to be handled by transmitters and receivers of radars. Even a small fraction of transmitted power may completely damage the receiver in case it leaks into it. Barring Doppler radars where leakage of a small fraction of transmitted power is used as a reference signal, the isolation between transmitters and receivers has to be as perfect as possible. This isolation can be achieved by using either some passive devices or active devices. The passive devices include hybrid junctions (viz. magic tee, rat race tee and short slot couplers), circulators and turnstile junctions. Hybrid junctions provide isolation to the tune of 20–30 dB in normal course and up to 60 dB with extreme precision but have 6 dB inherent loss at junctions on to-and-fro trips of the signal. Circulators provide 20–50 dB isolation without 6 dB inherent loss, and turnstile junctions provide 40–60 dB isolation. Small isolation can also be achieved by using separate polarisations. Isolation up to 80 dB is possible by separating two high gain antennas.

Duplexer falls under the category of active devices, which involve gas discharge tubes, ferrite devices and solid-state varactor diodes. It is normally used to provide a higher order of isolation by alternately switching a radar antenna to transmitter and receiver and thus protects the receiver from burnout or damage during transmission. Its various types include (i) branch-type duplexer using TR (transmit-receive) and ATR (anti-transmit-receive) tubes, (ii) balanced duplexer using ATR tube and (iii) non-reciprocal differential phase shift ferrite duplexers. In these, either TR or ATR tubes are used in conjunction with the short slot directional couplers with a broadband 3 dB coupling. The short slot hybrid junction provides an additional 20–30 dB isolation.

7.14 Diplexers

A diplexer is a circuit that separates signals by frequency. Filters are its essential constituents. It is effectively a three-port network. If two frequency bands are fed as inputs to a diplexer, each frequency band will emerge from a different port. The development of a diplexer and its working is illustrated in Figure 7.58.

7.15 Mode Suppressors

It is always desirable to propagate a single mode through the waveguide since the presence of higher-order modes results in power loss through dissipation. This loss increases with the order of the mode. Besides, the characteristics of waveguide components are adversely affected by their

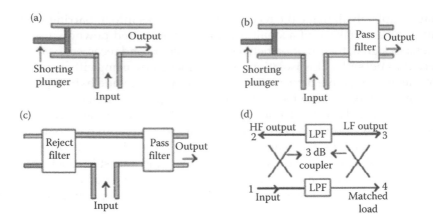

FIGURE 7.58
Diplexer and its formation. (a) Tee with shorting plunger, (b) with pass filter, (c) with reject filter in place of short and (d) balanced diplexer.

presence. Thus, all undesired modes should be effectively suppressed. For dominant mode, the waveguide dimensions are to be so selected that operation beyond cutoff for all but the dominant mode ceases to exist. If the dominant mode is not the desired one, as in rotary joints, other methods need to be employed.

The ideal mode filter will either totally reflect or absorb the undesired modes or totally transmit the desired mode. Conducting wires or sheets mounted parallel to E field of the undesired mode will result in a reflection of that mode. If these wires are normal to the desired mode's E field the mode will be passed unaffected. Thus, radial conducting wires in a circular waveguide (Figure 7.59a) allow the transmission of the TE_{10} mode. The concentric wires shown in Figure 7.59b supported by a thin dielectric allow the transmission of the TM_{01} mode.

Undesired modes can also be prevented if it is possible to cut slots in the waveguide walls in such a manner that the flow of undesired mode currents is prevented without interrupting the flow of current of the desired mode. Thus, longitudinal slots in the middle of (top and bottom) the broader walls of the rectangular guide will not affect the propagation of the TE_{10} mode

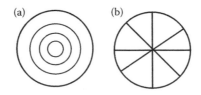

FIGURE 7.59
Mode filters. (a) Suppression of TE_{10} mode and (b) suppression of TM_{01} mode.

but will prevent propagation of any mode (such as TE_{01} mode) that would require currents to cross the slot axis.

EXAMPLE 7.1

Find the coupling coefficient of a directional coupler if the incident power in its main arm is 500 mW and power in its auxiliary arm is 400 μW.

Solution

The following data are given:

Incident power in the main arm $P_{in} = 500$ mW $= 500 \times 10^{-3}$ W
Output power from the auxiliary arm $P_{aux} = 400$ μW $= 400 \times 10^{-6}$ W
The coupling coefficient is given by
$C = 10 \log_{10}(P_{in}/P_{aux}) = 10 \log_{10} (500 \times 10^{-3}/400 \times 10^{-6}) = 10 \log_{10}$
$(1250) = 30.969$ dB

EXAMPLE 7.2

a. Find the coupling coefficient of a directional coupler if the incident power in its main arm is 500 mW and the power in its auxiliary arm is 400 μW.
b. If the incident power in its main arm is kept the same and the coupling coefficient obtained in part (a) is reduced to half, find the power in its auxiliary arm.
c. If the power in the auxiliary arm is halved and the coupling coefficient obtained in part (a) remains the same, find the power in its main arm.

Solution

a. The following data are given:

Incident power in the main arm $P_{in} = 500$ mW $= 500 \times 10^{-3}$ W
Output power from the auxiliary arm $P_{aux} = 400$ μW $= 400 \times 10^{-6}$ W
The coupling coefficient is given by

$$C = 10 \log 10(P_{in}/P_{aux}) = 10 \log 10 (500 \times 10^{-3}/400 \times 10^{-6})$$
$$= 10 \log 10 (1250) = 30.969 \text{ dB}$$

b. The following data are given:

Incident power in the main arm $P_{in} = 500$ mW $= 500 \times 10^{-3}$ W

$$C = 30.969/2 = 15.4845 \text{ dB}$$
$$C = 15.4845 = 10 \log_{10}(P_{in}/P_{aux}) = 10 \log_{10} (500 \times 10^{-3}/P_{aux})$$
$$(500 \times 10^{-3}/P_{aux}) = A \log 1.54845 = 35.3549$$
$$P_{aux} = (500 \times 10^{-3}/35.3549) = 0.01414 \text{ mW}$$

c. The following data are given:
Output power from the auxiliary arm $P_{aux} = 200$ μW $= 200 \times 10^{-6}$ W

$$C = 30.969 \text{ dB} = 10 \log_{10}(P_{in}/P_{aux}) = 10 \log_{10} (P_{in}/200 \times 10^{-6})$$
$$(P_{in}/200 \times 10^{-6}) = \text{Alog } 3.0969 = 1250$$
$$P_{in} = 1250 \times 200 \times 10^{-6} = 250 \text{ mW}$$

EXAMPLE 7.3

In a directional coupler, the output available from auxiliary arm is 200 mW with 15 W input and shorted output arm. This auxiliary input becomes 250 µW with the same power at the output port and shorted input arm. Find the directivity of the coupler.

Solution

P_{aux} (forward) = power coupled to the auxiliary arm when power is fed at input port of the main line = 200 mW = 200×10^{-3} W

P_{aux} (backward) = power coupled to the auxiliary arm when power is fed at output port of the main line = 250 µW = 250×10^{-6} W

The directivity is given by

$$D = 10 \log_{10}(P_3/P_4) = 10 \log_{10}\{P_{aux}(\text{forward})/P_{aux}(\text{backward})\} = 10 \log_{10}$$
$$(200 \times 10^{-3}/250 \times 10^{-6}) = 10 \log_{10} (800) = 29.031 \text{ dB}$$

EXAMPLE 7.4

a. Find the directivity of a directional coupler if the output readings from an auxiliary arm are –25 dBm in the forward direction and –45 dBm in the reverse direction.
b. If the directivity of the above coupler is 25 dB and if output readings from an auxiliary arm are –45 dBm in reverse direction, find the output reading of the auxiliary arm in the forward direction.

Solution

a. Given $\text{dBm}_1 = -25$ and $\text{dBm}_2 = -45$.
   The directivity (dB) in terms of dBm is $\text{dBm}_1 - \text{dBm}_2 = -25 - (-45) = 20$ dB.
b. Given $\text{dBm}_2 = -45$ and $D = 25$ dB.
   Here, $D = \text{dBm}_1 - \text{dBm}_2 = \text{dBm}_1 - (-45) = 25$ dB
   $\text{dBm}_1 - (-45) = 25 - 45 = -20$ dBm

EXAMPLE 7.5

A directional coupler with insertion loss of 0.5 dB is fed 15 mW power at the input port of the main arm. Find the power at the output port of this main line.

Solution

Given the power fed at the input port of main line = $P_1 = 15$ mW = 15×10^{-3} W.

Also given the insertion loss $I = 0.5$ dB. Let the power at the output port of this main line be P_4.

Then, the isolation is given by

$$I = 10 \log_{10}(P_1/P_4) = 10 \log_{10}(15 \times 10^{-3}/P_4) = 0.5$$
$$\log_{10}(15 \times 10^{-3}/P_4) = 0.05$$
$$15 \times 10^{-3}/P_4 = \text{Alog } 0.05 = 1.122$$
$$P_4 = 15 \times 10^{-3}/1.122 = 13.3687 \text{ mW}$$

EXAMPLE 7.6

The input power fed in a directional coupler is 1 mW. Calculate the power at the remaining ports if its directivity is 40 dB and coupling factor is 20 dB.

Solution

Given the power fed at the input port of the main line $= P_1 = 1$ mW $= 10^{-3}$ W
The directivity $D = 40$ dB and coupling factor $C = 20$ dB

$$C = 20 = 10 \log_{10}(P_1/P_3) \quad \log_{10}(10^{-3}/P_3) = 2$$
$$\text{Alog } 2 = 100 = 10^{-3}/P_3 \quad P_3 = 10^{-3}/100 = 10 \text{ μW}$$
$$D = 40 = 10 \log_{10}(P_3/P_4) \quad \log_{10}(P_3/P_4) = 4$$
$$\text{Alog } 4 = 10{,}000 = 10^{-5}/P_4 \quad P_4 = 10^{-5}/10{,}000 = 1 \text{ nW}$$

Power at port 2 (P_2) = Power input (P_1) − (power at coupled port (P_3) + power at isolated port (P_4)}
$$P_2 = 10^{-3} - 10^{-5} + 10^{-9} = 0.9989999 \text{ mW}$$

PROBLEMS

P-7.1 Find the coupling coefficient of a directional coupler if the incident power in its main arm is 300 mW and power in its auxiliary arm is 200 W.

P-7.2 Find the coupling coefficient of a directional coupler if the incident power in its main arm is 300 mW and power in its auxiliary arm is 200 W.

P-7.3 In a directional coupler, the output available from auxiliary arm is 100 mW with 15 W input and shorted output arm. This auxiliary input becomes 200 μW with the same power at the output port and shorted input arm. Find the directivity of the coupler.

P-7.4 a. Find the directivity of a directional coupler if the output readings from an auxiliary arm are −30 dBm in the forward direction and −40 dBm in the reverse direction.

b. If the directivity of the above coupler is 25 dB and if output readings from an auxiliary arm are −55 dBm in the reverse direction, find the output reading of the auxiliary arm in the forward direction.

P-7.5   A directional coupler with insertion loss of 0.3 dB is fed 10 mW power at the input port of the main arm. Find the power at the output port of this main line.

P-7.6   The input power fed in a directional coupler is 5 mW. Calculate the power at the remaining ports if its directivity is 50 dB and coupling factor of 10 dB.

Descriptive Questions

Q-7.1   Discuss the utility of ridged, dielectric loaded, bifurcated, trifurcated and flexible waveguides.

Q-7.2   Discuss the impact of dimensional change in waveguides.

Q-7.3   What are waveguide joints? Discuss their types and relative merits.

Q-7.4   Discuss waveguide bends and corners. How do these differ in terms of performance?

Q-7.5   What purpose is served by the waveguide twists and transitions?

Q-7.6   Discuss different types of input–output coupling methods in waveguides.

Q-7.7   Illustrate different feeding techniques in coaxial to waveguide adopter.

Q-7.8   In view of field distribution, explain the working of E- and H-plane tees.

Q-7.9   Why is EH tee called magic tee? Justify your answer with necessary illustrations.

Q-7.10 Explain the working of a hybrid ring. Why is it called rat race tee?

Q-7.11 Discuss the formation and working of multi-hole directional coupler.

Q-7.12 What are unidirectional and bidirectional couplers? Explain their modes of operation.

Q-7.13 Illustrate the terminations of different forms and explain their working.

Q-7.14 What are water loads and where are these used?

Q-7.15 Explain the working of different types of attenuators. Add necessary figures.

Q-7.16 Discuss types, application and properties of irises.

Q-7.17 What purpose is served by conducting posts and screws?

Q-7.18 Discuss the operation of various types of tuners.

Q-7.19 List the types of phase changers and their applicability in microwave systems.

Q-7.20 Discuss the purpose of diplexers and mode suppressors.

Further Reading

1. Adam, S. F., *Microwave Theory and Applications*. Prentice-Hall, Inc., New Jersey, 1969.
2. Alder, R. B. et al., *Electromagnetic Energy Transmission and Radiation*. MIT Press, Cambridge, MA, 1969.
3. Chatterjee, R., *Microwave Engineering*. East West Press, New Delhi, 1990.
4. Collin, R. E., *Foundations for Microwave Engineering*. McGraw-Hill, New York, 1966.
5. Das, A. and Das, S. K., *Microwave Engineering*. Tata McGraw-Hill, New Delhi, 2002.
6. Gandhi, O. P., *Microwave Engineering and Applications*. Pergamon Press, New York, 1981.
7. Gupta, K. C., *Microwaves*. Wiley Eastern, New Delhi, 1979.
8. Lance, A. L., *Introduction to Microwave Theory and Measurements*. McGraw-Hill, New York, 1964.
9. Lio, S. Y., *Microwave Devices & Circuits*, 3rd ed. Prentice-Hall of India, New Delhi, 1995.
10. Pozar, D. M., *Microwave Engineering*, 2nd ed. John Wiley & Sons, New Delhi, 1999.
11. Reich, H. J. et al. *Microwave Principles*, CBS Publishers and Distributors, New Delhi, 2004.
12. Rizzi, P. A., *Microwave Engineering—Passive Circuits*. Prentice-Hall, New Delhi, 2001.
13. Saad, T. and Hansen, R. C., *Microwave Engineer's Handbook*, Vol. 1. Artech House, Dedham, MA, 1971.
14. Wheeler, G. J., *Introduction to Microwaves*. Prentice-Hall, Englewood Cliffs, NJ, 1963.
15. Young, L., *Advances in Microwaves*. Academic Press, New York, 1974.

8

Scattering Parameters

8.1 Introduction

The scattering (or s) parameters are members of a family of parameters, which include Y, Z, h, g, T and ABCD parameters. At low frequencies, voltages and currents are measurable quantities and the characteristics of a network can be easily described in terms of Z and Y parameters. For the analysis of power frequency transmission lines, ABCD parameters are found to be more convenient. In case of high-frequency electronic circuits, hybrid (h and g) parameters are often used to characterise the behaviour of transistors. At UHF and above (generally beyond 1 GHz), the measurement of voltage (V) and current (I) becomes almost impossible as V and I vary with their physical locations. Besides, the measuring equipment at these frequencies is not readily available. In view of this non-availability, it becomes difficult to achieve short and open circuit conditions over a wide band of frequencies to be handled. The likelihood of instability for some of the active devices such as tunnel diode and power transistors mainly under short circuit condition also makes the measurement of conventional parameters difficult. Thus, the characteristics representation of devices at microwave range requires an entirely different set of parameters that are commonly referred to as scattering (or s) parameters. These differ from other parameters as these do not use open or short circuit conditions to characterise a linear electrical network; instead, the matched loads are used. These terminations are much easier to use at high frequencies than open and short circuit terminations. Moreover, the quantities are measured in terms of power.

Scattering parameters describe the electrical behaviour of linear electrical networks under the influence of electrical signals. These are useful for designing the electrical, electronic and communication systems. Although applicable at any frequency, the s-parameters are mostly used for networks operating at UHF and above, including the microwaves where signal power and energy considerations are more easily quantified than currents and voltages. As these parameters change with the frequency, the frequency needs to be specified for any stated s-parameter measurement, in addition to the characteristic impedance or system impedance. The scattering parameters are readily represented in matrix form and thus are the elements of a scattering (or S) matrix, which obey the rules of matrix algebra.

At microwave range, the wavelengths of the signals are dismally small and the dimensions of the components are drastically reduced. While propagating through such components or combination thereof, even a microscopic aberration in the surface of any device or mismatch at joints in the passage is seen by the wave as an obstacle. The unevenness and mismatch, bends, corners, twists and so on may (often) result in undesirable phenomena of reflection or scattering of energy. The s-parameters that represent the reflection coefficient of a port and transmission coefficients between different ports, when other ports are matched terminated, can account for any such discrepancy. As incident and reflected waves contain phases and amplitudes, s-parameters, which are closely related to these, they are complex quantities.

Scattering parameters are based on transmitted and reflected powers and their estimation does not require short circuit or open circuit. The short circuit and open circuit conditions are equally harmful to the active devices even at microwave frequencies. These parameters can readily be measured with the help of network analyser and some other indirect methods (Chapter 10). Thus, s-parameters are the most appropriate tools that characterise input and output ports of a device without harming the same. More specifically, s-parameters are the power-wave descriptors that permit to define the input–output relations of a network in terms of incident and reflected powers.

To understand the s-parameters, consider a generalised N-port network shown in Figure 8.1. Let a_1, a_2, ..., a_N be the inputs and b_1, b_2, ..., b_N be the outputs at ports 1, 2, ..., N. These inputs and outputs can be related by the matrix of Equation 8.1.

$$\begin{bmatrix} b_1 \\ b_2 \\ b_3 \\ \cdot \\ b_n \end{bmatrix} = \begin{bmatrix} s_{11} & s_{12} & s_{13} & \cdot & s_{1m} \\ s_{21} & s_{22} & s_{23} & \cdot & s_{2m} \\ s_{31} & s_{32} & s_{33} & \cdot & s_{3m} \\ \cdot & \cdot & \cdot & \cdot & \cdot \\ s_{n1} & s_{n2} & s_{n3} & \cdot & s_{nm} \end{bmatrix} \times \begin{bmatrix} a_1 \\ a_2 \\ a_3 \\ \cdot \\ a_m \end{bmatrix} \quad \text{or} \quad [b] = [S][a] \qquad (8.1)$$

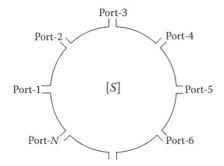

FIGURE 8.1
A generalised N-port network and its scattering matrix.

In Equation 8.1, [*b*] is a column vector of length '*n*', which characterises output ports, [*S*] is an '*n* × *m*' matrix, which contains s-parameters and [*a*] is a column vector of length '*m*', which characterises input ports. To understand the s-parameters that govern the behaviour of different microwave components, it is appropriate to study their properties. These properties are summarised in the following section.

8.2 Properties of Scattering Matrices

8.2.1 Order and Nature

The scattering matrix [*S*] is always a *square matrix* of an *order of* $(N \times N)$, where N represents the number of ports of microwave component or network. This S-matrix satisfies the following properties.

8.2.2 Symmetry Property

It states that if a network satisfies the reciprocity condition, then the s-parameters are equal to their corresponding transpose. To satisfy the reciprocity condition, the microwave network has to be linear and passive and should not contain any active device. Thus, if [*S*] is the scattering matrix and [*S*^T] is its transpose, [*S*] is symmetric provided it satisfies the condition of reciprocity, that is, [*S*] = [*S*^T]. Alternatively, if s_{ij} is an element of [*S*] and s_{ji} is an element of [*S*^T], then $s_{ij} = s_{ji}$. Thus, if a network is reciprocal, its [*S*] matrix is symmetric.

8.2.3 Unity Property

It states that the sum of the products of each term of any one column (or row) of [*S*] matrix with the complex conjugate of that column (or row) is unity.

$$\sum_{i=1}^{N} s_{ij} \cdot s_{ij}^{*} = 1 \quad \text{for } j = 1, 2, 3, \ldots, n \tag{8.2}$$

Since s_{ij} is symmetric for a lossless matched network, power input = power output.

That is, [*S*] [*S**] = [*I*].

8.2.4 Zero Property

The sum of the products of each term of any column (or row) of [*S*] with the complex conjugate of the corresponding terms of any other column (or row) is zero.

$$\sum_{i=1}^{N} s_{ik} \cdot s_{ij}^{*} = 0 \quad \text{for } j = 1, 2, 3, \ldots, n \text{ and } k = 1, 2, 3, \ldots, n \tag{8.3}$$

8.2.5 Phase Shift Property

If any terminal or reference plane is moved away from the junction by an electrical distance $\beta_k\, \ell_k$, then each scattering coefficients s_{ij} involving 'k' will get multiplied by the factor $e^{-\beta_k \ell_k}$.

8.3 Scattering Parameters for Networks with Different Ports

In practice, most of the microwave components have only 2, 3 or 4 ports. An N-port network will emerge only when some of these components are cascaded, which is actually done to form a complete microwave system. The following sub-sections deal with the network containing 2, 3 and 4 ports. The last of these sub-sections shows the configuration and its s-parameters of a cascaded network.

8.3.1 1-Port Network

The s-parameter for a 1-port network is a simple 1×1 matrix of the form $[S_{nn}]$, where n is the allocated port number. In case it is a passive load of some type, it complies the s-parameter definition of linearity.

8.3.2 2-Port Network

Figure 8.2 shows a simple 2-port network and its flow graph containing s-parameters. Such a network may represent a number of components, for example, waveguides, bends, corners, twists, attenuators, gyrators, isolators, phase changers, loads, terminations, slotted line and so on.

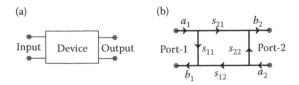

FIGURE 8.2
2-Port network and its flow graph. (a) Simple two-port network and (b) flow graph of a two-port network.

In this figure

- 'a' stands for incident node and a_1, a_2 are its terminals at which the wave incidents. a_1 and a_2 may also be considered as inputs (E contents of the wave) at these nodes.
- 'b' stands for reflected node and b_1, b_2 are its terminals from which the wave reflects. Here, b_1 and b_2 can also be considered as outputs (E contents of the wave) at these nodes.
- s_{11}, s_{12}, s_{21} and s_{22} are the scattering (s) parameters illustrated in the figure.
- In the symbols of the form 's_{mn}', the first subscript represents response or output and the second subscript indicates the input or excitation.

The inputs (a_1, a_2) and outputs (b_1, b_2) can be related to different s-parameters as follows:

$$b_1 = s_{11} a_1 + s_{12} a_2 \tag{8.4a}$$

$$b_2 = s_{21} a_1 + s_{22} a_2 \tag{8.4b}$$

Equation 8.4 can be written in the form of a matrix equation:

$$\begin{bmatrix} b_1 \\ b_2 \end{bmatrix} = \begin{bmatrix} s_{11} & s_{12} \\ s_{21} & s_{22} \end{bmatrix} \times \begin{bmatrix} a_1 \\ a_2 \end{bmatrix} \tag{8.5}$$

If port 2 is terminated in Z_0, there will be no reflection and thus $a_2 = 0$. This gives

$$b_1 = s_{11} a_1 \tag{8.6a}$$

$$b_2 = s_{21} a_1 \tag{8.6b}$$

If port 1 is terminated in Z_0, there will be no reflection and thus $a_1 = 0$. This gives

$$b_1 = s_{12} a_2 \tag{8.6c}$$

$$b_2 = s_{22} a_2 \tag{8.6d}$$

In view of the above conditions, the values of s-parameters can be obtained as follows:

$$s_{11} = \left. \frac{b_1}{a_1} \right|_{a_2=0} = \left. \frac{V_{\text{reflected-at-port1}}}{V_{\text{towards-port1}}} \right|_{a_2=0} \tag{8.7a}$$

$$S_{22} = \left.\frac{b_2}{a_2}\right|_{a_1=0} = \left.\frac{V_{\text{reflected-at-port2}}}{V_{\text{towards-port2}}}\right|_{a_1=0} \tag{8.7b}$$

$$S_{21} = \left.\frac{b_2}{a_1}\right|_{a_2=0} = \left.\frac{V_{\text{outof-port2}}}{V_{\text{towards-port1}}}\right|_{a_2=0} \tag{8.7c}$$

$$S_{12} = \left.\frac{b_1}{a_2}\right|_{a_1=0} = \left.\frac{V_{\text{outof-port1}}}{V_{\text{towards-port2}}}\right|_{a_1=0} \tag{8.7d}$$

In view of Equation 8.7 and Figure 8.3, different s-parameters can now be defined as follows:

- s_{11} is the input reflection coefficient measured when (output) port 2 is terminated by a matched load ($Z_L = Z_0$, Z_0 sets $a_2 = 0$) and ratio of reflected wave and incident wave is measured at port 1.

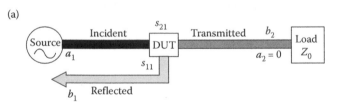

(a)

Load Z_0 is fully absorptive and all power is absorbed by the load; thus, $a_2 = 0$

$$s_{11} = \frac{\text{Reflected}}{\text{Incident}} = \left.\frac{b_1}{a_1}\right|_{a_2=0} \qquad\qquad s_{21} = \frac{\text{Transmitted}}{\text{Incident}} = \left.\frac{b_2}{a_1}\right|_{a_2=0}$$

(b)

Load Z_0 is fully absorptive and all power is absorbed by the load; thus, $a_1 = 0$

$$s_{12} = \frac{\text{Transmitted}}{\text{Incident}} = \left.\frac{b_1}{a_2}\right|_{a_1=0} \qquad\qquad s_{22} = \frac{\text{Reflected}}{\text{Incident}} = \left.\frac{b_2}{a_2}\right|_{a_1=0}$$

FIGURE 8.3
Illustration of meaning of scattering parameters due to device under test. (a) s_{11} and s_{21} and (b) s_{12} and s_{22}.

- s_{22} is the output reflection coefficient measured when (input) port 1 is terminated by a matched load ($Z_s = Z_0$, Z_s sets $a_1 = 0$) and ratio of reflected wave and incident wave is measured at port 2.
- s_{21} is the forward transmission (insertion) gain measured when (output) port 2 is terminated by a matched load ($Z_L = Z_0$ sets $a_2 = 0$) and signal is applied to port 1. It is then the ratio of the signals measured at b_2 and a_1 nodes in terms of voltages or currents.
- s_{12} is the reverse transmission (insertion) gain measured when (input) port 1 is terminated by a matched load ($Z_s = Z_0$ sets $a_1 = 0$) and signal is applied to port 2. It is then the ratio of the signals measured at b_1 and a_2 nodes in terms of voltages or currents.

Also, the square of these parameters are

- $|s_{11}|^2$ = power reflected from network input/power incident on network input
- $|s_{22}|^2$ = power reflected from network output/power incident on network output
- $|s_{21}|^2$ = power delivered to a Z_0 load/power available from Z_0 source = transducer power gain with both load and source having impedance Z_0
- $|s_{12}|^2$ = reverse transducer power gain with Z_0 load and source

In the above equations, either $a_1 = 0$ or $a_2 = 0$ indicates that the port is perfectly matched.

Equivalent circuit for a four-terminal network can be obtained by measuring input impedance (z_i) of the network, when it is terminated in (a) short circuit ($z_L = 0$), (b) open circuit ($z_L = \infty$) and (c) a matched load ($z_L = z_0$). The short circuit and open circuit conditions are obtained by sliding a movable short at output terminals. For the third condition, matched load replaces the movable short.

8.3.3 3-Port Network

A 3-port device includes E-plane and H-plane tees, 3-port circulator, bifurcated waveguide and a directional coupler if one of its ports is perfectly matched. All these result in a 3×3 matrix containing nine scattering parameters.

Figure 8.4a represents s-parameters of a 3-port network and Figure 8.4b shows its line diagram. These parameters marked in the figure can be written in the form of a matrix equation, which is as follows:

$$\begin{bmatrix} b_1 \\ b_2 \\ b_3 \end{bmatrix} = \begin{bmatrix} s_{11} & s_{12} & s_{13} \\ s_{21} & s_{22} & s_{23} \\ s_{31} & s_{32} & s_{33} \end{bmatrix} \begin{bmatrix} a_1 \\ a_2 \\ a_3 \end{bmatrix} \tag{8.8}$$

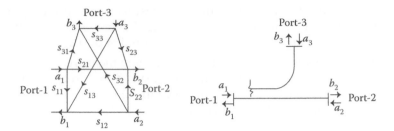

FIGURE 8.4
3-Port network. (a) *S*-parameters for a 3-port network and (b) line diagram of a multihole directional coupler with one port terminated.

8.3.4 *N*-Port Network

An *N*-port network results in an $(n \times n)$ matrix. As an example, a scattering matrix for $n = 4$, that is, a 4-port network obtained in the form of a (4×4) matrix equation as given below. The devices that fall under the category of 4-port network include magic tee, rat race tee, branch coupler, trifurcated waveguide and so on.

$$\begin{bmatrix} b_1 \\ b_2 \\ b_3 \\ b_4 \end{bmatrix} = \begin{bmatrix} s_{11} & s_{12} & s_{13} & s_{14} \\ s_{21} & s_{22} & s_{23} & s_{24} \\ s_{31} & s_{32} & s_{33} & s_{34} \\ s_{41} & s_{42} & s_{43} & s_{44} \end{bmatrix} \times \begin{bmatrix} a_1 \\ a_2 \\ a_3 \\ a_4 \end{bmatrix} \tag{8.9}$$

8.3.5 Cascaded Networks

Since microwave components are not used in isolation and are generally cascaded as shown in Figure 8.5, their scattering matrices are written after ascertaining the behaviour of each individual component. The figure illustrates s-parameters of two cascaded components.

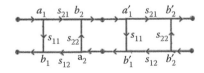

FIGURE 8.5
Two cascaded 2-port network.

8.4 Nature of Networks

Besides the classification in terms of the number of ports, a network can also be categorised as lossless and lossy. These networks can be classified in terms of their properties as follows.

8.4.1 Lossless Network

A lossless network is one that does not dissipate any power. In other words, the sum of the incident powers at all ports is equal to the sum of the reflected powers at all ports. In mathematical terms, $\Sigma |a_n^2| = \Sigma |b_n^2|$. This implies that the s-parameter matrix is unitary, that is, $[S^*] [S] = (I)$, where $[S^*]$ is the conjugate transpose of $[S]$ and $[I]$ is the identity matrix.

8.4.2 Lossy Network

A lossy passive network is one that dissipates power. In this case, the sum of the incident powers at all ports is greater than the sum of the reflected powers at all ports. In mathematical terms, $\Sigma |a_n^2| > \Sigma |b_n^2|$ and $(I) - [S^*] [S]$ is positive definite. The matrices $[S]$, $[S^*]$ and $[I]$ have the same meaning as in the case of lossless networks.

8.5 Types of s-Parameters

The s-parameters in general are described for the networks that operate at about single frequency. The receivers and mixers are not referred to as having s-parameters, although the reflection coefficients at each port can be measured and these parameters can be referred to as s-parameters. The problem arises when the frequency-conversion properties are to be described, which is not possible by using s-parameters.

8.5.1 Small-Signal s-Parameters

Small-signal s-parameters belong to the case when the signals involved are small enough to have only linear effects on the network and the gain compression does not take place. Passive networks, however, act linearly at any power level.

8.5.2 Large-Signal s-Parameters

In case of large signals, the s-parameters will vary with input signal strength and thus makes the analysis complicated.

8.5.3 Mixed-Mode s-Parameters

These refer to a special case of analysing balanced circuits.

8.5.4 Pulsed s-Parameters

These s-parameters are measured on power devices. They allow an accurate representation before heating up of the device.

8.6 Scattering Matrices for Some Commonly Used Microwave Components

This section includes S-matrices for H-plane, E-plane and *EH* tees, directional coupler, hybrid ring circulators, attenuator, gyrator and isolator. Some features of these components have been discussed in Chapters 5 and 7.

8.6.1 H-Plane Tee

Since H-plane tee (Figure 8.6) is a 3-port network, the order of its (3×3) S-matrix is

$$[S] = \begin{bmatrix} s_{11} & s_{12} & s_{13} \\ s_{21} & s_{22} & s_{23} \\ s_{31} & s_{32} & s_{33} \end{bmatrix} \qquad (8.10)$$

In view of the properties of the S-matrix
From symmetry of planes:

$$s_{13} = s_{23} \qquad (8.11)$$

From symmetry property:

$$s_{ij} = s_{ji} \qquad (8.12)$$

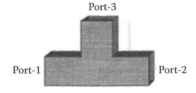

FIGURE 8.6
H-plane tee.

Thus

$$s_{13} = s_{31} \; s_{12} = s_{21} \quad s_{23} = s_{32} = s_{13} \tag{8.13}$$

The refection coefficient $s_{33} = 0$ as port 3 is perfectly matched.
In view of the above properties, $[S]$ matrix can be modified to

$$[S] = \begin{bmatrix} s_{11} & s_{12} & s_{13} \\ s_{12} & s_{22} & s_{13} \\ s_{13} & s_{13} & 0 \end{bmatrix} \tag{8.14}$$

In view of unity property

$$[s_{ij}] \, [s_{ij}^*] = [I] \tag{8.15}$$

$$\begin{bmatrix} s_{11} & s_{12} & s_{13} \\ s_{12} & s_{22} & s_{13} \\ s_{13} & s_{13} & 0 \end{bmatrix} \begin{bmatrix} s_{11}^* & s_{12}^* & s_{13}^* \\ s_{12}^* & s_{22}^* & s_{13}^* \\ s_{13}^* & s_{13}^* & 0 \end{bmatrix} = \begin{bmatrix} 1 & 0 & 0 \\ 0 & 1 & 0 \\ 0 & 0 & 1 \end{bmatrix} \tag{8.16}$$

This leads to the following set of equations:

$$s_{11} \, s_{11}^* + s_{12} \, s_{12}^* + s_{13} \, s_{13}^* = 1 \tag{8.17a}$$

$$s_{12} \, s_{12}^* + s_{22} \, s_{22}^* + s_{13} \, s_{13}^* = 1 \tag{8.17b}$$

$$s_{13} \, s_{13}^* + s_{13} \, s_{13}^* = 1 \tag{8.17c}$$

Using zero property,

$$s_{13} \, s_{11}^* + s_{13} \, s_{12}^* = 0 \tag{8.17d}$$

In view of unitary property, these equations lead to

$$|s_{11}|^2 + |s_{12}|^2 + |s_{13}|^2 = 1 \tag{8.18a}$$

$$|s_{12}|^2 + |s_{22}|^2 + |s_{13}|^2 = 1 \tag{8.18b}$$

$$|s_{13}|^2 + |s_{13}|^2 = 1 \tag{8.18c}$$

From Equation 8.18c,

$$2|s_{13}|^2 = 1 \quad \text{or} \quad s_{13} = \frac{1}{\sqrt{2}} \tag{8.19}$$

Thus, Equations 8.18a and 8.18b reduce to

$$|s_{11}|^2 + |s_{12}|^2 = \frac{1}{2} \tag{8.20a}$$

and

$$|s_{12}|^2 + |s_{22}|^2 = \frac{1}{2} \tag{8.20b}$$

Comparison of Equations 8.20a and 8.20b gives

$$|s_{11}|^2 = |s_{22}|^2 \quad \text{or} \quad s_{11} = s_{22} \tag{8.21}$$

Rewriting Equation 8.17d in the modified form

$$s_{13}\,(s_{11}^* + s_{12}^*) = 0 \tag{8.22a}$$

As

$$s_{13} \neq 0 \quad s_{11}^* + s_{12}^* = 0 \tag{8.22b}$$

This gives

$$s_{11}^* = -s_{12}^* \quad \text{or} \quad s_{11} = -s_{12} \quad \text{or} \quad s_{12} = -s_{11} \tag{8.22c}$$

From Equations 8.20a and 8.22c

$$2|s_{11}|^2 = \frac{1}{2} \quad \text{or} \quad s_{11} = \frac{1}{2} \tag{8.23a}$$

From Equation 8.22c

$$s_{12} = -\frac{1}{2} \tag{8.23b}$$

From Equation 8.21

$$s_{22} = \frac{1}{2} \tag{8.23c}$$

Substitution of values of s_{11} and s_{12} from Equation 8.23 and s_{13} from Equation 8.19 in Equation 8.14 leads to

$$[S] = \begin{bmatrix} 1/2 & -1/2 & 1/\sqrt{2} \\ -1/2 & 1/2 & 1/\sqrt{2} \\ 1/\sqrt{2} & 1/\sqrt{2} & 0 \end{bmatrix} \tag{8.24}$$

In view of Equation 8.8

$$\begin{bmatrix} b_1 \\ b_2 \\ b_3 \end{bmatrix} = \begin{bmatrix} 1/2 & -1/2 & 1/\sqrt{2} \\ -1/2 & 1/2 & 1/\sqrt{2} \\ 1/\sqrt{2} & 1/\sqrt{2} & 0 \end{bmatrix} \begin{bmatrix} a_1 \\ a_2 \\ a_3 \end{bmatrix} \qquad (8.25)$$

From Equation 8.25, we get

$$b_1 = \frac{1}{2}a_1 - \frac{1}{2}a_2 + \frac{1}{\sqrt{2}}a_3 \qquad (8.26a)$$

$$b_2 = -\frac{1}{2}a_1 + \frac{1}{2}a_2 + \frac{1}{\sqrt{2}}a_3 \qquad (8.26b)$$

$$b_3 = \frac{1}{\sqrt{2}}a_1 + \frac{1}{\sqrt{2}}a_2 \qquad (8.26c)$$

As power is fed through port 3, $a_3 \neq 0$, $a_2 = 0$ and $a_1 = 0$

$$b_1 = \frac{1}{\sqrt{2}}a_3 \qquad (8.27a)$$

$$b_2 = \frac{1}{\sqrt{2}}a_3 \qquad (8.27b)$$

and

$$b_3 = 0 \qquad (8.27c)$$

If P_3 is the power fed to port 3 (which corresponds to a_3), the power at port 1 and port 2 will be equally divided, provided these ports are properly terminated by matched loads. These powers will be P_1 and P_2 ($P_1 = P_2$) corresponding to a_1 and a_2.
As

$$P_3 = P_1 + P_2 = 2P_1 = 2P_2 \qquad (8.28)$$

Power gain $= 10 \log_{10} (P_1/P_3) = 10 \log_{10} (P_1/2P_1) = 10 \log_{10} (1/2) = -3$ dB
$$(8.29)$$

8.6.2 E-Plane Tee

E-plane tee (Figure 8.7) is a 3-port network and its (3 × 3) S-matrix is the same as Equation 8.10.

Assume that TE_{10} is the excited mode into E-arm, which corresponds to port 3 of E-plane tee. This port is assumed to be perfectly matched; thus, $s_{33} = 0$. Furthermore, since the outputs at port 1 and port 2 are 180° out of phase, therefore, $s_{23} = -s_{13}$. Equation 8.10 reduces to

$$[S] = \begin{bmatrix} s_{11} & s_{12} & s_{13} \\ s_{21} & s_{22} & -s_{13} \\ s_{31} & s_{32} & 0 \end{bmatrix} \tag{8.30}$$

From symmetry property

$$s_{12} = s_{21} s_{13} = s_{31} \quad \text{and} \quad s_{23} = s_{32} \tag{8.31}$$

Thus, S-matrix for E-plane tee becomes

$$[S] = \begin{bmatrix} s_{11} & s_{12} & s_{13} \\ s_{12} & s_{22} & -s_{13} \\ s_{13} & -s_{13} & 0 \end{bmatrix} \tag{8.32}$$

Using unitary property

$$[s_{ij}] [s_{ij}]^* = [I] \tag{8.33}$$

$$\begin{bmatrix} s_{11} & s_{12} & s_{13} \\ s_{12} & s_{22} & -s_{13} \\ s_{13} & -s_{13} & 0 \end{bmatrix} \begin{bmatrix} s_{11}^* & s_{12}^* & s_{13}^* \\ s_{12}^* & s_{22}^* & -s_{13}^* \\ s_{13}^* & -s_{13}^* & 0 \end{bmatrix} = \begin{bmatrix} 1 & 0 & 0 \\ 0 & 1 & 0 \\ 0 & 0 & 1 \end{bmatrix} \tag{8.34}$$

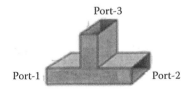

Port-3

Port-1   Port-2

FIGURE 8.7
E-plane tee.

In case of H-plane tee, Equation 8.34 leads to the following set of equations: In view of unitary property, these equations lead to

$$|s_{11}|^2 + |s_{12}|^2 + |s_{13}|^2 = 1 \tag{8.35a}$$

$$|s_{12}|^2 + |s_{22}|^2 + |s_{13}|^2 = 1 \tag{8.35b}$$

$$|s_{13}|^2 + |s_{13}|^2 = 1 \tag{8.35c}$$

From Equation 8.35c

$$|s_{13}|^2 = \frac{1}{2} \quad \text{or} \quad s_{13} = \frac{1}{\sqrt{2}} \tag{8.36}$$

Also, as in the case of H-plane tee, we again get from zero property

$$s_{13} \, s_{11}^* - s_{13} \, s_{12}^* = 0 \tag{8.37}$$

Equations 8.37 and 8.17d are the same.
In view of Equation 8.36, Equations 8.35a and 8.35b can be rewritten as

$$|s_{11}|^2 + |s_{12}|^2 = \frac{1}{2} \tag{8.38a}$$

and

$$|s_{12}|^2 + |s_{22}|^2 = \frac{1}{2} \tag{8.38b}$$

Equating Equations 8.38a and 8.38b to get

$$s_{11} = s_{22} \tag{8.39a}$$

Thus, from Equation 8.37

$$s_{11} = s_{12} \tag{8.39b}$$

In view of the above equivalence, we get from Equation 8.38

$$s_{11} = \frac{1}{2} \tag{8.40}$$

$$[S] = \begin{bmatrix} s_{11} & s_{12} & s_{13} \\ s_{12} & s_{22} & -s_{13} \\ s_{13} & -s_{13} & 0 \end{bmatrix} \tag{8.32}$$

Substitution of $s_{11} = 1/2$, $s_{13} = 1/\sqrt{2}$ and in view of Equations 8.39a and 8.39b, Equation 8.32 leads to

$$[S] = \begin{bmatrix} 1/2 & 1/2 & 1/\sqrt{2} \\ 1/2 & 1/2 & -1/\sqrt{2} \\ 1/\sqrt{2} & -1/\sqrt{2} & 0 \end{bmatrix} \tag{8.41}$$

In view of Equation 8.8
Thus,

$$\begin{bmatrix} b_1 \\ b_2 \\ b_3 \end{bmatrix} = \begin{bmatrix} 1/2 & 1/2 & 1/\sqrt{2} \\ 1/2 & 1/2 & -1/\sqrt{2} \\ 1/\sqrt{2} & -1/\sqrt{2} & 0 \end{bmatrix} \begin{bmatrix} a_1 \\ a_2 \\ a_3 \end{bmatrix} \tag{8.42}$$

From Equation 8.42, we get

$$b_1 = \frac{1}{2} a_1 + \frac{1}{2} a_2 + \frac{1}{\sqrt{2}} a_3 \tag{8.43a}$$

$$b_2 = \frac{1}{2} a_1 + \frac{1}{2} a_2 - \frac{1}{\sqrt{2}} a_3 \tag{8.43b}$$

$$b_3 = \frac{1}{\sqrt{2}} a_1 - \frac{1}{\sqrt{2}} a_2 \tag{8.43c}$$

As power is fed through port 3, $a_3 \neq 0$ and if $a_1 = a_2 = 0$

$$b_1 = \frac{1}{\sqrt{2}} a_3 \tag{8.44a}$$

$$b_2 = -\frac{1}{\sqrt{2}} a_3 \tag{8.44b}$$

and

$$b_3 = 0 \tag{8.44c}$$

Equation 8.44 indicates that an input power fed to port 3 will be equally divided into two halves at port 1 and port 2 but the two halves (b_1 and b_2) will be 180° out of phase, provided these ports are properly terminated by matched loads. Thus, E-plane tee also acts as a 3 dB power splitter.

8.6.3 *EH* Tee

EH tee (Figure 8.8), a 4-port device, results in 4×4 S-matrix, which is written as below:

$$[S] = \begin{bmatrix} s_{11} & s_{12} & s_{13} & s_{14} \\ s_{21} & s_{22} & s_{23} & s_{24} \\ s_{31} & s_{32} & s_{33} & s_{34} \\ s_{41} & s_{42} & s_{43} & s_{44} \end{bmatrix} \qquad (8.45)$$

As it is composed of E-plane and H-plane tees, we get
For H-plane tee

$$s_{23} = s_{13} \qquad (8.46a)$$

and for E-plane tee

$$s_{24} = -s_{14} \qquad (8.46b)$$

In view of its geometry, the input fed at port 3 will not emerge from port 4; thus

$$s_{34} = s_{43} = 0 \qquad (8.46c)$$

As port 3 and port 4 are assumed to be perfectly matched with regard to their input sources, there will be no reflections. This gives

$$s_{44} = s_{33} = 0 \qquad (8.46d)$$

In view of Equations 8.46a through 8.46d, the S-matrix becomes

$$s_{12} = s_{21}, \ s_{13} = s_{31}, \ s_{23} = s_{32}, \ s_{34} = s_{43}, \ s_{24} = s_{42} \quad \text{and} \quad s_{41} = s_{14} \qquad (8.47)$$

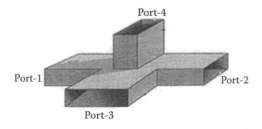

FIGURE 8.8
EH tee.

Thus, in view of Equations 8.46 and 8.47, the *S*-matrix of Equation 8.45 modifies to

$$[S] = \begin{bmatrix} s_{11} & s_{12} & s_{13} & s_{14} \\ s_{12} & s_{22} & s_{13} & -s_{14} \\ s_{13} & s_{13} & 0 & 0 \\ s_{14} & -s_{14} & 0 & 0 \end{bmatrix} \qquad (8.48)$$

Applying unitary property

$$\begin{bmatrix} s_{11} & s_{12} & s_{13} & s_{14} \\ s_{12} & s_{22} & s_{13} & -s_{14} \\ s_{13} & s_{13} & 0 & 0 \\ s_{14} & -s_{14} & 0 & 0 \end{bmatrix} \begin{bmatrix} s_{11}^* & s_{12}^* & s_{13}^* & s_{14}^* \\ s_{12}^* & s_{22}^* & s_{13}^* & -s_{14}^* \\ s_{13}^* & s_{13}^* & 0 & 0 \\ s_{14}^* & -s_{14}^* & 0 & 0 \end{bmatrix} = \begin{bmatrix} 1 & 0 & 0 & 0 \\ 0 & 1 & 0 & 0 \\ 0 & 0 & 1 & 0 \\ 0 & 0 & 0 & 1 \end{bmatrix} \qquad (8.49)$$

Equation 8.49 leads to the following set of equations:

$$|s_{11}|^2 + |s_{12}|^2 + |s_{13}|^2 + |s_{14}|^2 = 1 \qquad (8.50a)$$

$$|s_{12}|^2 + |s_{22}|^2 + |s_{13}|^2 + |s_{14}|^2 = 1 \qquad (8.50b)$$

$$|s_{13}|^2 + |s_{13}|^2 = 1 \qquad (8.50c)$$

$$|s_{14}|^2 + |s_{14}|^2 = 1 \qquad (8.50d)$$

From Equations 8.50c and 8.50d

$$s_{13} = 1/\sqrt{2} = s_{14} \qquad (8.51)$$

Substitution of s_{13} and s_{14} from Equation 8.51 in Equations 8.50a and 8.50b leads to

$$|s_{11}|^2 + |s_{12}|^2 = 0 \qquad (8.52a)$$

and

$$|s_{12}|^2 + |s_{22}|^2 = 0 \qquad (8.52b)$$

Comparing Equations 8.52a and 8.52b, we get

$$s_{11} = s_{22} \qquad (8.53)$$

In view of Equation 8.52

$$s_{11} = s_{12} = s_{22} = 0 \tag{8.54}$$

On substitution of the above values from Equations 8.51 and 8.54 in Equation 8.48

$$[S] = \begin{bmatrix} 0 & 0 & 1/\sqrt{2} & 1/\sqrt{2} \\ 0 & 0 & 1/\sqrt{2} & -1/\sqrt{2} \\ 1/\sqrt{2} & 1/\sqrt{2} & 0 & 0 \\ 1/\sqrt{2} & -1/\sqrt{2} & 0 & 0 \end{bmatrix} \tag{8.55}$$

Thus

$$\begin{bmatrix} b_1 \\ b_2 \\ b_3 \\ b_4 \end{bmatrix} = \begin{bmatrix} 0 & 0 & 1/\sqrt{2} & 1/\sqrt{2} \\ 0 & 0 & 1/\sqrt{2} & -1/\sqrt{2} \\ 1/\sqrt{2} & 1/\sqrt{2} & 0 & 0 \\ 1/\sqrt{2} & -1/\sqrt{2} & 0 & 0 \end{bmatrix} \begin{bmatrix} a_1 \\ a_2 \\ a_3 \\ a_4 \end{bmatrix} \tag{8.56}$$

From Equation 8.56

$$b_1 = \frac{1}{\sqrt{2}}(a_3 + a_4) \tag{8.57a}$$

$$b_2 = \frac{1}{\sqrt{2}}(a_3 - a_4) \tag{8.57b}$$

$$b_3 = \frac{1}{\sqrt{2}}(a_1 + a_2) \tag{8.57c}$$

$$b_4 = \frac{1}{\sqrt{2}}(a_1 - a_2) \tag{8.57d}$$

If $a_3 \neq 0\ a_1 = a_2 = a_4 = 0$, then from Equation 8.57

$$b_1 = \frac{1}{\sqrt{2}} a_3 \tag{8.58a}$$

$$b_2 = \frac{1}{\sqrt{2}} a_3 \tag{8.58b}$$

$$b_3 = b_4 = 0 \tag{8.58c}$$

Equation 8.58 indicates that power fed to port 3 will equally divide into two halves at port 1 and port 2.

If $a_4 \neq 0$, $a_1 = a_2 = a_3 = 0$, then from Equation 7.57

$$b_1 = \frac{1}{\sqrt{2}} a_4 \tag{8.59a}$$

$$b_2 = -\frac{1}{\sqrt{2}} a_4 \tag{8.59b}$$

$$b_3 = b_4 = 0 \tag{8.59c}$$

Equation 8.59 indicates that if power is fed to port 4, it will equally divide into two halves at port 1 and port 2 but these two powers will be 180° out of phase.

8.6.4 Directional Coupler

As directional coupler (Figure 8.9) is a 4-port device, its S-matrix of fourth order is the same as given by Equation 8.45. When ports are properly terminated with matched loads, all its diagonal elements become zero (i.e. $s_{11} = s_{22} = s_{33} = s_{44} = 0$) and the S-matrix reduces to

$$[S] = \begin{bmatrix} 0 & s_{12} & s_{13} & s_{14} \\ s_{21} & 0 & s_{23} & s_{24} \\ s_{31} & s_{32} & 0 & s_{34} \\ s_{41} & s_{42} & s_{43} & 0 \end{bmatrix} \tag{8.60}$$

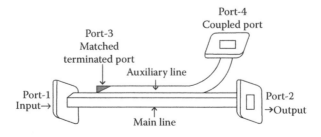

FIGURE 8.9
Multi-hole directional coupler.

As there is no coupling between port 1 and port 3 and between port 2 and port 4, we have

$$s_{13} = s_{31} = 0 \tag{8.61a}$$

and

$$s_{24} = s_{42} = 0 \tag{8.61b}$$

In view of Equation 8.61, Equation 8.60 becomes

$$[S] = \begin{bmatrix} 0 & s_{12} & 0 & s_{14} \\ s_{21} & 0 & s_{23} & 0 \\ 0 & s_{32} & 0 & s_{34} \\ s_{41} & 0 & s_{43} & 0 \end{bmatrix} \tag{8.62a}$$

Furthermore, in view of the symmetric property, Equation 8.62a gets modified to

$$[S] = \begin{bmatrix} 0 & s_{12} & 0 & s_{14} \\ s_{12} & 0 & s_{23} & 0 \\ 0 & s_{23} & 0 & s_{34} \\ s_{14} & 0 & s_{34} & 0 \end{bmatrix} \tag{8.62b}$$

From unity property of S-matrix

$$\begin{bmatrix} 0 & s_{12} & 0 & s_{14} \\ s_{21} & 0 & s_{23} & 0 \\ 0 & s_{32} & 0 & s_{34} \\ s_{41} & 0 & s_{43} & 0 \end{bmatrix} \begin{bmatrix} 0 & s_{12}^* & 0 & s_{14}^* \\ s_{21}^* & 0 & s_{23}^* & 0 \\ 0 & s_{32}^* & 0 & s_{34}^* \\ s_{41}^* & 0 & s_{43}^* & 0 \end{bmatrix} = \begin{bmatrix} 1 & 0 & 0 & 0 \\ 0 & 1 & 0 & 0 \\ 0 & 0 & 1 & 0 \\ 0 & 0 & 0 & 1 \end{bmatrix} \tag{8.63}$$

In view of unity property, the product of the first row of $[S]$ and first column of $[S^*]$ of Equation 8.63 gives

$$|s_{12}|^2 + |s_{14}|^2 = 1 \tag{8.64a}$$

In view of zero property, the product of the first row of $[S]$ and the third column of $[S^*]$ of Equation 8.63 gives

$$s_{12}\, s_{23}^* + s_{14}\, s_{34}^* = 0 \tag{8.64b}$$

In view of zero property, the product of the second row of [S] and the fourth column of [S*] of Equation 8.63 gives

$$s_{12}\, s_{14}^{*} + \left| s_{23}\, s_{34}^{*} \right| = 0 \tag{8.64c}$$

Equation 8.64b can be rewritten as

$$\left| s_{12} \right| \left| s_{23} \right| = \left| s_{14} \right| \left| s_{34} \right| \tag{8.65a}$$

Also, Equation 8.64c can be rewritten as

$$\left| s_{12} \right| \left| s_{14} \right| = \left| s_{23} \right| \left| s_{34} \right| \tag{8.65b}$$

Division of Equation 8.65a by Equation 8.65b gives

$$\left| s_{14} \right| = \left| s_{23} \right| \tag{8.66a}$$

Substitution of Equation 8.66a in Equation 8.65b gives

$$\left| s_{12} \right| = \left| s_{34} \right| \tag{8.66b}$$

For an ideal case, $\left| s_{14} \right| = \left| s_{23} \right|$ can be assumed to be an imaginary positive quantity $(j\alpha)$ and $\left| s_{12} \right| = \left| s_{34} \right|$ as a real positive quantity (β). Further, α and β are assumed to follow the relation $\alpha^2 + \beta^2 = 1$. Thus,

$$\left| s_{14} \right| = \left| s_{23} \right| = j\alpha \text{ (say)} \tag{8.66c}$$

and

$$\left| s_{12} \right| = \left| s_{34} \right| = \beta \text{ (say)} \tag{8.66d}$$

In view of Equations 8.66c and 8.66d, the S-matrix of Equation 8.62a becomes

$$[S] = \begin{bmatrix} 0 & \beta & 0 & j\alpha \\ \beta & 0 & j\alpha & 0 \\ 0 & j\alpha & 0 & \beta \\ j\alpha & 0 & \beta & 0 \end{bmatrix} \tag{8.67}$$

Thus,

$$\begin{bmatrix} b_1 \\ b_2 \\ b_3 \\ b_4 \end{bmatrix} = \begin{bmatrix} 0 & \beta & 0 & j\alpha \\ \beta & 0 & j\alpha & 0 \\ 0 & j\alpha & 0 & \beta \\ j\alpha & 0 & \beta & 0 \end{bmatrix} \begin{bmatrix} a_1 \\ a_2 \\ a_3 \\ a_4 \end{bmatrix} \qquad (8.68)$$

From Equation 8.68, we get

$$b_1 = \beta a_2 + j\alpha a_4 \qquad (8.69a)$$

$$b_2 = \beta a_1 + j\alpha a_3 \qquad (8.69b)$$

$$b_3 = j\alpha a_2 + \beta a_4 \qquad (8.69c)$$

$$b_4 = j\alpha a_1 + \beta a_3 \qquad (8.69d)$$

Equation 8.68 relates the input and output of an ideal directional coupler through its S-matrix. The coupling factor (C) of such a coupler may be given by

$$C = -20 \log \alpha = -10 \log \alpha^2 = -10 \log |s_{14}|^2 = -10 \log |s_{41}|^2 \qquad (8.70a)$$

The directivity (D) of the directional coupler is given as

$$D = 10 \log (P_{\text{aux(f)}}/P_{\text{aux(r)}}) = 10 \log (P_4/P_3) = 10 \log(|s_{41}^2|/|s_{31}^2|) \qquad (8.70b)$$

where $P_{\text{aux(f)}}$ and $P_{\text{aux(r)}}$ are the powers in auxiliary arms in forward and reverse directions, respectively. Proper matching of all the ports of the directional coupler is essential for the validity of the above equations.

8.6.5 Hybrid Ring

A hybrid ring (Figure 8.10) is a 4-port device. Its scattering matrix is given by

$$S = \begin{bmatrix} 0 & s_{12} & 0 & s_{14} \\ s_{21} & 0 & s_{23} & 0 \\ 0 & s_{32} & 0 & s_{34} \\ s_{41} & 0 & s_{43} & 0 \end{bmatrix} \qquad (8.71)$$

8.6.6 Circulators

Circulator is a non-reciprocal (3 or 4 port) device. The two cases are considered below.

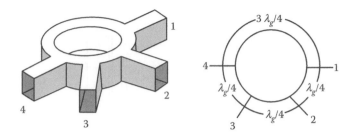

FIGURE 8.10
Hybrid ring.

8.6.6.1 4-Port Circulator

The scattering matrix for a perfectly matched lossless 4-port circulator shown in Figure 8.11 is

$$
S = \begin{bmatrix} 0 & 0 & 0 & s_{14} \\ s_{21} & 0 & 0 & 0 \\ 0 & s_{32} & 0 & 0 \\ 0 & 0 & s_{43} & 0 \end{bmatrix}
\tag{8.72}
$$

8.6.6.2 3-Port Circulator

3-Port (T- or Y-type) circulators are shown in Figure 8.12. Scattering matrices for perfectly matched lossless circulator having clockwise or anticlockwise circulation can be written separately. In clockwise-case, scattering parameters and scattering matrix can be written as

$$
s_{21} = \left. \frac{b_2}{a_1} \right|_{a_2=0, a_3=0}
\tag{8.73a}
$$

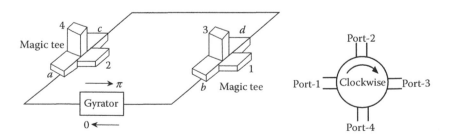

FIGURE 8.11
4-Port circulator.

(a)   (b)

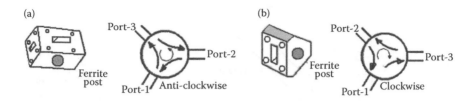

FIGURE 8.12
3-Port circulators. (a) T-type and (b) Y-type.

$$S_{32} = \left.\frac{b_3}{a_2}\right|_{a_1=0, a_3=0} \tag{8.73b}$$

$$S_{13} = \left.\frac{b_1}{a_3}\right|_{a_1=0, a_2=0} \tag{8.73c}$$

$$S = \begin{bmatrix} 0 & 0 & S_{13} \\ S_{21} & 0 & 0 \\ 0 & S_{32} & 0 \end{bmatrix} \tag{8.74}$$

In an ideal 3-port clockwise circulator, the values of surviving scattering parameters are

$$|S_{13}| = |S_{21}| = |S_{32}| = 1 \tag{8.75a}$$

All the other parameters assume zero values, that is

$$S_{11} = S_{12} = S_{22} = S_{23} = S_{31} = S_{33} = 0 \tag{8.75b}$$

For anticlockwise case, scattering matrix is as given below:

$$S = \begin{bmatrix} 0 & S_{12} & 0 \\ 0 & 0 & S_{23} \\ S_{31} & 0 & 0 \end{bmatrix} \tag{8.76}$$

8.6.7 Attenuator

An attenuator shown in Figure 8.13 is a 2-port reciprocal device. Its scattering matrix is

$$\begin{bmatrix} 0 & S_{12} \\ S_{21} & 0 \end{bmatrix} \tag{8.77}$$

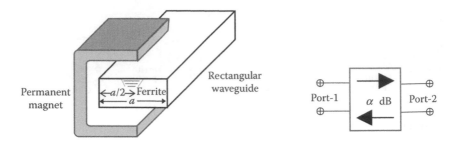

FIGURE 8.13
Attenuator.

Here, $s_{11} = s_{22} = 0$ represents that there is no reflection across the input and output ports. In general, $s_{12} = s_{21}$ = attenuation constant (α).

8.6.8 Gyrator

A gyrator given in Figure 8.14 is a 2-port non-reciprocal device. In the forward direction, it introduces 180° phase shift while in the backward direction, the phase shift introduced is 0°. Its scattering matrix can be written as

$$S = \begin{bmatrix} 0 & -s_{12} \\ s_{21} & 0 \end{bmatrix} \tag{8.78}$$

As expected, there are only two transmission coefficients $-s_{12}$ and s_{21}, which are 180° out of phase. For an ideal case, $s_{12} = 0$ and $s_{21} = \pi$ radians.

8.6.9 Isolator

An isolator (Figure 8.15) is a 2-port non-reciprocal device and its scattering matrix is given as

$$S = \begin{bmatrix} 0 & 0 \\ s_{21} & 0 \end{bmatrix} \tag{8.79}$$

Since it provides very high attenuation in one direction and zero in the other direction, only one of the two transmission coefficients s_{21} survives. In an ideal isolator, $s_{21} = 1$.

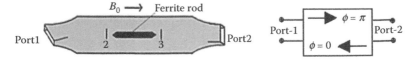

FIGURE 8.14
Gyrator.

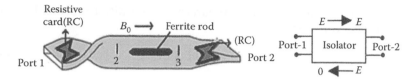

FIGURE 8.15
Isolator.

8.7 Electrical Properties of 2-Port Networks

s-Parameters can be used to express many electrical properties of networks comprising inductive, capacitive and resistive components. These may include gain, returnloss, VSWR, reflection coefficient and amplifier stability. A 2-port network may be reciprocal or non-reciprocal in nature. An amplifier operating under linear (small signal) conditions is a good example of a non-reciprocal network, whereas a matched attenuator is a reciprocal network. Assuming that the input and output connections to 2-port network are at port 1 and port 2, respectively, and further assuming that the nominal system impedance, frequency and any other factors that may influence the device, such as temperature, are fully specified, the various network properties can be given as in the following sub-sections.

8.7.1 Complex Linear Gain

It is simply the voltage gain as a linear ratio of output and input voltages provided all values are expressed as complex quantities. This complex linear gain is given by

$$G = s_{21} \qquad (8.80)$$

8.7.2 Scalar Linear Gain

It is the scalar voltage gain as a linear ratio of output and input voltages. This is given by

$$|G| = |s_{21}| \qquad (8.81)$$

8.7.3 Scalar Logarithmic Gain

It is measured in dBs and may have positive or negative values. In case of positive value, it is simply called a gain, and when negative, it is termed as negative gain or loss (i.e. –1 dB gain means a loss of 1 dB). The scalar logarithmic gain (g) is expressed as

$$g = 20 \log_{10} |s_{21}| dB \tag{8.82}$$

8.7.4 Insertion Loss

This refers to loss caused by a device that is introduced between two reference planes. This extra loss may be due to intrinsic loss in the device and/or mismatch. In case of extra loss, the IL is defined to be positive and is expressed in dBs. In case the two measurement ports use the same reference impedance, it is given as

$$IL = -20 \log_{10} |s_{21}| dB \tag{8.83}$$

8.7.5 Return Loss

Return loss is a positive scalar quantity implying the two pairs of magnitude (| |) symbols. The linear part $|s_{11}|$ is equivalent to the reflected voltage magnitude divided by the incident voltage magnitude.

8.7.5.1 Input Return Loss

It is a scalar measure of how close the actual input impedance of the network is to the nominal system impedance value. It is expressed in logarithmic magnitude. It is given by

$$RL_{in} = |20 \log_{10} |s_{11}|| dB \tag{8.84}$$

8.7.5.2 Output Return Loss

The output return loss has a similar definition as that of the input return loss but applies to the output (port 2) instead of input (port 1). It is given by

$$RL_{out} = |20 \log_{10} |s_{22}|| dB \tag{8.85}$$

8.7.6 Reverse Gain and Reverse Isolation

Scalar logarithmic (dB) expression for reverse gain is given by

$$g_{rev} = 20 \log_{10} |s_{12}| dB \tag{8.86}$$

In case it is expressed as reverse isolation, it becomes a positive quantity and is equal to the magnitude of g_{rev} and the expression becomes

$$I_{rev} = |g_{rev}| = |20 \log_{10} |s_{12}|| dB \tag{8.87}$$

8.7.7 Voltage Reflection Coefficient

The voltage reflection coefficients at input port (P_{in}) and output port (P_{out}) are given by

$$P_{in} = s_{11} \tag{8.88a}$$

and

$$P_{out} = s_{22} \tag{8.88b}$$

Since s_{11} and s_{22} are complex quantities, so are P_{in} and P_{out}. As voltage reflection coefficients are complex quantities, these may be represented on polar diagrams or Smith chart.

8.7.8 Voltage Standing Wave Ratio

The VSWR at a port, represented by 'ρ', is a similar measure of port match to return loss but is a scalar linear quantity. It relates to the magnitude of the voltage reflection coefficient and hence to the magnitude of either s_{11} for the input port or s_{22} for the output port.

At the input port, the VSWR is given by

$$\rho_{in} = \{(1 + |s_{11}|)/(1 - |s_{11}|)\} \tag{8.89a}$$

At the output port, the VSWR is given by

$$\rho_{out} = \{(1 + |s_{22}|)/(1 - |s_{22}|)\} \tag{8.89b}$$

The above relations are correct for reflection coefficients with magnitudes less than or equal to unity, which is usually the case. In case of tunnel diode amplifier, the magnitude of the reflection coefficient is greater than unity. In such a situation, these expressions will result in negative values. VSWR, however, from its definition, is always positive. A correct expression for port k of a multi-port network is given by

$$S_k = \{(1 + |s_{kk}|)/(1 - |s_{kk}|)\} \tag{8.90}$$

where S_k may represent ρ_{in} or ρ_{out}.

8.8 *s*-Parameters and Smith Chart

Any 2-port s-parameter may be located on a Smith chart using polar coordinates. The most meaningful representation belongs to those parameters (e.g., s_{11} and s_{22}) that can be converted into an equivalent normalised impedance (or admittance).

A 2-port network along with different s-parameters, input and output voltages, voltages at other nodes, and dependent and independent variables are illustrated in Figure 8.16. Also, Equation 8.5 contains independent variables a_1 and a_2 and dependent variables b_1 and b_2. The variables a_1, a_2, b_1, b_2, normalised incident voltages and the normalised reflected voltages are defined as

$$a_1 = \frac{V_1 + I_1 Z_0}{2\sqrt{Z_0}} = \frac{\text{voltage wave incident on port 1}}{\sqrt{Z_0}} = \frac{V_{i1}}{\sqrt{Z_0}} \tag{8.91a}$$

$$a_2 = \frac{V_2 + I_2 Z_0}{2\sqrt{Z_0}} = \frac{\text{voltage wave incident on port 2}}{\sqrt{Z_0}} = \frac{V_{i2}}{\sqrt{Z_0}} \tag{8.91b}$$

$$b_1 = \frac{V_1 - I_1 Z_0}{2\sqrt{Z_0}} = \frac{\text{voltage wave reflected from port 1}}{\sqrt{Z_0}} = \frac{V_{r1}}{\sqrt{Z_0}} \tag{8.91c}$$

$$b_2 = \frac{V_2 - I_2 Z_0}{2\sqrt{Z_0}} = \frac{\text{voltage wave reflected from port 2}}{\sqrt{Z_0}} = \frac{V_{r2}}{\sqrt{Z_0}} \tag{8.91d}$$

The square of the magnitudes of independent and dependent variables indicate the following:

$|a_1|^2$ = power incident on input of the network = power available from source impedance Z_0

$|a_2|^2$ = power incident on the output of the network = power reflected from the load

$|b_1|^2$ = power reflected from the input port of the network = power available from a Z_0 source minus the power delivered to the input of the network

$|b_2|^2$ = power reflected from the output port of the network = power incident on the load = power that would be delivered to a Z_0 load

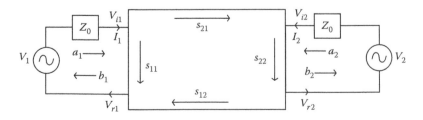

FIGURE 8.16
2-Port network showing different parameters.

While a and b represent transmission and reflection of power or voltage at the input and output of the 2-ports, the s-parameter coefficients defined by Equation 8.7 can be modified to

$$S_{11} = \left.\frac{b_1}{a_1}\right|_{a_2=0} = \left.\frac{V_{\text{reflected-at-port1}}}{V_{\text{towards-port1}}}\right|_{a_2=0} \tag{8.92a}$$

$s_{11} =$ input reflection coefficient with the output port terminated by a matched load

$$S_{22} = \left.\frac{b_2}{a_2}\right|_{a_1=0} = \left.\frac{V_{\text{reflected-at-port2}}}{V_{\text{towards-port2}}}\right|_{a_1=0} \tag{8.92b}$$

$s_{22} =$ output reflection coefficient with the input port terminated by a matched load

$$S_{21} = \left.\frac{b_2}{a_1}\right|_{a_2=0} = \left.\frac{V_{\text{outof-port2}}}{V_{\text{towards-port1}}}\right|_{a_2=0} \tag{8.92c}$$

$s_{21} =$ forward transmission (insertion) gain with the output port terminated by a matched load

$$S_{12} = \left.\frac{b_1}{a_2}\right|_{a_1=0} = \left.\frac{V_{\text{outof-port1}}}{V_{\text{towards-port2}}}\right|_{a_1=0} \tag{8.92d}$$

$s_{12} =$ reverse transmission (insertion) gain with the input port terminated by a matched load Squaring the s-parameters gives the following:

$|s_{11}|^2 =$ Power reflected from network input/Power incident
on network input (8.93a)

$|s_{22}|^2 =$ Power reflected from network output/Power incident
on network output (8.93b)

$|s_{21}|^2 =$ Power delivered to a Z_0 load/Power available from Z_0 source
$=$ Transducer power gain with both load and source having
an impedance of Z_0 (8.93c)

$|s_{12}|^2 =$ Reverse transducer power gain with Z_0 load and source (8.93d)

If Z_S and Z_L are the source and load impedances, respectively, and Z_0 is the reference (or characteristic) impedance, which is always defined to be positive and real, the s-parameters can be related to reflection coefficient, load impedance and so on of a transmission system.

It is to be noted that at very high frequencies, lumped elements no longer look lumped but instead begin to look like some kind of distributed elements. In addition, the other (2 port) parameters (Z, Y, ABCD, etc.) rely on the ability to create either an ideal short or an ideal open (or both) at one of the ports in order to measure the parameter. At high frequencies, this option is not available as inductance prevents the creation of a complete short, and capacitance prevents a complete open circuit.

The reflection coefficient (Equation 2.78) applied to a transmission line terminated in Z_L is

$$\Gamma_L = (Z_L - Z_0)/(Z_L + Z_0) \tag{8.94}$$

Γ_L ranges from -1 to $+1$ (when the load is either infinity or zero, respectively). When $Z_0 = Z_L$, $\Gamma_L = 0$ and there are no reflections allowing maximum power to be delivered to the load. In view of Equations 8.92 through 8.94, the following relations are obtained:

$$S_{11} = \frac{b_1}{a_1} = \frac{(V_1/I_1) - Z_0}{(V_1/I_1) + Z_0} = \frac{Z_1 - Z_0}{Z_1 + Z_0} = \Gamma_{in} \tag{8.95a}$$

$$S_{22} = \frac{b_2}{a_2} = \frac{(V_2/I_2) - Z_0}{(V_2/I_2) + Z_0} = \frac{Z_2 - Z_0}{Z_2 + Z_0} = \Gamma_{out} \tag{8.95b}$$

where V_1, V_2 are the voltages, I_1, I_2 are the currents, Z_1, Z_2 are the impedances and Γ_{in}, Γ_{out} are the reflection coefficients at port 1 and port 2, respectively. Using these values of s_{11} and s_{22}, the expressions for VSWR at the input and output ports are

$$VSWR\big|_{in} = \frac{1 + |s_{11}|}{1 - |s_{11}|} \tag{8.96a}$$

and

$$VSWR\big|_{out} = \frac{1 + |s_{22}|}{1 - |s_{22}|} \tag{8.96b}$$

As the Smith chart is developed for studying transmission lines, it can perfectly display s_{11} and s_{22}, which represent the transmission line reflection coefficients.

The rearrangement of the above relations gives the input and output impedances in terms of s_{11}, s_{22} and Z_0. These impedances are given as follows:

$$z_1 = z_0 \frac{1 + s_{11}}{1 - s_{11}} \tag{8.97a}$$

and

$$z_2 = z_0 \frac{1 + s_{22}}{1 - s_{22}} \tag{8.97b}$$

In view of the above relations, s-parameters are related to the parameters that form the basis of the Smith chart (viz. Γ, z, ρ, etc.). Thus, s-parameters can be mapped on the Smith chart.

8.9 Scattering Transfer (or *T*) Parameters

This is another class of scattering parameters and is referred to as scattering transfer parameters or simply *T*-parameters. Like s-parameters, *T*-parameters also have complex values. For a 2-port network, *T*-parameter matrix is closely related to the corresponding s-parameter matrix. In terms of the incident and reflected normalised waves at input and output ports, the *T*-parameter matrix can be given by two alternative relations as below:

$$\begin{pmatrix} b_1 \\ a_1 \end{pmatrix} = \begin{pmatrix} T_{11} & T_{12} \\ T_{21} & T_{22} \end{pmatrix} \begin{pmatrix} a_2 \\ b_2 \end{pmatrix} \tag{8.98a}$$

or

$$\begin{pmatrix} a_1 \\ b_1 \end{pmatrix} = \begin{pmatrix} T_{11} & T_{12} \\ T_{21} & T_{22} \end{pmatrix} \begin{pmatrix} b_2 \\ a_2 \end{pmatrix} \tag{8.98b}$$

T-parameters have an advantage over s-parameters as these can be readily used to determine the effect of cascading of 2 or more 2-port networks by simply multiplying the associated individual *T*-parameter matrices. If *T*-parameters of three different 2-port networks (1, 2 and 3) are given as (T_1), (T_2) and (T_3), then the *T*-parameter matrix (T_T) for the total cascaded network is given as

$$(T_T) = (T_1)\,(T_2)\,(T_3) \tag{8.99}$$

The conversion relations between S- and T-parameters for 2-port networks are given as follows.

8.9.1 Conversion from S-Parameters to T-Parameters

$$T_{11} = -\text{Determinant}(S)/S_{21} \tag{8.100a}$$

$$T_{12} = S_{11}/S_{21} \tag{8.100b}$$

$$T_{21} = -S_{22}/S_{21} \tag{8.100c}$$

$$T_{22} = 1/S_{21} \tag{8.100d}$$

8.9.2 Conversion from T-Parameters to S-Parameters

$$S_{11} = T_{12}/T_{22} \tag{8.101a}$$

$$S_{12} = \text{Determinant}(T)/T_{22} \tag{8.101b}$$

$$S_{21} = 1/T_{22} \tag{8.101c}$$

$$S_{22} = -T_{21}/T_{22} \tag{8.101d}$$

EXAMPLE 8.1

Find the power delivered through each port of a lossless H-plane tee when power input at one of the collinear port 1 is 30 mW and other two ports are matched terminated.

Solution

As power is fed at port 1 and port 2 and port 3 are matched terminated, $a_2 = a_3 = 0$.
  In view of Equation 8.13,

$$S_{13} = S_{31} \quad S_{12} = S_{21} \quad S_{23} = S_{32} = S_{13}$$

$S_{13} = 1/\sqrt{2}$ (8.19) $S_{11} = 1/2$ (8.23a) $S_{12} = -1/2$ (8.23b)

The effective power at port 1: $P_1 = |a_1|^2 (1 - |s_{11}|^2) = 30 (1 - 0.5^2)$
                                                                         $= 22.5$ mW.
Power transmitted to port 3: $P_3 = |a_1|^2 |s_{31}|^2 = 30 (1/\sqrt{2})^2 = 15$ mW.
Power transmitted to port 2: $P_2 = |a_1|^2 |s_{21}|^2 = 30 (-1/2)^2 = 7.5$ mW
  Thus, $P_1 = P_2 + P_3$.

EXAMPLE 8.2

Obtain the S-matrix for an 8 dB directional coupler having 40 dB directivity. The directional coupler is assumed to be lossless with VSWR $= 1$ at its all ports under matched conditions.

Solution

Given $C = 8$ dB, $D = 40$ dB and VSWR $= 1$.

The coupling factor is given by $C = -10 \log (P_1/P_4)$

$$C = 8 = -10 \log (P_1/P_4) \quad \text{or} \quad P_1/P_4 = A\log (-0.8) \quad \text{or} \quad P_1/P_4 = 0.1585 = |s_{41}|^2$$

$$s_{41} = \sqrt{(0.1585)} = 0.3981$$

The directivity is given by $D = 10 \log (P_4/P_3)$

$$40 = 10 \log (P_4/P_3) \quad \text{or} \quad P_4/P_3 = A\log (4) \quad \text{or} \quad P_4/P_3 = 10^4 = |s_{41}|^2/|s_{31}|^2$$

$$|s_{41}|/|s_{31}| = \sqrt{10^4} = 100 \quad s_{31} = |s_{41}|/100 = 0.3981/100 = 0.00398 = s_{13}$$

Since VSWR $= 1$ and $s_{11} = (\text{VSWR} - 1)/(\text{VSWR} + 1) = 0$.

As ports are properly terminated with matched loads, $s_{11} = s_{22} = s_{33} = s_{44} = 0$.

Also, in view of symmetry, $s_{13} = s_{31}$ $s_{12} = s_{21}$ $s_{23} = s_{32}$ $s_{42} = s_{24}$ and $s_{43} = s_{34}$

$$[S] = \begin{bmatrix} 0 & s_{12} & s_{13} & s_{14} \\ s_{12} & 0 & s_{23} & s_{24} \\ s_{13} & s_{23} & 0 & s_{34} \\ s_{14} & s_{24} & s_{34} & 0 \end{bmatrix}$$

Assuming that 1 and 2 are the main line ports and 3 and 4 are the auxiliary line ports and power P_1 is fed at port 1. If the power at ports 2, 3 and 4 is assumed to be P_2, P_3 and P_4, these can be related as

$$P_1 = P_2 + P_3 + P_4 \quad \text{or} \quad 1 = P_2/P_1 + P_3/P_1 + P_4/P_1 = |s_{21}|^2 + |s_{31}|^2 + |s_{41}|^2$$

$$1 = |s_{21}|^2 + (0.00398)^2 + (0.3981)^2$$

$$|s_{21}|^2 = 1 - (0.00398)^2 + (0.3981)^2 = 0.84151639 \quad s_{21} = 0.9173 = s_{12}$$

In view of identity property,

$$s_{12}^2 + s_{23}^2 + s_{24}^2 = 1 \quad s_{23}^2 + s_{24}^2 = 1 - s_{12}^2 = 1 - (0.9173)^2 = 0.15848$$

$$s_{13}^2 + s_{23}^2 + s_{34}^2 = 1 \quad s_{23}^2 + s_{34}^2 = 1 - s_{13}^2 = 1 - (0.00398)^2 = 0.99998$$

$$s_{14}^2 + s_{24}^2 + s_{34}^2 = 1 \quad s_{24}^2 + s_{34}^2 = 1 - s_{14}^2 = 1 - (0.398)^2 = 0.841596$$

From these equations,

$$s_{24}^2 - s_{34}^2 = 0.15848 - 0.99998 = -0.8415 \quad \text{and} \quad s_{24}^2 + s_{34}^2 = 0.841596$$

From this set of equations, $s_{24} = 0.00693$ and $s_{34} = 0.9174$

$$s_{23}^2 + s_{34}^2 = 0.99998 \quad \text{or} \quad s_{23}^2 = 0.99998 - s_{34}^2 = 0.99998 - (0.9174)^2 \quad \text{or}$$

$$s_{23} = 0.398$$

The final S-matrix can be written as

$$[S] = \begin{bmatrix} 0 & 0.9173 & 0.00398 & 0.398 \\ 0.9173 & 0 & 0.398 & 0.00693 \\ 0.00398 & 0.398 & 0 & 0.9174 \\ 0.398 & 0.00693 & 0.9174 & 0 \end{bmatrix}$$

EXAMPLE 8.3

Find the scattering coefficients of a matched non-ideal isolator, which has 0.4 dB isolation and 20 dB insertion loss.

Solution

The insertion loss (IL) and isolation (I) in terms of s-parameters are given as

$$IL = -10 \log|s_{21}|^2 = -20 \log|s_{21}| \quad 0.4 = -20 \log|s_{21}| \quad \text{or} \quad -0.02 = \log|s_{21}|$$
$$\text{or } s_{21} = 0.955$$

$$I = -10 \log|s_{12}|^2 = -20 \log|s_{12}| \quad 20 = -20 \log|s_{12}| \quad \text{or} \quad -1 = \log|s_{12}| \quad \text{or}$$
$$s_{12} = 0.2584$$

PROBLEMS

P-8.1 Find the power delivered through each port of a lossless E-plane tee when power input at one of the collinear port 1 is 20 mW and other two ports are matched terminated.

P-8.2 Obtain an S-matrix for 10 dB directional coupler having 25 dB directivity. This coupler is assumed to be lossless with VSWR = 1 at its all ports under matched conditions.

P-8.3 Find the scattering coefficients of a non-ideal isolator, which has 0.5 dB isolation and 30 dB insertion loss.

Descriptive Questions

Q-8.1 What are scattering parameters? How do s-parameters differ from other parameters used to analyse electrical and electronic networks?

Q-8.2 Discuss the various properties of scattering matrices.

Q-8.3 Discuss the s-parameters of a 2-port network and explain the meaning of each parameter.

Q-8.4 Obtain the *S*-matrix for H-plane tee.

Q-8.5 Explain the terms complex linear gain, scalar linear gain and scalar logarithmic gain.

Q-8.6 How are insertion loss, input return loss and output return loss related to *s*-parameters?

Q-8.7 Write the expressions for voltage reflection coefficient and VSWR in terms of *s*-parameters.

Q-8.8 How are *s*-parameters and Smith chart related?

Further Reading

1. Alder, R. B. et al., *Electromagnetic Energy Transmission and Radiation*. MIT Press, Cambridge, MA, 1969.
2. Baden Fuller, A. J. *An Introduction to Microwave Theory and Techniques*, II edn. Pergamon Press, Sao Paulo, 1979.
3. Bockelman, D. E. and Eisenstadt, W. R. Combined differential and common-mode scattering parameters: Theory and simulation. *MTT, IEEE Transactions* 43(7) part 1–2, 1530–1539, July 1995.
4. Chatterjee, R., *Microwave Engineering*. East West Press, New Delhi, 1990.
5. Collin, R. E., *Foundations for Microwave Engineering*, McGraw-Hill Book Company, New York, 1966.
6. Das, A. and Das, S. K., *Microwave Engineering*. Tata McGraw-Hill, New Delhi, 2002.
7. Gandhi, O. P., *Microwave Engineering and Applications*. Pergamon Press, New York, 1981.
8. Choma, J. and Chen, W. K., *Feedback Networks: Theory and Circuit Applications*. World Scientific, Singapore, 2007.
9. Rollett, J. M., Stability and power-gain invariants of linear two-ports, *IRE Transactions on Circuit Theory*, CT-9, 29–32, March 1962.
10. Kurokawa, K., Power waves and the scattering matrix, *IEEE Transactions MTT*, 194–202, March 1965.
11. Lance, A. L., *Introduction to Microwave Theory and Measurements*. McGraw-Hill, New York, 1964.
12. Lio, S. Y., *Microwave Devices & Circuits*, 3rd ed. Prentice-Hall, New Delhi, 1995.
13. Mavaddat. R., *Network Scattering Parameter*. World Scientific, Singapore, 1996.
14. Pozar, D. M., *Microwave Engineering*, 3rd edn. John Wiley & Sons, Inc., New Delhi, 2009.
15. William, E., Bob, S. and Bruce, T., *Microwave Differential Circuit Design Using Mixed-Mode S-Parameters*. Artech House, USA, 2006.

9

Microwave Antennas

9.1 Introduction

Antennas (or aerials) are devices that radiate (or couple) electromagnetic energy into space. Their effectiveness depends on their ability to concentrate (or direct) the energy into the desired direction. An efficient antenna is one in which very little or no energy is lost in the radiation process. The transmitting and receiving antennas have similar characteristics and in many applications, a single antenna is used to perform both the functions. As most of the radars and communication systems operate in microwave range, the antennas used in these applications are referred to as microwave antennas. Since the basic principles of operation of microwave antennas are similar to those used at lower frequencies, their basic characteristic parameters are also similar. The design of antennas relating to their sizes, shapes and types is mainly governed by the operating frequency, space occupied, weight and some other practical considerations. The electrical characteristics of antennas derived from the antenna theory are summarised in Section 9.2.

9.2 Antenna Theorems and Characteristic Parameters

9.2.1 Antenna Theorems

The three antenna theorems that correlate the properties of receiving and transmitting antennas, derived from network theorems, are as follows.

- *Equality of directional patterns:* The directional pattern of a receiving antenna is identical with its directional pattern as a transmitting antenna.

- *Equivalence of transmitting and receiving antenna impedances:* The impedance of an isolated antenna when used for receiving is the same as when it is used for transmitting.

- *Equality of effective length:* The effective length of an antenna for receiving is equal to its effective length as a transmitting antenna.

9.2.2 Antenna Characteristic Parameters

The characterising parameters of an antenna are briefly explained below.

Antenna efficiency: It is the ratio of total power radiated by the antenna to the total power fed to the antenna. This total power fed is the sum of radiated power and power loss. It is also called total efficiency. Since only a part of the power is radiated whereas the other part is lost due to the conditional status of the antenna and due to dielectric losses, the total efficiency becomes the sum of radiation efficiency, conditional efficiency and the dielectric efficiency.

Beam efficiency: It is the ratio of the main beam area to the total beam area.

Radiation pattern: It is the graphical representation of E (or H) field pattern of radiation from an antenna as a function of direction or space coordinates. The radiation pattern of an antenna may be very narrow (pencil beam), very wide in one of the planes (fan beam), omni shaped, shaped beam or a tilted beam type. In the case of array antennas, the radiation patterns may have a single beam or multiple beams, some of which may be electronically steered. When a radiation pattern is defined in terms of E, it is called E pattern and when it is defined in terms of H, it is called H pattern. These patterns are also termed as radiation density patterns or spatial variation of E field or H field. Power (radiation) pattern is the trace of received power per unit solid angle.

Beamwidth: It is the measure of directivity of an antenna or the angular measure of *half-power beamwidth* that is measured between the directions in which the antenna is radiating half of the maximum power. Also, the *beamwidth between the first two nulls* is the angular measurement between the directions radiating no power.

Solid angle (beam area): It is the angle subtended by an area expressed in steradians or square degrees.

Radiation intensity: It is the power radiated by an antenna per unit solid angle.

Gain: It is the measure of squeezed/concentrated radiations in a particular direction with respect to an omni (point source) antenna. *Power gain* is the ratio of radiation intensity in a given direction to the average total power. *Directive gain* is the ratio of power density in a particular direction at a given point to the power that would be radiated at the same distance by an omnidirectional antenna. *Directivity* is defined as the maximum directive gain.

Radiation resistance: It is the fictitious resistance, which when substituted in series with an antenna, will consume the same power as is actually radiated by the antenna.

Effective length: It is the ratio of induced voltage at the terminal of the receiving antenna under an open-circuit condition to the incident electric field strength.

Effective (collecting) aperture: It is the ratio of power radiated in watts to the Poynting vector (P) of the incident wave. The effective aperture accounts for the captured/collected power. The other two aperture terms accounting for scattered power and power loss in an antenna are referred as *scattering aperture* and the *loss aperture.*

Polarisation: It refers to the direction in which the electric field vector **E** is aligned during the passage of at least one complete cycle.

Antenna (equivalent noise) temperature: It is the fictitious temperature at the input of an antenna, which would account for noise ΔN at the output. ΔN is the additional noise introduced by the antenna itself.

Signal-to-noise ratio (SNR): It is the criterion of the detectability of a signal. It is applicable to any network representing a device or a system. If a network is given an input possessing certain characteristics and this signal emerges with some alterations (i.e. with reduced magnitude and/or phase deviation), it is presumed that these alterations are due to the noise introduced by the network itself. Therefore, the SNR is the measure of the quality of network vis-à-vis its imperfections. The concept of the SNR is equally applicable to an antenna since it may also introduce noise in the signal.

Front-to-back ratio: It is the ratio of power radiated in the desired direction to the power radiated in the opposite direction.

Driving point or terminal impedance: It is the impedance measured at the input terminals of an antenna. *Self-impedance* is the impedance of an isolated antenna, that is, when there is no other antenna in its vicinity. *Mutual impedance* is the measure of coupling between two antennas.

Antenna bandwidth: It is the range of frequency over which the antenna maintains its certain required characteristics, namely, gain, radiation resistance, polarisation, front-to-back ratio, standing wave ratio, impedance and so on.

It is to be noted that there are no hard-and-fast rules for choosing an antenna for a particular frequency range or application. Many electrical factors (characterising parameters discussed above), structural factors (shape, size, weight and appearance) and cost are the deciding factors. Besides, there is no difference in the referred factors relating to the transmitting and receiving antennas.

9.3 Types of Microwave Antennas

Theoretically, any antenna can be used in any frequency range. The commonly employed antennas at microwave range are discussed in the subsequent sections.

9.3.1 Reflector Antennas

Reflector antennas may comprise simple reflecting surfaces. These may be classified as active and passive reflectors. A passive reflector simply reflects the electromagnetic energy impinging on it and has no source or feed in its vicinity. These may be flat or of parabolic shapes. An active reflector is composed of a reflector and a feed. It modifies the radiation pattern of the feed. Active reflectors include corner reflectors and various versions of parabolic reflectors. The gain of corner reflectors is about 14 dB while of parabolic reflectors, it is more than 20 dB. The design of reflectors is relatively simple as they obey simple geometric laws of optics. These are frequently used in communication and radars.

Microwaves travel in a straight line and can be focussed and can get reflected just like light rays. While travelling, these form a wave front that is generally of spherical nature. As it moves away from the antenna, the energy spreads out in all directions in view of the nature of the wave front. Thus, the pattern produced by such a wave front is not very directional. For an antenna to be directive, a spherical wave front must change to a plane wave front. This task is performed by a reflector that transforms a spherical wave front into a plane wave front. A parabolic reflector is one of the most commonly used antennas through which high directivity can be obtained. Figure 9.1 illustrates a reflector with a microwave source placed at its focal point. The field leaves the feed as a spherical wave front, each part of which reaches the reflecting surface, with 180° phase shift. Each part of this field is then sent outwards at an angle that makes all parts of the field to travel in parallel paths. In view of the shape of a parabolic surface, all paths from F to the reflector and back to plane XY are of the same length. Therefore, the reflected field from the parabolic surface travels to XY plane at the same time.

The source at focus 'F' can be of many forms. In the case of a dipole, energy will be radiated from the antenna into space as well as towards the reflector. The energy directed into space will have a wide-beam pattern whereas that directed towards the reflector will form a narrow pattern. Away from the reflector, the patterns will get superimposed and the narrow pattern of

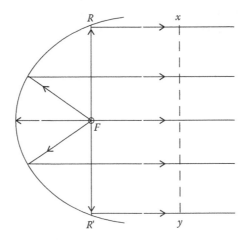

FIGURE 9.1
Reflection mechanism for a parabolic reflector.

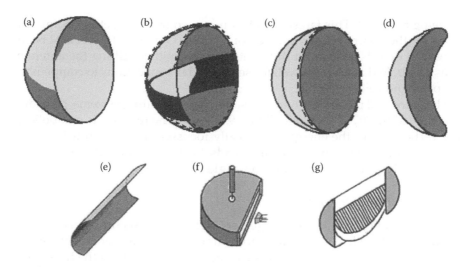

FIGURE 9.2
(a) Paraboloidal reflector, (b) truncated paraboloid (in horizontal plane), (c) truncated parabo-loid (in vertical plane), (d) orange peeled paraboloid, (e) cylindrical paraboloid, (f) pill box and (g) cheese antenna.

the parabolic reflector will get modified. To avoid such situation, some form of shield (usually hemispherical) may be used to direct most of the energy towards the parabolic surface. Thus, if direct radiation into space is somehow eliminated, the beam will be sharper with more concentration of power. The radiation pattern of a paraboloid reflector contains a major lobe and some minor lobes. The major lobe is directed along the axis of revolution. Very narrow beams are possible with this type of reflector.

Figure 9.2 illustrates the basic paraboloid reflector and its several versions used to produce different beam shapes required by special applications. The basic characteristics of these paraboloids are described in the following sections.

9.3.1.1 Paraboloidal Reflector

Figure 9.2a shows a conventional and the most commonly used paraboloidal reflector. It has a three-dimensional curved surface generated by rotating a parabola about its own axis. It is normally fed by a waveguide horn and generates a pencil beam. It bears different names such as a parabolic dish, dish reflector, microwave dish, parabolic reflector or simply a dish antenna.

9.3.1.2 Truncated Paraboloid

Figure 9.2b shows a truncated paraboloid. Since the reflector is parabolic in the horizontal plane, the energy is focussed into a narrow beam. As the reflector is truncated (vertically), the beam will spread out vertically instead of being focussed. This fan-shaped beam is used in radar detection applications for the accurate determination of bearing. Owing to the vertical beam spread, it detects aircrafts at different altitudes without tilting the antenna. This paraboloid also works well for search radar applications to compensate for pitch and roll of the ship.

The truncated paraboloid may be used in height-finding systems, provided it is rotated by 90°, as shown in Figure 9.2c. Since the reflector is parabolic in the vertical plane, the energy is vertically focussed into a narrow beam. In the horizontal direction, the beam spreads. Such a fan-shaped beam is used to accurately determine the elevation angle.

9.3.1.3 Orange-Peel Paraboloid

As shown in Figure 9.2d, an orange-peel reflector is a section of a complete circular paraboloid. Since the reflector is narrow in the horizontal plane and wide in the vertical plane, it produces a beam that is wide in the horizontal plane and narrow in the vertical plane. In its shape, the beam resembles a huge beaver tail. It is fed by a horn radiator. It intercepts almost the entire energy radiated by the horn and no energy escapes from its edges. These paraboloids are often used in height-finding equipment.

9.3.1.4 Cylindrical Paraboloid

A cylindrical paraboloidal section is used to generate a beam that is wider in one dimension than in the other. Figure 9.2e illustrates such an antenna. The cross section of a parabolic cylinder is just along one dimension that makes this reflector to be directive only in one plane. Such a reflector has a focal line instead of a single focal point and thus is fed by a linear array of dipoles, a slit on the side of the WG or by a thin WG radiator. If the radiator or an array of radiators is placed along this focal line, a directed beam is produced. Different beam shapes can be obtained by changing the width of the parabolic section. This type of antenna system is used in search radars and in ground-control approach systems.

9.3.1.5 Pillbox

Figure 9.2f shows a typical cylinder, called pillbox that is short in axial direction and is provided with conducting end plates. It can be simply fed by a probe or by extending the inner conductor of a coaxial cable to the space between the plates. It can also be fed by a WG horn or by an open-ended WG. Both of these feeds are illustrated in the figure. This antenna can be used to generate a fan beam.

9.3.1.6 Cheese Antenna

Figure 9.2g illustrates a cheese antenna, which is a combination of a pillbox and a parabolic cylinder. Another version of cheese antenna called parabolic

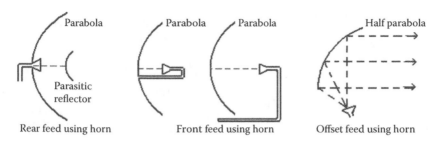

FIGURE 9.3
Different feed arrangements. (a) Rear feed, (b) front feed and (c) offset feed.

torus is generated by moving a parabolic contour over an arc of a circle whose centre is the axis of the parabola. It is illuminated by a moving feed and is used in surveillance radars with <120° scan angle.

Reflectors are generally fed by using point sources that are termed as feeds. A point source produces a spherical wave that is converted into a plane wave by the reflector. The feed may be a horn, a dipole or a helix. A feed may be regarded as a point source if it is located quite far from the reflector. Figure 9.3 shows different feed systems termed as front feed, rear feed and offset feed.

Parabolic structures may have continuous or discontinuous surfaces made of wire screen or perforated metal. The discontinuous surfaces have certain inherent advantages such as low wind pressure, light weight, low cost, easy shaping, easy fabrication and easy assembly. Their negative aspects include energy leakage, back lobe formation, decreased efficiency and increased lobe grating.

9.3.1.7 Spherical Reflector

In view of symmetry, these antennas have wider scanning angle than a paraboloid. A simple spherical reflector, however, does not produce an equiphase radiation pattern and thus, these have poor radiation patterns. The radiation pattern gets further degraded due to surface aberrations. These aberrations can be minimised by employing a reflector of sufficiently large radius so that a portion of the sphere can be approximated to a paraboloid, compensating these aberrations with special feeds or correcting lenses and by approximating a sphere by a stepped paraboloid.

9.3.1.8 Cassegrain Antenna

Figure 9.4 illustrates a Cassegrain antenna containing a parabolic reflector, a hyperbolic subreflector and a feed. As a rule of the thumb, the larger the subreflector the nearer it will be to the main reflector and shorter will be the axial dimension of the antenna assembly. But a larger subreflector results

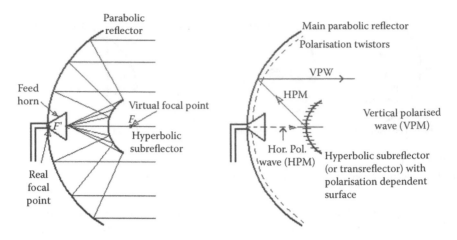

FIGURE 9.4
Cassegrain antenna.

in larger aperture blocking. Thus, a compromise in size and distance of the subreflector vis-a-vis the main reflector is to be made. To reduce aperture blocking, the subreflector can be made of horizontal grating of wires. Such a subreflector called transreflector passes a vertically polarised wave with negligible attenuation and reflects a horizontally polarised wave radiated by the feed. The main parabolic reflector is often coated with polarisation twisters (also called twist reflectors), each of which is equivalent to a quarter-wave plate. These produce 90° rotation of the plane of polarisation. The horizontally polarised wave reflected by the subreflector is rotated by the twist reflectors at the surface of the main dish and is thus transformed into a vertically polarised wave that passes through the subreflector with negligible attenuation. Such antennas are widely used in telescopes and monopulse tracking. These permit reduction in axial dimensions of the antenna, provide greater flexibility in the design of the feed system and eliminate the need for long transmission lines.

9.3.1.9 Corner Reflectors

A corner-reflector antenna comprises a half-wave radiator and a reflector. The reflector consists of two flat metal surfaces meeting at an angle immediately behind the radiator. It gives unidirectional radiation pattern. The construction of a corner reflector is shown in Figure 9.5. To obtain horizontal polarisation, these antennas are mounted in the horizontal position. In such cases, the radiation pattern is very narrow in the vertical plane, with maximum signal being radiated in line with the bisector of the corner angle. The directivity in the horizontal plane is approximately the same as for any half-wave radiator having a single-rod-type reflector behind it. If the antenna is mounted in the vertical position, as shown in Figure 9.5,

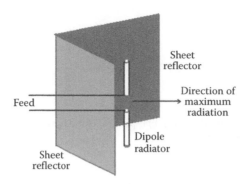

FIGURE 9.5
Corner-reflector antenna.

maximum radiation will be in a very narrow horizontal beam. Radiation in the vertical plane will be the same as for a similar radiator with a single-rod-type reflector behind.

9.3.2 Horn Antennas

A horn is a modified rectangular or circular WG. The mouth of a rectangular WG may be flared along a broader wall, narrower wall or both resulting in different forms of horns. Similarly, flaring of the opening of a circular WG will result in another form of horn. As the WG appears on the scene only at UHF and higher frequencies, fabrication of horn antennas at low frequencies is not practical.

Owing to its shape, the horn has good directional characteristics. The energy emitted from the flared end gets concentrated in a beam and effectively increases the power output in comparison to an unconcentrated beam. A horn may be used as an actual antenna, a standard for comparison, a feed to lenses and reflectors or as a pickup device for measuring the received energy from a source. One of the most common applications of the horn is the calibration of gain of other antennas of unknown characteristics. The standard gain horn is rated in dBs that indicates its effectiveness as compared to an isotropic antenna. Antennas used as feeds have a moderate gain of about 20 dBi. The standard gain horn's characteristics are described in terms of beam width in degrees and power gain is described in dBs. Both these characteristics vary with frequency as its gain usually increases with frequency. The specified gain is usually at some nominal frequency, or at the midfrequency of the operating range.

A pyramidal horn consists of a WG joined to a pyramidal section fabricated from a brass sheet. The variation in gain for this horn is about 1.5 dB over its entire band of operation. Horns may have different types in accordance with their shapes and sizes, some of which are illustrated in Figure 9.6.

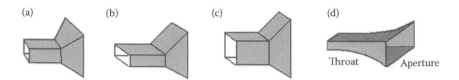

FIGURE 9.6
Rectangular horns. (a) Pyramidal horn, (b) E-plane sectoral horn, (c) *H*-plane sectoral horn and (d) exponentially tapered horn.

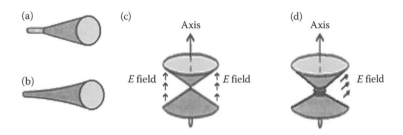

FIGURE 9.7
Circular WG horns. (a) Conical horn, (b) exponentially tapered horn, (c) TEM biconical horn and (d) TE10 biconical horn.

Similar to parabolic reflectors, horns can also be used to obtain directive radiation at microwave frequencies. Because they do not use resonant elements, horns have the advantage of being useful over a wide frequency band. Horn radiators are used with WGs because they serve both as an impedance-matching device and as a directional radiator. Horn radiators may be fed by coaxial and other types of lines. These are constructed in a variety of shapes, as illustrated in Figure 9.7. The shape of the horn determines the shape of the field pattern. The ratio of the horn length to the size of its mouth determines the beam angle and the directivity. In general, the larger the mouth, the more directive is the field pattern.

There are many versions of horn antennas. These include (i) pyramidal horns, (ii) sectoral *H*-plane and sectoral *E*-plane horns, (iii) conical and biconical horns, (iv) box horn, (v) hog horn, (vi) pointed WG and rounded WG horns, (vii) WG with a disc, (viii) WG with a dielectric, (ix) symmetrical or asymmetrical bevelled horn and (x) symmetrical or asymmetrical horn.

9.3.3 Slot Antennas

Figure 9.8 shows various shapes of slot antennas. Figure 9.9 illustrates slots cut in flat sheets, curved sheets, curved surface of a cylinder and WG walls. Figure 9.10 shows different forms of feed systems for slot antennas. Figure 9.11 represents vertically and horizontally polarised slots. The lengths of all slots are of the order of $\lambda/2$ and their widths are a small fraction of λ. A slot antenna

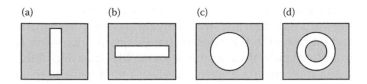

FIGURE 9.8
Shapes of slot antennas. (a) Vertical slot, (b) horizontal slot, (c) circular slot and (d) annular slot.

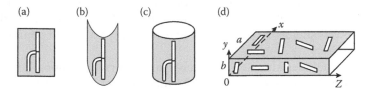

FIGURE 9.9
Slot in different surfaces. (a) Flat sheet, (b) curved sheet, (c) cylinder and (d) waveguide.

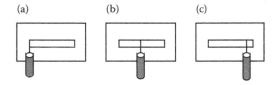

FIGURE 9.10
Different feed systems. (a) End feed, (b) center feed and (c) offset feed.

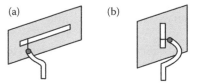

FIGURE 9.11
(a) Vertically and (b) horizontally polarised slots.

exhibits many of the characteristics of a conventional dipole antenna. When arranged in arrays, a high degree of directivity can be obtained. Their beams can scan a volume of space by changing either the frequency or phase of the signal driving the antenna elements.

9.3.4 Lens Antennas

Lenses can be used for radio waves just as for light. The structure of a lens antenna is a bit complex than that of the reflector antenna. Their gain is also 1 or 2 dB less than that of reflectors. These, however, can scan wider angles and their surface tolerance is more lenient than that of reflectors. Lenses have less rearward radiation and are relatively less lossy. These can be easily designed and fabricated in the desired shapes. Small microwave lenses can be constructed from polyethylene, polystyrene, Plexiglas and Teflon.

Lens antenna too converts a spherical wave into a plane wave in a given direction by using a feed that can be considered as a point source. This point source may again be an open-ended WG, a horn radiator or a simple dipole. The point source directs the electromagnetic energy towards the lens. All the radial segments of a spherical wave front are forced into parallel paths by a collimating lens. The effective lenses become small enough to be practical at microwaves. Fresnel lens reduces the size by using a stepped configuration called zoned lens.

Figure 9.12 includes almost all the necessary information about lens antennas. Accordingly, a lens antenna may be made of natural or artificial dielectrics. The dielectric lenses are normally wideband. The refractive index (n) of the dielectrics may be less than or greater than 1. These are referred to as dispersive ($n < 1$) and non-dispersive ($n > 1$) lenses. The bandwidth of the dispersive lens is much higher than that of the non-dispersive lens. As a rule of the thumb, if 'n' increases, the thickness of the lens and hence weight and size decreases. But a mismatch between the lens and free space increases and energy loss due to reflections increases. Thus, a compromised value of n is normally between 1.5 and 1.6.

Another classification of lenses is based on the number of surfaces encountered by the rays. It may be only one surface or the ray may come across two surfaces. In case of one surface, it may be the near/adjacent surface or the farther surface vis-à-vis the focal length. Although the two surface lenses are not frequently used they give better performance, there is no refocussing of energy into the feed and has a wide angle scanning ability for $\varepsilon_r \approx 2.5$.

Yet another important classification of lenses relates to the nature of their surfaces that may be continuous or discontinuous (zoned). The solid homogeneous

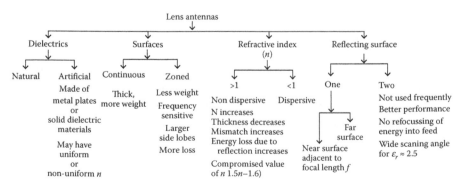

FIGURE 9.12
Classification of lens antennas.

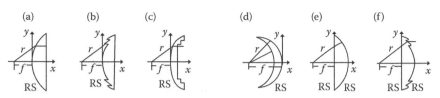

RS = Reflecting surface

FIGURE 9.13
(a) One surface, near end, (b) one surface, near end zoned, (c) one surface, far end zoned, (d) one surface, far end, (e) two surfaces and (f) two surfaces far end zoned.

dielectric lenses are heavier in weight and thicker in size. Stepping or zoning reduces their weight and thickness. The zoning is to be made in such a way that the optical path length through each of the zone is one wavelength less than the next outer zone. The zoning makes the lens frequency sensitive. Besides, due to zoning, there is loss of energy, increase in the side lobe level and the shadowing effect. Despite these drawbacks, zoned lenses are preferred due to their less weight and smaller size. Figures 9.13 and 9.14 illustrate some of the lenses with continuous and discontinuous surfaces for $n > 1$ and $n < 1$, respectively.

As in the case of reflector antennas, lens antennas also require feeds. Figure 9.15 shows some of the lenses along with their feeds. This figure also gives values of n and phase velocities.

In cases where the refraction occurs from the far (opposite) surface, lens surface is everywhere perpendicular to the incident ray and thus, the ray passes through this surface without refraction. When such an antenna is used as a transmitting antenna, the non-refracting surface lies in the constant wave front. Consequently, a reflected wave from this surface will converge at the focal point 'F' and will give rise to a reflection coefficient in the feed line that is approximately the same as the reflection coefficient of the surface. In view of the above, even moderate VSWR will often adversely affect the stability of the tube. To prevent the reflected signal from entering the source, the methods used are illustrated in Figure 9.16. These methods effectively reduce the VSWR and the secondary pattern of antennas remains

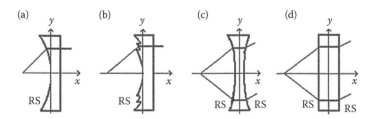

FIGURE 9.14
(a) One (near) surface, (b) one (near) surface zoned, (c) two (curved) surfaces and (d) two (flat) surfaces.

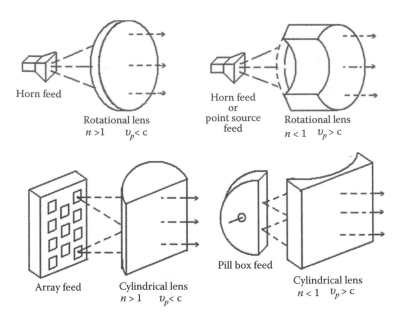

FIGURE 9.15
Lenses with their feeds, values of n and phase velocities (v_p).

by and large unaffected. These methods, however, do not eliminate other effects, namely, the loss of gain and increase in side lobe levels and demand some surface-matching techniques.

In view of the above discussion, different types of lenses can be constructed, a few of which are described below.

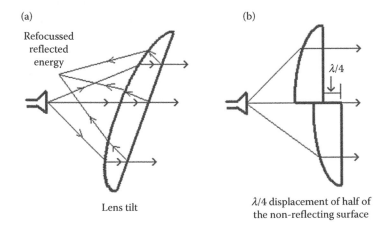

FIGURE 9.16
Prevention of reflected energy from entering the source by (a) tilted lens and (b) two displaced half lenses.

9.3.4.1 WG-Type Lens

The WG-type lens is sometimes referred to as a conducting lens. It consists of several parallel concave metallic strips that are placed parallel to the electric field of the radiated energy fed to the lens, as shown in Figure 9.17. These strips, displaced by slightly more than a half-wavelength, act as WGs in parallel for the incident wave.

The radiated energy consists of an infinite number of radial sections (rays). Each of these contains mutually perpendicular *E* and *H* lines and both are perpendicular to the direction of travel. Since each of these radial sections travels in a different direction, the point source, in itself, has poor directivity. The purpose of the lens is to convert the input spherical microwave segment (that consists of all the radial sections) into parallel (collimated) lines in a given direction at the exit side of the lens. The focussing action of the lens is accomplished by the refracting qualities of the metallic strips. The collimating effect of the lens is possible because the velocity of electromagnetic energy propagation through metals is greater than its velocity through air. Because of the concave construction of the lens, wave fronts arriving near the ends of the lens travel farther in the same amount of time than do wavefronts at the centre. Thus, the wave front emerging from the exit side of the lens appears as a plane wave. It consists of an infinite number of parallel sections (with both the *E* and the *H* field components) mutually perpendicular to the direction of travel.

9.3.4.2 Delay Lenses

Figure 9.18 illustrates the delay and non-delay construction. Another delay lens of convex form that is constructed of dielectric material is shown in

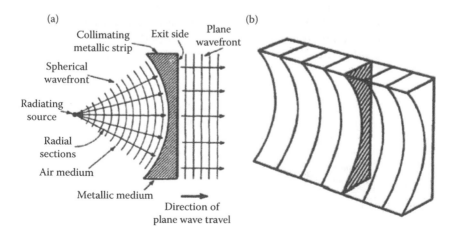

FIGURE 9.17
Parallel plate lens. (a) 2D view and (b) 3D view.

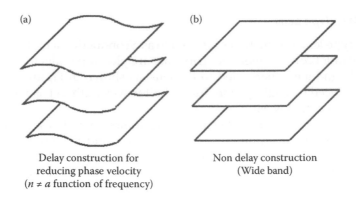

(a)

(b)

Delay construction for
reducing phase velocity
($n \neq a$ function of frequency)

Non delay construction
(Wide band)

FIGURE 9.18
Lenses. (a) Delay type and (b) non-delay type.

Figure 9.19. This lens delays or slows down the propagating wave as it passes through the lens. The phase delay in the passing wave is determined by the refractive index that is related to the dielectric constant ($n = \sqrt{\varepsilon}$) of the material. In most cases, artificial dielectrics, consisting of conducting rods or spheres that are small compared to the wavelength, are used to construct delay lenses. Owing to contours of the lens, the inner portion of the transmitted wave is decelerated for a longer interval of time than the outer portions. Such a non-uniform delay causes the radiated wave to be collimated.

9.3.4.3 Zoned Lens

The configurations of zoned lenses are shown in Figures 9.13b,c and f and 9.14b. Zoning is mainly done to reduce the weight of the lens. Zoning does have some adverse effects such as shadowing and refocussing of reflected energy into the feed antenna that ultimately reaches the source. These effects are to be nullified to save the generator. Some of the methods were discussed earlier.

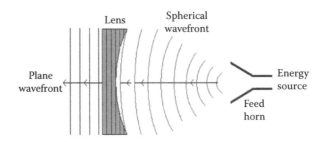

FIGURE 9.19
Delay-type lens.

9.3.4.4 Loaded Lens

A loaded lens is a multicellular array of thousands of cells. Each cell contains an element for slowing the wave (or causing delay). It is a serrated-metal, plastic-supported WG element that acts as a phase-controlling device. A loaded lens can focus microwave energy in much the same way as the WG type of lens does. This is caused since the speed of propagation is higher in the region between parallel plates than in free space. The parallel plates support the cells. The lens has an egg-crate appearance as it contains two lenses, one with vertical and the other with horizontal plates, in the same volume. The vertical plate lens focusses a vertically polarised beam, and the horizontal plate lens handles a horizontally polarised beam. This type of construction can be employed in multiple-beam applications where each beam has a different polarisation.

9.3.5 Frequency-Sensitive Antennas

A radar antenna uses a feed section to drive a horizontally stacked array section that radiates the applied RF pulses. The same array section receives the target returns. Each array contains slots to radiate and receive a particular frequency. Bearing data may be obtained by mechanically rotating the antenna by 360° and elevation data are obtained by electronically scanning the elevation. The radar antenna is frequency sensitive and radiates pulses at an elevation angle determined by the applied frequency. When the frequency is increased, the beam elevation angle decreases. Conversely, when the applied frequency is decreased, the beam elevation angle increases. The beam elevation angle is therefore selected by the application of a frequency corresponding to the desired angle of elevation. The physical length of the antenna feed section, called the serpentine section in relation to the wavelength of the applied energy determines the direction of the radiated beam. The beam shift occurs with a change in frequency because the positive and negative peaks of the energy arrive at adjacent slotted arrays at different times. The change in the field pattern is such that the angle of departure (angle at which the beam leaves the antenna) is changed. A change in phase of the applied RF energy causes the same effect.

9.4 Antenna Arrays

An array is a collection of radiating elements and may be 1D (linear), 2D (planar) or a 3D structure. Arrays are used where more directivity and steerable beam(s) are required. Its radiation pattern is controlled by relative amplitudes and phases of signals applied to each element. The scanning angle with (maintaining) reasonable gain depends on the pattern of the individual radiating element. An array can simultaneously generate several search or tracking beams within the same aperture.

2D planar arrays are more popular and are most commonly used. A planar array may be flat with square or rectangular shape or may be curved with cylindrical or spherical shape. The rectangular planar array generates a fan beam, whereas the square and circular planar arrays generate pencil beams. Planar arrays have lower side lobes than curved arrays due to their superior illumination. Besides, planar arrays require less maintenance. Since planar arrays can scan only a 60–70° angle (off-broadside), several faces are needed for full hemisphere coverage. A single curved-shaped array gives more coverage.

The radiating elements in an array may be arranged in different forms in terms of grid structures and cells therein. These elements may be dipoles, slots, polyrods, loops, horns, helix, spirals, log periodic structures and even parabolic dishes. Among these elements, dipoles give simple structures and are widely used for obtaining broadband patterns. Slots are easier to construct, give broadband pattern and are widely used due to the simple structure. Both dipoles and slots are used where a larger angular coverage with a single array is desired. Polyrods, helix, spirals and log periodic structures give more directive radiation patterns. Horns are also easier to construct and are thus used at microwave frequencies.

The element spacing in arrays governs grating lobes and limits the angle to which a beam can be steered. Elements are normally equispaced but non-equispaced configurations may also be used. The non-equispaced structures require lesser number of elements for comparable beamwidth that give broadband operation and patterns with lower side lobes. The number of elements in an array may be few to several thousands and are limited only by practical constraints.

Mutual coupling between elements is one of the major limitations of an array in relation to its performance. It causes input impedance of an individual element to be different from its own impedance in isolation. Large mutual coupling results in poor radiation pattern, changed radiation resistance, raised side lobes and in mismatching with the transmitter or receiver circuit.

To get the unidirectional radiation, a reflector can be placed behind an array as shown in Figure 9.20. Figure 9.21 illustrates horizontal array field patterns containing two, three and four elements.

9.4.1 Types of Arrays

As mentioned above, arrays can be classified in many ways, including the arrangements of elements and currents and phases therein. Such classification leads to broadside, end-fire, collinear and parasitic arrays. These are briefly described in the following sub-subsection.

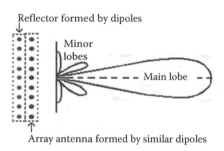

Reflector formed by dipoles

Minor lobes

Main lobe

Array antenna formed by similar dipoles

FIGURE 9.20
Field pattern of an array containing a reflector.

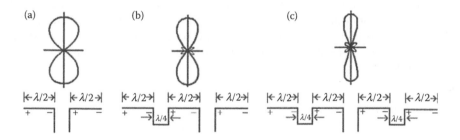

FIGURE 9.21
Horizontal arrays with field patterns. (a) Two element array, (b) three element array and (c) four element array.

9.4.1.1 Broadside Array

In broadside arrays (BSAs) (Figure 9.22), all elements are identical, equally spaced, lie along a line drawn perpendicular to their respective axis and are fed with currents of equal amplitude and same phase. Broadside antenna is bidirectional and may have a gain >30 dB.

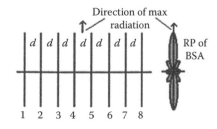

FIGURE 9.22
Broadside array.

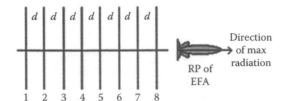

FIGURE 9.23
End-fire array.

9.4.1.2 End-Fire Array

In an end-fire array (EFA) (Figure 9.23), all elements are identical, equally spaced, along a line drawn perpendicular to their respective axis and are fed with currents of equal amplitudes but progressively increasing phases. An EFA in general is unidirectional but can be made bidirectional by feeding half of the elements with 180° out of phase. End fire is a narrowband with a gain >15 dB.

9.4.1.3 Collinear Array

As shown in Figure 9.24, its elements have a collinear arrangement. Antennas are mounted end to end in a straight line and their elements are fed with equal and in-phase currents. Its radiation pattern, perpendicular to the principal axis has (everywhere) a circular symmetry with the main lobe and it closely resembles that of the BSA. CLA is also called broadcast or omni-directional array. The gain of CLA is maximum when spacing between elements is between 0.3 λ and 0.5 λ. In view of its construction at these lengths, the elements are placed very close. The number of elements is generally not more than four. Further increase in number does not result in higher gain.

9.4.1.4 Parasitic Arrays

As shown in Figure 9.25, it contains one driven (feeding) element (DE) and some PEs. Yagi array falls to this category. The amplitude and phase of

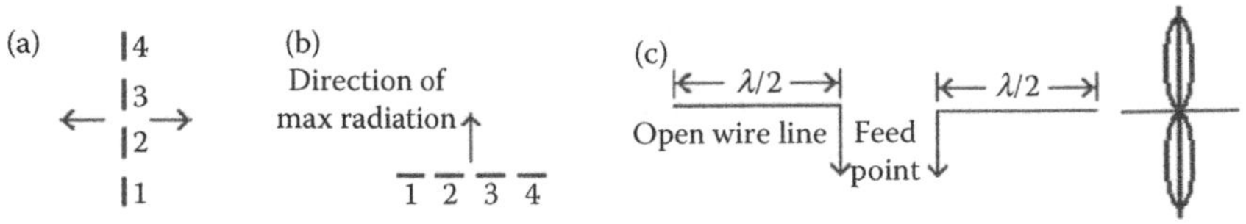

FIGURE 9.24
Collinear arrays. (a) Vertical antennas arranged colinearly, (b) horizontal antennas arranged colinearly and (c) two element colinear array and its radiation pattern.

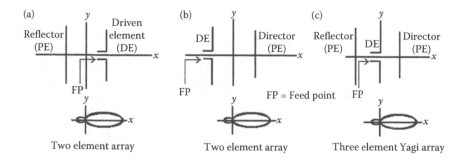

FIGURE 9.25
(a) Two element array PE > DE, (b) two element array PE < DE and (c) three element array.

the induced currents in PE depends on their tuning (length) and spacing between PEs and the DE. PE longer (by ≥5%) than DE is called the reflector and shorter to DE is called the director. The reflector makes the radiation maximum in the direction from PE to DE. The director makes the radiation maximum in the direction from DE to PE.

9.5 Microstrip Antennas

These antennas bear many names such as patch antenna, printed antenna, microstrip patch antenna and microstrip antenna (MSA). These are used where thickness and conformability to the host surfaces are the key requirements. MSAs are becoming increasingly popular in the mobile phone market. These can be directly printed onto a circuit board.

Patch antenna is basically a metal patch suspended over a ground plane (GP). Its assembly is usually contained in a plastic radome to protect the structure. These employ the same sort of lithographic patterning as used to fabricate Printed Circuit Boards (PCBs). In its most basic form, an MSA has a radiating patch on one side of a dielectric substrate and a GP on the other. It is fabricated by etching the antenna pattern in a metal trace bonded to an insulating dielectric substrate with a continuous metal layer bonded to the opposite side of the substrate that forms a GP. The simplest MSA uses a half-wavelength-long patch with a larger GP to give better performance but at the cost of increased antenna size. The GP is normally modestly larger than the active patch. The current flow is along the direction of the feed wire; so, the vector potential and thus E field follows the current. Such a simple patch antenna radiates a linearly polarised wave. The radiation can be regarded as being produced by the 'radiating slots' at the top and bottom, or equivalently as a result of the current flowing on the patch and the GP.

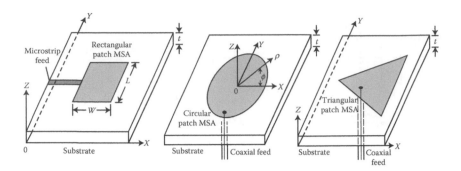

FIGURE 9.26
Configurations of MSAs.

MSAs can be of square, rectangular, circular, triangular, elliptical or theoretically of any other regular shape. The regular shape of a well-defined geometry simplifies analysis and helps in performance prediction. Rectangular and circular shapes are the most commonly used. Square patches generate a pencil beam and rectangular patches generate fan beams. Owing to the ease of fabrication, circular patches can also be used. The calculation of current distribution in circular patches is relatively more involved. Figure 9.26 shows configurations of rectangular, circular and triangular MSAs.

9.5.1 Feed Methods

The feed methods for MSAs are classified into contact and non-contact feeds. In the contacting method, the signal is directly fed to the radiating patch through a microstrip or coaxial cable. In the non-contacting case, power is fed through electromagnetic coupling that includes aperture coupling and proximity coupling. These methods are shown in Figures 9.27 through 9.29 and their comparison is given in Table 9.1.

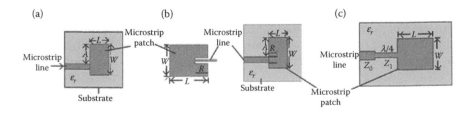

FIGURE 9.27
Contact feeds. (a) Offset feed, (b) inset feed and (c) feed with quarter-wave matching section with characteristic impedances Z_0 and Z_1 of line sections shown.

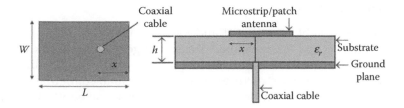

FIGURE 9.28
Coaxial or probe feed.

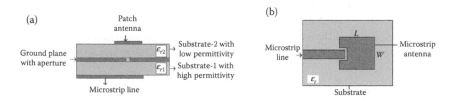

FIGURE 9.29
Non-contacting feeds. (a) Aperture coupled feed and (b) proximity coupled or indirect feed.

TABLE 9.1

Comparison of Different Feed Methods

	Microstrip	Coaxial	Aperture Coupled	Proximity Coupled
Spurious radiation from feed point	More		Less	Minimum
Fabrication	Easy	Requires soldering and drilling	Requires alignment	
Reliability	Better	Poor due to soldering	Good	
Impedance matching	Easy in all cases			
Bandwidth (with impedance matching)	2–5%		13%	

9.5.2 Characteristics of MSAs

All the characteristic parameters discussed in Section 9.2 are equally applicable to the MSAs. The return loss and radar cross section (RCS) are the two additional properties of MSAs. The return loss spells the ratio of the Fourier transforms of the incident pulse and the reflected signal, whereas RCS is the quantum of return of the incident energy. GPS guidance systems require low RCS platforms but the RCS of the conventional patch antenna is often too high. To reduce its RCS, the patch antenna is covered by a magnetic absorbing material. This, however, reduces the antenna gain by several dBs.

9.5.3 Merits of MSAs

These are lightweight, small-sized and low-cost antennas. These have less thickness and occupy lesser volume. These are good where thickness and conformability to the surface of mount or platform are the key requirements. In view of their conformal structures of low profile planar configuration, these can be easily moulded to any desired shape and hence, can be attached to any host surface. Besides, these antennas are simple to fabricate and are easy to produce, modify and customise. Their fabrication cost is low and thus can be manufactured in large quantities. Their fabrication process is compatible with microwave monolithic-integrated circuit (MMIC) and optoelectronic-integrated circuit (OEIC) technologies. These can support both linear and circular polarisations and are capable of dual- and triple-frequency operations. These are mechanically robust when mounted on rigid surfaces. Large arrays of half-wavelength or lesser spacing with MSAs can be easily formed.

9.5.4 Demerits of MSAs

These antennas have low bandwidth (usually 1–5%), low efficiency, low gain and low power-handling capacity. Owing to their smaller size, their design is complex that gets further enhanced when bandwidth enhancement techniques are adopted. These are resonant by inherent nature and suffer from radiations from feeds and junctions. These result in poor end-fire radiators except the tapered slot antennas. Surface wave excitation is an added limitation.

9.5.5 Applications

MSAs are commonly used in wireless applications such as cellular phones and pagers. These are immensely useful in telemetry, satellites, missiles, microwave and millimetric systems, airborne and spacecraft systems and in phased array radars. Many of these applications require a dielectric cover over the radiating element to provide protection from heat, physical damage and the environment. However, such coverage changes its resonance frequency and gain resulting in the degradation of performance. These changes can be minimised by taking some corrective measures in their design and fabrication.

Descriptive Questions

Q-9.1   Briefly describe antenna theorems and the characteristic parameters of antennas.

Q-9.2   Discuss the working of one reflector belonging to the active and passive category.

Q-9.3 Describe the reflection mechanism and radiation pattern of a parabolic reflector.

Q-9.4 Illustrate the different feed arrangements used in parabolic reflectors.

Q-9.5 Explain the working of a Cassegrain antenna.

Q-9.6 What is a horn antenna? Discuss the types and applications of horn antennas.

Q-9.7 Explain the working of vertically and horizontally polarised slot antennas.

Q-9.8 Discuss the relative merits and demerits of lens antennas.

Q-9.9 Discuss the parameters on the basis of which lens antennas are classified.

Q-9.10 Explain the working of sensitive antennas.

Q-9.11 Discuss the types, constructional features and radiation patterns of antenna arrays.

Q-9.12 List the applications where antenna arrays are preferred.

Q-9.13 Discuss the types, merits and demerits of MSAs.

Q-9.14 Discuss the relative merits of various feed methods of MSAs.

Further Reading

1. Brown, R. G. et al., *Lines, Waves and Antennas*, 2nd ed. John Wiley, New York, 1970.
2. Chang, K., *Microwave Ring Circuits and Antennas*. Wiley, New York, 1996.
3. Elliot, R. S., *Antenna Theory and Design*. Prentice-Hall, New Delhi, 1985.
4. Jesic, H. (ed.), *Antenna Engineering Handbook*. Tata McGraw-Hill, New York, 1961.
5. Jordon, E. C. and Balmain, K. G., *Electromagnetic Waves and Radiating Systems*. Prentice-Hall Ltd., New Delhi, 1987.
6. Kraus, J. D., Marhefka, R. J., and Khan, A. S., *Antennas and Wave Propagation*, 4th ed. Tata McGraw-Hill, New Delhi, 2010.
7. Sisodia, M. L. and Gupta, V. L., *Microwaves—Introduction to Circuits Devices and Antennas*. New Age International (Pvt) Ltd., New Delhi, 2006.
8. Skolnik, M. I., *Introduction to Radar Systems*. McGraw-Hill Publications, New York, 1962.
9. Weeks, W. L., *Antenna Engineering*. Tata McGraw-Hill Publishing Co., TMH Edition, New Delhi, 1974.

10

Microwave Measurements

10.1 Introduction

The concept of measurement in the microwave region is entirely different from that of low-frequency measurement of voltage, current, impedance, power or energy. Magnitudes of some of these quantities (e.g. voltage and current) reduce to such a dismal level that these are not even measurable in most of the cases. However, useful information can be gathered by measuring some other quantities. These include wavelength, frequency, voltage standing wave ratio (VSWR), attenuation, phase shift, quality factor, scattering parameters, dielectric constant, reflection and transmission coefficients, electric field strength, impedance and power. At microwave range the techniques and equipment used are mostly different from those used at lower frequencies. The signal sources, measuring devices and the detecting elements in most cases assume different names in view of their nature of operation. This chapter discusses the measurement setups, signal sources, measuring equipment and the components required in test benches for measuring these quantities.

Power source: In laboratories it is mainly the klystron power supply (KPS) along with klystron and tunable klystron mount.

Detectors: These include crystal detectors, thermistors and barreters.

Measuring devices: These may include VSWR meter, micro-ammeter, galvanometer and oscilloscope.

Components: These include slotted line with probe carriage, tunable probe, crystal detector mount, frequency meter, Bayonet Neill–Concelman (BNC) connectors, coaxial to waveguide adopter, waveguides with flanges or joints, bends or corners, waveguide stands, tees or hybrid junctions, directional couplers, circulators, isolators, phase shifters, movable shorts, waveguide twists or transitions, loads or terminations, tuners, attenuators, switches and solid or liquid cells.

10.2 Klystron Power Supply

KPS is a power source to operate low-voltage reflex klystron. The instrument normally provides a stable, well-regulated beam voltage continuously variable over a range of 250–350 V and capable of yielding current from 0 to 50 mA. The reflector voltage can be varied from –10 to –210 V with respect to the cathode. The internal square wave modulation variable in frequency and amplitude is provided for modulation of klystron. The klystron can be operated in carrier wave mode and can also be modulated from any external source through ultra high frequency (UHF) connector provided on the front panel. Connections to klystron are made through the octal base provided on the front as well as on rear panel for ease of operation. The instrument is normally wired for operation from 230 V and 50 Hz power supply. Figure 10.1 shows the probable locations of panel controls and terminals provided in most of the power supplies, the description of which is given below.

Power switch (1): It is used to put the power supply in ON or OFF positions.

Output (2): It is an octal socket, which mates with the octal plug of klystron mount and provides all the voltages required for operating the reflex klystron.

Beam voltage (3): When turned in the clockwise direction the beam voltage increases which can be read on the meter scale.

Meter function switch (4): It controls meter function for reading beam voltage or current. It is normally kept at current. Under *OFF* position HT is disconnected from the output socket and is not applied to the klystron. In *VOLTAGE* position HT is applied to klystron and the beam voltage is read on the upper scale of the meter. And lastly, in *CURRENT* position the meter measures beam current of the klystron which can be read on the lower scale.

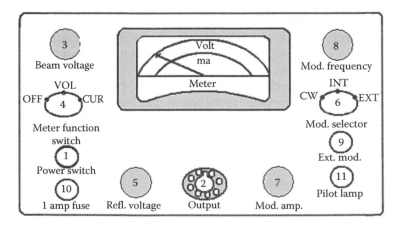

FIGURE 10.1
Front panel of a KPS.

Reflector voltage (5): It varies the reflector voltage which can be adjusted to obtain different modes of klystron or to optimise the power output.

Modulation selector (6): This switch selects the mode of operation. In *CW position* no modulation is applied and klystron operates at carrier wave. At *INT position* the internal square-wave modulator is switched-on and modulation is applied to the reflector whose frequency and amplitude are adjusted with relevant controls. In *EXT* position internal modulator is switched off. External modulation is applied through UHF connector on the panel.

Mod. amp (7): It adjusts amplitude of modulating signal (both internal and external) fed to the reflector of the klystron.

Mod. freq. (8): This controls the frequency of the internal modulating signal approximately from 800 to 2000 Hz.

Ext. mod (9): This control provides input for external modulation.

Besides the above, a *1A fuse (10)* and a *Pilot lamp (11)* are also provided on the front panel

10.2.1 Operating Procedure

For proper functioning of a KPS it has to be handled as per the instructions supplied by the manufacturer. Some basic common points to be adopted are as follows.

The power supply is to be properly earthed and the klystron mount leads are to be properly connected to the output socket (2). Initially the meter function switch is to be kept in OFF position, the voltage potentiometer is maintained in an extreme anti-clockwise position and the reflector voltage potentiometer in an extreme clockwise position.

1. *Carrier wave operation:* For carrier wave operation:

    a. Set the modulation selector switch to CW.

    b. Throw power switch to ON position and allow klystron about 2-min warm-up time.

    c. Turn meter switch to voltage position. The meter will yield beam voltage. This beam voltage may be adjusted with beam voltage potentiometer to the specified value. This value may differ from klystron to klystron, and is to be adjusted accordingly (e.g., 300 V for 2k25 or 723 A/B).

    d. Turn the meter function switch to current position. Adjust the reflector voltage till microwave power is obtained from the klystron. At the juncture of generation of microwave power the beam voltage will show a slight increase.

    e. Obtain the required frequency by tuning the klystron mechanically through a tuning screw or by any other method provided in a particular device. Small adjustments can also be made by readjusting the reflector voltage.

2. *Operation with internal modulation*
   a. Set the modulation selector switch to INT and repeat steps 1a–1e. Adjust modulation amplitude and reflector voltage potentiometers to obtain maximum power.
   b. Adjust modulation frequency potentiometer to get the modulation frequency. When a tuned amplifier is connected at the output the square wave modulation is normally used. Usually, this amplifier is a VSWR meter tuned to a frequency of 1 kHz so that the modulation frequency of modulator is tuned for maximum output in the VSWR meter. An oscilloscope with appropriate specifications can also be used.
3. *Operation with external modulation*
   a. Set external modulation selector switch to EXT.
   b. Connect external modulating signal to external modulation knob and repeat steps 1a–1e.
   c. Adjust the frequency and amplitude of the modulation at the generator.
   d. Attenuate the signal with modulation amplitude potentiometer (if required).
   e. If external modulating signal is a saw-tooth voltage, usually from an oscillator, mode of klystron (variation of output power with reflector voltage) can be seen on CRO.

10.2.2 Reflex Klystron

Reflex klystron is the most commonly used signal source for test benches. Its constructional features, working and other details are discussed in Chapter 12.

10.2.3 Klystron Mount

The klystron mount is illustrated in Figure 10.2. It contains a rectangular waveguide with a movable short on one end and a socket in which the reflex

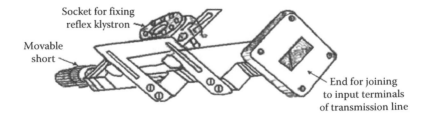

FIGURE 10.2
Klystron-mount with movable short.

klystron is to be inserted. Just beneath the socket a small hole is cut in the upper wall of the waveguide for insertion of a probe. This probe is a thin wire through which signal output of klystron is coupled to the waveguide. As can be seen the socket contains a number of small holes around its periphery into which different terminals of klystron get inserted. All these holes are connected to a multi-wire lead to be attached to KPS. The reflex klystron will generate microwave power only when its different terminals are applied with appropriate potentials.

10.3 VSWR Meter

The standing wave and VSWR are discussed in detail in Chapter 2 in connection with the plane waves and the transmission lines. The two important features of interest are (i) the distance between the maxima and the adjacent minima is a quarter of the guide wavelength and (ii) the height or depth of the wave from the crest to the trough denotes the degree of mismatch. The frequency of operation can now be obtained from this guide wavelength. Also by measuring VSWR ($=E_{max}/E_{min} = 1/M$) the degree of mismatch (M) can be determined. If the circuit is perfectly matched M is unity. The VSWR can be measured by a device commonly called VSWR meter. It is a highly sensitive, high gain tuned electronic instrument operating at a single (fixed) frequency of 1 kHz with tolerance of $\pm2\%$ and BW $\cong 40$ Hz. It is calibrated to indicate directly both voltage and power, standing-wave ratios when used with square law detectors such as crystal diodes and barreters. It operates on 230 V, 50 Hz ac and consumes about 45 W of power. In normally available equipment the different controls provided are described below.

1. *ON/OFF power switch:* It connects or disconnects the device to the power supply.
2. *Range switch:* It permits range-to-range readings (up to 0.1 μV with noise levels <0.03 μV) without loss of accuracy and sensitivity.
3. *Gain control:* It provides adjustment of sensitivity over a ($\cong 8.5$ dB) meter scale.
4. *Expand-normal + 5 dB switch:* It selects an expanded meter scale when desired and in the +5 dB position shifts the meter reading up by 5 dB.
5. *Meter:* It is calibrated to indicate VSWR directly in decibels (with an accuracy of $\approx \pm0.1$ dB).
6. *Selector switch:* It connects the input circuit of the equipment to the crystal diode (for low-level signal), barreter or another high impedance signal source (for medium or high-level signals).

a. For barreters, the correct bias current is automatically supplied through the input jack, the current value being selected by a selector switch for either of the two different barreter resistances in common use.

b. The internal impedance seen at the input jack is either 200 Ω or 200 kΩ.

c. When set to Bolo, a dc bias $\approx$8.75 mA or 4.5 mA is automatically applied to a 200 Ω barreter connected to the input jack.

7. *Input:* The signal to be measured is to be applied.

10.4 Travelling Wave Detection

In Figure 10.3a and b two types of travelling wave detectors are shown. The travelling wave detector shown in Figure 10.3a) as such is not a separate entity but forms part of the transmission line. It contains a rectangular waveguide with a slit in the middle of its top wall. A sliding platform with the provision to accommodate a probe is placed on the top wall. With the movement of this platform the thin probe moves in the slit, senses the field present in the guide and passes it to the detector end of the probe.

10.4.1 Slotted Section

Slotted section (Figure 10.4) or slotted line is a plane transmission line with a slot. It is usually mounted on the probe carriage assembly. The slot in line section is cut in such a way that no appreciable power leaks from the line. This slotted section is the same as discussed above.

In a waveguide, the slot is cut in the centre of a broader wall and parallel to the flow of currents for the same reason. The slotted section has dimensions

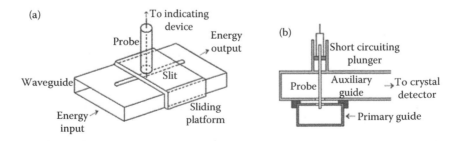

FIGURE 10.3
Travelling-wave detector. (a) Basic travelling detector and (b) improved travelling detector.

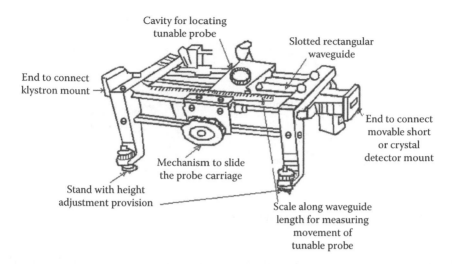

FIGURE 10.4
Slotted transmission line with a probe carriage.

suitable for the frequency range in use. The exploring probe can move along the interior of the waveguide through the slot. In this way, the electric field due to the transmitted and reflected energy can be sampled. Useful information such as impedance, standing wave ratio, wavelength, frequency and so forth can be obtained by using the slotted section.

In the coaxial line the slot is cut in the outer conductor parallel to the direction of the propagation or to the flow of current in the outer conductor to cause negligible interference with the current.

The probe carriage, which accepts tunable probe, moves over the slotted section with the help of a helical rack and pinion. The movement of the carriage can be read to the accuracy of fraction of millimetres with the overlapping vernier scale. For more accurate measurements, facilities can be provided by attaching a dial gauge to the carriage.

10.4.2 Tunable Probes

This can be considered as a pickup device to sample the energy under examination and to apply it to other devices (e.g. VSWR or oscilloscope) for measurement and observation. The probe is mounted atop a slotted line and moves along it. In a probe a crystal detector is mounted which is connected to the input of a galvanometer or standing wave amplifier. The tip (control wire like projection of the probe) goes into the waveguide through the slot. As the tip moves in the electric field, current flows in the probe. The probe current remains the same as long as the energy level in the waveguide is constant. Owing to standing waves in the guide, the

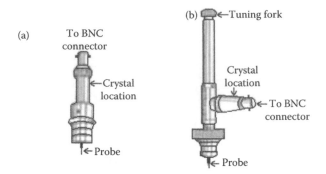

FIGURE 10.5
Tunable probes. (a) Broadband (fixed) tuned probe and (b) tunable probe.

current in the probe varies with the field variations. A probe can be broadband or narrowband and can be fixed or tunable. Two of such probes are shown in Figure 10.5.

10.4.3 Movable Shorts

Adjustable waveguide shorts are used for terminating a waveguide with a short circuit. It consists of a sliding short inside the waveguide housing. The movement of the plunger is screw controlled. The reflection coefficient close to unity is obtained with the non-contacting-type plungers. Figure 10.6 shows two movable shorts one of which has a calibrated micrometer and thus referred to as precision movable short.

10.4.4 Frequency (or Wave) Meters

The microwave frequency can be measured by either electronic or mechanical technique. The electronic method includes frequency counters or high-frequency heterodyne systems. In these the unknown frequency is compared with harmonics of a known lower frequency by use of a low-frequency generator, a harmonic generator and a mixer. The accuracy of the

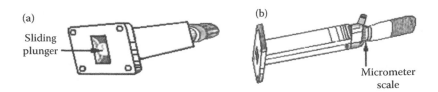

FIGURE 10.6
Movable shorts. (a) Movable short and (b) precision movable short.

measurement is limited by the accuracy and stability of the reference source. The mechanical methods require a tunable resonator such as an absorption wave meter, which has a known relation between a physical dimension and the frequency.

In a laboratory, lecher lines can be used to measure the wavelength on a parallel wire transmission line. This wavelength can be used to calculate the frequency. A slotted waveguide (Section 10.4.1) or slotted coaxial line can also be used to directly measure the wavelength. These devices consist of a probe (Section 10.4.2) introduced into the line through a longitudinal slot, so that the probe is free to move up and down the line. Slotted lines are primarily intended for measurement of the voltage standing wave ratio on the line. If a standing wave is present on the line it may also be used to measure the distance between the nodes (equal to half wavelength). Once the wavelength is known it leads to the determination of frequency. The precision of this method is limited by the determination of the nodal locations.

Figure 10.7 illustrates two frequency meters one of them is a direct reading type while the second is of micrometer type. In direct reading the signal frequency can be directly read after its proper adjustment. In case of micrometer frequency meter a calibrated chart is supplied by the manufacturer. The frequency can be read from the calibration chart. One of such chart is given in Table 10.1. This chart contains values of frequencies corresponding to some selected micrometer readings. The frequencies corresponding to

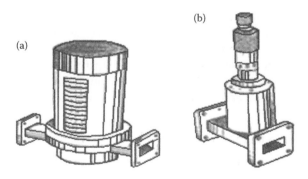

FIGURE 10.7

Frequency (wave) meters. (a) Direct reading type and (b) micrometer type.

TABLE 10.1

Calibration Chart for a Typical Frequency Meter

Frequency (GHz)	8.2	8.6	9.0	9.4	9.8	10.2	10.6	11.0	11.4	11.8	12.2
Micrometer reading (mm)		18.88	16.20	14.09	12.36	10.94	9.69	8.63	7.72	6.88	6.15	5.45

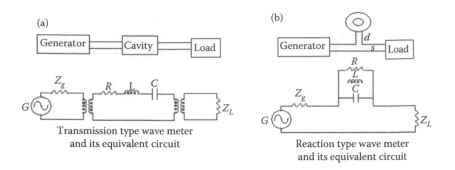

FIGURE 10.8
(a) Transmission and (b) reaction type of wave meters.

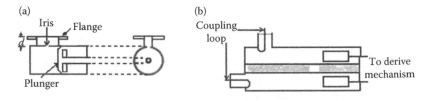

FIGURE 10.9
Rectangular and coaxial cavity wave meters. (a) Cavity wave meter and (b) coaxial wave meter.

any other micrometer reading not given in the chart can be worked out by interpolation.

The frequency meters can be made either by using a transmission type of cavity or a reaction type of cavity. In the first case its equivalent circuit amounts to a series resonant circuit, whereas for the second it is a parallel resonant circuit. The two cases are shown in Figure 10.8.

The cavity wave meter (CWM) can be carved out from a rectangular waveguide or from a coaxial cable. Figure 10.9 shows two such cavities.

10.4.5 Microwave Detectors

These devices are used to convert microwave signal to dc or low-frequency signal, which can be measured. A crystal or bolometer element in a transmission line should receive the microwave power with very little reflection, since any reflection represents a loss and a potential source of error. In a waveguide mount one end of the crystal or bolometer is usually grounded and the other end is attached to the inner conductor of a coaxial line. To prevent the microwave signal from entering the measuring apparatus a large bypass capacitor is to be located at the junction of the coaxial line and the waveguide. In a coaxial line mount the crystal is usually put in series with the inner conductor and a bypass capacitor is used on the output end.

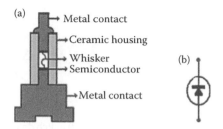

FIGURE 10.10
Crystal diode. (a) Construction and (b) symbol.

To have a closed dc circuit it is sometimes necessary to add a dc return on the input side of the crystal. In a waveguide system, it is often convenient to use a waveguide to coaxial transition and a coaxial crystal mount. Germanium or silicon is commonly used as the material in construction for the crystal. The silicon type IN-21 and IN-23 crystals are suitable for use at the X-band. This crystal is a low-voltage device. The output voltage (or current) is not directly proportional to the input voltage but follows a square law. The output of the crystal is fed to an indicating instrument such as a galvanometer or VSWR meter. Figure 10.10 illustrates the construction and its equivalent representation.

10.4.6 Tunable Crystal Detector Mounts

Tunable detector mount is designed for detecting of microwave signals and monitoring microwave power in conjunction with a suitable detector (IN-23 for X-band). It is mounted with a crystal detector in a waveguide section. A non-contacting shorting plunger is used for matching purpose. Figure 10.11 illustrates three detector mounts. Each mount has a provision for a BNC connector so that a lead can take the detected signal to a measuring device.

In a tunable waveguide mount the distance between the diode and the waveguide is adjustable with the aid of the plunger. Thus, these are capable of broad band operation with small reflections. Some of the details of these mounts are shown in Figure 10.12. The waveguide crystal mount, Figure 10.12a, uses half-wavelength short-circuited line to provide an RF short circuit through the crystal. It still allows rectified current to be measured. Figure 10.12b shows a coaxial crystal mount with tuning elements which not only tune but provide path to direct current.

10.4.7 Bolometer

The term bolometer is referred to both the barrater and the thermistor. Both of these are temperature-sensitive devices.

FIGURE 10.11
Different types of detector mounts. (a) Waveguide (fixed) match detector mount, (b) waveguide tunable detector mount and (c) coaxial detector mount.

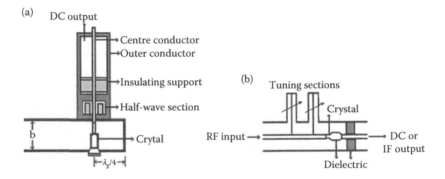

FIGURE 10.12
Crystal mount. (a) Waveguide type and (b) coaxial type.

10.4.7.1 Barraters

Barraters (Figure 10.13a) have maximum power dissipation from 50 mW to 2 W and cold resistance of 6–120 Ω. These are of two types.

1. A short length ($< \lambda/10$, for keeping the current distribution uniform) of thin (3–10 μm thickness for minimising skin effect), tungsten or platinum wire characterised by a *positive temperature coefficient of resistance*, mounted in a glass envelope with metal end caps or leads. To increase power dissipation, the envelope is filled with an inert gas. These are used up to 1 GHz.

FIGURE 10.13
Bolometer configurations. (a) Barrater and (b) thermistor bead.

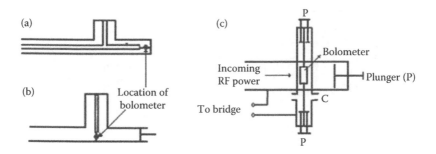

FIGURE 10.14
Bolometer mounts. (a) Coaxial line type, (b) waveguide type and (c) details of (b).

2. Thin film bolometer consists of a thin platinum film of a very small (few μm) thickness, vacuum deposited on a glass or mica substrate for frequency up to 50 GHz. It has a simple design and can readily be connected in and matched to a waveguide transmission line.

10.4.7.2 Thermistor

The thermistor (Figure 10.13b) consists of a small (~0.5 mm diameter) bead of a semiconductor material (usually a mixture of metallic oxides such as manganese, nickel and cobalt) with platinum leads (25 μm dia.). The bead is given an extremely thin coat of glass and enclosed in glass envelope with stiffer leads sealed into it. It has negative temperature coefficient of resistance, wider (few thousand Ω to hundreds of kΩ) range of resistance, faster rate of resistance change with absorbed power and thus higher sensitivity (from 10 to 100 Ω/mW). Its operating point is usually set at 100–300 Ω by preheating, most often with dc power, used to measure power from few μW to several mW at 16 GHz.

10.4.7.3 Bolometer Mounts

Figure 10.14 shows the location of the bolometer in a mount. This mount may be made of coaxial cable or a waveguide. A few details of waveguide type of mount are given in Figure 10.14c. Figure 10.15 illustrates some of the basic characteristics of a thermistor and a barrater.

10.5 Qualities of Microwave Components and Devices

For effective and efficient working of signal sources and measuring equipment the microwave components must conform to some basic minimum standards. Besides, all the components to be connected in a test bench have

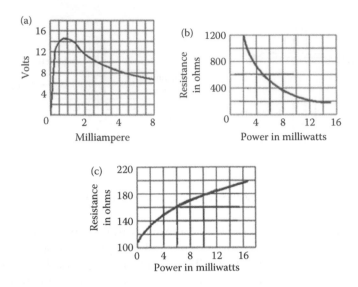

FIGURE 10.15
Characteristics of thermistor and barrater. (a) Static voltage current curve for a thermistor bead, (b) power versus resistance curve for a thermistor bead and (c) resistance versus power absorbed by a barrater.

TABLE 10.2

Required Qualities of Microwave Devices and Components

Sources	Measuring Instruments	Components
Power stability	High sensitivity	Low attenuation
Frequency stability	Minimum errors	Low VSWR or
Mode stability	Short warm-up time	Low reflection coefficient
Waveform stability	High reliability	
Modulation stability	Insignificant loading effect	
No excessive heating	on measuring circuit	
Small in size	Small in size	
Light in weight	Light in weight	
Low in cost	Low in cost	

to be good in terms of their electrical properties. The required properties for each category noted above are listed in Table 10.2.

10.6 Precautions

Waveguides are made from soft conducting material such as copper or aluminium. They are therefore easy to dent or deform. Even the slightest damage of the inner surface of a waveguide will result in the formation of standing waves and may cause in internal arcing. Any such arcing causes further damage to the waveguide. A continuous occurrence of such phenomena may render a waveguide unusable. The use of a waveguide with damaged inner surfaces can greatly reduce its efficiency. Thus, the use of a waveguide on a damaged surface will make the system not only inefficient but also unworkable.

In some applications, the copper/aluminium waveguides are used in conjunction with other equipment/platforms made of steel or other materials. In view of electrolytic corrosion of the metals at the points of contacts between two dissimilar metals, the waveguides can be completely destroyed and that too in a relatively short period of time. Thus, any direct contact between dissimilar metals must be avoided.

All practical waveguide systems are exposed to weather. To avoid natural corrosion waveguides are to be properly painted. The dielectric in waveguides is air which is an excellent media as long as it is free from moisture. The moment the air is wet it becomes a poor dielectric and can result in internal arcing. To prevent the moisture from entering the waveguides all joints are to be properly sealed.

At microwave range the quantum of power involved is quite low. Efforts are to be made to avoid loss of the same. To get better results a number of precautions, encompassing plumbing, handling and selection of equipment, are listed below.

1. All the components are to be properly connected to prevent leakage/reflection of microwave energy.
2. The entire setup is to be perfectly levelled.
3. The klystron is to be properly placed in its mount and leads to the KPS.
4. The leads to klystron mount are to be connected accurately.
5. Ensure microwave generation by proper biasing of electrodes of klystron.
6. RF source must be followed by an isolator (for low signals) or an attenuator (for medium or high signals) to prevent reflections reaching the source.
7. Matched load is to be discarded if the device under test is an absorbing element.
8. The KPS, klystron, VSWR meter and other devices and components are to be operated strictly in accordance with the instructions provided for the equipment.
9. Power output and frequency of signal source are dependent on load, discontinuities and mismatched terminations. Change in their values may result in erroneous readings.

10. Reflections from excessive penetration of detector probe may alter position of standing wave or even result in uneven spacing between nulls.

11. Presence of higher-order modes of propagation in the vicinity of waveguide discontinuities may also result in uneven spacing between nulls.

12. Excessive attenuation introduced may necessitate excessive probe penetration for adequate output, which may alter positioning of standing waves.

10.7 Some Standard Norms

The block diagrams for experimental setups involve several devices and components and some of these bear more than one name, a brief description of the same is listed below.

1. Microwave devices and components are to be connected in the sequences shown.

2. Source, signal source and signal generator bear the same meaning.

3. Klystron, KPS and klystron mount are part of the signal source.

4. Power supply and KPS bear the same meaning.

5. Pad is referred to the isolator or (fixed/variable) attenuator.

6. Wave meter, CWM and frequency meter bear the same meaning.

7. Absorption-type CWM is more commonly used in laboratories.

8. Transmission line, slotted section and slotted transmission line bear the same meaning.

9. Probe carriage forms the part of a slotted section and generally not shown separately.

10. Probe contains a diode detector and no separate detector is needed.

11. Crystal detector mount and detector mount have the same meaning.

12. Detector mount is loaded with detector and no separated block for detector is shown.

13. Movable shorting plungers are normally part of the detector mounts.

14. An indicator/measuring device bears the same meaning.

Section 10.8 includes measurement of some basic quantities and a few practical applications are discussed in Section 10.9. All quantities referred to as 'basic' are not basic in the true sense and some of them are interrelated with each other.

10.8 Measurement of Basic Quantities

This section includes measurement of wavelength, frequency, VSWR, phase, quality factor, attenuation constant or insertion loss, scattering parameters, dielectric constant, impedance, power, reflection coefficient and transmission coefficient. Some of these quantities are measured in more than one way.

10.8.1 Measurement of Wavelength

The resonant cavity is a microwave analogue of tuned circuit. Earlier it was mentioned that these cavities may be of transmission or absorption types. Transmission cavity allows maximum transmission of signal at the resonance frequency. The absorption type absorbs maximum signal and allows minimum signal to pass through at the resonance frequency.

The experimental setup is shown in Figure 10.16. To measure the wavelength first the power level is adjusted to obtain the full-scale reading on the (output) meter. The movable short is adjusted to obtain maximas and minimas as sharp as possible. The probe carriage is moved and the distance between locations of maximas and minimas is noted. The process is repeated to take a number of readings in order to get an average value of λ. The guide wavelength (λ_g) is evaluated by using the relation:

$$\lambda_g = \lambda/\sqrt{[1 - (\lambda/\lambda_c)^2]} \tag{10.1}$$

where λ is the free space wavelength, λ_c (= $2a/c$) is the cutoff wavelength, a is the dimension of the broader wall of the waveguide and c (= 3×10^8 m/s) is

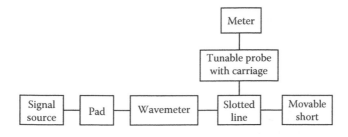

FIGURE 10.16
Experimental setup to measure wavelength.

the speed of light. Once the wavelength is determined the frequency can also be obtained by using the relation:

$$f = c/\lambda_g \tag{10.2}$$

It is to be mentioned that in case of wavelength measurement the insertion of probe is generally kept a little deeper than in the case of VSWR measurement. This method can also be employed to measure the frequency taking some additional steps for which CWM is slowly tuned to get a dip in the power level. The frequency is read either from direct reading CWM or from micrometer CWM. In case of micrometer type the micrometer reading is converted into frequency by using the given calibrated chart (such as Table 10.2). The two frequencies obtained are compared with that obtained from Equation 10.2.

10.8.2 Measurement of Frequency

10.8.2.1 Transmission Method

The experimental setup to be used is shown in Figure 10.17.

In this the CWM is placed between the source and the detector. CWM is slowly tuned until the detected signal reaches maximum. At this point the frequency read from CWM is the frequency of RF signal.

10.8.2.2 Reaction Method

This experimental setup is shown in Figure 10.18.

Here, CWM placed between the detector and the source is tuned until a dip in the power level is observed. When CWM is tuned to produce resonance

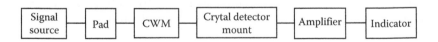

FIGURE 10.17
Experimental setup to measure frequency (transmission method).

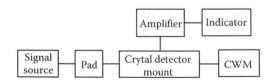

FIGURE 10.18
Experimental setup to measure frequency (reaction method).

the impedance coupled into the system changes rapidly near the resonance and produces a reaction upon the detected signal. The frequency of the greatest reaction is the frequency of the signal. For calibration of a device whose frequency is affected by temperature, humidity and so forth, more accurate methods are required such as direct and interpolation methods.

10.8.3 Measurement of VSWR

10.8.3.1 Transmission Line Method

The experimental setup for transmission line method is shown in Figure 10.19.

This method is preferred for measuring low VSWR. In this the probe is moved through probe carriage and the detected output is read from the attached meter. The VSWR (= V_{max}/V_{min}) is obtained by noting the maximum and minimum values of voltages.

In case of deeper insertion of the probe, errors due to reflections may get introduced and VSWR obtained may be lower than the actual. These errors become appreciable for VSWR < 10. Thus, the probe insertion has to be as less as possible for low VSWR since it represents discontinuity in itself. The insertion has to be a little more for high VSWR. For low signals ferrite isolator should be used as pad, whereas for higher signals an attenuator may be used. Matched load is not required if the device under test (DUT) is of absorbing nature. The meter may be a VSWR meter or any other voltage measuring device in this frequency range.

10.8.3.2 Twice Minimum Method

This method is also preferred for measuring low VSWR since it eliminates errors due to insertion. The setup for this method is the same as given in Figure 10.19. Here, first a probe is inserted and a minimum located. The probe is moved in one direction to a point, where power is twice the minimum. The position of the probe at this power location is noted as d_1. The probe is again moved but in other direction to a point, where power is twice the minimum, this position is noted as d_2. The VSWR [= $\lambda_g/(d_1 - d_2)$] can now be obtained.

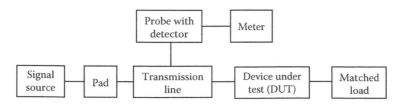

FIGURE 10.19
Experimental setup to measure VSWR (transmission line method).

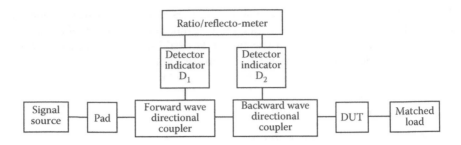

FIGURE 10.20
Experimental setup to measure VSWR (ratiometer method).

10.8.3.3 Reflectometre Method

The experimental setup is shown in Figure 10.20.

This method can be used for low, medium and high VSWR as there is no probe insertion or no problem of discontinuity. Figure 10.21 shows the use of two directional couplers to get the ratio of readings (D_1 and D_2) of the two detector indicators. D_1 is the measure of signal travelling in the forward direction, whereas D_2 is the measure of the reflected signal. The ratio D_2/D_1 is a measure of the reflection coefficient (Γ) that is related to VSWR by the relation

$$\text{VSWR} = \frac{E_{max}}{E_{min}} = \frac{[1+|\Gamma|]}{[1-|\Gamma|]} \tag{10.3}$$

If D_1 and D_2 are so combined that their ratio can directly be read this setup can be used to measure the reflection coefficient. Such an instrument is called ratiometer or reflectometer.

10.8.4 Measurement of Phase Shift

Knowing the phase shift (ϕ) is equivalent to knowing the electrical length of a network. Since ($m\lambda + \lambda/4$) for ($0 \le m < \infty$) will have the same effect in terms of phase, prior knowledge of approximate electrical length is necessary.

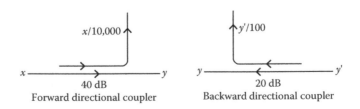

FIGURE 10.21
Combination of (a) forward and (b) backward directional couplers.

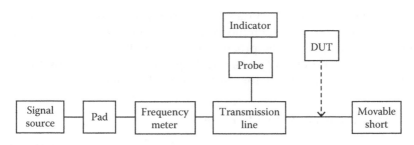

FIGURE 10.22
Experimental setup 1 for measuring phase shift.

10.8.4.1 Method I

The experimental setup is shown in Figure 10.22. First, the location of minima (say L_1) is noted without locating the DUT in the setup. Later, the device is inserted and the new location of minima (say L_2) is noted. The network insertion will introduce a phase shift [$\cong (L_2 \sim L_1) \lambda$]. Since one λ is equivalent to the phase shift (ϕ) of $360°$ or 2π radians the value of ϕ can be calculated. The change of phase of a component or network due to variation of some parameter is sometimes desired. As an example consider the movement of a dielectric rod inside a waveguide from one end to the mid-position or the other end. For such a case a graph of ϕ as a function of position can be plotted.

10.8.4.2 Method II

The experimental setup shown in Figure 10.23 can be used to measure the phase of a non-reciprocal device (e.g., ferrite phase shifter).

In this case power is fed into power splitter (E-plane tee) for dividing the signal into two halves. For proper matching and to prevent interaction appropriate pads may be used. In slotted line two equal or nearly equal signals of the same frequency will travel in opposite directions resulting in deep nulls $\lambda/2$ apart. The minimas with zero current are noted. The current is to be

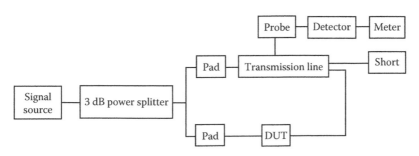

FIGURE 10.23
Experimental setup 2 for measuring phase shift.

varied in small steps and the positions of new minimas are to be recorded. If the component is $\lambda/2$ long, then the minima will move $\lambda/4$ towards the component. Thus, the path of the probe is lengthened by $\lambda/4$ and that through the component is shortened by $\lambda/4$, resulting in a total of $\lambda/2$. Thus, the phase shift using this setup is twice the shift in the null position.

10.8.5 Measurement of Quality Factor

The experimental setup is given in Figure 10.24. The steps to be adopted are as follows:

The frequency of the signal is varied and the output at different frequencies is recorded. In view of these observations the frequency response can be plotted. The nature of the likely response is shown in Figure 10.25. The Q $\{=f_0/(f_2-f_1)\}$ can now be calculated.

In case of unloaded cavity matched load is not to be connected. Q can also be obtained using the method for measuring attenuation. Other more accurate methods, not commonly used due to non-availability of required components in labs include (i) dynamic method, (ii) comparison method and (iii) impedance method.

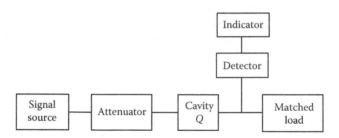

FIGURE 10.24
Experimental setup to measure quality factor 'Q'.

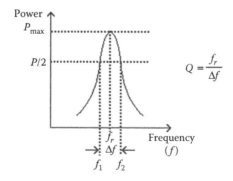

FIGURE 10.25
Frequency response of the cavity.

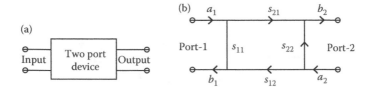

FIGURE 10.26
Illustration of (a) a two-port network and (b) its signal flow graph.

10.8.6 Measurement of Scattering Parameters

Scattering parameters have been discussed in Chapter 8, where the relations of s_{11}, s_{12}, s_{21} and s_{22} are given by Equation 8.7. To understand the measurement process of these parameters consider Figure 10.26, which illustrates a simple two-port network and its signal flow graph.

The equivalent circuit for a four-terminal network can be obtained by measuring input impedance (z_i) of the network, (i) when it is short circuited ($z_L = 0$), (ii) open circuited ($z_L = \infty$) and (iii) when a matched load ($z_L = z_0$) is connected. The short-circuit and open-circuit conditions are obtained by sliding movable short at output terminals. For third condition, matched load (ML) replaces the movable short. If the three output impedances cause the reflection coefficients $\Gamma_{oc} = -1$, $\Gamma_{sc} = 1$ and $\Gamma_{ML} = \Gamma_0$ and are measured, one gets

$$s_{11} = \Gamma_0 \qquad (10.4a)$$

$$s_{22} = (2\Gamma_0 - \Gamma_{sc} - \Gamma_{oc})/(\Gamma_{sc} - \Gamma_{oc}) \qquad (10.4b)$$

$$s_{11}s_{22} = s_{12}^2 = \{\Gamma_0(\Gamma_{oc} - \Gamma_{sc}) - 2\Gamma_{sc}\Gamma_{oc}\}/(\Gamma_{sc} - \Gamma_{oc}) \qquad (10.4c)$$

If network and the reference planes are physically symmetrical

$$s_{11} = s_{22} = (\Gamma_{sc} + \Gamma_{oc})/(2 - \Gamma_{sc} + \Gamma_{oc}) \qquad (10.5a)$$

$$s_{11}s_{22} - s_{12}^2 = (2\Gamma_{sc}\Gamma_{oc} + \Gamma_{sc} - \Gamma_{oc})/(2 - \Gamma_{sc} + \Gamma_{oc}) \qquad (10.5b)$$

In view of the above expressions, the problem now reduces to the measurement of reflection and transmission coefficients the measurements of which are discussed in subsequent subsections.

10.8.7 Measurement of Insertion Loss/Attenuation

The insertion loss of a network is given by

$$IL = 10 \log_{10} (P_1/P_2) = L_D + L_R + L_i \qquad (10.6)$$

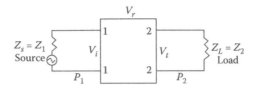

FIGURE 10.27
Voltage contents of waves and impedances at different locations.

where P_1 is the power delivered by source to the load in the absence of the network and P_2 is the power delivered in the presence of the network. These P_1 and P_2 may differ due to dissipation loss (L_D) in network, reflection loss (L_R) due to mismatch and the intrinsic loss (L_i) of network for one-way propagation. These losses are related to scattering parameters as

$$L_D = 10 \log_{10}\left[|s_{12}|^2/(1 - |s_{12}|^2)\right] \tag{10.7a}$$

$$L_R = 10 \log_{10}(1 - |s_{11}|^2) \tag{10.7b}$$

$$L_i = 10 \log_{10}|s_{11}|^2 \tag{10.7c}$$

The efficiency of the system can also be calculated by using the relation:

$$\eta = |s_{12}|^2/(1 - |s_{11}|^2) \tag{10.8}$$

If P_1 is the maximum power available from the source and P_2 is the maximum power available at the load end, then the insertion loss is given by

$$\text{Insertion loss} = 10 \log_{10}(P_1/P_2) \tag{10.9}$$

If V_i, V_t and V_r are the voltage contents of incident, transmitted and reflected wave at respective locations and Z_1 and Z_2 are the impedances shown in Figure 10.27, the insertion loss is

$$\text{Insertion loss} = 10 \log_{10}(V_i^2/V_t^2)(Z_2/Z_1) \tag{10.10}$$

10.8.7.1 Insertion or Power Ratio Method

The block diagram of experimental setup is shown in Figure 10.28. The block shows a setup without a device wherein the device whose insertion loss is to be measured is inserted. This method is fairly accurate and most commonly used. The steps to be followed are as follows:

The readings of meters M_1 (without insertion) and M_2 (with insertion) are noted, the insertion loss ($= M_2 - M_1$) is calculated and converted into decibels as given below

$$\text{Loss in dB} = 20 \log_{10}(M_1/M_2) = 20 \log_{10}|1/\Gamma| \tag{10.11}$$

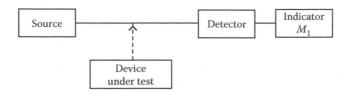

FIGURE 10.28
Experimental setup for measurement of insertion loss (power ratio method).

If matching is not perfect at input or output or both ends one of the following methods is used.

10.8.7.2 Substitution Methods

Three arrangements based on this principle are shown. In all these cases the signal source, unknown network, precision variable attenuator and detecting device are to be connected in series. Power level at detector is to be adjusted to some convenient reference level by standard attenuator. The attenuator readings are then noted. The network is removed and the same power level is adjusted again by changing the attenuation. The attenuator readings are again noted. Difference between the two-attenuator readings will give the insertion loss.

If the detector is a bolometer, the problems of low response and linearity are important. The range of measurement depends on the signal power available, sensitivity of the receiver system and the range of the attenuator. The signal power is limited by the ability of the network to withstand it. The sensitivity of the receiver has a theoretical limit. If narrow bandwidths are used extremely low level of signals can be measured.

R. F. substitution: This method (Figure 10.29) may be used upto 50 dB attenuation.

A. F. or D. C. substitution: This method (Figure 10.30) may be used upto 100 dB. It gives better accuracy upto 50 dB.

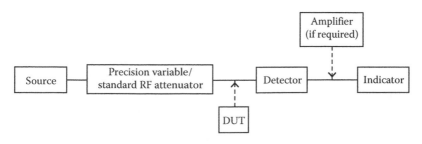

FIGURE 10.29
Experimental setup for measurement of insertion loss (RF substitution method).

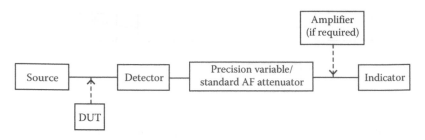

FIGURE 10.30
Experimental setup for measurement of insertion loss (AF/DC substitution method).

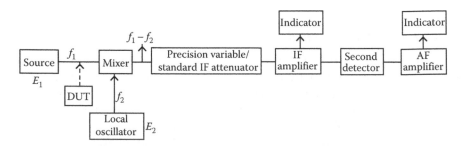

FIGURE 10.31
Experimental setup for measurement of insertion loss (IF substitution method).

If substitution using linear detection: This method is good upto 100–150 dB.

In Figure 10.31, frequencies of signal source (f_1) and local oscillator (f_2) and their respective field strengths E_1 and E_2 are shown. As $E_1 \ll E_2$, $E_1 \ll E_2, E_1^2 \lll E_2^2$, these field strengths give

$$(E_1 + E_2)^2 = E_1^2 + E_2^2 + 2E_1E_2 \approx E_2^2 + 2E_1E_2 \qquad (10.12)$$

Thus, the output is linearly related with E_1.

10.8.7.3 Cavity Resonance Method

Cavity resonators have been discussed at length in Chapter 4. A long resonator can be formed by shortening a transmission line (say waveguide) at both ends. If L is the length of the line, R is the resistance of shorting device at its ends, λ_g is the guide wavelength, λ_0 is the free space wavelength in dielectric medium filling the waveguide, Z_0 is the characteristic impedance and α the attenuation constant in nepers per unit length, then the quality factor Q is given by

$$Q = (\lambda_g/\lambda_0)^2 \, \pi L/[\alpha L + (2R/Z_0)] \, \lambda_g \qquad (10.13a)$$

$$= (\pi/\alpha\lambda_g) \, (\lambda_g/\lambda_0)^2 \quad \text{(when R approaches zero)} \qquad (10.13b)$$

$$1/Q = [(\lambda_0/\lambda_g)^2 \, (\alpha\lambda_g/\pi) + (2R\lambda_g/(\pi L Z_0))] \tag{10.14a}$$

$$= (\lambda_0/\lambda_g)^2 \, (\alpha\lambda_g/\pi) \quad \text{(when R approaches zero)} \tag{10.14b}$$

If λ_0 for TEM mode, λ_g and Q are known α can be determined as

$$\alpha = (\pi/Q \, \lambda_g) \, (\lambda_g/\lambda_0)^2 \quad \text{in nepers per unit length} \tag{10.15}$$

10.8.7.4 Scattering Coefficient Method

The setup is shown in Figure 10.32. Insertion loss L_{12} of the network shown (for terminal 1-1 to 2-2) or L_{21} (for 2-2 to 1-1) is given by

$$L_{12} = -10 \log_{10} (1 - |s_{11}|^2) - 10 \log_{10} [|s_{12}|^2/(1 - |s_{11}|^2)] \tag{10.16a}$$

$$L_{21} = -10 \log_{10} (1 - |s_{22}|^2) - 10 \log_{10} [|s_{21}|^2/(1 - |s_{22}|^2)] \tag{10.16b}$$

If s_{11}, s_{12}, s_{21} and s_{22} are known L_{12} and L_{21} can be determined.

10.8.8 Measurement of Dielectric Constant

The input impedance of a homogeneous non-magnetic dielectric filled semi-infinite waveguide carrying a TE mode is given by

$$Z = \{1 + (\lambda_0/\lambda_c)^2\}^{1/2}/\{\varepsilon - (\lambda_0/\lambda_c)^2\}^{1/2} \tag{10.17}$$

where λ_0 is the wavelength in free space, λ_c is the cutoff wavelengths for empty waveguide and $\varepsilon \, (= \varepsilon' - \theta\varepsilon'')$ is the complex permittivity of the (solid or liquid) material. The real and imaginary parts of permittivity are given by

$$\varepsilon' = (\lambda_0/\lambda_c)^2 + [\{1 - (\lambda_0/\lambda_c)^2\} \{\rho^2 \sec^4 \beta d - (1 - \rho^2)^2 \tan^2 \beta d\}/(1 + \rho^2 \tan^2 \beta d)^2] \tag{10.18a}$$

$$\varepsilon'' = [\{1 - (\lambda_0/\lambda_c)^2\} \{2\rho (1 - \rho^2) \sec^2 \beta d - \tan^2 \beta d\}/(1 + \rho^2 \tan^2 \beta d)^2] \tag{10.18b}$$

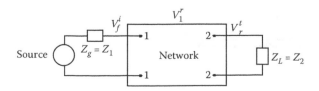

FIGURE 10.32
Experimental setup for measuring scattering coefficients.

In these expressions ρ is the VSWR, d is the separation of first minima from the load and $\beta\ (=2\pi/\lambda)$ is the propagation constant in the feeding guide. In view of the above the problem of measuring dielectric constant reduces to the problem of measurement of ρ and d.

10.8.9 Measurement of Impedance

10.8.9.1 Bridge Method

There are many bridges used to measure the impedance at microwave frequencies including Wheat-Stone Bridge, Byrene Impedance Bridge Pseudo Bridge and Magic Tee Bridge. Here, Magic Tee Bridge is the only one described.

The configuration of Magic Tee Bridge is shown in Figure 10.33a and its equivalent line diagram in Figure 10.33b. It contains a magic tee, a variable attenuator and a movable short. Items connected at different ports are also shown. This configuration follows the principle of basic bridge circuit composed of four arms. One of these arms is connected to a source, the second arm to a detector, the third arm to known impedance and the fourth one to unknown impedance. The bridge is balanced by adjusting the known (standard) impedance and the detector reading is noted. The procedure is shown by an illustration in Figure 10.33b. In this case the bridge is balanced by varying the impedance S. For null detection $X = S$. Since the construction of variable standard impedance is difficult this limits the usefulness of this method.

10.8.9.2 Slotted Line Method

For this method, the setup for measuring the VSWR is to be used. Transmission line terminated in an unknown impedance z_L is shown in Figure 10.34. z_s and z_L can be written as

$$z_s = -(z_L + j \tan \beta l)/(1 + j\, z_L\, \beta l) \tag{10.19a}$$

$$y_s = -(y_L + j \tan \beta l)/(1 + j\, y_L\, \beta l) \tag{10.19b}$$

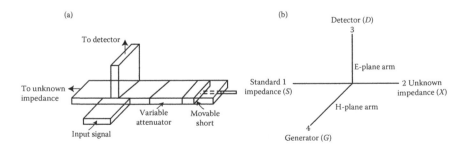

FIGURE 10.33
(a) Magic tee bridge and (b) its equivalent line diagram.

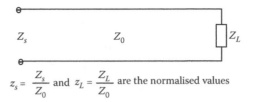

$$z_s = \frac{Z_s}{Z_0} \text{ and } z_L = \frac{Z_L}{Z_0} \text{ are the normalised values}$$

FIGURE 10.34
Terminated transmission line.

Here, z_L can be calculated from VSWR and distance L of minimum position from the unknown impedance. At minimum position the voltage and current both are real, thus z_L is also real. At minimum Z_s is reciprocal of VSWR and at maximum Z_s = VSWR. Since Z_s is a complex quantity, Smith chart may be used to avoid calculations. Move the probe on the transmission line (with movable short attached to it) and measure the minima. This minimum is called reference minima and position of the short is the reference plane. Replace movable short by the unknown impedance, and measure the minima. The minima will shift in its location. The direction of shift in respect of the reference minima can be moved on the Smith chart and z_L can be determined.

10.8.10 Measurement of Power

Microwave power levels for the purpose of measurement are generally divided into three categories as given below:

1. *Low power*: in this category the power is generally less than 10 mW
2. *Medium power*: this varies between 10 mW and 10 W and
3. *High power*: the power level greater than 10 W

At microwave range the power can be measured in the following ways.

(a) By absorbing all the available power in the measuring device. Since the maximum power dissipation capability of device will limit the absorption and hence the range of measurement. (b) By sampling a known fraction of power from transmission path going to the load. This method extends the range of measurement. Under this category the methods commonly employed for power measurement are (i) bolometer bridge method, (ii) unbalanced bridge method, (iii) automatically balanced power ratio Bridge method and (iv) microwave (static or circulating) calorimetric method.

10.8.10.1 Bolometer Bridge Method

The bolometer element is placed into the bolometer mount or holder (shown in Figure 10.35a). It forms one of the arms of the bridge (shown in Figure 10.35b)

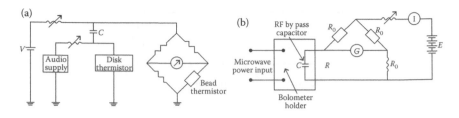

FIGURE 10.35
Power measurement. (a) Bolometer bridge circuit and (b) experimental setup.

and is subjected to heating by incoming microwave power. There should not be any leakage of power. The resistance of bolometer changes with the heating.

Initially the bridge is balanced by supplying dc power through an amme-ter and an adjustable resistor. Next, the microwave power is applied, which results in the change of resistance of bolometer and unbalances the bridge. The bridge is again balanced by changing the dc power. This change is a measure of the microwave power. This method is useful for low and medium power measurements. For high power measurement calorimetric method is preferred.

10.8.10.2 Calorimetric Method

Calorimetric method (Figure 10.36) is based on the conversion of a fraction of radio frequency energy into heat, which is absorbed by a liquid (most often flow of water at microwave frequencies), filling the power detector (calorimeter). The difference in temperature between the water entering and leaving the chamber is proportional to the RF energy applied. Water may absorb heat from a terminating resistor or serve as load itself. If Q is the quantity of heat, W is the volume of water flow through a calorimeter, ΔT is the difference between the inlet (T_1) and outlet (T_2) temperatures, Q is given by

$$Q = W \, \Delta T \tag{10.20a}$$

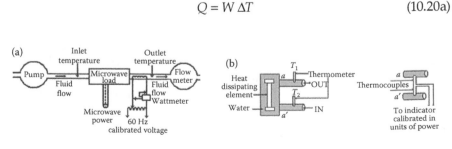

FIGURE 10.36
Calorimeter–wattmeter. (a) Experimental setup and (b) details of (a).

From Joules–Lenz's law

$$Q = 0.24 \ I^2R \ t = 0.24 \ Pt \qquad (10.20b)$$

$$\text{Thus, } Q = W \ \Delta T = 0.24 \ Pt \qquad (10.20c)$$

$$P = W \ \Delta T/0.24 \ t = 4.17 \ W \ \Delta T/t \qquad (10.20d)$$

where P is the unknown power (to be measured) and t is the measurement time. The volume W is determined as the difference between readings of total flow meter at the beginning and at the end of the observation.

The temperature difference is measured by thermometers or by thermocouples after a steady state is reached. With thermocouples connected in series opposition, the micro-ammeter may be calibrated in unit of ΔT. If, in addition, the rate of flow of water ($u = W/t$) is maintained constant, the calibration may be directly obtained in the units of power ($P = 4.1 \ u \ \Delta T = a \ \Delta T$). The matching of generator and line to load (for applied microwave energy to get fully absorbed) is of paramount importance. This method is used for measuring medium and high power range. The accuracy of the measurement depends on the measurement of ΔT, u, quality of match and leakage of microwave power.

10.8.11 Measurement of Electric Field Intensity

The pointing vector $\mathbf{P}$ in W/m² is given by $\mathbf{P} = \mathbf{E} \times \mathbf{H}$, where $\mathbf{E}$ is the electric field intensity and $\mathbf{H}$ is the magnetic field intensity. Since $\mathbf{E}$ and $\mathbf{H}$ are perpendicular $\mathbf{P}$ will be perpendicular to both $\mathbf{E}$ and $\mathbf{H}$. If one component of each, $\mathbf{E}$ and $\mathbf{H}$, exists $\mathbf{P}$ vector will also have one component. P ($=EH$) can simply be written in scalar form. Since $E/H = \eta = 120\pi$, the expression of power can also be written as

$$P = E^2/120\pi \quad \text{or} \quad E = [120\pi \ P]^{1/2} \qquad (10.21a)$$

Now if, W ($=PS$) is the total power dissipated by a surface area (S in m²) in Watts

$$E = [120\pi \ W/S]^{1/2} \qquad (10.21b)$$

Thus, the problem of measurement of E reduces to the measurement of power. The block diagram as shown in Figure 10.37 can be used to measure W and hence E.

When an antenna (with effective length h_e) is placed in the electromagnetic field of plane wave parallel to $\mathbf{E}$, an emf is induced in it, which can be given by the relation $V = E \ h_e$. The problem of measurement of E reduces to the measurement of voltage, which can be measured by a voltmeter or by other methods.

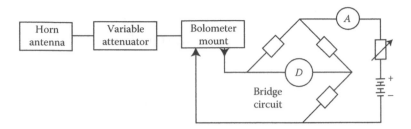

FIGURE 10.37
Setup to measure electric field intensity.

Loop antennas are used for measurement in LF, MF and HF band, half-wave dipoles in VHF and UHF band and horn antennas in microwave band.

10.8.12 Measurement of Reflection Coefficient

Reflection coefficient can be obtained if VSWR is known. It can also be obtained by using directional coupler through which forward and reflected wave quantities can be measured.

10.8.12.1 Comparison Method

The setup for this method is shown in Figure 10.38. The steps to be followed are as follows:

The power generated by source is to be suitably attenuated. This power from the attenuator enters E arm and divides into two halves in arms A and B. The power of arm B is reflected by the network connected at reference plane $X–X$. A suitable adjustment of the tuner reflects power of arm A. For standard reflector at $X–X$, adjust the tuner such that the reflected powers in two branches are equal in amplitude and phase. At this point, the detector will

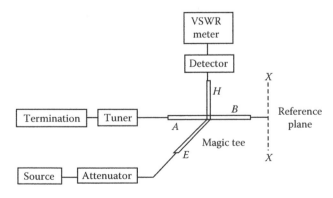

FIGURE 10.38
Setup for comparison method.

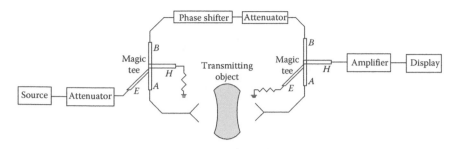

FIGURE 10.39
Setup for measurement of transmission coefficient by comparison method.

display a null. Replace standard reflector by an unknown reflector. The two reflecting waves will differ in amplitude. The difference is detected, amplified and displayed on a calibrated meter.

10.8.13 Measurement of Transmission Coefficient

Since the transmission coefficient $T = 1 - |\Gamma|$, if reflection coefficient is known the transmission coefficient can be obtained.

10.8.13.1 Comparison Method

The setup to be used for this method is given in Figure 10.39. The steps are as follows:

A signal is fed to arm E, which divides into two halves appearing at arms A and B. H arm is terminated into a matched load. Signal from arm B travels through a phase shifter and a variable attenuator to arm B'. Signal from arm-A travels through a transmitting load placed between two horns to arm-A'. Initially a known reference load is taken. The phase shifter and attenuator are so set that the signals arriving at B' and A' are equal in phase and amplitude. Thus, there is no signal at arm H and detector gives a null. Change the load. The detector at A' changes, signals at A' and B' differ, the difference is detected, amplified and displayed on a calibrated meter.

10.9 Some Practical Applications

10.9.1 Measurement of Gain of an Aerial Horn

Setup for measurement of gain is shown in Figure 10.40. The steps are as follows:

Two power meters, one in the transmitting circuit and another in the receiving circuit, are connected to obtain gain. A correction factor to account

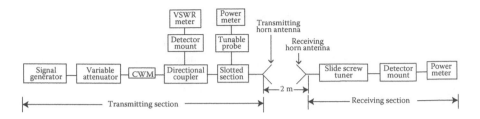

FIGURE 10.40
Setup for measurement of gain of an aerial horn.

for reflections is needed, which is obtained by using the VSWR meter to calculate reflection coefficient. For improving the accuracy of measurements
special care is required to ensure total interception of power radiated by the
transmitting antenna. This demands exact alignment, proper tuning and
orientation of horns to the same plane.

10.9.2 *E* and *H* Plane Radiation Patterns

The setup is given in Figure 10.41. The procedure to be followed is as follows:
Waveguide twist is used only for *H*-plane measurement. For *E*-plane measurement twist is replaced by a plane waveguide. Suitable components are
connected to both horns for power measurements. Once the power level at
the transmitting end is set the output power meter reading can be noted by
rotating the transmitting antenna (through rotating table) to various angular
positions. Points are marked on a polar graph and joined to get radiation
pattern. The direction of maximum radiation and the beam width can be
obtained from the graph.

10.9.3 Measurement of Thickness of a Metallic Sheet

Figure 10.42a gives the method for measuring the thickness. The procedure
is described as follows:

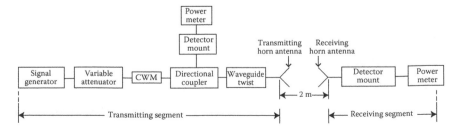

FIGURE 10.41
Setup for the measurement of *E* and *H* plane radiation patterns.

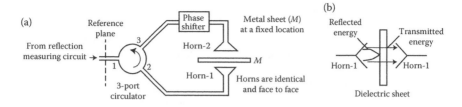

FIGURE 10.42
Setup for thickness measurement of (a) metallic sheet and (b) dielectric sheet.

Power enters port 1, appears at port 2, travels to horn 1, gets reflected, arrives at port 2, appears at port 3, travels to horn 2 through a phase shifter, gets reflected, travels to port 3 again passing through phase shifter, enters port 3, appears at port 1 and finally travels to reflection measuring circuit. As a first step, take M of standard (known) thickness, adjust the tuner and phase shifter to obtain null deflection. Replace standard sheet by sheet of unknown thickness. A change in thickness will result in change of detected signal, which can be brought to null either by changing the phase or by changing tuning. The thickness can be calibrated either in terms of phase or detector reading.

10.9.4 Measurement of Thickness of a Dielectric Sheet

In this case also the setup for measuring transmission coefficient (Figure 10.39) is to be used. As the sheet in the present case is made of a dielectric material there will be partial reflection and partial transmission. It is illustrated in Figure 10.42b. Thus, the reflected energy to be measured will be relatively less. In case of change of thickness the ratio of transmitted and reflected power will change. Indicating meter can thus be calibrated in terms of thickness.

10.9.5 Measurement of Wire Diameter

In the setup for measurement of transmission coefficient (Figure 10.39) the sheet is to be replaced by a wire (Figure 10.43a). Since the wire will diffract power radiated by horn 1, a change in wire diameter will alter the amplitude and phase of the signal reaching horn 2. These changes in phase and amplitude can be detected and displayed on a calibrated meter. Such a setup can also be used in drawing operation of wires.

10.9.6 Measurement of Moisture Content

Measurement of moisture content is equivalent to the measurement of attenuation and phase shift of electromagnetic wave passing through the material and reflected from it. Thus, the setups used for attenuation and phase

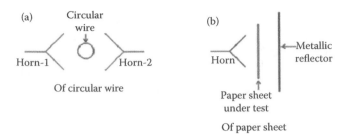

FIGURE 10.43
Location between horns of (a) circular wire and (b) paper sheet.

shift can be employed to monitor the moisture in paper and textile industry. Figure 10.43b illustrates the location of a paper sheet in between a horn and a metallic reflector.

The measurement techniques may be modified in accordance to the structure and shape of the test object. Setups for sheet materials (i.e. paper, textile, laminated wood) will differ from those for testing granulated or porous materials (sand, clay, bricks, etc.) or liquids.

10.9.7 Measurement/Monitoring Moisture Content in Liquids

Moisture content changes the effective permittivity of a liquid. Thus, the technique used to measure the resonance frequency of the cavity can be used to monitor the moisture. For optimum results E should be maximum along the axis of tube (Figure 10.44). If both liquids are moisture free there will be a null detection. If liquids differ in moisture contents, the meter will indicate a reading. The meter, therefore, can be calibrated for different moisture contents.

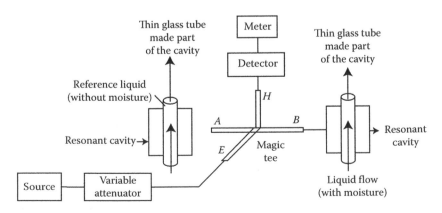

FIGURE 10.44
Monitoring of moisture content in liquids.

Descriptive Questions

Q-10.1 Discuss the working of a KPS. List the steps required for its proper operation in different modes.

Q-10.2 Discuss the working of a VSWR meter. Why it is preferred over other devices?

Q-10.3 How is slotted transmission line used in the measurement of frequency and wavelength?

Q-10.4 Discuss the working of (i) tunable probe, (ii) movable short and (iii) frequency meter.

Q-10.5 Illustrate the construction of a microwave detector and explain its working.

Q-10.6 Discuss the salient features and characteristics of barrater and thermistor.

Q-10.7 List the qualities to be possessed by a source, a measuring device and components to yield proper results at microwave frequencies.

Q-10.8 Discuss the precautions to be taken in microwave measurements.

Q-10.9 Describe a method which simultaneously yields values of wavelength and frequency.

Q-10.10 Discuss various methods for measuring phase shift.

Q-10.11 Illustrate the experimental setup to measure the quality factor.

Q-10.12 Discuss various ways in which the insertion loss or attenuation can be measured.

Q-10.13 Describe various methods of impedance measurement.

Q-10.14 Why are power levels categorised into low, medium and high power. How are these different power levels measured?

Q-10.15 How are E and H plane radiation patterns of a horn antenna measured?

Further Reading

1. Ginzton, E. L., *Microwave Measurements*. McGraw-Hill, New York, 1957.
2. Gupta, K. C., *Microwaves*. Wiley Eastern, New Delhi, 1979.
3. Kellijan, R. and Jones, C. L., *Microwave Measurement Manual*. McGraw-Hill, University of Wisconsin–Madison, 1965.
4. Kneppo, I., Comparison method of measurement Q of microwave resonators. *Trans. IEEE*, MTT-25, 423–426, 1977.
5. Lance, A. L., *Introduction to Microwave Theory and Measurements*. McGraw-Hill, New York, 1964.

6. Misra, D. K., Permittivity measurement of modified infinite samples by a directional coupler and sliding load. *Trans. IEEE*, 29, 65–67, 1981.
7. Reich, H. J. et al. *Microwave Principles*, D. Van Nostrand, Reinhold Co., New York, 1957 (an East West edition).
8. Saad, T. and Hansen, R. C., *Microwave Engineer's Handbook*, Vol. 1. Artech House, Dedham, MA, 1971.
9. Thomas, H. E., *Hand Book of Microwave Techniques and Equipment*. Prentice-Hall, Inc., Englewood Cliffs, NJ, 1968.
10. Thomas, H. L., *Microwave Measurement and Techniques*. Artech House, USA, 1976.

11

Basics of Microwave Tubes

11.1 Introduction

Before the microwave energy is transmitted, radiated, propagated or processed to achieve the desired goal, it has to be generated and amplified. At the dawn of the electronic era, this task was carried out by vacuum tubes. In view of the explosion of knowledge, the use of higher frequencies became a necessity. As the frequency range was enhanced, it was noticed that these conventional (vacuum) tubes utterly fail to yield the desired results beyond certain limits. In the quest to overcome these limitations, a new class of tubes, referred to as microwave tubes, emerged. Before discussing the construction, principles of operation and other aspects of microwave tubes, it is necessary to understand the following aspects:

1. Causes of failure of the conventional tubes at higher frequencies
2. Behaviour of charged particles in the presence of different fields since the working of all the tubes, particularly those used at microwave range, is based on the movement of electrons
3. Concept of velocity modulation, a key phenomenon in almost all microwave tubes

In the following sections, these three aspects that can be termed as basics of microwave tubes are discussed. Finally, the classification of microwave tubes is also included.

11.2 Frequency Limitations of Conventional Tubes

Figure 11.1 illustrates some of the vacuum tubes that remained at the helm of affair in the domain of electronics and are still used in high-power RF transmitters. In the given figures, *A*, *K*, *G*, *CG* and *SG* represent anode (or plate), cathode, grid, control grid and suppressor grid, respectively.

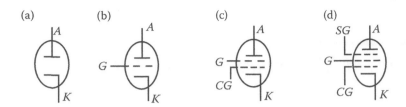

FIGURE 11.1
Four conventional tubes. (a) Diode, (b) triode, (c) tetrode and pentode.

These tubes perform well at lower frequencies. As the frequency is raised, the wavelengths become shorter and a circuit designer starts facing problems. By the time the designer enters the microwave domain, the situation becomes almost hopeless. The same tubes that operate well vis-à-vis generation and amplification remain no longer useful. The efficiency of a conventional tube remains largely independent of frequency up to a certain range and declines rapidly thereafter. The key factors that influence the performance of ordinary tubes at higher frequencies include inter-electrode capacitance, lead inductance and the electron transit time.

11.2.1 Inter-Electrode Capacitances

As per basic definition, the configuration of two electrodes separated by a dielectric possesses a property of storing electrostatic energy and the capacity of storage is termed as capacitance. As a tube may comprise several electrodes that are maintained at different potentials, there are capacitances between these electrodes. Depending on their surface areas and separation, these capacitances, termed as inter-electrode capacitances, may assume different values. Figure 11.2 shows inter-electrode capacitances in a triode-based circuit. These include the capacitance between plate and grid (C_{pg}), plate and cathode (C_{pk}) and grid and cathode (C_{gk}). Besides, L and C are the inductance and capacitance of the tuned circuit.

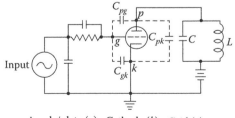

Anode/plate (p)   Cathode (k)   Grid (g)

FIGURE 11.2
Inter-electrode capacitances in a vacuum tube.

At low or medium frequencies, the capacitive reactances of these inter-electrode capacitances are quite large and tube operation is not noticeably affected. As the frequency increases, the reactances of these capacitances start reducing. At a stage, these reactances become small enough and greatly affect the circuit performance. Thus, the higher the frequency, or the larger the inter-electrode capacitance, the higher will be the current through capacitance.

The circuit in Figure 11.2 shows that C_{gk} lies in parallel with the signal source. As the frequency of the input signal increases, the effective grid-to-cathode impedance of the tube decreases due to the decrease in the reactance of the inter-electrode capacitance. At about 100 MHz, the reactance of C_{gk} becomes quite small and the signal is short circuited within the tube. In view of Figure 11.2, some of the inter-electrode capacitances are effectively in parallel with the tuned circuit and thus, will affect the resonance frequencies of the tuned circuits to which they belong.

11.2.2 Lead Inductances

Inductance is defined as flux linkage per unit ampere of the current. Wherever there is flow of current, it results in a magnetic flux. Its linkage with the source current or with those flowing in another branch or circuit, in the vicinity, results in inductance. At lower frequencies, these inductances remain almost sleeping elements and become active as the frequency is raised to a sufficiently higher level. Since lead inductances within a tube are effectively in parallel with the inter-electrode capacitance, their net effect is to increase the limit of the frequency. As the inductance of the cathode lead is common to both the grid and plate circuits, it provides a path for degenerative feedback that reduces the overall circuit efficiency.

11.2.2.1 Impact of Lead Inductance and Inter-Electrode Capacitance

Figure 11.3 illustrates a circuit that contains a triode tube and an equivalent circuit wherein the triode is replaced by its equivalent current source and inter-electrode capacitances. In obtaining the equivalent circuit, it is assumed

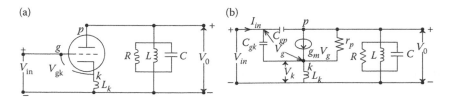

FIGURE 11.3
(a) Triode circuit and (b) equivalent circuit.

that inter-electrode capacitances (C_{gp} and C_{gk}) and cathode inductance (L_K) are the only parasitic elements present. These bear the following relations:

$$C_{gp} \ll C_{gk} \text{ and } \omega L_K \ll 1/\omega C_{gk} \tag{11.1}$$

Thus, the input voltage (V_{in}) and input current (I_{in}) can be written as

$$V_{in} = V_g + V_k = V_g + j\omega L_k g_m V_g = V_g[1 + j\omega L_k g_m] \tag{11.2a}$$

$$I_{in} = j\omega C_{gk} V_g \text{ or } V_g = I_{in}/j\omega C_{gk} \tag{11.2b}$$

The potentials V_g and V_k are shown in Figure 11.3b. Also, g_m is the transconductance of the tube.

From Equations 11.2a and 11.2b

$$V_{in} = I_{in}(1 + j\omega L_k g_m)/(j\omega C_{gk}) \tag{11.3}$$

The input admittance ($Y_{in} = I_{in}/V_{in}$) of the tube can be written as

$$Y_{in} = j\omega C_{gk}/(1 + j\omega L_k g_m) = \omega^2 L_k C_{gk} g_m + j\omega C_{gk} = A + jB \tag{11.4}$$

The cathode lead is usually short in length and is quite large in diameter and $g_m \ll 1$ m-mhos. Thus, the inequality ($\omega L_k g_m \ll 1$) is almost always true and the input impedance (Z_{in}) at very high frequencies is

$$Z_{in} = 1/Y_{in} = 1/(A + jB) = [1/(A + B^2/A)] - [j/(B + A^2/B)] \tag{11.5}$$

From Equations 11.4 and 11.5

$$Z_{in} = 1/[(\omega^2 L_k C_{gk} g_m) + (C_{gk}/(L_k g_m))] - j/(\omega C_{gk} + \omega^3 L_k^2 C_{gk} g_m^2) \tag{11.6}$$

At high frequencies, terms involving ω^2 and ω^3 are much higher than the associated terms in the denominator, that is, $[(\omega^2 L_k C_{gk} g_m) \gg \{C_{gk}/(L_k g_m)\}$ and $\omega^3 L_k^2 C_{gk} g_m^2 \gg \omega C_{gk}$. Thus, Equation 11.6 reduces to

$$Z_{in} \approx \{1/(\omega^2 L_k C_{gk} g_m)\} - j\{1/(\omega^3 L_k^2 C_{gk} g_m^2)\} \tag{11.7}$$

In Equation 11.7, Re[Z_{in}] $\propto 1/\omega^2$ and Im[Z_{in}] $\propto 1/\omega^3$.

For $f > 1$ GHz, Re[Z_{in}] tends to zero, the signal source is merely short circuited and thus, output power decreases rapidly.

Y_{in} and Z_{in} for a pentode can be similarly obtained. For a pentode also, Re[Z_{in}] $\propto 1/\omega^2$ and Im[Z_{in}] $\propto 1/\omega^3$ and the output power decreases as before.

By reducing the lead length and the electrode area, the effect of L and C can be minimised. Such minimisation, however, limits the power-handling capacity of the tube.

11.2.2.2 Gain: BW Product Limitation

Figure 11.4 illustrates an output-tuned circuit of a pentode. In such vacuum tubes, the maximum gain is generally obtained by resonating the output circuit. From this equivalent circuit, the following expressions can be obtained.

In Figure 11.4, r_p is the plate resistance, R is the lead resistance, L and C are the tuning elements, G is the conductance and V_L is the load voltage. Also, f_r is the resonant frequency, A_m is the maximum voltage gain at f_r and g_m is the trans-conductance of the tube. These parameters for $r_p \gg \omega L_k$ are related as

$$G = 1/r_p + 1/R \tag{11.8a}$$

$$V_L = (g_m V_g)/[G + j(\omega C - 1/\omega L)] \tag{11.8b}$$

$$f_r = 1/[2\pi \sqrt{(LC)}] \tag{11.8c}$$

$$A_m = g_m/G \tag{11.8d}$$

Since the bandwidth (BW) is measured at half-power points, the denominator of the load voltage expression can be related by

$$G = (\omega C) - (1/\omega L) = (\omega^2 LC - 1)/\omega L \tag{11.9a}$$

$$\text{or } \omega^2 - \omega(G/C) - 1/LC = 0 \tag{11.9b}$$

Equation 11.9b is a quadratic equation and its roots are

$$\omega_{2,1} = (G/2C) \pm \sqrt{[(G/2C)^2 + (1/LC)]} \tag{11.10a}$$

$$\text{BW} = \omega_2 - \omega_1 = G/C \quad \text{for } (G/2C)^2 \gg (1/LC) \tag{11.10b}$$

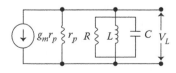

FIGURE 11.4
Output-tuned circuit of a pentode.

In view of Equations 11.8d and 11.10b, the gain BW product is

$$A_m\text{BW} = g_m/C \tag{11.11}$$

This product is not a function of frequency. Thus, for high gain, BW has to be narrow. This restriction is applicable to resonant circuits only. In megawatt devices, reentrant cavities and slow wave structures help in getting a higher gain over a wider BW.

There are several ways to overcome the limitations of conventional tubes. Inter-electrode capacitance can be reduced by increasing the spacing between electrodes or by reducing the size of electrodes of the tube. The increase in spacing of electrodes increases the problems associated with transit time. The reduction of tube size lowers the power-handling capability. Thus, none of these methods work well beyond 1 GHz. An effort to overcome one particular limitation will have a contrary effect on the other. Beyond an upper limit of approximately 1 GHz, the conventional tubes are rendered almost useless and may be used only after incorporating certain modifications and that too, up to a certain specified frequency range. For still higher frequencies, specialised (microwave) tubes are now readily available that can generate and amplify signals to high-power levels more cheaply than solid-state devices over a wide range of microwave frequencies.

11.2.3 Transit Time Effect

Transit time is the time taken by electrons to travel from the cathode to the anode. At lower frequencies, this time is quite insignificant as compared to the signal time period and can safely be assumed that electrons leave the cathode and arrive at the anode instantaneously. At microwave frequencies, the transit time becomes an appreciable fraction of signal time period and begins to affect the efficiency. As an example, a transit time of 1 ns is only 0.001% of (1 μs) the time corresponding to 1 MHz. At 1 GHz, the same transit time becomes equal to the time required for an entire cycle. The delay due to the transit time becomes comparable to the time period and therefore, the phase difference between the input and its response is too large.

The transit time depends on the spacing and the potential difference between the electrodes in a tube. If the transit time exceeds 10% of a cycle, it results in a significant decrease in tube efficiency. This decrease is caused, in part, by a phase shift between the plate current and grid voltage. The plate current must be in phase with the grid-signal voltage and must be 180° out of phase with the plate voltage for effective operation of the tube. When the transit time approaches 25% of a cycle, this phase relationship does not hold true. The electrons leave the cathode and travel to the plate under the influence of a positive swing of a high-frequency grid signal. The current due to this flow of electrons is initially in phase with the grid voltage. As the transit time becomes an appreciable fraction of a cycle, this current arriving at the plate lags behind the grid-signal voltage. As a result, the anode and grid signals are no longer 180° out of phase. This causes design problems especially with feedback in oscillators. The grid begins to take power from the deriving source. This power is absorbed by the grid and results in heating. Thus, it results in the decrease of output of the tube and in an increase of the plate power dissipation.

In mathematical terms, the electron transit angle θ_g is given by

$$\theta_g = \omega\tau_g = \omega d/v_0 \qquad (11.12)$$

where ω is the angular frequency, d is the separation between the grid and the cathode, τ_g ($=d/v_0$) is the electron transit time across the gap, v_0 ($=0.593 \times 10^{-6}\sqrt{V_0}$) is the velocity of the electrons and V_0 is the dc voltage.

In view of the above discussion, one can conclude that below microwave frequencies, the transit time is almost negligible. At microwave frequencies, it is large compared to the period of the microwave signal and the potential between the grid and the cathode may alternate from 10 to 100 times during the electron transit. Thus, during the negative half-cycle, the grid potential removes energy that was given to electrons during the positive half-cycle. Hence, the electron may oscillate back and forth in the grid–cathode space or may even return to the cathode. It results in the reduction of operating efficiency of the tube. This degenerative effect becomes more significant for frequencies much higher than 1 GHz. Once electrons pass the grid, they are quickly accelerated to the anode by the high plate voltage. For lesser frequencies ($f < 1$ GHz), output delay is negligible in comparison to the phase of the grid voltage. This indicates that the transadmittance is a large real quantity that is the usual trans-conductance g_m. At microwaves, the transit angle is not negligible and the trans-conductance becomes a complex quantity with relatively small magnitude. As a result, output power decreases.

Besides the above key factors, there are some more effects that also influence the tube performance up to some extent. These include (i) skin effect, (ii) electrostatic induction and (iii) dielectric losses. These are briefly described in the following section.

11.2.4 Skin Effect

As discussed in Section 2.9, at high frequencies, the current in conductors tries to flow entirely on the skin of conductors. In view of the inverse relation of resistance R ($=\rho L/A$), a decrease in the area (A) of current carrying the segment of R increases. Thus, with the increase of frequency, skin effect causes a significant increase in series resistance of leads at UHF and above.

11.2.5 Electrostatic Induction

It is another cause of power loss in tubes. As the electrons forming the plate current move, they electrostatically induce a potential in the grid. This induction causes currents of positive charges to move back and forth in the grid structure. This action resembles the action of hole–current in semiconductor devices. At low frequencies, the current induced on one side of the grid by the approaching electrons is equal to the current induced on the other side by the receding electrons. The net effect is zero as the two currents are in opposite directions and cancel each other. At higher frequencies, when the transit time becomes an appreciable part of a cycle, the number of electrons approaching the grid is not always equal to the number of electrons going away. As a result, the induced currents do not cancel. This uncancelled current results in power loss in the grid and is considered to be of resistive nature. This loss amounts to the presence of an imaginary resistor between the grid and the cathode. The resistance of this resistor decreases rapidly with the increase of frequency. Its value may ultimately become so low that the grid is short circuited to the cathode. Thus, the proper operation of the tube is prevented.

11.2.6 Dielectric Loss

With the increase in frequency, the dielectric loss increases and tube efficiency decreases. Thus, for proper amplification, the base of the tube is to be made of low loss dielectric.

11.3 Influence of Fields on Motion of Charged Particles

Irrespective of the frequency of operation, the working of all the tubes revolves around the movement of electrons. These electrons get initially liberated under the influence of heat energy from the cathode, move forward and try to reach the anode, their assigned destination, under the influence of field setup by the potential difference maintained between different electrodes. There are situations (e.g. in vacuum tubes) wherein the electrons move only under the influence of the electric (E) field. In some other cases (e.g. tubes with focussing magnets), electrons are influenced by the magnetic (B) field. Since most of the tubes are used for the amplification of weak ac signals, the field due to ac signals also affects the behaviour of electrons. In view of the alternating nature of this field, the electrons may get accelerated or decelerated. Thus, the behaviour of electrons may be affected by one (E or B), two (E and B) or all the three (E, B and ac) fields, simultaneously. The three cases of influence of the fields on electrons are considered below.

1. In the case of a potential difference (V) between two electrodes, electrons will come under the influence of an electric field intensity E

(= V/d, where d is the separation between two electrodes). The force (F) exerted by E on an electron (with charge $-e$) is given by

$$\mathbf{F} = -e\,\mathbf{E} \tag{11.13}$$

Equation 11.13 imposes no restriction on the charge whether it is moving or stationary.

2. In most of the tubes, the liberated electrons are to be focussed on a narrow beam. The task is accomplished by the magnetic field. If only this field is present and the electrons move with a velocity $\mathbf{U}$, the resulting force can be given by

$$\mathbf{F} = -e\,(\mathbf{U} \times \mathbf{B}) \tag{11.14}$$

According to this equation, $\mathbf{B}$ exerts force only if the particle is in motion (i.e. $\mathbf{U} \neq 0$). Also, this force is perpendicular to both **(U)** and **(B)** vectors.

3. If both $\mathbf{E}$ and $\mathbf{B}$ are simultaneously present, the net force is the sum of Equations 11.13 and 11.14. The resulting expression referred to as *Lorentz force equation* is given below:

$$\mathbf{F} = -e\,[\mathbf{E} + (\mathbf{U} \times \mathbf{B})] \tag{11.15}$$

4. If a tube is used to amplify, the presence of an ac field ($E'\cos\omega t$) is also to be accounted. Assuming this field to be along the direction of $\mathbf{E}$, Equation 11.15 may be modified to

$$\mathbf{F} = -e\,[(\mathbf{E} + \mathbf{E}'\cos\omega t) + (\mathbf{U} \times \mathbf{B})] \tag{11.16}$$

Equation 11.16 gives a force on a charge due to the combined effect of E-, B- and ac fields. In this, E' is the magnitude of the field intensity of the ac field and ω is the angular frequency.

The journey from Equations 11.13 through 11.16 indicates the gradual increase of the complexity of force expressions. From the point of view of the solution, the situation drifts from bad to worst. To simplify the problem, certain assumptions to be made are given below:

1. The particle on which the force is exerted is an isolated point charge.

2. In no case, other field(s) except that involved in a particular equation are present.

3. Since the motion of charges constitutes current and the presence of any current causes the magnetic field, the impact of all such fields on electrons is ignored.

Practically, none of the above conditions may be truly met but these are required for solving the given equations. Besides the assumptions, the physical configurations from which these equations emerge are to be accommodated in the appropriate coordinate systems. This accommodation will depend on the shape of the tube and the coordinates along which different fields may align. Also, **E** or **B** may be functions of only one, two or all the three coordinates. In general, tube configurations can be accommodated in cylindrical or Cartesian systems and in most of the cases, both E and B can be assumed to have only one component.

11.3.1 Motion in Electric Field

In the electric field, the force 'F' results in time rate of the change of momentum and thus, Equation 11.13 can be rewritten as

$$F = -eE = \frac{d}{dt}(mU) = m\frac{dU}{dt} \tag{11.17}$$

Equation 11.17 is of the general form and may be related to any coordinate system. Its simplified solution is obtained in rectangular and cylindrical coordinate systems.

11.3.1.1 Cartesian Coordinate System

In this case, if both E and U are functions of all the three coordinates, Equation 11.17 can be written as

$$-eE_x = m\frac{dU_x}{dt} = m\frac{d^2x}{dt^2} \tag{11.18a}$$

$$-eE_y = m\frac{dU_y}{dt} = m\frac{d^2y}{dt^2} \tag{11.18b}$$

$$-eE_z = m\frac{dU_z}{dt} = m\frac{d^2z}{dt^2} \tag{11.18c}$$

Figure 11.5 shows two parallel plates located at $x = 0$ and $x = d$ and are maintained at $V = 0$ and $V = V_0$ voltages, respectively. It is assumed that a particle (electron 'e') enters the E-field at time $t = 0$ at a physical location $x = y = z = 0$ with initial velocities $U_x = U_{x0}$, $U_y = U_{y0}$ and $U_z = 0$. Since $E_x = -V_0/d$ and $E_y = E_z = 0$, from Equation 11.18

$$\frac{dU_x}{dt} = \frac{d^2x}{dt^2} = -\frac{e}{m}E_x = -\frac{e}{m}\left(-\frac{V_0}{d}\right) = \frac{eV_0}{md} = k \tag{11.19a}$$

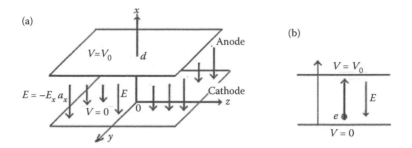

FIGURE 11.5
Parallel plate diode having the E-field. (a) Geometry and field distribution and (b) movement
of electron.

$$\frac{dU_y}{dt} = \frac{d^2y}{dt^2} = 0 \tag{11.19b}$$

$$\frac{dU_z}{dt} = \frac{d^2z}{dt^2} = 0 \tag{11.19c}$$

In view of *Appendix A2*, Equation 11.19 leads to

$$U = \sqrt{(2eV/m)} \tag{11.20}$$

where $V = E \cdot x = V_0 \cdot x/d$ is the potential at any point.
   The substitution of the values of e ($= 1.602 \times 10^{-19}$ C) and m ($= 9.1091 \times 10^{-31}$ kg)
in Equation 11.20 gives

$$U = 5.932 \times 10^5 \sqrt{\{V_0(x/d)\}} = 5.932 \times 10^5 \sqrt{(x/d)}\sqrt{V_0} \tag{11.21a}$$

At $x = d$, the expression for velocity reduces to

$$U = 0.5932 \times 10^6 \sqrt{V_0} \text{ m/s} \tag{11.21b}$$

   As the force is directly proportional to E ($F = QE$), the electron will travel
from the cathode to the anode along a straight line. Figure 11.5b illustrates
the movement of an electron that adopts a straight line path between two
parallel plates of a diode under the influence of the E-field.

11.3.1.2 Cylindrical Coordinate System

Figure 11.6 illustrates the geometry of a cylindrical diode along with
the potentials of the anode and the cathode. This also shows the E-field

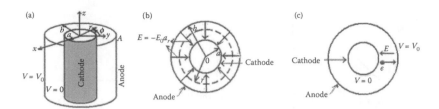

FIGURE 11.6
Cylindrical diode having the E-field. (a) Geometry and potentials, (b) cross section and E field and (c) electron movement.

distribution along its cross section. For this case, Equation 11.17 can be written as

$$-\frac{e}{m}E = \frac{d}{dt}(U_r a_r + U_\phi a_\phi + U_z a_z) = \frac{d}{dt}(U_r a_r) + \frac{d}{dt}(U_\phi a_\phi) + \frac{d}{dt}(U_z a_z) \quad (11.22a)$$

Equation 11.22a is obtained from Equation 11.17 by putting:

$$U = U_r a_r + U_\phi a_\phi + U_z a_z \quad (11.23b)$$

where $\qquad U_r = dr/dt, \ U_\phi = rd\phi/dt \quad \text{and} \ U_z = dz/dt \quad (11.22c)$

In Equation 11.22, U_r, U_ϕ and U_z are the velocities and a_r, a_ϕ and a_z are the unit vectors along r-, ϕ- and z-axes, respectively. The evaluation of the three terms yields the following expressions:

$$\frac{d}{dt}(U_r a_r) = U_r \frac{d\phi}{dt} a_\phi + \frac{dU_r}{dt} a_r \quad (11.23a)$$

$$\frac{d}{dt}(U_\phi a_\phi) = -U_\phi \frac{d\phi}{dt} a_r + \frac{dU_\phi}{dt} a_\phi \quad (11.23b)$$

and

$$\frac{d}{dt}(U_z a_z) = \frac{dU_z}{dt} a_z \quad (11.23c)$$

In view of Equation 11.23, Equation 11.22 can be written as

$$-\frac{e}{m}E_r = \frac{dU_r}{dt} - U_\phi \frac{d\phi}{dt} \quad (11.24a)$$

$$-\frac{e}{m}E_\phi = U_r \frac{d\phi}{dt} + \frac{dU_\phi}{dt} \quad (11.24b)$$

and

$$-\frac{e}{m}E_z = \frac{dU_z}{dt} \tag{11.24c}$$

In view of *Appendix A3*, Equation 11.24 leads to the following expression:

$$U_r = \frac{dr}{dt} = \sqrt{\frac{2eV_0 \ln(r/a)}{m \ln(b/a)}} \tag{11.25a}$$

By substituting the values of e and m, Equation 11.25a gives

$$U_r = \left[5.932 \times 10^5 \sqrt{\frac{\ln(r/a)}{\ln(b/a)}} \right] \sqrt{V_0} \tag{11.25b}$$

At $r = b$, Equation 11.25b reduces to

$$U = 5.932 \times 10^5 \sqrt{V_0} \tag{11.26}$$

Equations 11.26 and 11.21b are exactly the same and will yield the same value of velocity for a given voltage. In the case of a cylindrical diode also, the electron 'e' emerging from the cathode will follow a straight line path to arrive at the anode as illustrated in Figure 11.6c.

11.3.2 Motion in Magnetic Field

Here, Equation 11.14 is the starting point and Equation 11.17 can be modified to

$$F = -e(U \times B) = \frac{d}{dt}(mU) = m\frac{dU}{dt} \tag{11.27}$$

The solution of Equation 11.27 can again be obtained in two coordinate systems as given below.

11.3.2.1 Cartesian Coordinate System

This configuration is illustrated in Figure 11.7a. If $U = U_x a_x + U_y a_y + U_z a_z$ and $B = B_x a_x + B_y a_y + B_z a_z$, Equation 11.27 can be written in terms of its components as

$$\frac{-e}{m}(U_y B_z - U_z B_y) = \frac{dU_x}{dt} \tag{11.28a}$$

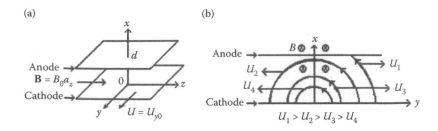

FIGURE 11.7
Parallel plate diode having the B-field. (a) Geometry and field and (b) motion of electron with different velocities.

$$\frac{-e}{m}(U_zB_x - U_xB_z) = \frac{dU_y}{dt} \tag{11.28b}$$

$$\frac{-e}{m}(U_xB_y - U_yB_x) = \frac{dU_z}{dt} \tag{11.28c}$$

Equation 11.28 can be written in the following form:

$$\frac{d^2x}{dt^2} = \frac{-e}{m}(B_z\frac{dy}{dt} - B_y\frac{dz}{dt}) \tag{11.29a}$$

$$\frac{d^2y}{dt^2} = \frac{-e}{m}(B_x\frac{dz}{dt} - B_z\frac{dx}{dt}) \tag{11.29b}$$

$$\frac{d^2z}{dt^2} = \frac{-e}{m}(B_y\frac{dx}{dt} - B_x\frac{dy}{dt}) \tag{11.29c}$$

The solution to Equation 11.29 (*Appendix A4*) leads to the following relations:

$$x = (U_{y0}/\omega_0)(\cos \omega_0 t - 1) \tag{11.30a}$$

$$y = (U_{y0}/\omega_0)\sin \omega_0 t \tag{11.30b}$$

Equations 11.30a and 11.30b are the parametric equations of a circle with radius '*a*' given by

$$a = U_{y0}/\omega_0 = U/\omega_0 = mU/(eB_0) \tag{11.31}$$

The centre of the circle lies at $x = -a/2$ and $y = a/2$.

The energy of the particle shall remain unchanged as expected in the case of a constant (steady) magnetic field. The linear velocity of the particle bears a relation with the angular velocity that can be obtained from Equation 11.28. The other relations are

The linear velocity of the particle is

$$U = a\,\omega_0 = a\,e\,B_0/m \qquad (11.32a)$$

The radius of the path adopted by the particle is

$$a = m\,U/eB_0 \qquad (11.32b)$$

The cyclotron angular frequency of the circular motion of the particle is

$$\omega_0 = U/a = e\,B_0/m \qquad (11.32c)$$

Period (T) for one complete revolution is

$$T = 2\pi/\omega_0 = 2\pi m/eB_0 \qquad (11.32d)$$

In view of the above relations, the following conclusions can be drawn:

- The force exerted on the electron by the magnetic field is normal to the motion on every instant and thus, no work is done on the electron and its velocity remains constant.
- The force remains constant in magnitude but changes in the direction of the motion because the electron is pulled by the magnetic force in a circular path.
- The radius of the circular path of motion is directly proportional to the velocity of the electron but the angular velocity and the period are independent of the velocity or radius. Thus, faster moving electrons traverse the circle of a larger radius at the same time that a slower moving electron moves in a smaller circle.

Figure 11.7b illustrates that the emerging electrons may return to the cathode for lower velocities and will fall into the cathode if the velocity is sufficiently higher.

11.3.2.2 Cylindrical Coordinate System

If the velocity $\mathbf{U} = U_r\,a_r + U_\phi\,a_\phi + U_z\,a_z$ and the magnetic flux density $\mathbf{B} = B_r\,a_r + B_\phi\,a_\phi + B_z\,a_z$, Equation 11.14 can be written in terms of its components as

$$\frac{-e}{m}(U_\phi B_z - U_z B_\phi) = \frac{dU_r}{dt} - U_\phi\frac{d\phi}{dt} \qquad (11.33a)$$

$$\frac{-e}{m}(U_z B_r - U_r B_z) = U_r \frac{d\phi}{dt} + \frac{dU_\phi}{dt} \qquad (11.33b)$$

$$\frac{-e}{m}(U_r B_\phi - U_\phi B_r) = \frac{dU_z}{dt} \qquad (11.33c)$$

As shown in Figure 11.8a, it may be assumed that $\mathbf{B} = B_0 a_z$ and $B_r = B_\phi = 0$. Further, an electron ($-e$) is assumed to enter the field at $t = 0$ at $r = a$, $\phi = 0$ and $z = 0$ with an initial velocity $U = U_{r0} a_r$. As $U_z = dz/dt = 0$ and the particle starts at $z = 0$, z remains 0 for all times. Also, $U_\phi = 0$ at $t = 0$, $d\phi/dt = \omega$ and $U_\phi = r \omega$. Thus, Equation 11.33 may be written as

$$\frac{-e}{m} B_0 \omega r = \frac{d^2 r}{dt^2} - r\omega^2 \qquad (11.34a)$$

$$\frac{e}{m} B_0 \frac{dr}{dt} = \frac{1}{r}\frac{d}{dt}(r^2 \omega) \qquad (11.34b)$$

$$\frac{dU_z}{dt} = 0 \qquad (11.34c)$$

These equations can be solved by following the procedure used in the case of the rectangular coordinate system. This case will lead to almost the same results as obtained in the case of Section 11.3.2.1. The electron emerging from the cathode will follow a circular path. In case the velocity is varied, the radii of the circles will also vary. As can be seen in Figure 11.8b, the emerging electrons may return to the cathode for lower velocities or may fall into the anode if the velocity is sufficiently higher.

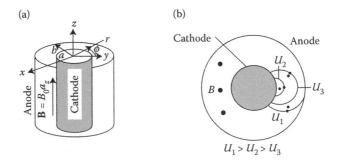

(a)   (b)

FIGURE 11.8
Cylindrical diode having the B-field. (a) Geometry and field and (b) movement of electrons.

11.3.3 Motion in Combined Electric and Magnetic Field

Here, Equation 11.15 is the starting point and Equation 11.17 (in the rectangular coordinate system) can be modified to the following form:

$$\frac{d^2x}{dt^2} = -\frac{e}{m}\left(E_x + B_z\frac{dy}{dt} - B_y\frac{dz}{dt}\right) \tag{11.35a}$$

$$\frac{d^2y}{dt^2} = -\frac{e}{m}\left(E_y + B_x\frac{dz}{dt} - B_z\frac{dx}{dt}\right) \tag{11.35b}$$

$$\frac{d^2z}{dt^2} = -\frac{e}{m}\left(E_z + B_y\frac{dx}{dt} - B_x\frac{dy}{dt}\right) \tag{11.35c}$$

In Figure 11.7, assume that an electron enters the combined field ($E_x = -V_0/d$, $E_y = E_z = 0$ and $B = B_0 a_z$, $B_x = B_y = 0$) at time $t = 0$ at a physical location $x = y = z = 0$ with an initial velocity $U = U_{y0}\, a_y$. As before, if $U_z = 0$, $z = 0$ for all times. Also, $U_x = 0$ at $t = 0$. In view of these assumptions, Equation 11.35 reduces to

$$\frac{d^2x}{dt^2} = -\frac{e}{m}\left(E_x + B_z\frac{dy}{dt} - B_y\frac{dz}{dt}\right) \tag{11.36a}$$

$$\frac{d^2y}{dt^2} = -\frac{e}{m}\left(B_x\frac{dz}{dt} - B_z\frac{dx}{dt}\right) \tag{11.36b}$$

$$\frac{d^2z}{dt^2} = -\frac{e}{m}\left(B_y\frac{dx}{dt} - B_x\frac{dy}{dt}\right) \tag{11.36c}$$

In the cylindrical coordinate system:

$$\frac{d^2r}{dt^2} - r\omega^2 = \frac{e}{m}\left(E_r - B_z\omega r - B_\phi\frac{dz}{dt}\right) \tag{11.37a}$$

$$\frac{1}{r}\frac{d}{dt}(r^2\omega) = -\frac{e}{m}\left(E_\phi + B_r\frac{dz}{dt} - B_z\frac{dr}{dt}\right) \tag{11.37b}$$

$$\frac{d^2z}{dt^2} = -\frac{e}{m}\left(E_z + B_\phi\frac{dr}{dt} - B_r\omega r\right) \tag{11.37c}$$

Equations 11.36 and 11.37 can be solved by using similar steps as adopted in Sections 11.3.1.1 and 11.3.2.1. The resulting solution will indicate the behaviour of electrons in the presence of the combined (electric and magnetic) field.

In the next section, the motion of the electron is studied in the presence of E-, B- and ac fields. In the resulting solution, if an ac field is eliminated, this case will yield the same result as will be obtained from Equation 11.36.

11.3.4 Motion in Electric, Magnetic and an AC Field

In this case, Equation 11.16 is the starting point. This equation is reproduced below:

$$\mathbf{F} = Q[(\mathbf{E} + E' \sin \omega t) + (\mathbf{U} \times \mathbf{B})] \tag{11.38}$$

11.3.4.1 Cartesian Coordinates

Figure 11.9 shows a parallel plate magnetron with d as the separation between the two plates. These plates are maintained at 0 and V_0 voltages. Thus, E has only the E_x component. This magnetron also contains a magnetic field along its z-axis (i.e. $B = B_0 a_z$). Furthermore, this magnetron is also applied at a time-varying potential ($V_1 \cos \omega t$), the resulting E-field of which is aligned with the E-field of the dc potential. Since any time-varying electric field will be associated with a time-varying magnetic field or vice versa, all such associated fields are neglected. It is to be noted that in this case, E and B will not satisfy Maxwell's equations. In view of the above and Figure 11.9, we can write

$$E = E_z = -V/d = (-V_0/d)\ \{1 + (V_1/V_0)\cos\omega t\}\ a_x \quad \text{and}\ E_y = E_z = 0 \tag{11.39}$$

$$\text{Thus, } E = (-V_0/d)\ \{1 + \alpha \cos \omega t\}a_x, \quad \text{where } \alpha = V_1/V_0 \tag{11.40}$$

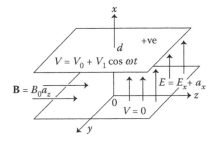

FIGURE 11.9
Parallel plate magnetron with E-, B- and ac field.

$$\text{Also,} \quad B = B_0 a_z \quad \text{and} \quad B_y = B_x = 0 \tag{11.41}$$

The resulting equations are

$$\frac{d^2x}{dt^2} = -\frac{e}{m}\left[-\frac{V_0}{d}(1 + \alpha \cos \omega t) + B_0 \frac{dy}{dt}\right] \tag{11.42a}$$

$$\frac{d^2y}{dt^2} = -\frac{e}{m}\left(-B_0 \frac{dx}{dt}\right) \tag{11.42b}$$

and

$$\frac{d^2z}{dt^2} = 0 \tag{11.42c}$$

The solution of Equation 11.42 given in *Appendix A5* finally leads to the following expressions:

$$U_x = \frac{k}{\omega_0}\left(1 - \frac{\alpha\omega_0^2}{\omega^2 - \omega_0^2}\right)\sin \omega_0 t + \frac{\alpha k \omega}{(\omega^2 - \omega_0^2)}\sin \omega t = \frac{dx}{dt} \tag{11.43a}$$

$$U_y = \frac{k}{\omega_0}\{1 - \left(1 - \frac{\alpha\omega_0^2}{\omega^2 - \omega_0^2}\right)\cos \omega_0 t)\} - \frac{\alpha k \omega^2}{(\omega^2 - \omega_0^2)}\cos \omega t = \frac{dy}{dt} \tag{11.43b}$$

$$x = \frac{k}{\omega_0^2}\left[\left(1 - \frac{\alpha\omega_0^2}{\omega^2 - \omega_0^2}\right)\cos \omega_0 t - \frac{\alpha\omega_0^2}{(\omega^2 - \omega_0^2)}\cos \omega t\right] \tag{11.44a}$$

$$y = \frac{k}{\omega_0^2}[\omega_0 t - (1 - \frac{\alpha\omega_0^2}{\omega^2 - \omega_0^2})\sin \omega_0 t) - \frac{\omega_0}{\omega}\frac{\alpha\omega_0^2}{(\omega^2 - \omega_0^2)}\sin \omega t] \tag{11.44b}$$

Equations 11.43 and 11.44 cannot be interpreted in their present forms. Thus, two simplified cases are considered below.

ac field absent: If there is no ac field, $\alpha = 0$. Thus, Equations 11.43 and 11.44 reduce to

$$U_x = \frac{k}{\omega_0}\sin \omega_0 t \tag{11.45a}$$

$$U_y = \frac{k}{\omega_0}\{1 - \cos \omega_0 t\} \tag{11.45b}$$

$$x = \frac{k}{\omega_0^2}(1 - \cos\omega_0 t) \qquad (11.46a)$$

$$y = \frac{k}{\omega_0^2}(\omega_0 t - \sin\omega_0 t) \qquad (11.46b)$$

Equations 11.46a and 11.46b are parametric equations of a cycloid. This curve may be traced by a point on the circumference of a rolling wheel. The velocity components of such a wheel movement are (a) angular velocity = ω_0 rad/s and (b) forward translational velocity = k/ω_0 m/s. The radius of a wheel may be found from the linear velocity of a point on the circumference ignoring the translation of k/ω_0 and producing a radius of $2k/\omega_0^2$ m. The maximum velocity attained will be $2k/\omega_0$ measured nearest to the anode leading to the maximum kinetic energy.

In view of Figure 11.10, there can be the following two possibilities:

1. The separation $d \le 2k/\omega_0^2$: In this case, the electron will fall into the plate (anode).
2. The separation $d > 2k/\omega_0^2$: In this case, the electron will not enter the anode and will return in $2\pi/\omega_0$ s to the cathode at a distance $2\pi k/\omega_0^2$ from the point of origination in the cathode. This distance is equal to twice the radius of the wheel.

The particle will attain the maximum velocity of $2k/\omega_0$ m/s measured nearest to the anode leading to the maximum kinetic energy. This kinetic energy (from the relation $1/2\ mv^2$) is

$$KE_{max} = \frac{1}{2}m(2k/\omega_0)^2 = (2mV_0^2)/(d^2B_0^2) \qquad (11.47)$$

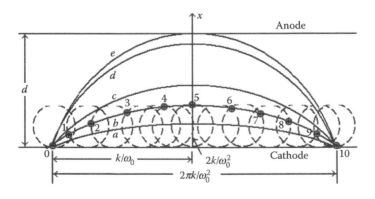

FIGURE 11.10
Movement of a spot on the wheel forming a cycloidal path.

This energy will completely return to the electric field by the time the electron returns to the cathode. As noted at the end of Section 11.3.3, this case is similar to that depicted by Equation 11.36 since only E and B are present. In this case, the electron emerging from the cathode follows a cycloidal path. It returns to the cathode or may fall into the anode as per the conditions noted above.

ac field present: In the presence of an ac field, let

$$\frac{\alpha \omega_0^2}{\omega^2 - \omega_0^2} = \frac{1}{2} \tag{11.48a}$$

and

$$\omega = 1.1\omega_0 \tag{11.48b}$$

The substitution of Equation 11.48b in Equation 11.48a gives

$$\alpha = (V_1/V_0) = 0.105 \tag{11.48c}$$

Equation 11.48c obtained from Equations 11.48a and 11.48b indicates that the magnitude of an ac field (V_1) is much smaller than the magnitude of the dc field (V_0). In this case, it is only about 10%. This situation is normally met in the devices wherein a small ac signal is to be amplified at the cost of a large applied dc source.

In view of the above relations, Equation 11.44 reduces to

$$
\begin{aligned}
x &= \frac{k}{\omega_0^2}(1 - \frac{1}{2}\cos\omega_0 t - \frac{1}{2}\cos 1.1\omega_0 t) \\
&= \frac{k}{\omega_0^2}(1 - \cos 0.05\omega_0 t \cos 1.05\omega_0 t) \\
&= \frac{k}{\omega_0^2} - \frac{k}{\omega_0^2}\cos 0.05\omega_0 t \cos 1.05\omega_0 t
\end{aligned} \tag{11.49a}
$$

$$y = \frac{k}{\omega_0^2}\left[(\omega_0 t - \frac{1}{2}\sin\omega_0 t) - \frac{1}{1.1}\frac{1}{2}\sin 1.1\omega_0 t\right] \tag{11.49b}$$

The oscillatory behaviour of a charged particle obtained in view of Equation 11.49 is illustrated in Figure 11.11. The electron starts from the cathode to a maximum distance of almost $2k/\omega_0^2$. If this electron is not captured by the anode, it returns nearer to the cathode, turns around and goes towards the anode but this time with a slightly lesser distance towards the plate. It continues oscillating back and forth with decreasing amplitudes until it comes

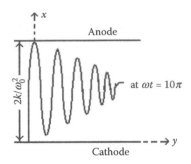

FIGURE 11.11
Motion of a charged particle in the combined field.

to rest at $x = k/\omega_0^2$ at $\omega_0 = 10\pi$. Owing to the non-zero y component of the velocity, it also progresses a little in the y direction.

11.3.4.2 Cylindrical Coordinates

The problem of a cylindrical magnetron with a cathode and an anode with the number of cavities requires a cylindrical coordinate system. Such a magnetron involves electric, magnetic and ac fields. Thus, Equation 11.38 is to be solved to account for the motion of the charged particle in the presence of all these fields. The mathematical analysis of this problem is given in Section 12.6.8. In view of the analogy of the behaviour of the electron in combined E-, B- and ac fields, obtained in Section 11.3.4.1, some fruitful conclusions can be drawn. Accordingly, the electron emerging from the cathode will follow a cycloidal path provided the ac field is absent and will appear to be as shown in Figure 11.12a. In the presence of an ac field, the electron will keep on oscillating between the cathode and the anode along with its progress in

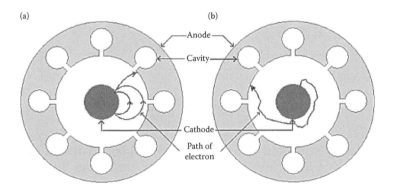

FIGURE 11.12
Electron movement in the cylindrical cavity magnetron. (a) In E and B fields and (b) in E, B and ac field.

the forward direction. As noted earlier, its capture by the anode or return to the cathode will depend on the amplitude and polarity of the ac signal and relative values of the electric and magnetic field. The likely path of an electron is shown in Figure 11.12b.

11.4 Velocity Modulation

An electron in motion possesses kinetic energy ($KE = 1/2\, mv^2$ where m is its mass and v is the velocity of motion). The relation that energy is directly proportional to the square of velocity spells the key principle of energy transfer and amplification in microwave tubes. In view of Section 11.3, the velocity of a charged particle is governed by the potential difference between the electrodes and an electron can be accelerated or decelerated by an electrostatic field. Figure 11.13 shows a moving electron in an electrostatic field.

11.4.1 Influence of Electric Field on Electrons

Figure 11.13 illustrates the configurations of two oppositely charged electrodes. It also shows the orientation of the E-field and an electron travelling therein. This figure has two parts. In Figure 11.13a, the electron is travelling in the opposite direction to that of the E-field, while in Figure 11.13b, the direction of travel of the electron is same as that of the E-field. These two cases vis-à-vis influence of E on electrons can be summarised as in the following paragraphs.

In the case of Figure 11.13a, as electrons are negatively charged particles, they will be attracted by the positively charged electrode. As a result of this attractive force, the electrons will get accelerated and their velocity will

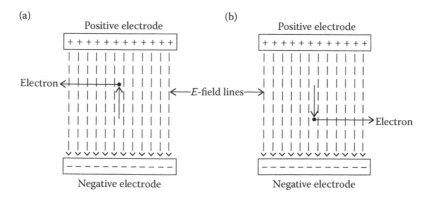

FIGURE 11.13
Moving electron in the electrostatic field. (a) Electron movement opposite to E-field and (b) electron movement along E-field.

increase. As a consequence, the energy level of electrons will also increase. The electrons will acquire this additional energy from the electrostatic field.

In the case of Figure 11.13b, the negatively charged electrode will repel the electron. This repulsive force will result in deceleration of the electron. When the velocity of the electron is reduced, its energy level also reduces. The energy lost by the electron due to deceleration is gained by the electrostatic field.

The acceleration and deceleration of electrons in view of their direction of travel is termed as velocity modulation. The velocity of a moving charge will also get influenced by the change in the magnitude of the field. Thus, it is not only the direction of travel but also the magnitude of the field that will result in velocity modulation.

11.4.2 Operation of Velocity-Modulated Tubes

The operation of a velocity-modulated tube depends on a change in velocity of the electrons passing through its electrostatic field. This change in electron velocity, during its travel over a section of the tube, causes bunching of electrons. These bunches are separated by spaces in which there are relatively fewer electrons. Velocity modulation is then defined as the variation in the velocity of a beam of electrons caused by the alternate speeding up and slowing down of the electrons in the beam. This variation is usually caused by a voltage signal applied between the grids through which the beam passes.

Figure 11.14a shows a simplified version of an electron gun containing a cathode. This electron gun liberates a stream of electrons. The electrons emitted from the cathode will be attracted by the positive accelerator grid. Some of these electrons will form a beam that will pass through a pair of closely spaced grids, referred to as buncher grids. Each of these grids is connected to one side of a tuned circuit. The tuned circuit in the illustration represents the doughnut-shaped resonant cavity surrounding the electron stream. As shown in Figure 11.14b, the buncher grids are the dashed lines at the centre of the cavity and are at the same dc potential as the accelerator

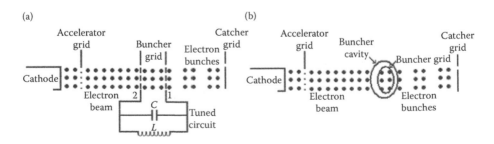

FIGURE 11.14
An electron gun. (a) With buncher grid and tuned circuit and (b) with buncher cavity and buncher grid.

grid. The alternating voltage that exists across the resonant circuit causes the velocity of the electrons leaving the buncher grids to differ from the velocity of the electrons arriving at the buncher grids. The amount of difference depends on the strength and direction of the electrostatic field within the resonant cavity as the electrons pass through the grids.

11.4.2.1 Bunching Process

For a better understanding of the bunching process, consider the motions of individual electrons illustrated in Figure 11.15. Figure 11.15a shows three electrons nearer to the left of the cavity that is maintained at zero potential. When the voltage across the grids becomes negative (Figure 11.15b), electron 1 crossing the gap gets slower. Figure 11.15c shows that the potential across the gap is 0 V; the velocity of electron 2 passing the cavity remains unaffected. Electron 3 enters the gap (Figure 11.15d) when the voltage across the gap is positive and thus, its velocity is increased. The combined effect (shown in Figure 11.15e) is that all the three electrons arrive on the other side of the cavity almost at the same time and thus, these are referred to as a bunch of electrons and the process is called the bunching process.

11.4.2.2 Current Modulation

The velocity modulation of the beam in itself does not serve any useful purpose as no useful power is produced at this juncture. The energy gained by the accelerated electrons is compensated by the energy lost by the decelerated electrons. If the velocity-modulated electrons are allowed to drift into a region where there is no electrostatic field, it will result in a new and useful beam distribution. In this field-free area beyond the buncher cavity, there will be continued formation of bunches in view of the new velocity relationships between the electrons. Unless the beam is subjected to some other force, these bunches will tend to form and disperse until the original beam distribution is eventually reformed. The net effect of velocity modulation

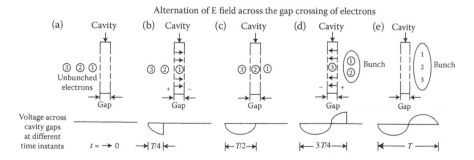

FIGURE 11.15
Bunching of electrons at (a) $t = 0$, (b) $t = T/4$, (c) $t = T/2$, (d) $t = 3T/4$ and (e) $t = T$.

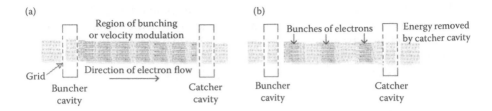

FIGURE 11.16
(a) Velocity-modulation and (b) energy removal.

is thus to result in current-density-modulated beam that varies at the same rate as the grid-signal frequency. Thus, a useful power from the beam can be extracted. Figure 11.16a and b illustrates the current-modulated (bunched) electron beam in various stages of formation and dispersion.

11.4.2.3 Energy Extraction

As shown in Figure 11.16, a second cavity, called a catcher cavity is placed at a point of maximum bunching to take useful energy out from the beam. The physical location of this cavity is determined by the frequency of the buncher-grid signal as this signal determines the transit time of the electron bunches. Both these cavities are resonant at the buncher-grid frequency. The electron bunches induce an RF voltage in the grid gap of the second cavity causing it to oscillate. The proper location of the catcher cavity causes the induced grid-gap voltage to decelerate the electron bunches as and when they arrive at the gap. Since the largest electron concentration is in the form of bunches, slowing of these bunches results in a transfer of energy to the output cavity. The balance of energy gets disturbed due to the placement of this second cavity. As bunches arrive at the cavity, the voltage has to be of correct polarity to increase the velocity of the electrons and the beam absorbs energy. The gaps between these bunches contain very few electrons. The energy removed from the beam is much greater than the energy required to speedup these scattered electrons. Thus, the location of the second cavity has to be such that a useful energy can be removed from a velocity-modulated electron beam.

11.5 Classification of Microwave Tubes

Microwave tubes perform the same functions of generation and amplification in the microwave segment of the frequency spectrum as that performed by vacuum tubes at lower frequencies. There are a number of tubes, for example, klystrons, TWTs, backward wave oscillators, magnetrons and other crossed-field devices, which operate in microwave range. Their construction, the basic operating principles and other related matters are discussed in Chapter 12. The variations of microwave tubes for specific applications are so numerous that all of them cannot be encompassed in one or two chapters. Also, since the general principles of operation for most of these tubes are similar, only a few of the most commonly used tubes are discussed in Chapter 12.

Microwave tubes can be broadly classified into two categories, namely, linear beam (or O-type) tubes and crossed-field (or M-type) tubes. Further classification of these tubes is given in the following section.

11.5.1 Linear Beam (or O-Type) Tubes

The classification of linear beam tubes is based on the types of structures employed. These structures may be of resonant or non-resonant (periodic) nature. The first category includes different forms of resonant cavities and the second category encompasses slow wave structures that may be in the form of helices, coupled cavities, ring loops and ring bars. The two main tubes that involve resonant cavities are the klystron and reflex klystron. The tubes with slow wave structures may utilise forward or backward waves. The tubes with forward waves in slow wave structures include TWTs and forward wave amplifiers (FWAs), whereas with backward waves, slow wave structures are referred to as backward wave amplifiers (BWAs) and BWOs. A tube that is a combination of klystron and TWT is referred to as twystron.

Like klystron, TWT also operates on the principle of interaction between the electron beam and the RF field. Both the klystrons and TWTs are linear beam tubes with a focussed electron beam similar to that in cathode ray tube (CRT). In klystron, the RF field is stationary; thus, the interaction between the beam and the RF field is not continuous. In TWTs, the interaction between the beam and the RF field is made continuous by using the slow wave structures.

The klystron and TWT are the most commonly used tubes. Their peak power outputs may be of the order of 30 MW with beam voltages of 100 kV at about 10 GHz. At these frequencies, an average power output of 700 kW can also be obtained. The gains of these tubes may lie between 30 and 70 dB and the efficiency may lie between 15% and 16%. The BW for klystrons is of the order of 1–8% and TWT is of the order of 10–20% of its operating frequencies.

11.5.2 Crossed-Field (or M-Type) Tubes

These tubes can be classified into the following two broad categories. In the first category of the tubes, maser action is used. The gyration is one of the tubes that fall in this class. The second category employs resonant or non-resonant cavities. Some of the tubes belonging to the second category are given in the following section.

11.5.2.1 Tubes Using Resonant Cavities

Magnetron: All magnetrons consist of some form of anode and cathode. These operate with a dc magnetic field normal to a dc electric field. In view of the crossed-field nature, these devices are referred to as crossed-field devices. The magnetron uses reentrant resonant cavities that are located all around the anode, each of which represents one-half-wave period. The RF field circulates from one cavity to the next. Since wave velocity approximately equals electron velocity, there is continued interaction between the RF field and the electron beam. Thus, the magnetron also operates on the principle of continuous interaction between the electron beam and the RF field.

Like klystrons and TWTs, the magnetrons also have a number of versions that are based on their operational and constructional features. The magnetrons can thus be classified as the following:

A. Classification based on operational features
   a. *Split anode magnetron:* It uses a static negative resistance between two anode segments; thus, these are also called negative resistance magnetrons. Such magnetrons generally operate below microwave range.
   b. *Cyclotron-frequency magnetron:* It operates under the influence of synchronism between an alternating component of the electric field and a periodic oscillation of electrons in a direction parallel to the field. Its frequency of operation covers the microwave range. Normally, its output power is very small (about 1 W at 3 GHz) and the efficiency is quite low (about 10% for the split anode type and 1% for the single anode type).
   c. *Travelling-wave magnetrons:* It depends on the interaction of electrons with a travelling field of linear velocity. It is often referred to simply as magnetron.

B. Classification based on constructional features
   a. *Linear magnetron:* These cavities are linearly arranged along the cathode of planar shape.
   b. *Coaxial magnetron:* This magnetron is formed by involving coaxial lines and is thus termed as coaxial magnetron. The inside-out coaxial magnetron is also a modified version of this class. In these, the principle of integrating a stabilising cavity into the magnetron geometry is introduced.

c. *Voltage-tunable magnetron (VTM):* It contains the cathode–anode geometry of the conventional magnetron but its anode is made easily tunable.

d. *Inverted magnetron:* It contains an inverted geometry of the conventional magnetron wherein the cathode is placed outside to surround the anode and the microwave circuit.

e. *Cylindrical magnetron:* This type of magnetron is also called a conventional magnetron. These contain several reentrant cavities that are connected to the gaps.

f. *Frequency agile or dither-tuned magnetron:* This is specifically designed for frequent frequency changes.

g. *Electron-resonance magnetron:* Its operation is based on the electron transit time.

A magnetron is sometimes also named in accordance with the shape of the cavity. A few of such electrons are referred to as (a) hole and slot cavity magnetron, (b) vane cavity magnetron, (c) slot cavity magnetron and (d) rising sun cavity magnetron.

11.5.2.2 Tubes with Non-Resonant Cavities

a. *Forward wave crossed-field amplifier (FWCFA):* It is also called M-type forward wave amplifier (or M-type FWA). It uses reentrant non-resonant cavities and forward waves.

b. *Dematron:* Another device that uses forward wave for its operation but uses non-reentrant non-resonant cavities is referred to as dematron.

c. *Backward wave crossed-field amplifier (BWCFA):* It is also called amplitron. This also uses reentrant non-resonant cavities. It is a broadband, high-power, high-gain and high-efficiency tube. *BWCFA* is widely used in airborne radars and space-borne applications.

d. *Backward wave crossed-field oscillator (BWCFO):* It is also called M-type forward wave oscillator (or M-type FWO). Its trade name is carcinotron. It uses non-reentrant non-resonant cavities.

Descriptive Questions

Q-11.1 Discuss the frequency limitations of conventional tubes.

Q-11.2 Explain the terms inter-electrode capacitance, lead inductance and transit time effect.

Q-11.3 Illustrate the path of a charge under the influence of the (a) *E*-field, (b) *B*-field, (c) *E*- and *B*-fields and (d) *E*-, *B*- and ac fields between two parallel plates.

Q-11.4 Illustrate the path of a charge under the influence of (a) *E*-field, (b) *B*-field, (c) *E*- and *B*-fields and (d) *E*-, *B*- and ac fields between two cylindrical planes.

Q-11.5 Use the necessary illustrations to describe velocity modulation.

Q-11.6 Discuss the classification of microwave tubes.

Further Reading

1. Artsimovich, L. A. and Lukyanov, S. Y., *Motion of Charged Particles in Electric and Magnetic Fields*. Mir Publishers, Moscow, 1980.
2. Chodorow, M. and Susskind, C., *Fundamentals of Microwave Electronics*. McGraw-Hill, New York, 1964.
3. Dalman, G. C., *Microwave Devices, Circuits and Their Interaction*. John Wiley & Sons Inc, New York, 1994.
4. Gewartosky, J. W. and Watson, H. A., *Principles of Electron Tubes*. D. Van Nostrand Company, Princeton, NJ, 1965.
5. Gilmour, A. S. Jr., *Microwave Tubes*. Artech House, Dedham, MA, 1986.
6. Hayt, W., *Engineering Electromagnetics*, 1st ed. McGraw-Hill Publishing Co. Ltd., New York, 1958.
7. Hutter, R. G. E., *Beam and Wave Electrons in Microwave Tubes*. D. Van Nostrand Company, Princeton, NJ, 1960.
8. Terman, F. E., *Electronics and Radio Engineering*. McGraw-Hill, Tokyo, 1955.
9. Warnecke, R. R. et al., *Velocity Modulated Tubes, Advances in Electronics*, Vol. 3. Academic Press, New York, 1951.

12

Microwave Tubes

12.1 Introduction

The classification of microwave tubes was briefly discussed in Chapter 11. This chapter describes some of the most commonly used tubes. To assess their strength and weakness relating to the generation and amplification of signals, this chapter includes their constructional features, working principles and performance parameters.

12.2 Klystron

The word *klystron* is derived from a Greek word that means to wash or break over, such as waves wash a beach. Although invented in 1937 by the Varian brothers, it is still widely used. It finds applications in radars and communication equipment as an oscillator and amplifier. The three basic phenomena, namely, velocity modulation, current modulation and transit time effect, which govern its operation, were discussed in Chapter 11. The velocity modulation in itself is the result of interaction between the signal and the electron beam. Since the available length for this interaction is generally short, a strong electrostatic field becomes a necessity to yield the desired results.

12.2.1 Types of Klystron

Klystron tubes have many versions and can be classified as follows.

12.2.1.1 Reflex Klystron

Reflex klystron is a low-power low-efficiency oscillator used to generate microwave energy. It contains only one cavity followed by a reflector that returns the beam through the cavity. By optimising the phase of reflected beam, the device can be made to oscillate at the resonant frequency of the cavity.

12.2.1.2 Two-Cavity Klystron

This klystron contains two cavities called buncher and catcher cavities. The buncher cavity acts as the input cavity and the catcher cavity as the output cavity. Thus, the input signal is coupled into the buncher cavity and the output is taken from the catcher cavity. The RF voltage across the input cavity gap velocity modulates the electron beam by accelerating and retarding the electrons. The separation between input and output cavities referred to as drift space determines the length of the tube. This length is designed to yield optimum bunching of electrons at the output cavity. These bunches induce higher RF current in the output cavity. This current is then coupled out.

The two-cavity klystron can also be constructed as an oscillator. The main difference between a two-cavity klystron amplifier and the oscillator relates to the construction of their cavities. Since in an oscillator a feedback is required, the cavities are to be modified to allow feedback from the output cavity to the input cavity. These can generate higher power than reflex klystrons.

12.2.1.3 Multi-Cavity Klystron

It contains additional 'idler' cavities between the input and output cavities to enhance the velocity modulation by causing further acceleration or retardation of electrons.

12.2.1.4 Extended Interaction Klystron

In this klystron, one or more of the single cavities of the conventional klystron are replaced by structures containing two or more interaction gaps. These are electromagnetically coupled both to the beam and also to each other.

It is to be mentioned that except the reflex klystron, wherein generally no focussing is required, almost all klystrons employ magnetic focussing, which may be in the form of permanent magnets or electromagnets.

12.3 Two-Cavity Klystron

This klystron contains an electron gun, a collector and buncher and catcher cavities. It is a more commonly used device. For proper operation, each of its electrodes is to be maintained at different potentials. This also contains control and accelerator grids. The control grid is maintained at a low positive voltage and the accelerator grid at a high positive dc potential.

12.3.1 Constructional Features

The components of a two-cavity klystron, shown in Figure 12.1, include the following:

1. *Electron gun*: It contains a heater, a cathode, an anode, focussing electrodes and an insulator. It generates electron beam and radiates the same.

2. *Input cavity*: It is also called the *buncher cavity*. Its velocity modulates the electron beam. Its main components include input window, drift tube and cavity.

3. *Output cavity*: It is also called the *catcher cavity*. It catches the signal amplified by bunching of electrons. Its main components include output window, drift tube and cavity.

4. *Collector*: It collects the electron beam and releases the energy as heat. Its main components are collector and insulator.

5. *Focussing magnet*: It concentrates electrons into a (narrow) beam through the length of the tube. It normally comprises permanent magnet.

6. *Cavity tuner*: It changes the resonant frequency by changing the shape of the cavity. Its main constituents include tuner board and bellows.

The two-cavity klystron may be used either as an oscillator or an amplifier. The configuration shown in Figure 12.2 depicts the oscillator operation.

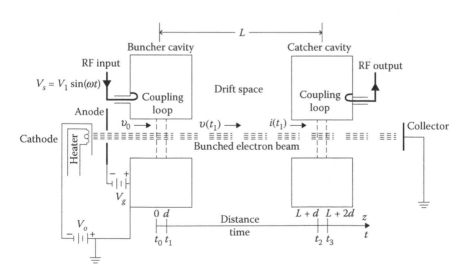

FIGURE 12.1
Two-cavity klystron amplifier.

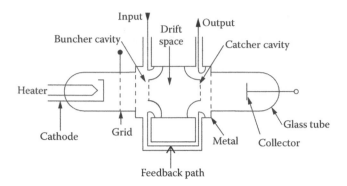

FIGURE 12.2
Two-cavity klystron with feedback arrangement.

The feedback path shown provides energy of the proper delay and phase relationship to sustain oscillations. These can generate higher power levels than reflex klystrons. A signal applied at the buncher grids gets amplified if the feedback path is removed.

12.3.2 Operation

The operation of a two-cavity klystron revolves around velocity modulation and bunching process, which were earlier discussed in Section 11.4 and can further be explained in view of its functional diagram as illustrated in Figure 12.3. It indicates that when a klystron is energised, the electrons emitted from the cathode move towards the anode. These get focussed and form

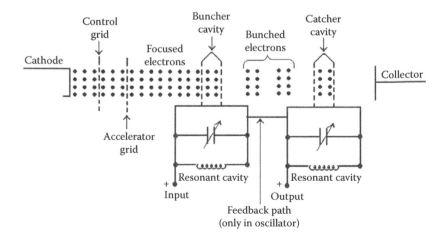

FIGURE 12.3
Functional view of two-cavity klystron.

a beam in view of the low positive voltage of the control grid. This beam travels towards the buncher cavity initially with uniform velocity. It gets accelerated by a very high positive dc potential applied in equal amplitude to the accelerator and the buncher grid. The buncher grid is also connected to a cavity resonator wherein an ac potential (RF signal) is superimposed on the dc voltage. These ac potentials are produced by spontaneous oscillations within the cavity that begin when the tube is energised. These initial spontaneous oscillations are due to random fields and circuit imbalances that are present at the time the circuit is energised. The oscillations within the cavity produce an oscillating electrostatic field at the same frequency as the natural frequency of the cavity. The direction of this field changes with the frequency at which the cavity is tuned. These changes alternately accelerate and decelerate electrons passing through the grids. The area beyond the buncher grid called drift space is the area wherein electrons form bunches in view of alternate acceleration and deceleration of electrons. Another grid called catcher grid is placed along the beam at a point where bunches are fully formed. Its function is to absorb energy from the electron beam. Its location is determined by the transit time of bunches at the natural resonant frequency, which is the same for both the catcher and buncher cavities. The location is chosen because maximum energy transfer to the catcher cavity occurs when the electrostatic field is of correct polarity to slow down the electron bunches.

12.3.3 Mathematical Analysis

The mathematical analysis of velocity modulation requires certain simplified assumptions, which are given below:

1. Cross section of the beam is assumed to be uniform in terms of concentration of electrons.
2. There is negligible or no effect of space charge.
3. The magnitude (V_1) of microwave input signal (V_s) is much smaller than that of the dc accelerating voltage (V_0).
4. The velocity of electrons (v_0) at the instant of liberation from the cathode is zero.

12.3.3.1 Velocity Modulation

After liberation, the excited electrons come under the influence of the dc voltage (V_0) and move forward with uniform velocity v_0 given by Equation 11.21b, which is reproduced as follows:

$$v_0 = 0.5932 \times 10^6 \sqrt{V_0} \text{ m/s} \tag{12.1}$$

When a microwave signal is applied to the input terminals of the buncher cavity, its gap voltage in grid space will appear as

$$V_s = V_1 \sin(\omega t) \tag{12.2}$$

As assumed, $V_1 \ll V_0$, the average transit time (τ) through the buncher gap (d) is

$$\tau \approx d/v_0 = t_1 - t_0 \tag{12.3}$$

The voltages $(V_s$ and $V_1)$, times $(t_1$ and $t_0)$ and buncher gap d are shown in Figures 12.1 and 12.4.

The average gap transit angle can be written in terms of average transit time (τ) as

$$\theta_g = \omega\tau = \omega (t_1 - t_0) = \omega d/v_0 \tag{12.4}$$

The average microwave voltage in buncher gap can be obtained by integrating V_s, dividing the outcome by average transit time (τ), and on using Equation 12.4, the expression becomes

$$\langle V_s \rangle = \frac{1}{\tau} \int_{t_0}^{t_1} V_1 \sin \omega t \, dt = -\frac{V_1}{\omega\tau}(\cos \omega t_1 - \cos \omega t_0)$$

$$= -\frac{V_1}{\omega\tau}[\cos \omega t_0 - \cos\{\omega t_0 + (\omega d/v_0)\}] \tag{12.5a}$$

In Equation 12.5a, if A is substituted for $\omega t_0 + \omega d/2v_0 = \omega t_0 + (\theta_g/2)$ and B for $\omega d/2v_0 = \theta_g/2$, the involved trigonometric terms can be written as

$$\cos \omega t_0 = \cos [\{\omega t_0 + (\theta_g/2)\} - (\theta_g/2)] = \cos (A - B)$$

$$\cos (\omega t_0 + \omega d/v_0) = \cos (\omega t_0 + \omega d/2v_0 + \omega d/2v_0)$$

$$= \cos \{\omega t_0 + (\theta_g/2) + (\theta_g/2)\} = \cos (A + B)$$

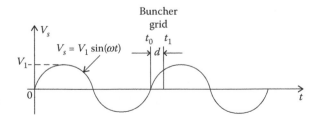

FIGURE 12.4
Parameters V_S and V_1 and time t_1 and t_0.

Thus, $\cos(A - B) - \cos(A + B) = 2 \sin A \sin B = 2 \sin(\omega t_0 + \theta_g/2) \sin(\theta_g/2)$. Equation 12.5a can be written as

$$\langle V_s \rangle = -\frac{2V_1}{\theta_g} \sin\left(\frac{\theta_g}{2}\right) \sin\left(\omega t_0 + \frac{\theta_g}{2}\right) \tag{12.5b}$$

Beam coupling coefficient: The parameter β_i called beam coupling coefficient of input cavity gap is defined as

$$\beta_i = -\sin(\theta_g/2)/(\theta_g/2) \tag{12.6}$$

As θ_g increases, the coupling between the electron beam and the buncher cavity decreases or the velocity modulation of beam for a given microwave signal decreases.

The exit velocity $v(t_1)$ from the buncher gap is given by

$$v(t_1) = \sqrt{[(2e/m)\ \{V_0 + \beta_i V_1 \sin(\omega t_0 + \theta_g/2)\}]}$$

$$= \sqrt{[(2e/m)\ V_0\ \{1 + (\beta_i V_1/V_0) \sin(\omega t_0 + \theta_g/2)\}]} \tag{12.7}$$

Depth of velocity modulation: The factor $\beta_i V_1/V_0$ involved in Equation 12.7 is referred to as the depth of velocity modulation. If it is assumed that $\beta_i V_1 \ll V_0$ and bracketed term $\{1 + (\beta_i V_1/V_0) \sin(\omega t_0 + \theta_g/2)\}^{1/2}$ is expanded using binomial expansion and using relation $v_0 = \sqrt{[(2e/m)\ V_0]}$, Equation 12.7 becomes

$$v(t_1) = v_0\ \{1 + (\beta_i V_1/2V_0) \sin(\omega t_0 + \theta_g/2)\} \tag{12.8}$$

Equation 12.8 can also be written as

$$v(t_1) = v_0\ \{1 + (\beta_i V_1/2V_0) \sin(\omega t_1 - \theta_g/2)\} \tag{12.9}$$

12.3.3.2 Process of Bunching

The electrons leaving the buncher cavity drift with a velocity are given by Equation 12.8 or 12.9. They now move in the field free space between two cavities. The impact of velocity modulation on these electrons is quite significant as electrons in the beam now travel in the form of bunches. The current carried by this beam becomes non-uniform in accordance with the bunching. This current is referred to as modulated current and the process is called *current modulation*.

At par with Figure 11.15, Figure 12.5 illustrates the essence of bunching wherein three electrons pass through the buncher cavity at different time instants corresponding to different voltages. The electrons that pass the buncher cavity at $V_s = 0$ travel through with unchanged velocity v_0 and may be considered as the bunching centre or the reference. Those electrons that

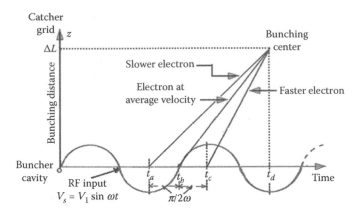

FIGURE 12.5
The bunching process.

pass the buncher cavity during the positive half cycle ($V_s > 0$) get accelerated and travel faster than the reference electrons. Similarly, those electrons that pass the buncher cavity during the negative half cycle ($V_s < 0$) get decelerated and travel slower than the reference electrons. At a distance ΔL, electrons join each other to form a bunch and travel further. This figure also illustrates the trajectories of electrons travelling with slower, average and faster speeds. It is to be noted that the bunching occurs from the later part of the negative half cycle to the first half of the positive cycle. From the second half of the positive cycle to the first half of the negative cycle, there is no bunching simply because the electron leaving earlier and travelling faster can only be joined by those leaving later and moving slower.

In view of Figure 12.5, the distance from the buncher grid to the location of the electron bunching for the electrons at t_a, t_b and t_c can be written as

$$\Delta L = v_0 \, (t_d - t_b) \quad \text{for } t_b \tag{12.10a}$$

$$= v_{min} \, (t_d - t_a) = v_{min} \, (t_d - t_b + \pi/2\omega) \quad \text{for } t_a \tag{12.10b}$$

$$= v_{max} \, (t_d - t_c) = v_{max} \, (t_d - t_b - \pi/2\omega) \quad \text{for } t_c \tag{12.10c}$$

From Equations 12.8 and 12.9

$$v_{min} = v_0 \, [1 - (\beta_i V_1/2V_0)] \tag{12.11a}$$

and

$$v_{max} = v_0 \, [1 + (\beta_i V_1/2V_0)] \tag{12.11b}$$

Substitution of Equations 12.11a and 12.11b in Equations 12.10b and 12.10c yields

$$\Delta L = v_0 \, (t_d - t_b) + [v_0 \, (\pi/2\omega) - v_0 \, (\beta_i V_1/2V_0) \, (t_d - t_b) - v_0 \, (\beta_i V_1/2V_0) \, (\pi/2\omega)]$$

$$(12.12a)$$

$$\Delta L = v_0 \, (t_d - t_b) + [-v_0 \, (\pi/2\omega) + v_0 \, (\beta_i V_1/2V_0) \, (t_d - t_b) + v_0 \, (\beta_i V_1/2V_0) \, (\pi/2\omega)]$$

$$(12.12b)$$

The necessary condition for meeting of electrons at t_a, t_b and t_c at the same distance ΔL is

$$[v_0 \, (\pi/2\omega) - v_0 \, (\beta_i V_1/2V_0) \, (t_d - t_b) - v_0 \, (\beta_i V_1/2V_0) \, (\pi/2\omega)] = 0 \quad (12.13a)$$

$$[v_0 \, (\pi/2\omega) + v_0 \, (\beta_i V_1/2V_0) \, (t_d - t_b) + v_0 \, (\beta_i V_1/2V_0) \, (\pi/2\omega)] = 0 \quad (12.13b)$$

From which

$$(t_d - t_b) \approx \pi \, v_0/(\omega \, \beta_i V_1) \tag{12.14a}$$

and

$$\Delta L = v_0 \, (t_d - t_b) = v_0 \, [\pi v_0/(\omega \, \beta_i V_1)] \tag{12.14b}$$

The distance given by Equation 12.14b is not the one for maximum bunching. Figure 12.6 shows the distance time plot. This plot is often called the Applegate diagram.

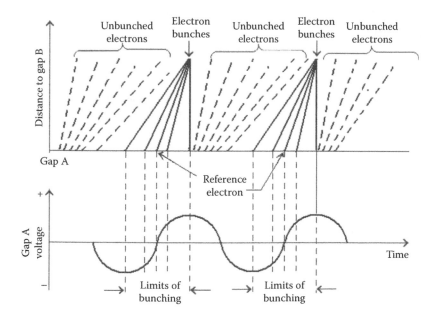

FIGURE 12.6
Applegate diagram (repetitive formation of bunches).

The transit time for an electron to travel distance L, between the buncher and catcher cavities, is

$$T = t_2 - t_1 = L/v(t_1) \tag{12.15a}$$

And in view of Equation 12.9

$$T = (L/v_0) \{1 + (\beta_i V_1/2V_0) \sin (\omega t_1 - \theta_g/2)\}^{-1} \tag{12.15b}$$

Replacing the binomial expansion of $(1 + x)^{-1}$ for x {= $(\beta_i V_1/2V_0)$ sin $(\omega t_1 - \theta_g/2)$} $\ll 1$ and the dc transit time L/v_0 by T_0, Equation 12.15b can be written as

$$T = T_0\{1 - (\beta_i V_1/2V_0) \sin (\omega t_1 - \theta_g/2)\} \tag{12.15c}$$

Equation 12.15 can be written in terms of radians after simplification

$$\omega T = \omega t_2 - \omega t_1 = \theta_0 - X \sin (\omega t_1 - \theta_g/2) \tag{12.16}$$

In Equation 12.16

$$\theta_0 = \omega L/v_0 = 2\pi N \tag{12.17}$$

where θ_0 is the dc transit angle and N is the number of electron transit cycle in drift space.

Bunching parameter: Comparison of Equations 12.15c and 12.16 gives a parameter called the bunching parameter of klystron and is denoted by X. It is given by

$$X \equiv \frac{\beta_i V_1}{2V_0} \theta_0 \tag{12.18}$$

The bunching is maximum during its retarding phase; thus, kinetic energy is transferred from electrons to the field of second cavity. The electrons emerge from the second cavity with reduced velocity and finally terminate at the collector. In case of more number of intermediate cavities, the bunching is likely to be more complete. As the unbunched electrons can also be captured by the catcher cavity, their presence in the output signal is regarded as noise.

The impact of bunching on signal output can be seen in Figure 12.7 in terms of input/output voltages of a tube. Accordingly, the weak input signal ($V_s = V_1$ sin ωt) enhances to higher signal ($V_s = V_2$ sin ωt) at the output. Besides the parameters, β_i (Equation 12.6), T (Equation 12.15) and X (Equation 12.18), the other useful relations for two-cavity klystron are given below:

Beam current (i_2)

$$i_2 = I_0 + \sum_{n=1}^{\infty} 2I_0 J_n(nX) \sin\{n\omega(t_2 - \tau - T_0)\} \tag{12.19}$$

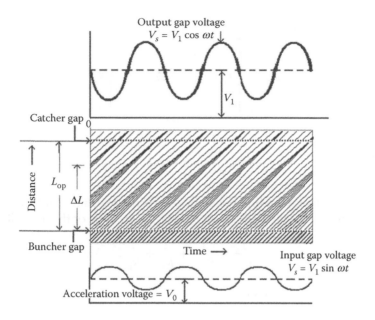

FIGURE 12.7
Input and output voltages in view of bunching.

The value of $2I_0 J_n(nX)|_{n=1} = 2I_0 J_1(X) = I_f$ is the fundamental component of beam current (i_2), which has its maximum value at $X = 1.841$. The optimum distance (L_{opt}) at which this fundamental component (I_f) becomes maximum is given by

$$L_{opt} = \frac{3.682 v_0 V_0}{\omega \beta_i V_1} \qquad (12.20)$$

In Equations 12.19 and 12.20, I_0 is the dc current, $J_n(nX)$ is the Bessel's function of order n and argument (nX), and β_i is the beam coupling coefficient. Parameters X, V_0, V_1, v_0, T_0, τ and t_2 were given earlier.

Induced current ($i_{2\,ind}$) in catcher cavity

$$i_{2\,ind} = \beta_0 i_2 = \beta_0 2I_0 J_1(X) \sin\{\omega(t_2 - \tau - T_0)\} \qquad (12.21)$$

where β_0 is the beam coupling coefficient of the catcher cavity.

When buncher and catcher cavities are identical, $\beta_i = \beta_0$ and the fundamental component of induced current is

$$I_{2\,ind} = \beta_0 I_2 = \beta_0 2I_0 J_1(X) \qquad (12.22)$$

Output power delivered to the catcher cavity and load: Figure 12.8 shows an output equivalent circuit in which R_{sho} is the wall resistance of the catcher

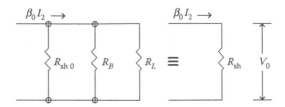

FIGURE 12.8
Output equivalent circuit.

cavity, R_B is the beam loading resistance, R_L is the resistance of external load and R_{sh} is the total equivalent shunt resistance. In view of this circuit

$$\frac{1}{R_{sh}} = \frac{1}{R_{sh0}} + \frac{1}{R_B} + \frac{1}{R_L}$$
(12.23)

The output power is given by

$$P_{out} = \frac{\beta_0 I_2^2}{2} R_{sh} = \frac{\beta_0 I_2 V_2}{2}$$
(12.24)

Efficiency of klystron:

$$\text{Efficiency} = \frac{P_{out}}{P_{in}} = \frac{\beta_0 I_2 V_2}{2 I_0 V_0}$$
(12.25)

where V_2 is the fundamental component of catcher gap voltage.

Mutual conductance (G_m) of a klystron amplifier: The equivalent mutual conductance $|G_m|$ is given as

$$|G_m| = \frac{i_{2ind}}{V_1} = \frac{2\beta_0 I_0 J_1(X)}{V_1}$$
(12.26)

In view of Equation 12.18 and with the assumption of $\beta_i = \beta_0$

$$V_i = \frac{2V_0}{\beta_0 \theta_0} X$$
(12.27)

The substitution of V_1 from Equation 12.27 in Equation 12.26 gives

$$\frac{|G_m|}{G_0} = \frac{\beta_0^2 \theta_0 J_1(X)}{X}$$
(12.28)

where $G_0 = I_0/V_0$ is the dc beam conductance.

Voltage gain of klystron amplifier: This is given as

$$A_v \equiv \frac{|V_2|}{|V_1|} = \frac{\beta_0 I_2 R_{sh}}{V_1} = \frac{\beta_0^2 \theta_0}{R_0} \frac{J_1(X)}{X} R_{sh} \qquad (12.29)$$

where $R_0 = V_0/I_0 \ (= 1/G_0)$ is the dc beam resistance.

Substitution of Equation 12.28 and the value of R_0 in Equation 12.29 gives

$$A_v = G_m R_{sh} \qquad (12.30)$$

Power required for bunching: According to the relation derived by Feenberg

$$\frac{P_B}{P_0} = \frac{V_1^2}{2V_0} F(\theta_g) \qquad (12.31)$$

P_B is the power given by the buncher cavity to produce the beam bunching:

$$P_B = \frac{V_1^2}{2} G_B \qquad (12.32a)$$

The supplied dc power P_0 is

$$P_0 = V_0^2 G_0 \qquad (12.32b)$$

and

$$F(\theta_g) = \frac{1}{2}\left[\beta_i^2 - \beta_i \cos\left(\frac{\theta_g}{2}\right)\right] \qquad (12.32c)$$

In Equation 12.32b, $G_0 = I_0/V_0$ is the equivalent conductance of the electron beam and G_B is the equivalent beam conductance. Substitution of Equations 12.32a and 12.32b in Equation 12.31 gives

$$\frac{G_B}{G_0} = \frac{R_0}{R_B} = F(\theta_g) \qquad (12.33a)$$

$$G_B = \frac{G_0}{2}\left[\beta_i^2 - \beta_i \cos\left(\frac{\theta_g}{2}\right)\right] \qquad (12.33b)$$

Thus, the power delivered (P_d) by the electron beam to the catcher cavity can be expressed as

$$P_d = V_2^2/2R_{sh} \qquad (12.34)$$

where R_{sh} is given by Equation 12.23.

Quality factor: The loaded quality factor of catcher cavity at resonant frequency is given by

$$\frac{1}{Q_L} = \frac{1}{Q_0} + \frac{1}{Q_B} + \frac{1}{Q_{ext}} \tag{12.35}$$

where Q_L, Q_0, Q_B and Q_{ext} are the quality factors of the whole catcher circuit, the catcher wall, the beam loading and the external circuit, respectively.

12.3.4 Multi-Cavity Klystron Amplifiers

In two-cavity klystron, the bunching of electrons remains incomplete and a large number of out-of-phase electrons arrive at catcher cavity between bunches. Thus, more than two cavities are employed in practical klystron amplifiers. Partially bunched current pulses excite oscillations in intermediate cavities, and these cavities in turn set up gap voltages, which results in completion of bunching. These extra cavities result in improved amplification, power output and efficiency. Additional cavities serve to further velocity-modulate the electron beam and thus result in more energy at the output.

A representative three-cavity klystron is illustrated in Figure 12.9. For safety reasons, the entire drift-tube assembly, the cavities and the collector of the three-cavity klystron are operated at ground potential. As in two-cavity case, the electron beam is formed and accelerated towards the drift tube by

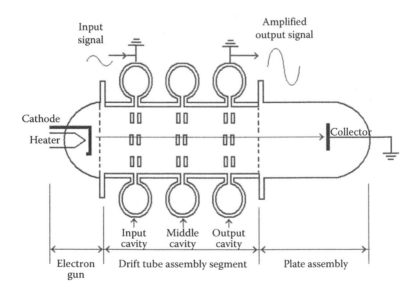

FIGURE 12.9
Three-cavity klystron.

a large negative pulse applied to the cathode. Here also, the magnetic focussing concentrates the electrons into a tight beam and keeps it away from the walls of the tube. The high-powered klystrons use concave-shaped cathode for focussing of the beam.

The output of any klystron, regardless of the number of cavities used, is developed by velocity modulation of the electron beam. The electrons that are accelerated by the cathode pulse are acted upon by RF fields developed across the input and middle cavities. Some electrons are accelerated, some are decelerated and some are unaffected. Electron reaction depends on the amplitude and polarity of the fields across the cavities when the electrons pass the cavity gaps. During the time the electrons are travelling through the drift space between the cavities, the accelerated electrons overtake the decelerated electrons to form bunches. As a result, bunches of electrons arrive at the output cavity at the proper instance during each cycle of the RF field and deliver energy to the output cavity. Only a small degree of bunching takes place within the electron beam during the interval of travel from the input cavity to the middle cavity. The amount of bunching is sufficient, however, to cause oscillations within the middle cavity and also to maintain a large oscillating voltage across the input gap; most of the velocity modulation produced in the three-cavity klystron is caused by the voltage across the input gap of the middle cavity. The high voltage across the gap causes the bunching process to proceed rapidly in the drift space between the middle and the output cavity. The electron bunches cross the gap of the output cavity when the gap voltage is at maximum negative. Maximum energy transfer from the electron beam to the output cavity occurs under these conditions. The energy released by the electrons is the kinetic energy that was originally absorbed from the cathode pulse.

Klystron amplifiers with as many as five intermediate cavities in addition to the input and output cavities are available. The intermediate cavities improve electron bunching and enhance amplifier gain. The overall efficiency of the tube is also improved but to a lesser extent. Adding more cavities can be considered equivalent to the addition of more stages to a conventional amplifier.

12.3.5 Tuning

A klystron may have the following two types of tuning.

12.3.5.1 Synchronous Tuning

In synchronous tuning, all cavities of a klystron are tuned to the same frequency. In this case, the overall amplifier gain increases but the bandwidth reduces. A klystron amplifier will thus deliver high power over a narrow bandwidth. Thus, synchronous tuning may be used for narrow-band operations.

12.3.5.2 Staggered Tuning

In this case, different cavities are tuned to different frequencies. This tuning is employed for broadband operations. The transmission of TV signal may be such an example where broadband operation is required. In this tuning, intermediate cavities are tuned to either side of the central frequency. If the cavities are tuned to slightly different frequencies, the gain of the amplifier will reduce but the bandwidth will appreciably increase. Since Q of involved cavities is quite high, the stagger tuning always requires a bandwidth exceeding 1% of the central frequency.

12.3.6 Two-Cavity Klystron Oscillator

As stated earlier, a klystron can also be used as an oscillator. For oscillations, a portion of signal in the catcher cavity is to be coupled back to the buncher cavity. This feedback must have the correct polarity and sufficient magnitude. Oscillations in two cavity klystrons behave as in any other feedback oscillator, having been started by a switching or noise impulse. These oscillations continue as long as dc power is present.

12.3.7 Performance

Klystrons can operate over a frequency range of 250 MHz to 95 GHz. Their efficiency ranges from about 35% to 50% and power gain is approximately 30–35 dB at UHF and 60–70 dB at microwave range. These tubes can operate in CW and pulse mode and can deliver an average CW power up to 700 kW and pulsed power up to 30 MW at 10 GHz. A 50 kW UHF klystron may be over 2 m long with nearly 250 kg weight. About two-thirds of this weight is contributed by the focussing magnets. This weight can be reduced to nearly one-third by employing periodic permanent magnet (PPM) focussing.

12.3.8 Applications

Klystrons are extensively used in communication and radar services, including UHF TV transmitters, long-range pulsed radars, linear particle accelerators, troposcatter links and in earth station transmitters.

12.4 Reflex Klystron

In Section 12.3.6, it was noted that a klystron can work as an oscillator, if a fraction of its output power is fed back to its input cavity. It will oscillate if its loop gain has unity magnitude with a phase shift multiple of 2π. In general,

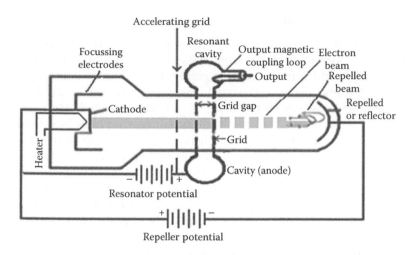

FIGURE 12.10
Constructional features of reflex klystron.

an oscillator is not constructed of a two-cavity klystron mainly because when an oscillation frequency is varied, the resonant frequency of each cavity and the feedback path phase shift is to be readjusted for a positive feedback. The reflex klystron is a single-cavity device that overcomes the above drawback.

Reflex klystron is a low-power tube of 10–500 mW output in the frequency range of 1–25 GHz. Its efficiency is low and may lie between 20% and 30%. It is commonly used in applications wherein the low power can serve the purpose, for example, in laboratory test benches, in microwave receivers as local oscillator, in commercial, military and airborne Doppler radars and in missiles. Its constructional features are shown in Figure 12.10.

12.4.1 Constructional Features

It contains a small electron gun comprising a heater and an electrode called the cathode, a reflector plate referred to as repeller, an oscillating resonant cavity (called anode) and focussing electrodes. If the device has a short beam, no focussing electrodes are required. After electrons are liberated from the cathode, they travel in the form of a narrow beam, which gets accelerated towards the cavity under the influence of an electric field. The existence of this field is due to a high positive potential at which the cavity (anode) is maintained. The excited electrons overshoot the cavity gap and continue their journey to the next electrode called the repeller. These electrons never reach the repeller as it is maintained at a high negative potential. Electrons in the beam reach some point in the repeller space and then turn back, eventually to fall into the anode cavity. Figure 12.10 illustrates the constituents of reflex klystron and the repelled electron beam as described above.

12.4.2 Power Sources Required

The reflex klystron operation requires three power sources. The first source is required to liberate electrons from the cathode. This liberation is due to the heat energy, which is caused by a supply to the heater filament. Now, if these liberated and thermally excited electrons come under the influence of a suitable (positive) potential field, they get accelerated. Thus, the second source, a positive resonator voltage (often referred to as beam voltage), is used to accelerate these electrons through the grid-gap of the resonant cavity. As stated earlier, these electrons are to be repelled. This task is performed by the third source, that is, a (negative) repeller voltage. In addition to the above, if the liberated electrons spread and are to be focussed into a narrow beam, it may require an electrostatic field, which may be set up by the resonator (positive) potential in the body of the tube. This resonator potential may be common with the resonator cavity, the accelerating grid and the entire body of the tube. Two of the three power sources are shown in Figures 12.10 and 12.11.

12.4.3 Velocity Modulation

The phenomenon of velocity modulation has the same relevance in reflex klystrons as in klystrons. In klystrons, the electron beam is modulated by passing it through an oscillating resonant cavity. In reflex klystron, the

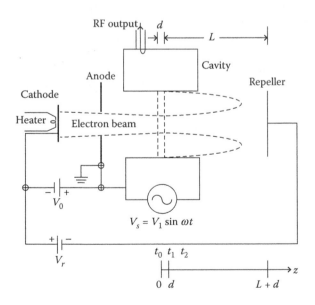

FIGURE 12.11
Functional diagram of reflex klystron.

modulation is accomplished by using the repeller. The feedback required to maintain the oscillations within the cavity is obtained by reversing the direction of the beam that is, by returning it towards the cavity. The electrons in the beam are velocity-modulated before the beam passes through the cavity the second time. This reversal is due to the negatively charged electrode that repels the beam. The returning beam gives up the energy required to maintain oscillations. In view of this reflex action, the klystron gets a new name, that is, the reflex klystron.

The functional diagram of a reflex klystron shown in Figure 12.11 illustrates a number of parameters required in the analysis. These parameters include t_0 (time for an electron entering the cavity gap at $z = 0$), t_1 (time for the same electron leaving cavity gap at $z = d$) and t_2 (time for the same electron returned by the retarding field at $z = d$ and collected by the cavity). The analysis of reflex klystron is almost similar to that of a two-cavity klystron. As in the case of the two-cavity klystron, the space-charge effect on the electron motion is neglected and it is assumed that the electrons enter the cavity gap at $z = 0$ at time t_0 to have a uniform velocity given by Equation 12.1. The same electron leaves the cavity gap at $z = d$ at time t_1 with the velocity

$$v(t_1) = v_0\left[1 + \frac{\beta_i V_1}{2V_0}\sin\left(\omega t_1 - \frac{\theta_g}{2}\right)\right] \qquad (12.36)$$

This expression for the electron moving in the forward direction is identical to that of Equation 12.7. The parameters involved also have the same meaning as before. In reflex klystron, the same electron is repelled to the cavity at $z = d$ at time t_2 by the retarding electric field E. This electric field assumed to be constant in the z-direction is given by

$$E = \frac{V_r + V_0 + V_1\sin(\omega t)}{L} \qquad (12.37)$$

The parameters involved in Equation 12.37 are shown in Figure 12.11). As discussed in Section 11.3, the force equation for one electron in the repeller region can be written as

$$m\frac{d^2z}{dt^2} = -eE = -e\frac{V_r + V_0}{L} \qquad (12.38)$$

where the gradient relation $E = -\nabla V$ is used only in the z-direction. In Equations 12.37 and 12.38, V_r is the magnitude of the repeller voltage. Equation 12.38 is obtained in view of Equation 12.37 with the assumption that $|V_1\sin(\omega t)| \ll (V_r + V_0)$.

Integration of Equation 12.38 gives the velocity term in the z-direction:

$$\frac{dz}{dt} = -\frac{e}{m}\frac{V_r + V_0}{L}\int_{t_1}^{t} dt = -\frac{e}{m}\frac{V_r + V_0}{L}(t - t_1) + k_1 \tag{12.39a}$$

Arbitrary constant k_1 is evaluated as at $t = t_1$, $dz/dt = v(t_1) = k_1$. Equation 12.39a now becomes

$$\frac{dz}{dt} = -\frac{e}{m}\frac{V_r + V_0}{L}(t - t_1) + v(t_1) \tag{12.39b}$$

Integration of Equation 12.39b gives the distance term:

$$z = -\frac{e}{m}\frac{V_r + V_0}{L}\int_{t1}^{t}(t - t_1)\,dt + v(t_1)\int_{t1}^{t} dt + k_2 \tag{12.40a}$$

Arbitrary constant k_2 is evaluated since at $t = t_1$, $z = d = k_2$. Equation 12.40a becomes

$$z = -\frac{e}{m}\frac{V_r + V_0}{2L}(t - t_1)^2 + v(t_1)(t - t_1) + d \tag{12.40b}$$

In view of the assumption that the electron leaves the gap at $z = d$, $t = t_1$ with a velocity $v(t_1)$ and returns to the gap at $z = d$ and time $t = t_2$, then at $t = t_2$ and $z = d$, Equation 12.40b reduces to

$$0 = -\frac{e}{m}\frac{V_r + V_0}{2L}(t_2 - t_1)^2 + v(t_1)(t_2 - t_1)$$

or

$$\frac{e}{m}\frac{V_r + V_0}{2L}(t_2 - t_1) = v(t_1) \tag{12.41}$$

On substituting $v(t_1)$ from Equation 12.36 the time taken by the electron in leaving and returning to the same spot ($z = d$), the round-trip transit time $\{(T' = (t_2 - t_1)\}$ in the repeller region is

$$T' = (t_2 - t_1) = \frac{2mL}{e(V_r - V_0)}v(t_1) = T_0'\left[1 + \frac{\beta_i V_1}{2V_0}\sin\left(\omega t_1 - \frac{\theta_g}{2}\right)\right] \tag{12.42a}$$

$$T_0' = \frac{2mL}{e(V_r - V_0)}v_0 \tag{12.42b}$$

where T_0' is the round-trip dc transit time of the centre-of-the-bunch electron. The multiplication of Equation 12.42a with ω gives

$$\omega T' = \omega(t_2 - t_1) = \omega T_0' \left[1 + \frac{\beta_i V_1}{2V_0} \sin\left(\omega t - \frac{\theta_g}{2} \right) \right] = \theta_0' + X' \sin\left(\omega t_1 - \frac{\theta_g}{2} \right) \quad (12.43)$$

In Equation 12.43

$$\theta_0' = \omega T_0' \quad (12.44a)$$

and

$$X' = \frac{\beta_i V_1}{2V_0} \theta_0' \quad (12.44b)$$

where θ_0' is the round-trip dc transit angle of the centre-of-the-bunch electron and X' is the bunching parameter of the reflex klystron oscillator.

12.4.4 Process of Bunching

Equation 12.44b gives the mathematical expression for the bunching parameter X', which has almost the same meaning as that of X given by Equation 12.18 for klystron. Thus, the bunching process also has the same physical implication. The process in reflex klystrons, however, differs since reflex klystron has only one cavity and a repeller, whereas klystron has more than one cavity but no repeller.

In reflex klystron, when the tube is energised, the resonator potential causes the resonant cavity to begin oscillations at its natural frequency. These oscillations cause an electrostatic field across the grid-gap of the cavity the direction of which changes with the frequency of the cavity. The changing electrostatic field affects the electrons in the beam as they pass through the gap. Some are accelerated and some are decelerated, depending on the polarity of the field as they pass through the gap. Figure 12.12 illustrates the three possible ways in which an electron can be affected as it passes through the gap. The velocity of passing electron may increase, decrease, or remain constant. Since the resonant cavity is oscillating, the grid potential is of the alternating nature that causes the electrostatic field between the grids to follow a sine-wave. As a result, the velocity of electrons passing through the gap is affected uniformly as a function of that sine-wave. The amount of the velocity change is dependent on the strength and polarity of the grid voltage.

With properly adjusted voltages, returning electrons give more energy to the cavity than they take on the forward journey and oscillations result.

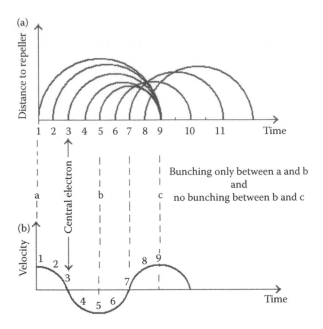

FIGURE 12.12
Electron bunching diagram. (a) Distance–time diagram and (b) velocity–time diagram.

Figure 12.12 termed as electron bunching diagram explains the behaviour of electrons in terms of (a) distance–time and (b) velocity–time diagrams. Their combination is referred to as Applegate diagram. In this diagram, consider a reference electron (marked as 3) that passes the gap when the gap voltage is zero and is going negative. It overshoots the gap unaffected, penetrates some distance into repeller space and returns to the cavity in its backward journey. An electron (marked as 2) passing the gap slightly earlier encounters slightly positive voltage at the gap. The resulting acceleration would propel it farther into repeller space. Since it will be reflected by the repeller, it will take a little more time to return to the cavity. Another electron (marked as 4) passing the gap slightly later encounters negative voltage at the gap. This electron faces retardation, shortened journey in repeller space and thus takes a little shorter time in returning to the gap. In view of the normal, longer and shorter times taken by the three electrons (2, 3 and 4) leaving at different instances, all the three are likely to arrive at the same time at the gap. This results in bunching of returning electrons. It is to be noted that the behaviour of electron-1 will be similar to that of electron-2 and that of electron-5 will be similar to that of electron-4. Thus, all electrons between locations '*a*' and '*b*' on time scale will get bunched. It can be further seen from Figure 12.12 that the conditions met by electrons between locations '*b*' and '*c*' are just reverse of that met by electrons between '*a*' and '*b*'; hence there will be no bunching between '*b*' and '*c*'.

12.4.4.1 Current Modulation

In view of Sections 11.4 and 12.4.4, the velocity-modulated electrons get bunched before arriving at the cavity. In view of this bunching, the amount of current varies from instant to instant. The process of this current variation is referred to as current modulation. Thus, the process of velocity modulation gets converted to the current modulation. The returning beam can now be referred to as current-modulated electron beam.

12.4.4.2 Round-Trip Transit Angle

The retuning electron beam will generate a maximum amount of energy to the oscillation provided it crosses the cavity gap when the gap field is maximum retarding. In view of Figure 12.11, for a maximum energy transfer, the round-trip transit angle with reference to the centre of bunch is given by

$$\omega(t_2 - t_1) = \omega T_0' = (n - 1/4)2\pi = N2\pi = 2n\pi - \pi/2 \qquad (12.45)$$

where n is any positive integer for cycle number and $N = (n - 1/4)$ is the number of mode. Also, it is assumed that $V_1 \ll V_0$.

12.4.4.3 Beam Current

The beam current injected into the cavity gap from the repeller region flows in the negative z-direction. Thus, the beam current of a reflex klystron can be written as

$$i_{2i} = -I_0 - \sum_{n=1}^{\infty} 2I_0 J_n(nX')\cos\{n(\omega t_2 - \theta_0' - \theta_g)\} \qquad (12.46)$$

The fundamental component of induced current in the cavity by the modulated electron beam is obtained by neglecting θ_g (as $\theta_g \ll \theta_0$):

$$i_2 = -\beta_i I_2 = 2I_0\beta_i J_1(X')\cos(\omega t_2 - \theta_0') \qquad (12.47)$$

The magnitude of the fundamental component is

$$I_2 = 2I_0\beta_i J_1(X') \qquad (12.48)$$

Equations 12.46 through 12.48 involve the bunching parameter X' for reflex klystron. The expressions of X' and X (for two-cavity klystron) yield values of opposite signs.

12.4.4.4 Power Output and Efficiency

DC Power: The dc power supplied by the beam voltage V_0 is

$$P_{dc} = V_0 I_0 \tag{12.49}$$

AC Power: The ac power delivered to the load is

$$P_{ac} = V_1 I_1 / 2 = V_1 I_0 \beta_i J_1(X') \tag{12.50}$$

In view of Equations 12.44b and 12.45, the ratio of V_1/V_0 is

$$\frac{V_1}{V_0} = \frac{2X'}{\beta_i(2n\pi - \pi/2)} \tag{12.51}$$

Substitution of Equation 12.51 in Equation 12.50 gives

$$P_{ac} = \frac{2V_0 I_0 X' J_1(X')}{(2n\pi - \pi/2)} \tag{12.52}$$

Efficiency: The electronic efficiency of the reflex klystron oscillator is defined as the ratio of ac power delivered to the load to the dc power supplied by the beam voltage V_0. Thus, in view of Equations 12.49 and 12.52

$$\text{Efficiency} \equiv \frac{P_{ac}}{P_{dc}} = \frac{2X' J_1(X')}{(2n\pi - \pi/2)} \tag{12.53a}$$

As shown in Figure 12.13, the factor $X' J_1(X')$ involved in Equation 12.53 attains maximum value of 1.25 at $X' = 2.408$. Further, from Figure 3.16 $J_1(X') = 0.52$ at $X' = 2.408$. As $n = 2$ has the most power output for $n = 2$, $N = (n - \frac{1}{4}) = 1\frac{3}{4}$ mode. For this case, the maximum electronic efficiency becomes

$$\text{Efficiency} = \frac{2 \times 2.408 \times 0.52}{3.5\pi} = 0.2277 \tag{12.53b}$$

Thus, the maximum theoretical efficiency of a reflex klystron oscillator ranges between 20% and 30%. In view of Equations 12.1, 12.42b and 12.43, the relationship between the repeller voltage and the cycle number n is

$$\frac{V_0}{(V_r + V_0)^2} = \frac{(2\pi n - \pi/2)^2}{8\omega^2 L^2} \frac{e}{m} \tag{12.54a}$$

The power output expressed in terms of the repeller voltage is

$$P_{ac} = \frac{V_0 I_0 X' J_1(X')(V_r + V_0)}{\omega L} \sqrt{\frac{e}{2mV_0}} \tag{12.54b}$$

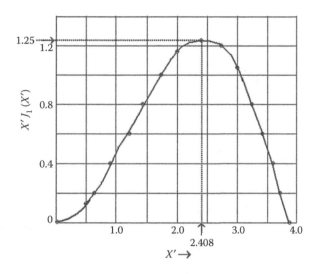

FIGURE 12.13
Variation of factor $X'J_1(X')$ with X'.

12.4.4.5 Electronic Admittance

The induced current given by Equation 12.47 can be written as

$$i_2 = 2I_0 \beta_i J_1(X') e^{-j\theta_0'} \tag{12.55}$$

The voltage across the gap at $t = t_2$ in phasor form is

$$V = V_2 e^{-j\pi/2} \tag{12.56}$$

The ratio of i_2 to V_2 is the electronic admittance of the reflex klystron. This is given as

$$Y_e = \frac{i_2}{V_2} = \frac{I_0}{V_0} \frac{\beta_i^2 \theta_0'}{2} \frac{2J_1(X')}{X'} e^{j(\pi/2 - \theta_0')} \tag{12.57}$$

In view of Equation 12.57, the amplitude of phasor admittance Y_e is a function of dc beam admittance (I_0/V_0), the dc transit angle (θ_0') and the second transit of the electron beam through the cavity gap. The electronic admittance is a non-linear quantity and is proportional to the factor $2J_1(X')/X'$. The parameter X' in itself is proportional to the signal voltage. This factor approaches zero when the signal voltage becomes zero.

12.4.4.6 Equivalent Circuit

The equivalent circuit of a reflex klystron is shown in Figure 12.14. In this circuit, L and C represent the energy storage elements of the cavity, G_C is the

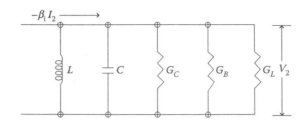

FIGURE 12.14
Equivalent circuit of a reflex klystron.

conductance representing the copper losses, G_B is the beam loading and G_L represents the load conductance.

Effective conductance of this circuit is related to the effective shunt resistance R_{sh} as

$$G = G_C + G_B + G_L = 1/R_{sh} \qquad (12.58)$$

The necessary condition for oscillations in the cavity is

$$|-G_e| \geq G \qquad (12.59)$$

The electronic admittance Y_e given by Equation 12.57 can be written in the rectangular form as

$$Y_e = G_e + jB_e \qquad (12.60)$$

where G_e and B_e are the real and imaginary parts of Y_e.

The phase of Y_e is $\pi/2$ when θ_0' is zero. The rectangular plot of Y_e is a spiral as shown in Figure 12.15. Any value of θ_0' for which the spiral lies in the area to the left of the line $(-G - jB)$ will lead to the oscillations. This value may be given in terms of mode number N as

$$\theta_0' = (n - 1/4)2\pi = N\, 2\pi \qquad (12.61)$$

12.4.5 Modes of Operation

Figure 12.16 illustrates the two modes of operations (for $n = 1$ and $n = 2$ or for $N = \frac{3}{4}\pi$ and $1\frac{3}{4}\pi$) of a reflex klystron. The physical interpretation of modes is given as follows:

In oscillators, it is common to assume that the oscillations start by noise or switching transients. For oscillations to be maintained, the transit time in the repeller space or the time taken by the reference electron from the instant it leaves the gap to the instant it returns must have the correct value. The most suitable departure time is centred on the reference electron (3) at 90° point of the cosine-wave across the resonator gap. It needs to be noted that ideally no

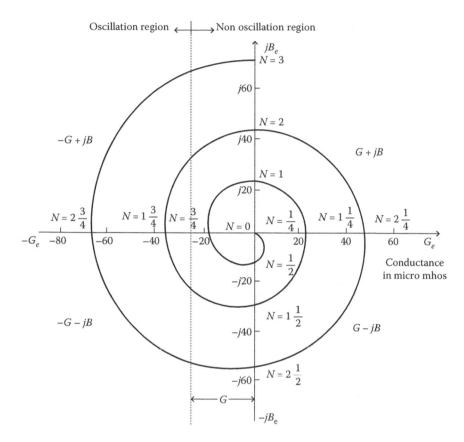

FIGURE 12.15
Electronic admittance spiral of a reflex klystron.

energy goes into velocity modulation of beam. The amount of energy taken by accelerating electrons is given by the retarded ones since over one complete cycle as many electrons are retarded as accelerated by the gap voltage. The best possible time for electrons to return to the gap is when the voltage then existing across the gap will apply maximum retardation to them. This is the time when the gap voltage is maximum positive. Electrons then fall through the maximum negative voltage between the gap-grid, thus giving maximum amount of energy to the gap. The return of electrons after time T ($= n + \frac{3}{4}$, $n = 0, 1, 2, \ldots$) satisfies this requirement, where 'T' is the transit time of electrons in repeller space, and n is an integer.

The transit time depends on repeller and anode voltages and thus both need careful adjustment. When the cavity is tuned to correct frequency both of these voltages are adjusted to the correct value of T. Each combination of acceptable anode–repeller voltages will permit oscillations for a particular value of n. In turn, each n is said to correspond to a different reflex klystron mode.

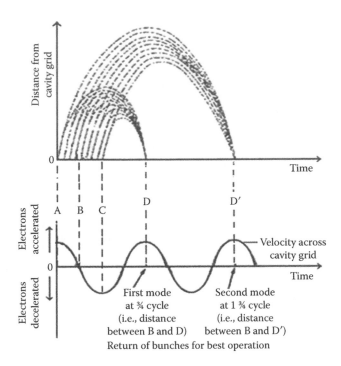

FIGURE 12.16
Two modes of operations of a reflex klystron.

Figure 12.16 shows the effect of the repeller field on the electron bunch for ¾ and 1¾ cycle. If the electrons remain in the field for longer than ¾ cycles, the difference in electron transit time changes the tube performance characteristics. The differences in operating characteristics are identified by the modes of operation. The reflex klystron operates in a different mode for each additional cycle the electrons remain in the repeller field. Mode 1 is obtained when the repeller voltage produces an electron transit time of ¾ cycles. Mode 2 has an electron transit time of 1¾ cycles, mode 3 has an electron transit time of 2¾ cycles and so on. The physical design of the tube limits the number of modes possible in practical applications. A range of four modes of operation is normally available. The actual mode used (1¾ cycles to 4¾ cycles, 2¾ cycles to 6¾ cycles, etc.) depends on the application. The choice of mode is determined by the difference in power available from each mode and the band of frequencies over which the circuit can be tuned.

12.4.6 Debunching

Debunching is the opposite of the bunching process. In the bunching process, the electrons join each other while travelling forward, whereas in debunching, they spread out before they enter the electrostatic fields across the cavity grid on their return journey. The lower concentration of electrons in the returning bunches delivers lesser power to the oscillating cavity. As a result, the amplitude of cavity oscillations reduces and thus output power is also reduced. In higher-mode operation, the bunching process is slower than that in lower-order modes. In view of mutual repulsion between the negatively charged electrons, the slower bunching is more likely to be affected by the debunching process. Further, the long drift time in higher-order modes allows more time for electron interaction and hence the debunching effect is more severe. As mutual repulsion changes the relative velocities, the electrons within bunches become more prone to spreading out. Thus, the output power differs for different modes of operation.

12.4.7 Tuning

The resonant frequency of reflex klystrons can be adjusted in many ways. These may include mechanical methods involving screws, bellows or dielectric inserts or electronic method in which the repeller voltage is changed. Mechanical tuning may give a frequency variation of ±20 MHz at X-band to ±4 GHz at 200 GHz. The electronic tuning by adjustment of repeller voltage has the range of ±8 MHz at X-band and ±80 MHz for submillimeter klystrons.

In electronic tuning, accomplished by altering the repeller voltage, the centre frequency of the cavity is kept constant at some specified value. The frequency is then varied around this centre frequency within the mode of operation. Such variations, above or below the centre frequency, are confined within the half-power points of the mode. In case the centre frequency is to be changed, it can be done either by grid-gap tuning method or by using the paddles or slugs. In grid-gap tuning, the central frequency is varied by altering the distance between the grids, which results in the change of physical size of cavity. To change the physical distance between the grids, a tuning screw can be used to mechanically move one of the grids. The change of distance results in the variation of the cavity capacitance. In case of paddles or slugs, the inductance of the cavity changes and by appropriate adjustment of paddles or slugs, the cavity can be tuned.

Figure 12.17 illustrates the electronic tuning range and output power of a reflex klystron. Each mode has a centre frequency which is predetermined by the physical size of the cavity. The output power increases as the repeller voltage is made more negative. This is because the transit time of the electron bunches is decreased.

12.4.8 Applications

Despite the fact that the reflex klystron in most of the applications is replaced by the semiconductor devices, it is still used as a signal source in microwave generators, local oscillator in microwave receiver frequency-modulated

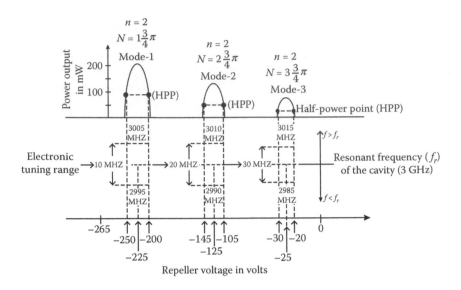

FIGURE 12.17
Electronic tuning and output power of a reflex klystron.

oscillator in portable microwave links and pump oscillator for parametric amplifiers. It is a useful millimetric and submillimetric oscillator producing more power (with very low noise) at higher frequency than semiconductor devices. Reflex klystrons with integral cavities from 4 to 200 GHz are available. Demountable cavities are also possible in amplifier klystrons. The reflex klystrons have typical overall power ranges from 3 W in X-band to 10 mW at 220 GHz. Their efficiencies are normally less than 10%.

12.5 Travelling-Wave Tube

The travelling-wave tube (TWT) is a high-gain, low-noise, wide-bandwidth microwave amplifier. It is capable of gains greater than 40 dB with bandwidths exceeding an octave where one octave bandwidth spells that the upper-to-lower-frequency ratio is two. TWTs have been designed for frequencies as low as 300 MHz and as high as 50 GHz. The TWT is primarily a voltage amplifier. The wide-bandwidth and low-noise characteristics make it ideal for use as an RF amplifier in microwave equipment. Its major elements include a vacuum tube, an electron gun, a slow wave structure, a focussing structure, an RF input, an RF output and a collector.

Like klystron, the TWT also operates on the principle of interaction between electron beam and RF field. Both the klystrons and TWTs are linear-beam tubes with a focussed electron beam similar to that in CRT. In klystron, the RF field is stationary; thus, interaction between the beam and the RF field is not continuous. In TWT also, the RF field travels with the speed of light, while the velocity of electron beam does not exceed 10% of the speed of light, even with very high anode voltage. Continuity of interaction between beam and RF field demands that these two travel in the same direction and with almost the same velocity. TWT uses such a structure to make the interaction continuous.

12.5.1 Simplified Model of TWT

Imagine that an electron beam emerging from an electron gun is passing along a straight wire. This wire represents a (terminated) non-resonant transmission line and is fed an RF signal as input. This input causes an electromagnetic wave progressing at the speed of light along the wire towards the output. The electric field of this wave may interact with the electron beam provided the field and the beam have the same velocity and the same direction of travel. If the interaction is proper, it will modulate the velocity of the beam, which in turn will result in bunching of electrons. But since the velocity of electron beam is not at par with that of the velocity of the wave, the probability of inter-action is much less. At present, there is no known method to accelerate an electron beam to the speed of light. Since the electron beam and wave cannot travel together, bunching will not take place and such a simplified model will not work. The TWT, therefore, requires some delay structure, which can slow down the travelling wave to the speed of electrons in the beam.

12.5.2 Slow Wave Structures

There are many structures that can be used to slow down the wave. Besides the most commonly used helical structure, these structures may be of the shape of (a) folded back line, (b) zigzag line, (c) interdigital line and (d) cor-rugated waveguide. These are shown in Figure 12.18. Some other structures, namely, coupled-cavity, ring-loop and ring-bar structures confirming to these shapes are also briefly described.

12.5.2.1 Helical Structure

A common TWT delay structure is a wire, wound in the form of a long coil or helix. The shape of the helix slows the effective velocity of the wave along its common axis with the tube to about one-tenth of the speed of light. The wave still travels down the helix wire at the speed of light, but the coiled shape causes the wave to travel a much greater total distance than that travelled by the electron beam. The speed at which the wave travels can be varied by changing either the number of turns (i.e., by changing the pitch p) or the diameter of turns in the helix wire.

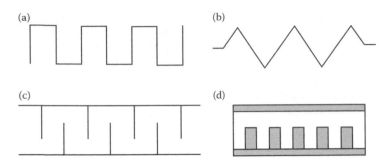

FIGURE 12.18
Slow wave structures. (a) Folded back line, (b) zigzag line, (c) interdigital and (d) corrugated waveguide.

The geometry of a helix is illustrated in Figure 12.19. In the given figure, 'p' is the pitch, 'a' is the radius, d is the diameter, C (= πd) is the circumference and ψ is the pitch angle of the helix. It is inherently a non-resonant structure and gives very large bandwidth. The turns of the helix are in close proximity. Oscillations caused by feedback occur at high frequencies. The helix diameter is to be reduced at higher frequencies to contain a high RF field at its centre. With the reduction of diameter, the focussing of the beam becomes difficult and if a high-power beam is intercepted by the helix, it is likely to melt.

The helical delay structure works well because it has the added advantage of causing a large proportion of electric fields that are parallel to the electron beam. The parallel fields provide maximum interaction between the fields and the electron beam. Figure 12.20 shows the electric fields that are parallel to the electron beam inside the helix.

The basic principle of operation of a TWT lies in the interaction between electron beam and RF signal. Equation 11.20b, reproduced below, gives the velocity of an electron beam:

$$v = \sqrt{\frac{2eV_a}{m}} = 0.5932 \times 10^6 \sqrt{V_a} \text{ m/s} \tag{12.1}$$

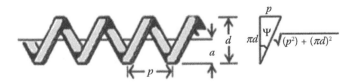

FIGURE 12.19
Geometry of helix.

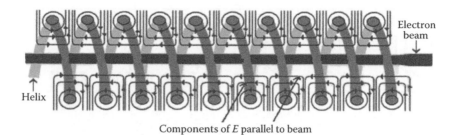

Components of E parallel to beam

FIGURE 12.20
Electric field parallel to the electron beam inside the helix.

where V_a is the (accelerating) anode voltage, e is the electron charge
($= 1.6 \times 10^{-19}$ C) and m is the mass of electron ($= 9.1 \times 10^{-31}$ kg). With an anode
voltage of 5 kV, an electron velocity can be calculated from Equation 12.1. It
comes out to be about 4.2×10^7 m/s, whereas the signal in general travels with
the velocity of light ($c = 3 \times 10^8$ m/s). Thus, there is much difference between
the two velocities. As noted earlier, there is no method by which the velocity of electron beam is increased to that of light. Even if it is assumed to be
possible, then in view of the theory of relativity, the electron mass increases
with the velocity and approaches infinity as its velocity approaches c. Thus,
achieving electron velocities approaching c makes the situation complicated.
If, however, the signal velocity is brought to the level of velocity of the electron beam, it is possible to amplify the signal by virtue of its interaction with
the beam.

This interaction is usually achieved by using a slow wave helical structure.
Without the helix, the signal would travel at velocity c, whereas with the
helix, the axial signal velocity approximately equals $c \times (p/2\pi a)$, where a and
p are shown in Figure 12.19. Thus, the signal is slowed down by the factor
$p/2\pi a$, which is independent of the signal frequency. The signal travelling
along the helix is known as a slow wave, and the helix is referred to as a slow
wave structure. The condition for equal slow wave and electron beam velocity is therefore approximately obtained from the relation

$$v = \frac{c \cdot p}{2\pi a} = \sqrt{\frac{2eV_a}{m}} \tag{12.62}$$

Helix is the most commonly used slow wave structure in TWTs. This can,
however, be replaced by some other forms shown in Figure 12.18 or by ring-
bar, ring-loop, or coupled-cavity structure. The selection of the structure
depends on the requirements of the desired gain/bandwidth and power
characteristics.

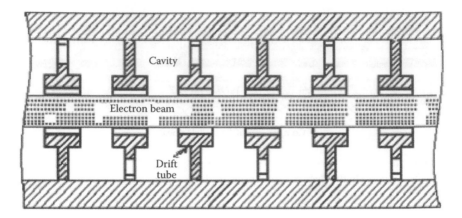

FIGURE 12.21
Coupled-cavity slow wave structure.

12.5.2.2 Coupled-Cavity Structure

The coupled-cavity TWT (Figure 12.21) uses a slow wave structure consisting of a large number of cavities coupled to one another. These resonant cavities together are coupled with a transmission line. Such an arrangement acts as a distributed network, with its principle of operation identical to that of pulse forming networks. The electron beam is velocity modulated by a RF signal at the first resonant cavity. This RF energy travels along the cavities and induces RF voltages in each subsequent cavity. If the spacing between these cavities is correctly adjusted, the voltages at each cavity induced by the modulated beam are in phase and travel along the transmission line to the output, with an additive effect. Thus, the output is much greater than the power input. Coupled-cavity structure gives good bandwidth but less than that obtained from the helix. Its applications are limited to the frequencies of less than 100 GHz beyond which ring-bar and other structures are used.

12.5.2.3 Ring-Loop Structure

This slow wave structure consists of loops of rings. A larger number of such rings are tied together (Figure 12.22). These devices have the capability to

FIGURE 12.22
Ring-loop slow wave structure.

operate at higher power levels than conventional helix TWTs. These, however, have significantly less bandwidth which ranges between 5% and 15% of the operating frequency. Also, these have a lower cutoff frequency of about 18 GHz. The ring-loop slow wave structure has high coupling impedance and low harmonic components. Thus, these tubes have the advantages of high gain (between 40 and 60 dBs), small dimensions, higher operating voltage and lesser backward-wave oscillations.

12.5.2.4 Ring-Bar Structure

The characteristics of ring-bar TWT (Figure 12.23) are similar to those of ring-loop TWT. This slow wave structure is easier to construct as it can be made by cutting a copper tube.

12.5.3 Construction

The physical construction of a typical TWT is shown in Figure 12.24. It contains an electron gun that produces an electron beam similar to those produced in CRTs, although the beam current is relatively much larger than that in CRTs. Electrons from a heated cathode are accelerated towards the anode along the axis of the tube. The anode is held at a high positive potential with respect to the cathode. A significant proportion of electrons liberated from

FIGURE 12.23
Ring-bar slow wave structure.

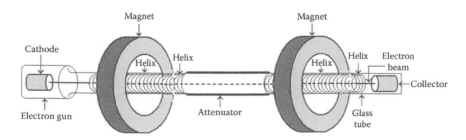

FIGURE 12.24
Physical construction of a TWT.

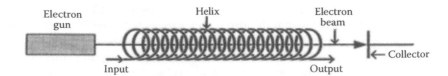

FIGURE 12.25
Input–output locations in TWT of Figure 12.24.

the cathode passes through a hole in the anode to produce the beam. In some of the tubes a grid is inserted between the cathode and anode. It is maintained at an adjustable negative potential of few tens of volts with respect to the cathode. The function of this grid is to control the beam current. The electron beam travels down the tube to the collector, which is maintained at a higher potential with respect to the cathode as well as the anode of the electron gun. The beam remains confined inside the helix, which is also held at a high potential with respect to the cathode. The helix current is low because of the beam focussing.

Some of the basic components of TWT shown in Figure 12.24 are re-illustrated in Figure 12.25 in a simplified form. These include the electron gun, input and output locations, helix, electron beam and collector.

The surrounding magnets shown in Figure 12.24 provide a magnetic field along the axis of the tube to focus the electrons into a tight beam. The helix, at the centre of the tube, is a coiled wire that provides a low-impedance transmission line for RF energy within the tube. The RF input and output are coupled into and removed from the helix by the respective waveguide directional couplers that have no physical connection to the helix.

Figure 12.26 further illustrates the essential features of a TWT including the location of an attenuator, which is provided to prevent any reflected waves from travelling back.

12.5.4 Focussing

As TWTs have sufficient lengths, the electron beam needs to be focussed for its proper operation. The possible focussing methods include a permanent

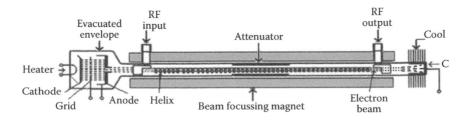

FIGURE 12.26
Essential features of TWT with wave propagation from left to right.

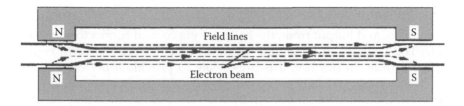

FIGURE 12.27
Focussing of an electron beam by magnetic field.

magnet, an electromagnet, a solenoid, an integral solenoid and periodic permanent magnet. The magnetic field required for good focussing will be very large, which would mean a bulky and heavy magnet. Thus, the use of permanent magnets is somewhat awkward, whereas the electromagnets become bulky and cannot be used where space is a premium. The solenoids provide excellent magnetic field and are often employed on high-power tubes used in ground-based radars. The integral solenoids make the assembly lighter and thus are quite suitable for airborne applications. This type of focussing is shown in Figure 12.27.

Figure 12.28 illustrates PPM focussing wherein a number of small toroidal permanent magnets of alternating polarity are arranged along the tube with spacing. This figure also shows the contour of beam. Defocussing (if any) that may occur between any two magnets gets corrected by the next magnet. The PPM configuration is often used.

As illustrated in Figures 12.27 and 12.28, the focussing of beam is achieved by a magnet around the outside of the tube. An electron with a component of velocity perpendicular to magnetic field lines experiences a restoring force tending to bring back its direction parallel to field lines as was shown in Figure 12.20.

The weight of magnet greatly reduces by using PPM arrangement. Under ideal conditions, this is reduced by a factor $1/N^2$, where N is the number of magnets used. The alternative method of solenoid focussing is used only in high-power earth station TWTs, where size and weight are not of much importance.

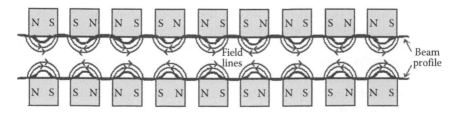

FIGURE 12.28
Periodic permanent magnetic (PPM) focussing.

12.5.5 Input–Output Arrangements

The input to, and output from, the helix are via coaxial connectors or waveguide. The locations of input and output in a TWT were shown in Figures 12.25 and 12.26. In practice, it is impossible to provide a perfect match at these transitions, especially over a wide bandwidth, so an attenuator is used to prevent the reflection of energy through the helix, which may result in instability. This attenuator (Figures 12.24 and 12.26) is usually in the form of a resistive coating on the outside of the central portion of the tube. A physical discontinuity in the helix is also used in some cases. The attenuator reduces the RF signal as well as any reflected signal, but the electron bunches set up by the signal remain unaffected.

The helix is a fairly delicate structure, and must be provided with adequate thermal dissipation to prevent damage. Figure 12.26 illustrates some cooling fins provided for this purpose. In medium-power tubes, the helix is often supported on a beryllia or alumina substrate. For high-power TWTs, alternative slow wave structures may be employed (e.g., coupled cavities), though usually at the expense of bandwidth. In this form, the TWT resembles a klystron amplifier.

12.5.6 Basic Operating Principle

TWT employs an electron gun to produce a very narrow electron beam. This beam travels through the centre of the helix without touching the helix itself. The helix is made positive with respect to the cathode and the collector is made still more positive. Narrow and correctly focussed beam attracted by the collector acquires high velocity. A dc magnetic field prevents beam spreading. Input signal is applied to the input end of the helix via a waveguide or a coaxial line. The field starts progressing along the helix with almost the speed of light in free space.

The speed with which the field advances axially is equal to the velocity of light multiplied by the ratio of the helix pitch to the helix circumference. Thus, this can be made (relatively) quite slow and approximately equal to the electron beam velocity. Axial RF field and beam now interact continuously, with the beam bunching and energy transfer from the electron beam to the RF field. Almost complete bunching results in high gain and minimum noise. TWTs are capable of high BW and are used as medium- and high-power (CW/pulsed) amplifier at microwave range.

Electrons leaving the cathode at random encounter a weak RF field at the input of the helix, which is due to input signal. As in klystron and magnetron, the interaction between the RF field and the beam results in velocity modulation and bunching of electrons takes place. This bunching tends to be more complete as the field progresses in the forward-axial direction along the helix. Simultaneously, the RF field starts growing (exponentially) from

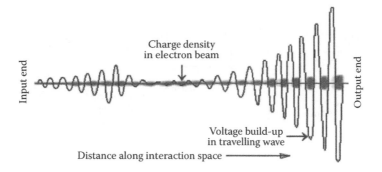

FIGURE 12.29
Growth of signal and bunching along TWT.

turn to turn. As shown in Figure 12.29, the process continues as both the wave and the beam travel towards the output end.

The interaction between the beam and the slow wave results in the 'velocity modulation' of the beam. In essence, the term 'velocity modulation' indicates that some of the electrons are accelerated, whereas certain others are retarded. Such a phenomenon results in the formation of bunches within the beam. The beam current, therefore, becomes modulated by the RF signal. The bunches react with the RF fields associated with the slow wave travelling down the helix, resulting in a net transfer of energy from the beam to the signal, and consequent amplification. Since there are no resonant structures involved in this interaction, amplification is obtained over a wide bandwidth. In fact, the principal factors that limit bandwidth are the input/output coupling arrangements.

12.5.7 Power Supply

The power supply arrangements along with the voltages and currents for a typical 10 W (output) TWT are shown in Figure 12.30. Although the alignment details apply to almost all tubes, the manufacturer's data regarding electrode voltages and tube operating conditions need to be referred to before operating a particular tube. As the power reflected from any mismatch at the output is dissipated in the helix and can burn it out, it is essential that the output of the tube is suitably matched to the load. For the same reason, the antenna must also be appropriately matched.

The beam current is controlled by the grid-cathode voltage. Since all modern TWTs are provided with the beam focussing arrangement, no further adjustments are necessary. In case the focussing itself is adjustable, the tube should initially be run at low beam (collector) current, and the beam focussing magnets adjusted for minimum helix current. The helix voltage should

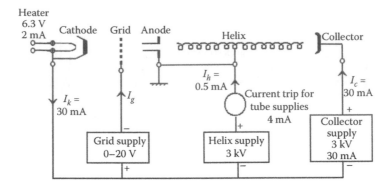

FIGURE 12.30
Power supply arrangement at different electrodes of a TWT.

also be set for minimum helix current. With the tube running at its specified collector current, the RF drive can be applied. The collector current will
hardly change, but the helix voltage should be set for maximum output, con-
sistent with not exceeding the tube voltage or helix current ratings. If the
focussing is adjustable, this should be readjusted for minimum helix current,
since the RF drive will defocus the beam slightly. As the helix is a fragile
structure and will not dissipate more than a certain amount of power without getting damaged, the helix current should be metered, and a current
trip incorporated to cut the power supplies to the tube if the helix current
exceeds the limit. The extra high tension (EHT) supplies to the tube should
also be well smoothed because ripples present will phase-modulate the output and give a rough note. If the collector dissipates more than about 100 W,
it may be necessary to use a blower to cool the collector end of the tube.
Typical efficiency of the TWTA is about 10%, though some modern tubes
have reportedly attained up to 40% efficiency.

12.5.8 Mathematical Analysis

The phase velocity of a wave in an air-filled waveguide is greater than c. The
slow wave structures are the special circuits used in microwave tubes to
reduce this phase velocity to the tune of the velocity of the beam. In view of
the geometry of the helix (Figure 12.19), the ratio of phase velocity v_p along
the pitch to the phase velocity along the coil can be given by

$$\frac{v_p}{c} = \frac{p}{\sqrt{p^2 + (\pi d)^2}} = \sin\psi \qquad (12.63)$$

Since the helical coil, in general, may be within a dielectric-filled cylinder,
Equation 63 can be re-written as

$$v_{p\varepsilon} = \frac{p}{\sqrt{\mu\varepsilon\{p^2 + (\pi d)^2\}}} \tag{12.64}$$

A slow wave structure, with a large dielectric constant, is more lossy and thus less efficient. For small pitch angle, the phase velocity along the coil in free space is approximately given by

$$v_p \approx \frac{c \cdot p}{\pi d} = \frac{\omega}{\beta} \tag{12.65}$$

12.5.8.1 Brillouin Diagram

Figure 12.31 shows the ω–β (or Brillouin) diagram for a helical slow wave structure. This curve is very useful in the design of a helix. Once β is known, v_p can be computed from Equation 12.65 for a given dimension of a helix. The group velocity (v_{gr}) of the wave is simply the slope of the curve and is given by

$$v_{gr} = \partial\omega/\partial\beta \tag{12.66}$$

A slow wave structure has the property of periodicity in axial direction. In helical slow wave structure, a back or forth translation through a distance of one pitch length results in the same identical structure. Thus, the pitch is the period of slow wave helical structure. Its field must be distributed according to the Floquet's theorem for periodic boundaries. Thus, if $E(x, y, z)$ is a periodic function of z with period L, β_0 is the phase constant of average electron

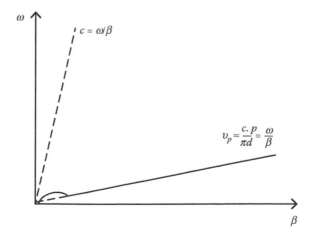

FIGURE 12.31
ω–β (or Brillouin) diagram.

velocity in the axial direction, then this theorem can be stated in terms of a mathematical relation

$$E(x,y,z-L) = E(x,y,z)e^{j\beta_0 L} \tag{12.67}$$

where $E(x, y, z)$ is postulated to be the solution to Maxwell's equations in a periodic structure, which can be written as

$$E(x,y,z) = f(x,y,z)e^{-j\beta_0 z} \tag{12.68}$$

In Equation 12.68, $f(x, y, z)$ is a periodic function of z with period L of slow wave structure.

Equation 12.67 can now be written as

$$E(x,y,z-L) = f(x,y,z-L)e^{-j\beta_0(z-L)} \tag{12.69}$$

Since $f(x, y, z - L)$ is a periodic function with period L, thus

$$f(x, y, z - L) = f(x, y, z) \tag{12.70}$$

In view of Equations 12.68 through 12.70, we get

$$E(x,y,z-L) = f(x,y,z)e^{j\beta_0 L} \tag{12.71}$$

Equation 12.71, the mathematical statement of Floquet's theorem, is satisfied by Equation 12.68.

Since any function that is periodic, single-valued, finite and continuous can be expanded in the Fourier series, the function $E(x, y, z)$ can be written as

$$E(x,y,z) = \sum_{n=-\infty}^{\infty} E_n(x,y)e^{-j(2n\pi/L)z}e^{j\beta_0 z} = \sum_{n=-\infty}^{\infty} E_n(x,y)e^{-j\beta_n z} \tag{12.72}$$

where

$$E_n(x,y) = \frac{1}{L}\int_0^L E(x,y,z)e^{-j(2n\pi/L)z}\, dz \tag{12.73a}$$

$$\beta_n = \beta_0 + (2n\pi/L) \tag{12.73b}$$

In Equation 12.73, $E_n(x, y)$ is the amplitude of the nth harmonic and β_n is its corresponding phase constant. In Equation 12.72, quantities $E_n(x,y)e^{-j\beta_n z}$ are known as spatial harmonics by the analogy with the time domain Fourier series.

The substitution of Equation 12.72 in wave equation for lossy media (Equation 2.8a) gives

$$\nabla^2 \left[\sum_{n=-\infty}^{\infty} E_n(x,y)e^{-j\beta_{nz}} \right] - \gamma^2 \left[\sum_{n=-\infty}^{\infty} E_n(x,y)e^{-j\beta_{nz}} \right] = 0 \qquad (12.74)$$

In view of the linear nature of wave equation, Equation 12.74 can be written as

$$\sum_{n=-\infty}^{\infty} \left[\nabla^2 E_n(x,y)e^{-j\beta_{nz}} - \gamma^2 E_n(x,y)e^{-j\beta_{nz}} \right] = 0 \qquad (12.75)$$

This equation indicates that if spatial harmonic for each value of n is a solution of the wave equation, the summation of all space harmonics also satisfies the wave equation. Thus, only the complete solution of Equation 12.72 can satisfy the boundary conditions of a periodic structure. Equation 12.72 also shows that in a periodic structure, the field can be expanded in the form of an infinite series, each component of which has the same frequency but different phase velocities v_{pn}. Thus,

$$v_{pn} = \frac{\omega}{\beta_n} = \frac{\omega}{\beta_0 + (2n\pi/L)} \qquad (12.76)$$

The group velocity, which comes out to be independent of n, can be obtained in view of Equations 12.66 and 12.76. Its value is as given as follows:

$$v_{gr} = \left[\frac{d\{\beta_0 + (2n\pi/L)\}}{d\omega} \right]^{-1} = \frac{\partial \omega}{\partial \beta_0} \qquad (12.77)$$

In view of Equation 12.76, the phase velocity v_{pn} in the axial direction decreases with the increase of n and β_0. Thus, for some *n*, the value of v_{pn} may become less than the velocity of light. For still higher *n*, the phase velocity may further reduce and the interaction between microwave signal and electron beam may become possible, which may ultimately lead to the amplification process.

Figure 12.32 illustrates another ω–β (or Brillouin) diagram but this time with several spatial harmonics. From this diagram, it can be noted that the second quadrant indicates the negative phase velocity that corresponds to the negative n. It spells that the electron beam moves in the positive z-direction while the beam velocity coincides with the negative spatial harmonic's phase. This property is exhibited by the backward-wave oscillator (BWO). The shaded areas shown are referred to as forbidden regions for propagation. When the

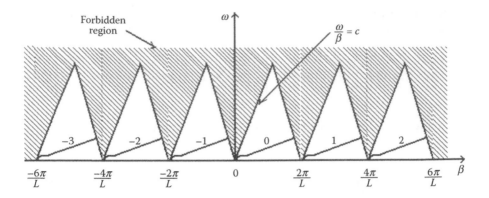

FIGURE 12.32
ω–β (or Brillouin) diagram of spatial harmonics for helical structure.

axial phase velocity of any spatial harmonic exceeds the velocity of light, the structure radiates energy.

12.5.8.2 Amplification Process

The schematic diagrams of a helix-type TWT are shown in segments in Figures 12.24 through 12.26 and the Brillouin diagram characterising a slow wave helix structures is given in Figure 12.32. The phase shift per period of the fundamental wave on the helix is

$$\theta_1 = \beta_0 L \tag{12.78}$$

where $\beta_0 = \omega/v_0$ is the phase constant of average beam velocity and L is the period or pitch.

The dc transit time of an electron is given by

$$T_0 = L/v_0 \tag{12.79}$$

The phase constant of the nth space harmonic is

$$\beta_n = \frac{\omega}{v_0} = \frac{\theta_1 + 2n\pi}{v_0 T_0} = \beta_0 + \frac{2n\pi}{L} \tag{12.80}$$

In deriving Equation 12.80, which is identical to Equation 12.73b, it is assumed that the axial space-harmonic phase velocity and the beam velocity are synchronised. Such synchronisation is necessary for interaction between the electron beam and electric field. Thus,

$$v_{np} = v_0 \tag{12.81}$$

According to Equation 12.81, theoretically two velocities have to be equal but in practice the proper transfer of energy requires that the dc velocity of the electron beam is kept slightly greater than the axial velocity of the wave.

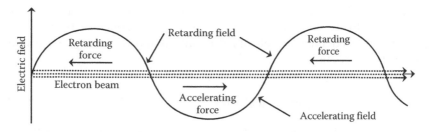

FIGURE 12.33
Interaction between electron beam and electric field.

When an RF signal is coupled into the helix, it results in an electric field. Its axial component exerts force on the electrons in view of the relations $F = -eE$ and $E = -\nabla V$. In view of this force, the electrons entering the retarded field are decelerated and those in the accelerating field get accelerated. Thus, the electrons start bunching around those electrons, which enter the helix at the instant when field is zero. The process is illustrated by Figure 12.33.

Since the dc velocity of the electron beam is slightly greater than the axial wave velocity, there are more electrons in the retarded field than in the accelerating field where more energy is transferred from the beam to the electric field. As a result, bunches start becoming increasingly compact. The process of amplification begins, and by the time bunches arrive at the output end, there is a larger amplification of the signal. The axial magnetic field due to focussing magnets prevents beam spreading as it travels down the tube. The attenuator attenuates reflected waves due to any mismatch and prevents them from reaching the input to cause oscillation. The bunched electrons induce a new electric field of the same frequency, which in turn induces a new amplified microwave signal on the helix. For the travelling wave progressing in the z-direction, the z component of the electric field (E_z) can be expressed as

$$E_z = E_1 \sin (\omega t - \beta_p z) \tag{12.82}$$

where E_1 is the magnitude of the field E_z, β_p ($= \omega/v_p$) is the axial phase constant of the microwave signal and v_p is the axial phase velocity of the wave. The electric field is assumed as maximum if $t = t_0$ at $z = 0$.

The equation of motion of the electron is given as

$$m \frac{dv}{dt} = -eE_1 \sin(\omega t - \beta_p z) \tag{12.83}$$

If the velocity of the electron (v) is assumed to be given by

$$v = v_0 + v_e \cos (\omega_e t + \theta_e) \tag{12.84}$$

Then,

$$dv/dt = -v_e \, \omega_e \sin (\omega_e t + \theta_e) \tag{12.85}$$

In Equations 12.84 and 12.85, v_0 is the dc electron velocity, v_e is the magnitude of the velocity fluctuation in the velocity-modulated electron beam, ω_e is the angular frequency of velocity fluctuation and θ_e is the phase angle of the velocity fluctuation.

Substitution of Equation 12.85 in Equation 12.83 gives

$$m\, v_e\, \omega_e \sin{(\omega_e t + \theta_e)} = e\, E_1 \sin(\omega t + \beta_p z) \tag{12.86}$$

There will be interaction between the electric field and the electron beam, provided the velocity of the modulated electron beam and the dc electron velocity are approximately the same, that is, $v \approx v_0$. Thus, the distance travelled by the electrons is

$$z = v_0\, (t - t_0) \tag{12.87}$$

Substitution of z from Equation 12.87 in Equation 12.86 gives

$$m\, v_e\, \omega_e \sin{(\omega_e t + \theta_e)} = e\, E_1 \sin\{\omega t - \beta_p v_0\, (t - t_0)\} \tag{12.88a}$$

Rearrangement of the terms after substitution of $\beta_p = \omega/v_p$ or $\omega = \beta_p\, v_p$ at $t = t_0$ gives

$$v_e \sin{(\omega_e t + \theta_e)} = \{eE_1/(m\omega_e)\} \sin\{\beta_p\, v_p\, t - \beta_p v_0 t + \beta_p v_0 t_0\} \tag{12.88b}$$

Comparison of the left and right sides of Equation 12.88b yields the following relations:

$$v_e = eE_1/(m\omega_e) \tag{12.89a}$$

$$\omega_e = \beta_p\, (v_p - v_0) \tag{12.89b}$$

$$\theta_e = \beta_p v_0 t_0 \tag{12.89c}$$

Equation 12.89a shows that the magnitude of the velocity fluctuation of the electron beam is directly proportional to the magnitude of the axial field.

12.5.8.3 Convection Current

The convection current (i) in the electron beam induced by the axial electric field is given by the equation called *electronic equation* as given below:

$$i = j\frac{\beta_e I_0}{2V_0(j\beta_e - \gamma)^2} E_1 \tag{12.90}$$

where $\beta_e \equiv \omega/v_0$ is the phase constant of the velocity-modulated electron beam, $\gamma (= \alpha_e + j\beta_e)$ is the propagation constant of the axial wave, I_0 is the dc current, V_0 is the dc voltage and E_1 is the magnitude of electric field in the z-direction.

12.5.8.4 Axial Electric Field

The axial electric field (E_1) of the slow wave helix is determined by the equation called the *circuit equation* as given below:

$$E_1 = -i\frac{\gamma^2\gamma_0 Z_0}{\gamma^2 - \gamma_0^2} \tag{12.91}$$

where $\gamma_0 \{\equiv j\omega\sqrt{(LC)}\}$ is the propagation constant when $i = 0$ and $Z_0 \{= \sqrt{(L/C)}\}$ is the characteristic impedance of the lossless line with distributed L and C parameters representing the slow wave helix.

12.5.8.5 Wave Modes

Substituting Equation 12.91 in Equation 12.90 gives

$$(\gamma^2 - \gamma_0^2)(j\beta_e - \gamma)^2 = -j\frac{\gamma^2\gamma_0 Z_0 \beta_e I_0}{2V_0} \tag{12.92}$$

Equation 12.92 is of fourth order in γ and thus has four roots. Its exact solution can be obtained by using numerical methods and digital computer. Its approximate solution can however be obtained by equating the dc electron beam velocity to the axial phase velocity of the travelling wave. The resulting values of the four propagation constants are

$$\gamma_1 = -\beta_e C(\sqrt{3}/2) + j\beta_e(1 + C/2) \tag{12.93a}$$

$$\gamma_2 = \beta_e C(\sqrt{3}/2) + j\beta_e(1 + C/2) \tag{12.93b}$$

$$\gamma_3 = j\beta_e(1 - C) \tag{12.93c}$$

$$\gamma_4 = -j\beta_e(1 - C^3/4) \tag{12.93d}$$

In Equation 12.93, C is the gain parameter of the TWT and is defined as

$$C \equiv \left(\frac{I_0 Z_0}{4V_0}\right)^{1/3} \tag{12.94}$$

In O-type helical TWTs, these four propagation constants represent the following four different modes of wave propagation:

1. The propagation constant γ_1 corresponds to a forward wave whose amplitude grows exponentially with distance. It propagates at a phase velocity slightly greater than the electron beam velocity and the energy is transferred from the beam to the wave.

2. The propagation constant γ_2 corresponds to a forward wave with exponentially decaying amplitude with distance. It propagates at the same velocity as that of the growing wave but the energy flows from the wave to the beam.

3. Parameter γ_3 corresponds to a forward wave whose amplitude remains constant. It travels at a velocity slightly higher than that of the electron beam. In this case, there is no net energy flow between the wave and the beam.

4. Propagation constant γ_4 corresponds to a backward wave whose amplitude remains constant. It progresses in the negative z-direction with a velocity slightly higher than that of the electron beam provided the value of C is about 0.02.

12.5.8.6 Output Power Gain

With the assumption of perfectly matched structure, the total circuit voltage is the sum of the voltages of three forward travelling waves, which can be written as

$$V(z) = V_1 e^{-\gamma_1 z} + V_2 e^{-\gamma_2 z} + V_3 e^{-\gamma_3 z} \tag{12.95a}$$

$$\text{As } V(0) \{= V_1 + V_2 + V_3\} = V(z) \quad \text{at } z = 0 \tag{12.95b}$$

$V(l) = V(z)$ at $z = 1$, is the voltage at output end, which can be obtained as

$$V(l) = \{V(0)/3\}\exp(\sqrt{3}\,\pi\,NC) \tag{12.95c}$$

The output power gain (A_p in dBs) is defined as

$$A_p \equiv 10\log\left|\frac{V(l)}{V(0)}\right|^2 = -9.54 + 47.3NC \tag{12.96}$$

where NC is a numerical number.

Equation 12.96 shows that there is an initial loss of 9.54 dB at the circuit input. This loss is the result of the split of input voltage into three waves of equal amplitudes. Thus, the growing voltage is only one-third of the input voltage. It can be further noted that gain is proportional to the length N in electronic wavelength of the slow wave structure and gain parameter C of the circuit.

12.5.9 Characteristics

Figure 12.34 shows the transfer characteristic of TWT, which is linear up to a certain limit. As the input is increased, the amplifier saturates. Its linear range permits the use of TWT to amplify single side band (SSB) signals. It

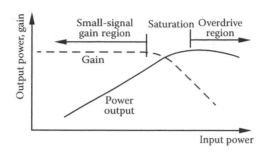

FIGURE 12.34
Characteristics of TWT amplifier.

is one of its great advantages in an amateur context. Although the tube can also be used at saturated output, the amplification is no longer linear. Also, if appreciable harmonic power is generated, this may be reflected at the output transition and might damage the helix through over dissipation.

The output from the amplifier can also be amplitude-modulated by a signal on the grid, but the attendant phase modulation is quite high. This method is normally not used to produce a great depth of modulation, except when the TWT is operated in pulsed mode. This is mainly because between maximum and minimum output voltages, beam may be intercepted by the helix, which may result in excessive dissipation unless the transitions are quite rapid. Phase modulation is obtained by varying the helix voltage over a small range. Typically, ±100 V deviation from 2 kV helix voltages will give 2 radian phase shift, with 1–2 dB reduction in output. This is due to the fact that the gain is very sensitive to the cathode-helix voltage. In view of the above, it is useful to include some permanent form of power monitor of the output from the amplifier. This can be in the form of a directional coupler and a diode detector.

12.5.10 Prevention of Oscillations

The gain in TWTs can exceed even 80 dB. In case of poor load matching reflections are likely to occur. This problem aggravates if the turns of cavities are close enough. Thus, in all practical tubes, some form of attenuator is used, which in turn reduces the gain. Though the attenuator attenuates both the forward and backward waves, the forward wave continues to grow past the attenuator, because bunching remains unaffected.

12.5.11 Applications

TWTs are highly sensitive, low-noise, wideband amplifiers. These can be used as low-power devices in radar and other microwave receivers. These are also used as medium- and high-power devices in communication, electronic counter measure (ECM) and space vehicles. These can be used as CW tubes to generate or amplify AM and FM signals. In case of amplitude modulation, the tube starts saturating at about 70% of the maximum output. TWT's are also employed as pulsed tubes in airborne, ship-borne and high-power ground-based radars.

12.6 Magnetron

Magnetrons are the high-power microwave sources. These may operate up to 70 GHz frequency range. Military radars generally employ conventional TWT magnetrons to generate high-peak-power RF pulses. The conventional magnetron has an edge over any other microwave device in terms of its size, weight, voltage and efficiency. A magnetron can deliver output peak power of the order of 40 MW with 50 kV dc voltage at 10 GHz. The average power outputs obtained from magnetrons are up to 800 kW and their efficiency may range between 40% and 70%. As stated earlier, the magnetrons employ re-entrant resonant cavities, which are located all around the anode. The RF field circulates from one cavity to the next. Magnetrons operate on the principle of continuous interaction between electron beam and the RF field as the wave velocity approximately equals electron velocity. In Section 11.5.2, the classification of magnetrons was discussed at length. In the subsequent section, the conventional cylindrical magnetron with hole-and-slot cavities is discussed in detail, whereas some others are only briefly described.

12.6.1 Construction of Conventional Cylindrical Magnetron

Magnetron is an oscillator that converts dc pulse power into microwave power by means of a standing wave structure. Pulsed operation enables the very high peak power required by the accelerator guide to be generated while limiting the average power demanded. It is in the form of a diode usually of cylindrical construction. Magnetron is commonly used in radars, and in microwave ovens. It employs radial (biasing) electric field, axial (applied) magnetic field and an anode structure (generally) with permanent cavities. It contains a cylindrical cathode that is surrounded by a cylindrical anode as shown in Figure 12.35. This anode contains a number of cavities.

In view of the cylindrical nature of the structure, the dc electric field is along the radial direction and the magnetic field is along the axial direction. The magnetic force due to the (applied) dc axial magnetic field passes through the cathode and the surrounding interaction space. Since the axial magnetic field and the radial electric field are perpendicular to each other, the magnetron is called a *crossed-field device*. In magnetron, the output lead is usually a *probe* or a *loop*. Depending on the frequency of operation, the loop or probe extends into one of the cavities and is coupled to the coaxial line or waveguide.

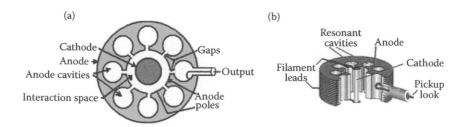

(a)   (b)

FIGURE 12.35
Hole-and-slot magnetron. (a) Cross-sectional view and (b) 3D view.

12.6.2 Working Mechanism

The working of a magnetron requires an understanding of the impact of different fields on an electron. In a magnetron, three types of fields are simultaneously present. These include electric, magnetic and RF fields. When an electron comes out from a cathode, all these fields try to influence it in their own way. To understand the combined impact of all these fields, the roles played by each individual field are to be considered separately. Some of the aspects of these effects were discussed in detail in Section 11.3. These effects are reconsidered below with special reference to magnetron.

12.6.2.1 Effect of Electric Field

In Chapter 11, it was noted that an electric field (**E**) exerts a force (**F**) on a charge (Q). This force (**F** $= Q$**E**) imposes no restriction on the charge whether it is moving or stationary. Thus, if an electron emerges from the cathode, it will travel straight to the anode, along E-lines, in view of the linear relation between **F** and **E**. This is shown in Figure 12.36.

12.6.2.2 Effect of Magnetic Field

As discussed in Chapter 11, a moving charge represents a current and the magnetic field (**B**) exerts a force on this moving charge (Q). This force is given

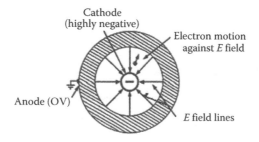

FIGURE 12.36
Electron motion in E field.

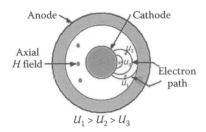

$$U_1 > U_2 > U_3$$

FIGURE 12.37
Electron motion in *B* field.

by $F = Q(U \times B)$, where U is the velocity of the moving charge. This relation imposes restriction on Q, which has to be a moving charge and no force is exerted on a charge at rest. Further, the exerted force on Q is perpendicular to both B and U. For a non-zero resulting force, B and U must also be perpendicular to each other. If this charge is an electron (e) and the velocity is represented by v, the magnitude of force is proportional to Bev. In this relation, B is the component of the magnetic field in a plane perpendicular to the direction of travel of electron. Thus, if the electron is moving in the horizontal direction and B acts vertically downward, the path of electron will be curved to the left. Since B is constant, the force exerted by B (or radius of curvature followed by e) will depend on the forward radial velocity of electron. Figure 12.37 illustrates the path of an electron in B field across the cross section of a cavity magnetron.

12.6.2.3 Effect of Combined Field

When both E and B fields act simultaneously on an electron, the force exerted is given by $F = e(E + U \times B)$. The path of the electron gets altered due to the exertion of two forces. The new path is the combination of linear path due to the E field and the circular path due to the B field. The curvature of this path will depend on the relative amplitudes of electric and magnetic fields. Figure 12.38 illustrates the electron motion in the combined E and B field.

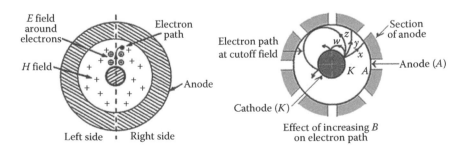

FIGURE 12.38
Electron motion in combined E and B field.

Figure 12.38 also shows a number of paths adopted by an electron under the influence of constant radial electric field and variable axial magnetic field. When $B = 0$, electron goes straight from the cathode (K) to the anode (A) under the influence of E (path x). When B is slightly increased, it will exert a lateral force on the electron, bending its path to the left (path y). The electron motion no more remains rectilinear. As the electron approaches A, its velocity continues to increase rapidly as it is accelerating. Thus, as the effect of B increases, the path curvature becomes sharper as electron approaches A. If B increases further, the electron may not reach the anode (path z). B required for the return of the electron to the cathode after just grazing the anode is called *cutoff field*. Since the cutoff field reduces the anode current to zero, the value of B is quite important to reckon with. If B is still stronger, the path of the electron will be more curved, and the electron will return to the cathode (path w) even sooner (only to be re-emitted). Since B is greatly in excess of the cutoff field, in the absence of any other (RF) field, all electrons will return to the cathode, overheat it and ruin the tube.

As the electrons emitted by the hot cathode travel towards the anode along curved paths, oscillating electric and magnetic fields are produced in the cavities. The accumulation of charges at the ends of cavities gives rise to the capacitance and the flow of current around cavities results in inductance. Thus, each cavity is equivalent to a parallel resonant circuit, which tunes to a particular frequency, provided the cavity volume remains constant. The resulting parallel resonant circuit is shown in Figure 12.39.

12.6.2.4 RF Field

Figure 12.40 shows the RF field inside each of the cavities. It further illustrates that this field extends into the interaction space and each gap corresponds to a maximum voltage point. According to this figure, the field alternates, making these maximas of opposite polarity.

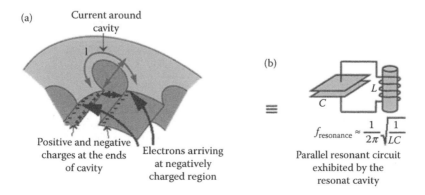

FIGURE 12.39
(a) Cavity and (b) its equivalent parallel resonant circuit.

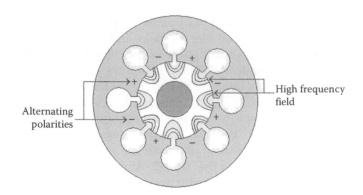

FIGURE 12.40
High-frequency E field.

12.6.2.5 Effect of Combined (E, B and RF) Fields

In the absence of the RF electric field, electrons will follow the paths shown in Figure 12.38 under the influence of E and B fields. Since the RF field exists inside each of the cavities, in its presence, these paths will get modified. Each cavity acts in the same way as a short-circuited quarter wave section of transmission line. Each gap corresponds to a maximum voltage point in the resulting standing wave pattern, with electric field extending into the interaction space. To understand the behaviour of charges under the influence of E, B and RF fields, consider Figure 12.41. It shows different charges emerging from the cathode at different time instants and may be from different locations. These charges are named a, b, c and so on. The behaviour of each charge is studied separately as given below.

Electron a: The presence of the RF field results in a tangential component of electric field. When electron 'a' is situated at point 1, this tangential component opposes the tangential velocity of the electron. Electron 'a' is thus

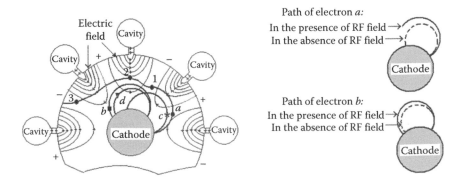

FIGURE 12.41
Paths traversed by an electron in magnetron under π-mode of oscillations.

retarded and gives energy to the RF field. In view of retardation, the force of the dc magnetic field is diminished; as a result, 'a' moves closer to the anode. If conditions are so arranged that by the time electron 'a' arrives at point 2, the polarity of the field is reversed, 'a' will again be in a position to give energy to the RF field, through it being retarded. Magnetic force on 'a' diminishes once more and another interaction of this nature occurs, this time at 3. This assumes that E field has reversed polarity each time this electron arrives at a suitable interaction position. Thus, favored electrons spend a considerable time in interaction space and are capable of orbiting the cathode several times before arriving at the anode.

Electron b: Electron *b* is so located that it gets accelerated by the RF field and it extracts energy from the RF field. The force exerted on it by the dc magnetic field increases. It spends much less time in interaction space than 'a'. Its interaction with the RF field takes as much energy as it was supplied by 'a'. It returns to K even sooner than it would have in the absence of the RF field. For maintaining oscillations, more energy must be given to the RF field than extracted from it. It is apparently unlikely since there are as many electrons of type 'a' as that of 'b'. However, charges 'a' and 'b' encounter entirely different situations. Electron 'a' spends more time in the RF field than 'b' as 'b' returns to the cathode only after one or two interactions. Thus, 'a' gives more energy to the RF field than 'b' takes from it. Besides charge 'a' gives energy repeatedly and 'b' extracts energy once or twice. Since more energy is given to the RF field than taken from, it results in sustained oscillations. The unfavorable electrons of type 'b', which fall into the cathode result in its heating (termed as *back heating*) to the tune of 5% of anode dissipation. It is not actually a total loss since it is possible to maintain the required temperature by relying on back heating after switching off the filament. The bunching process is also present in the magnetron but it is called the *phase focussing effect*. It is quite important since without phase focussing, favored electrons would fall behind the phase change of E-field across gaps, as electrons are retarded at each interaction with the RF field.

Electron c: Consider another electron 'c', which contributes some energy to the RF field. Since the tangential component is not much strong at this point, it does not give as much energy as 'a'. At first it appears to be less useful. It however encounters not only a diminishing tangential RF field but also a component of the radial RF field. This has the effect of accelerating it radially outward. At this juncture, the dc magnetic field exerts a strong force on 'c' tending it back to K but also accelerating it somewhat in the counter-clockwise direction. This in turn gives 'c' a fair chance of catching 'a'. In a similar manner, electron *d* will be tangentially retarded by the dc magnetic field. It will be caught up by the favored electrons. This process leads to bunching. It is to be noted that the favored position of an electron means it is in a position of equilibrium. If an electron slips back or forth, it will quickly return to the correct position with respect to the RF field, by phase focussing effect.

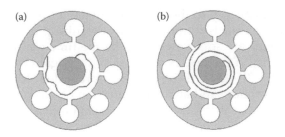

FIGURE 12.42
Paths traversed by a single electron for falling into anode. (a) After about one round and (b) after several rounds.

Figure 12.42a illustrates a fluctuating path traversed by a single electron between the cathode and the anode, whereas as Figure 12.42b shows that the electron may take several rounds before it falls into the anode.

Figure 12.43 shows the joining of a number of electrons to form bunches. These bunches now travel in the form of spokes of a wheel in cavity magnetron. These spokes of bunches rotate counter-clockwise with the correct velocity to keep up with the RF phase changes between adjoining anode poles. Thus, a continued interchange of energy takes place with an RF field receiving much more energy than it gives.

Since the RF field changes its polarity, each favored electron, by the time it arrives opposite the next gap, meets the same situation of there being a positive anode pole above it and to the left, and a negative anode pole above it and to the right. It can be imagined that the *E* field itself is rotating counter-clockwise at the same speed as the electron bunches. The cavity magnetron is also called the *travelling-wave magnetron* precisely because of these rotating fields.

12.6.3 Modes of Oscillation

Since magnetron has a number of resonant cavities, it has a number of resonant frequencies and/or modes of oscillation. The modes have to be

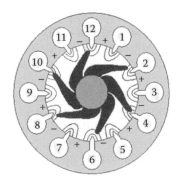

FIGURE 12.43
Space-charge wheels in a 12-cavity magnetron with RF field shown.

self-consistent. To understand the consistency of modes, assume that in the eight-cavity magnetron, the phase difference between adjacent anode pieces is 30°. In this case, the total phase shift around the anode comes out to be 240° (= 8 × 30°). This would mean that first pole piece will be 120° out of phase with itself. If the phase difference between adjacent anode pieces is 45°, the discrepancy noticed in case of 30° is removed and 45° becomes the smallest usable phase difference for the eight-cavity magnetron. This does not yield suitable characteristics and the π-mode is preferred for practical reasons.

The self-consistent oscillations can exist only if the phase difference between adjoining anode poles is $n\pi/4$, where n is an integer. For best results, $n = 4$ is used in practice. The resulting π-mode oscillations are shown in Figure 12.44 at an instant of time when RF voltage on the top left-hand anode is maximum positive. In view of the alternative nature of the RF field, this anode may be maximum negative at some other instant of time, while the RF voltage between that pole and the next may be zero at other instant. This figure also shows the connections of cavities for π-mode and $\pi/2$-mode and corresponding polarities at one particular time instant.

A magnetron may have eight or more coupled-cavity resonators that it may operate in several different modes. The oscillating frequencies corresponding to different modes are not the same. Thus, it is quite likely that a 3-cm π mode may become 3.05, 3/4π mode. Such change of mode is referred to as *mode jumping*. Adjusted magnetic and electric fields for π mode may still support spurious mode up to some extent, since its frequency is not too far distant.

12.6.4 Strapping

Mode jumping is an undesirable phenomena occurrence, which is quite common in identical cavity magnetrons. This can be prevented by using strapping, which consists of two rings of heavy gauge wire. These wires connect alternate anode poles. Since the number of cavities used in magnetron is

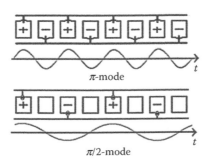

FIGURE 12.44
Waveforms for π-mode and $\pi/2$-mode.

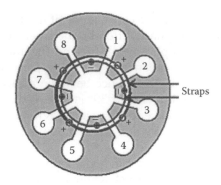

FIGURE 12.45
Strapping rings in magnetron with hole-and-slot cavities.

generally even, such alternate connection ensures that the connected anodes have identical polarities. With the use of strapping, the frequency of π-mode is separated from the frequency of other modes. For the π-mode, all parts of each strapping ring are at the same potential, but the two adjacent rings have alternately opposing potentials. These are shown in Figure 12.45.

The stray capacitance between the strapping rings adds capacitive loading to the resonant mode. For other modes, however, a phase difference exists between the successive segments connected to a given strapping ring, which causes the current to flow in the straps. Also, the straps contain inductance, and an inductive shunt is placed in parallel with the equivalent circuit. This lowers the inductance and increases the frequency at modes other than the π-mode. Strapping may become unsatisfactory due to losses in straps in very high power magnetrons, particularly at high frequencies. At high frequencies, the cavities are small in size and large in number (16 or 32 more common) to ensure suitable field in the interaction space. Thus, at high frequencies, the likelihood of mode jumping is more.

Figure 12.46 illustrates four types of cavities that may be used in magnetrons. These include the conventional hole-and-slot type, vane type, slot type and rising sun type of cavities. Anode block with a pair of cavity system of

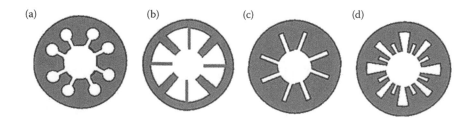

FIGURE 12.46
Common types of cavity cross-sectional views. (a) Hole-and-slot type, (b) vane type, (c) slot type and (d) rising sun type.

dissimilar shape and resonant frequency is one of the solutions. A rising sun anode structure has the effect of isolating mode frequency from others. No strapping is needed in case of rising sun cavity structure.

12.6.5 Frequency Pushing and Pulling

Change in anode voltage affects the orbital velocity of the electron cloud. This in turn alters the rate at which energy is given to the anode resonators. This changes the oscillating frequency, permitted by the bandwidth. This process is called *frequency pushing*. Also, the magnetron is susceptible to the frequency variations due to change of load impedance. This effect is more significant in case of reactive load. Such a frequency variation is called *frequency pulling*.

12.6.6 Coupling Methods

The RF energy can be removed from a magnetron by means of a *coupling loop*. At frequencies lower than 10 GHz, the coupling loop is made by bending the inner conductor of a coaxial line into a loop. The loop is then soldered to the end of the outer conductor so that it projects into the cavity as shown in Figure 12.47a. The loop may also be located at the end of the cavity shown in Figure 12.47b. It causes the magnetron to obtain sufficient pickup at higher frequencies. The *segment-fed loop method* is shown in Figure 12.47c wherein the loop intercepts the magnetic lines passing between cavities. The *strap-fed loop method* is shown in Figure 12.47d. The energy is intercepted between the strap and the segment. On the output side, the coaxial line feeds another coaxial line directly or feeds a waveguide through a choke joint. The vacuum seal at inner conductor helps to support the line. *Aperture* or *slot coupling* is shown in Figure 12.47e. Energy is coupled directly to the waveguide through an iris.

12.6.7 Magnetron Tuning

A magnetron can be tuned by using the inductive or capacitive tuning method. Both these tuners are symmetrical in nature, that is, in each, the cavity is affected in the same manner.

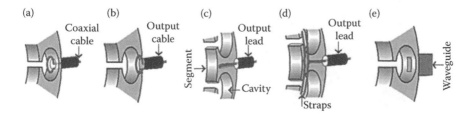

FIGURE 12.47
(a) Loop projected into cavity, (b) loop located at the end of cavity, (c) segment-fed loop method, (d) strap fed loop method and (e) aperture or slot coupling.

12.6.7.1 Inductive Tuning

In inductive tuning, the resonant frequency is changed by varying the inductance of the resonant cavity. An inductive tuning element is inserted into the hole of the hole-and-slot cavities. Since the ratio of the surface area to the cavity volume in a high-current region changes, the inductance of the resonant circuits is changed. The tuner illustrated in Figure 12.48a is called a *sprocket tuner* or *crown-of-thorns tuner*. All its tuning elements are attached to a frame, which is positioned by a flexible bellows arrangement. The insertion of the tuning elements into each anode hole decreases the inductance of the cavity and therefore increases the resonant frequency. The inductive tuning lowers the unloaded Q of the cavities and thus reduces the efficiency of the tube.

12.6.7.2 Capacitive Tuning

In capacitive tuning, the resonant frequency is changed by varying the capacitance of the resonant cavity. In this case, a ring element is inserted into the cavity slot. This increases the slot capacitance and decreases the resonant frequency. Since the gap is narrowed in width, the breakdown voltage is lowered. Therefore, capacitive tuned magnetrons must be operated with low voltages and at low-power outputs. The tuner illustrated in

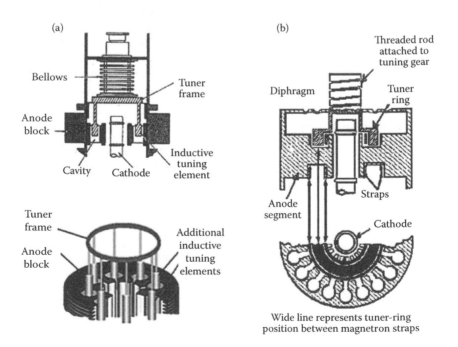

FIGURE 12.48
Magnetron tuning (a) inductive and (b) capacitive.

Figure 12.48b is called a *cookie-cutter tuner*. It consists of a metal ring that is inserted between the two rings of a double-strapped magnetron. It increases the strap capacitance. Because of the associated mechanical and voltage breakdown problems, cookie-cutter tuner is more suitable for use at longer wavelengths.

Both the above methods allow frequency variation of up to 10% of the central frequency. For obtaining larger tuning range, the two tuning methods may be used in combination.

12.6.8 Mathematical Analysis

12.6.8.1 Electron Motion

In cylindrical (conventional) magnetron, a number of cavities are connected to the gaps. A dc voltage (V_0) is applied in the radial direction (between the cathode and the anode) and the magnetic field (B_0) along the axial (positive z) direction. With proper adjustment of V_0 and B_0, the electrons will follow cycloidal paths in the cathode anode space.

Since $E = E_r a_r$ and $B = B_z a_z$, only Equations 11.38a and 11.38b reduce to

$$\frac{d^2r}{dt^2} - r\omega^2 = \frac{e}{m}(E_r - B_z\omega r) \tag{12.97a}$$

$$\frac{1}{r}\frac{d}{dt}(r^2\omega) = \frac{e}{m}\left(B_z\frac{dr}{dt}\right) \tag{12.97b}$$

where

$$\omega = d\varphi/dt \tag{12.97c}$$

Equation 12.97b can be rearranged in the following form:

$$\frac{d}{dt}(r^2\omega) = \frac{e}{m}rB_z\frac{dr}{dt} = \frac{1}{2}\omega_c\frac{d}{dt}(r^2) \tag{12.98}$$

where $\omega_c = (e/m)B_z$ is the cyclotron angular frequency. Integration of Equation 12.98 gives

$$r^2\omega = \frac{1}{2}\omega_c r^2 + k_1 \tag{12.99}$$

At the radius of the cathode cylinder $r = a$, and $\omega = 0$, the constant $k_1 = -\omega_c a^2/2$. Thus, ω can be expressed as

$$\omega = \frac{1}{2}\omega_c\left(1 - \frac{a^2}{r^2}\right) \tag{12.100}$$

The kinetic energy (KE) of the electron is given by

$$\text{KE} = \frac{1}{2}mv^2 = eV \tag{12.101}$$

In view of Equation 12.101, the electron velocity v can be obtained. This velocity has r and ϕ components as given below:

$$v^2 = \frac{2e}{m}V = v_r^2 + v_\phi^2 = \left(\frac{dr}{dt}\right)^2 + \left(r\frac{d\phi}{dt}\right)^2 \tag{12.102}$$

At the radius $(r = b)$ of the inner edge of the anode, $V = V_0$ and $dr/dt = 0$, when the electrons just graze the anode, Equation 12.100 becomes

$$\left(\frac{d\phi}{dt}\right) = \frac{1}{2}\omega_c\left(1 - \frac{a^2}{b^2}\right) \tag{12.103}$$

Equation 12.103 can also be manipulated as

$$b^2\left(\frac{d\phi}{dt}\right)^2 = \frac{2e}{m}V_0 \tag{12.104}$$

Substitution of Equation 12.103 in Equation 12.104 gives

$$b^2\left(\frac{1}{2}\omega_c\left(1 - \frac{a^2}{b^2}\right)\right)^2 = \frac{2e}{m}V_0 \tag{12.105}$$

12.6.8.2 Hull Cutoff Equations

Substitution of $\omega_c = (e/m)B_z$ in Equation 12.105 and the appropriate manipulation gives the following two relations:

$$B_{0c} = \frac{(8V_0m/e)^{1/2}}{b(1 - a^2/b^2)} \tag{12.106}$$

$$V_{0c} = \frac{e}{8m}B_0^2b^2(1 - a^2/b^2)^2 \tag{12.107}$$

Equation 12.106 is referred to as the *Hull cutoff magnetic equation*, which gives the cutoff magnetic field B_{0c}. This equation spells that for a given V_0 if $B_0 < B_{0c}$, the electron will not reach the anode. Similarly, Equation 12.107 is referred to as the *Hull cutoff voltage equation*, which gives the cutoff voltage

V_{0c}. This equation indicates that for a given B_0, if $V_0 < V_{0c}$, the electron will not reach the anode.

12.6.8.3 Cyclotron Angular Frequency

As noted earlier, the magnetic field is perpendicular to the cycloidal path adopted by the moving electrons. The outward centrifugal force on the electron is equal to the pulling force. Thus,

$$mv^2/R = evB \tag{12.108}$$

where R is the radius of the cycloidal path and v is the tangential velocity of the electron.

The cyclotron angular frequency of the circular motion of electron is

$$\omega_c = v/R = eB/m \tag{12.109}$$

The period of one complete revolution is

$$T = 2\pi/\omega = 2\pi m/eB \tag{12.110}$$

Oscillations in a closed slow wave structure are possible only if the total phase shift around the structure is an integral multiple of 2π radians. Thus, the phase shift between two adjacent cavities in N re-entrant cavity structure for the nth mode of the oscillation is

$$\phi_n = 2\pi n/N \tag{12.111}$$

The oscillations in the slow wave structure will be produced when the average rotational velocity of electrons corresponds to the phase velocity of the field. This can be achieved by adjusting the anode dc voltage. Magnetron oscillators generally operate in π-mode. The phase shift requirement for such operation is given by

$$\phi_n = \pi \tag{12.112}$$

Figure 12.49 shows the lines of force in the π-mode for eight-cavity magnetron wherein the excitation is largely in the cavities with opposite phases in successive cavities. The consecutive rise and fall of field in adjacent cavities may be considered as a travelling wave along the slow wave structure. The moving electrons will transfer energy to this travelling wave when the electrons are decelerated by a retarded field when they pass through each anode cavity. For the mean separation L between cavities, the phase constant (β_0) of the fundamental-mode field is given by

$$\beta_0 = 2\pi n/NL \tag{12.113}$$

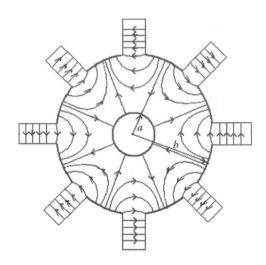

FIGURE 12.49
Lines of force in π-mode for eight-cavity magnetron.

The solution for the fundamental ϕ component of the travelling-wave electric field in slow wave structures can be obtained by solving Maxwell's equations and with the application of appropriate boundary conditions. This is of the form

$$E_{\phi 0} = jE_1 e^{j(\omega t - \beta 0 \phi)} \tag{12.114}$$

The travelling field of fundamental mode travels around the structure with the angular velocity

$$d\phi/dt = \omega/\beta_0 \tag{12.115}$$

From this relation, it can be concluded that the interaction between the field and the electrons occurs and the energy gets transferred when the cyclotron and the angular frequencies are equal.

$$\omega_c = \beta_0 \, (d\phi/dt) \tag{12.116}$$

12.6.8.4 Equivalent Circuit

Figure 12.50 illustrates an equivalent circuit for resonator of a magnetron. In this figure, Y_e is the electronic admittance, V is the RF voltage across the vane tips, C is the capacitance at these vane tips, L is the inductance, G_r is the conductance of the resonator and G is the load conductance per resonator. Each resonator of slow wave structure is to comprise a similar separate resonant circuit.

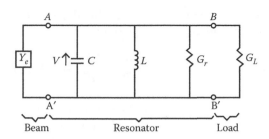

FIGURE 12.50
Equivalent circuit for resonator of a magnetron.

12.6.8.5 Quality Factor

In view of the resonant circuit, the unloaded quality factor (Q_{un}) is given by

$$Q_{un} = \omega_0 C/G_r \qquad (12.117a)$$

The external quality factor (Q_{ext}) is given by

$$Q_{ext} = \omega_0 C/G_L \qquad (12.117b)$$

Also, the loaded quality factor (Q_L) is given by

$$Q_L = \omega_0 C/(G_r + G_L) \qquad (12.117c)$$

In Equations 12.117a, b and c, ω_0 ($= 2\pi f_0$) is the angular resonant frequency.

12.6.8.6 Power and Efficiency

The term efficiency encompasses circuit efficiency and electronic efficiency. In mathematical terms, these are given as follows:
Circuit efficiency

$$\eta_c = G_L/(G_L + G_r) = G_L/G_{ext} = 1/(1 + Q_{ext}/Q_{un}) \qquad (12.118)$$

In view of Equation 12.118, η_c is maximum when $G_L \gg G_r$. The large G_L amounts to heavy loading of magnetron. It is undesirable in some cases since it makes the tube sensitive to load.
Electronic efficiency

$$\eta_e = P_{gen}/P_{dc} \qquad (12.119)$$

where P_{gen} ($= P_{dc} - P_{loss}$) is the RF power induced into the anode circuit, P_{dc} ($= V_0 I_0$) is the power from dc supply, P_{loss} is the power loss in the anode circuit, V_0 is the (dc) anode voltage and I_0 is the (dc) anode current. Thus, the RF power generated by electrons can be given by

$$P_{\text{gen}} = V_0 I_0 - P_{\text{loss}} = V_0 I_0 - I_0 \frac{m}{2e} \frac{\omega_0^2}{\beta^2} + \frac{E_{\text{max}}^2}{B_z^2} = \frac{1}{2} N |V|^2 \frac{\omega_0 C}{Q_L} \qquad (12.120)$$

In Equation 12.120, N is the total number of resonators, V is the RF voltage across the resonator gap, E_{max} ($= M_1 |V|/L$) is the maximum electric field, β is the phase constant, B_z is the magnetic flux density, L is the centre-to-centre distance between the vane tips and M_1 is the gap factor for a π-mode operation. The *gap factor* M_1 is given by

$$M_1 = \sin(\beta_n \delta/2)/(\beta_n \delta/2) \qquad (12.121)$$

$M_1 \approx 1$ for small values of δ.

Equation 12.120 may be further simplified to the following form:

$$P_{\text{gen}} = \frac{NL^2}{2M_1^2} \frac{\omega_0 C}{Q_L} E_{\text{max}}^2 \qquad (12.122)$$

In view of the above, the electronic efficiency can be written as

$$\eta_e = \frac{P_{\text{gen}}}{P_{\text{dc}}} = \left(1 - \frac{m\omega_0^2}{2eV_0\beta^2}\right) \bigg/ \left(1 + \frac{I_0 m M_1^2 Q_L}{B_z e NL^2 \omega_0 C}\right) \qquad (12.123)$$

12.6.9 Other Magnetrons

In Section 11.5.2, the classification of magnetrons was briefly discussed. Thereat it was noted that magnetrons can be classified in more than one way. Accordingly, a number of magnetrons were listed, some of which are briefly described below.

12.6.9.1 Linear Magnetron

In this magnetron, the cavities are linearly arranged along the cathode of planar shape. Figure 12.51 illustrates a linear magnetron wherein an anode with a number of cavities is parallel to a flat cathode. This figure also shows the coordinate system and the directions of electric and magnetic fields.

In view of the shape of the configuration, its mathematical analysis can be carried in a Cartesian coordinate system and the problem can be tackled by following the procedure adopted in Section 11.3.4. Accordingly, the Hull cutoff voltage for a linear magnetron is obtained as

$$V_{0c} = \frac{1}{2} \frac{e}{m} B_0^2 d^2 \qquad (12.124)$$

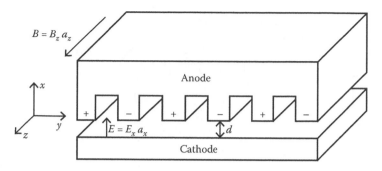

FIGURE 12.51
Schematic diagram of linear magnetron.

It spells that for a given B_z (= B_0), if $V_0 < V_{0c}$, the electron will not reach the anode.

Similarly, the Hull cutoff magnetic flux density for a linear magnetron can be obtained as

$$B_{0c} = \frac{1}{d}\sqrt{2\frac{m}{e}V_0} \qquad (12.125)$$

Thus, for a given voltage, if $B_0 > B_{0c}$, the electron will not reach the anode.

12.6.9.2 Coaxial Magnetron

It comprises an anode resonator structure, which is surrounded by an inner single high-Q cavity. The structure shown in Figure 12.52 operates in TE_{011} mode. There are slots in the back walls of alternate cavities of the anode structure. These slots tightly couple the electric fields in resonators to the surrounding cavity. In the π-mode operation, the electric fields in every other cavity are in phase; thus, the coupling in the surrounding cavity is in the same direction. This field in the surrounding coaxial cavity stabilises the magnetron in the desired π-mode operation.

In the TE_{011} mode, the path of the electric fields in the cavity is circular. This field reduces to zero at the cavity walls. The path of the resulting current in the cavity walls is also circular about the tube axis in accordance with the field. The device also contains an attenuator within the inner cylinder near the ends of the coupling slots. This attenuator damps out an undesirable mode. In this magnetron, the tuning mechanism is simple and reliable. Coaxial magnetron does not require straps. In coaxial magnetron, the anode resonator can be larger and less complex than that required in conventional strapped magnetron. Thus, it has lower cathode loading and reduced voltage gradients.

A typical X-band coaxial magnetron has 400 kW minimum peak power in the 8.9–9.6 GHz range, with 0.0013 duty cycle, 32 kV nominal voltage and 32 A peak anode current.

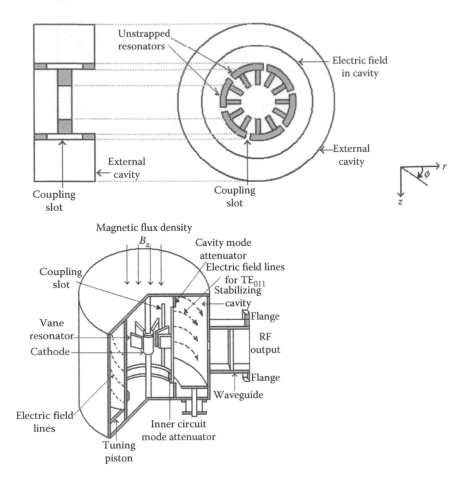

FIGURE 12.52
Structure of coaxial magnetron.

12.6.9.3 Inverted Coaxial Magnetron

In this device, the locations of the anode and the cathode are interchanged or inverted; thus, it is called inverted coaxial magnetron. In this, the cavity is located inside a slotted cylinder and an array of resonator vanes is arranged on the outside. Its cathode is in the form of a ring around the anode. Its schematic diagram is shown in Figure 12.53.

In this case, the Hull cutoff voltage equation is

$$V_{oc} = \frac{e}{8m} B_0^2 a^2 \left(1 - \frac{b^2}{a^2}\right)^2 \tag{12.126}$$

Thus, for a given B_0, if $V_0 < V_{oc}$, the electron will not reach the anode.

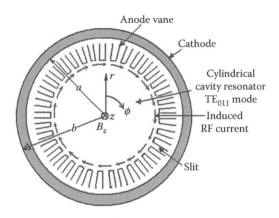

FIGURE 12.53
Schematic diagram of an inverted coaxial magnetron.

Similarly, the Hull cutoff magnetic equation is

$$B_{0c} = -\left(8V_0 \frac{m}{e}\right)^{1/2} \bigg/ a\left(1 - \frac{b^2}{a^2}\right) \tag{12.127}$$

In this case, for a given V_0, if $B_0 > B_{oc}$, the electron will not reach the anode.

12.6.9.4 Voltage Tunable Magnetron

It is a broadband oscillator in which the frequency is changed by varying the applied voltage between the anode and the sole. In this device, the electrons are emitted from a short cylindrical cathode located at its one end. Near to the cathode, these electrons are formed into a hollow beam by the electric and magnetic forces. The beam gets accelerated radially outward from the cathode and ultimately enters in the region between the sole and the anode. This beam rotates around the sole. The rate of its rotation is controlled by the axial magnetic field and the dc voltage applied between the anode and the sole. The cross-sectional view of this magnetron is shown in Figure 12.54.

This magnetron uses low Q resonator. Its bandwidth may exceed 50% at low-power levels. In its π-mode operation with hollow beam, the bunching process occurs in the resonator. Its oscillation frequency depends on the rotational velocity of the beam and can be controlled by varying the applied dc voltage between the anode and the sole. The control electrode in the electron gun can be used to coarsely adjust the power output. The high-power and high-frequency voltage tunable magnetron has only limited bandwidth. The bandwidth may approach 70% in low-power low-frequency case.

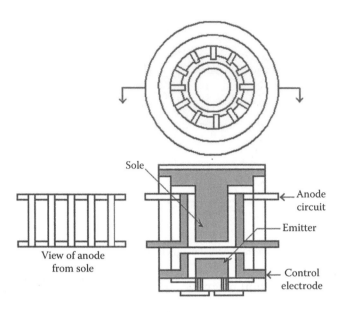

FIGURE 12.54
Cross-sectional view of voltage tunable magnetron.

12.6.9.5 Frequency-Agile Coaxial Magnetron

The frequency agility of a coaxial magnetron refers to the capability to tune the output frequency with sufficiently high speed to produce a pulse-to-pulse frequency change greater than the amount required to effectively obtain decorrelation of adjacent radar echoes. The reduced target scintillation, increased detectability of targets in a clutter environment and improved resistance to electronic countermeasures are some of the key requirements of radar, particularly that used in electronic warfare. The increase of the pulse-to-pulse frequency separation improves the radar system performance. Greater pulse-to-pulse frequency separation makes it difficult for a jamming transmitter to centre on the radar frequency and thus the effective interference with system operation also becomes difficult. The frequency-agile magnetron along with appropriate integrated circuit serves this purpose. These magnetrons are classified into three types.

Dither magnetron: In this, the output frequency varies periodically with a constant excursion, a constant rate and a fixed centre frequency.

Tunable/dither magnetron: In this, the output frequency varies periodically with a constant excursion, and constant rate but the centre frequency can be manually or mechanically tuned.

Accutune magnetron: In this case, the output frequency variations are determined by the waveforms of an externally generated low-level voltage signal. With properly selected waveform, this magnetron has the features of dither

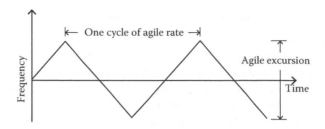

FIGURE 12.55
Agile rate and agile excursion.

and tunable/dither magnetrons, together with a capability for varying the excursion, rate and tuning waveform.

Figure 12.55 illustrates the agile rate and agile excursion of a typical X-band frequency–agile coaxial magnetron with pulse voltage 15 kV, pulse current 15 A, maximum duty cycle 0.0011 accutune range 1 GHz, centre frequency 9.10 GHz and peak power 90 kW.

This figure leads to the following definitions:

Agile rate is the number of times per second that the transmitter frequency traverses the agile excursion and returns to its starting frequency.

Agile excursion is the total frequency variation of the transmitter during the agile operation.

As the number of effectively integrated pulses cannot exceed the number of pulses placed on the target during one antenna scan, the integration period of radar is determined by the antenna beamwidth and scan rate. The design value for agile excursion can thus be expressed in terms of radar operating parameters.

$$\text{Agile excursion} = N/\tau \tag{12.128}$$

where N is the number of pulses placed on the target during one radar scan and τ is the shortest pulse duration used in the system.

The duty cycle is defined as

$$\text{Duty cycle} \equiv \tau/T = \tau f \tag{12.129}$$

The agile rate can thus be written as

$$\text{Agile rate} = 1/2T \tag{12.130}$$

As two excursions through the agile frequency occur during each cycle of agile rate, a factor 2 appears in the denominator.

12.6.9.6 Negative-Resistance or Split-Anode Magnetron

This is also called split-anode, negative-resistance magnetron and is operated by a static negative resistance between its electrodes. This oscillator has a frequency equal to the frequency of the tuned circuit connected to the tube. This is a variation of the basic magnetron that operates at a higher frequency. It is capable of greater power output than the basic magnetron. Its general construction is similar to the basic magnetron except that it has a split plate, as shown in Figures 12.56a and 12.56b. The two half plates are operated at different potentials to provide an electron motion as shown in Figure 12.56c. The electron progressing from the cathode to the high-potential plate is deflected by the magnetic field and follows the path shown in Figure 12.56c. After passing through the split between the two plates, the electron enters the electrostatic field set up by the lower-potential plate.

Here, the magnetic field has more effect on the electron and deflects it into a tighter curve. The electron then continues and makes a series of loops through B and E fields until it finally arrives at the low-potential plate. Oscillations are started by applying the proper magnetic field to the tube. The field value required is slightly higher than the critical value. In the split-anode tube, the critical value is the field value required to cause all the electrons to miss the plate when its halves are operating at the same potential. The alternating voltages impressed on the plates by the oscillations generated in the tank circuit will cause electron motion and current will flow. Since a very concentrated magnetic field is required for the negative-resistance magnetron oscillator, the length of the tube plate is limited to a few centimeters to keep the magnet at reasonable dimensions. In addition, a small-diameter tube is required to make the magnetron operate efficiently at microwave frequencies. A heavy-walled plate is used to increase radiating properties of the tube. Artificial cooling methods, such as forced-air or water-cooling are used to obtain still higher dissipation in these high-output tubes.

The output of a magnetron is reduced by the bombardment of the filament by electrons, which travel in loops. This increases the filament temperature under conditions of a strong magnetic field and high plate voltage and sometimes leads to instability of the tube. This effect can be reduced by operating

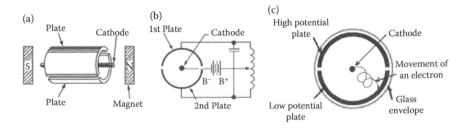

FIGURE 12.56
Split-anode magnetron. (a) Its components, (b) its biasing and (c) electron movement.

the filament at a reduced voltage. In some cases, the plate voltage and field strength are also reduced to the destruction of the filament.

12.6.9.7 Electron-Resonance Magnetron

The operation of this oscillator is based on electron transit time. This is capable of generating very large-peak-power outputs at frequencies in thousands of megahertz. Although its average power output over a period of time is low, it can provide very high-powered oscillations in short bursts of pulses.

In the electron-resonance magnetron, the plate is constructed to resonate and to function as a tank circuit. Thus, this magnetron has no external tuned circuits. Power is delivered directly from the tube through transmission line as shown in Figure 12.57a. In this case, the constants and operating conditions of tube are such that the electron paths differ from those of Figure 12.56c. Instead of closed spirals or loops, the path is a curve with a series of sharp points. Ordinarily, this magnetron has more than two segments in the plate. For example, Figure 12.57b illustrates an eight-segment plate. This magnetron is most widely used for microwave frequencies in view of its reasonably high efficiency and relatively high output. Its average power is limited by the amount of cathode emission, and its peak power is limited by the maximum voltage rating of the tube components.

Some common types of anode blocks used in electron-resonance magnetrons were shown in Figure 12.46. The anode block of Figure 12.46a has cylindrical cavities and that of Figure 12.46b has trapezoidal cavities. These two anode blocks are strapped, that is, their alternate segments are so connected that each segment is opposite in polarity to the segment on either side. Both these blocks require an even number of cavities. The rising-sun type anode block shown in Figure 12.46d contains alternate large and small trapezoidal cavities. These trapezoidal cavities in fact are modified form of slotted cavities shown in Figure 12.46c. This cavity arrangement of Figure 12.46d results in a stable frequency between the resonant frequencies of the cavities of large and small sizes.

Figure 12.58a shows a hole-and-slot cavity, which consists of a cylindrical hole in the copper anode and a slot that connects the cavity to the interaction

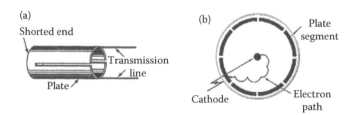

FIGURE 12.57
Electron-resonance magnetron. (a) Plate tank circuit and (b) electron path.

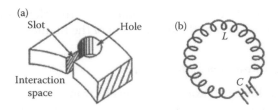

FIGURE 12.58
A hole-and-slot cavity and its electrical circuit. (a) Configuration and (b) equivalent circuit.

space. Its equivalent circuit is shown in Figure 12.58b. As described earlier in connection with Figure 12.39, the parallel sides of the slot form the plates of a capacitor while the walls of the hole act as an inductor. The combination of hole and slot thus forms a high-Q, resonant LC circuit. As shown in Figure 12.46, the anode of a magnetron contains a number of these cavities. An analysis of the anodes in the hole-and-slot block reveals that LC tanks of each cavity are in series in case the straps are removed and anode block after strapping of alternate segments the cavities are connected in parallel. Figure 12.59 shows equivalent circuits for the two cases.

12.6.10 Performance Parameters and Applications

- Pulsed radar and linear particle accelerators, duty cycle typically 0.1, power range 10 kW at 100 GHz, 2 MW at X-band and 10 MW at UHF, efficiency around 50%, cathode temperature up to 1800°C.
- VTMs 200 MHz to X-band with CW power up to 10 W, efficiency up to 75%, used as sweep oscillators in telemetry and missile applications.
- Fixed-frequency CW magnetrons used in industrial heating and microwave ovens over 900 MHz to 2.5 GHz range giving 300 W to 10 kW at 70% efficiency.

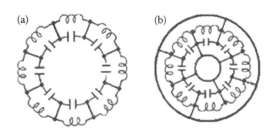

FIGURE 12.59
Equivalent of hole–slot cavities. (a) Series connection without straps and (b) parallel connection with straps.

12.6.11 Arcing

Arcing in magnetrons is a common phenomenon that generally occurs in new tubes or when it remains idle for long period. One of its prime causes is the release of gas from tube elements during idle periods. It may also occur due to the presence of sharp surfaces within the tube, mode shifting and by drawing excessive current. While the cathode can withstand considerable arcing for a short period, its regular occurrence shortens the life of the magnetron or may completely destroy it. Thus, after each excessive arcing, the tube must be baked-in until the arcing ceases and the tube is stabilised.

12.6.11.1 Baking-In Procedure

Raise the magnetron voltage from a low value until arcing occurs several times a second. Leave the voltage at that value until arcing dies out. Furthermore, raise the voltage until arcing again occurs and leave the voltage at that value until the arcing again ceases. When arcing becomes violent and resembles a continuous arc, the applied voltage is to be reduced to allow the recovery of the magnetron. When normal rated voltage is reached and the magnetron stabilises at the rated current, the baking-in is complete.

12.7 Crossed-Field Amplifier

The crossed-field amplifiers (CFA) can be grouped in accordance with their mode of operation as forward wave and backward wave types. In this, the direction of the phase and group velocity of the energy on the microwave circuit are of prime concern. The behaviour of phase velocity is of prime importance since the amplification results only when the electron beam interacts with the RF electric field. The process of interaction was discussed earlier in connection with the ω–β diagram shown in Figure 12.31. In forward-wave CFAs, the slow wave structures are generally of the helix type, whereas the strapped bar line structure is an appropriate choice for backward-wave mode.

12.7.1 Forward-Wave Crossed-Field Amplifier

The forward-wave crossed-field amplifier (FWCFA) is also referred to as simply crossed-field amplifier. CFAs can further be grouped by their electron stream source as emitting-sole or injected-beam types. In this group, the emphasis is laid on the method through which the electrons are made to reach the interaction space and on their mode of control. In the emitting-sole tube, the electric field forces the current to emanate from the cathode. This current is a function of the cathode dimensions, its emission properties and

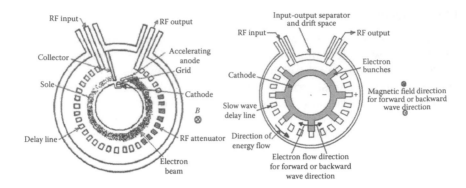

FIGURE 12.60
Configuration of FWCFA.

the applied voltage. The high perveance of the interaction geometry results in a high current and high power capability at relatively low voltage. In the injected-beam tube, the electron beam is generated in a separate gun assembly and then injected into the interaction region. The beam–circuit interaction features in emitting sole and injected beam tubes are similar. In both, the favorably phased electrons travelling towards the positively polarised anode are collected and unfavorably phased electrons are directed towards the negatively polarised electrode. The configuration comprising different components of FWCFA is shown in Figure 12.60.

The linear beam interaction was discussed earlier in connection with the TWT. There it was noted that the electron stream emerging out of the cathode is first accelerated by an electric gun to the full dc velocity. This dc velocity approximately equals the axial phase velocity of the RF filed in the slow wave structure. After interaction, the spent electron beam leaves the interaction region with reduced average velocity. The difference between the velocities of incoming and outgoing beams accounts for the RF energy generated. In CFA, the electron travels under the influence of the force due to the dc electric field, magnetic field and RF electric field. It tends to travel along equipotential lines in spiral trajectories.

12.7.2 Backward-Wave Crossed-Field Amplifier

The backward-wave crossed-field amplifier is also called Amplitron. It is a broadband microwave amplifier and can also be used as an oscillator. It is similar in operation to the magnetron and is capable of providing relatively large amounts of power with high efficiency. Its bandwidth at any given instant is roughly ±5% of the rated centre frequency. It amplifies any incoming signals within this bandwidth. CFAs can deliver peak power of many MWs and average power of tens of kilowatts, with efficiencies greater than

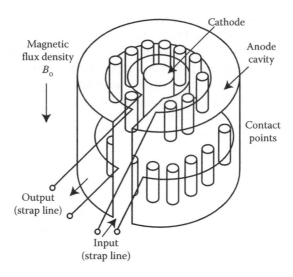

FIGURE 12.61
Configuration of Amplitron.

70%. CFA is used in many microwave electronic systems in view of its desirable characteristics, wide bandwidth, high efficiency and the ability to handle large amounts of power. CFAs are quite suitable for space-telemetry applications due to its high efficiency. Also, its high power and stability aspects have made it useful in high-energy, linear atomic accelerators. Such tubes are commonly used in air surveillance radar and military pulse radar. These may also be used as the intermediate or final stage in high-power radar systems.

Figure 12.61 shows the schematic diagram of Amplitron. Its resonator circuit comprises the anode cavity and pins. A pair of its pins and the cavity are excited in opposite phase by the strap line. The electron beam interacts with the electromagnetic waves in the resonant circuit.

12.8 Backward-Wave Oscillators

If instead of forward wave, the backward wave is utilised with the TWT structures, the resulting tubes are called backward-wave oscillators (BWOs). Such tubes have the advantage of a very wide tunable range, which may be greater than or equal to an octave. These tubes have extensive applications in swept frequency sources (sweepers). With the advances in semiconductor devices, these are rapidly being replaced by Gunn diodes and, more recently by transistor sources.

The BWO is a microwave-frequency velocity-modulated tube. It operates on TWT principles of electron RF field interaction. It is like a shorter and

thicker TWT. If the presence of initial oscillations is assumed, BWO operation becomes similar to that of TWT. Electrons are ejected from the cathode
of the electron gun, focussed by axial magnetic field and collected at the far
end of glass tube. During the travel through a helical slow wave structure,
these electrons get bunched. Bunching becomes more complete as they arrive
nearer to the collector. As in TWT, interchange of energy occurs with radio
frequency field along the helix. The signal grows as it progresses towards
the helix end. Oscillation may be thought to be occurring because of reflections from an imperfectly terminated collector end of the helix. Thus, there is
feedback and the output is collected from the cathode end of helix, towards
which reflection took place.

Helix is essentially a non-resonant structure. The bandwidth of BWO is
very high and is limited by interaction between the beam and the slow wave
structure. The operating frequency is determined by collector voltage along
with its associated cavity system. BWO with ring cathode generating a hollow beam with maximum near helix increases the interaction. In BWOs, permanent magnets are normally used for focussing, which result in simplest
magnets and smallest tube. The use of solenoids at higher frequencies provides best penetration and distribution of axial magnetic field. These oscillators can be either *O*-type or *M*-type. Both these are briefly described in the
following section.

12.8.1 *O*-Type Backward-Wave Oscillator

It consists of an electron gun and a tape helix-type slow wave structure.
The tube is surrounded by an axial helix to focus the electron beam. The
RF power is taken near the gun end of the helix. Its principle of operation
is based on the phenomenon of interaction of the RF field and the electron
beam as the phase velocity of the RF wave on the helix is comparable to the
velocity of the electrons. As the electron moves towards the collector, it alternately experiences accelerating and decelerating fields between two successive gaps of the helix and as a result bunching takes place.

The electron beam is slightly faster than the RF field. As a result, the electron beam gets advanced compared to the axial component of the electric
field by 90° at the collector end. Thus, the electron beam gets retarded and
transfers energy to the wave on helix. Owing to the decelerating effect, the
RF signal is amplified continuously as the wave on the helix progresses in
the backward direction. Thus, the maximum energy given by the electron
beam to the RF field will be at the collector end and the maximum amplitude
of this field will be near the gun end at the place of signal output.

A typical BWO constructed from a folded transmission line or waveguide
that winds back and forth across the path of the electron beam is shown in
Figure 12.62. The folded waveguide serves the same purpose as that served
by the helix in a TWT. The fixed spacing of the folded waveguide, however,
limits its bandwidth. The frequency of a given waveguide is constant. In

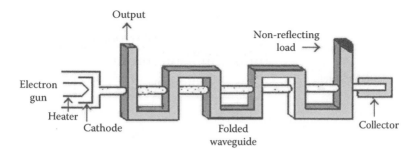

FIGURE 12.62
A typical backward-wave oscillator.

such a BWO, it is controlled by the transit time of the electron beam. The transit time in turn is controlled by the collector potential. The output frequency of this BWO can thus be changed by varying the collector voltage. This aspect is an added advantage. Like TWT, the electron beam in BWO is also focussed by a magnet placed around the body of the tube.

12.8.2 Backward-Wave *M*-Type Cross-Field Oscillator

It contains an electrode called sole, which may have a linear or circular structure. In the first case, it is referred to as linear *M*-carcinotron oscillator, whereas in the second case, it is called the circular *M*-carcinotron oscillator.

12.8.2.1 Linear M-Carcinotron Oscillator

In the linear *M*-carcinotron oscillator, the slow wave helix structure is kept parallel to the sole and a dc electric field is maintained between these. The configuration is also applied with a dc magnetic field perpendicular to the electric field to deflect the electrons emitted by the cathode through 90°. The electrons interact with a backward-wave space harmonic as the energy in the circuit flows opposite to the direction of electron motion. Because of its cross-field nature, this type of oscillator has high efficiency ranging between 30% and 60%. Its configuration is illustrated in Figure 12.63.

12.8.2.2 Circular M-Carcinotron Oscillator

The construction of this tube is shown in Figure 12.64. Its slow wave structure and sole are circular. This structure is terminated at the collector end. The slow wave structure here is an inter-digital line. The output is taken from the gun as mentioned before. Its basic principle of operation also depends on the interaction between the electron beam and the RF field in the slow wave structure.

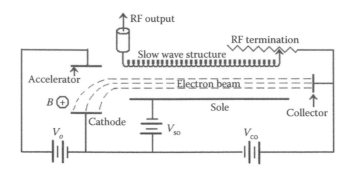

FIGURE 12.63
Linear *M*-carcinotron oscillator.

12.8.3 Applications and Performance Parameters

Linear BWOs do not have an attenuator like the TWTs. These have good tuning and frequency range and are used as signal source in instruments and transmitters. These tubes can be made equivalently broadband noise sources similar to TWTs. Their amplified wideband output is transmitted as jamming signal. These can operate from 1 GHz to over 1000 GHz. Thomas-CSF BWO operates over 320–400 GHz, 50 mW and at 2000 GHz, 0.8 mW CW power. Normal BWO peak power between 10 and 100 mW and CW power of 20 W is reported at high frequencies. A pulsed power of 250 kW with duty

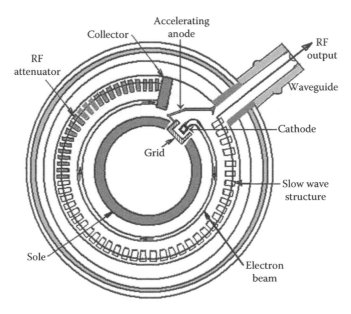

FIGURE 12.64
Circular M-carcinotron oscillator.

cycle <1 s has also been reported. Their typical tuning range is reported to be of the order of an octave at around 40 GHz.

As the beam velocity of this tube is controlled by varying the voltage, its frequency of oscillation can be widely varied. This property makes the BWO suitable to be used as a sweep oscillator. Further, by changing its beam current, the amplitude of oscillation can be greatly reduced. This tube therefore can also be used for amplitude modulation. Besides, it can be used as a voltage tunable bandpass amplifier if its beam current does not exceed the oscillation threshold level.

EXAMPLE 12.1

Find (a) the electron velocity leaving the cathode, (b) the gap transit angle, (c) the beam coupling coefficient, (d) the dc transit angle between cavities and (e) the maximum input gap voltage to get maximum voltage V_2, if the data for two cavity klystron are $V_0 = 2$ kV, $R_0 = 50$ kΩ, $I_0 = 25$ mA, $f = 5$ GHz, gap spacing $= d = 1$ mm, spacing between two cavities $= L = 4$ cm and effective shunt impedance without beam loading $= R_{sh} = 25$ kΩ.

Solution

$f = 5$ GHz $\quad \omega = 2\pi f = \pi \times 10^{10}$

a. The electron velocity leaving the cathode is given by Equation 12.1:

$$v_0 = \sqrt{(2 \text{ eV/m})} = 0.5932 \times 10^6 \sqrt{V_0} = 0.5932 \times 10^6 \sqrt{2000}$$
$$= 0.5932 \times 10^6 \times 44.72136 = 2.653 \times 10^7 \text{ m/s}$$

b. The gap transit angle is given by Equation 12.4:

$$\theta_g = \omega d/v_0 = \pi \times 10^{10} \times 10^{-3}/2.653 \times 10^7 = 1.1842 \text{ rad}$$

c. For identical buncher and catcher cavities, the beam coupling coefficient (β_i) of the buncher cavity is the same as that of the catcher cavity (β_0), β_i is defined by Equation 12.6:

$$\beta_i = \beta_0 = \sin(\theta_g/2)/(\theta_g/2) = \sin(1.1842/2)/(1.1842/2)$$
$$= \sin(0.592)/(0.592) = 0.9426$$

d. The dc transit angle between the cavities is given by Equation 12.17:

$$\theta_0 = \omega L/v_0 = \pi \times 10^{10} \times 4 \times 10^{-2}/2.653 \times 10^7 = 47.3666 \text{ rad}$$

e. The relation for maximum input voltage can be obtained from Equation 12.18, which defines the bunching parameter X {$= (\beta_i V_1 \theta_0)/2V_0$}. For maximum V_2, $J_1(X)$ must be maximum. In view of Figure 7.13, this corresponds to $J_1(X) = 0.582$ at $X = 1.841$. Thus,

$$V_{1max} = (2V_0 X)/(\beta_i \theta_0) = (2 \times 2000 \times 1.841)/(0.9426 \times 47.3666)$$
$$= 164.935 \text{ V}$$

EXAMPLE 12.2

Use the data of Example 12.1 to calculate (a) the depth of velocity modulation, (b) the minimum and maximum velocities attained by electrons, (c) the fundamental component of the beam current and (d) the optimum distance at which this fundamental component becomes maximum.

Solution

a. The depth of velocity modulation (δ) is given by the factor

$$\delta = \beta_i V_1 / V_0 = 0.9426 \times 164.935/2000 = 0.0777$$

b. The minimum and maximum velocities attained by the electrons are given by Equations 12.11a and 12.11b:

$$v_{min} = v_0 [1 - (\beta_i V_1/2 \, V_0)] = 2.653 \times 10^7[1 - 0.0777/2]$$
$$= 2.653 \times 10^7 \times 0.9611 = 2.55 \times 10^7 \text{ m/s}$$

$$v_{max} = v_0 [1 + (\beta_i V_1/2 \, V_0)] = 2.653 \times 10^7[1 + 0.0777/2]$$
$$= 2.653 \times 10^7 \times 1.03885 = 2.756 \times 10^7 \text{ m/s}$$

c. The expression for beam current (I_2) is given by Equation 12.19. It has fundamental component $\{2 \, I_0 J_1(X)\}$, which is maximum at $X = 1.841$.

$$I_2 = 2 \, I_0 J_1(X) = 2 \times 25 \times 10^{-3} \times 0.582 = 0.0291 \text{ A}$$

d. The optimum distance at which the fundamental component becomes maximum is given by Equation 12.20:

$$L_{opt} = \frac{3.682 v_0 V_0}{\omega \beta_i V_1} = \frac{3.682 \times 2.653 \times 10^7 \times 2000}{\pi \times 10^{10} \times 0.9426 \times 164.935} = 0.04 \text{ m}$$

EXAMPLE 12.3

Find (a) the voltage gain for the case of no beam loading of the output cavity, (b) V_2, (c) the amplifier efficiency without beam loading, (d) the beam loading conductance, (e) the power required for bunching and (f) the supplied dc power for the data given in Example 12.1.

Solution

a. Voltage gain for the case of no beam loading of the output cavity is given by Equation 12.29:

$$A_v = \frac{\beta_0^2 \theta_0}{R_0} \frac{J_1(X)}{X} R_{sh}$$

$$= (0.9426)^2 \times 47.3666 \times 0.582 \times 25 \times 10^3)/(50 \times 10^3 \times 1.841) = 6.65$$

b. In view of Equation 12.24

$$V_2 = I_2 R_{sh} = 0.0291 \times 25 \times 10^3 = 727.5 \text{ V}$$

c. The amplifier efficiency without beam loading is given by Equation 12.25:

Efficiency $= (\beta_0 I_2 \, V_2)/(2 \, I_0 \, V_0)$
$= (0.9426 \times 0.0291 \times 727.5)/(2 \times 25 \times 10^{-3} \times 2000) = 19.955\%$

d. The beam loading conductance is given by Equation 12.33b:

$$G_B = (G_0/2)[\beta_i^2 - \beta_i \cos(\theta_g/2)]$$

where
$G_0 = I_0/V_0$ is the dc beam conductance
$G_0 = I_0/V_0 = 25 \times 10^{-3}/2000 = 12.5 \times 10^{-6}$ mho
$\beta_i = \beta_0 = 0.9426$, $\theta_g = 1.1842$ rad, cos $(1.1842/2) = 0.83$
$G_B = (12.5 \times 10^{-6}/2)[(0.9426)^2 - 0.9426 \times 0.83] = 0.6625 \times 10^{-6}$

The beam loading resistance

$$R_B = 1/G_B = 1/(0.6625 \times 10^{-6}) = 1.51 \times 10^6 \; \Omega.$$

e. The power required for bunching given by Equation 12.32a is as given below:

$$P_B = (V_0^2/2) \, G_B = \{(164.935)^2/2\} \times 0.6625 \times 10^{-6} = 9.011177 \text{ mW}$$

f. The supplied dc power is given by Equation 12.32b:

$$P_0 = V_0^2 G_0 = (2 \times 10^3)^2 \times 12.5 \times 10^{-6} = 50 \text{ W}$$

EXAMPLE 12.4

The operating conditions for a reflex klystron are specified by the following data: $V_0 = 500$ V, $V_1 = 1$ V, gap spacing $d = 1$ mm and $R_{sh} = 20$ kΩ. The tube oscillates at $f_r = 10$ GHz and operates in $n = 2$ (1¾) mode. Find (a) the gap transit angle θ_g, (b) the beam coupling coefficient β_0, (c) the value of the repeller voltage V_r, (d) the round-trip dc transit time of the centre-of-the-bunch electron, (d) the round-trip dc transit time T_0', (e) the round-trip dc transit angle and (f) the Bunching parameter of the reflex klystron. Neglect the transit time through the gap and the beam loading.

Solution

Given $V_0 = 500$ V, $V_1 = 1$ V, $L = 1$ mm, $R_{sh} = 20$ kΩ, $n = 2$, $d = 1$ mm
$f_r = 10$ GHz, $\omega = 2\pi, f = 2\pi \times 10 \times 10^9$
$v_0 = 0.5932 \times 10^6$, $\sqrt{V_0} = 0.5932 \times 10^6 \sqrt{500} = 13.264 \times 10^6$ m/s
$e = 1.602 \times 10^{-1}$ C $\quad m = 9.109 \times 10^{-31}$ kg $\quad e/m = 1.759 \times 10^{11}$
$m/e = 5.686 \times 10^{-12}$

a. The gap transit angle is given by Equation 12.4:

$$\theta_g = \omega d/v_0 = 2\pi \times 10^{10} \times 10^{-3}/13.264 \times 10^6 = 4.737 \text{ rad}$$

b. In the reflex klystron, the buncher cavity is also the catcher cavity; thus, the beam coupling coefficient $\beta_i = \beta_0$, which is defined by Equation 12.6:

$$\beta_i = \beta_0 = \sin(\theta_g/2)/(\theta_g/2) = \sin(4.737/2)/(4.737/2)$$
$$= \sin(2.3685)/(2.3685) = 0.2948$$

c. The value of the repeller voltage (V_r) can be evaluated from Equation 12.54a:

$$\frac{V_0}{(V_r + V_0)^2} = \frac{(2\pi n - \pi/2)^2}{8\omega^2 L^2}\frac{e}{m} = \frac{(2\pi \times 2 - \pi/2)^2}{8 \times (2\pi \times 10^{10})^2 (10^{-3})^2}(1.759 \times 10^{11})$$

$$= 0.6734 \times 10^{-3}$$

or $V_r = 361.7$ V

d. The round-trip dc transit time (T_0) of the centre-of-the-bunch electron is given by Equation 12.42b:

$$T_0' = 2(m/e) L\ v_0/(V_r + V_0)$$
$$= \{2 \times 5.686 \times 10^{-12} \times 10^{-3} \times 13.264 \times 10^6\}/(361.7 + 500)$$
$$= 1.75 \times 10^{-10}\ \text{s}$$

e. The round-trip dc transit angle is given by Equation 12.44a:

$$\theta_0' = \omega T_0' = 2\pi \times 10^{10} \times 1.75 \times 10^{-10} = 11°$$

f. The bunching parameter of the reflex klystron is given by Equation 12.44b:

$$X' = (\beta_i V_1 \theta_0')/2V_0 = (0.2948 \times 1 \times 11)/2 \times 500 = 0.0032428$$

EXAMPLE 12.5

In Example 12.4, if $\beta_0 = 0.8$, calculate (a) the direct current required to give a microwave gap voltage of 250 V and (b) the electronic efficiency.

Solution

a. In view of Equations 12.24 and 12.48, the direct current required to give a microwave gap voltage $V_2 = 250$ V for $\beta_0 = 0.8 = \beta_i$

$$V_2 = I_2\ R_{sh} = 2\ I_0\ \beta_0\ J_1(X')\ R_{sh}$$

$$I_0 = V_2/\{2\ \beta_0\ J_1(X')\ R_{sh}\} = 250/\{2 \times 0.8 \times 0.582 \times 20 \times 10^3\} = 13.4\ \text{mA}$$

b. The electronic efficiency (η) is given by Equation 12.53a:

$$\eta = 2X'J1(X')/(2n\pi - \pi/2)$$
$$= (2 \times 1.841 \times 0.582)/(2 \times 3.154 \times 2 - 3.154/2) = 19.4123\%$$

EXAMPLE 12.6

Calculate (a) the pitch of a helix used in TWT if its radius is 5 cm and accelerating anode voltage is 5 kV, (b) the phase velocity along coil (in air) and (c) the phase velocity along the coil located in dielectric ($\varepsilon_r = 3.250$ filled cylinder). Discuss the results.

Solution

Given radius $= 0.06$ m, diameter $d = 0.1$ m and anode voltage $V_a = 5000$ V
   From Equation 12.1,

$$v = 0.5932 \times 10^6\sqrt{V_a}$$
$$= 0.5932 \times 10^6\sqrt{(5 \times 10^3)} = 0.5932 \times 10^6 \times 70.71$$
$$= 0.4195 \times 10^8 \text{ m/s}$$

In view of Equation 12.62, $v = c\,p/(2\pi a)$ for air-filled coil
Thus, pitch

$$p = v(\pi d)/c = 41.95 \times 10^6\,(3.1415 \times 0.1)/3 \times 10^8 = 0.0439 \text{ m}$$

In view of Equation 12.63, v_p for air coil in air is given by

$$v_p = \frac{pc}{\sqrt{p^2 + (\pi d)^2}} = \frac{0.0439 \times 3 \times 10^8}{\sqrt{(0.0439)^2 + (0.1\pi)^2}} = \frac{0.1317 \times 10^8}{0.3172} = 0.4152 \times 10^8$$

In view of Equation 12.64, $v_{p\varepsilon}$ for coil in dielectric-filled cylinder is

$$v_{p\varepsilon} = \frac{p}{\sqrt{\mu\varepsilon\{p^2 + (\pi d)^2\}}} = \frac{1}{\sqrt{\mu\varepsilon}}\frac{v_p}{c}$$

$$\sqrt{\mu\varepsilon} = \sqrt{\mu_0\varepsilon_0\varepsilon_r} = \sqrt{(4\pi \times 10^{-7} \times 8.854 \times 10^{-12} \times 3.25)} = 6.0133 \times 10^{-9} \quad \text{and}$$
$$1/\sqrt{\mu\varepsilon} = 1.663 \times 10^8$$

$$v_{p\varepsilon} = 1.663 \times 10^8 \times 0.4152 \times 10^8/3 \times 10^8 = 0.23 \times 10^8$$

The phase velocity of a wave in an air-filled waveguide is greater than c (i.e., $>3 \times 10^8$ m/s). In air-filled helix, it reduces to 0.4152×10^8 m/s and in helix within a dielectric-filled cylinder with $\varepsilon_r = 3.25$, it reduces to $= 0.23 \times 10^8$ m/s.

EXAMPLE 12.7

A travelling-wave tube having a helix of characteristic impedance $Z_0 = 10$ ohms operates at 10 GHz. Its beam voltage $V_0 = 5$ kV, beam current $I_0 = 40$ mA and circuit length $N = 40$. Find its (a) gain parameter, (b) output power gain, (c) phase constant of velocity-modulated electron beam β_e and (d) propagation constants.

Solution

Given $Z_0 = 10$ ohms, $V_0 = 5$ kV, $I_0 = 40$ mA, $f = 10$ GHz and $N = 40$

a. The gain parameter from Equation 12.94 is

$$C \equiv (I_0 Z_0/4 \, V_0)^{1/3} = \{(40 \times 10^{-3} \times 10)/(4 \times 5 \times 10^3)\}^{1/3} = 2.7144 \times 10^{-2}$$

b. The output power gain from Equation 12.96 is

$$A_p = -9.54 + 47.3 \, NC = -9.54 + 47.3 \times 40 \times 2.7144 \times 10^{-2} = 41.816448$$

c. Phase constant of velocity modulated electron beam is

$$\beta_e = \omega/\upsilon_0 = (2\pi f)/(0.5932 \times 10^6 \sqrt{V_a})$$
$$= (2 \times \pi \times 10^{10})/(0.5932 \times 10^6 \sqrt{5000})$$
$$= 1497.94 \text{ rad/m}$$

d. The four propagation constants are given by Equation 12.93:

$$\gamma_1 = -\beta_e C(\sqrt{3}/2) + j\beta_e(1 + C/2)$$

$$\gamma_2 = \beta_e C(\sqrt{3}/2) + j\beta_e(1 + C/2)$$

$$\gamma_3 = j\beta_e(1 - C)$$

$$\gamma_4 = -j\beta_e(1 - C^3/4)$$

$$\beta_e C \sqrt{3}/2 = 1.49794 \times 10^3 \times 2.7144 \times 10^{-2} \times 0.866 = 35.213$$

$$\beta_e(1 + C/2) = 1.49794 \times 10^3 \times (1 + 2.7144 \times 10^{-2}/2) = 1518.27$$

$$1 - C = 1 - 2.7144 \times 10^{-2} = 0.972856$$

$$1 - C^3/4 = 1 - (2.7144 \times 10^{-2})^3/4 = 0.9999993214$$

$$\gamma_1 = -\beta_e C(\sqrt{3}/2) + j\beta_e(1 + C/2) = -35.213 + j1518.27$$

$$\gamma_2 = \beta_e C(\sqrt{3}/2) + j\beta_e(1 + C/2) = 35.213 + j1518.27$$

$$\gamma_3 = j\beta_e(1 - C) = j1.49794 \times 10^3 \times 0.972856 = j1457.2799$$

$$\gamma_4 = -j\beta_e(1 - C^3/4) = -j1.49794 \times 10^3 \times 0.9999993214 = -j1497.94$$

EXAMPLE 12.8

A cylindrical magnetron with outer cathode radius of 4 cm, inner radius of anode of 8 cm has anode voltage of 20 kV, beam current 25 A and magnetic flux density of 0.35 Wb/m^2. Calculate (a) the cyclotron angular frequency, (b) the cut of flux density for a fixed anode voltage and (c) the cutoff voltage for a fixed flux density.

Solution

Given $V_0 = 20$ kV, $I_0 = 25$ A, $B_0 = 0.35$ Wb/m^2, $a = 4$ cm and $b = 8$ cm
As $e = 1.6 \times 10^{-19}$ C and $m = 9.1 \times 10^{-31}$ kg,
$e/m = 1.6 \times 10^{-19}/9.1 \times 10^{-31} = 0.1758 \times 10^{12}$ and $m/e = 5.688 \times 10^{-12}$

$a/b = 4/8 = 0.5$ $(a/b)^2 = 0.25$ $1 - (a/b)^2 = 0.75$

a. The cyclotron angular frequency is given as

$$\omega_c = (e/m)\, B_z = 0.1758 \times 10^{12} \times 0.35 = 0.06154 \times 10^{12}$$

b. The cutoff flux density (B_{0c}) for a fixed anode voltage (V_0) given by Equation 12.106 is

$$B_{0c} = \frac{(8V_0 m/e)^{1/2}}{b(1 - a^2/b^2)} = \frac{(8 \times 20 \times 10^3 \times 5.688 \times 10^{-12})^{0.5}}{8 \times 10^{-2} \times 0.75}$$

$$= \frac{9.54 \times 10^{-4}}{6 \times 10^{-2}} = 15.9 \, \text{mWb/m}^2$$

c. The cutoff anode voltage (V_0) for a fixed flux density (B_0) given by Equation 12.107 is

$$V_{0c} = \frac{e}{8m} B_0^2 b^2 (1 - a^2/b^2)^2 = (0.1758 \times 10^{12}/8) \times 0.1225$$

$$\times 64 \times 10^{-4} \times (0.75)^2 = 9690.975 \, \text{kV}$$

EXAMPLE 12.9

A conventional magnetron operates at an anode voltage (V_0) of 5 kV, beam current (I_0) of 5 A and frequency (f) of 10 GHz. Its resonator conductance (G_r) is 0.25 milli-mhos, loaded conductance (G_l) is 0.03 milli-mhos and vane capacitance (C) is 3 pF. The power loss (P_{loss}) in the device is 15 kW. Calculate (a) the angular resonant frequency, (b) the unloaded Q, (c) the loaded Q, (d) the external Q, (e) the circuit efficiency and (f) the electronic efficiency.

Solution

Given $V_0 = 5$ kV, $I_0 = 5$ A, $f = 10^{10}$ Hz, $G_r = 2.5 \times 10^{-4}$ mho, $G_l = 0.3 \times 10^{-4}$, $C = 3$ pF, power loss $P_{loss} = 15$ kW

a. The angular resonant frequency $\omega_r = \omega_0 = 2\pi f = 2\pi \times 10^{10} = 6.2832 \times 10^{10}$ rad/s

b. The unloaded quality factor is given by Equation 12.117a:

$$Q_{un} = \omega_0 C/G_r = 6.2832 \times 10^{10} \times 3 \times 10^{-12}/2.5 \times 10^{-4} = 754$$

c. The loaded quality factor is given by Equation 12.117c:

$$Q_L = \omega_0 C/(G_r + G_l)$$
$$= 6.2832 \times 10^{10} \times 3 \times 10^{-12}/(2.5 \times 10^{-4} + 0.3 \times 10^{-4}) = 673.2$$

d. The external quality factor is given by Equation 12.117b:

$$Q_{ext} = \omega_0 C/G_l = 6.2832 \times 10^{10} \times 3 \times 10^{-12}/0.3 \times 10^{-4} = 6283.1853$$

e. Circuit efficiency is given by Equation 12.118:

$$\eta_c = 1/(1 + Q_{ext}/Q_{un})$$
$$= 1/(1 + 6283.1853/754) = 0.107145 = 10.7145\%$$

f. The electronic efficiency is given by Equation 12.119 as

$$\eta_e = P_{gen}/P_{dc}$$

$$P_{dc} = V_0 I_0 = 5 \times 10^3 \times 5 = 25 \text{ kW}$$

$$P_{gen} = P_{dc} - P_{loss} = 25 - 15 = 10 \text{ kW}$$

Thus,

$$\eta_e = 10/25 = 0.4 \quad \text{or} \quad 40\%.$$

PROBLEMS

P-12.1 Find (a) the velocity of an electron just leaving the cathode, (b) the gap transit angle, (c) the beam coupling coefficient, (d) the dc transit angle between cavities and (e) the maximum input gap voltage to get maximum voltage V_2, if the data for two cavity klystron are $V_0 = 3$ kV, $R_0 = 60$ kΩ, $I_0 = 30$ mA, $f = 6$ GHz, gap spacing $= d = 1.5$ mm, spacing between two cavities $= L = 5$ cm and effective shunt impedance without beam loading $= R_{sh} = 30$ kΩ.

P-12.2 Use data of P-12.1 to calculate (a) the depth of velocity modulation, (b) the minimum and maximum velocities attained by electrons, (c) the fundamental component of beam current and (d) the optimum distance at which this fundamental component becomes maximum.

P-12.3 Find (a) the voltage gain for the case of no beam loading of the output cavity, (b) V_2 (c) the amplifier efficiency without beam loading, (d) the beam loading conductance, (e) the power required for bunching and (f) the supplied dc power for the data given in P-12.1.

P-12.4 The operating conditions for a reflex klystron are specified by the following data: $V_0 = 400$ V, $V_1 = 1$ V, gap spacing $d = 1.5$ mm and $R_{sh} = 25$ kΩ. The tube oscillates at $f_r = 8$ GHz and operates in $n = 2$ (1¾) mode. Find (a) the gap transit angle θ_g, (b) the beam coupling coefficient β_0, (c) the value of the repeller voltage V_r, (d) the round-trip dc transit time of the centre-of-the-bunch electron, (e) the round-trip dc transit time T_0', (f) the round-trip dc transit angle and (g) the bunching parameter of the reflex klystron. Neglect the transit time through the gap and the beam loading.

P-12.5 In P-12.4, if $\beta_0 = 0.9$, calculate (a) the direct current required to give a microwave gap voltage of 350 V and (b) the electronic efficiency.

P-12.6 Calculate (a) the pitch of a helix used in TWT if its radius is 6 cm and accelerating anode voltage is 6 kV, (b) the phase velocity along coil (in air) and (c) the phase velocity along the coil located in dielectric ($\varepsilon_r = 2.25$ filled cylinder). Discuss the results.

P-12.7 A travelling-wave tube having a helix of characteristic imped-ance $Z_0 = 15$ ohms operates at 8 GHz. Its beam voltage $V_0 = 4$ kV, beam current $I_0 = 30$ mA and circuit length $N = 50$. Find its (a) gain parameter, (b) output power gain, (c) phase constant of velocity-modulated electron beam β_e and (d) propagation constants.

P-12.8 A cylindrical magnetron with outer cathode radius of 4 cm and inner radius of anode of 10 cm has an anode voltage of 30 kV, beam current 20 A and magnetic flux density 0.3 Wb/m². Calculate the (a) cyclotron angular frequency, (b) cutoff flux density for a fixed anode voltage and (c) cutoff voltage for a fixed flux density.

P-12.9 A conventional magnetron operates at an anode voltage (V_0) of 6 kV, beam current (I_0) of 6 A and frequency (f) 8 GHz. Its resona-tor conductance (G_r) is 0.2 milli-mhos, loaded conductance (G_l) is 0.02 milli-mhos, and vane capacitance (C) is 4 pF. The power loss (P_{loss}) in the device is 10 kW. Calculate (a) the angular resonant fre-quency, (b) unloaded Q, (c) loaded Q, (d) external Q, (e) circuit effi-ciency and (f) electronic efficiency.

Descriptive Questions

Q-12.1 List the linear beam and crossed-field tubes and further classify these in accordance with the involvement of resonant and non-resonant cavities.

Q-12.2 Give the constructional features of two-cavity klystron and explain its operation.

Q-12.3 With the aid of necessary figures, explain the process of bunch-ing in two-cavity klystron.

Q-12.4 Discuss the multi-cavity klystron amplifier. How does it differ from the two-cavity klystron?

Q-12.5 Discuss the process of bunching in the reflex klystron. Add fig-ures if required.

Q-12.6 Explain the terms velocity modulation, current modulation and modes of operation in reference to reflex klystrons.

Q-12.7 Discuss the tuning process in reflex klystrons.

Q-12.8 Discuss the basic operating principle of travelling-wave tubes.

Q-12.9 Give the constructional features, focussing and input/output arrangement of TWT.

Q-12.10 List the various slow wave structures. Illustrate the geometry of helix and give the factor by which the wave gets slowed down.

Q-12.11 Explain the working of O-type backward-wave oscillator.

Q-12.12 Discuss the salient features of a circular M-carcinotron oscillator.

Q-12.13 Explain the working of a magnetron by including the effect of electric field, magnetic field, combined E and B fields and combined E, B and RF fields.

Q-12.14 Explain the modes of oscillation, strapping and frequency pushing and pulling.

Q-12.15 Discuss various coupling methods used in magnetrons.

Q-12.16 Discuss various tuning methods employed in magnetrons.

Q-12.17 Explain the working of the negative-resistance magnetron.

Q-12.18 Explain the operating principle of the electron-resonance magnetron.

Further Reading

1. Brillouin, L., *Wave Propagation on Periodic Structures*, 2nd ed. Dover, NY, 1953.
2. Brown, W. C., The microwave magnetron and its derivatives. *IEEE Trans. Electron Devices*, ED-31, 11, 1595–1605, November 1984.
3. Chodorow, M. and Susskind, C., *Fundamentals of Microwave Electronics*. McGraw-Hill Book Company, New York, 1964.
4. Chodorow, M. and Wessel-berg, T., A high-efficiency klystron with distributed interaction. *IRE Trans. Electron Devices*, ED-8, 44–55, January 1961.
5. Chodorow, M. and Craig, R. A., Some new circuits for high-power travelling-wave tubes. *Proc. IRE*, 45, 1106–1118, August 1957.
6. Curnow, H. J., A general equivalent circuit for coupled-cavity slow-wave structures. *IEEE Trans. Microwave Theory Tech.*, MTT-13, 671–675, September 1965.
7. Dalman, G. C., *Microwave Devices, Circuits and Their Interaction*, John Wiley & Sons Inc, New York, 1994.
8. Gewartowski, J. W. and Watson, H. A. *Principles of Electron Tubes*. D. Van Nostrand Company, Princeton, NJ, 1965.
9. Gilmour, A. S., Jr., *Microwave Tubes*. Artech House, Dedham, MA, 1986.
10. Hutter, R. G. E., *Beam and Wave Electrons in Microwave Tubes*. D. Van Nostrand Company, Princeton, NJ, 1960.

11. *IEEE Proceedings*, 61, No. 3, March 1973. Special issue on high power microwave tubes.

12. Kennedy, G., *Electronic Communication Systems*, 3rd ed. McGraw-Hill, New Delhi, 1996.

13. Larue, A. D. and Rubert, R. R., Multi-Megawatt Hybrid TWT's at *S*-Band and *C*-Band. Presented to the IEEE Electron Devices Meeting, Washington, DC, October, 1964.

14. Liao, S. Y., *Microwave Electron Tubes*. Prentice-Hall, Inc., NJ, 1988.

15. Mendel, J. T., Helix and coupled-cavity travelling-wave tubes. *Proc. IEEE*, 61(3), 280–298, March 1973.

16. Pierce, J. R., *Traveling-Wave Tubes*, D. Van Nostrand Company. Princeton, NJ, 1950.

17. Ruetz, A. J. and Yocom, W. H., High-power travelling-wave tubes for radar systems. *IRE Trans. Mil. Electron*, MIL-5, 39–45, April 1961.

18. Skowron, J. F., The continuous-cathode (emitting-sole) crossed-field amplifier. *Proc. IEEE*, 61(3), 330–356, March 1973.

19. Soohoo, R. F., *Microwave Electronics*. Addison-Wesley Publishing Company, MA, 1971.

20. Staprans, A. et al., High-power linear beam tubes. *Proc. IEEE*, 61(3), 299–330, March 1973.

21. Warnecke, R. R. et al., Velocity modulated tubes. *Advances in Electronics*, Vol. 3. Academic Press, New York, 1951.

13

Microwave Diodes

13.1 Introduction

Chapter 12 encompassed different tubes to generate and amplify microwave energy. These tasks can also be performed by semiconductor devices which include different types of diodes and transistors. The tubes are suitable for higher powers, higher frequencies of operation, higher operating voltages and for higher degree of amplification linearity. On the contrary, the semiconductor devices have smaller size, lesser weight, low cost, higher reliability and greater energy efficiency particularly with CMOS technology. The highly automated manufacturing processes and easy availability of their complementary devices are the added advantages.

The entire matter related with the semiconductor devices is divided into two chapters. This chapter deals with the microwave diodes whereas Chapter 14 is devoted to microwave transistors. This chapter includes brief description and limitations of some conventional diodes and detailed discussion about different types of microwave diodes.

13.2 Basics of Semiconductor Devices

13.2.1 Properties of Semiconducting Materials

In microwave solid state devices the electrical behaviour of solids plays the pivotal role. The transport of charges through a semiconductor is not only the function of properties of electrons but strongly linked to the arrangement of atoms in solids. The electrical conductivities of semiconducting materials fall in between those of conductors and insulators and can be widely varied by changing their temperature, optical excitation and impurity contents. Table 13.1 includes the properties of important semiconductors. These include bandgap energy, mobility and the relative dielectric constants.

TABLE 13.1

Properties of Important Semiconductors

Semiconductor Substance	Bandgap Energy (eV)		Mobility at 300 K (cm²/V s)		Relative Dielectric Constant (ε_r)
	0 K	330 K	Holes	Electrons	
C (Carbon)	5.51	5.47	1600	1800	5.5
Ge (Germanium)	0.89	0.803	1900	3900	16
Si (Silicon)	1.16	1.12	450	1600	11.8
AlSb (Aluminium antimony)	1.75	1.63	420	200	11
GaSb (Germanium antimony)	0.80	0.67	1400	4000	15
GaAs (Galium arsenide)	1.52	1.43	400	8500	13.1
GaP (Galium phosphide)	2.40	2.26	75	110	10
InSb (Indium antimonide)	0.26	1.80	750	78,000	17
InAs (Indium arsenide)	0.46	0.33	460	33,000	14.5
InP (Indium phosphide)	1.34	1.29	150	4600	14
CdS (Cadmium sulphide)	2.56	2.42	50	300	10
CdTe (Cadmium telluride)	1.85	1.70		800	10
ZnO (Zinc oxide)		3.20		200	9
ZnS (Zinc sulphide)	3.70	3.60		165	8

The energy band plays a key role vis-à-vis the electrical properties of conductors, semiconductors and dielectrics. In semiconductors there is a forbidden energy region between conduction and valence bands. Within the forbidden band no allowable states can exist. The energy level at the bottom of the conduction band is designated as E_c and that at the top of valence band as E_v. The difference between these two energy levels E_g ($= E_c - E_v$) is the most important parameter in semiconductors. As shown in Figure 13.1 this energy is conventionally taken to be positive when measured upward for electrons and downward for holes.

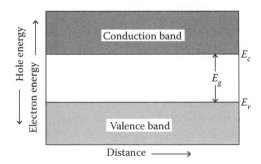

FIGURE 13.1
Energy band diagram.

13.2.2 Mechanism Involved

The operation of microwave semiconductor diodes is based on a number of mechanisms. In tunnel diodes the carriers cross over from one (filled) energy state to another (empty) through tunnelling. This tunnelling mechanism is a quantum mechanical phenomenon with no analogy with the classical semiconductor physics. Another class of microwave semiconductor diodes operates on bulk effect of semiconductors. These devices operate on the principle of transfer of electrodes from one energy level to another and thus are referred to as transferred electron devices. These include Gunn diode, LSA diode, indium phosphide diode and cadmium telluride diode. There is another category of diodes which are specifically meant to operate at microwave frequencies and include Read diode, IMPATT diode, TRAPATT diode and BARITT diode. This class is referred to as Avalanche transit time devices. These devices by virtue of their inherent nature operate with reversed bias and use the avalanche effect. Beside the above the microwave diodes also include point-contact diode, step-recovery diode (SRD), PIN diode, Schottky diode and Varactor or varicap diode. These diodes have their own mechanism of operation and do not fall under any of the above three categories.

13.2.3 Significance of Negative Resistance

Some of the devices discussed in the subsequent sections exhibit negative resistance. To understand the significance of negative resistance two simple RLC circuits are shown in Figure 13.2. In Figure 13.2a the circuit contains a positive resistance. In this circuit the moment switch S is turned on at $t = 0$ the switching transients will result in oscillations but these oscillations will quickly die out due to the presence of positive resistance. In Figure 13.2b the circuit also contains a negative resistance. In this case oscillations will start in a similar manner at $t = 0$. These oscillations will be sustained due to the presence of negative resistance. The negative resistance in fact gives power to the circuit, part of which compensates the losses due to positive

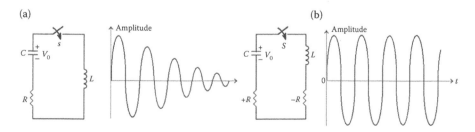

FIGURE 13.2
Illustration of the significance of negative resistance. (a) Simple RLC circuit, (b) damped oscillations, (c) RLC circuit with negative resistance and (d) sustained oscillations.

resistance and the remaining part sustains oscillations. Thus, the study of active devices which exhibit negative resistance property attains a special significance.

13.3 Conventional Diodes

The diodes referred to here as conventional are those which are not used at microwave range. These fall into either of the two categories. The first category includes those diodes which are generally used in low RF frequency applications. The second category deals with the diodes that are meant to operate in optical range. The circuit symbols of some of the diodes belonging to these categories are illustrated in Figure 13.3.

13.3.1 Low-Frequency Conventional Diodes

There are many types of diodes which are used in electronic circuits mainly at lower frequency range. They perform different functions in accordance with their properties. Some of the available conventional diodes belonging to the first category are briefly described below.

13.3.1.1 PN Junction Diodes

These are normal or standard type of diodes. These can be further categorised as signal diode and rectifier diode. *Signal diodes* are used for applications where small values of currents are drawn. These can also be used for other functions within circuits where the 'one way' effect of a diode is required. *Rectifier diodes* are used in power supplies for rectifying ac power inputs, that is, for high-current and high-voltage applications. These are generally PN junction diodes, although Schottky diodes may also be used if low-voltage drops are needed. They are able to rectify current levels that may range from an ampere upwards.

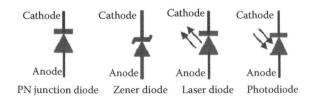

FIGURE 13.3
Circuit symbols of some conventional diodes.

13.3.1.2 *Zener Diode*

It operates under reverse bias conditions and breaks down when a specified voltage limit is crossed. If the current is limited through a resistor, it produces a stable voltage. It is widely used to provide a reference voltage in power supplies. The two modes of reverse breakdown include Zener breakdown (predominates below 5.5 V) and impact ionisation (predominates above 5.5 V). Irrespective of mode of breakdown this diode is named as Zener diode.

13.3.1.3 *Avalanche Diode*

It operates in reverse bias and uses the avalanche effect for its operation. In general it is used for photo-detection where the avalanche process enables high levels of sensitivity to be obtained, even if there are higher levels of associated noise.

13.3.2 Characteristic Parameters of Conventional Diodes

The semiconducting material used in diodes is of paramount importance because it affects their characteristics and properties. For low-frequency devices silicon is the most widely used material followed by germanium. For more specialised (microwave) diodes other materials (e.g. GaAs, InP, CdTe, etc.) are used. The important parameters of these diodes include forward voltage drop, peak inverse voltage, maximum forward current, leakage current and junction capacitance.

13.3.2.1 *Forward Voltage Drop (V_f)*

It is the result of flow of forward current in any electronics device. In power rectification losses will be higher for a high forward voltage drop. In RF diodes the forward voltage drop has to be small as signals themselves may be small and will have to overcome this drop.

13.3.2.2 *Peak Inverse Voltage*

It is the maximum voltage a diode can withstand in the reverse direction. This voltage must not be exceeded by the peak-to-peak value of the waveform otherwise the device may fail.

13.3.2.3 *Maximum Forward Current*

It is the maximum allowable forward current level for a diode. As the current exceeds beyond this level additional heat is generated which needs way for its effective removal.

13.3.2.4 Leakage Current

It is the flow of small current (of μA or pA order) in diode when it is reverse biased. In PN junction diode flow of such current is due to the minority carriers in the semiconductor. The level of leakage current depends on the reverse voltage, temperature and the type of material.

13.3.2.5 Junction Capacitance

All PN junction diodes exhibit junction capacitance. This capacitance exists between the two plates assumed to be formed at the edges of depletion region. Width of depletion region is the dielectric spacing between these plates. The value of this capacitance depends on the reverse voltage which alters the width of depletion region. This property of formation of capacitance is fully utilised in varactor or varicap diodes.

13.3.3 High-Frequency Limitations of Conventional Diodes

At higher frequencies there are three factors which pose problems. The first factor is junction capacitance. Its existence affects the gain and speed of the device. For most RF applications this capacitance is to be greatly minimised. The *lead inductance* is another factor which limits the performance of conventional devices. Its presence may result in degenerative feedback which in turn reduces the overall performance of device. The *transit time* is the third important factor. It attains more significance in microwave transistors. If its value exceeds 10% of a cycle the transistor efficiency decreases which in turn lowers the transconductance and speed.

13.3.4 Optical Frequency Diodes

These diodes operate at optical frequencies. These include light-emitting diodes (LEDs), laser diodes (LDs) and photodiodes (PDs). These are briefly described below.

13.3.4.1 Light-Emitting Diodes

These produce (incoherent) light when forward biased with current flowing through the junction. These use component semiconductors, and can produce light of a variety of colours. The development of high output LEDs and OLEDs has altogether changed the concept of displays. The different LED characteristics include colours, light/radiation wavelength and light intensity.

13.3.4.2 Laser Diodes

These produce coherent light and are widely used in many applications from DVD and CD drives to laser light pointers for presentations. Although LDs are much cheaper than other forms of laser generator, they are considerably expensive than LEDs and have limited life.

13.3.4.3 Photodiodes

These are used for detecting the light. When light strikes a PN junction these liberate electrons and holes. These are typically operated under reverse bias conditions where even small amount of current flow resulting from the light can be easily detected. PDs can also be used to generate small signals. A PD may be of PIN, PN, Avalanche or Schottky form. In some applications, PIN diodes work very well as photo-detectors.

13.4 Microwave Diodes

The requirement of small-sized microwave devices led to the solid-state devices with higher and higher frequency ranges. These devices are predominantly active and two-terminal diodes. This section deals with only six types of diodes, namely (i) point-contact diodes, (ii) SRD, (iii) PIN diode, (iv) tunnel diode, (v) Schottky diode and (vi) Varactor diodes. Some other microwave diodes, namely transferred-electron devices, and avalanche transit-time diodes are discussed in the subsequent sections.

13.4.1 Point-Contact Diode

Point-contact diodes, also called crystals, are the oldest microwave semiconductor devices. These were developed during the Second World War and are still used as receiver mixers and detectors.

13.4.1.1 Construction

It is one of the most basic forms of diodes in terms of its construction. It consists of a piece of N-type semiconductor, onto which a sharp point of a specific type of (group III) metal wire is placed. As this physical junction is formed, some of the metal from the wire migrates into the semiconductor and produces a PN junction. Point-contact diodes have a very low junction capacitance because the resulting junction is very small.

Unlike the PN-junction diode, this diode depends on the pressure of contact between a point and a semiconductor crystal for its operation. Figure 13.4 illustrates a point-contact diode. Its constituents are shown in Figure 13.4a and formation of p region around the contact point in Figure 13.4b. One section of the diode consists of a small rectangular crystal of n-type silicon. A fine beryllium-copper, bronze-phosphor, or tungsten wire

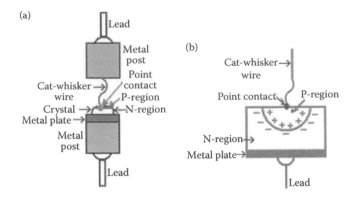

FIGURE 13.4
Point-contact diode. (a) Construction and constituents and (b) p-region around point contact.

called the cat whisker presses against the crystal and forms the other part of the diode.

During formation of point-contact diode, a relatively large current is passed through the cat whisker to the silicon crystal. It results in the formation of a small region of p-type material around the crystal in the vicinity of the point contact. Thus, a pn junction is formed which behaves in the same way as a normal pn junction. The pointed wire is used instead of a flat metal plate to produce a high-intensity electric field at the point of contact without using a large external source voltage. The application of large voltages across the average semiconductor is avoided to prevent excessive heating. The end of cat whisker is one of the terminals of the diode. It has a low-resistance contact to the external circuit. A flat metal plate on which the crystal is mounted forms the lower contact of the diode with the external circuit. Both contacts with the external circuit are low-resistance contacts. Figure 13.5a illustrates a cutaway view of the point-contact diode, whereas its schematic symbol is shown in Figure 13.5b.

13.4.1.2 Operation

Its operation is similar to that of a PN junction diode. Its *characteristics* under forward and reverse bias are somewhat different from those of the junction diode. With forward bias, its resistance is higher than that of the junction diode. With reverse bias, its current flow is not as independent of the applied voltage as it is in the junction diode. The point-contact diode has an *advantage* over the junction diode because the capacitance between the cat whisker and the crystal is less than the capacitance between the two sides of the junction diode. As such, the capacitive reactance existing across the point-contact diode is higher and the capacitive current that flows in

(a)

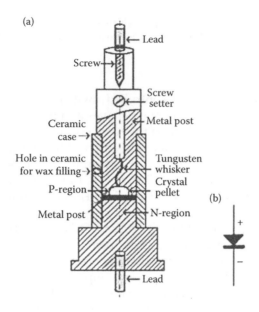

FIGURE 13.5
Cutaway view and symbol of point-contact diode. (a) Cutaway view and (b) symbol.

the circuit at high frequencies is smaller. In view of its very low capacitance it is ideal for many RF applications. The main *limitation* of this diode is its inability of handling high currents due to its small junction. These diodes, however, are very cheap.

13.4.2 Step-Recovery Diode

SRD characterised by very fast switching times is a form of semiconductor diode that can be used as a charge-controlled switch. It has the ability to generate very sharp pulses at very high frequencies. These rely on a very fast turn-off characteristic of the diode for their operation. These are doped differently than other types of diodes, with less doping at the pn junction than away from it. In view of its method of operation, it is also called the *'Snap-off' diode, 'charge storage diode'* or *'memory varactor'*.

13.4.2.1 Construction

The SRD is fabricated with the doping level gradually decreasing as the junction is approached or as a direct PIN structure. This profile reduces the switching time because there are fewer charge carriers in the region of the junction and hence less charge is stored in this region. This allows these stored charges to be released more rapidly when biasing is changed from forward to reverse. The rapid establishment of forward current in SRDs is an added advantage in comparison to ordinary junction diode.

13.4.2.2 Operation

The SRD is used as a charge-controlled switch. When SRD is forward biased and when the charge enters it behaves as a normal diode. When it is switched from forward conduction to reverse cut-off a reverse current flows for a while as stored charge is removed. When all the charges are removed it suddenly turns-off or snaps-off. It is the abruptness with which the reverse current ceases that enables the SRD to be used for the generation of micro-wave pulses and also for waveform shaping. The elaboration of its operation is as under.

Under the normal forward bias conditions SRD conducts normally. When it is quickly reverse biased it initially appears as low impedance, typically less than an ohm. Once the stored charge depletes, the impedance abruptly increases to its normal (very high) reverse impedance. This transition occurs very quickly, typically well within a nanosecond. This property allows the SRD to be used in pulse shaping (sharpening) and pulse generator circuits. The high harmonic content of the signal produced by any repetitive wave-forms from SRD circuits enables them to be used as comb generator where a comb of harmonically related frequencies is generated.

13.4.2.3 Applications

SRDs are primarily used in communication circuits above 1 GHz. It finds applications in microwave radio frequency (RF) electronics as very short pulse generator or parametric amplifier, ultra-fast waveform generator, comb generator and high-order frequency multiplier. It is also capable of working at moderate power levels. This gives it a distinct advantage over some other available RF technologies.

13.4.3 PIN Diode

PIN diode operates as variable resistor at RF frequencies. It may be referred to as voltage-controlled resistor with low noise. Its resistance value is deter-mined only by the forward bias dc current. It is able to control large RF sig-nals with much smaller level of dc excitation.

13.4.3.1 Construction

PIN diode consists of a heavily doped p-type semiconducting material (P^+) and a heavily doped N-type semiconducting material (N^+). Between these two there is a thicker layer of intrinsic (I) semiconductor which has little or no dop-ing and hence represents extremely high resistivity. The same semiconductor

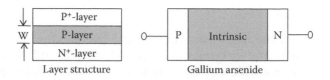

FIGURE 13.6
Layer structure of PIN diode.

material is used for all the three areas. Silicon is the most commonly used material in PIN diodes. It has better power-handling capability and provides a highly resistive intrinsic region. For improving the speed of response GaAs is an alternative. The intrinsic semiconductor increases the area of depletion region which is useful for switching applications as well as for use in photodiodes, among others. The layer structure of PIN diode is shown in Figure 13.6.

The semiconductor sandwich of p-type, intrinsic and n-type materials gives this diode its name. Thus, it may be written as P-type—Intrinsic—N-type. As no practical semiconductor is absolutely free from impurities a PIN diode may be considered as a semiconductor diode which consists of two heavily doped P and N regions separated by a substantially higher resistivity P or N region. This leads to two types of PIN diode structures. The first structure (referred to as *π-type*) contains heavily doped P and N regions separated by an unusually lightly doped P-type intrinsic layer. In this case, semiconductor junction occurs at N⁺ interface. Second structure (referred to as *V-type*) contains heavily doped P and N regions separated by a lightly doped N-type intrinsic layer. In this case, semiconductor junction occurs at P⁺ interface. The structure, impurity distribution and field distribution of a PIN diode are shown in Figure 13.7.

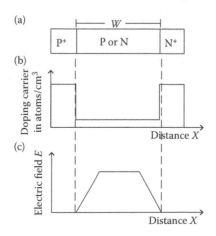

FIGURE 13.7
Doping profile and field distribution in PIN diode. (a) Structure, (b) doping profile and (c) electric field distribution.

13.4.3.2 Operation

PIN diode can operate in forward and reverse bias modes. When it is forward biased, holes and electrons are injected from P and N region into I-region. These charges do not immediately combine. A finite quantity of charges always remains stored resulting in lower resistivity of I-region. This quantity of charges depends on recombination/carrier-life time and forward bias current. A forward bias PIN diode is a current-controlled resistor which is useful in low distortion attenuator and amplitude modulation applications. Reverse bias equivalent circuit consist of PIN diode capacitance (C_T), a shunt loss element (R_s) and a parasitic inductance (L_s). The reverse biased PIN diode is easier to impedance matching.

Pin diode acts as an ordinary diode up to about 100 MHz, but beyond this frequency its operational characteristics change. The large intrinsic region increases the transit time of electrons crossing the region and electrons begin to accumulate in the intrinsic region. This carrier storage in intrinsic region causes the diode to stop acting as a rectifier and it begins to act as a variable resistance. The equivalent circuit of a pin diode at microwave frequencies is shown in Figure 13.8a and a resistance versus voltage characteristic curve is shown in Figure 13.8b.

In view of Figure 13.8b the microwave resistance of a pin diode changes with the variation of its bias voltage. It reduces from 6 kΩ under negative bias to about 5 Ω under positive bias. Thus, when a diode is mounted across a transmission line or waveguide, the loading effect is insignificant while the diode is reverse biased, and the diode presents no interference to power flow. When the diode is forward biased, the resistance drops to approximately 5 Ω and most power is reflected. Thus, the diode acts as a switch when mounted in parallel with a transmission line or waveguide. Several diodes in parallel can switch peak power in excess of 150 kW. The upper power limit is determined by the ability of diode to dissipate power. The upper frequency limit is determined by the shunt capacitance of the pn junction, shown in Figure 13.8a. PIN diodes with upper limit frequencies in excess of 30 GHz are available. Performance of PIN diode primarily depends on the chip geometry and nature of semiconductor material particularly in the I–region. Its characteristics may be controlled

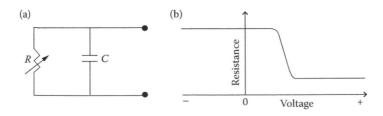

FIGURE 13.8
Equivalent circuit and characteristic of a pin diode. (a) Equivalent circuit and (b) resistance versus voltage curve.

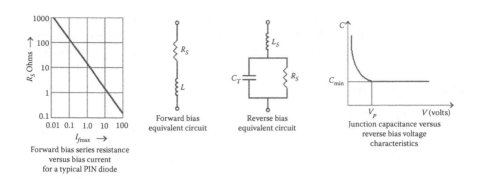

FIGURE 13.9
Characteristics and equivalent circuits of PIN diode under different biasing.

by thickness of I-region. Figure 13.9 shows the forward bias (FB) and reverse bias (RB) characteristics and corresponding equivalent circuits.

13.4.3.3 Advantages

A forward bias PIN diode is useful in low distortion attenuator and amplitude modulation applications. The reverse-biased PIN diode is easier to impedance matching. A PIN diode can have long carrier life time and very high resistivity. These factors enhance its ability to control RF signal with minimum distortion and low dc power requirement.

13.4.3.4 Applications

PIN diodes can be used as variable attenuator, as switch and as microwave phase shifter. It can be used in mixers, detectors and power control devices. It can be used over a frequency range of 100 MHz to 100 GHz.

13.4.4 Tunnel Diode

Tunnel diode is a semiconductor p–n junction diode. It follows a tunnelling mechanism.

13.4.4.1 Construction

It is constructed either from Ge or GaAs. In both cases p and n regions are heavily doped with impurity concentration of 10^{19}–10^{20} atoms/cm^3. At the junction the depletion layer barrier is very thin which makes possible for majority carriers to cross the potential barrier through tunnelling. Since it is a majority carrier effect, the operation of tunnel diode is very fast. The time of tunnelling is governed by quantum transition probability/unit volume and not by classical transit time concept (i.e., transit time = barrier width/carrier velocity). If the barrier thickness is less than 3 Å the probability that a particle will tunnel through the potential barrier even though it does not possess enough kinetic energy to pass over the same barrier, is quite appreciable.

13.4.4.2 Operation

Figure 13.10 illustrates various bands and corresponding energy levels of P and N semiconductors. These levels are termed as E_c (conduction band), E_v (valence band), E_g (energy gap) and E_F (Fermi level) on both the sides. It also shows empty state on the P-side and filled state on the N-side. The lowest point of empty state and upper-most point of filled state are assumed to be aligned.

13.4.4.2.1 Open-Circuit Condition

Tunnelling will take place if the filled state at one side and empty state on another side are at same energy level. Since filled state of N-side and empty state of P-side are not at the same level, there will be no tunnelling and hence no flow of current in either direction across the junction. In case of ordinary diode, Fermi level is in forbidden band. In tunnel diode, both P and N sides are heavily dropped and so at P-side Fermi level is in valence band (VB) and at N-side Fermi level is in conduction band (CB). This condition is also illustrated in Figure 13.10.

13.4.4.2.2 Forward Bias Condition

This is illustrated in Figure 13.11 for different values of forward voltages. These voltages corresponding to peak, valley and Fermi level currents are named as V_p, V_v and V_F, respectively. Figure 13.11a shows the same state as that of Figure 13.10 since $V = 0$. When forward bias is applied, the energy level of N-side is increased (Figure 13.11b). Some of the carriers present in filled band of N-side come at the same level as that of empty state of P-level.

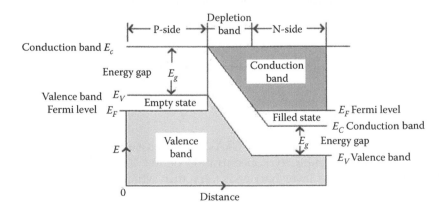

FIGURE 13.10
Tunnel diode under open circuit or zero-bias condition.

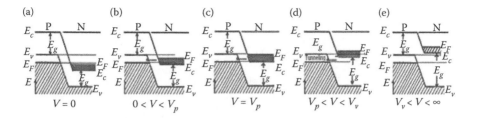

FIGURE 13.11

Tunnel diode under following forward bias conditions: (a) $V = 0$, (b) $0 < V < V_p$, (c) $V = V_p$, (d) $V_p < V < V_v$ and (e) $V_v < V < \infty$.

Electrons start tunnelling from N- to P-side resulting in flow of current. If forward voltage is further increased (Figure 13.11c), the filled state (FS) and empty state (ES) appear at the same level. Maximum electrons will tunnel through barrier from N to P side. At this juncture at a particular voltage V_p peak current I_p is obtained. If forward bias is increased beyond V_p (Figure 13.11d), energy level of N-side goes above that of P-side. Numbers of tunnelling electrons start reducing and current starts decreasing. Still more forward biasing will bring the two sides totally at different levels (Figure 13.11e). It will result in no tunnelling and hence no current or very low current flow. If forward voltage is increased beyond the valley voltage, ordinary injection current starts flowing and increases exponentially. The ratio of I_p to I_v is different in different materials. In GaAs this ratio is 15 in Ge 8 and in Si it is only 3. Also GaAs offers maximum voltage swing between V_v and V_p of 1 V while in Ge it is only 0.45 V.

As illustrated in Figure 13.12 the total current flowing through the tunnel diode is the sum of tunnel current and injection current.

13.4.4.3 V–I Characteristics

Figure 13.13 illustrates $V–I$ characteristics of a tunnel diode. The (combined action) curve is superimposed by a load line which intersects the characteristic curve at points a, b and c. If voltage and current vary about b, the final values

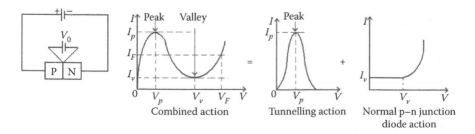

FIGURE 13.12

Current flow through tunnel diode.

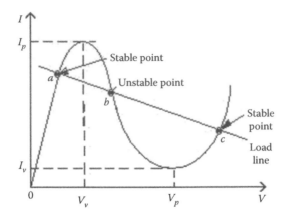

FIGURE 13.13
V–I characteristics of a tunnel diode.

of V and I will be given by points a and c and not by point b. Thus, points a and c are called the stable state and b the unstable state. Since tunnel diode has two stable states for this load line the circuit is called bistable and it can be used as a binary device in switching circuits. The circuit with load line crossing point b in negative resistance (NR) region is called astable. Another load line crossing point a in positive resistance region indicates monostable circuit. If R_n is the magnitude of negative resistance the negative conductance is given as

$$-g = \left.\frac{\partial I}{\partial V}\right|_{V_p} = -\frac{1}{R_n} \tag{13.1}$$

13.4.4.4 Advantages

Tunnel diode exhibits high speed, requires low voltage for operation and generates low noise. Its peak current-to-valley current ratio is high, cost is low and weight is light. It is used in forward-bias state and exhibits negative resistance (NR).

13.4.4.5 Applications

These are used in microwave mixers, detectors, low-noise amplifiers and oscillators. It is useful in microwave oscillators and amplifiers since it exhibits negative resistance characteristics in the region between peak and valley currents. It can also be used as binary memory.

13.4.5 Schottky Diode

The Schottky barrier diode is a variation of point-contact diode in which the metal semiconductor junction is a surface rather than a point contact.

In view of the large contact area, or barrier, between the metal and the semiconductor these have lower forward resistance, lower capacitance due to metal–semiconductor contact and lower noise generation. The Schottky diode or Schottky barrier diode is sometimes also referred to as *surface barrier diode*. These are also called the *hot carrier diode* or *hot electron diode* because the electrons flowing from the semiconductor to the metal have a higher energy level than the electrons in the metal. The effect is the same as it would be if the metals were heated to a higher temperature than normal.

13.4.5.1 Construction

The Schottky barrier diode can be manufactured in a variety of forms. The simplest form is the point-contact diode (Figure 13.14a) wherein a metal wire is pressed against a clean semiconductor surface. Vacuum-deposited Schottky diode structure (Figure 13.14b) is another form in which the metal is vacuum deposited. Figure 13.14c shows its structure and Figure 13.14d its symbol.

13.4.5.2 Fabrication

The most critical element in its fabrication is to ensure a clean surface for an intimate contact of metal with the semiconductor surface. This requirement is achieved chemically. The metal is normally deposited in a vacuum either by evaporation or by sputtering process. In some cases chemical deposition is also employed. In certain other cases instead of a pure metal contact silicides are used. This is normally done by depositing the metal and then heat treating to give the silicide. This process is advantageous since the reaction uses the surface silicon, and the actual junction propagates below the surface, where the silicon is not exposed to any contaminants. Schottky structure is fabricated by using relatively low-temperature techniques. It generally does not require high-temperature steps needed in impurity diffusion.

Breakdown effect around the edge of the metalised area is one of the problems with the simple deposited metal diode. This arises from the presence of high electric fields around the edge of the plate. Leakage effect is yet another problem. To overcome these problems a guard ring of P⁺ semiconductor

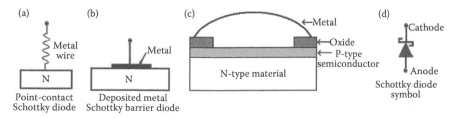

FIGURE 13.14
Schottky diode. (a) Point-contact type, (b) deposited metal type, (c) structure of (b) and (d) symbol.

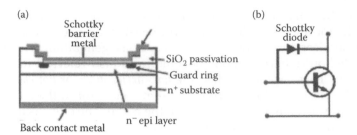

FIGURE 13.15
Schottky diode. (a) Structure with guard ring and (b) NPN transistor with Schottky diode clamp.

along with an oxide layer around the edge is used. This guard ring is fabricated by using diffusion process. In some cases metallic silicides may also be used in place of the metal. The guard ring in this form of Schottky diode structure operates by driving this region into avalanche breakdown before the Schottky junction is damaged by large levels of reverse current flow during transient events. The structure with guard rings is shown in Figure 13.15a. This form of structure is used in many forms of Schottky. It is however particularly suitable in rectifier diodes where the voltages may be high and breakdown problem may be more.

13.4.5.3 Characteristics

Schottky diode is called a majority carrier device. It does not rely on holes or electrons recombination when they enter the opposite type of regions as in a conventional diode. This gives it tremendous advantages in terms of speed. By making the devices small the normal RC time constant can be reduced which makes them faster than the conventional PN diodes. This factor makes them suitable for RF applications. This diode also has a much higher current density than an ordinary PN junction resulting in lower forward voltage drops. This aspect makes it ideal for use in power rectification applications. Its main drawback is its relatively high reverse current. In some more exacting applications this aspect needs proper monitoring.

13.4.5.4 Advantages

The advantages of Schottky diodes include (i) lower forward resistance, (ii) lower noise generation, (iii) low turn on voltage, (iv) fast recovery time and (v) low junction capacitance.

13.4.5.5 Limitations

These have a high reverse leakage current. This creates problems with sensing circuits. Leakage paths into high impedance circuits give rise to false readings. This aspect needs to be addressed in the circuit design.

13.4.5.6 Applications

The applications of the Schottky barrier diode are the same as those of the point-contact diode. Besides high switching speed and high-frequency capability its low noise level makes it especially suitable for microwave receiver detectors and mixers. These can also be used in high-power applications, as rectifiers due to its high current density and low forward voltage drop leading to low power loss compared to ordinary PN junction diodes. Its other applications include (i) power OR circuits where a load is driven by two separate power supplies, (ii) solar cell systems since solar cells do not like the reverse charge applied and require low-voltage drop diodes and (iii) clamp diode in a transistor circuits to speed-up the operation when used as a switch. Figure 13.15b illustrates a Schottky diode which is inserted between the collector and base of the driver transistor to act as a clamp.

13.4.6 Varactor or Varicap Diode

The term varactor is derived from the words variable reactor. Its reactance is varied in a controlled manner with a bias voltage. It is a two-terminal device which acts as a variable capacitor.

In varactor diode, the capacitance is varied by changing its (reverse) bias. This bias varies the width of the depletion region and variation of this width changes the capacitance. Thus, the varactor or varicap diode acts as a capacitor with the depletion region being the insulating dielectric and the capacitor plates formed by the extent of the conduction regions.

Junction capacitance exists in all reverse biased diodes because of the depletion region. In a varactor diode this junction capacitance is optimised and used for many RF and switching applications. Varactor diodes are often used for electronic tuning applications in FM radios and televisions. These are often called voltage-variable capacitor diodes and used as voltage-variable capacitor at lower frequencies. By taking the advantage of its nonlinear *V–I* characteristics these are also used as frequency multiplier at microwave frequencies.

13.4.6.1 Operation

When reverse bias is applied a depletion layer is formed which acts as a dielectric between two materials. Thus, the junction acts as a capacitor (C). If reverse bias is increased, width of depletion layer increases and C decreases. Thus, junction capacitance can be varied by varying reverse bias voltage. A reverse biased p–n junction is shown in Figure 13.16 wherein C varies in non-linear fashion which can easily follow MW signal. This C depends on applied reverse bias voltage and the junction design. This junction C is denoted by C_j and is proportional to the reverse bias voltage {i.e., $C_j \propto (V_{rev})^{-n}$}. In this relation parameter n decides the type of junction.

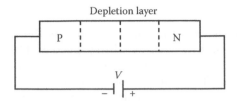

FIGURE 13.16
Reverse biased p–n junction.

In some cases applied reverse bias voltage is kept constant to give some fixed value of junction capacitance C_0. Sensitivity (S) is another important parameters of this diode. It is related to its doping profile m. Junctions of varactor diode are assigned different names in accordance with the values of m. These are given below.

$$S = \frac{1}{m + 2} \tag{13.2}$$

For $m = 0$, $S = 1/2$ it is called one-sided abrupt junction.

For $m = 1$, $S = 1/3$ it is called one-sided linearly graded junction

For $m < 0$ (say $m = -1$) $S = 1$ it is called hyper-abrupt junction and has the highest sensitivity.

If the value of C is larger, large capacitance variation is obtained with the applied voltage.

13.4.6.2 Characteristics

Figure 13.17 shows the symbol of varactor diode and junction capacitance Vs voltage characteristics. The graph of depletion layer capacitance Vs reverse bias voltage is shown in Figure 13.18a) with log scale on both the axis. Its equivalent circuit is shown in Figure 13.18b.

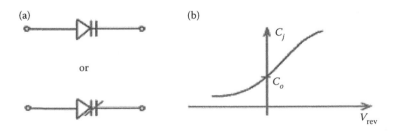

FIGURE 13.17
(a) Symbol of varactor diode and (b) the junction capacitance versus voltage characteristics.

(a)   (b)

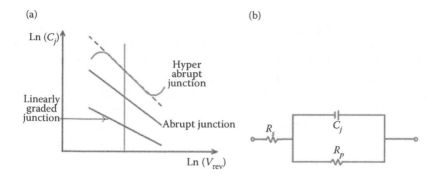

FIGURE 13.18
Varactor diode. (a) Capacitance–voltage relationship and (b) equivalent circuit.

In these figures C_j is the junction capacitance, R_s is series resistance and R_p is the parallel resistance. This R_p is the equivalent resistance of generation-recombination current, diffusion current and surface leakage current. With the increase in reverse bias voltage both C_j and R_s decrease but R_p increases. The efficiency of a varactor diode is expressed in terms of its quality factor Q. Its maximum value of Q is given by

$$Q_{max} = \left(\frac{R_p}{4R_s} \right)^{1/2} \tag{13.3}$$

13.4.6.3 Applications

Varactor diodes are used in harmonic generation, voltage variable tuning, mixing, detection and modulation of microwave signals, active filters and parametric amplifiers. Varactor is an active element in parametric amplifiers.

13.5 Transferred Electron Devices

There is a class of microwave devices which operates on the principle of transfer of electrons between different energy levels. This class of devices is referred to as transferred electron devices (TEDs). These include Gunn diode, limited space-charge accumulation (LSA) diode, indium phosphide (InP) diode and cadmium telluride (CdTe) diode. Since all these devices operate on the principle of transfer of electrons it is essential to understand the process of transfer in accordance with the band structures of the materials and formation of field domains therein.

Most of the semiconductors contain two major bands termed as valence band (VB) and conduction band (CB). The valence band is the lower-most and conduction band is the upper-most band. Conduction band contains free electrons which contribute towards the flow of current, whereas the valence band contains bound electrons which take no part in conduction process. The band gap between valence and conduction bands is called forbidden band or forbidden energy gap. In some semiconductors conduction bands contain high and low mobility electrons at different energy levels. In TEDs electrons transfer between these energy levels, that is, conduction electrons shift from high-to-low mobility state under the influence of strong electric field.

13.5.1 Valley Band Structure

The conduction band may be divided into a number of energy levels which are referred to as valleys. The conduction band with lower energy level is termed as lower (or central) valley, whereas the conduction bands with higher energy levels are termed as upper (or satellite) valleys. The number of satellite valleys is more as compared to the lower (or central) valleys.

Figure 13.19a illustrates the two-valley structure of GaAs. In this the central valley is above 1.43 eV from valence band. There are 6-secondary or satellite valleys above this level. The energy level of last satellite valley is 1.79 eV higher from the valence band. Similarly, InP has three-valley band structure as shown in Figure 13.19b. In this the lower valley (LV) is 1.33 eV from valence band, mid-valley (MV) is 0.6 eV from LV and upper valley (UV) is 0.8 eV from LV. The figure also illustrates that there is weak coupling between MV and LV and also between UV and MV but there is strong coupling between UV and LV.

Properties which characterise electrons in each band include the density of electrons (n) effective mass (m) and the total mobility (μ). Electrons present in the lowest valley (n_ℓ) have small effective mass (m_l) and higher mobility (μ_ℓ).

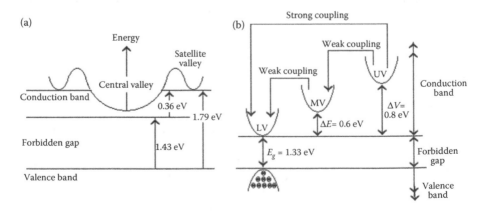

FIGURE 13.19
Band (or valley) structures. (a) Two valleys for GaAs and (b) three valleys for InP.

Electrons present in the upper valley (n_u) have higher effective mass (m_u) and lower mobility (μ_u). Under the influence of electric field the drift velocity (v_d) of electrons (with charge e) is given by

$$v_d = \mu E \qquad (13.4)$$

And the current density is given by

$$J = nev_d = ne\mu E \qquad (13.5)$$

To understand the transfer of electrons, consider the two-valley model of GaAs semiconductor. At room temperature when applied E-field is low (i.e., $E < E_\ell$) almost all electrons are present in the lower valley, that is, at the bottom level of the conduction band. When E is increased (i.e., $E_\ell < E < E_u$) photons are generated. These photons strike the electrons present in the valley. This collision results in transfer of energy from photon to electrons. As a result, kinetic energy of electrons increases. Electrons start shifting in upward direction into one of the secondary valleys. This shift is referred to as electron transfer. The mode of electron transfer between the two valleys under different field conditions is shown in Figure 13.20. In this figure, E is the applied field, E_ℓ is the lower valley field and E_u is the upper valley field.

If E is further increased electrons start moving towards satellite valley. The effective mass of electrons at higher valley is almost eight times more than that in the lower valley. With further increase (i.e., $E_u < E$) energy of electrons increases but their velocity decreases. Thus, current reduces in this range of increasing E. This gives negative differential coefficient ($-dI/dE$). Thus, whenever there is an increase in energy, electrons shift towards the higher valley. The mobility of electrons gradually starts reducing. There is no sudden change in the mobility. Almost similar phenomenon of transfer is observed in InP semiconductor. However in this case the maximum drift velocity is reached for relatively higher E. Figure 13.21 illustrates the variation of drift velocities for GaAs and InP semiconductors with electric field intensity.

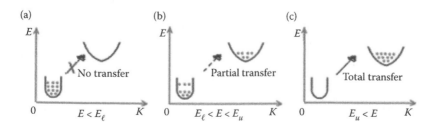

FIGURE 13.20
Electron transfer between two valleys when (a) $E < E_\ell$, (b) $E_\ell < E < E_u$ and (c) $E_u < E$.

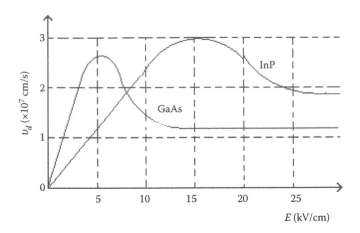

FIGURE 13.21
Electron drift velocity versus *E*-field curve.

Table 13.2 shows data for gap energy at 300 K, separation energy between two valleys, threshold field and peak velocity for different semiconductors having two and three valley structures, whereas Table 13.3 includes the data effective mass and mobility for GaAs.

13.5.1.1 Conductivity and Current Density

Figure 13.22 illustrates the division of drift velocity (v_d) versus *E* curve into Ohmic and (differential) negative resistance (NR) regions. In Ohmic region electrons are present in the central valley. With the increase in *E*, v_d also increases and can be given by

$$v_d = \mu_l E \tag{13.6}$$

TABLE 13.2

Data for Two-Valley Semiconductors

Semiconductor	Gap Energy at 300 K E_g (eV)	Separation Energy between Two Valleys ΔE (eV)	Threshold Field E_{th} (kV/cm)	Peak Velocity v_p (10^7 cm/s)
Ge	0.80	0.18	2.3	1.4
Ga As	1.43	0.36	3.2	2.2
In P (3-Valley SC)	1.33	0.60, 0.80	10.5	2.5
Cd Te	1.44	0.51	13.0	1.5
In As	0.33	1.28	1.6	3.6
In Sb	0.16	0.41	0.60	5.0

TABLE 13.3

Data for Two-Valley GaAs

Valley	Effective Mass M_e	Mobility μ
Lower	$M_{el} = 0.068$	$\mu_l = 8000$ cm²/V s
Upper	$M_{eu} = 1.2$	$\mu_u = 180$ cm²/V s

Increase in v_d in this region is upto a certain limit of E-field which is referred to as threshold electric field (E_{th}). When E is increased beyond E_{th}, v_d decreases. The region in which v_d decreases with increasing E, is referred to as differential negative resistance region. At this stage,

$$v_d = \mu'Ev_d \tag{13.7}$$

where μ' is the mean mobility and is given by

$$\mu' = \frac{n_l\mu_l + n_u\mu_u}{n_l + n_u} \tag{13.8}$$

The current density is given as

$$J = n\mu'eE = (n_l + n_u)\mu'eE = \sigma E \tag{13.9}$$

where $n = (n_l + n_u)$. The conductivity (σ) is given by

$$\sigma = e(n_l\mu_l + n_u\mu_u) \tag{13.10}$$

The time required by electrons to transfer from one valley to another is called *inter-valley transfer time* and is of the order of picoseconds. The existence of drop in mobility with an increase in E creates the negative conduction.

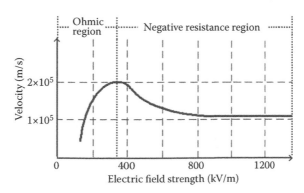

FIGURE 13.22
Division of drift velocity curve for GaAs.

This mechanism discovered by *Riddley, Watking* and *Hilsum* is named after them. This effect called *negative differential conductivity effect* depends on bulk properties of SC and not on the junction.

13.5.1.2 Ridley–Watkins–Helsum Theory

It revolves around the concept of differential negative resistance (NR) developed in bulk solid state, in III–V group compound, when a voltage (or current) is applied to the terminals of a sample. The negative resistance has two modes called as (i) voltage-controlled negative resistance (VCNR) mode in which the current density can be multi-valued and (ii) current-controlled negative resistance (CCNR) mode wherein the voltage can be multi-valued.

The appearance of a differential NR region in current density curve renders the sample electrically unstable. As a result, initially homogeneous sample becomes electrically heterogeneous in an attempt to reach stability. In VCNR mode high field domains are formed separating two low field regions. The interface separating low and high field domains lie along equipotentials, thus they are planes perpendicular to the current directions. In CCNR mode, sample splitting results in high current filaments running along the direction of field. The negative resistance of the sample at a particular region can be expressed as

$$\frac{dI}{dV} = \frac{dJ}{dE} \tag{13.11}$$

The phenomena relating to VCNR mode and CCNR mode are shown in Figure 13.23.

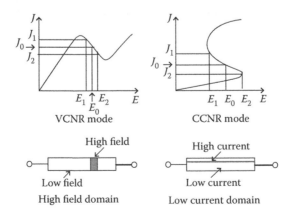

FIGURE 13.23
VCNR and CCNR modes.

13.5.1.3 *Differential Negative Resistance*

In the expression of the conductivity (σ) of n-type GaAs given by Equation 13.10 'e' is the electron charge, n_l and n_u are the electron densities, μ_l and μ_u are the electron mobilities in the lower and upper valleys, respectively. When applied field is sufficiently large electrons get accelerated and their effective temperature rises above the lattice temperature. Thus, electron density n ($= n_l + n_u$) and mobility μ are both functions of E.

Differentiation of Equation 13.10 [$\sigma = e\,(n_l\,\mu_l + n_u\,\mu_u)$] with respect to E gives

$$\frac{d\sigma}{dE} = e\left\{\left(\mu_l\frac{dn_l}{dE} + \mu_u\frac{dn_u}{dE}\right) + \left(n_l\frac{d\mu_l}{dE} + n_u\frac{d\mu_u}{dE}\right)\right\} \tag{13.12}$$

If μ_l and μ_u are proportional to E^p where p is a constant

$$\frac{d(n_l + n_u)}{dE} = \frac{dn}{dE} = 0 \tag{13.13}$$

$$\frac{dn_l}{dE} = -\frac{dn_u}{dE} \tag{13.14}$$

and

$$\frac{d\mu}{dE} \propto \frac{dE^p}{dE} = pE^{p-1} = p\frac{E^p}{E} \propto p\frac{\mu}{E} = \mu\frac{p}{E} \tag{13.15}$$

Substitution of Equations 13.13 through 13.15 into Equation 13.12 results in

$$\frac{dn}{dE} = e(\mu_l - \mu_n)\frac{dn_l}{dE} + e(n_l\mu_l + n_u\mu_u)\frac{p}{E} \tag{13.16}$$

Differentiation of ohms law ($J = \sigma E$) with respect to E yields:

$$\frac{dJ}{dE} = \sigma + \frac{d\sigma}{dE}E \tag{13.17}$$

Or

$$\frac{1}{\sigma}\frac{dJ}{dE} = 1 + \frac{d\sigma}{dE}\frac{\sigma}{E} \tag{13.18}$$

The condition for negative resistance is

$$\frac{d\sigma}{dE}\frac{\sigma}{E} > 1 \tag{13.19}$$

Substitution of Equation 13.10 into Equation 13.16 with $f = n_u/n_l$ yields

$$\frac{\mu_l - \mu_u}{\mu_l + \mu_u f}\left(-\frac{E}{n_l} \cdot \frac{dn_l}{dE} - p\right) > 1 \qquad (13.20)$$

From the above equations the following conclusions can be drawn.

1. The exponent p is a function of scattering mechanism and should be negative and large. This factor makes impurity quite undesirable.
2. When it is dominant, positive mobility rises with increasing E. When lattice scattering is dominant, p is negative and must depend on the lattice and carrier temperature.

To satisfy Equation 13.19 the un-bracketed term in Equation 13.20 must be positive or $\mu_l > \mu_u$, electrons must begin in a low mass valley and transfer to high mass valley when they get heated by E-field. The maximum value of this term is unity that is for $\mu_l \gg \mu_u$. The factor dn_l/dE is the rate of change of the carrier density with field at which electrons transfer to the upper valley. This is a function of $(n_u - n_l)$, electron temperature and energies of the two valleys. Thus, to exhibit negative resistance, the band structure of a semiconductor must satisfy the following criterion:

1. $\Delta E > KT$ or $\Delta E > 0.36$ eV at room temperature. Here, ΔE is the separation energy between the lower valley and the bottom of upper valley.
2. This requires that $\Delta E < E_g$, otherwise the semiconductor will breakdown and will become highly conducting before the electrons begin to transfer from lower-to-upper valley because hole–electron pair formation shall begin.
3. The electrons in the lower valley must have high mobility, small effective mass and low density of state, whereas those in the upper valley must have low mobility, large effective mass and high density of state.

Neither Si and Ge nor InAs, GaP, InSb satisfy these conditions. This criterion is satisfied by semiconductor compounds such as GaAs, CdTe and InP. Figure 13.24 illustrates the positive E versus J characteristics of a two-valley semiconductor.

In view of Equations 13.7 and 13.9 {with $(n_l + n_u)$ replaced by n and v_d by v} the current density can be written as

$$J = e \cdot n \cdot v \qquad (13.21)$$

Thus,

$$\frac{dJ}{dE} = en\frac{dv}{dE} \qquad (13.22)$$

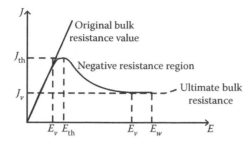

FIGURE 13.24
Positive E versus J characteristics.

The condition for negative resistance may be written as

$$\frac{dv_d}{dE} = \mu n < 0 \tag{13.23}$$

where v is the average velocity and v_d is the drift velocity of electrons.

13.5.2 High Field Domain

In connection with the RWH theory it was noted that in VCNR mode high field domains are formed. To understand the formation of such domains consider a sample of n-type of material (say GaAs wherein electrons are the majority carriers) across which an E-field (or voltage) is applied. As long as the applied voltage is small both the electric field (E) and current density (J) remain uniform throughout and GaAs remains Ohmic since the drift velocity (v_d) is proportional to E ($J = \sigma E_x$ $a_x = \sigma(V/L)$ $a_x = \rho$ v_x $a_x = \rho U$). When the applied voltage (V) exceeds the threshold voltage (V_{th}) a high field domain is formed near the cathode (Figure 13.25a–c). As a result, E reduces in the rest of the material and current (I) drops to about two-third of its maximum value (Figure 13.25e). This high-field domain drifts with the stream of carriers and disappears at the anode. When E (or V) further increases the drift velocity increases and GaAs exhibits negative resistance (Figure 13.25e). As illustrated in Figure 13.25d if the diode is biased at E_A electrons entering the location A are more than those leaving A. This results in the increase of excess negative space charge at A. The field on the left side of A decreases and on its right side increases (Figure 13.25c) and results in greater space charge. This accumulation of negative charges is caused either by random noise fluctuations or possibly by non-uniformity in the doping. The process continues until both low and high fields attain values outside the differential NR region and settle at 1 and 2 (Figure 13.25e) where currents in two regions are equal. As a result a travelling space-charge accumulation is formed. The process continues as long as enough charges to form the space charge are available.

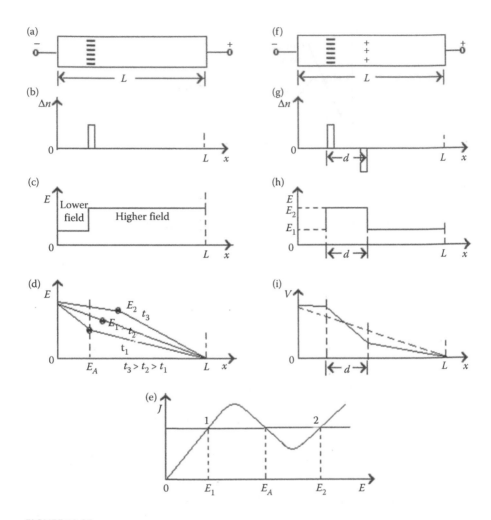

FIGURE 13.25
a) Formation of high field domain near cathode for $V > V_{th}$, (b) differential electron density, (c) E-field distribution, (d) variation of E for different time instants, (e) E vs J variation, GaAs exhibit NR for $V \gg V_{th}$, (f) drift of high field domain, (g) new differential electron density, (h) E-field distribution and (i) voltage variation.

When positive and negative charges are separated by a small distance d a dipole domain is formed (Figure 13.25f and g). The E field inside the dipole domain becomes greater than on either side of the dipole (Figure 13.25h). The current in the low field side becomes greater than that in the high field side due to differential NR. The two field values tend towards equilibrium conditions outside the differential NR region where the low and high currents are the same. The dipole field reaches a stable state and moves through the specimen towards anode. When high field domain disappears at the anode formation of a new field starts at the cathode and the process continues.

13.5.2.1 Properties of High Field Domain

It is to be noted that the high field domain follows certain set norms in its formation and extinction. These norms referred to as properties of high field domain are as below.

1. A domain starts to form near the cathode whenever in the region of a sample the applied E-field exceeds threshold value (i.e., $E > E_{th}$). This domain drifts with carrier stream through the device towards anode. When the increase of E results in decrease of drift velocity (v_d) the (GaAs) sample exhibits negative resistance.

2. If a sample (or device) already contains domains then any additionally applied E-field results in increase of the domain size due to absorption of more energy from the field. This results in the decrease of current.

3. A domain, in a sample with uniform doping area, does not disappear before reaching the anode end unless voltage is reduced appreciably below the threshold value.

4. In a non-resonant circuit the formation of a new domain can be prevented by decreasing the voltage slightly below the threshold voltage.

5. A domain, as it passes through the regions of different doping and cross-sectional area, either modulates the current through a device, or disappears. The effective doping may be varied in the regions along the drift path by additional contacts.

6. The domain's length L is generally inversely proportional to the doping level, thus the devices with the same product of doping and length ($n \times L$) behaves similarly in terms of product of frequency and length ($f \times L$), the ratio of voltage and length (V/L) and the efficiency.

7. As a domain crosses a point in the device there is sudden voltage change. This change can be detected by a capacitive contact. The presence of domain anywhere in a device can be detected by the decrease in current or by change in the differential impedance.

8. The properties listed in 6 and 7 are valid only when L is much greater than the thermal diffusion length for the carriers. This length for GaAs is about 1 μm for doping of 10^{16} per cubic-cm and about 10 μm for a doping of 10^{14} per cubic-cm.

13.5.3 Modes of Operation

In view of the above properties a number of modes of operation are possible which can be defined in view of the relative values of products $n_0 \times L$, $f \times L$ and some other related parameters. In these products, f is the frequency of operation, L is the length of the specimen and n_0 is the doping concentration. The defined modes of operation are as under.

13.5.3.1 Gunn Diode Mode

It is defined where $f \cdot L$ is about 10^7 cm/s and $n_0 \cdot L > 10^{12}/\text{cm}^2$. The device is unstable because of cyclic formation of either accumulation layer or high field domain. In a circuit with relatively low impedance the device operates in high field domain mode and frequency of oscillation is near to intrinsic frequency (f_i). When device is operated in relatively high Q and properly coupled to the load, the domain is quenched or delayed (or both) before nucleating. In this case, the oscillating frequency is almost entirely determined by the resonant frequency (f_r) of the cavity and has a frequency several times than intrinsic frequency (f_i).

If the transit time (T_t), oscillation period (T_0), drift velocity (v_d) and sustaining velocity (v_S) known as the Gunn mode is again divided into the following sub-modes.

13.5.3.1.1 Transit-Time Mode

In this mode $v_d = v_S$, high-frequency domain is stable, $v_d = 10^7$ cm/s, $T_0 = T_t$, current is collected only when the domain arrives at the anode. The efficiency of this mode is about 10%.

13.5.3.1.2 Delayed Mode

This mode is also called inhibited mode. The domain is collected when $E < E_{th}$. A new domain cannot form until $E > E_{th}$ again. In this case, $T_0 > T_t$, $10^6 < f \cdot L < 10^7$ cm/s and efficiency is 20%.

13.5.3.1.3 Quenched Mode

For this mode, $f \cdot L > 2 \times 10^7$ cm/s and $T_0 = T_t$. If bias field drops below E_s during negative half-cycle, domain collapses before reaching the anode. If bias field swings again above threshold voltage (E_{th}) new domain forms and the process restarts. The efficiency of this mode can be upto 13%.

13.5.3.2 Stable Amplification Mode

It is defined where $T_0 < T_t$, $T_0 = 3T_d$ (where T_d is the dielectric relaxation time), $fL \approx 10^7$ cm/s, $n_0 \cdot L$ is between 10^{11} and $10^{12}/\text{cm}^2$. This mode is not used in microwave range.

13.5.3.3 LSA Mode

It is defined where $f \cdot L > 10^7$ cm/s (generally $> 2 \times 10^7$), and n_0/f is between 2×10^4 and 2×10^5.

13.5.3.4 Bias-Circuit Oscillation Mode

It occurs only when there is either Gunn or LSA oscillation and is usually at the region where $f \cdot L$ is too small. When a bulk diode is biased to threshold,

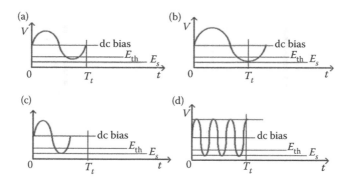

FIGURE 13.26
Wave shapes for different modes. (a) Transit time mode, (b) delayed mode, (c) quenched mode and (d) LSA mode.

average current suddenly drops as Gunn oscillations begin. This drop can lead to oscillations in bias circuit typically at 1 kHz to 100 MHz.

Figure 13.26 illustrates Gunn modes given at a, b, c and the LSA mode in d. Figure 13.27 shows different modes of operation for Gunn diode listed above.

13.5.4 Criterion for Mode's Classification

At early stage of space-charge accumulation the time rate of growth of space-charge layers is

$$Q(x,t) = Q\{(x - v \cdot t), 0\}\exp(t/\tau_d) \tag{13.24}$$

where τ_d {$= \varepsilon/\sigma = \varepsilon/(en_0\mu_n)$} is the magnitude of the negative dielectric relaxation time. In this relation, e is the electron charge, n_0 is the doping

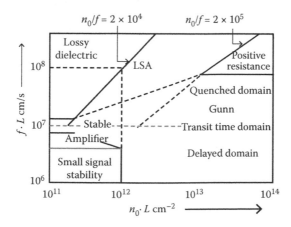

FIGURE 13.27
Different modes of operation for Gunn diode.

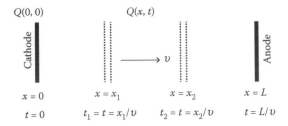

FIGURE 13.28
Drifted layers along with their timings and locations.

concentration and μ_n is the negative mobility. Equation 13.24 is valid throughout the transit time in the entire space-charge layer.

The factor for maximum growth of space-charge accumulation (called growth factor) is given by

$$\frac{Q(L, L/v)}{Q(0,0)} = \exp\left(\frac{L}{v\tau_d}\right) = \exp\left(\frac{Ln_0 e \left|\mu_n\right|}{\varepsilon v}\right) \tag{13.25}$$

In Equation 13.25, the drift of layer is assumed to begin at the cathode at $x = 0$, $t = 0$ and arrive at the anode at $x = L$ and $t = L/v$ (Figure 13.28). For large growth of space charge this factor must be greater than unity. This requires

$$n_0 L > \frac{\varepsilon v}{e \left|\mu_n\right|} \tag{13.26}$$

Equation 13.26 spells criterion for classifying the modes of operation for Gunn-effect diodes. For n-type GaAs the value of $(\varepsilon v / e|\mu_n|)$ is about $10^{12}/\text{cm}^2$, where $|\mu_n|$ is assumed to be 150 cm²/Vs. Table 13.4 gives the time relations, doping levels and nature of circuits corresponding to the different modes.

TABLE 13.4

Different Modes and Their Corresponding Data

Mode	Time Relation	Doping Level	Nature of Circuit
Stable amplifier	$T_0 \geq T_t$	$n_0 L < 10^{12}$	Non-resonant
Gunn domain	$T_g \leq T_t$, $T_0 = T_t$	$n_0 L > 10^{12}$	Non-resonant and constant voltage
Quenched domain	$T_g \leq T_t$, $T_0 = T_t$	$n_0 L > 10^{12}$	Resonant and finite impedance
Delayed domain	$T_g \leq T_t$, $T_0 > T_t$	$n_0 L > 10^{12}$	Resonant and finite impedance
LSA	$T_0 \leq T_g$, $T_0 > T_d$	$2 \times 10^4 < n_0/f$ $< 2 \times 10^5$	Multiple resonances, high impedances, high dc bias

The transferred electron devices include Gunn diode, LSA diode, indium phosphide diode and cadmium telluride diode. These are discussed in the following sections.

13.5.5 Gunn Diode

The Gunn oscillator is a source of microwave energy that uses the bulk effect of gallium–arsenide semiconductor. It operates on the principle of transfer of electrons from one energy level to another and thus it is referred to as transferred electron device. The process of electron transfer, high field domain formation and possible modes of operation have already been discussed. The working of Gunn diode and other TEDs can now be described as below.

13.5.5.1 Gunn Effect

The working of Gunn diode is based on Gunn effect discovered by J. B. Gunn in 1963. According to the Gunn's observation above some critical voltage corresponding to the electric field of 2–4 kV/cm current in every specimen becomes a fluctuating function of time. In GaAs specimen these fluctuations are in the form of a periodic oscillation superimposed upon the pulse current. The frequency of these fluctuations is inversely proportional to the transit time of a domain across the semiconductor. The transit time in itself is proportional to the length of semiconductor material, and to some extent the applied voltage. Thus, the frequency of such oscillations is determined mainly by the specimen and not by external circuit. The period of oscillation is usually inversely proportional to the specimen length and closely equals transit time (slightly above 10^7 cm/s) of electrons between the electrodes. According to the Gunn the peak pulse microwave power delivered by GaAs specimen to a matched load was as high as 0.5 W at 1 GHz and 0.15 W at 3 GHz corresponding to 1–2% of the pulse input power. From Gunn's observation carrier drift velocity v_d linearly increases from zero to maximum, when E varies from zero to a threshold value. For E greater than threshold value (3 kV/cm for n-type GaAs), v_d decreases and diode exhibits negative resistance. The variation of drift velocity is shown earlier in Figures 13.21 and 13.22. This phenomenon observed by Gunn is referred to as *Gunn effect*. Microwave devices that operate on the principle of Gunn effect are called Gunn diodes.

Figure 13.29a illustrates a GaAs semiconductor specimen while Figure 13.29b which is analogous to Figure 13.28 shows the movement of domains therein. Each domain causes a pulse of current at the output. Its output is determined by the physical length of the semiconductor chip. This oscillator can deliver continuous power up to about 65 mW and pulsed (peak) power up to about 200 W. The power output of a solid chip is limited by the heat generated and its effective removal from this small chip area. Much higher power outputs have been achieved using wafers of gallium–arsenide as a single source.

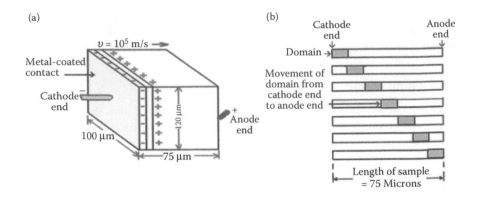

FIGURE 13.29
n-type GaAs diode. (a) Schematic diagram and (b) domain movement.

13.5.5.2 Current Fluctuations

The fluctuation of current is shown in Figure 13.30. This current waveform was produced by applying a voltage pulse of 16 V amplitude and 10 ns duration to a specimen of n-type GaAs, 2.5 cm in length. Oscillation frequency was 4.5 GHz. Lower trace had 2 ns/cm in the horizontal axis and 0.23 A/cm in the vertical axis. Gunn found that the period of these oscillations was equal to the transit time of electrons through the specimen, calculated from the threshold current. Gunn also discovered that threshold electric field E_{th} varies with length and type of material.

13.5.6 LSA Diode

As discussed above the LSA mode is defined where $f \cdot L > 2 \times 10^7$ cm/s, and n_0/f is between 2×10^4 and 2×10^5. Now consider that T_t is the domain transit

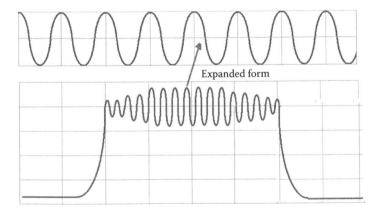

FIGURE 13.30
Current fluctuations.

time, T_d is the dielectric relaxation time at low field, T_g is the domain growth time, T_0 is the natural period of oscillation of high Q external electric circuit, n_o is the doping concentration, L is the length of the sample and f is the frequency of operation. If conditions $n_0L < 10^{12}$ and $2 \times 10^4 < n_0/f < 2 \times 10^5$ are met high field domain and space charge do not have sufficient time to buildup. Magnitude of RF voltage has to be large enough to drive the diode below threshold during each cycle to dissipate space charge. The portion of each cycle during which RF voltage exceeds threshold has to be short enough to prevent domain formation and space-charge accumulation. Only the primary accumulation layer forms near the cathode, the rest of the sample remains fairly homogeneous. Thus, with limited space-charge formation the remainder of sample appears as a series of negative resistance that increases with frequency of oscillation in the resonant circuit.

In LSA mode the diode is placed in a resonator tuned to an oscillator frequency of $f_0 = 1/T_0$. The device is biased to several times of the threshold voltage. As RF voltage swings beyond threshold, space charge starts building-up at the cathode. Since T_0 of RF signal is less than T_g the total voltage swings below the threshold before the domain can form. Further, since $T_0 \gg T_d$ the accumulated space charge is drained in a very small fraction of RF cycle. Thus, the device spends most of the RF cycle in NR region and the space charge is not allowed to build-up. The frequency of oscillation in LSA mode is independent of T_t of the carriers and is determined solely by circuit external to the device. Also the power-impedance product does not fall off as $1/f_0^2$ and thus the power output in LSA mode can be greater than that in the other modes.

LSA mode is very sensitive to load conditions and temperature fluctuations and doping. RF circuit must allow quick field build-up in order to prevent domain formation.

The power output of an LSA oscillator is given by the relation

$$P = \eta V_0 I_0 = \eta(ME_{th}L)(n_0ev_0A) \tag{13.27}$$

where V_0 is operating voltage, I_0 is operating current, η is dc to RF field conversion efficiency (which depends on the material and the circuit considerations) M is the multiple of operating voltage above negative resistance threshold voltage, E_{th} is the threshold field (~3400 V/cm), L is the device length (~10–200 μm), n_0 is the donor concentration (~10^{15} e/cm³), e is the electron charge (1.6×10^{-19} C), v_0 is the average carrier drift velocity (~10^7 cm/s) and A is the device area (~3×10^{-4} to 20×10^{-4} cm²). Figure 13.31 shows the operation of the LSA mode.

For an LSA oscillator n_0 is primarily determined by the desired operating frequency f_0, so that for a properly designed circuit, the peak power output is directly proportional to active volume (length $L \times$ area A) of the device. Active volume cannot be increased indefinitely. Theoretically it is limited by electrical length and skin depth. Practically it is limited by available bias, thermal dissipation capability and technical aspects associated with material

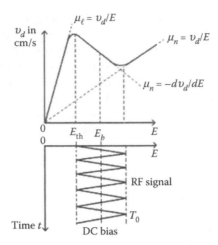

FIGURE 13.31
LSA mode of operation.

uniformity. Power capability of LSA oscillator varies from 6 kW (pulse) at 1.75 GHz to 400 W (pulse) at 51 GHz.

13.5.7 Indium Phosphide Diodes

Indium phosphide (InP) diode uses an N-type material. Its basic operation is at par with that of GaAs diode. This too relies on Gunn effect. Its working however is based on the three-valley structure (Figure 13.19b). InP oscillators operate through transit time phenomenon and do not oscillate in bulk mode of LSA. Frequencies are found to be dependent on active layer thickness. These oscillators are tunable over large frequency range by adjusting the cavity size. This range is bounded only by thickness. Further details of GaAs

TABLE 13.5

Comparison of GaAs, InP and CdTe

Assessment Parameter	GaAs	InP	CdTe
Electrical behaviour explained by:	Two-valley model	Three-valley model	Two-valley model
Threshold level (kV/cm)	Lower (~3)	Midway (~8)	Higher (~13)
Velocity of electrons transfer from LV to UV	Slower	Much faster	Similar to GaAs
Coupling between LV and UV	Strong	Weak	Strong
Current contribution	More by LV than UV	MV supplies additional energy	More by LV than UV
Peak current-to-valley current ratio	Low	Large	Similar to GaAs

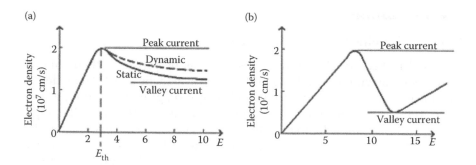

FIGURE 13.32
Electron density versus electric field curves. (a) For GaAs and (b) for InP.

and InP materials are given in Table 13.5. The electron density versus electric field curves for GaAs and InP diodes are shown in Figure 13.32.

13.5.8 Cadmium Telluride Diodes

Cadmium telluride (CdTe) is an N-type material. Similar to GaAs the basic operation of CdTe diode is based on Gunn effect and the two-valley model theory. CdTe compound has substantially higher threshold (~13 kV/cm) as compared to that of GaAs (~3 kV/cm). This higher threshold can be thought of as associated with relatively strong coupling of electrons to longitudinal phonons. This limits the mobility and hence the energy acquisition from the field. It also provides an efficient mechanism for energy transfer to lattice, thereby minimises kinetic energy in electron distribution. In CdTe (as in GaAs) the spike amplitude can be as large as 50% of the maximum total current. A similar maximum efficiency for CdTe and GaAs can be expected. Domain velocities in CdTe and GaAs are approximately equal. Samples of same lengths will operate at about the same frequency in transit time mode. Higher threshold field for CdTe combined with its poor thermal conductivity creates heating problem. This requires sufficiently short pulse to be used for dissipation of heat. High operating field of the sample can be an advantage.

13.5.9 Advantages

TEDs require low power supply (12 V). These devices are of low cost, lightweight, are reliable and have low noise and high gain.

13.5.10 Applications

TEDs are used in C, X and Ku-band for ECM amplifiers in wideband systems. These may also be used in X and Ku band transmitters for radar systems such as in air traffic control.

13.6 Avalanche Transit Time Devices

As noted in Section 13.2.1.3 avalanche devices by virtue of their inherent nature operate with reversed bias and use the avalanche effect. These devices may have many structures which include n^+-p-i-p^+, p^+-n-i-n^+, p^+-n-n^+, n^+-p-p^+, p-n-p, p-n-i-p, p-n-metal and metal-n-metal. In these structures the superscript ($^+$) indicates high doping and the letter i refers to an intrinsic material. In these devices if the structure description begins with n^+, holes are the carriers, whereas if it begins with p^+ electrons are the carriers. When such a device is reverse biased and its voltage exceeds the breakdown limit, the space-charge region extends from one junction through the intermediate region to another junction and the carriers start moving. These moving carriers result in the rising field of appropriate polarity. This field may attain a maximum value of about several hundred kV/cm at a particular junction. While moving in a high field these carriers nearer to a junction may acquire sufficient energy to knock other carriers into conduction band and produce the hole–electron pairs. This pair production is referred to as *avalanche multiplication*. The rate of this pair production is a sensitive non-linear function of the field. With proper doping the field may have a relatively sharp peak and as a result the avalanche multiplication can be confined to a very narrow region at a junction.

The electrons move into n^+ region and the holes to p^+ region. For silicon, this movement is at a constant drift velocity (v_d) of about 10^7 cm/s and field throughout the space charge is about 5 kV/cm. The transit time of carriers across the drift i-region of length L is given as

$$\tau = \frac{L}{v_d} \tag{13.28}$$

And the avalanche multiplication factor is given by

$$M = \frac{1}{1 - (V/V_b)^n} \tag{13.29}$$

where V is the applied voltage, V_b is the avalanche breakdown voltage, n is the numerical factor which depends on doping at p^+ n or n^+ p junction. For silicon the value of n varies from 3 to 6. The breakdown voltage for silicon p^+-n junction is given as

$$|V_b| = \frac{\rho_n \mu_n \varepsilon_s |E_{max}|_b^2}{2} \tag{17.30}$$

where ρ_n is the resistivity, μ_n is the electron mobility, ε_s is the semiconductor permittivity and E_{max} is the maximum breakdown electric field. Figure 13.33

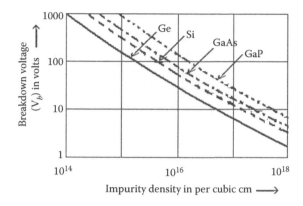

FIGURE 13.33
Avalanche breakdown voltage versus impurity density.

illustrates the variation of avalanche breakdown voltages with impurity at a p^+-n junction for different semiconductors.

In view of the above description the devices which involve avalanche multiplication and transit time are referred to as avalanche transit time devices. These include Read, IMPATT, TRAPATT and BARITT diodes.

13.6.1 Read Diodes

The Read diode is one of the avalanche transit time devices which involve avalanche multiplication and transit time effect and may have n^+-p-i-p^+ or p^+-n-i-n^+ structure.

13.6.1.1 Operation

Figure 13.34a shows an n^+-p-i-p^+ reverse biased Read diode structure. In this structure, the holes are generated from the avalanche multiplication. This device consists of (a) a high field (or avalanche) thin p-region in which the avalanche multiplication occurs, (b) the intrinsic i region through which the generated holes drift toward p^+ contact and (c) the space between n^+-p and i-p^+ termed as space-charge region. In case of p^+-n-i-n^+ structure electrons generated from avalanche multiplication drift through the i-region.

The impedance of this diode is capacitive. It is mounted in a microwave cavity having mainly inductive impedance. The combination of diode and cavity forms a resonant circuit. The device produces a negative ac resistance which in turn delivers power from dc bias to the oscillations.

When applied reverse-biased voltage is well above the breakdown voltage, the space-charge region extends from n^+-p junction through the p and i-region to i-p^+ junction. Fixed charges in various regions are shown in Figure 13.34b.

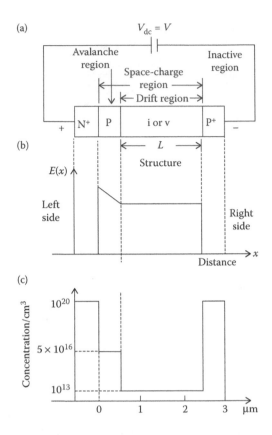

FIGURE 13.34
Field distribution and doping profile in a Read diode. (a) Structure, (b) field distribution and (c) doping profile.

A positive charge, while moving from left to right, gives a rising field. The maximum value (of about several hundred kV/cm) of this field occurs at the n^+ p junction. The hole carriers moving in the high field near n^+ p junction acquire sufficient energy to knock the valence electrons into conduction band resulting in generation of hole–electron pairs and hence in avalanche multiplication. This multiplication, a sensitive non-linear function of the field (having sharp peak with proper doping) is confined to a very narrow region at the n^+ p junction. As noted earlier the electrons and holes move at a constant drift velocity (of about 10^7 cm/s) in a field (of about 5 kV/cm) throughout the space charge into n^+ and p^+-regions, respectively. The transit time (τ) of a hole across the drift space (L), avalanche multiplication factor (M) and the breakdown voltage (V_b) for silicon p^+-n junction are given by Equations 13.28 through 13.30. Figure 13.34c illustrates the doping profile in the diode.

Figure 13.35a shows the Read diode mounted in a microwave resonant circuit to which an ac signal (Figure 13.35c) is applied at a given frequency. With

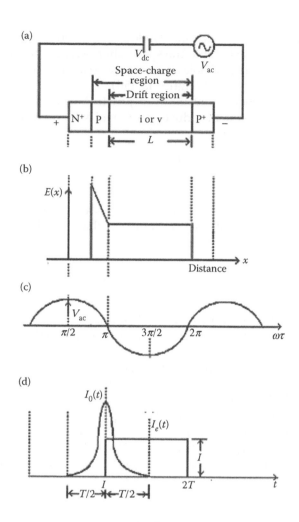

FIGURE 13.35
Applied voltage and currents in Read diode. (a) Structure, (b) field distribution, (c) applied ac voltage and (d) currents $I_0(t)$ and $I_e(t)$.

this application the total field across the diode becomes the sum of dc and ac fields. It results in larger spike at n^+ p junction as shown in Figure 13.35b. This field during positive half of ac cycle will cause breakdown at n^+ p junction provided it exceeds the breakdown voltage. The carrier current $I_0(t)$ generated at n^+ p junction by the avalanche multiplication grows exponentially with time. During the negative half-cycle of ac voltage if the field is below the breakdown voltage the carrier current $I_0(t)$ decays exponentially to a small steady-state value. The carrier current $I_0(t)$ is the current only at the junction and is of the form of a pulse of very short duration (Figure 13.35d). This current attains its maxima either in the mid of ac voltage cycle or a quarter cycle later than from the voltage maxima. Under the influence of electric field the generated holes are injected into the space-charge region towards the negative terminal. As the injected holes traverse the drift space they induce a current $I_e(t)$ in the external circuit (Figure 13.35d).

When the holes generated at the n⁺-p junction drift through the space-charge region they result in reduction of field which in accordance with Poisson's equation is given by

$$\frac{dE}{dx} = -\frac{\rho}{\varepsilon_s} \tag{13.31}$$

where ρ is the volume charge density and ε_s is the semiconductor permittivity. Since drift velocity of holes in the space-charge region is constant, the induced current in external circuit is

$$I_e(t) = \frac{Q}{\tau} = \frac{Q v_d}{L} \tag{13.32}$$

where Q is the total charge of moving holes, v_d is drift velocity of holes and L is the length of the drift (intrinsic) region. $I_e(t)$ is equal to the average current in space-charge region.

When pulse of hole current $I_0(t)$ is suddenly generated at n⁺-p junction a constant current ($I_e(t)$) starts flowing in the external circuit. This current continues to flow during time τ in which the holes are moving across the space-charge region. Thus, on the average $I_e(t)$ is delayed by $\tau/2$ or 90° relative to $I_0(t)$ generated at n⁺-p junction. Since the carrier current $I_0(t)$ is delayed by 90° relative to the ac voltage, $I_e(t)$ is then delayed by 180° relative to voltage (Figure 13.35d). Therefore, the cavity should be tuned to give a resonant frequency as

$$f = 1/2\tau = v_d/2 L \tag{13.33}$$

Since applied ac voltage and $I_e(t)$ are 180°, out-of-phase negative conductance occurs and the Read diode can be used for microwave amplifiers and oscillators. For $v_d = 10^7$ cm/s for silicon the optimum operating frequency for a Read diode with an L-region length of 2.5 μm is 20 GHz.

13.6.1.2 Output Power and Quality Factor

As illustrated in Figure 13.35d $I_e(t)$ is a square wave. It spreads over a very small duration of positive half cycle of ac voltage and almost constant during negative half cycle. Since the direct current (I_d) supplied by dc bias is the average external current or conduction current it follows that the amplitude of variation of $I_e(t)$ is approximately equal to I_d. If V_0 is the amplitude of ac voltage the ac power delivered is

$$P = 0.707 \, V_0 \, I_d \text{ W/unit area} \tag{13.34}$$

And Q of the circuit is given by

$$Q = \omega \text{ (maximum energy stored/average dissipated power)} \quad (13.35)$$

Since Read diode supplies ac energy it has a negative Q in contrast to the positive Q of cavity. At the stable operating point $-Q = +Q$ if the magnitude of ac voltage is increased the stored energy or energy of oscillation increases faster than the energy delivered per cycle. This is required for stable oscillations to persist.

13.6.2 IMPATT Diode

IMPATT stands for IMPact Avalanche Transit Time. It uses avalanche and transit time properties of semiconductors. It produces negative resistance at microwave frequencies and operates in reverse-breakdown (avalanche) region. Applied voltage causes momentary breakdown once per cycle. This starts a pulse of current moving through the device. For ac operation, current passing through the diode is 180° out of phase with the voltage. Its frequency of operation depends on the thickness of the device.

The Read diode can be referred to as the basic type of IMPATT diode. The power generation requires sustained oscillations which in turn require the property of differential negative resistance. IMPATT diodes too exhibit this property due to (i) the impact ionisation avalanche effect which causes the carrier current and ac supply to be phase-out by 90° and (ii) the transit-time effect which further delays the external current relative to ac voltage by 90°. Thus, the net phase shift between current through diode is 180° out of phase with the voltage.

13.6.2.1 *Principle of Operation*

The operation of IMPATT diode mainly depends on avalanche and saturation processes. The avalanche process takes place in reverse biased p-n junction wherein the leakage current is due to minority carriers. When high voltage is applied, minority carriers are accelerated. They acquire maximum velocity and strike other crystal lattices leading to the generation of electron–hole pairs which may take part in conduction. The current produced in this process is given by

$$I = M I_0 \quad (13.36)$$

where I_0 is the leakage current and M is the multiplication factor given by

$$M = \frac{1}{1 - (V_a/V_b)^n} \quad (13.37)$$

where V_b is the applied voltage, V_b is the breakdown voltage and parameter n varies between 3 and 6 for silicon. In the saturation process the saturation velocity for carriers generally exists in all semiconductor devices. In p-n junction the motion of carriers takes place with carrier drift velocity. This velocity becomes constant under the influence of E. The threshold value of E for silicon is 10^4 V/m.

13.6.2.2 Construction

IMPATT diode illustrated in Figure 13.36a is a solid-state power device made from a highly doped semiconductor layer. This is called n+ layer. On this layer another n⁻ layer is deposited epitaxially. At extreme side of this layer p impurities are diffused. This resulting layer is referred to as p+ layer. This diode is stuck on a support called a post. The anode and cathode are made from copper with gold wires for contacts. These diodes may be made either from Si, GaAs or InP. Although silicon is cheaper and makes the fabrication simple, its IMPATTs provide higher output at higher frequencies but GaAs is preferred as its IMPATTs have higher efficiency and lower noise.

13.6.2.3 Working

When steady bias voltage is applied i-layer (Figure 17.36b) is formed between the anode and the cathode. An avalanche current multiplication takes place. Due to this multiplication process 180° phase shift is introduced between applied voltage and the current through diode. This phase shift is the result of time delay provided for the avalanche process. This delay is required for the generation of avalanche current multiplication and the time required for carriers to pass through the drift space. This time is called the transit time. This diode operates in negative conduction mode. An ac signal is superimposed on the steady bias voltage. This voltage provides the higher voltage gradient across the diode. The required value of this bias voltage is about

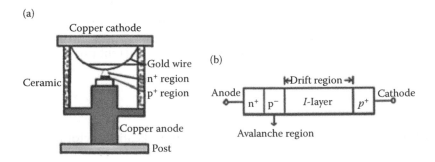

FIGURE 13.36
(a) Construction and (b) schematic diagram of IMPATT diode.

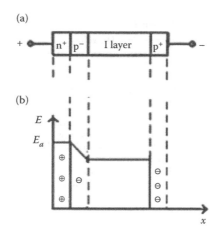

FIGURE 13.37
Layers and field distribution in IMPATT diode. (a) Silicon layers and (b) field distribution.

400 kV/cm. Whenever there is a positive cycle of ac voltage, negative ac current is obtained and vice versa. Assume that avalanche multiplication takes place in a thin region near n+-P junction. An electric field distribution in the device under the reverse bias is shown in Figure 13.37 wherein E_a is its critical value. As applied ac voltage goes positive, more holes are generated in n+ region. Similarly, more electrons are generated in p+ region.

As positive voltage goes on increasing, more electrons and holes are generated. These minority carriers strike the atoms of crystal lattice resulting in more number of electron–hole pairs. Thus, multiplication of carriers takes place through the process termed as *avalanche multiplication*. The current generated is called *avalanche current*. This process occurs at the critical value of electric field. Since the sum of critical value of dc and ac voltages during the positive half-cycle is always $>E_a$ avalanche multiplication process continues. Avalanche multiplication is not an instantaneous process but requires some time delay. Thus, current pulse becomes maximum at the junction whenever voltage is zero and going to the negative side. Drift of current pulse, with respect to applied dc voltage plus RF voltage superimposed, takes place from the anode to the cathode. This is shown in Figure 13.38.

The current pulse reaches its peak not at $\pi/2$ when the voltage is maximum but at π when voltage is zero. Thus, there is an inherent delay in avalanche multiplication process. Additional delay in view of time taken by current pulses in drifting from anode to cathode is also present. This results in 90° phase delay between current and voltage. Waveforms for diode voltage, junction current and diode current for p+-n⁻ n+ structure of IMPATT are shown in Figure 13.39.

Figure 13.40 illustrates the time dependence of the growth of signal and the drift of holes. The pulse of holes is shown by dotted line on the field diagram.

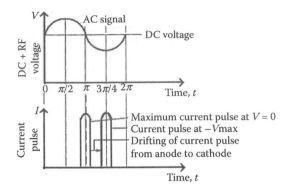

FIGURE 13.38
Drift of current pulses.

13.6.2.4 Performance

Output power and efficiency are given in Table 13.6. The maximum operating fundamental frequency of 100 and 400 GHz with harmonic generation is reported. In case of avalanche multiplication noise signal is produced. Besides, the high operating currents generate shot noise. The IMPATT-based amplifiers have a noise figure of about 30.

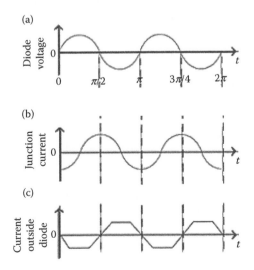

FIGURE 13.39
Waveforms of voltage and currents. (a) Diode voltage, (b) junction current and (c) current outside diode.

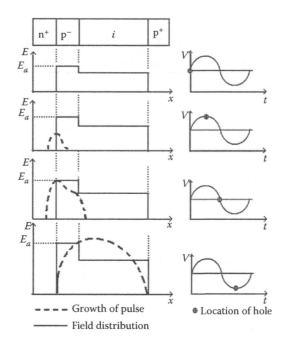

FIGURE 13.40
Time dependence of the growth and drift of holes.

13.6.2.5 Advantages and Limitations

IMPATT diodes are the most powerful CW solid-state microwave power sources. These can be fabricated from Ge, Si and GaAs and can be operated at very high frequencies. These in comparison to Gunn diode are more efficient, generate more power and require low power supply. These are reliable, compact, inexpensive, lightweight and moderately efficient. Their basic limitation is that they produce more noise and require higher supply voltage.

TABLE 13.6
Output and Efficiency at Different Frequencies for IMPATT Diodes

Frequency (GHz)	Output Power	Efficiency (%)	Type
10	10 W	20	GaAs
50	1.5 W	10	GaAs
70	0.13 W	5	Si
100	60 mW	1	GaAs
220	50 mW	1	Si

TABLE 13.7

Different Relative Aspects of Gunn Diode and IMPATT Diode

Assessment Parameter	Gunn Diode	IMPATT Diode
Possibility of broadband operation	Yes	No
Noise	Low	More
Power output	Less	More
Efficiency	Low	High
Possible use in pump oscillators	Yes	No, due to noise problem
Supply voltage required	Low	High
Frequency of operation	Low	High

13.6.2.6 Applications

These are used in several types of radars and can be used in power tubes and in microwave oscillators at highest frequencies. These are used as transmitters for millimeter-wave communication. Table 13.7 gives the relative aspects of Gunn diode and IMPATT diode for various assessment parameters.

13.6.3 TRAPATT Diode

TRAPATT is a short form of TRApped Plasma Avalanche Trigger Transit-time. Its structure is shown in Figure 13.41. It is also a microwave avalanche diode. Its operation is limited to the frequencies below a few GHz. It is much noisier than IMPATT. IMPATT has p^+-n-n^+ structure, whereas TRAPATT has n^+-p-p^+. Here, the cathode is connected to positive potential and anode to negative. Its junction is called trapped avalanche region. In this diode the doping level changes gradually and not abruptly. These diodes are the planar silicon diodes.

13.6.3.1 Operation

Its operation can be explained with reference to IMPATT. Let an IMPATT be mounted in a coaxial cavity and is so located that there is a short circuit (SC) at operating frequency. This SC is present at $\lambda/2$ distance from the diode.

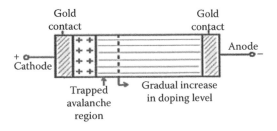

FIGURE 13.41
Structure of TRAPATT diode.

As an ac (RF) signal is superimposed on dc signal, whenever oscillations from this diode start most of the power gets reflected due to SC at operating frequency. This reflected power will get added to RF signal resulting in its enhancement. Thus, the total enhanced voltage across diode is the sum of RF signal and dc bias voltage. If this increased voltage exceeds the threshold voltage avalanche process begins. There is transition of more electrons and holes. This is termed as plasma of electrons and holes. Plasma will create a large potential across the junction, which will oppose the dc potential and reduce the total voltage. Thus, the current pulse is trapped, which will try to move towards n+ region (or cathode). Thus, net potential of TRAPATT is much less than that of IMPATT resulting in less dissipation and improved efficiency. For a given thickness the TRAPATT diode possesses much lower operating frequency as compared to IMPATT. This is because in TRAPATT drift velocity is low and hence transition time is less.

13.6.3.2 Performance

Good results from TRAPATT are obtained for frequencies below 10 GHz. Since it is harmonic sensitive the entry of second, third and fourth harmonics in circuit needs to be blocked. In a proper resonant circuit plasma is built up cyclically as shown in Figure 13.42. The discharge of this plasma leads to efficient generation of microwaves.

13.6.3.3 Advantages and Limitations

TRAPATT has higher efficiency than IMPATT. It has low power dissipation due to involvement of lower potential. The rich harmonic pulses generated by these allow wide range tuning of oscillators. It has high peak and average power. TRAPATTs too require low power supply, are reliable and are of low cost. Its basic limitations are its low transit time and much lower operating frequency than that of IMPATT diode.

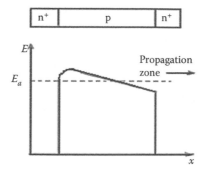

FIGURE 13.42
TRAPATT cycle.

TABLE 13.8

Comparison between TRAPATT and IMPATT

Assessment Parameter	TRAPATT Diode	IMPATT Diode
Structure type	n^+-p-p^+	p^+-n-n^+
Transit time	Less	More
Operating frequency	Low	High
Efficiency	High	Low
Power dissipation	Low	High
Sensitiveness to harmonics	More	Less

13.6.3.4 Applications

These find application in radar beacons, low power Doppler radar as local oscillator and in landing systems. These are used in S-band pulsed transmitter for phased-array radar systems.

Table 13.8 compares different aspects of TRAPATT and IMPATT diodes, whereas Figure 13.43 compares three diodes in terms of their achieved peak powers.

13.6.4 BARITT Diode

The word BARITT stands for BARrier Injected Transit Time. It is basically a back-to-back pair of diodes. This diode has long drift region like IMPATTs. In these the carriers, traversing the drift region, are not extracted from the plasma of an avalanche region. These are generated by the minority carrier's injection from forward-biased junction.

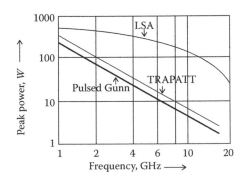

FIGURE 13.43
Peak power levels achieved by pulsed Gunn, TRAPATT and LSA diodes.

13.6.4.1 Structures

These diodes may have different structures including p-n-p, p-n–i-p, p-n-metal and metal-n-metal. A structure of a metal-n-metal BARITT diode is illustrated in Figure 13.44. It contains a crystal, n-type silicon 10 μm thin sliced wafer, with 11 Ω cm resistivity and 4×10^{14} per cubic cm doping. It is sandwiched between two PtSi Schottky barrier contacts of about 0.1 μm thickness.

13.6.4.2 Operation

Consider the p-n–i-p structure of BARRIT diode. The forward-biased p-n junction emits holes into the i region. These holes drift through the i-region with saturation velocity. Finally, these carriers are collected at the p-contact. These diodes exhibit a negative resistance for transit angle between π and 2π the optimum of which is around 1.6π.

As noted above, the BARITT diode can be fabricated simply by sandwiching n-type silicon of 1 Ω cm resistivity and 10 μm thickness between two pt-Si Schottky barrier contacts of about 0.01 μm thickness. In this case, the avalanche multiplication is not the cause for increase of current. It is the thermionic emission of holes injected from the forward bias contact which

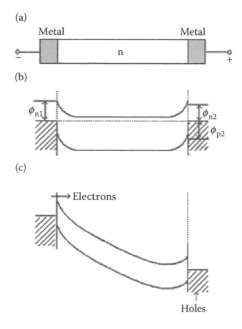

FIGURE 13.44
M-n-M diode. (a) Structure, (b) energy band in thermal equilibrium and (c) bands under bias condition.

results in such an increase. The microwave oscillations in a BARITT diode are mainly governed by the following two factors.

- There is a rapid increase of carrier injection process due to decreasing potential barrier of the forward biased metal–semiconductor contact.
- The injected carriers traversing the semiconductor depletion layer has an apparent transit angle of $3\pi/2$.

The rapid increase in terminal current with applied voltage results due to thermionic-hole injection into the semiconductor as the depletion layer under reverse-biased spreads through the entire thickness of the device. The critical voltage (V_c) is given as

$$V_c \cong qn\frac{L^2}{2\varepsilon_s} \tag{13.38}$$

As before q is the charge of carrier, n is doping concentration, L is thickness of semiconductor and ε_s is permittivity of the semiconductor.

13.6.4.3 Advantages and Limitations

It is less noisy compared to other transit time devices. For silicon BARITT amplifier the noise figure at C-band is as low as 15 dB. It is low cost and reliable and requires low power supply. Its bandwidth is narrow, power output is low (a few mW) and conversion efficiency is also low.

13.6.4.4 Applications

These are used in local oscillators in communication and radar receivers.

13.6.4.5 Comparison of Power Levels

In the end Figure 13.45 illustrates variation of power with frequency for Gunn (pulse) diode, LSA diode and TRAPATT. According to this figure the power for Gunn diode and TRAPATT linearly reduces with the frequency. In case of LSA this reduction is not very steep.

13.7 Parametric Devices

Word parametric is derived from the term parametric excitation. This excitation can be produced by using non-linear or time-varying reactances. These reactances may be capacitive or inductive in nature and can be obtained through time-varying capacitance $C(t)$ or inductance $L(t)$. This parametric excitation is subdivided into (*i*) parametric amplification and (ii) parametric oscillation. Solid-state varactor diode is the most commonly used element in parametric devices. As in quantum amplifier lasers and masers, parametric amplifiers employ ac instead of dc power supply. In view of ac power supply the non-linear capacitance $C(t)$ or inductance $L(t)$ can be regarded as a function of voltage (*v*) or current (*i*) as both of these are functions of time. Similarly, the electric charge Q and the magnetic flux ϕ also become functions of time in view of ac supply. Thus, the non-linear capacitance and inductance can be defined in terms of Q, ϕ and the instantaneous values of v and I, respectively. These defining equations are

$$C(v) = \frac{\partial Q}{\partial v} \tag{13.39a}$$

$$L(i) = \frac{\partial \phi}{\partial i} \tag{13.39b}$$

When the voltages of two or more frequencies are fed to a non-linear reactance, these signals get mixed. This mixing forms the basis of operation for parametric devices. Let the two voltages v_s (signal voltage with frequency ω_s) and v_p (pump voltage with frequency ω_p) be impressed to a non-linear device. With the assumption that $v_s \ll v_p$ the total (resulting) voltage (*v*) across the non-linear capacitance becomes

$$v = v_s + v_p = V_s \, \text{Cos}(\omega_s t) + V_p \cos(\omega_p t) \tag{13.40}$$

Since Q is the function of instantaneous voltage (*v*) it can be written as

$$Q(v) = Q(v_s + v_p) \tag{13.41}$$

The expression for $Q(v)$ can be expanded in Taylor's series about the point $v_s = 0$. Retaining the first two terms of the expanded series to get

$$Q(v) = Q(v_p) + \frac{dQ(v_p)}{dv}\bigg|_{v_s=0} v_p \tag{13.42}$$

Assuming

$$C(v_p) = \frac{dQ(v_p)}{dv} = C(t) \tag{13.43}$$

where $C(v_p)$ is periodic with fundamental frequency ω_p. Its expansion in Fourier series gives

$$C(v_p) = \sum_{n=0}^{\infty} C_n \cos(n\omega_p) \tag{13.44}$$

Since v_p is a function of time $C(v_p)$ is also function of time. Thus,

$$C(t) = \sum_n C_n \cos(n\omega_p t) \tag{13.45}$$

In Equation 17.45 the coefficient C_n represents the magnitude of each harmonic of time-varying C. Also coefficients C_n are non-linear functions of v_p. Since $C(t)$ is non-linear, principle of superposition does not apply for arbitrary ac signal amplitudes. The current through $C(t)$ is

$$i = \frac{dQ}{dt} = \frac{dQ(v_p)}{dt} + \frac{d\{C(t)v_s\}}{dt} \tag{13.46}$$

If $v_s \ll v_p$, nonlinear capacitance behaves like a time-varying linear capacitance. The first term of current i yields a current at f_p and not related to f_s. If $v_s \nless v_p$ the Taylors series can be expanded about a dc bias voltage V_0 in a junction diode, wherein C is proportional to

$$(\phi_0 - V)^{-1/2} = V_0^{-1/2} \tag{13.47}$$

where ϕ_0 is the junction barrier potential and V is the negative voltage supply. Since

$$\{V_0 + v_p \cos(\omega_p t)\}^{-1/2} \approx V_0^{-1/2}\{1 - \frac{v_p}{3V_0}\cos(\omega_p t)\} \quad \text{for } v_p \ll 3V_0 \tag{13.48}$$

$C(t)$ can be expressed as

$$C(t) = C_0[1 + 2\gamma\cos(\omega_p t)] \tag{13.49}$$

Parameter γ is proportional to v_p and indicates the coupling effect between voltages at f_s and f_0, where f_0 is the output frequency.

13.7.1 Manley–Rowe Power Relations

These relations indicate power flow into and out of an ideal non-linear reactance and predict the possibility of power gain in a parametric amplifier. Figure 13.45 shows an equivalent circuit obtained from Manley–Rowe relation. The circuit reveals that it has a signal generator operating at frequency f_s and a pump generator operating at frequency f_p. Together these are associated with series resistances and band-pass filters. Signals at f_p

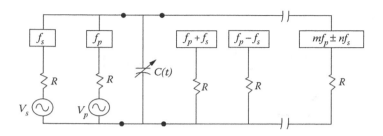

FIGURE 13.45
Equivalent circuit obtained from Manley–Rowe relation.

and f_s are applied to a non-linear capacitance $C(t)$. These resonating filter circuits are designed to reject power at all frequencies other than their respective frequencies. In the presence of f_s and f_p an infinite number of resonant frequencies of $mf_p \pm nf_s$ are generated (m & $n = 0, 1, 2,..., \infty$). Each resonant circuit is assumed to be ideal. Thus, the power loss by non-linear susceptance is negligible, that is, input to $C(t)$ at f_p = output of $C(t)$ due to non-linear interaction.

In view of Equation 13.40

$$v = v_s + v_p = v_s \cos(\omega_s t) + v_p \cos(\omega_p t)$$
$$= v_s \{\exp(j\omega_s t) + \exp(-j\omega_s t)\} + v_p \{\exp(j\omega_p t) + \exp(-j\omega_p t)\} \quad (13.50)$$

The general expression for charge Q deposited on C is given by

$$Q = \sum_{m=-\infty}^{\infty} \sum_{n=-\infty}^{\infty} Q_{m,n} \exp^{j(m\omega_p t + n\omega_s t)} \quad (13.51)$$

For Q to be real

$$Q_{m,n} = \{Q_{-m,n}\}^* \quad (13.52)$$

The total voltage v can be expressed as a function of Q. Taylor series expansion of $v(Q)$ gives

$$v = \sum_{m=-\infty}^{\infty} \sum_{n=-\infty}^{\infty} v_{m,n} \exp^{j(m\omega_p t + n\omega_s t)} \quad (13.53)$$

For v to be real

$$v_{m,n} = \{Qv_{-m,-n}\}^* \quad (13.54)$$

The current (i) through $C(t)$ is

$$i = dQ/dt = \sum_{m=-\infty}^{\infty} \sum_{n=-\infty}^{\infty} j(m\omega_p + n\omega_s) Q_{m,n} \exp^{j(m\omega_p t + n\omega_s t)}$$

(13.55)

$$= \sum_{m=-\infty}^{\infty} \sum_{n=-\infty}^{\infty} I_{m,n} \exp^{j(m\omega_p t + n\omega_s t)}$$

where,

$$I_{m,n} = j(m\omega_p + n\omega_s) Q_{m,n} = I_{-m,-n}$$

(13.56)

Since $C(t)$ is assumed to be pure reactance, the average power at frequencies $mf_p + nf_s$ is

$$P_{m,n} = V_{m,n} \{I_{m,n}\}^* + \{V_{m,n}\}^* I_{m,n} = V_{-m,-n} \{I_{-m,-n}\}^* + \{V_{-m,-n}\}^* I_{-m,-n} = P_{-m,-n}$$

(13.57)

In view of power conservation:

$$\sum_{m=-\infty}^{\infty} \sum_{n=-\infty}^{\infty} P_{m,n} = 0$$

(13.58)

On multiplying and dividing Equation 13.58 by $(m\omega_p + n\omega_s)$ and rearranging the terms:

$$\omega_p \sum_{m=-\infty}^{\infty} \sum_{n=-\infty}^{\infty} \frac{m P_{m,n}}{m\omega_p + n\omega_s} + \omega_s \sum_{m=-\infty}^{\infty} \sum_{n=-\infty}^{\infty} \frac{n P_{m,n}}{m\omega_p + n\omega_s} = 0$$

(13.59)

Since $(I_{m,n}/m\omega_p + n\omega_s) = jQ_{m,n}$, $P_{m,n}/m\omega_p + n\omega_s$ becomes $-jv_{m,n} Q_{m,n} - jv_{-m,-n} Q_{-m,-n}$ and is independent of ω_p and ω_s for any f_p and f_s, the resonant circuit external to that of $C(t)$ the currents keeps all the voltage amplitudes ($v_{m,n}$) unchanged. The charges $Q_{m,n}$ also remains unchanged as these are functions of $v_{m,n}$. The frequencies f_p and f_s can thus be arbitrarily adjusted to make

$$\sum_{m=-\infty}^{\infty} \sum_{n=-\infty}^{\infty} \frac{m P_{m,n}}{m\omega_p + n\omega_s} = 0$$

(13.60a)

$$\sum_{m=-\infty}^{\infty} \sum_{n=-\infty}^{\infty} \frac{n P_{m,n}}{m\omega_p + n\omega_s} = 0$$

(13.60b)

Equation 13.59 can now be expressed as

$$\sum_{m=0}^{\infty}\sum_{n=-\infty}^{\infty}\frac{mP_{m,n}}{m\omega_p + n\omega_s} + \sum_{m=0}^{\infty}\sum_{n=-\infty}^{\infty}\frac{-mP_{m,n}}{m\omega_p + n\omega_s} = 0 \qquad (13.61)$$

In view of the relation $P_{m,n} = P_{-m,-n}$ (Equation 13.57) and on replacing ω_p and ω_s by f_p and f_s Equation 13.60 becomes

$$\sum_{m=0}^{\infty}\sum_{n=-\infty}^{\infty}\frac{mP_{m,n}}{mf_p + nf_s} = 0 \qquad (13.62a)$$

$$\sum_{m=-\infty}^{\infty}\sum_{n=0}^{\infty}\frac{nP_{m,n}}{mf_p + nf_s} = 0 \qquad (13.62b)$$

Equations 13.62a and b are the standard forms of Manley–Rowe power relation. The term $P_{m,n}$ represents the real power flow into or leaving the non-linear capacitor at frequencies of $mf_p + nf_s$. The terms f_p and f_s stand for the fundamental frequencies of pumping and signal voltage oscillators, respectively. $P_{m,n}$ is positive if power flows into $C(t)$ or emerges out of two (signal and pump) voltage generators and negative if power leaves or flows into the load resistance.

If power flow is allowed at $f_p + f_s$ and all other harmonics are short circuited the currents will exist only corresponding to f_p, f_s and $f_p + f_s$. Thus, m and n will have values –1, 0 and +1 and Equation 13.62 will reduce to

$$\frac{P_{1,0}}{f_p} + \frac{P_{1,1}}{f_p + f_s} = 0 \qquad (13.63a)$$

$$\frac{P_{0,1}}{f_s} + \frac{P_{1,1}}{f_p + f_s} = 0 \qquad (13.63b)$$

13.7.2 Classification of Parametric Devices

The parametric devices can be classified in view of Manley–Rowe relations. Equation 13.63 indicates that $P_{1,0}$ and $P_{0,1}$ are the powers supplied by two generators at f_p and f_s. These powers are considered to be positive. Also $P_{1,1}$ the power flowing from reactance into the resistive load at f_s and f_p is considered to be negative. The power gain (G) is defined as the ratio of power delivered by capacitance at $f_s + f_p$ to the power absorbed by it at f_s. This gives the gain for an amplifier

$$G = (f_p + f_s)/f_s \qquad (13.64a)$$

Sum-frequency parametric amplifier or up converter: In this the maximum power gain is simply the ratio of output frequency (f_0) to the input frequency (f_s). It is given as

$$G = f_0/f_s \qquad (13.64b)$$

where $f_0 = (f_p + f_s) > f_s$ and this up converter is nothing but a *modulator*.

Down converter: In this case $f_s = (f_p + f_0)$. Equation 13.64 predicts that parametric device will have a gain and device acts as *demodulator*. This power gain is in fact a loss.

$$G = f_s/(f_p + f_s) \qquad (13.64c)$$

Negative resistance parametric amplifier: In this case $f_p = f_s + f_0$, the power $P_{1,1}$ supplied at f_p is positive. Both $P_{1,0}$ and $P_{0,1}$ are negative. The capacitor delivers power to the signal generator at f_s instead of absorbing it. The power gain may be infinite, which is an unstable condition. The circuit may oscillate at both f_s and f_0.

13.7.3 Parametric Amplifiers

Parametric amplifier is a low-noise microwave amplifier that uses variable reactance to amplify microwave signals. Figure 13.46 illustrates an example of a non-degenerative parametric amplifier.

In superheterodyne receivers an RF is mixed with local oscillator (LO) signal in a non-linear circuit (called mixer) to generate sum and difference frequencies. In parametric amplifiers the local oscillator is replaced by a pump generator (e.g. reflex klystron) and non-linear element by a time-varying capacitor such as varactor diode or time-varying inductor.

In Figure 13.47 f_s and f_p are mixed in non-linear capacitor. Accordingly, a voltage of f_p and f_s and sum and difference $mf_s \pm nf_p$ appears across the

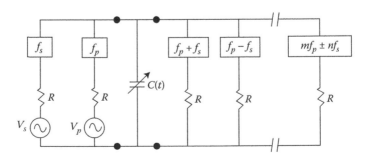

FIGURE 13.46
Non-degenerative parametric amplifier.

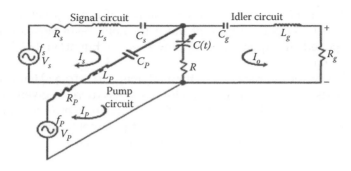

FIGURE 13.47
Equivalent circuit of parametric amplifier.

capacitor. The output circuit which does not require external excitation is called *idler circuit*. If a resistive load is connected across the terminals of idler circuit an output voltage can be generated across load at output frequency f_0. Output frequency f_0 of idler circuit is expressed as

$$f_0 = mf_s \pm nf_{p\prime} \tag{13.66}$$

where $m, n = 0, 1, 2, \ldots , \infty$.

If $f_0 > f_s$ the device is called a parametric up converter (PUC) and if $f_0 < f_s$ it is called parametric down converter (PDC).

13.7.3.1 Parametric Up Converter

As given in Section 13.7.2 in this case f_0 $(>f_s)$ is given by $f_0 = mf_s + nf_p$ and there is no power flow in parametric device at frequencies other than f_s, f_p and f_0. Its maximum power gain is

$$G = \frac{f_0}{f_s} \frac{x}{(1 + \sqrt{1 + x^2})} \tag{13.67a}$$

where,

$$x = \frac{f_s}{f_0}(\gamma Q)^2 \tag{13.67b}$$

$$Q = \frac{1}{2\pi f_s C R_d} \tag{13.67c}$$

$$G' = \frac{x}{1 + \sqrt{1 + x^2}} \tag{13.67d}$$

In Equation 13.67c R_d is the series resistance of a p–n junction diode in Equation 13.67b γQ is the figure of merit for non-linear capacitor and the factor G' spelled by Equation 13.67d involved in Equation 13.67a may be regarded as gain.

If y is taken as the degradation factor as R_d tends to zero, γQ tends to infinity and y tends to γ. Also, the gain (G) for a lossless diode is given by f_0/f_s. In a typical case $\gamma Q = 10$. Thus, if $f_0/f_s = 15$, $G = 7.3$ dB. Since pure reactance does not add thermal noise, the noise figure (F) of parametric amplifier is quite low in comparison to the transistor amplifier. It is given as

$$F = 1 + \frac{2T_d}{T0}\left[\frac{1}{\gamma Q} + \frac{1}{(\gamma Q)^2}\right] \tag{13.68}$$

where T_d is the diode temperature in K and $T_0 = 300$ K is the ambient temperature.

In microwave diode $\gamma Q = 10$, if $f_0/f_s = 10$ and $T_d = 300$ K, $F_{min} = 0.9$ dB, bandwidth (BW) of parametric amplifier up converter is given by

$$BW = 2\gamma\sqrt(f_0/f_s) \tag{13.69}$$

If $f_0/f_s = 10$, $\gamma = 0.2$, $BW = 1.264$.

It is to be noted that the PUC circuit is not practical at high frequencies.

13.7.3.2 Parametric Down Converter

As noted in Section 13.7.2, $f_s = f_0 + f_p$. Here, input power is fed to idler circuit and output is taken from the signal circuit. For PDC the gain (G) is given by

$$G = \frac{f_s}{f_0}\frac{x}{(1 + \sqrt{1 + x^2})} \tag{13.70}$$

This circuit is used at HF with circulators.

13.7.3.3 Negative Resistance Parametric Amplifiers

In view of Section 13.7.2, if a significant segment of power flows only at f_s, f_p and f_i a regenerative condition with possibility of oscillation both at signal frequency f_s and idler frequency f_i will occur. The idler frequency is related with f_s and f_p by

$$f_i = f_p - f_s \tag{13.71}$$

For the mode below the oscillation threshold the device behaves as a bilateral negative resistance parametric amplifier. For NRPA the power gain (G) is given by

$$G = \frac{4f_i}{f_s} \frac{R_g R_i}{R_{Ts} R_{Ti}} \frac{a}{(1-a)^2} \tag{13.72}$$

where R_i is the output resistance of idler generator, R_{Ts} is the total series resistance at f_s, R_{Ti} is the total series resistance at f_i and a is the ratio R/R_{Ts}, where R representing equivalent negative resistance is given as

$$R = \frac{\gamma^2}{(\omega_s \omega_i C^2 R_{Ti})} \tag{13.73}$$

The noise figure (F) and bandwidth (BW) are:

$$F = 1 + \frac{2T_d}{T_0}\left[\frac{1}{\gamma Q} + \frac{1}{(\gamma Q)^2}\right] \tag{13.74a}$$

$$BW = \frac{\gamma}{2}\sqrt{\frac{f_i}{f_s \text{gain}}} \tag{13.74b}$$

The expression of noise figure F is identical to that of up converter. If gain = 20 dB, $f_i = 4f_s$, $\gamma = 0.3$, BW = 0.03.

13.7.3.4 Degenerate Parametric Amplifiers or Oscillator

It is defined as non-linear amplifier with $f_s = f_i$, since $f_i = f_p \sim f_s$, $f_s = f_p/2$. These are used for low noise figures. Its power gain and BW are exactly the same as for PAUC. Its noise figures are separately defined for single-side band (F_{SSB}) and double-side band (F_{DSB}). These are

$$F_{SSB} = 2 + \frac{2T_d^* R_d}{T_0 R_g} \tag{13.75a}$$

$$F_{DSB} = 1 + \frac{T_d^* R_d}{T_0 R_g} \tag{13.75b}$$

where T_d^* is the average diode temperature in K, R_d is the diode series resistance and R_g is the external output resistance for signal generator.

13.7.4 Relative Merits

PUC has positive input impedance. Its power gain is not a function of changes in source impedance. It does not require circulator and its typical BW is of the order of 5%. Also PUC is a unilateral, stable device with wide bandwidth and low gain. NRPA is a bilateral and unstable device with narrow BW and high gain. These are used with circulators at high frequencies. DPA does not require separate signal and idler circuit coupled by the diode and is the least complex type of PA. DPAs are used for low noise figures.

13.8 Masers

Maser is an acronym for 'Microwave Amplification by Stimulated Emission of Radiation'. It is used for creation, amplification and transmission of an intense, highly focused beam of high-frequency radio waves. It is an oscillator in which the basic frequency is controlled through an atomic resonance rather than a resonant electronic circuit. It produces a coherent wave which is much closer to an ideal single-frequency source. Thus, the output of a maser can be transmitted over fairly large distances with relatively little loss.

In maser, electromagnetic radiation is produced by stimulated emission. An atom or molecule in an excited (or enhanced energy) state emits photon of a specific frequency when struck by a second photon of the same frequency. The emitted photon and the striking photon emerge in phase and in the same direction. For sufficient intensity of such emissions, to become a steady source of radiation, many atoms are to be pumped to the higher energy state.

The first maser used a beam of ammonia molecules. The beam was passed along the axis of a cylindrical cage of metal rods, with alternate rods having positive and negative electric charge. The non-uniform electric field from the rods sorts-out the excited and unexcited molecules, focus the excited molecules through a small hole into the resonator. The output was less than 1 μW. The wavelength, being determined primarily by the ammonia molecules, was so constant and reproducible that it could be used to control a clock that would gain or lose no more than a second in several hundred years. This maser can also be used as a microwave amplifier. Paramagnetic ions in crystals have also been used as the source of coherent radiation for a maser.

Maser amplifiers have the advantage that they are much quieter than those comprising vacuum tubes or transistors. Thus, masers add very little or almost no noise to the signals during amplification process. In view of this significant aspect it allows utilisation of very weak signals.

A maser oscillator requires a source of excited atoms or molecules and a resonator to store their radiation. The excitation must force more atoms or molecules into the upper energy level than in the lower, in order for amplification by stimulated emission to predominate over absorption. For wavelengths of a few millimetres or longer, the resonator can be a metal box whose dimensions are chosen so that only one of its modes of oscillation coincides with the frequency emitted by the atoms; that is, the box is resonant at a particular frequency. The losses of such a resonator can be made quite small, so that radiation can remain stored for adequate time to stimulate emission from successive atoms as they are excited. Thus, all the atoms are forced to emit in such a way as to augment this stored wave. Output is obtained by allowing some radiation to escape through a small hole in the resonator.

13.8.1 Principle of Operation

The materials required for masers may be solid or gaseous. These materials must have some distinct energy levels referred to as E_1, E_2, E_3 and so forth. Each level is populated by different numbers of atoms or molecules. These numbers are referred to as N_1, N_2, N_3 and so on. The lowest level (E_1) is termed as the ground level. The maser is referred to as two- or three-level maser depending on the energy levels of the material used.

In natural state of the materials the lower levels are more populated than higher levels. The masing action requires an unnatural state wherein the population at the higher level is more than that at the lower level. This unnatural state is referred to as population inversion. This inversion can be achieved by feeding (pumping) sufficient energy to the ground level atoms so as to raise these to the higher level. During this process, these atoms absorb the pumped energy resulting in excitation of electrons. When these excited electrons jump from higher level to the lower (ground) level energy is released. If the pump source and the released energy have the same frequency it is related to the energy levels as under:

$$f = (E_3 - E_1)/h \qquad (13.76)$$

where E_1 and E_3 are the two energy levels, h is the Planck's constant and f is the natural frequency. Under this condition the energy will add up. With only E_1 and E_3 levels it is a two-level maser. In case there is an intermediate energy level E_2 atoms may give some small energy to this level first and then return to ground level. Energy released in this case will correspond to

$$f' = (E_2 - E_1)/h \qquad (13.77)$$

Most of the atoms follow the second process and they are simulated by the presence of cavity surrounding the maser material. These are called three-level masers.

13.8.2 Types of Masers

These can be classified into (i) atomic beam masers which include ammonia maser, free electron maser and *hydrogen maser,* (ii) gas masers which include rubidium maser and (iii) solid-state masers which include Ruby maser. Some of these masers are described below:

13.8.2.1 Ammonia Maser

These masers use ammonia (NH_3) gas as working material. Since ammonia molecules are already distributed in two different (E_1 and E_3) levels, no pumping is required. With the help of a dc magnetic field the higher energy molecules are focused into a microwave cavity resonator. The frequency of this resonator corresponds to natural frequency of ammonia molecules. Inside this cavity there will be downward transition of higher energy molecules. If a signal is coupled to the cavity it will get amplified. After this transition the low-energy molecules are pumped from the resonant cavity. The amplification process in the maser cavity will regenerate since the amplified signal energy will stimulate further transitions.

Ammonia maser is generally operated with beam strength and cavity coupling conditions. Under these conditions the gain increases to infinity and oscillations result in the cavity. Thus, the device becomes a high stability and pure spectral density oscillator. Its thermal noise is reduced by placing the whole system into liquid helium bath. The power output of this maser is small. Its frequency stability is of the order of 1 part in 10^6 in a minute period which can be compared to that of an atomic clock.

The advantages of ammonia maser include (i) low noise, (ii) high-frequency stability and (iii) high spectral purity. Its main limitation is that it can amplify only a very narrow band of frequencies. Also, this maser is not tunable. In view of these negative aspects it has largely been superseded by other kinds such as solid-state ruby maser.

13.8.2.2 Ruby Maser

Ruby is a crystal of silica with a slight natural doping of chromium. The crystal acquires paramagnetic properties due to the presence of chromium. Figure 13.48a illustrates the energy levels of a three-level ruby maser. Energy at the correct frequency is added to the chromium atoms to raise them to the uppermost level. In this laser, first there is transition from upper-most layer to intermediate layer and then to the ground level. The cavity surrounding the ruby, resonates at the second transition frequency. The ruby atoms emit more energy than the energy applied which amounts to amplification. In this maser, the variation of dc magnetic field alters the energy levels of chromium atoms and thus a form of tuning can be obtained by adjusting the dc magnetic field. A sectional view of ruby

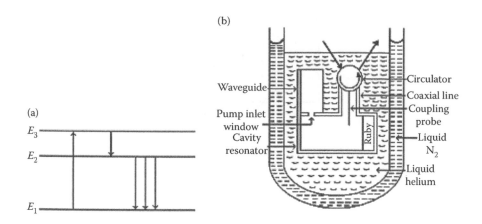

FIGURE 13.48
Ruby maser. (a) Energy levels and (b) sectional view.

maser is shown in Figure 13.48b. As it is a single port device it requires a circulator to separate the input from the output. It also requires tuned circuits for input and the pump signals.

The disadvantage of ruby maser is its narrow bandwidth which is governed by the cavity itself. Although there is possibility of tradeoff between gain and bandwidth, one of these has to be limited. One of the solutions to this problem is the travelling wave maser wherein a slow-wave structure is employed as in the case of TWT. In this structure, the signal is allowed to grow at the expense of the pump signal. The TW masers provide larger bandwidth, greater stability and easy tunability. Such maser is a four-port device which does not require circulator and thus losses caused due to circulator are eliminated. A typical maser may have a gain of 25 dB, bandwidth of 25 MHz at 1.6 GHz.

13.8.2.3 Hydrogen Maser

It is one of the most important types of masers currently used as an atomic frequency standard. Its oscillation relies on stimulated emission between two hyperfine levels of atomic hydrogen. With reference to Figure 13.49 its working can be briefly described below.

First, a beam of atomic hydrogen is produced by submitting the low-pressure gas to an RF discharge. To get some stimulated emission, it is necessary to create a population inversion of the atoms. It is done after passing the beam of atomic hydrogen through an aperture and a magnetic field. During its passage many of the atoms in the beam are left in the upper energy level of the lasing transition. From this state, the atoms can decay to the lower state and emit some microwave radiation. This step is termed as 'state selection'. A cavity of high-quality factor confines the microwaves and re-injects

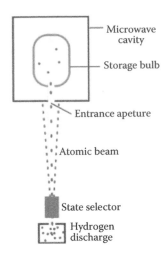

FIGURE 13.49
A hydrogen maser.

them repeatedly into the beam of atomic hydrogen. The stimulated emission amplifies microwaves on each passage of the beam. The combination of amplification and feedback results in oscillations. The resonant frequency of the microwave cavity is exactly tuned to the hyperfine structure of hydrogen: 1420 405 751.768 Hz. A small fraction of the signal in the microwave cavity is coupled into a coaxial cable and then sent to a coherent receiver. This microwave signal emerging from the maser is very weak (a few pW). The frequency of the signal is fixed and extremely stable. The coherent receiver is used to amplify the signal and change the frequency. This is done using a series of phase-locked loops and a high-performance quartz oscillator.

13.8.3 Maser Applications

A maser may be used as an amplifier or as an oscillator. The latter application requires a higher power level. One of the most useful types of maser is based on atomic hydrogen in which the transitions occur at about 1.421 GHz. The hydrogen maser with very sharp, constant oscillating signal serves as a time standard for an atomic clock. Thus, these are used as high-precision frequency references. These 'atomic frequency standards' are one of the many forms of atomic clocks. Due to low noise aspect these are widely used as electronic amplifiers in radio telescopes, space probe receivers, radio astronomy and extra terrestrial communications. Masers are also developed for applications in directed-energy weapons.

EXAMPLE 13.1

Determine the conductivity of an n-type Gunn diode if the charge of electron $e = 1.6 \times 10^{-19}$ C, electron density at lower valley $n_\ell = 3 \times 10^{10}$ cm^{-3}, electron density at upper valley is $n_u = 5 \times 10^8$ cm^{-3}, mobility of charges in lower valley $\mu_\ell = 8000$ cm^2/V s and that in upper valley $\mu_u = 180$ cm^2/V s.

Solution

Given $e = 1.6 \times 10^{-19}$ C, $n_\ell = 3 \times 10^{10}$ cm$^{-3} = 3 \times 10^{16}$ m^3, $n_u = 5 \times 10^8$ cm$^{-3} =$
5×10^{14} m^3, $\mu_\ell = 8000$ cm^2/V s $= 8000 \times 10^{-4}$ $\mu_u = 180$ cm^2/V s $= 180 \times 10^{-4}$
In view of Equation 13.10

$$\sigma = e\,(n_\ell\,\mu_\ell + n_u\,\mu_u) = 1.6 \times 10^{-19}\,(3 \times 10^{16} \times 8000 \times 10^{-4} + 5 \times 10^{14} \times 180 \times 10^{-4})$$
$$= 1.6 \times 10^{-19}\,(2.4 \times 10^{16} + 0.0009 \times 10^{16})$$
$$= 3.84144 \text{ milli-mhos}$$

EXAMPLE 13.2

Determine the mean mobility (μ') and drift velocity (v_d) of electrons if the electron density at lower valley $n_\ell = 10^{10}$ cm^{-3}, at upper valley is $n_u = 10^8$ cm^{-3}, mobility of charges in lower valley $\mu_\ell = 6000$ cm^2/V s, in upper valley $\mu_u = 200$ cm^2/V s and $E = 3000$ V.

Solution

Given $n_\ell = 10^{10}$ cm$^{-3} = 10^{16}$ m^{-3} $\quad n_u = 10^8$ cm$^{-3} = 10^6$ m^{-3}

$$\mu_\ell = 6000 \text{ cm}^2/\text{V s} = 6000 \times 10^{-4} \text{ m}^2/\text{V s},$$
$$\mu_u = 200 \text{ cm}^2/\text{V s} = 200 \times 10^{-4} \text{ m}^2/\text{V s}$$

In view of Equation 13.8

$$\mu' = (n_\ell\,\mu_\ell + n_u\,\mu_u)/(n_\ell + n_u) = (10^{16} \times 6000 \times 10^{-4} + 10^{14} \times 200 \times 10^{-4})$$
$$/(10^{16} + 10^{14}) - 0.59426 \text{ m}^2/\text{V s}$$

In view of Equation 13.7

$$v_d = \mu'E = 0.59426 \times 3000 = 1782.77 \text{ m/s}$$

EXAMPLE 13.3

Calculate (a) electron drift velocity, (b) current density, (c) negative electron mobility and (d) magnitude of the negative dielectric relaxation time in an n-type GaAs Gunn diode having an applied field $E = 3000$ V, threshold field $E_{th} = 2500$ V, device length $L = 12$ μm, doping concentration $n_0 = 2 \times 10^{14}$ cm^{-3}, operating frequency $f = 10$ GHz and $\varepsilon_r = 13.1$.

Solution

Given $E = 3000$ V, $E_{th} = 2500$ V, $L = 12$ μm $= 12 \times 10^{-6}$, $n_0 = 2 \times 10^{14}$ cm$^{-3} =$
2×10^{20} m^{-3}, $f = 10$ GHz $= 10 \times 10^9 = 1/T$ and $\varepsilon_r = 13.1$.
Since $v_d = L/T = Lf = 10 \times 10^9 \times 12 \times 10^{-6} = 12 \times 10^4$ m/s $= 1.2 \times 10^7$ cm/s
In view of Equation 13.21

$$J = q\,n\,v_d = e\,n_0\,v_d = 1.6 \times 10^{-19} \times 2 \times 10^{20} \times 12 \times 10^4 = 3.84 \times 10^6 \text{ A/m}^2$$

In view of Equation 13.4

$$\mu_n = -v_d/E = -12 \times 10^6/3000 = -4000 \text{ cm}^2/\text{V s}$$

$$\varepsilon = \varepsilon_0\varepsilon_r = 13.1 \times 8.854 \times 10^{-12} = 1.159874 \times 10^{-10}$$

$$\sigma = e\,n_0\,|\mu_n| = 1.6 \times 10^{-19} \times 2 \times 10^{20} \times 40 = 1.28 \times 10^3$$

$$\tau_d = \varepsilon/\sigma = \varepsilon_0\varepsilon_r/\sigma = 1.159874 \times 10^{-10}/1.28 \times 10^3 = 9.06 \times 10^{-14} \text{ s}$$

EXAMPLE 13.4

Calculate the product of the doping concentration and the device length to spell the criterion for classifying the modes of operation in an n-type GaAs Gunn diode having electron drift velocity $v_d = 2.25 \times 10^5$ m/s, negative electron mobility $|\mu_n| = 0.015$ m^2/V s and relative dielectric constant $\varepsilon_r = 13.1$. Also, calculate the growth factor if L (= 10 µm) is the length of the device and the doping concentration $n_0 = 2 \times 10^{14}$ cm^{-3}

Solution

Given $v_d = 2.25 \times 10^5$ m/s, $|\mu_n| = 0.01$ m^2/V · s, $\varepsilon_r = 5.25$, $L = 10$ µm $= 10^{-5}$ m and $n_0 = 2 \times 10^{14}$ cm$^{-3} = 2 \times 10^{20}$ m^{-3}

In view of Equation 13.26 the criterion for classifying modes of operation is given by

$$n_0L > \varepsilon v_d/\{e|\mu_n|\}$$

Right-hand side of the expression $= \varepsilon v_d/\{e|\mu_n|\}$
$$= (8.854 \times 10^{-12} \times 13.1 \times 2.25 \times 10^5)$$
$$/(1.6 \times 10^{-19} \times 0.01)$$
$$= 1.6311 \times 10^{14}/\text{m}^2 = 1.6311 \times 10^{10}/\text{cm}^2$$

Left-hand side $= n_0L = 2 \times 10^{14} \times 10^{-3} = 2 \times 10^{11}$
Thus, the criterion for classifying modes of operation is satisfied.
In view of Equation 13.25 the growth factor is given as
Growth factor $= \exp(L\,n_0\,e|\mu_n|/\varepsilon v)$
$$= \exp\{(10^{-5} \times 2 \times 10^{20} \times 1.6 \times 10^{-19} \times 0.01)$$
$$/(8.854 \times 10^{-12} \times 5.25 \times 2.25 \times 10^5)\}$$
$$= 0.3059$$

EXAMPLE 13.5

Calculate the output power of an LSA oscillator if its threshold field, $E_{th} = 3.3$ kV/m, device length $L = 15$ µm, donor concentration $n_0 = 2 \times 10^{21}$/m^3, electron charge $e = 1.6 \times 10^{-19}$ C, average carrier drift velocity $v_0 = 2 \times 10^5$ m/s, device area $A = 4 \times 10^{-8}$ m^2, dc to RF field conversion efficiency $\eta = 0.05$ and multiple of operating voltage above negative resistance threshold voltage $M = 4$.

Solution

Given $E_{th} = 3.3$ kV/m, $L = 15$ μm, $n_0 = 2 \times 10^{21}/m^3$, $e = 1.6 \times 10^{-19}$ C, $v_0 = 2 \times 10^5$ m/s, $A = 4 \times 10^{-8}$ m², $\eta = 0.05$ and $M = 4$

In view of Equation 13.27

$$P = \eta\,(M\,E_{th}\,L)\,(n_0\,e\,v_0\,A) = 0.05\,(4 \times 3.3 \times 10^3 \times 15 \times 10^{-6})$$
$$\times (2 \times 10^{21} \times 1.6 \times 10^{-19} \times 2 \times 10^5 \times 4 \times 10^{-8})$$
$$= 25.344 \text{ mW}$$

EXAMPLE 13.6

Calculate the avalanche multiplication factor (M) if the applied voltage $V = 100$ V, avalanche breakdown voltage $V_b = 200$ V and numerical factor for silicon is $n = 5$.

Solution

Given $V = 100$ V, $V_b = 200$ V, $n = 4$

In view of Equation 13.29

$$M = 1/\{1 - (V/V_b)^n\} = 1/\{1 - (100/200)^4\} = 1/\{1 - (0.5)^4\} = 1.067$$

EXAMPLE 13.7

Calculate the critical voltage of a BARITT diode if the relative dielectric constant of silicon $\varepsilon_r = 11.8$, donor concentration $N = 3 \times 10^{21}$ m⁻³ and specimen length $L = 5$ μm.

Solution

Given $\varepsilon_r = 11.8$, $n_0 = 3 \times 10^{21}$ m⁻³ and $L = 5$ μm

In view of Equation 13.38

$$V_c = qnL^2/2\varepsilon_s = 1.6 \times 10^{-19} \times 3 \times 10^{21} \times (5 \times 10^{-6})^2/(2 \times 11.8 \times 8.854 \times 10^{-12})$$
$$= 57.4288 \text{ V}$$

EXAMPLE 13.8

A parametric amplifier has ratio of its output to input frequency of 30, figure of merit of 12, factor of merit figure of 0.5 and diode temperature 300 K. Calculate its power gain, noise figure and bandwidth.

Solution

Given $f_0/f_s = 30$, $\gamma Q = 12$, $\gamma = 0.5$ and $T_d = 300$ K

In view of Equation 13.67

$$x = (f_s/f_0)\,(\gamma Q)^2 = (12)^2/30 = 4.8\ 1 + x^2 = 24.04\ \sqrt{(1 + x^2)} = 4.903\ 1 + \sqrt{(1 + x^2)}$$
$$= 5.903$$

$$G = (f_0/f_s)\,[x/\{1 + \sqrt{(1 + x^2)}\}] = 30 \times 4.8/5.903 = 24.39 \text{ dB}$$

In view of Equation 13.68 the noise figure is given by

$$F = 1 + \frac{2T_d}{T_0}\left[\frac{1}{\gamma Q} + \frac{1}{(\gamma Q)^2}\right] = 1 + (2 \times 300/273)\{(1/12) + (1/12)^2\}$$

$$= 1 + 2.1978(0.083 + 0.00694) = 1.19767\,\text{dB}$$

In view of Equation 13.69 the bandwidth is given by

$$\text{BW} = 2\gamma\sqrt{(f_0/f_s)} = 2 \times 0.5\sqrt{30} = 5.477$$

PROBLEMS

P-13.1 Determine the conductivity of an n-type Gunn diode if the charge of electron $e = 1.6 \times 10^{-19}$ C, electron density at lower valley $n_\ell = 3.5 \times 10^{10}$ cm^{-3}, electron density at upper valley is $n_u = 5.5 \times 10^8$ cm^{-3}, mobility of charges in lower valley $\mu_\ell = 7000$ cm^2/V s and that in upper valley $\mu_u = 150$ cm^2/V s.

P-13.2 Determine the mean mobility (μ') and drift velocity (v_d) of electrons if the electron density at lower valley $n_\ell = 2 \times 10^{10}$ cm^{-3}, at upper valley is $n_u = 1.5 \times 10^8$ cm^{-3}, mobility of charges in lower valley $\mu_\ell = 5000$ cm^2/V s, in upper valley $\mu_u = 300$ cm^2/V s and $E = 4000$ V.

P-13.3 Calculate (a) electron drift velocity, (b) current density, (c) negative electron mobility and (d) magnitude of the negative dielectric relaxation time in an n-type GaAs Gunn diode having an applied field $E = 3500$ V, threshold field $E_{th} = 3000$ V, device length $L = 10$ μm, doping concentration $n_0 = 3 \times 10^{14}$ cm^{-3}, operating frequency $f = 9$ GHz and $\varepsilon_r = 13.1$.

P-13.4 Calculate the product of the doping concentration and the device length to spell the criterion for classifying the modes of operation in an n-type GaAs Gunn diode having electron drift velocity $v_d = 3.25 \times 10^5$ m/s, negative electron mobility $|\mu_n| = 0.025$ m^2/V · s and relative dielectric constant $\varepsilon_r = 5.7$. Also calculate the growth factor if L (= 8 μm) is the length of the device and the doping concentration $n_0 = 3 \times 10^{14}$ cm^{-3}.

P-13.5 Calculate the output power of an LSA oscillator if its threshold field, $E_{th} = 3.5$ kV/m, device length $L = 10$ μm, donor concentration $n_0 = 2.5 \times 10^{21}$/m^3, electron charge $e = 1.6 \times 10^{-19}$ C, average carrier drift velocity $v_0 = 2.5 \times 10^5$ m/s, device area $A = 3 \times 10^{-8}$ m^2, dc to RF field conversion efficiency $\eta = 0.06$ and multiple of operating voltage above negative resistance threshold voltage $M = 3.5$.

P-13.6 Calculate the avalanche multiplication factor (M) if the applied voltage $V = 90$ V, avalanche breakdown voltage $V_b = 150$ V and numerical factor for silicon is $n = 5.5$.

P-13.7 Calculate the critical voltage of a BARITT diode if the relative dielectric constant of silicon $\varepsilon_r = 11.8$, donor concentration $N = 4 \times 10^{21}$ m^{-3} and specimen length $L = 6$ μm.

P-13.8 A parametric amplifier has ratio of its output to input frequency of 35, figure of merit of 10, factor of merit figure of 0.4 and diode temperature 320°K. Calculate its power gain, noise figure and bandwidth.

Descriptive Questions

Q-13.1 Briefly explain the construction and working of point-contact diode.

Q-13.2 Discuss the structure and operation of SRD.

Q-13.3 Illustrate the layer structure, impurity distribution and E-field distribution of PIN diode.

Q-13.4 Discuss tunnelling phenomena and draw V–I curve of a tunnel diode under various bias conditions.

Q-13.5 Explain the working of Schottky barrier diode and illustrate its structure.

Q-13.6 Discuss the working of varactor diode and draw its junction capacitance V_s voltage characteristics.

Q-13.7 Explain the process of electron transfer between two valleys under different field conditions and show its electron drift velocity Vs E-Field curve for GaAs and InP.

Q-13.8 Discuss Ridley–Watkins–Helsum (RWH) theory.

Q-13.9 In view of high field domain properties discuss various possible modes of operations.

Q-13.10 Discuss the Gunn effect and the working of a Gunn diode.

Q-13.11 Discuss the working of limited space-charge accumulation (LSA) diode.

Q-13.12 Explain the operation of InP Diodes and compare it with that of (CdTe) diode.

Q-13.13 Explain the meaning of avalanche multiplication.

Q-13.14 Write expression for avalanche multiplication factor and draw avalanche breakdown voltage Vs impurity density curves for different materials.

Q-13.15 Explain the working of Read diode and include its structure, field distribution and doping profile.

Q-13.16 Discuss construction and principle of operation of IMPATT diode.

Q-13.17 Discuss the construction and working of BARITT diode.

Q-13.18 Discuss and classify parametric devices in view of Manley–Rowe relations.

Q-13.19 Discuss types of parametric amplifiers. What is degenerate parametric amplifier?

Further Reading

1. Bahl, I., ed. *Microwave Solid-State Circuit Design*. John Wiley & Sons, New York, 1988.
2. Blackwell, L. A. and K. L. Kotzebue, *Semiconductor-Diode Parametric Amplifiers*. Prentice-Hall, Inc., Englewood Cliffs, NJ, 1961.
3. Chang, K. K. N., *Parametric and Tunnel Diodes*. Prentice-Hall, Inc., Englewood Cliffs, NJ, 1964.
4. Chang, K., *Microwave Solid-State Circuits and Applications*. Wiley, New York, 1994.
5. Chodorow, M. and C. Susskind, *Fundamentals of Microwave Electronics*. McGraw-Hill, New York, 1964.
6. Clorfeine, A. S. et al., A theory for the high efficiency mode of oscillation in avalanche diodes. *RCA Rev.*, 30, 397–421, September 1969.
7. Coleman, D. J., JR. and S. M. Sze, A low-noise metal–semiconductor–metal (MSM) microwave oscillator. *Bell System Tech. J.*, 50, 1695–1699, May–June 1971.
8. Dalman, G. C. *Microwave Devices, Circuits and Their Interaction*. John Wiley & Sons Inc, New York, 1994.
9. Colliver, D. and B. Prew, Indium phosphide: Is it practical for solid state microwave sources. *Electronics*, 110–113, April 10, 1972.
10. Conwell, E. M. and M. O. Vassell, High-field distribution function in GaAs. *IEEE Trans. Electron Devices*, ED-13, 22, 1966.
11. Copeland, J. A., Bulk negative-resistance semiconductor devices. *IEEE Spectrum*, No. 5, 40, May 1967.
12. Copeland, J. A., CW operation of LSA oscillator diodes – 44 to 88 GHz. *Bell System Tech. J.*, 46, 284–287, January 1967.
13. Copeland, J. A., Characterization of bulk negative-resistance diode behaviour. *IEEE Trans. Electron Devices*, ED-14(9), 436–441, September 1967.
14. Copeland, J. A., Stable space-charge layers in two valley-semiconductors. *J. Appl. Phys.*, 37(9), 3602, August 1966.
15. Deloach, B. C. and D. L. Scharfetter, Device physics of TRAPATT oscillators. *IEEE Trans. Electron Devices*, ED-17(1), 9–21, January 1970.
16. Deloach, B. C., JR., Recent advances in solid state microwave generators. *Advances in Microwaves*, Vol. 2. Academic Press, New York, 1967.
17. Eastman, L. F., *Gallium Arsenide Microwave Bulk and Transit-Time Devices*. Artech House, Dedham, MA, 1973.

18. Elliott, B. J., J. B. Gunn and J. C. Mcgroddy, Bulk negative differential conductivity and travelling domains in n-type germanium. *Appl. Phys. Lett.*, 11, 253, 1967.

19. Foyt, A. G. and A. L. Mcwhorter, The Gunn effect in polar semiconductors. *IEEE Trans. Electron Devices*, ED-13, 79–87, January 1966.

20. Gilden, M. and M. E. Hines, Electronic tuning effects in the Read microwave avalanche diode. *IEEE Trans. Electron Devices*, ED-13, 5–12, January 1966.

21. Gunn, J. B., Instabilities of current in III–V semiconductors. *IBM J. Res. Dev.*, 8, 141–159, April 1964.

22. Gunn, J. B., Instabilities of current and of potential distribution in GaAs and In *7th Int. Conf. on Physics of Semiconductor Plasma Effects in Solid*, pp. 199–207, Tokyo, 1964.

23. Gunn, J. B., Microwave oscillations of current in III–V semiconductors. *Solid-State Communications*, 1, 89–91, September 1963.

24. Gunn, J. B., Effect of domain and circuit properties on oscillations in GaAs. *IBM J. Res. Dev.*, 10, 310–320, July 1966.

25. Haddad, G. I., ed., *Avalanche Transit-Time Devices*. Artech House, Dedham, MA, 1973.

26. Haddad. G. I. et al., Basic principles and properties of avalanche transit-time devices. *IEEE Trans.* MTT-18(11), 752–772, November 1970.

27. Hess, K., *Advanced Theory of Semiconductor Devices*. Prentice-Hall, Inc., Englewood Cliffs, NJ, 1988.

28. Hieslmair, H. et al., State of the art of solid-state and tube transmitters. *Microwave J.*, 26(10), 46–48, October 1983.

29. Hilsum, C., Transferred electron amplifier and oscillators. *Proc. IEEE*, 50, 185–189, February 1962.

30. Kroemer, H., Theory of Gunn effect. *Proc. IEEE*, 52, 1736, 1964.

31. Kroemer, H., Negative conductance in semiconductors. *IEEE Spectrum*, 5(1), 47, January 1968.

32. Liao, S. Y., *Microwave Solid-State Devices*. Prentice-Hall, Inc., Englewood, NJ, 1985.

33. Liao, S. Y., *Microwave Devices & Circuits*, 3rd ed. Prentice Hall of India, New Delhi, 1995.

34. Liao, S. Y., *Semiconductor Electronic Devices*. Prentice-Hall, Englewood Cliffs, NJ, 1990.

35. Ludwig, G. W., Gunn effect in CdTe. *IEEE Trans.* ED-14(9), 547–551, Sept. 1967.

36. Ludwig, G. W., R. E. Haslsted and M. Aven, Current saturation and instability in CdTe and ZnSe. *IEEE Trans.* ED-13, 671, August–September 1966.

37. Milnes, A. G., *Semiconductor Devices and Integrated Electronics*. Van Nostrand Reinhold Company, New York, 1980.

38. Nanavati, R. P., *Semiconductor Devices*. Intext Education Publishers, Scranton, PA, 1975.

39. Navon, D. H., *Semiconductor Micro Devices and Materials*. Holt, Rinehart and Winston, New York, 1986.

40. Oliver, M. R. and A. G. Foyt, Gunn effect in n-CdTe. *IEEE Trans.*, ED-14(9), 617–618, September 1967.

41. Parker, D., TRAPATT oscillations in a p-i-n avalanche diode. *IEEE Trans. Electron Devices*, ED-18(5), 281–293, May 1971.

42. Prager, H. J. et al., High-power, high-efficiency silicon avalanche diodes at ultra high frequencies. *Proc. IEEE (Letters)*, 55, 586–587, April 1967.

43. Read, W. T., A proposed high-frequency negative-resistance diode. *Bell System Tech. J.*, 37, 401–446, 1958.
44. Ridley, B. K. and T. B. Watkins, The possibility of negative resistance effects in semiconductors. *Proc. Phys. Soc.*, 78, 293–304, August 1961.
45. Ridley, B. K., Specific negative resistance in solids. *Proc. Phys. Soc. (London)*, 82, 954–966, December 1963.
46. Ruch, J. G. and G. S. Kino, Measurement of the velocity-field characteristics of gallium arsenide. *Appl. Phys. Lett.*, 10, 50, 1967.
47. Shockley, W., Negative resistance arising from transit time in semiconductor diodes. *Bell System Tech. J.*, 33, 799–826, July 1954.
48. Shur, M., *Physics of Semiconductor Devices*, Prentice Hall of India, New Delhi, 1990.
49. Sobol, H. and F. Sterzer, Solid-state microwave power sources. *IEEE Spectrum*, 4, 32, April 1972.
50. Sze, S. M., *Physics of Semiconductor Devices*, 2nd ed. John Wiley & Sons, New York, 1981.
51. Sze, S. M., Microwave avalanche diodes. *IEEE Proc.*, 59(8), 1140–1171, August 1971.
52. Sze, S. M., *Semiconductor Devices: Physics and Technology*. John Wiley, New York, 1985.
53. Taylor, B. C. and D. J. Colliver, Indium phosphide microwave oscillators. *IEEE Trans. Electron Devices*, ED-18, No. 835–840, October 1971.
54. Thim, H. W., Computer study of bulk GaAs devices with random one-dimensional doping fluctuations. *J. Appl. Phys.*, 39, 3897, 1968.
55. Thim, H. W. and M. R. Barber, Observation of multiple high-field domains in n-GaAs. *Proc. IEEE*, 56, 110, 1968.
56. Thim, H. W., Linear microwave amplification with Gunn oscillators. *IEEE Trans. Electron Devices*, ED-14(9), 520–526, September 1967.
57. Tyagi, M. S. *Introduction to Semiconductor Materials and Devices*. Wiley, New York, 1999.
58. Weiss, M. T., A solid-state microwave amplifier and oscillator using ferrites. *Phys Rev.*, 107, 317, July 1957.
59. Wilson, W. E., Pulsed LSA and TRAPATT sources for microwave systems. *Microwave J.*, 14, 33–41, August 1971,

14

Microwave Transistors

14.1 Introduction

This chapter deals with microwave transistors used to generate and amplify microwave signals. These include bipolar junction transistors (BJTs), hetero-junction bipolar transistors (HBTs), junction field effect transistors (JFETs), metal semiconductor field effect transistors (MESFETs), high electron mobility transistors (HEMTs) and metal-oxide-semiconductor field effect transistors (MOSFETs), including CMOS.

14.2 Transistors and Vacuum Tubes

Vacuum tubes remained the key players for long in the field of electronics and are still used in some applications. These have higher operating voltage, higher degree of amplification linearity and are suitable for high power and high operating frequencies. The transistors, on the contrary, have smaller size, lesser weight, higher efficiency and reliability, and are preferred where these aspects are of concern. The manufacturing process of transistors is highly automated and their complementary devices are readily available. Their applications, however, are limited only to the low frequency range. The causes of their limitations are briefly described under the heading of conventional transistors.

14.2.1 Conventional Transistor

An ordinary or conventional (bipolar) transistor is an electronic component with three semiconductor regions called emitter (E), base (B) and collector (C). These three appropriately but differently doped segments are arranged either in p–n–p or n–p–n form. Further, any of these three segments may be made common to input and output circuits. In view of this commonality, the three resulting circuits are referred to as common emitter (CE), common

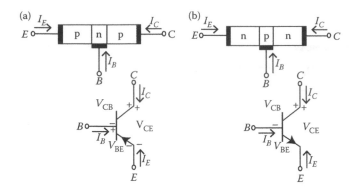

FIGURE 14.1
Transistors having E, B and C terminals. (a) p–n–p form and (b) n–p–n form.

base (CB) and common collector (CC) configurations. Each of these has its own characteristic parameters, relative merits and suitability of application.

Figure 14.1 shows p–n–p and n–p–n transistors (along with their circuit symbols) with positive current convention wherein the arrow on the emitter lead specifies the direction of current when the emitter–base junction is forward-biased. In both, the assumed positive direction of terminal currents I_E, I_B and I_C are into the transistor. However, the actual directions of currents I_B and I_C in p–n–p and I_E in n–p–n will be opposite to those shown in Figure 14.1.

14.2.2 Limitations of Conventional Transistors

Like vacuum tubes, the performance of ordinary BJTs deteriorates at higher frequencies due to the presence of interelectrode capacitances. The reactance of these capacitances, particularly those between base and emitter and base and collector junctions, starts decreasing with the increase of frequency and ultimately tends to behave as a short circuit at very high frequencies. As a result, the signal gets short circuited and appears directly at the collector terminal; thus the gain of transistor is grossly affected. The values of these capacitances depend on the width of the depletion layer. This width in itself is a function of transistor biasing. Although the lead inductance in a transistor also affects its performance, it can be minimised by using shorter leads. Besides, in transistors, the transit time plays the same role as in vacuum tubes. Here too the transit time limits the frequency of operation. For reducing the transit time, the device length can be reduced but this reduction affects the electric field strength. As a result, the speed of charges reduces and the transit time again increases. In view of these high-frequency limitations, ordinary transistors fail to serve the purpose. This failure led to the development of an entirely new class of devices termed as microwave transistors.

14.3 Microwave Transistors

Like a conventional transistor, a microwave transistor is also a non-linear device. Its principle of operation is similar to that of a low-frequency transistor. However, at microwaves, the requirement for dimensions, process control, heat sinking and packaging becomes much more stringent. The major concerns for a microwave transistor include transit time, parasitic capacitance and parasitic resistance.

Microwave transistors can be broadly classified into *bipolar* and *unipolar* categories. The bipolar transistors include (a) BJTs and (b) HBTs. The unipolar transistors encompass (a) MOSFETs, (b) MESFETs and (c) HEMTs. Both BJTs and field effect transistors (FETs) are of low-cost, small-size, lightweight and reliable devices and both require low power supply and give high CW power output.

14.3.1 Bipolar Junction Transistors

As illustrated in Figure 14.2, a BJT can be visualised as a combination of two back-to-back intimately coupled p–n junctions wherein the current is conveyed by the slow process of diffusion. Figure 14.2a shows the movement of electrons from emitter to collector, whereas Figure 14.2b shows the movements of electrons and holes in the same n–p–n configuration. Different resulting currents are also illustrated. The corresponding circuit symbol is the same as shown in Figure 14.1b.

Currently, the majority of BJTs fabricated from silicon dominate amplifier applications at lower microwave range. These are cheap, durable and integrative. Also, these have higher gains than those of FETs. Their noise figure in RF amplifiers is moderate and I/f noise characteristic is 10–20 dB higher to that of GaAs MESFETs. These are the main devices for local oscillators. Traditionally, BJTs are the choice of analogue circuit designers. It is mainly due to their higher transconductance and higher output impedance.

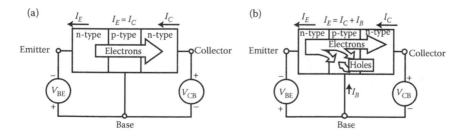

FIGURE 14.2
Bipolar (n–p–n) junction transistor. Movement of (a) electrons and (b) holes.

14.3.1.1 Transistor Structures

Almost all microwave transistors are of planar form and of silicon n–p–n type. Silicon BJTs dominate for frequency range below S band (i.e. about 3 GHz). These are inexpensive, durable and integrative. This structure is preferred as the *electron mobility* μ_n (=1500 cm²/V·s) and is much higher than the *hole mobility* μ_p (=450 cm²/V·s). Besides, the *carrier densities* of electrons (n_n) and holes (p_n) in different segments of n–p–n transistor are different. The values are $n_n = 1.7 \times 10^{17}$, $p_n = 3.7 \times 10^9$ in emitter, $n_n = 1.7 \times 10^{11}$, $p_n = 3.7 \times 10^{15}$ in base, and $n_n = 1.7 \times 10^{14}$, $p_n = 3.7 \times 10^{12}$ in collector.

In view of Figure 14.3, the geometry of BJTs can be categorised into (i) interdigitated, (ii) overlay and (iii) matrix, mesh or emitter grid. The *figures-of-merit* (M) for interdigitated is given by $M = 2(\ell + w)/\{\ell(w + s)\}$ and for overlay and matrix $M = 2(\ell + w)/\{(\ell + p)(w + s)\}$. The dimensional parameters (ℓ, w, s and p) relate to the fingers and cells shown in Figure 14.4.

Although a transistor can be fabricated in many ways, generally the diffusion and ion implantation processes are preferred. The fabrication begins with a lightly doped n-type epitaxial layer to form a collector. The base region is formed by counter doping this layer by p-type semiconductor through diffusion. The emitter is formed by a shallow heavily doped n-type diffusion or by ion implantation. In an interdigital planar arrangement, contacts for emitter and base are generally located on the semiconductor surface.

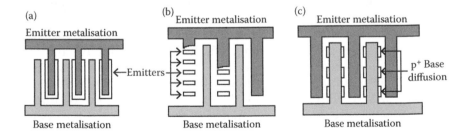

FIGURE 14.3
Three geometrical structures of BJTs. (a) Interdigitated, (b) overlay and (c) matrix.

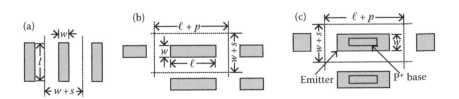

FIGURE 14.4
Dimensional parameters of fingers and unit cells. (a) Interdigitated, (b) overlay and (c) matrix.

The p–n–p structure is simply the complementary of the n–p–n structure wherein n and p are interchanged by p and n. The interdigital structure contains n + ℓ base fingers for n emitter fingers. This number (n) up to a certain limit enhances the power-handling capability of transistor beyond which the parasitic effects degrade the noise figure and upper frequency capability of the device.

14.3.1.2 Transistor Configurations

In general, a transistor can be connected in *CB, CE* or *CC* configurations and the bias voltages therein are connected to terminals in accordance with the operational requirements.

In *common base configuration* (Figure 14.5), the base is made common to the input and output circuits. In p–n–p transistors, the holes are the major contributors towards current. The flow of holes is from the emitter to the collector and down towards the grounded bases terminal. In n–p–n case, all current and voltage polarities are opposite to those of p–n–p. For CB configuration, the input voltage V_{EB} and output current I_C are functions of output voltage V_{CB} and input current I_E and can be expressed accordingly. Currents I_E and I_C are shown in Figure 14.1. The current gain of CB configuration of a transistor is slightly less than one but its voltage gain may be very high. Thus, CB configuration is usually employed in amplifier applications.

In *common emitter configuration* (Figure 14.6), the emitter terminal is made common to input and output circuits. This is the most commonly used configuration. In this configuration, the input current (I_B) and output voltage (V_{CE}) are independent variables. The input voltage (V_{BE}) and output current (I_C) are the functions of output voltage (V_{CE}) and input current (I_B). As the transistor in the CE configuration remains open in the cutoff mode and closed in the saturation mode, it is commonly used as switch or pulse amplifier.

In the *common collector configuration* (Figure 14.7), the collector is made common to input and output circuits. In this configuration, the output voltage of the load (R_L) is taken from emitter terminals instead of the collector as done

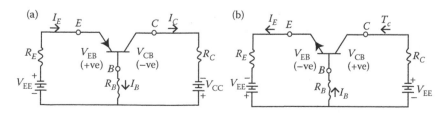

FIGURE 14.5
Common base configurations. (a) p–n–p and (b) n–p–n.

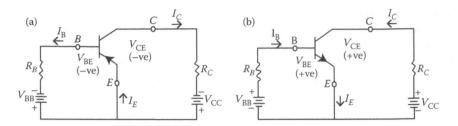

FIGURE 14.6
Common emitter configurations. (a) p–n–p and (b) n–p–n.

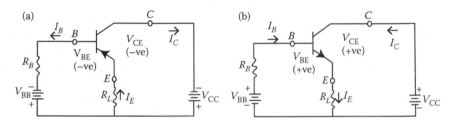

FIGURE 14.7
Common collector configurations. (a) p–n–p and (b) n–p–n.

in CB and CE configurations. There is no current flow in the emitter terminal at load when the transistor operates in the cutoff mode. In the saturation mode, the load current approaches its maximum value. This configuration can also be used as a switch or pulse amplifier. It however differs from the CE configuration as its voltage gain is slightly less than one but its current gain is very high ($\approx \beta + 1$). This aspect makes the common collector configuration useful for power amplification.

14.3.1.3 Biasing and Modes of Operation

For the proper operation of a transistor, its different junctions need to be connected to voltages of different polarities. Such supply connections are referred to as *biasing*, which may be in the form of forward or reversed bias. The term *forward bias* for n–p–n transistor spells that the p side of a junction is connected to the negative polarity of the voltage and the n side to the positive polarity of a p–n junction. In *reversed bias*, these connections are opposite to that of the forward bias. Depending on biasing, a BJT can operate in four different modes, which are referred to as normal (or forward-active), saturation, cutoff and inverse or inverted (or reverse-active) modes. These modes of operations are shown in Figure 14.8.

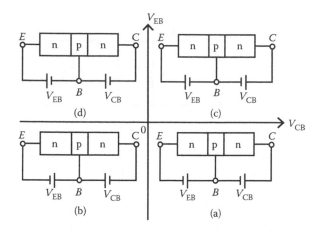

FIGURE 14.8
Different operational modes of an n–p–n transistor. (a) normal mode, (b) saturation mode, (c) cuttoff mode and (d) inverse mode.

In the *normal mode*, the base–emitter junction of an n–p–n transistor is forward-biased and its base–collector is reverse-biased. Most transistor amplifiers operate in the normal mode. Its common base current gain (α) is called the normal α_N.

In the *cutoff mode*, both base–emitter and base–collector junctions are reverse-biased and the transistor acts like an open circuit. In both the saturation and the cutoff modes, a transistor can be used as an ON–OFF device referred to as switching device.

In the *saturation mode*, both transistor junctions are forward-biased. In this mode, the transistor has very low resistance and acts like a short circuit.

In the *inverse (or inverted) mode*, the biasing of junctions is opposite to that of the normal mode, that is, base–emitter junction is reverse-biased and its base–collector junction is forward-biased. Its current gain is designated as the inverse alpha (α_I). In symmetric transistors, $\alpha_N \cong \alpha_I$. In view of unequal doping, these two gains are not actually equal. The use of this mode is not common, except in TTL logic gates.

14.3.1.4 Equivalent Model

In the normal active mode of the CE configuration, for small signal operation of the transistor, normally the hybrid-pi equivalent model is used. This model is illustrated in Figure 14.9. The non-linear or ac parameters of this hybrid-pi model can be expressed as

$$V_{be} = h_{ie} i_b + h_{re} v_{ce} \tag{14.1a}$$

$$i_c = h_{fe} i_b + h_{oe} v_{ce} \tag{14.1b}$$

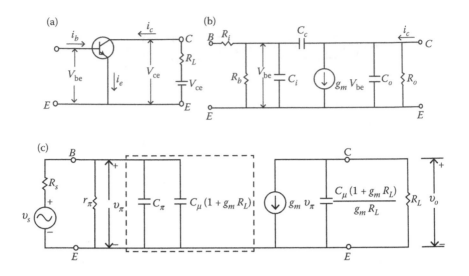

FIGURE 14.9

Hybrid-pi equivalent model of transistor in CE configuration. (a) CE configuration of n–p–n transistor, (b) hybrid π model and (c) simplified equivalent circuit of hybrid π model.

where h_{ie}, h_{re}, h_{fe} and h_{oe} called hybrid parameters can be written as

$$h_{ie} = \left.\frac{\partial v_{be}}{\partial i_b}\right|_{v_{ce}=\text{constant}} \tag{14.2a}$$

$$h_{re} = \left.\frac{\partial v_{be}}{\partial v_{ce}}\right|_{i_b=\text{constant}} \tag{14.2b}$$

$$h_{fe} = \left.\frac{\partial i_c}{\partial i_b}\right|_{v_{ce}=\text{constant}} \tag{14.2c}$$

$$h_{oe} = \left.\frac{\partial i_c}{\partial v_{ce}}\right|_{i_b=\text{constant}} \tag{14.2d}$$

In Equations 14.1 and 14.2, h_{ie} is the dynamic resistance of the emitter–base circuit, h_{oe} is the dynamic conductance of the collector–base circuit and h_{re} represents the relative effectiveness of the emitter and collector voltages in influencing emitter current. Lastly, h_{fe} is the negative of the current gain parameter (α). These parameters hold under the specified conditions in the respective expressions. Also, i_b, i_e and i_c are the base, emitter and collector currents, respectively, V_{be} is the voltage between emitter and base and V_{ce}

is the voltage between the emitter and collector. R_L is the load connected between the emitter and the collector. It also includes R_o, the resistance between the base and emitter junction. $R_i (= r_b + r_\pi)$ is the sum of base spreading resistance r_b and the incremental resistance of the emitter base diode r_π, C_i is the capacitance between the base and emitter junction and C_o is the value of capacitance between the collector and the emitter. The parameter g_m called mutual conductance (or transconductance) is a key transistor parameter defined for small signals. It is the ratio of incremental change of the collector current (∂i_c) at the output terminals owing to the incremental change of the emitter voltage (∂V_{be}) at the input terminals. The resistances R_b, R_e and R_o are, in fact, the parasitic resistances and C_i, C_o and C_c are the parasitic capacitances. The parameter r_π is related to an incremental resistance of the emitter base diode, C_π is the diffusion capacitor, C_μ is the depletion region capacitance of the reverse-biased collector–base junction, $v_\pi = r_\pi i_b$ and $i_c = g_m$ v_π. All these parameters are shown in Figure 14.9.

14.3.1.5 Transistor Characteristics

The current–voltage (I–V) characteristics of an ideal n–p–n transistor connected in CB configuration are shown in Figure 14.10. These characteristics contain three regions, namely, active region, saturation region and cutoff region.

In the *active region*, the base–emitter junction is forward-biased and the base–collector junction is reverse-biased. The collector current (I_c) is independent of the collector voltage and depends only on the emitter current (I_E). When the emitter current becomes zero, the collector current is equal to the reverse saturation current (I_{co}).

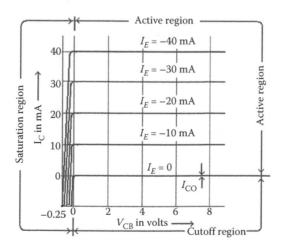

FIGURE 14.10
I–V characteristics of an n–p–n transistor in CB configuration.

In the *saturation region*, both emitter and collector junctions are forward-biased. In this region, the collector current sharply increases in view of the flow of the electron current from the n-side across the collector junction to the p-side base.

In the **cutoff region**, since both the emitter and the collector junctions are reverse-biased, the emitter current is cut off to zero.

14.3.1.6 Transistor Assessment Parameters

As BJTs are used for the amplification of signals, their assessments are to be made in terms of their gains. Since the common collector configuration has no voltage gain, the amplification phenomena mainly relates to common base or common emitter configurations.

Common base n–p–n transistor is normally assessed in terms of its current gain parameter (α), which represents the ratio of the output current ($I_C + I_{C0}$) to the input current (I_E) for a small signal. It can also be defined by the current components crossing the emitter and collector junctions. Thus,

$$\alpha = -(I_C + I_{C0})/I_E \tag{14.3}$$

where I_C is the collector current, I_E is the emitter current and I_{C0} is the collector junction reverse saturation current when the emitter current is zero. These two currents (I_C and I_E) are of opposite signs. Since α represents gain, it has to be positive and thus a negative sign is placed in the expression. Its numerical value varies between 0.9 and 0.995.

The other important assessment parameters include the emitter efficiency (or injection efficiency) and the transport factor. The *emitter efficiency* (γ) is defined as the ratio of the current of the injected carriers at the emitter junction to the total emitter current. The *transport factor* (β^*) is defined as the ratio of injected carrier current reaching the collector junction to the injected carrier current at the emitter junction. The *current gain* (α) can be expressed as the product of the *emitter efficiency* and the *transport factor* (i.e. $\alpha = \gamma\beta^*$).

For *common emitter configuration* in the active region of the n–p–n transistor, the emitter junction is forward-biased and the collector junction is reverse-biased. In this case, the *emitter current gain* (β) is defined as

$$\beta = (I_C + I_{C0})/I_B \tag{14.4}$$

Under the condition $I_E + I_B + I_C + I_{C0} = 0$, the two current gains (α and β) can be related as

$$\beta = \alpha/(1 - \alpha) \tag{14.5}$$

14.3.1.7 Limitations of Transistors

Johnson studied the limitations of transistors in view of the following assumptions:

- The maximum possible velocity of carriers in a semiconductor is the saturated drift velocity (v_s), which is on the order of 6×10^6 cm/s for electrons and holes in silicon and germanium.
- The maximum electric field (E_m) sustained in a semiconductor without dielectric breakdown is about 10^5 V/cm in germanium and 2×10^5 V/cm in silicon.
- The maximum current of a microwave power transistor is limited by its base width.

In view of these assumptions, the four derived relations specify (i) voltage–frequency limitation, (ii) current–frequency limitation, (iii) power–frequency limitation and (iv) power gain–frequency limitation. These relations are as follows:

Voltage–frequency limitation: This limitation is spelled by the following relation:

$$V_m f_t = (E_m v_s)/2\pi = 2 \times 10^{11} \text{V/s for silicon} \quad \text{and}$$

$$1 \times 10^{11} \text{V/s for germanium} \tag{14.6}$$

In Equation 14.6, $V_m = E_m L_{\min}$, $f_t = (1/2\pi\tau)$ and $\tau = L/v$, where V_m is the maximum allowable applied voltage, E_m is the maximum electric field intensity, τ is the charge carrier transit time or the average time for charge carrier moving at an average velocity (v) to traverse the emitter–collector distance (L), v_s is the maximum possible saturated velocity, and $L_{\min}$ is the minimum value of distance (L).

With $v_s = 6 \times 10^6$ m/s, the transit time can be reduced by reducing L. The lower limit of L can be reached when the electric field becomes equal to dielectric breakdown field. L is presently limited to about 25 µm for overlay and matrix devices and 250 µm for interdigitated devices. These values of L set an upper limit on cutoff frequency. In view of non-uniformity of v_s and E field, the cutoff frequency is considerably less than the maximum possible frequency obtained from Equation 14.6.

Current–frequency limitation: This limitation is spelled by the relation

$$(I_m X_c) f_t = (E_m v_s) / 2\pi \tag{14.7}$$

where I_m is the maximum current of the device, X_c ($= 1/\omega_\tau C_o = 1/2\pi f_\tau C_o$), is the capacitive reactance and C_o is the collector–base capacitance. The other parameters have the same meaning as defined earlier. In practice, the area of device cannot be increased indefinitely. This limits the value of capacitance and hence the reactance, which in turn limits the maximum current for a maximum attainable power.

Power–frequency limitation: This limitation is specified by the relation

$$\left(P_m X_c\right)^{0.5} f_\tau = \left(E_m v_s\right)/2\pi \tag{14.8}$$

This equation is obtained by multiplying Equations 14.6 and 14.7 and by replacing $V_m I_m$ with P_m. Accordingly, as the device cutoff frequency increases, the power capacity decreases for a given device impedance. For a given value of the product $E_m v_s$ (for a material), the maximum power that can be delivered to the carriers traversing the transistor is infinite if the transistor cross section is made infinitely large or X_c approaches zero. Figure 14.11 illustrates the results predicted by this equation and the experimental results reported by the manufacturers.

Power gain frequency limitation: This last limitation is given by

$$(G_m V_{th} V_m)^{0.5} f = (E_m v_s)/2\pi \tag{14.9}$$

In this, $G_m \{ = (f_\tau/f)^2(Z_{out}/Z_{in}) = (f_\tau/f)^2(C_{in}/C_{out})\}$ is the maximum available power gain, V_{th} ($= kT/e$) is the thermal voltage, k ($= 1.38 \times 10^{-23}$ J/K) is the Boltzmann's constant, T is the absolute temperature in kelvin (K) and e ($= 1.6 \times 10^{-19}$ C) is the electron charge. In the expression of G_m, Z_{in} and Z_{out} are the input and output impedances, respectively. These impedances are reciprocal to C_{in} (input capacitance) and C_{out} (output capacitance). Parameters V_m and E_m are the same as defined in Equation 14.6.

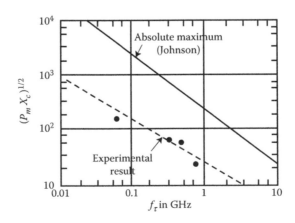

FIGURE 14.11
Power–frequency limitation.

BJTs can be fabricated by using silicon–germanium and only silicon. These have entirely different performances vis-à-vis their utility and gain. Silicon–germanium n–p–n BJTs have higher gain (21 dB at 1.8 GHz), lower noise and higher current gain (h_{fe} = 110 for I_C = 50 mA, V_{CE} = 1.5 V). An n–p–n silicon RF transistor gives higher power, lower gain (15 dB at 13.56 MHz) and lower current gain (h_{fe} = 23 with I_C = 10 A, V_{CE} = 6 V).

In Section 13.5, transferred electron devices (TEDs) were discussed in detail. While comparing these with BJTs, it can be noted that the theory and technology of TEDs is not applicable to transistors. TEDs are bulk devices having no junctions or gates and are fabricated from compound semiconductors such as GaAs, InP and CdTe. These operate with hot electrons whose energy is much greater than the thermal energy. In contrast, microwave transistors operate with either junctions or gates and are fabricated mainly from elemental semiconductors, such as silicon or germanium. These operate with warm electrons whose energy is not much greater than the thermal energy (0.026 eV at room temperature) of electrons in semiconductors.

14.3.2 Heterojunction Bipolar Transistors

Section 14.3.1 was devoted to BJTs, which use the same material for its segments of base, emitter and collector with different degrees of doping. The junctions formed therein are called homojunctions. HBT is an improved version of BJT wherein two different semiconducting materials for emitter and base regions are used. The junction formed between two different materials with different bandgap energy is referred to as heterojunction. The effect of such a junction is to limit the injection of holes from the base into the emitter region since the potential barrier in the valence band is higher than in the conduction band. This allows a high doping density to be used in the base, which results in small base resistance without affecting the gain. Also, a low emitter doping reduces the base emitter junction capacitance, which is one of the key requirements for high-frequency operation.

14.3.2.1 Materials and Energy Band Diagrams

In HBTs, the materials that may be used for the substrate include silicon, gallium arsenide and indium phosphide. Also, the epitaxial layer materials may include silicon, silicon–germanium alloys, aluminium gallium arsenide, gallium arsenide, indium phosphide and indium gallium arsenide. Wide-bandgap semiconductors, for example, gallium nitride and indium gallium nitride are especially promising.

The basic requirement for the formation of heterojunction is the matching of lattice constants (a's) of two semiconductor materials. Any mismatch of these constants may introduce a large number of interface states and degrade the operation. The two commonly used materials include Ge and GeAs as their lattice constants (a = 5.646 Å for Ge and a = 5.653 Å for GaAs) are closely

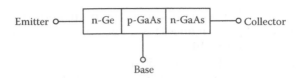

FIGURE 14.12
Heterojunction transistor diagram.

matched. Since each of these materials may be n or p type, there are four possibilities for the formation of a heterojunction. These include (i) p-Ge to p-GaAs junction, (ii) p-Ge to n-GaAs junction, (iii) n-Ge to p-GaAs junction and (iv) n-Ge to n-GaAs junction. A heterojunction transistor diagram having n-Ge, p-GaAs and n-GaAs semiconductors is shown in Figure 14.12.

Figure 14.13 illustrates the energy band diagrams for n-type germanium (Ge) and p-type gallium arsenide (GaAs) drawn side by side. In this figure, E_{V1} and E_{C1} represent the energy level of valence and conduction bands for n-type germanium and E_{V2} and E_{C2} represent the energy level of valence and conduction bands for p-type gallium arsenide. The Fermi level energy for the two materials is shown as E_{F1} and E_{F2}. The difference between two valence band and conduction band energy levels is given as Δ_{EV} and Δ_{EC}. Besides the difference, the energy gap for the two semiconductors is shown as E_{g1} and E_{g2}. In this figure, the vacuum level is used as the reference. The work functions are denoted by y_1 and y_2 for n-Ge and p-GaAs, respectively. As can be seen from Figure 14.13, when an n-Ge and a p-GaAs are isolated, their Fermi levels are not aligned. The different energies of the conduction band edge and the valence band edge are

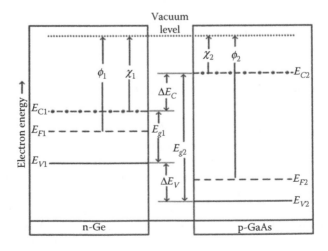

FIGURE 14.13
Energy band diagram for isolated n-Ge and p-GaAs.

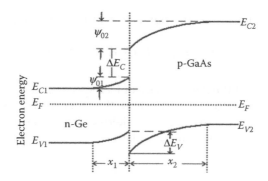

FIGURE 14.14
Energy band diagram for n-Ge and p-GaAs jointed together.

$$\Delta E_V = E_{g2} - E_{g1} - \Delta E_C \qquad (14.10a)$$

$$\Delta E_C = \chi_1 - \chi_2 \qquad (14.10b)$$

where χ is the electron affinity (in eV) and E_g is the bandgap energy. Suffix 1 is used for n-Ge and 2 for p-GaAs in different mathematical terms.

Figure 14.14 shows the energy band diagram when the two semiconductors are joined. This joining results in the alignment of the Fermi energy levels of two materials. Also their energy bands get depleted at the junction. As a result, the electrons from n-Ge are injected into p-GaAs and the holes get transferred from p-GaAs to n-Ge until their Fermi levels are aligned. This leads to the bending or depletion of energy bands at the junction. This energy bending creates a built-in voltage in both sides of the junction shown as ψ_{01} (in n-Ge) and ψ_{02} (in GaAs). The total built-in voltage is given by

$$\psi_0 = \psi_{01} + \psi_{02} \qquad (14.11a)$$

The built in voltages in acceptor and donor regions are given by
$\psi_{01} = V_T \ln |N_a/n_i|$ and $\psi_{02} = V_T \ln |N_d/n_i|$
Thus,

$$\psi_0 = V_T \ln \left| N_a N_d / n_i^2 \right| \qquad (14.11b)$$

In Equation 14.11b, $V_T = KT/q \approx 26 \times 10^{-3}$, N_a and N_d are acceptor and donor densities, n_i is the intrinsic charge density ($n_i \approx 1.5 \times 10^{10}$ for silicon and 1.8×10^6 for G_aA_s, both at 300 K), k is the Boltzmann's constant, T is the temperature in kelvin (K) and q is the electron charge.

The electric flux (D) at the junction is continuous and is given by

$$D = \varepsilon_0 \varepsilon_{r1} E_1 (= \varepsilon_1 E_1) = \varepsilon_0 \varepsilon_{r2} E_2 (= \varepsilon_2 E_2) \qquad (14.12)$$

In Equation 14.12, ε_0 is the permittivity of free space, ε_{r1} and ε_{r2} are the relative permittivities of the two materials; thus, their permittivities are ε_1 ($= \varepsilon_0 \varepsilon_{r1}$) and ε_2 ($= \varepsilon_0 \varepsilon_{r2}$). E_1 and E_2 are the electric fields on the two sides. Space charges on the two sides follow the equality relation

$$x_1 N_d = x_2 N_a \qquad (14.13)$$

where x_1 and x_2 are the depletion widths for n-Ge and p-GaAs, respectively. Also, E fields for the two sides can be written as

$$E_1 = (\psi_{01} - V_1)/x_1 \qquad (14.14a)$$

$$E_2 = (\psi_{02} - V_2)/x_2 \qquad (14.14b)$$

where V_1 and V_2 are the portions of biased voltages in n-Ge and p-GaAs, respectively. In view of Equations 14.12 through 14.14, the resulting expression becomes

$$(\psi_{01} - V_1) \varepsilon_{r1} N_{d1} = (\psi_{02} - V_2) \varepsilon_{r2} N_{d2} \qquad (14.15)$$

As the potential ($\psi_{01} + \psi_{02} + \Delta E_C/q$) between E_{c1} and E_{c2} (Figure 14.14), that is, across the junction for electron injection is very high, the electron current from n-Ge to p-GaAs is very small. In contrast, the hole current from p-GaAs side to n-Ge is dominant due to the low potential barrier for the injection of holes. Now, if A is the cross section, q is the charge, D_p is the hole diffusion constant, L_p is the hole diffusion length, V is the bias voltage, V_T is the voltage equivalent of temperature and p_{n0} is the minority equilibrium hole density in n-Ge, the current is given by

$$I = \{(A\, q\, D_p p_{n0})/L_p\}\{\exp(V/V_T) - 1\} \qquad (14.16)$$

Tapering of a bandgap leads to a field-assisted transport in the base. It speeds up the transport through the base and increases the frequency response.

14.3.2.2 Advantages

HBT has lower forward transit time (τ_f) and much lower base resistance (R_b). It has the ability to turn off devices with a small change of base voltage. For a typical (AlGaAs/GaAs) HBT, τ_f is 4 ps whereas for silicon BJT, it is of the order of 12 ps. Similarly, R_b of this HBT in terms of equivalent power is of the order of 70 W, whereas it is 200 W for silicon BJT. Thus, HBTs can handle signals of several hundred gigahertz.

14.3.2.3 Applications

These are used in (i) modern ultra-fast circuits in RF systems, (ii) high-power efficiency applications such as power amplifiers in cellular phones and (iii) high-reliability applications such as laser drivers. It is a potential candidate for high-speed switching devices such as GaAs MESFETs.

14.4 Field Effect Transistors

The FETs mainly include (i) JFET, (ii) MESFET and (iii) MOSFET. These are described as below.

14.4.1 Junction Field Effect Transistor

FET is a unipolar device. It may be in the form of a p–n junction gate or a Schottky-barrier gate. FETs of the first category are called JFETs while the second category is referred to as MESFETs. JFETs can be further classified as n-channel JFET and p-channel JFET. In n-channel JFET, n-type of material is sandwiched between two highly doped layers of p-type material designated as p^+. In p-channel JFETs, the sandwiched part is a p-type semiconductor.

14.4.1.1 JFET Structure

Figure 14.15 illustrates an n-channel JFET with the two p-type regions. These regions are referred to as *gates*. Each end of the n-channel is joined by a metallic contact. The left-side contact is referred to as the *source* as the electrons emerge from this terminal. The right-side contact is the *drain* contact as it drains out the electrons. This designation of contacts is in accordance with the directions of biasing voltages. This figure also shows its circuit symbol. The drain current flows from the drain to the device.

In p-channel JFET, the polarities of both the biasing voltages are to be reversed. Thus, the direction of the current flow also gets reversed. As the mobility of the electrons is higher than that of the holes, n-channel JFETs are preferred in view of their higher conductivity and faster speed.

JFET (often referred to simply as FET) is a three-terminal device capable of both amplification and switching at microwave range. The three terminals of FET are named as gate, source and drain. With respect to a BJT, the gate of a FET corresponds to the base, the drain to the collector and the source to the emitter terminal.

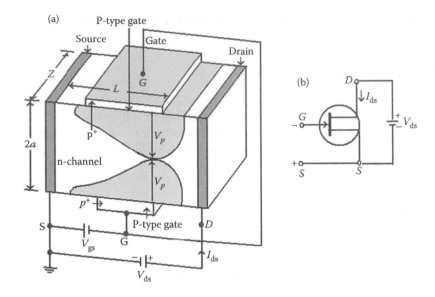

FIGURE 14.15
n-channel JFET. (a) Structure and (b) circuit symbol.

When JFET is used as an amplifier, its gate is most often configured as the input terminal, the source is grounded and the drain is the output terminal. Its output current (I_d) is controlled by the input voltage (V_g). This is called *common source configuration* since the source is common to the input and output ground connections. Although unusual, it is possible to have a *common gate configuration* in which the gate is grounded. Such amplifier configuration provides lesser voltage gain than that of the common source amplifier. This configuration, however, is easier to impedance match than the normal common source amplifier.

14.4.1.2 Types of Microwave FETs

FTEs can also be classified as (i) low-noise FETs, (ii) power FETs and (iii) switch FETs. *Low-noise FETs* are optimised to provide the lowest possible noise figure at very low voltage and power (about 1.5 V and 10 ma). *Power FETs* have higher breakdown voltage than low-noise FETs and can operate at higher voltages. These have much larger periphery than that of low noise FETs. *Switch FETs* operate passively where there is no drain current and no gain. In these, the gate voltage is used only to switch the device from resistive to small capacitive element. These can be configured with a transmission line in series or in shunt, with the source grounded. In series case, the drain and the source act as the input or the output. There are many design possibilities to the gate structure. One of these is called a *meandered gate* wherein multiple FET gates are hooked up in series, rather than through a single bus-bar.

FETs can also be divided into depletion mode and enhancement mode. A *depletion mode FET* is one where the gate is mainly used to reduce the current within the channel. Most common microwave FETs are depletion mode FETs. An *enhancement mode FET* does not conduct drain to source until the gate is slightly forward-biased. The enhancement mode FETs can be considered as a depletion mode FET with a zero-volt pinch-off voltage. One of the limitations of enhancement mode FETs is that the turn-on voltage (0.7 V) of the Schottky contact cannot be exceeded. Thus, the gate-etching process has to be well controlled to produce a FET with pinch-off of 0 V or nearer to a few tenths of a positive volt.

14.4.1.3 Principles of Operation

With reference to the parameters shown in Figure 14.15, the steps involved in the operation of a JFET are as given below:

1. Under a normal operating condition, when there is no gate voltage (i.e. $V_g = 0$), the drain current (I_d) is also zero and the channel between the gate junctions is entirely open.
2. When a small drain voltage (V_d) is applied between the drain and the source, the n-type semiconductor bar acts as a simple resistor and the current I_d increases linearly with V_d.
3. With the application of a reverse gate voltage (V_g) across the p–n gate junctions, the majority of the free electrons get depleted from the channel and space charge regions get extended into the channel.
4. As V_d is further raised, the space charge regions expand and join together and all free electrons get completely depleted into the joined region. This condition is called pinched-off.
5. Under pinched-off condition, I_d remains almost constant, but V_d is continuously increased.

Pinch-off voltage (V_p): The pinch-off voltage is the gate reverse voltage (V_g) that removes all the electrons from the channel. This is where the drain–source terminals start to look like an open circuit, and no appreciable current flows even at high drain–source potentials. In practice, there is always some residual current and the actual V_p measurement must make an allowance for this. It is given by the relation

$$V_p = (q\, N_d a^2)/(2\varepsilon_s)\text{V} \tag{14.17}$$

Equation 14.17 spells that V_p is a function of charge q in coulombs, doping concentration N_d in electrons/m³, channel height a and permittivity ε_s of the material in farad/metre. This relation basically emerges from the Poisson's equation for the voltage in the n channel with volume charge density ρ (in coulombs/m^3) and on application of appropriate boundary conditions. If doping is increased to the limit set by the gate breakdown voltage and V_p is made large enough, the drift saturation effect becomes dominant. The pinch-off voltage (V_p) under saturation condition can be written as

$$V_p = V_d + |V_g| + \psi_o \tag{14.18}$$

where $|V_g|$ is the absolute value of the gate voltage and ψ_o is the built-in or barrier voltage at the junction. In view of Equation 14.18, the saturation drain voltage is given by

$$V_{d\,sat} = \{(q\,N_d a^2)/(2\varepsilon_s)\} - |V_g| - \psi_o \tag{14.19}$$

When the gate bias voltage is zero, the JFET has a conducting channel between the source and the drain. This state is referred to as ON state and the device is called a normally *ON JFET*. For OFF state, a gate voltage is to be applied to deplete all carriers in the channel. The device is now referred to as depletion mode JFET or *D-JFET*.

Channel resistance: The drain current of n-channel JFET is dependent on the drain and gate voltages and the channel resistance. The channel resistance for n-channel is expressed as

$$R = \frac{L}{2q\mu_n N_d Z(a - W)} \tag{14.20}$$

where μ_n is the electron mobility, L is the length in the x direction, Z is the distance in the z direction, W is the width of the depletion layer and 'a' is the width between the p–n junction. The parameters q and N_d were defined earlier.

Drain current: The expression for the drain current (I_d) is given as follows:

$$I_d = I_p \left[\frac{V_d}{V_p} - \frac{2}{3}\left(\frac{V_d + |V_g| + \psi_0}{V_p} \right)^{1/2} + \frac{2}{3}\left(\frac{|V_g| + \psi_0}{V_p} \right)^{3/2} \right] \tag{14.21}$$

where the pinch-off current (I_p) is given as

$$I_p = \frac{\mu_n q^2 N_d^2 Z a^3}{L\varepsilon_s} \tag{14.22}$$

Figure 14.16 illustrates the normalised ideal I–V characteristics of a typical JFET for a pinch-off voltage of 3.2 V.

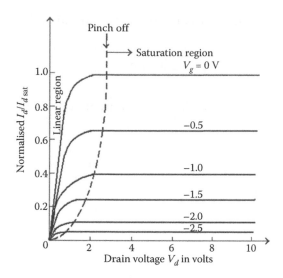

FIGURE 14.16
Normalised ideal I–V characteristics of a typical JFET.

In the linear region, $V_d \ll (|V_g| + \psi_0)$; thus, the expression of drain current reduces to

$$I_d = \frac{I_p V_d}{V_p} \left[1 - \left(\frac{|V_g| + \psi_0}{V_p} \right)^{1/2} \right] \tag{14.23}$$

Drain conductance: The drain conductance (g_d) is given by

$$g_d = \left. \frac{\partial I_d}{\partial V_d} \right|_{V_g = \text{constant}} = \frac{I_p}{V_p} \left[1 - \left(\frac{|V_g| + \psi_0}{V_p} \right)^{1/2} \right] \tag{14.24}$$

Transconductance: The transconductance (g_m) is expressed by

$$g_m = \left. \frac{\partial I_d}{\partial V_g} \right|_{V_d = \text{constant}} = \frac{I_p V_d}{2 V_p^2} \left[1 - \left(\frac{V_p}{|V_g| + \psi_0} \right)^{1/2} \right] \tag{14.25}$$

Drain saturation current: The drain current gets saturated at the pinch-off voltage, thus by putting $V_p = V_d + |V_g| + \psi_0$, the expression for the drain saturation current (I_d) becomes

$$I_d = I_p \left[\frac{1}{3} - \left(\frac{|V_g| + \psi_0}{V_p} \right) + \frac{2}{3} \left(\frac{|V_g| + \psi_0}{V_p} \right)^{1/2} \right] \tag{14.26}$$

Saturation drain voltage: The corresponding saturation drain voltage ($V_{d\ sat}$) is given by

$$V_{d\ sat} = V_p + |V_g| + \psi_0 \tag{14.27}$$

Mutual conductance at saturation: The mutual conductance in saturation region is

$$g_m = \left.\frac{\partial I_d}{\partial V_g}\right|_{V_d=\text{constant}} = \frac{I_p}{V_p}\left[1 - \left(\frac{|V_g|+\psi_0}{V_p}\right)^{1/2}\right] \tag{14.28}$$

where

$$\frac{I_p}{V_p} = \frac{2\mu_n q N_d a Z}{L} \tag{14.29}$$

Drain resistance: For small signal, the drain or output resistance (r_d) is given by

$$r_d = \left.\frac{\partial V_d}{\partial I_d}\right|_{V_g=\text{constant}} \tag{14.30}$$

Amplification factor: The amplification factor (μ) is given by

$$\mu = \left.\frac{\partial V_d}{\partial V_g}\right|_{I_d=\text{constant}} = r_d g_m \tag{14.31}$$

Cutoff frequency: The cutoff frequency (f_c) for JFET is

$$f_c = \frac{g_m}{2\pi C_g} = \frac{2\mu_n q N_d a^2}{\pi \varepsilon_s L} \tag{14.32}$$

where C_g represents the capacitance between the gate and the source and is given by

$$C_g = LZ\varepsilon_s/2a \tag{14.33}$$

Pinch-off region: When the E field appears along the x-axis, the drain end of the gate is more reverse-biased than the source end. The boundaries of the

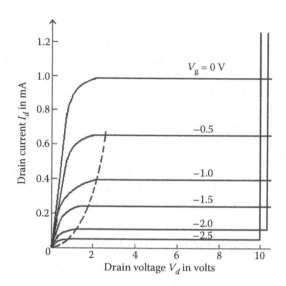

FIGURE 14.17
I–V characteristics with breakdown voltages of a typical JFET.

depletion region do not remain parallel to the centre of the channel. As the drain voltage and drain current are further increased, the channel is finally pinched-off.

Breakdown voltage: For a constant gate voltage, the increase in the drain voltage beyond a certain limit results in avalanche breakdown across the gate junction. Consequently, the drain current rises very sharply. Thus, it is the gate–drain breakdown voltage, which is measured on the I–V curves. At high drain–source potential and near pinch-off, the I–V curves tend to bend up. The breakdown voltage (combination of $V_g = -4$ V and $V_d = 6$ V) is approximately 10 V. This breakdown voltage (V_b) is related to V_d and V_g as

$$V_b = V_d + |V_g| \qquad (14.34)$$

The I–V characteristics of Figure 14.17 give V_b for different constant values of V_g.

Knee voltage: It is the voltage at which the curves transition from linear to saturation. In the linear region, I_d depends on both V_g and V_d (from $V_d = 0$ to ≈ 2 V). In the saturation region, I_d depends mainly on V_g and not on V_d. This is the right side of the curve, beyond $V_d = 2$ V.

14.4.1.4 Comparison between FETs and BJTs

As discussed earlier, BJTs are bipolar devices, whereas FETs are unipolar. In case of BJTs, the conduction involves both holes and electrons. In FETs, it involves holes for p-channel and electrons for n-channel. Both are three terminal devices that are named as source, drain and gate in FETs and base, emitter and collector in BJTs. The output in FETs is determined by the field effect of applied voltage, whereas in BJTs it is determined by input current. Thus, FETs are referred to as voltage-controlled and BJTs as current-controlled devices. The change in output current for the same change in applied voltage in FETs is less as compared to BJTs. In case of FETs, input impedance, efficiency and operating frequency are higher and the temperature stability is more than that of BJTs. However, FETs have less ac gain as an amplifier, low noise figure and small relative size in comparison to BJTs. The IC chip stability of FETs is better than BJTs and these can have voltage gain in addition to the current gain, which is not possible in BJTs.

14.4.2 Metal Semiconductor Field Effect Transistors

MESFET is quite similar to JFET in construction and terminology. It differs from JFET since it uses a Schottky (metal-semiconductor) junction instead of a p–n junction for a gate and is sometimes referred to as Schottky-barrier field effect transistor. MESFETs may have n-channel or p-channel. It utilises the advantage of high-speed characteristics of GaAs as the base semiconductor material. Thus, these are faster but more expensive than silicon-based JFETs or MOSFETs. These can operate up to about 45 GHz and are commonly used in microwave communication and radar circuits. As the scale of integration increases, it is all the more difficult to use MESFETs as the basis for digital ICs in comparison to CMOS silicon-based fabrication. The current in the channel causes a voltage drop along its length, which results in a charge-depletion region.

MESFETs are usually constructed by using compound semiconductor (such as GaAs, InP or SiC) technologies. Thus, these lack in high-quality surface passivation. The GaAs MESFET has higher electron mobility, higher electric field and higher electron saturation drift velocity than the silicon devices. As a result, the GaAs MESFETs have larger output power and low noise figure and are commonly used in microwave integrated circuits wherever high power, low noise and broadband amplification is desired. The maximum reported cutoff frequency for GaAs MESFETs is 168 GHz.

14.4.2.1 Structure

Figure 14.18 shows diagram of a GaAs MESFET wherein its terminals, different layers and regions are illustrated. This also illustrates the coordinates to show the alignments of different components along with its circuit symbol.

Chromium (Cr), which has an energy level near the centre of GaAs bandgap, is used as a substrate in GaAs MESFETs. As Cr is the dominant impurity, the Fermi level is pinned near the centre of the bandgap. This process results in a semi-insulating GaAs substrate with a very high resistivity

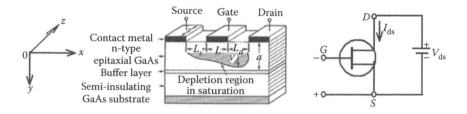

FIGURE 14.18
Schematic diagram of a GaAs MESFET.

(near to 10^8 ohm-cm). A thin layer of lightly doped n-type GaAs is epitaxially grown on this non-conducting substrate to form the channel region of FET. In many cases, a high resistivity GaAs epitaxial layer is grown between the n-type GaAs layer and the substrate. Such a layer is called the buffer layer. To define the patterns in the metal layers such as Au–Ge for source and drain ohmic contacts and in Al layer for Schottky-barrier gate contacts, the photolithographic process may be employed. The GaAs MESFET can also be grown by ion implantation method. A thin n-type layer can be formed at the surface of the substrate by implanting Si or other suitable donor impurity. This process, however, requires annealing to remove the radiation damage. In these methods, the source and drain contacts may be improved by further n^+ implantation in these regions.

The MESFET devices have interdigitated structure, which are fabricated by using an n-type GaAs epitaxial film of about 0.15–0.35 µm thick on a 100 µm semi-insulating substrate. For doping of the n-channel layer, either sulphur or tin is used with doping concentration N (between $8 \times 10^{16}/cm^3$ to $2 \times 10^{17}/cm^3$). The range of electron mobility in the layer is between 3000 and 4500 $cm^2/V \cdot s$. The Schottky gate is formed by evaporated aluminium. The source and drain contacts are formed of Au–Ge, Au–Te or Au–Te–Ge alloys. The metallisation pattern of gold is used to bring the source, drain and gate contacts out of bonding pads over the semi-insulating surface. A buffer layer is often fabricated between the active n-type epitaxial layer and semi-insulating layer. Its thickness is kept around 3 µm with a doping concentration of $10^{15}–10^{16}$ per cm.

14.4.2.2 Principles of Operation

When voltages are applied to reverse bias the p–n junction between the source and the gate and to forward bias the junction between the source and the drain, the majority carriers (electrons) flow in the n-type epitaxial layer from the source electrode, through the channel beneath the gate to drain electrode. This current causes voltage drop in the channel along its length. The Schottky-barrier-gate electrode becomes more reverse-biased towards the drain. A charge depletion region is formed in the channel. This gradually pinches off the channel against the semi-insulating substrate towards the drain end. With the increase of reverse bias between the source and the gate, the height of the charge depletion region also increases. The decrease in channel height in the pinch-off region increases the channel resistance. The drain current I_d is modulated by the gate voltage V_g. This process leads to a family of curves shown in Figure 14.19.

Transconductance: The transconductance of a FET is given as

$$g_m = \frac{dI_d}{dV_g}\bigg|_{V_g = \text{constant}} \tag{14.35}$$

For a fixed drain-to-source voltage V_d, the drain current I_d is a function of reverse biasing gate voltage V_g. As I_d is controlled by the field effect of V_g, the device is referred to as field effect transistor. Any increase in I_d results in the voltage drop between the source and the channel and in further reverse biasing of p–n channel. Its continuous increase eventually results in pinch-off of the channel. When the channel gets pinched-off, the drain I_d is almost constant and does not vary with any increase in the drain to source voltage V_d.

Pinch-off voltage: It is the gate reverse bias voltage, which removes all the free charges from the channel. The expression for pinch-off voltage is given as

$$V_p = (qN_d a^2)/2\varepsilon_s \tag{14.36}$$

where the involved parameters are the same as were defined for JFET.

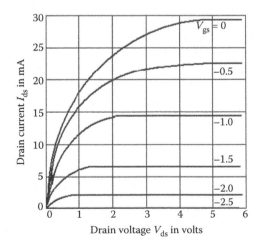

FIGURE 14.19
I–v characteristics of a GaAs MESFET.

14.4.2.3 Small-Signal Equivalent Circuit

In MESFETS, the channel is of very short length. The velocity saturation in its channel occurs before reaching the pinch path. Its electronic characteristics depend on many intrinsic and extrinsic parameters. The intrinsic parameters include transconductance (g_m), drain conductance (G_d), input resistance (R_i), gate–source capacitance (C_{gs}) and the gate–drain (or feedback) capacitance (C_{gd}). The extrinsic parameters include gate metallisation resistance (R_g), source–gate resistance (R_s), drain–source capacitance (C_{ds}), gate bonding-pad parasitic resistance (R_p), gate bonding-pad parasitic capacitance (C_p) and the load impedance (Z_L). Figure 14.20 illustrates the cross section of GaAs MESFET with its various parameters and equivalent circuit.

The values of the intrinsic and extrinsic parameters depend on the type of channel, materials used, structure of the device and the dimensions of the Schottky-barrier-gate FET. The power gain and efficiency are closely related to the extrinsic resistances and will decrease with the increase in these resistances. The noise figure however improves as these resistances become high. Since the increase in the channel doping N decreases the relative influence of the feedback capacitance, increases the transconductance and the dc open circuit gain, doping should be made as high as possible. This increase, however, reduces the breakdown voltage of the gate. A value of 10^{18} per cubic cm may be the upper limit set.

Drain current: The drain current (I_d) of a Schottky-barrier-gate FET is given by

$$I_d = I_p \frac{3(u^2 - \rho^2) - 2(u^3 - \rho^3)}{1 + \eta(u^2 - \rho^2)} \tag{14.37}$$

where I_p the saturation current for the Shockley case at $V_g = 0$ is given by

$$I_p = (q\, N_d \mu\, a\, Z\, V_p)/3L \tag{14.38a}$$

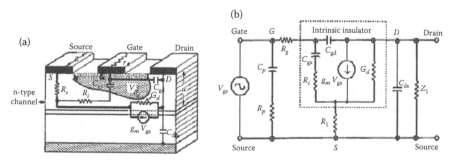

FIGURE 14.20
GaAs MESFET. (a) Different R and C parameters and (b) equivalent circuit.

u, the normalised sum of drain and gate voltage with respect to pinch-off voltage, is

$$u = \left[\{V_d + |V_g|\}/V_p\right]^{0.5} \tag{14.38b}$$

ρ, the normalised gate voltage with respect to pinch-off voltage, is given by

$$(\rho = |V_g|/V_p)^{0.5} \tag{14.38c}$$

η, the velocity ratio, is given as

$$\eta = (\mu|V_p|/v_sL) = v/vs \tag{14.38d}$$

In Equations 14.37 and 14.38, μ is the low field mobility in m²/V·s, Z is the channel depth or width, L is the gate length, v_s is the saturation drift velocity and V_p is the pinch-off voltage. The symbols q, N_d and a are defined earlier.

In practice, the drain current and the mutual conductance for a GaAs MESFET are given by

$$I_{ds} = I_{dss}\left(1 + \frac{|V_g|}{V_p}\right) \tag{14.39}$$

$$g_m = \frac{2I_{dss}}{|V_p|}\left(1 + \frac{|V_g|}{V_p}\right)^{1/2} \tag{14.40}$$

14.4.2.4 MESFET Characteristics

Figure 14.21 illustrates a GaAs MESFET and its various curves in the saturation region. In MESFET, the narrowest channel cross section is under the drain end of the gate. In the electric field distribution curve, the peak of the field appears near the drain. The drift velocity rises to peak at x_1, close to the centre of the channel, and falls to the low saturated value under the gate edge. Since the channel cross section is getting narrowed down, the region requires heavy accumulation of electron in order to preserve the continuity of the current. Also, since the electrons are becoming progressively slower with increasing x, between x_2 and x_3 the process is just the reverse. The channel widens and electrons move faster resulting in a strong depletion layer. The charges in the accumulation and depletion layers are nearly equal. Also, most of the drain voltage drops in this stationary dipole layer.

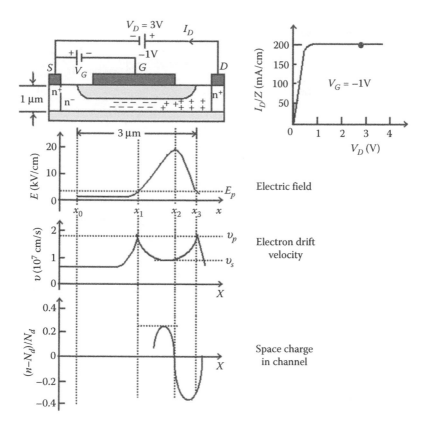

FIGURE 14.21
A GaAs MESFET and its various curves in the saturation region.

The drain current for a Si MESFET is given by

$$I_d = Z\,q\,n(x)\upsilon(x)\mathrm{d}(x) \tag{14.41}$$

where Z is the channel width or depth, $n(x)\,(=N_d)$ is the density of conduction electrons, $\upsilon(x)$ is the drift velocity, $\mathrm{d}(x)$ is the thickness of conductive layer and x represents the coordinate of the direction of the electron drift.

The current–voltage characteristics of a Si MESFET are shown in Figure 14.22. It is to be noted that any segment given in the earlier part of the figure exists even if it is not shown in the latter part.

As shown in Figure 14.22a, there is no metal gate electrode. The source and drain contacts are made at the two ends of the surface of the conducting layer. When a positive voltage $V\,(=V_{ds})$ is applied, electrons will flow from the source to the drain. Figure 14.22b shows the addition of a metal gate, which is also shorted (connected) to the source terminal. The application of a

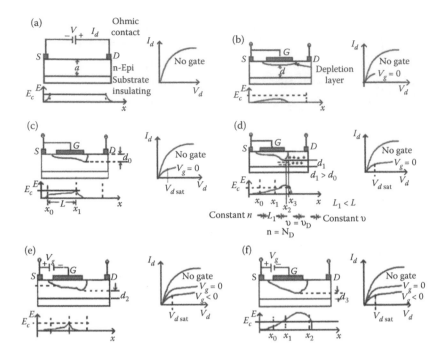

FIGURE 14.22
I–V characteristics of a Si MESFET. (a) No metal gate, (b) metal gate added, (c) increase in V_{ds} widens depletion layer, (d) V_{ds} exceeds V_{dsat}, (e) gate to channel junction reverse-biased and (f) V_{gs} becomes negative.

small drain voltage creates a depletion layer just below the gate. The flow of a current ($I_d = I_{ds}$) between the drain and the source is given by

$$I_{ds} = I_{dss}(1 + |V_s|/V_p) \tag{14.42}$$

Figure 14.22c indicates that the increase in drain voltage (V_{ds}) widens the depletion layer. As a result, the conductive cross section decreases. In order to maintain the current through the channel, this decrease in cross section is compensated by an increase of electron velocity. With further increase of drain voltage, the electron attains their saturation velocity (v_s).

As indicated in Figure 14.22d, when the drain voltage is further increased and exceeds $V_{d\,sat}$, the depletion layer further widens but towards the drain. Figure 14.22e shows that the gate-to-channel junction is reverse-biased. It further widens the depletion layer. Figure 14.22f illustrates that when the gate voltage V_{gs} becomes more negative, the channel is almost pinched-off and the drain current I_{ds} is nearly cut off.

Cutoff frequency (f_c): The cutoff frequency of a Schottky-barrier-gate FET depends on the formation of the transistor. In a wide-band lumped circuit, it is given by

$$f_c = g_m/(2\pi C_{gs}) = v_s/4\pi L \tag{14.43}$$

where g_m is the transconductance, L is the gate length, v_s is the saturation drift velocity and C_{gs} is the gate–source capacitance. The C_{gs} is given by

$$C_{gs} = \left.\frac{dQ}{dV_{gs}}\right|_{V_{gd}=\text{constant}} \tag{14.44}$$

From the charge carrier transit time (τ) discussed in connection with the power frequency limitation of BJTs, the cutoff frequency is obtained to be

$$f_c = 1/2\pi\tau = v_s/2\pi L \tag{14.45}$$

This value is just half of the value obtained in Equation 14.43.

Maximum oscillation frequency (f_{max}): The maximum frequency of oscillation, in a distributed circuit, depends on the transconductance and the drain resistance. It is given by

$$f_{max} = \frac{f_c}{2}(g_m R_d)^{1/2} = \frac{f_c}{2}\left[\frac{\mu E_p(\mu_m - \rho)}{v_s(1 - u_m)}\right]^{1/2} \text{ Hz} \tag{14.46}$$

where R_d is the drain resistance, g_m is the transconductance of the device, E_p is the electric field at the pinch-off region in the channel, u_m is the saturation normalisation of u and v_s, μ and ρ have been defined earlier.

For $\rho = 0$ (i.e., $V_{gs} = 0$) and $\eta \gg 1$, so that $f_c = v_s/4\pi L$ and $u_m = (3/\eta)^{1/3} \ll 1$, f_{max} becomes

$$f_{max} = \gamma(v_s/L)(3/\eta)^{1/6} \text{ Hz} \tag{14.47}$$

where in case of GaAs, $\gamma = 0.14$ for $\mu E_p/v_s = 13$ and $\gamma = 0.18$ for $\mu E_p/v_s = 20$.

In view of the experimental results

$$f_{max} = (33 \times 10^3)/L \text{ in Hz} \tag{14.48}$$

The highest value of f_{max} for maximum power gain with matched input and output networks is

$$f_{max} = \frac{f_c}{2}\left(\frac{R_d}{R_s + R_g + R_i}\right)^{1/2} \text{ Hz} \tag{14.49}$$

where R_d is the drain resistance, R_s is the source resistance, R_g is the gate metallisation resistance and R_i is the input resistance.

14.4.2.5 Advantages and Limitations of Using Schottky Barrier Diode

The relative values of voltage drops for normal diode are between 0.7 and 1.7 V, whereas for the Schottky diode, these are between 0.15 and 0.45 V. Also, the relative values of reverse recovery time for normal diode are about 100 ns, whereas for the Schottky diode, this value is almost zero. With the Schottky diode, lower resistance devices are possible and fabrication is greatly simplified. Such devices have relatively low reverse voltage (~50 V), high reverse leakage current and higher system efficiency.

14.4.2.6 Applications

MESFET's main application areas include military communications, satellite communications, commercial optoelectronics, power amplifier for output stage of microwave links, as front-end low noise amplifier of microwave receivers and as a power oscillator.

14.4.3 Metal-Oxide-Semiconductor Field Effect Transistors

MOSFET is a four-terminal device and belongs to the class of FETs. Its terminals are referred to as the source, gate, drain and substrate wherein the source may be made common to the substrate. It may be constructed with both n-type and p-type channels. It is used for amplifying or switching electronic signals. Currently, it is the most commonly used transistor in both digital and analog circuits. Its distinguishing features include the presence of an insulator between the gate and the remainder of the device.

Its current flow is controlled by an applied vertical electric field. When the gate bias voltage is zero, the two back-to-back p–n junctions between the source and the drain prevent the flow of current in either direction. When a positive voltage is applied between the gate and the source, negative charges are induced in the channel to cause a flow of current. Since the current is controlled by the electric field, this device is called the JFET.

14.4.3.1 Structure

In MOSFETs, a voltage on the oxide-insulated gate electrode induces a conducting channel between the source and the drain. This channel can be of n- or p-type and in accordance the device is termed as nMOSFET (or nMOS) and pMOSFET (or pMOS). Figure 14.23 illustrates the physical structure of an nMOSFET along with coordinates of alignments and circuit symbol.

In *n-channel MOSFETs*, two highly doped n⁺ sections are diffused into a slightly doped p-type of substrate. These n⁺ sections act as the source and the drain and are separated by about 0.5 µm. A thin layer of insulating silicon

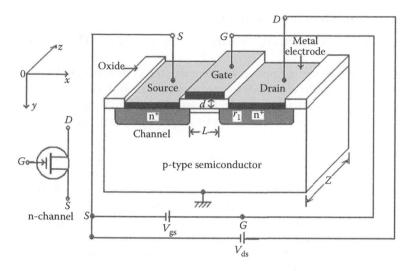

FIGURE 14.23
Physical structure of an nMOSFET.

dioxide (SiO_2) is grown over the surface of the structure. The insulator is also provided with a metal contact, which is called the gate.

In *p-channel MOSFETs*, two highly doped p^+ sections are diffused into a slightly doped n-type of substrate. These sections are for the source and the drain. For gate electrode, the heavily doped polysilicon or a combination of silicide and polysilicon can be used.

In microwave integrated circuits, a MOSFET is commonly surrounded by a thick layer of oxide to isolate it from the adjacent devices. Its basic parameters include the length of the channel (L), depth of the channel (z), insulator thickness (d) and the junction thickness of n^+ section (r_1). The channel length refers to the distance between two $n^+ - p$ junctions just beneath the insulator. The orders of these parameters L, z, d and r_1 may be about (say) 0.5, 5, 0.1 and 0.2 µm, respectively.

In earlier MOSFETs, metal (aluminium) was used as the gate material. Later it got replaced by a layer of polysilicon (polycrystalline silicon). Thus the word 'metal' involved in its name became almost a misnomer. This replacement was mainly due to the capability of polysilicon to form self-aligned gates. Since with polysilicon it is difficult to increase the speed of operation of transistors, the metallic gates are again gaining ground. A related term almost a synonym of the MOSFET is the insulated-gate field effect transistor (IGFET). This term is more inclusive, since many MOSFETs use a gate that is not metal, and a gate insulator that is not oxide. Metal-insulator-semiconductor FET (MISFET) is yet another synonym.

14.4.3.2 Operation of n-Channel MOSFET

In view its physical structure, the functioning of an nMOSFET is as follows.

1. When no voltage is applied to the gate, the connection between the source and drain electrodes corresponds to a link of two p–n junctions connected back-to-back. Under this biasing, no current will flow from the source to the drain except a small reverse leakage current.

2. When a positive voltage is applied at the gate with respect to the source (wherein the substrate is either grounded or connected to the source), positive charges are deposited on the gate metal. It leads to the induction of negative charges in the p-type semiconductor at the insulator–semiconductor interface. As a result, a depletion layer within a thin surface region containing mobile electrons is formed. These induced electrons form the n-channel and allow the flow of current from the drain to the source.

3. For a given gate voltage (V_g), the drain current (I_d) gets saturated for some drain voltage (V_d). The minimum gate voltage (V_g) required to induce the channel is called threshold voltage (V_{th}). Before a conducting n-channel (mobile electrons) is induced, the gate voltage must exceed the threshold voltage (i.e. $V_g > V_{th}$) in n-channel MOSFET. In case of p-channel MOSFETs, the gate voltage must be more negative than the threshold voltage (i.e. $V_g < V_{th}$) before p-channel (mobile holes) is formed.

The modes of operation for both n-channel and p-channel MOSFETs include the enhancement mode and depletion mode. These modes are shown in Figure 14.24.

- *The n-channel enhancement mode* (Figure 14.24a): When the gate voltage is zero, the channel conductance is very low and the device does not operate. A positive voltage must be applied to the gate to form

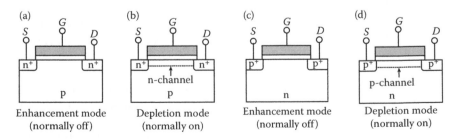

FIGURE 14.24
Modes of operation of MOSFETs (a and b) n-channel (c and d) p-channel.

an n-channel for conduction. The drain current is enhanced by the positive voltage. Such a device is referred to as enhancement mode (normally off) n-channel MOSFET.

- *The n-channel depletion mode* (Figure 14.24b): If an n-channel exists at equilibrium (or zero bias), a negative gate voltage must be applied to deplete the carriers in the channel. As a result, the channel conductance is reduced and the device is turned off. Such a device is called as the depletion mode (normally on) n-channel MOSFET

- *The p-channel enhancement mode* (Figure 14.24c): A negative voltage must be applied to the gate to induce a p-channel for conduction. Such a device is referred to as the enhancement mode (normally off) p-channel MOSFET.

- *The p-channel depletion mode* (Figure 14.24d): A positive voltage must be applied to the gate to deplete the carriers in the channel for non-conduction. Such a device is called the depletion mode (normally on) p-channel MOSFET.

Figure 14.25 illustrates the symbols, output I–V characteristics and the transfer characteristics of the modes of MOSFETs.

14.4.3.3 Circuit Symbols

In Figure 14.26, circuit symbols for n- and p-channels for enhancement and depletion modes are shown. In these symbols, a line represents a channel. The source and the drain leave this line at right angles and then bend back at right angles into the same direction as the channel. Sometimes, three line segments are used to represent the enhancement mode and a solid line for the depletion mode. Another line is drawn parallel to the channel to represent the gate. The bulk connection, if shown, is shown connected to the back of the channel with an arrow indicating PMOS or NMOS. These arrows always point from P to N. Thus, an NMOS (N-channel in P-well or P-substrate) has the arrow pointing in (from the bulk to the channel). If the bulk is connected to the source (as is generally the case with discrete devices), it is sometimes angled to meet up with the source leaving the transistor. If the bulk is not shown (as is often the case in IC design as they are generally common bulk) an inversion symbol is sometimes used to indicate PMOS, alternatively an arrow on the source may be used in the same way as for bipolar transistors (i.e. out for nMOS, in for pMOS). In general, the MOSFET is a four-terminal device, and in ICs many of the MOSFETs share a body connection, not necessarily connected to the source terminals of all the transistors. These symbols with indication of channels and modes are again shown in Figure 14.26.

Drain current: The drain current I_d of MOSFET depends on V_d. It first increases linearly with V_d in the linear region and thereafter it levels off to a

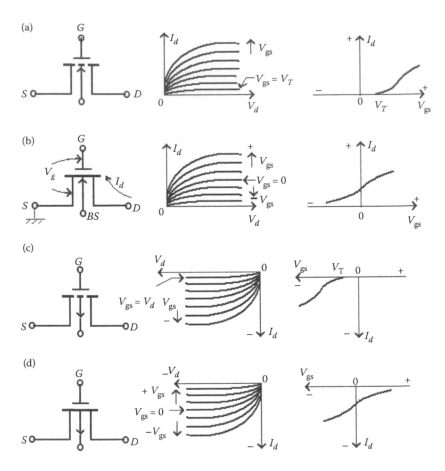

FIGURE 14.25
Symbols, output I–V and transfer characteristics of MOSFET. (a) n-channel enhancement mode
(normally off), (b) n-channel depletion mode (normally on), (c) p-channel enhancement mode
(normally off) and (d) p-channel depletion mode (normally on).

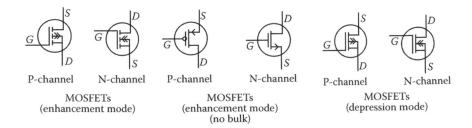

FIGURE 14.26
MOSFET symbols.

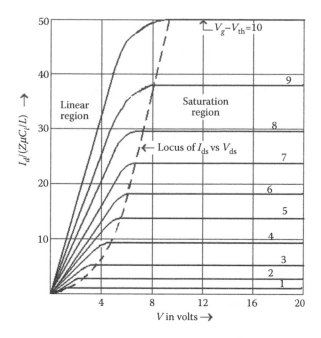

FIGURE 14.27
Current–voltage characteristics of an n-channel MOSFET.

saturated value in the saturation region. The current–voltage characteristics of an n-channel MOSFET are shown in Figure 14.27.

The mathematical expression for drain current is given as

$$I_d = \frac{Z}{L}\mu_n C_i \left[\left(V_g - 2\psi_b - \frac{V_d}{2} \right) V_d - \frac{2}{3C_i}(2\varepsilon_s q N_a)^{1/2}\left\{ (V_d + 2\psi_b)^{1/2} - (2\psi_b)^{1/2} \right\} \right]$$

$$(14.50)$$

where μ_n is the electron carrier mobility, C_i $(= \varepsilon_i/d)$ is the insulator capacitance per unit area, ε_i is the insulator permittivity, V_g is the gate voltage, V_d is the drain voltage, V_{th} is the threshold voltage, ψ_b $\{= (E_i - E_F)/q\}$ is the potential difference between the Fermi level E_F and the intrinsic Fermi level E_i, ε_s is the semiconductor permittivity, q is the carrier charge and N_a is the acceptor concentration. The parameters L, Z and d are shown in Figure 14.23. In the linear region, the drain voltage is small and the expression of I_d reduces to

$$I_d \approx \frac{Z}{L}\mu_n C_i \left[(V_g - V_{th})V_d - \left(\frac{1}{2} + \frac{\sqrt{\varepsilon_s q N_a/\psi_b}}{4C_i} \right) V_d^2 \right] \qquad (14.51)$$

$$I_d \approx \frac{Z}{L}\mu_n C_i (V_g - V_{th})V_d \quad \text{for } V_d \ll (V_g - V_{th}) \qquad (14.52a)$$

where

$$V_{th} = 2\psi_b + (2/C_i)(\varepsilon_s q\, N_a\psi_b)^{1/2} \tag{14.52b}$$

Drain current in saturation region: In the saturation region, the drain current is

$$I_{d\,sat} \approx (mZ/L)\mu_n C_i(V_g - V_{th})^2 \tag{14.53}$$

where m = 0.5 is the low doping factor. It may be assumed that $V_{d\,sat} \approx (V_g - V_{th})$.

Transconductance: In the linear region, the trans- (or mutual) conductance g_m is given by

$$g_m = \left.\frac{\partial I_d}{\partial V_g}\right|_{V_d=\text{constant}} = (Z/L)\mu_n C_i V_d \tag{14.54}$$

In saturation region, it becomes

$$g_{m\,sat} = (2mZ/L)\mu_n C_i(V_g - V_{th}) \tag{14.55}$$

Channel conductance: The channel conductance is given as

$$g_d = \left.\frac{\partial I_d}{\partial V_g}\right|_{V_g=\text{constant}} = (Z/L)\mu_n C_i(V_g - V_{th}) \tag{14.56}$$

All the equations from Equations 14.49 through 14.56 are for idealised n-channel MOSFET. For p-channel, the voltages V_g, V_d and V_{th} are negative and I_d flows from the source to the drain.

Non-idealised n-channel: For non-idealised n-channel, the saturation current is given by

$$I_{d\,sat} = Z C_i(V_g - V_{th})v_s \tag{14.57}$$

The transconductance becomes

$$g_m = Z C_i v_s \tag{14.58}$$

where $v_s = L/\tau$ is the carrier drift velocity.

The threshold voltage is given by

$$V_{th} = (\Phi_{ms}/q) - (Q_f/C_i) + 2\psi_b + (2/C_i)(\varepsilon_s q\, N_a\psi_b)^{1/2} \tag{14.59}$$

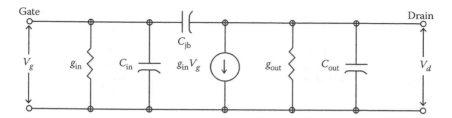

FIGURE 14.28
Equivalent circuit of common source MOSFET.

where Φ_{ms} $(= \Phi_m - \Phi_s)$ is the difference of work functions (in eV) of metal (Φ_m) and the semiconductor (Φ_s) and Q_f is the fixed oxide charges.

Maximum operating frequency: It is determined by the circuit parameters given in the equivalent circuit shown in Figure 14.28. In these parameters, g_m is the input conductance due to the leakage current, which is negligible as the leakage current is very small (about 10^{-10} A/cm²), g_{out} $(= g_d)$ is the output conductance, C_{in} $(= ZLC_i)$ is the input capacitance, C_{jb} is the feedback capacitance, C_{out} $\{ = C_i\,C_s/(C_i + C_s)\}$ is the sum of two p–n junction capacitances in series with the semiconductor capacitance/unit area.

The maximum operating frequency in the linear region is given as

$$f_{max} = \omega_m/2\pi = g_m/(2\pi C_m) = \mu_n V_d/(2\pi L^2) \qquad (14.60a)$$

In saturation, it is given by a simple relation

$$f_{max} = v_s/2\pi L \qquad (14.60b)$$

As $V_g \gg V_{th}$ in the saturation region, the transconductance reduces to

$$g_{m\,sat} = (Z/L)\mu_n C_i V_g = Z\,C_i v_s \qquad (14.61)$$

14.4.3.4 Sub-Division of Enhancement Mode

The *enhancement mode* can be further divided into three sub-modes. These include cutoff mode, triode mode and saturation or active mode. These sub-modes for n-channel MOSFETs are described below. These are shown in Figure 14.29.

1. *Cutoff mode:* The cutoff mode is also called *sub-threshold or weak-inversion mode.* For this operation, V_{gs} is less than V_{th} and according to the basic threshold model, the transistor is turned off. There is no conduction between the drain and the source. However, the Boltzmann

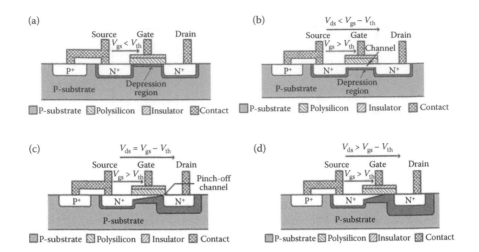

FIGURE 14.29
Sub-modes of enhancement modes. (a) Sub-threshold mode, (b) linear operating region (operating mode), (c) active mode at onset of pinch-off and (d) active mode well into pinch-off.

distribution of electron energies allows some of the more energetic electrons at the source to enter the channel and flow to the drain. It results in a sub-threshold current, which is an exponential function of gate–source voltage. Thus, when the transistor is turned off, the current between the drain and the source should ideally be zero except a weak inversion current, referred to as sub-threshold leakage. In weak inversion, the drain current (I_d) varies exponentially with the difference of gate-to-source voltage (V_{gs}) and the threshold voltage (V_{th}). It is given by an approximate relation:

$$I_d \approx I_{d0}\exp\{(V_{gs} - V_{th})/nV_T\} \tag{14.62}$$

where I_{d0} is the current when $V_{gs} = V_{th}$, $V_T (= kT/q)$ is the thermal voltage and $n (= 1 + C_d/C_{ox})$ is the slope factor. In the expressions of V_T and n, K is the Boltzmann's constant, T is the absolute temperature, q is the charge, C_d is the capacitance of depletion layer and C_{ox} is the capacitance of the oxide layer.

Figure 14.29a shows a cross section of an nMOSFET. When $V_{gs} < V_{th}$, there is little or no conduction between the source and the drain and the switch is off. When the gate becomes more positive, it attracts electrons. An n-type conductive channel is induced in the substrate below the oxide. This channel allows electrons to flow between the n-doped terminals and the switch is on.

2. *Triode mode (or linear region):* It is also called the *ohmic mode*. As illustrated in Figure 14.29b, for this mode, $V_{gs} > V_{th}$ and $V_{ds} < (V_{gs} - V_{th})$.

The transistor is on, and a channel is created that allows current to flow between the drain and the source. The MOSFET operates like a resistor, controlled by the gate voltage relative to both the source and drain voltages. The current from the drain to the source is given as

$$I_d = \mu_n C_{ox}(Z/L)\{(V_{gs} - V_{th})\, V_{ds} - (V_{ds})^2/2)\} \tag{14.63}$$

where μ_n is the charge-carrier effective mobility, Z is the gate width, L is the gate length and C_{ox} is the gate oxide capacitance per unit area. The transition from the exponential sub-threshold region to the triode region is not sharp.

3. *Saturation or active mode:* This case is shown in Figure 14.29c wherein $V_{gs} > V_{th}$ and $V_{ds} > (V_{gs} - V_{th})$. The switch is turned on, and a channel is created, that allows the current to flow between the drain and the source. Since the drain voltage is higher than the gate voltage, the electrons spread out, and conduction is not through a narrow channel but through a broader, two- or three-dimensional current distribution extending away from the interface and deeper in the substrate. The onset of this region is also known as pinch-off to indicate the lack of channel region near the drain. The pinch-off channel is illustrated in Figure 14.29d.

The drain current is now weakly dependent on drain voltage and controlled primarily by the gate–source voltage, and modelled approximately as

$$I_d = \{(\mu_n C_{ox})/2\}(Z/L)(V_{gs} - V_{th})^2\{1 + \lambda(V_{ds} - V_{ds\,sat})\} \tag{14.64}$$

The factor λ is the channel length modulation parameter. It models the current dependence on drain voltage due to the 'early effect', or channel length modulation. $V_{ds\,sat}\ (= V_{gs} - V_{th})$ represents the saturation value of V_{ds}. The transconductance (g_m) is a key design parameter of MOSFET, which is given by

$$g_m = 2\,I_d/(V_{gs} - V_{th}) = 2I_d/V_{ov} \tag{14.65}$$

where $V_{ov}\ (= V_{gs} - V_{th})$ is called the overdrive voltage and $V_{ds\,sat}$ accounts for a small discontinuity in I_d, which would otherwise appear at the transition between the triode and saturation regions. The output resistance $r_{out}\ (= 1/\lambda I_d)$ is another key design parameter of the MOSFET and is the inverse of $g_{ds}\ (= \partial I_{ds}/\partial V_{ds})$. When λ is zero, it results in an infinite output resistance of the device and leads to unrealistic circuit predictions, particularly in analog circuits.

Depletion mode: The depletion-mode MOSFETs are less commonly used than the enhancement-mode MOSFETs. These devices are so doped that a channel exists even with zero voltage from the gate to the source. In order to control the channel, a negative voltage is applied to the gate (for an n-channel device). The channel is depleted and as a result the current flow through the device is reduced. The depletion-mode device (DMD) is equivalent to a normally closed (ON) switch, whereas the enhancement-mode device (EMD) is equivalent to a normally open (OFF) switch. A PMOS switch has about 3 times the resistance of an NMOS device of equal dimensions because electrons have about 3 times the mobility of holes in silicon.

Body effect: The body effect or back-gate effect describes the changes in the threshold voltage by the change in source-bulk voltage (V_{SB}) and is given by the approximate relation

$$V_{th} = V_{T0} + \gamma \left\{ \sqrt{(V_{SB} + 2\phi)} - \sqrt{(2\phi)} \right\} \tag{14.66}$$

where V_{th} is the threshold voltage with substrate bias present, and V_{T0} is the zero-V_{SB} value of the threshold voltage, γ is the body effect parameter and 2ϕ is the surface potential parameter. The body can be operated as a second gate, sometimes referred to as the back gate.

14.4.3.5 Materials

Silicon is the most commonly used semiconductor in MOSFETs. The use of silicon germanium (Si–Ge) compound has also been reported. The increase in power consumption due to gate current leakage is greatly checked on replacing silicon dioxide by high-ε dielectric for the gate insulator and on replacing polysilicon by metal gates. The gate is separated from the channel by a thin insulating layer of silicon dioxide and later of silicon oxy-nitride. In some devices, a high-ε dielectric + metal gate combination in the 45 nm mode is used. Though not an ideal conductor, the use of highly doped polycrystalline silicon as gate material is quite common. The silicon oxide is the most commonly used insulating material.

14.4.3.6 Advantages and Limitations

Its fabrication technology is quite mature. Its high integration levels and capability for low-voltage operation are the added advantages. Its limitations include high resistive channels, lower transconductance and increased capacitance due to channel widening.

14.4.3.7 Applications

The MOSFET transistor is one of the most important devices used in the design and construction of ICs for digital computers. Its thermal stability and other general characteristics make it extremely popular in computer circuit design.

14.4.3.8 Comparison of MOSFETs with BJTs

- Transconductance of MOSFETs in switching region is lower than that of BJTs.

- Output impedance of MOSFETs in switching region is significant as compared to BJTs.

- Impact on characteristics and performance due to dimensional change in analog circuits is significant in MOSFETs as compared to BJTs.

- A MOSFET acts as nearly ideal switching element whereas a BJT does not.

- In the linear region, MOSFET works as a precision resistor with much higher controlled resistance than BJT.

- In high-power circuits, MOSFETs do not suffer from thermal runaway as BJTs do.

- MOSFETs can be formed into capacitors and gyrator circuits. This allows op-amps made from them to appear as inductors, thereby allowing all of the normal analog devices (except diodes) to be built entirely out of MOSFETs. Thus, a complete analog circuit can be made on a single silicon chip in a much smaller space than that made from BJTs.

- Some ICs combine analog and digital MOSFET circuitry on a single mixed signal IC, making the required space smaller. In such circuits, analog and digital circuits are to be isolated by using isolation rings and silicon-on-insulator (SOI).

- BJTs and MOSFETs can be incorporated into a single device during the fabrication process as BJTs in the analog design process can handle a larger current in a smaller space. Mixed BJT-MOSFET devices are called *Bi-FETs* if they contain just one BJT-FET and BICOMS if they contain complementary BJT-FETs. Such devices have the advantages of both insulated gates and higher current density.

14.4.4 High Electron Mobility Transistors

These devices involve two materials with different bandgaps. The joining of these materials results in a heterojunction structure. Thus, these devices are also known as heterostructure FETs (HFETs). The semiconductor devices usually employ impurities to generate the electrons that take part in the conduction process. These electrons are slowed down through collisions with impurities that are used to generate them. The impact of this slow down can be compensated by using high-mobility electrons. These high-mobility electrons are generated by creating a heterojunction between a highly doped wide-bandgap n-type donor–supply layer and a non-doped narrow-band-gap channel layer with no dopant impurities. Since the generation of high-mobility electrons is related to the appropriate doping, these FETs are also known as modulation-doped FETs (MODFETs).

The electron mobility in MESFET channel with typical donor concentrations of about 10^{17} cm^{-3} ranges from 4000 to 5000 cm^2/V·s at room temperature. Owing to the ionised impurity scattering, the figure of mobility does not go significantly higher at 77 K. The mobility in undoped GaAs of 2 to 3×10^5 cm^2/V·s has, however, been achieved at 70 K. It was found that the mobility of GaAs with high electron concentration can be increased through the modulation doping method. This aspect was demonstrated in GaAs–AlGaAs super-lattices. A HEMT based on modulation-doped GaAs–AlGaAs single-heterojunction structure was developed. HEMTs have exhibited lower noise figure and higher power gain at microwave frequencies. HEMT amplifiers are constructed for 40–70 GHz. Amplifier gains of 4.5–6 dB for frequency band 56–62 GHz are exhibited with associated noise figure of 6 dB at 57.5 GHz. A 72 GHz amplifier achieved a gain of 4–5 dB with a BW of 2.5 GHz. The HEMTs have an edge over MESFETs in terms of shorter gate lengths, reduced gate and source contact resistances and optimised doping profile.

14.4.4.1 HEMT Structure

A basic HEMT structure is illustrated in Figure 14.30. It uses selectively doped GaAs–AlGaAs heterojunction. An undoped GaAs layer and a Si-doped n-type AlGaAs layer are grown on a semi-insulating (S.I.) GaAs substrate. It also illustrates a two-dimensional electron gas (2DEG) created between the undoped and n-type layers and a buffer sandwiched between the undoped GaAs layer and the S.I. GaAs substrate.

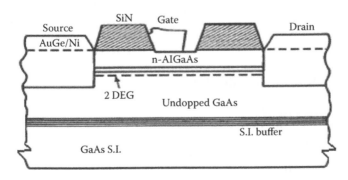

FIGURE 14.30
HEMT structure.

The materials (AlGaAs and the GaAs) involved in the structure shown in Figure 14.30 have different bandgaps. The heterojunction created by these materials forms a quantum well in the conduction band on the GaAs side. The electrons generated in the thin n-type AlGaAs layer fall completely into the GaAs layer and form a depleted AlGaAs layer. Thus, the electrons can now move quickly (through this depleted layer) without colliding with any impurities because the GaAs layer is undoped. It results in the creation of a very thin layer of highly mobile conducting electrons with very high concentration, giving the channel very low resistivity. These highly mobile conducting electrons are referred to as high-mobility electrons and this thin layer is called a 2DEG.

14.4.4.2 Operation

Since GaAs has higher electron affinity, free electrons from the AlGaAs layer are transferred to the undoped GaAs layer. Here they form a two-dimensional high-mobility electron gas within 100 Å of the interface. The n-type AlGaAs layer of HEMT is completely depleted through two depletion mechanisms. These include (i) the trapping of free electrons by surface states causing surface depletion and (ii) the transfer of electrons into the undoped GaAs layer bringing about interface depletion.

The Fermi energy level of the gate metal is matched to the pinning point, which is 1.2 eV below the conduction band. With the reduced AlGaAs layer thickness, the electrons supplied by donors in the AlGaAs layer are insufficient to pin the layer. As a result, band bending is moving upward and the 2DEG does not appear. When the applied voltage to the gate is more positive than the threshold voltage, electrons accumulate at the interface and a 2DEG is formed.

HEMTs may be of enhancement mode HEMT (E-HEMT) and of depletion mode HEMT (D-HEMT). The operations of these can be controlled by the concentration of electrons. The electron mobility increases with the decrease in temperature due to reduced phonon scattering. It is reported that it increases from 8000 cm^2/V·s at 300 K to 2×10^5 cm^2/V·s at 77 K to 1.5×10^6 cm^2/V·s at 50 K to 2.5×10^6 cm^2/V·s at 4.5 K.

14.4.4.3 Current–Voltage Characteristics

The drain current can be given by the relation

$$I_{ds} = q\,n(z)\,W\upsilon(z) \tag{14.67}$$

where q is the electron charge, $n(z)$ is the concentration of 2DEG, W is the gate width and $\upsilon(z)$ is the electron velocity.

Figure 14.31 illustrates the I–V characteristics of the HEMT amplifier with $W = 150$ μm and L_g (gate length) $= 0.6$ μm at 30 GHz.

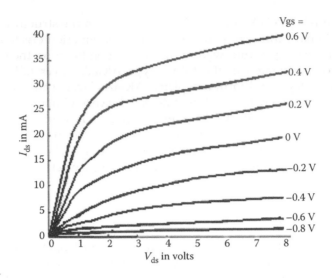

I–V characteristics of HEMT amplifier.

14.4.4.4 HEMTs Fabrication Using IC Technology

HEMTs can be fabricated by using IC technology. Figure 14.32 gives the sequence for the self-aligned gate procedure in the fabrication of large-scale integration HEMTs, including the enhancement mode HEMT (E-HEMT) and depletion mode HEMT (D-HEMT).

1. *Ohmic contact formation:* In this step, the active region is isolated by a shallow mesa step (180 nm). This isolation, in general, is achieved

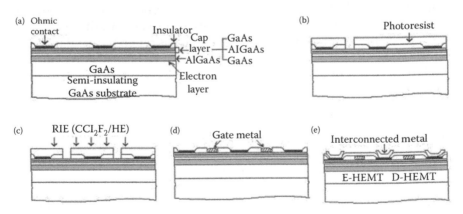

FIGURE 14.32
Sequence for fabrication of n-channel depletion-type MOSFET. (a) Ohmic contact formation, (b) opening gate windows, (c) selective dry etching, (d) gate metallisation and (e) interconnect metallisation.

in one go, and can be made nearly planar. The source and drain for E-HEMT and D-HEMT are metallised with an AuGe eutectic alloy and an Au overlay alloy to produce ohmic contacts with the electron layer. The step is shown in Figure 14.32a.

2. *Opening gate windows:* In this step, fine gate patterns are formed by E-HEMTs as the top GaAs layer. Later, the thin ($Al_{0.3}Ga_{0.7}As$) stoppers are etched off by non-selective chemical etching. This step is shown in Figure 14.32b.

3. *Selective dry etching:* With the same photoresist process for the formation of gate patterns in D-HEMTs, selective dry etching is performed to remove the top GaAs layer for D-HEMTs and also to remove the GaAs layer under the thin ($Al_{0.3}Ga_{0.7}As$) stopper for E-HEMTs. This step is given in Figure 14.32c.

4. *Gate metallisation:* Schottky contacts for the E- and D-HEMT gate are provided by depositing aluminium. The Schottky gate contacts and GaAs top layer for ohmic contacts are then self-aligned to achieve high-speed performance. The step is shown in Figure 14.32d.

5. *Interconnect metallisation:* Electrical connections from the interconnecting metal composed of Ti, Pt and Au to the device thermals are provided through the contact holes etched in the crossover insulator film. The last step is shown in Figure 14.32f.

Figure 14.33 shows the cross-sectional view of a self-aligned structure. It represents E- and D-HEMT forming an inverter for the direct coupled FET logic (DCFL) circuit configuration.

14.4.4.5 Materials Used

The material selection for HEMTs, out of a wide variation of material combination, depends on its applications. A commonly used combination includes GaAs with AlGaAs. Devices incorporating indium show improved high-frequency

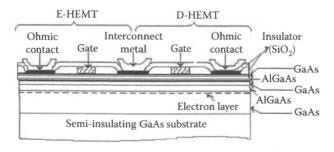

FIGURE 14.33
Cross-sectional view of a self-aligned HEMT structure.

TABLE 14.1

Comparison of HEMT with Other Devices

Device	Frequency	Noise Figure	Power	Speed
HEMT	Up to 70 GHz	Very good	Very good	Excellent
GaAs MESFET	40 GHz	Good	Good	Good
GaAs–AlGaAs HBT	20 GHz	Good	Good	Excellent
Si MOSFET	10 GHz	Poor	Very good	Very poor
Silicon BJTs	1 GHz	Poor	Poor	Good

performance. In recent years, gallium nitride has also attracted the attention of designers.

14.4.4.6 Advantages

It exhibits low noise figure and high gain up to 70 GHz. Its switching speed is about 3 times as fast as that of a GaAs MESFET. Its power dissipation is reported at about 100 pW. It can be fabricated by using integrated circuit technology. Table 14.1 compares HEMT with other devices used at microwave frequencies in view of different assessment parameters.

14.4.4.7 Applications

It is used in mm-wave analog and high-speed digital applications. It is also used in systems where high gain and low noise at high frequencies are required. It is used in cellular communications, direct broadcast receivers, warfare systems and radio astronomy.

14.5 Metal-Oxide-Semiconductor Transistors

MOS is an abbreviation for metal-oxide-semiconductor. In technical terminology, it stands for metal-oxide-semiconductor field effect transistors (or MOSFETs). The source, the channel and the drain of the MOS transistors are encompassed by a depletion region. Thus, its individual components do not require further isolation at the time of fabrication. The elimination of isolation leads to greater packaging density on a semiconductor chip. This advantage of MOS is not feasible with BJTs.

MOS transistors can be classified into three categories. These are named as N-channel MOSFETs (or NMOS), P-channel MOSFETs (or PMOS) and complementary MOSFETs (or CMOS). The CMOS provides n-channel and p-channel MOSFETs on the same chip.

Currently, CMOS has largely replaced NMOS and PMOS. The fabrication of NMOS is much simpler than that of BJTs. Also, the power dissipation in CMOS circuits is less than that in BJTs and NMOS. Thus, both NMOS and CMOS devices are quite useful for high-density ICs.

14.5.1 NMOS Logic

NMOS logic uses MOSFETs to implement logic gates and other digital circuits. NMOS transistors have four modes of operation. These modes are referred to as cutoff (or sub-threshold) mode, triode mode, saturation (or active) mode and velocity saturation mode. The n-type MOSFETs are arranged in 'pull-down network' (PDN) between the logic gate output and the negative supply voltage, while a resistor is placed between the logic gate output and the positive supply voltage. The circuit is designed in such a way that if the desired output is low, then PDN will be active, creating a current path between the negative supply and the output.

A MOSFET can be made to operate as a resistor, so the whole circuit can be made with n-channel MOSFETs only. The major problem with NMOS is that a dc current must flow through a logic gate even when the output is low. Thus there is power dissipation even when the circuit is not switching. NMOS circuits are slow to transition from low to high. During high to low transition, the transistors provide low resistance and capacitative charge at the output drains away very quickly. Since the resistance between the output and the positive supply rail is much greater, the low to high transition takes longer time. Using a resistor of lower value will speed up the process but also increases static power dissipation. However, a better and most common way to make the gates faster is to use depletion mode instead of enhancement mode transistors as loads. This is called depletion-load NMOS logic. The asymmetric input logic levels make NMOS circuits susceptible to noise.

14.5.1.1 NMOS Structures

Figure 14.34 shows a depletion-type structure and its I–V characteristics. Figure 14.35 illustrates an induction-type structure of MOSFETs and its characteristics. In case of depletion type, when the terminals of the device are open, there is a conducting channel that links the drain to the source. With these open-circuited terminals, the induced type has a channel of opposite type to that of the drain and the source linking the two regions. In PMOS, the device has p-type source and drain regions.

The structures of enhancement and depletion modes basically refer to relative increase or decrease of the majority carrier density in the channel connecting the source and the drain. In case of enhancement (normally off) mode, a given gate bias tends to increase the majority carrier density in the channel. Thus, an NMOS is said to be operating in the enhancement mode if its gate is biased by a positive voltage with respect to the substrate, which

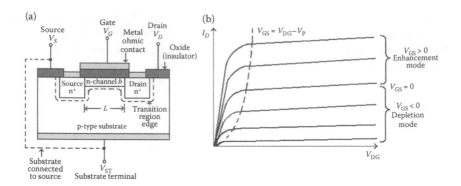

FIGURE 14.34
n-Channel depletion-type MOSFET. (a) Structure and (b) I–V characteristics.

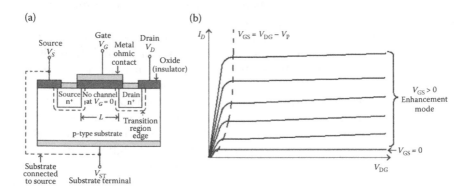

FIGURE 14.35
n-Channel induced-type MOSFET. (a) Structure and (b) I–V characteristics.

tends to drain more electrons into the n-type channel. An NMOS is said to be operating in the depletion mode if its gate is supplied a negative voltage with respect to the substrate, which diminishes the electron density in the n-channel. In case of PMOS, the situation is the just reverse to that of NMOS. In the enhancement mode, the gate is negatively biased, which drains more electrons into the p-type channel, and in the depletion mode, the gate is positively biased, which diminishes the electron density in the p-channel. When a MOS is operated in the enhancement mode, the higher majority carriers leads to higher drain current, whereas in the depletion mode, the lower majority carrier density results in lower drain current.

14.5.1.2 NMOS Operation

The operation is described as a logic gate shown in Figure 14.36. It contains two EMDs in series with a DMD. The three resistors are connected

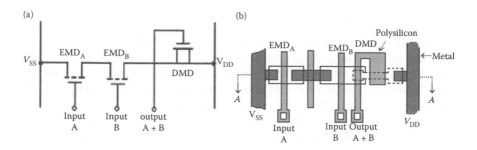

FIGURE 14.36
NMOS (a) operation and (b) layout.

between the positive power supply V_{DD} and a reference ground potential V_{SS}. The DMD at $V_{GS} = 0$ is normally on and acts as a current source for two EMDs. The gates A and B of the two EMDs act as input to the logic circuit. DMD's gate source connection is the output. The output voltage of the two-input NAND circuit is low only when both EMDs are turned on at their logic high level.

14.5.2 P-Type Metal-Oxide-Semiconductor Logic

PMOS logic uses p-type MOSFETs to implement logic gates and other digital circuits. This logic also has the same four modes of operation as that for NMOS. These are arranged in pull-up network (PUN) between the logic gate output and positive supply voltage, while a resistor is placed between the logic gate output and the negative supply voltage. The circuit is designed such that if the desired output is high, then PUN will be active, creating a current path between the positive supply and the output.

PMOS logic is easy to design and manufacture but has several limitations. The foremost of these is that a dc current flows through a PMOS logic gate when the PUN is active that is whenever the output is high. This leads to static power dissipation even when the circuit is idle. Also, PMOS circuits are slow to transition from high to low. When transitioning from low to high, the transistors provide low resistance, and the capacititative charge at the output accumulates very quickly. But the resistance between the output and the negative supply rail is much greater, so the high to low transition takes longer time. The use of resistors of lower value speeds up the process but static power dissipation increases. Also, the asymmetric input logic levels make PMOS circuits susceptible to noise.

14.5.3 Complementary Metal-Oxide-Semiconductor

CMOS uses complementary and symmetrical pairs of p-type and n-type MOSFETs for logic functions. It is sometimes also referred to as complementary-symmetry metal-oxide-semiconductor (COS-MOS).

14.5.3.1 CMOS Structures

CMOS structures are classified into three categories shown in Figure 14.37. These are referred to as n-tub, p-tub and twin tub. In n-tub, a tub or well is formed in p-type substrate, whereas in p-tub, a tub is formed in n-type substrate. In twin tub, both n-tub and p-tub are combined on the same substrate. A tub can be produced by an extra diffusion step.

CMOS circuits implement logic gates by using p-type and n-type MOSFETs to create paths to the output from either the voltage source or the ground. When a path is created from the voltage source, the circuit is said to be pulled up. When a path is created from the ground, the output is pulled down to the ground potential.

Figure 14.38 shows both PMOS and NMOS logics together in the same wafer of CMOS. CMOS circuits are constructed in such a way that all PMOS transistors are fed an input either from the voltage source or from another PMOS transistor. Similarly, all NMOS transistors are given an input either from the ground or from another NMOS transistor. The composition of a PMOS transistor creates low resistance between source and drain contacts for low gate voltage and high resistance for high gate voltage. The composition of the NMOS transistor creates high resistance between the source and

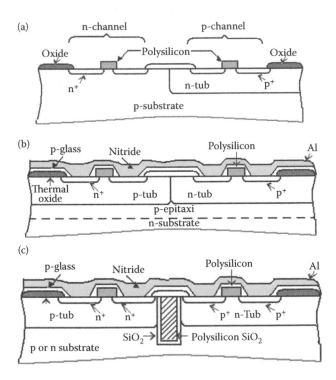

FIGURE 14.37
Three types of CMOS structures. (a) n-tub, (b) p-tub and (c) twin-tub.

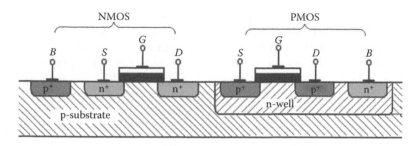

FIGURE 14.38
Cross section of two transistors in a CMOS gate.

the drain with low gate voltage and low resistance when a high gate voltage is applied. To accomplish the current reduction in CMOS, every nMOSFET is complemented with a pMOSFET and both gates and drains are connected together. A high voltage on the gates makes nMOSFET to conduct and pMOSFET not to conduct while a low voltage causes the reverse on both the gates. Owing to this arrangement, there is a large reduction in power consumption and heat generation. Since both MOSFETs briefly conduct during switching as the voltage changes the state, there are brief spikes in power consumption. This becomes a serious issue at high frequencies.

14.5.3.2 CMOS Operation

Figure 14.39a illustrates a static CMOS inverter. It contains a PMOS transistor and an NMOS transistor. Both of these are fed with an input voltage at A, which may be low or high. When the voltage at A is low, the NMOS transistor's channel is in a high resistance state and limits the flow of current from Q to the ground. PMOS transistor's channel is in low resistance state, which allows much more current from supply to the output. As the resistance between the supply voltage and Q is low, the

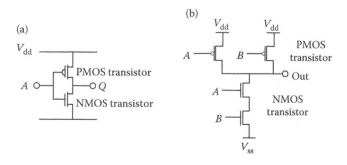

FIGURE 14.39
Devices in CMOS logic. (a) Static inverter and (b) NAND gate.

voltage drop between the supply voltage and Q due to a current drawn from Q is small. Thus, the output registers a high voltage. Similarly, when the voltage of input A is high, the PMOS transistor is in OFF (high resistance) state, which limits the current flow from positive supply to the output. Also, when the NMOS transistor is in an ON (low resistance) state, it allows the output to drain to the ground. As the resistance between Q and the ground is low, the voltage drop due to a current drawn into Q placing Q above the ground is small. This low drop results in the output registering a low voltage. In view of the above, it can be concluded that the outputs of PMOS and NMOS transistors are complementary. Thus, when the input is low, the output is high, and when the input is high, the output is low. Alternatively, it can be stated that the output of a CMOS circuit is the inversion of the input.

The duality between PMOS and NMOS transistors of CMOS is its important characteristic. A CMOS circuit is created to allow a path from the output to either the power source or the ground. This can be accomplished if the set of all paths to the voltage source is the complement of the set of all paths to the ground. This can be easily done by defining one in terms of the NOT of the other. In view of the De Morgan's laws-based logic, PMOS transistors in parallel have correspondence with NMOS transistors in series while the PMOS transistors in series have correspondence with NMOS transistors in parallel. Complex logic functions such as those involving AND and OR gates require manipulation of the paths between gates to represent the logic. When a path consists of two transistors in series, both transistors must have low resistance to the corresponding supply voltage, modelling an AND gate. When a path consists of two transistors in parallel, either one or both of the transistors must have low resistance to connect the supply voltage to the output, modelling an OR logic.

Figure 14.39b shows a NAND gate in CMOS logic. If both A and B inputs are high, then both NMOS transistors will conduct, neither of the PMOS transistors will conduct and a conductive path will be established between the output and V_{ss} (ground), bringing the output to be low. If either A or B input is low, one of the NMOS transistors will not conduct, one of the PMOS transistors will conduct and a conductive path will be established between the output and V_{dd} (voltage source), bringing the output to be high. The N device is manufactured on a P-type substrate (p-well) while the P device is manufactured on an n-well. A P-substrate 'tap' is connected to V_{ss} and an n-well tap is connected to V_{dd} to prevent latchup.

14.5.3.3 Power Dissipation

CMOS logic dissipates less power than NMOS logic circuits and BJTs. It dissipates power only during switching between on and off states. Such dissipation is referred to as dynamic dissipation. Static NMOS logic dissipates power whenever the output is low because there is a current path from V_{dd}

to V_{ss} through the load resistor and the n-type network. This dissipation is referred to as static dissipation. It dissipates nearly zero power when idle.

14.5.3.4 Analog CMOS

Analog CMOS are used in analog applications such as operational amplifiers, transmission gates (in lieu of signal relays) and in RF circuits all the way to microwave frequencies. Currently, the conventional CMOS devices work over a range of $-55°C$ to $+125°C$.

14.5.3.5 Advantages

CMOS devices have high noise immunity and low static power consumption. It draws significant power only when the transistors in the device are switching between on and off states. These devices dissipate lesser power than TTL or NOMS logic. It gives high density of logic functions on a chip. CMOS has an advantage over NMOS since both low-to-high and high-to-low output transitions are fast since the pull-up transistors have low resistance when switched on, unlike the load resistors in NMOS logic. Besides, in CMOS, there is full voltage swing between the low and high rails. This strong, more nearly symmetric response makes CMOS more resistant to noise.

14.5.3.6 Applications

The CMOS technology is used for constructing integrated circuits in microprocessors, microcontrollers, static RAM, digital logic circuits and analog circuits (viz. image sensors, data convertors and integrated transceivers) for many types of communication.

14.6 Memory Devices

A device used to store data or information in terms of binary digits is referred to as a memory device. This involves semiconductor devices such as flip-flops, capacitors and ferroelectric materials exhibiting polarisation with applied electrical field. These are classified as in the following sections.

14.6.1 Memory Classification

Memory devices can be classified in many ways. These include functionality, nature of storage mechanism and access pattern. On the basis of functionality, these can be classified as read-only memory (ROM) and read-write memories (RWM). RWMs are the most flexible memories.

In regard to the storage mechanism, the data in a memory may be stored either in flip-flop or as a charge on a capacitor. These memory cells are called static and dynamic, respectively. In static memory, data are retained as long as the supply voltage is retained. The dynamic memory requires periodic refreshing to compensate for the loss of charge due to leakage. The static RAM (SRAM) and dynamic RAM (DRAM) memories belong to the above two categories. RWMs use active circuitry to store information; thus, they are referred to as volatile memories. In these, the data are lost when the supply is removed. SRAM and DRAM memories fall into this category. ROMs encode information into the circuit topology, for example, by adding or removing transistors. In view of the hard-wired nature of topology, data can only be read but cannot be modified. Because in ROM structures the removal of supply voltage does not result in loss of stored data, these are referred to as non-volatile memories. This class includes resistance RAM (RRAM), ferroelectric RAM (FeRAM) and phase change RAM (PCRAM) memories.

There is a class of memories that are non-volatile but still perform both read and write functions. These are named as non-volatile read-write memories (NVRWM). These include erasable programmable RPM (EPROM), electrically erasable programmable ROM (EEPROM) and flash memories. EPROM and EEPROM belong to the RWM category, but in these, the read operation is much faster than the write operation; thus, the word 'write' is dropped from their names.

The access pattern, which spells the order in which data are accessed in a memory, forms another basis for memory classification. The memories wherein data are read or written in random order are called random access memories.

14.6.2 Semiconductor Memories

The major semiconductor memories include the following.

14.6.2.1 Read-Only Memory

This memory shown in Figure 14.40a is also called mask ROM. In ROM, information is inscribed in the form of presence or absence of a link between the word (access) line and bit (sense) line. This information inscribed at the

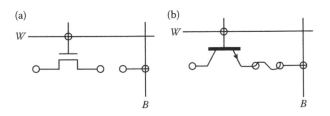

FIGURE 14.40
Semiconductor memories. (a) ROM and (b) PROM.

fabrication stage cannot be altered. When the word line (*W*) is activated, this causes the presence and absence of a readout signal on the bit line (*B*). Its link determines the cell size and the turnaround time. These have high speed and high density.

14.6.2.2 Programmable Read-Only Memory

This memory shown in Figure 14.40b is one of the forms of ROM that can be programmed. It lacks the erase capability. It uses cells either with a fuse that can be electrically blown open or a p–n diode that can be short circuited by an avalanche pulse. It can be of bipolar or of MOS form. The latter has higher density but lower speed.

14.6.2.3 Erasable Programmable Read-Only Memory

This memory shown in Figure 14.41a consists of a floating gate of a double-poly-gate MOSFET. The presence or absence of charge in this gate determines the state of the logic. For programming, it is injected with energetic carriers generated by drain p–n junction avalanche breakdown. This increases the threshold voltage of the memory transistor. For erasing the memory, it is packaged with a glass window through which an ultraviolet light is thrown on the floating gate to release charges. This device is also referred to as floating-gate avalanche injection MOS (FAMOS).

14.6.2.4 Electrically Erasable Programmable Read-Only Memory

This is shown in Figure 14.41b. In order to become electrically erasable, it requires some means to inject and extract charge carriers into and from a floating gate. Its latest version uses a floating gate separated from the silicon by an oxide. Programming and erasing is achieved by forcing the channel current to flow between gate and substrate with the control biased and negative supply.

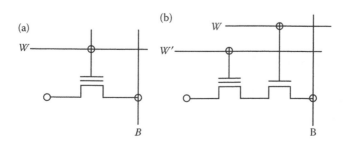

FIGURE 14.41
Semiconductor memories. (a) EPROM and (b) EEPROM.

Both EPROM and EEPROM store charge on a conductive region in the middle of a MOS gate oxide. Thus, these are critically dependent on the MOS structure, especially high-field carrier transport in both silicon and oxide. Both these memories are of non-volatile nature as carriers in the floating island remain there even after the removal of the power supply.

14.6.2.5 Static Random Access Memory

SRAM is a matrix of static cells using a bistable flip-flop structure to store the logic state. The flip-flop consists of two cross-coupled CMOS inverters (T_1, T_3 and T_2, T_4). The output of the inverter is connected to the input node of the other inverter. This configuration is called 'latch'. The operation of SRAM is static since the logic state is sustained as long as the power is applied. SRAM does not have to be refreshed. These require low power and battery back-up system. Their speed is high but their cost is also high. As shown in Figure 14.42a, its configuration contains a CMOS SRAM cell, wherein T_1 and T_2 are (p-channel) load transistors, T_3 and T_4 are (n-channel) drive transistors and T_5 and T_6 are (n-channel) access transistors.

14.6.2.6 Bipolar Static Random Access Memory

It is the fastest of all semiconductor memories, whereas MOS SRAM is the fastest among the MOS memories. Its circuit representation is given in Figure 14.42b.

14.6.2.7 Dynamic Random Access Memories

Modern DRAM technology consists of a cell array using a storage cell structure shown in Figure 14.43. The cell includes a MOSFET and a MOS capacitor (i.e. one transistor (T)/one capacitor (C) cell). The MOSFET acts as a switch

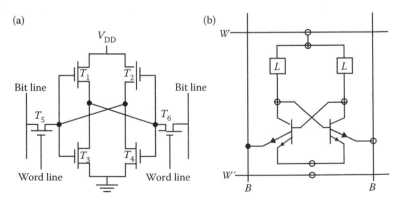

FIGURE 14.42
Semiconductor memories. (a) SRAM and (b) BSRAM.

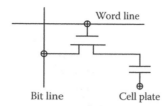

FIGURE 14.43
Basic configuration of a DRAM cell.

to control the writing, refreshing and reading-out actions of the cell. The capacitor is used for charge storage. During the write cycle, the MOSFET is turned on so that the logic state in the bit line is transferred to the storage capacitor. The operation of DRAM is 'dynamic', since the data need to be 'refreshed' periodically within a fixed interval, typically 2–50 ns. It has a simple structure, high density, low cost and small area.

DRAM has been developed to reduce the cell area and power consumption. The stored charge will be removed typically in a few milliseconds mainly because of the leakage current of the capacitor; thus, dynamic memories require periodic 'refreshing' of the stored charge. Figure 14.44 illustrates a single-transistor DRAM cell with storage capacitor. Its cell layout is shown in Figure 14.44a and its cross section through A–A' is given in Figure 14.44b.

Figure 14.45 illustrates the information storage capacity of Si-chips (DRAM) over four decades.

To meet the requirements of high-density DRAM, the DRAM structure has been extended to the third dimension with stacked or trench capacitors. The advantage of the trench type (a) is that the capacitance of the cell can be increased by increasing the depth of the trench without increasing the surface area of the silicon occupied by the cell. The storage capacitance increases as a result of stacking the storage capacitor on top of the access transistor. The dielectric is formed using the thermal oxidation or CVD nitride methods. Hence, the stacked cell process is easier than the trench-type process.

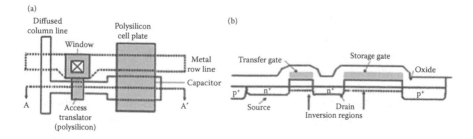

FIGURE 14.44
Single-transistor DRAM cell with storage capacitor. (a) Cell layout and (b) cross section through A–A'.

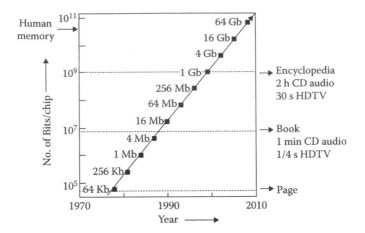

FIGURE 14.45
Information storage capacity of Si-chips (DRAM).

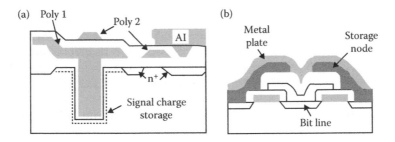

FIGURE 14.46
DRAM. (a) Trench cell structure and (b) single-layer stacked-capacitor cell.

Figure 14.46 shows a DRAM with a trench cell structure (a) and with single-layer stacked-capacitor cell (b).

14.6.2.8 Flash Memories

These were developed by Toshiba in 1987 and include both the analog and the digital memories. By nature these are non-volatile, of high density and of low cost. Most of the devices used for present-day non-volatile memories have a floating gate structure.

Figure 14.47 shows the floating-gate non-volatile memory and its equivalent circuit, which is represented by two capacitors in series for the gate structure. The stored charge gives rise to a threshold voltage shift, and the device is switched to a high-voltage state (logic 1). For a well-designed memory device, the charge retention time can be more than 10 years. To erase the stored charge and return the device to a low threshold voltage state (logic 0),

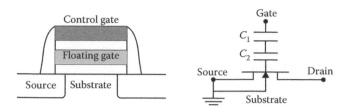

FIGURE 14.47
Non-volatile memory. (a) Floating-gate structure and (b) equivalent circuit.

an electrical bias or other means, such as UV light, can be used. In floating-gate devices, the programming can be done by either hot carrier injection or a tunnelling process. The two cases are shown in Figures 14.48 and 14.49. The hot electrons are 'hot' because they are heated to a high-energy state by the high field near the drain. Some of the hot electrons with energy higher than the barrier height of SiO_2/Si conduction band (~3.2 eV) can surmount the barrier and get injected into floating gate.

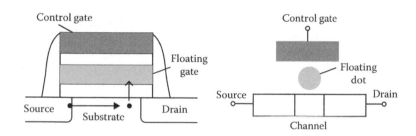

FIGURE 14.48
Programming by hot carrier injection.

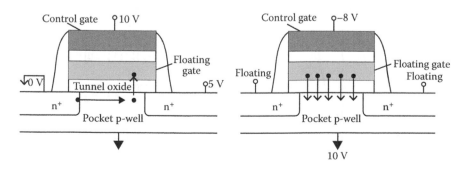

FIGURE 14.49
Programming by tunnelling process.

14.6.3 Some Other Memories

These memories, different from semiconductor memories, include ferroelectric RAM, resistance RAM, phase change RAM and ovonics unified memory. These are discussed below.

14.6.3.1 Ferroelectric RAM

Ferroelectric materials exhibit spontaneous polarisation with applied electrical field due to atomic displacement of the body-centred atom in the perovskite (ABO_3) structure. The remnant polarisation state is maintained after the removal of the electric field. The two stable states of ferroelectric materials are the basis of memory application. Figure 14.50 illustrates the hysteresis Vs the domain movement where the domain is the region that has the same polarity. In this figure, V_c is the coercive voltage where the net polarisation is zero. P_r and P_s are polarisations at the remnant and saturation points, respectively.

14.6.3.2 Resistance RAM

RRAM is a promising candidate for the next-generation non-volatile memories. Its technology can be integrated with that of CMOS. With RRAM it is possible to have stable high-temperature programming up to 300°C. In RRAM, the memory resistor can be switched between high and low resistance states over a large number of cycles without memory degradation (10^6 times of set/reset and 10^{12} times of reading cycles confirmed). The cell resistance can be read without affecting the stored data. Figure 14.51 illustrates typical V–I characteristics of a material (say NiO with 60 nm slice). In this figure, arrow 1 indicates high-resistance state (means a lower current flow in the wafer), arrow 2 indicates set transition ($V = V_{set}$), arrows 3 and 4 indicate low

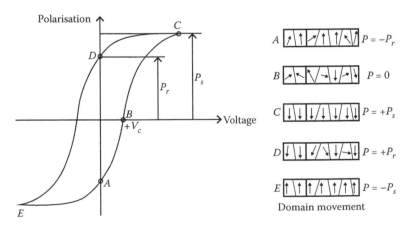

FIGURE 14.50
Hysteresis Vs domain movement.

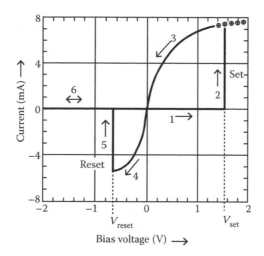

FIGURE 14.51
V–I characteristics.

resistance states (means a higher currents flow in the wafer), arrow 5 indicates reset transition ($V = V_{reset}$) and arrow 6 again indicates high resistance state.

14.6.3.3 Phase Change RAM

This memory uses chalcogenide material, which is the general class of switching media in CD-RW and DVD-RW. In this, laser beam energy is used to control the switching between crystalline and amorphous phases. In this material, the higher energy represents amorphous phase and medium energy the crystalline phase. Low-energy laser beam is used to read the information. When produced in high volume, it results in low-cost devices. Figure 14.52 illustrates the material characteristics. The amorphous and crystalline phases of the material and the corresponding diffraction patterns are

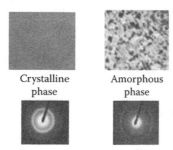

Crystalline phase   Amorphous phase

Electron diffraction patterns

FIGURE 14.52
Material characteristics.

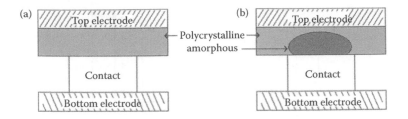

FIGURE 14.53
Joule heating by cell current. (a) Low resistance (set state = 0) and (b) high resistance (reset state = 1)

also shown. The amorphous phase has short-range atomic order, low free electron density, high activation energy and high resistivity. The crystalline phase contains long-range atomic order, high free electron density, low activation energy and low resistivity.

14.6.3.4 Ovonics Unified Memory

In this case, instead of laser beam, electric current is used to heat the material to switch between amorphous and crystalline phases. One of the methods of heating referred to as joule heating by cell current is illustrated in Figure 14.53. Here, the high current leads to high temperature, amorphous phase and high resistance, whereas medium current results in lower temperature, crystalline phase and low resistance. Low current is used to sense the resistance.

14.7 Charge-Coupled Devices

Charge-coupled device (CCD), a MOS diode structure, was invented in 1969 by Boyle and Smith (of Bell Labs) as an alternative to magnetic bubble memory storage. In CCD, charges can be moved along a predetermined path. The movement is controlled by clock pulses. In view of transfer of charges, it is also referred to as charge transfer device (CTD). In CCDs, the motion of charge packets is transversely controlled by the applied gate voltage. As this process is similar to the motion of carriers in MESFET or MOSFET, the CCD can be referred to as field effect CCD.

14.7.1 Structure and Working

Figure 14.54 shows the basic structure of a CCD. In CCD, the charge storage and transfer element is the MOS capacitor shown in Figure 14.55. If a suitably large voltage (+12 or +15 V) is applied to the gate electrode, it will set up an electric field (Figure 14.55a). This field will repel the holes from the region below the gate electrode and a depletion region will be formed. The E-lines

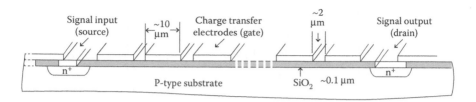

FIGURE 14.54
Basic structure of a CCD.

emanating from the gate electrode will terminate on the negatively charged acceptor ions in the depletion region, which will extend for several micrometers into the highly doped p-type substrate. This depletion region is referred to as a potential well. This well will get gradually filled due to the flow of thermally generated electron (Figure 14.55b). The electrons will accumulate near the silicon surface and form a surface (Figure 14.55c), which is called n-type inversion layer. Figure 14.55d illustrates the situation of a completely filled potential well. At this juncture, the inward diffusion of electrons into the well will be balanced by the outward flow of electron, so that there is no net flow of electrons into the potential well.

The depletion region, potential well and the situation of a completely filled potential well is also illustrated in Figure 14.56.

The filling of the potential well by thermally generated electrons is not an instant phenomena; it takes some time usually of the order of 0.1 s. This sets the lower limit of the speed of operation of CCD. Also, the signal passing through this potential well takes much lesser time. In all practical applications, a large number of CCDs are used, which are arranged in the form of arrays. In an array, a potential well is produced under the gate at the source end and a given amount of charge is injected from the source region into this potential well. All such injected charges are in the form of a given number of packets of electrons. These packets can then be transferred along the array from gate to gate in successively formed potential wells from the source end of the CCD array to the drain end. These charges are then extracted from the drain end.

Figure 14.57 illustrated three different situations. This figure illustrates three electrodes named ϕ_1, ϕ_2 and ϕ_3. These electrodes are maintained at

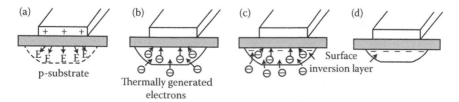

FIGURE 14.55
Energy band diagram of MIS structure. (a) E-field begins, (b) gradual filling of well, (c) n-type of inversion layer and (d) complete filling of potential well.

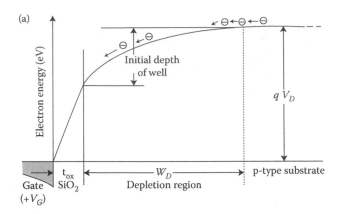

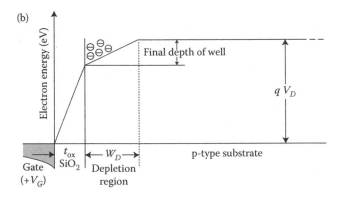

FIGURE 14.56
Location of SiO$_2$, depletion region, p-substrate and (a) initial depth and (b) final depth of potential well.

different potentials. In Figure 14.57a, the charge packet is localised in the potential well underneath the ϕ_1 electrode. As the electrode applied to electrode ϕ_2 increases (from 0 to 15 V) and at the same time, the voltage of ϕ_1 decreases from 15 V to zero, the electron will move from the collapsing potential well underneath ϕ_1 electrode to now expanding well underneath the ϕ_2 electrode (Figure 14.57b). It may be noted that due to very close electrode spacing (of about 2 µm (Figure 14.57c)), the depletion regions and potential wells of the two electrodes will overlap during this transfer.

The electrons are injected into the CCD array at the source or input end. In this injection process (Figure 14.58), the input gate electrode increases, which produces an n-type surface inversion layer between the n$^+$ source region and the potential well underneath the signal input gate electrode. This potential well will immediately be filled with electrons from the source region to a level corresponding to a net charge $Q = C_{ox}(V_A - V_t)$, where C_{ox} is the MOS capacitance, V_A is the analog signal level and V_t is the threshold level for surface inversion.

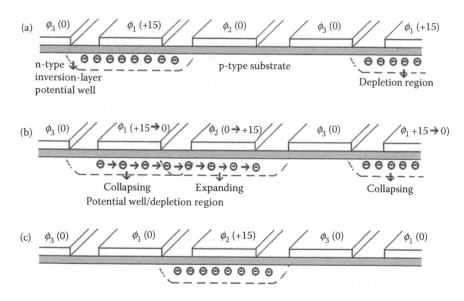

FIGURE 14.57
Electrodes potentials and transfer of charges. (a) Charge packet is localized in potential-well underneath ϕ_1 electrode, (b) charge packet moves to ϕ_2 electrode and (c) overlapped depletion regions and potential wells.

The input gate then decreases, which now isolates the signal input potential well from n⁺ source region and the electron packet in the input gate potential well can then be propagated along the CCD array to the drain end.

Next, as ϕ_2 decreases and simultaneously ϕ_3 increases, the electron packet will get transferred to ϕ_3. Further, as ϕ_1 increases and simultaneously ϕ_3 decreases, the electron packet will get transferred to ϕ_1. Thus, the three-phase charge transfer cycle is completed. This cycle is repeated in a similar fashion and the charge packet will be propagated along the CCD array from the source to the drain end. At the output or drain end, the electron packet is collected by the n⁺ drain region (Figure 14.59).

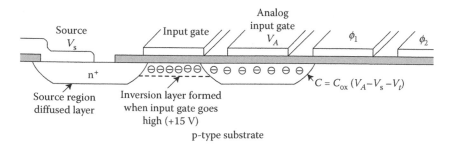

FIGURE 14.58
Charge injection into a CCD array.

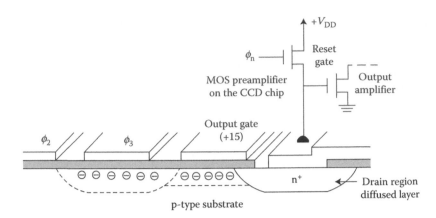

FIGURE 14.59
Charge extraction at the end of a CCD array.

14.7.2 Types of CCDs

CCDs can be classified as surface channel CCD (SCCD), buried channel CCD (BCCD) and junction CCD (JCCD). In SCCD and BCCD, the charges are stored and transferred along the surface or interior of the semiconductor. In JCCD, this store or transfer occurs at the p–n junction. CCDs can also be categorised in accordance with their mode of illumination. Thus, these can be referred to as front-illuminated CCD (FI CCD), back-illuminated CCD (BI CCD) and deep depletion CCD (DD CCD). CCDs can also be classified in terms of number of phases. Figure 14.60 illustrates 2, 3 and 4 phase CCDs.

14.7.3 Dynamic Characteristics

These include charge transfer efficiency, frequency response and power dissipation.

14.7.3.1 Charge Transfer Efficiency (η)

It is defined as the fraction of charge transferred from one well of the CCD to the next. The leftover fraction is termed the transfer loss (ε). Thus, charge transfer efficiency can be written as

FIGURE 14.60
Two, three and four phase CCDs. (a) Two phase (b) three phase and (c) four phase.

$$\eta = 1 - \varepsilon \tag{14.68}$$

Consider that a single pulse with initial amplitude of P_0 transfers down the CCD register. If there are n numbers of transfers (or phases), the amplitude P_n of the nth transfer can be given as

$$P_n = P_0\eta^n = P_0(1 - \varepsilon)^n = P_0(1 - n\varepsilon) \quad \text{for } \varepsilon \ll 1 \tag{14.69}$$

14.7.3.2 Frequency Response

In CCDs, the potential well does not remain unfilled indefinitely and gets eventually filled with thermally generated electron (or holes) at some instant of time. These charges remain stored for some time. This storage time has to be much lesser than the thermal relaxation time of CCD's capacitor. As this relaxation time depends on the channel length L, the maximum frequency is limited by L.

14.7.3.3 Power Dissipation

The power dissipation (P) per bit is simply the product of number of transfers (n), frequency of operation (f), voltage (V) and the maximum value of quality factor (Q_{max}). Thus

$$P = n f V Q_{max} \tag{14.70}$$

14.7.4 Other Assessment Parameters

Some other assessment parameters of CCD include its (i) quantum efficiency, (ii) spectral range, (iii) linearity and (iv) dynamic range. These are illustrated in the figures given below.

Several quantum efficiency curves of different types of CCDs as a function of the wavelength are given in Figure 14.61a. Figure 14.61b shows the spectral sensitivities for different CCDs.

Spectral ranges for FI CCD, BI CCD and DD CCD are shown in Figure 14.61c. The dynamic range, which is defined as the ratio of brightest and faintest detectable signal is shown in Figure 14.61c. It can be noted that CCDs are extremely linear detectors, that is, the received signal increases linearly with the exposure time. Therefore, CCDs enable the simultaneous detection of both very faint and very bright objects. In contrast photographic plates have a very limited linear range. There is a minimum required time for which an object needs to be exposed for the formation of its proper image. Further on during the exposure, the image gets saturated quickly. The dynamic range of CCDs is about 100 times larger compared to films.

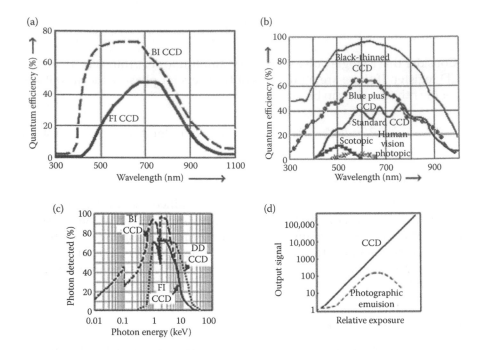

FIGURE 14.61
CCDs characterising parameters. (a) Quantum efficiency, (b) spectral sensitivities, (c) spectral ranges and (d) linearity and photographic emulsion.

14.7.5 Advantages

The advantages of CCDs include good spatial resolution, high quantum efficiency, large spectral window, very low noise, large dynamic range, high photometric precision, very good linearity and reliable rigidity

14.7.6 Applications

CCDs can be used for the detection and imaging of infrared light from targets. In scanned IR system, time delay and integration are its most important functions. These can perform many analog and digital-processing functions such as delay, multiplexing, de-multiplexing and transversal and recursive filtering, integration, analog and digital memory and that of digital logic. Thus, these are widely used in special applications involving VLSI circuits. There is a well-established use of CCDs in x-ray astronomy. Also, CCDs may lead to excellent spectroscopic resolution.

EXAMPLE 14.1

A microwave silicon transistor has a saturation drift velocity $v_s = 5 \times 10^5$ cm/s, transit time cutoff frequency $f_T = 5$ GHz, reactance $X_C = 1$ ohm and $L = 30$ μm. Calculate the maximum electric field intensity

E_m, maximum allowable applied voltage V_m, minimum value of distance L (L_{min}), collector–base capacitance C_o and maximum current of the device I_m.

Solution

Given $v_s = 5 \times 10^5$ cm/s, $f_T = 5$ GHz, $X_C = 1\ \Omega$ and $L = 30\ \mu m$.
In view of Equation 14.6, $V_m f_\tau = (E_m\ v_s)/2\pi = 2 \times 10^{11}$ V/s for silicon.

$$E_m\ v_s/2\pi = 2 \times 10^{11}\ \text{V/s} \quad \text{or} \quad E_m = 4\pi \times 10^{11}/5 \times 10^5 = 2.51 \times 10^6\ \text{V/cm}$$

$$V_m f_\tau = V_m \times 5 \times 10^9 = 2 \times 10^{11} \quad \text{or} \quad V_m = 2 \times 10^{11}/5 \times 10^9 = 40\ \text{V}$$

$$V_m = E_m\ L_{min} \quad \text{or} \quad L_{min} = V_m/E_m = 40/2.51 \times 10^6 = 15.92\ \mu m$$

In view of Equation 14.7, $(I_m X_c)\ f_\tau = (E_m v_s)/2\pi = 2 \times 10^{11}$

$$X_c = 1/2\pi f_\tau C_o$$

Or

$$C_o = 1/2\pi f_\tau\ X_c = 1/(2\pi \times 5 \times 10^9 \times 1) = 31.83\ \text{pF}$$

$$(I_m\ X_c)\ f_\tau = 2 \times 10^{11}$$

Or

$$I_m = 2 \times 10^{11}/(X_c f_\tau) = 2 \times 10^{11}/(5 \times 10^9) = 40\ \text{A}$$

EXAMPLE 14.2

A silicon transistor has $v_s = 5 \times 10^5$ cm/s, $f_T = 4$ GHz, reactance $X_C = 1$ ohm and $E_m = 1.5 \times 10^5$ V/cm. Calculate the maximum power carried by the transistor.

Solution

Given $v_s = 5 \times 10^5$ cm/s, $f_T = 4$ GHz, $X_C = 1$ ohm and $E_m = 1.5 \times 10^5$ V/cm.
In view of Equation 14.8, $(P_m\ X_c)^{0.5} f_\tau = (E_m v_s)/2\pi$

$$P_m = \{(E_m v_s)/2\pi\}^2/(X_c f_\tau^2) = (1.5 \times 10^5 \times 5 \times 10^5/2\pi)^2/\{1 \times (4 \times 10^9)^2\}$$
$$= 8.875\ \text{W}$$

EXAMPLE 14.3

In the Ge–GaAs heterojunction transistor, the given parameters for Ge are marked with suffix 1 and for GaAs with suffix 2. These include electron affinity $\chi_1 = 4.01$, $\chi_2 = 4.08$ and energy gap $E_{g1} = 0.82$, $E_{g2} = 1.44$. Calculate the energy difference between Ge and GeAs for conduction and valence bands.

Solution

Given $\chi_1 = 4.01$, $\chi_2 = 4.08$, $E_{g1} = 0.82$ and $E_{g2} = 1.44$.

In view of Equations 14.10a and 14.10b

$$\Delta E_C = \chi_1 - \chi_2 = 4.01\text{--}4.08 = -0.07 \text{ eV}$$

$$\Delta E_V = E_{g2} - E_{g1} - \Delta E_C = 1.44 - 0.82 + 0.07 = 0.69 \text{ eV}$$

EXAMPLE 14.4

Calculate the pinch-off voltage (V_p) of a Si-JFET if its channel height $a = 0.15 \text{ μm}$, relative permittivity $\varepsilon_r = 11.8$ and electron concentration $N_d = 6 \times 10^{17} \text{ cm}^{-3}$.

Solution

Given $a = 0.15 \text{ μm}$, $\varepsilon_r = 11.8$ and $N_d = 6 \times 10^{17} \text{ cm}^{-3}$.
In view of Equation 14.17

$$V_p = (q \, N_d \, a^2)/(2\varepsilon_s) \text{ V}$$
$$= \{1.6 \times 10^{-19} \times 6 \times 10^{17} \times (0.15 \times 10^{-4})^2\}/(2 \times 8.854 \times 10^{-14} \times 11.8) = 10.34 \text{ V}$$

EXAMPLE 14.5

Calculate the (a) pinch-off voltage, (b) pinch-off current, (c) built-in voltage, (d) drain current, (e) saturation drain voltage and (f) cutoff frequency if the given parameters for a silicon JEFET at 300 K include electron density $N_d = 3 \times 10^{17} \text{ cm}^{-3}$, hole density $N_a = 2 \times 10^{19} \text{ cm}^{-3}$, $\varepsilon_r = 11.8$, channel length $L = 10 \times 10^{-4} \text{ cm}$, channel width $Z = 40 \times 10^{-4} \text{ cm}$, channel height $a = 0.15 \times 10^{-4} \text{ cm}$, drain voltage $V_d = 9 \text{ V}$, gate voltage $V_g = -1.2 \text{ V}$ and electron mobility $\mu_n = 750 \text{ cm}^2/\text{V} \cdot \text{s}$

Solution

Given $N_d = 3 \times 10^{17} \text{ cm}^{-3}$, $N_a = 2 \times 10^{19} \text{ cm}^{-3}$, $\varepsilon_r = 11.8$, $L = 10 \times 10^{-4} \text{ cm}$, $Z = 40 \times 10^{-4} \text{ cm}$, $a = 0.1 \times 10^{-4} \text{ cm}$, $V_d = 9 \text{ V}$, $V_g = -1.2 \text{ V}$ and $\mu_n = 800 \text{ cm}^2/\text{V} \cdot \text{s}$. Also, $q = 1.6 \times 10^{-19} \text{ C}$ and $n_i = 1.5 \times 10^{10}$.

a. Pinch-off voltage (V_p) is given by Equation 14.17

$$V_p = (q \, N_d \, a^2)/(2\varepsilon_s)$$
$$= \{1.6 \times 10^{-19} \times 3 \times 10^{17} \times (0.1 \times 10^{-4})^2\}/(2 \times 8.854 \times 10^{-14} \times 11.8)$$
$$= 2.3 \text{ V}$$

b. Pinch-off current (I_p) is given by Equation 14.22

$$I_p = \{\mu_n q^2 N_d^2 Z a^3\}/(L\varepsilon_s)$$
$$= 0.08 \times (1.6 \times 10^{-19})^2 \times (3 \times 10^{23})^2$$
$$\quad \times 40 \times 10^{-6} \times (10^{-7})^3/(10^{-5} \times 8.854 \times 10^{-12} \times 11.8)$$
$$= 7.068 \text{ mA}$$

c. Built-in voltage (ψ_0) is given by Equation 14.11b
$\psi_0 = V_T \ln|N_a N_d / n_i^2|$, where V_T (= 26 mV at 300 K) is the voltage equivalent of the temperature

$$= 26 \times 10^{-3} \ln|(3 \times 10^{17} \times 2 \times 10^{19})/(1.5 \times 10^{10})^2| = 0.983 \text{ V}$$

d. Drain current (I_d) is given by Equation 14.21

$$I_d = I_p\left[\frac{V_d}{V_p} - \frac{2}{3}\left(\frac{V_d + |V_g| + \psi_0}{V_p}\right)^{1/2} + \frac{2}{3}\left(\frac{|V_g| + \psi_0}{V_p}\right)^{3/2}\right]$$

$|V_g| + \psi_0 = 1.2 + 0.983 = 2.183$ $\{|V_g| + \psi_0\}/V_p = 2.183/2.3 = 0.949$

$V_d + |V_g| + \psi_0 = 9 + 1.2 + 0.983 = 11.183$

$\{V_d + |V_g| + \psi_0\}/V_p = 11.183/2.3 = 4.862$ $[\{|V_g| + \psi_0\}/V_p]^{1/2} = 0.974$

$[\{|V_g| + \psi_0\}/V_p]^{3/2} = 0.9245$ $[\{V_d + |V_g| + \psi_0\}/V_p]^{1/2} = 2.2$

Id $= 2.35 \times 10^{-3} \times \{(9/2.3) - (2/3) 2.2 + (2/3) 0.974\} = 7.274$ mA

e. Saturation drain voltage ($V_{d\,sat}$) is given by Equation 14.27

$$V_{d\,sat} = V_p + |V_g| + \psi_0 = 2.3 + 1.2 + 0.983 = 4.48 \text{ V}$$

f. Cutoff frequency (f_c) is given by Equation 14.32

$f_c = 2\,\mu_n\,q\,N_a\,a^2/(\pi\,\varepsilon_s\,L^2)$

$= 2 \times 800 \times 1.6 \times 10^{-19} \times 3 \times 10^{17} \times (10^{-5})^2/(\pi \times 8.854 \times 10^{-14} \times 11.8 \times 10^{-6})$

$= 2.34$ GHz

EXAMPLE 14.6

Using the data of Example 14.5, calculate the (a) drain conductance, (b) transconductance, (c) and the capacitance between the gate and the source.

Solution

Given in Example 14.5: $\varepsilon_r = 11.8$, $L = 10 \times 10^{-4}$ cm, $Z = 40 \times 10^{-4}$ cm, $a = 0.1 \times 10^{-4}$ cm and $V_d = 9$ V. Also, the calculated values are $I_p = 2.35$ mA, $V_p = 2.3$ V, $|V_g| + \psi_0 = 2.183$ and $[\{|V_g| + \psi_0\}/V_p]^{1/2} = 0.974$.
   In view of the above values

$$I_p/V_p = 2.35 \times 10^{-3}/2.3 = 1.022 \times 10^{-3} \quad \text{and} \quad \{V_p/(|V_g| + \psi_0)\}^{1/2}$$

$$= (2.3/2.183)^{1/2} = 1.026$$

a. The drain conductance (g_d) is given by Equation 14.24

$$g_d = \frac{I_p}{V_p}\left[1 - \left(\frac{|V_g| + \psi_0}{V_p}\right)^{1/2}\right] = 1.022 \times 10^{-3} \times (1 - 0.974)$$

$$= 26.572\mu\text{-mho}$$

b. The transconductance (g_m) is expressed by Equation 14.25

$$g_m = \frac{I_p V_d}{2V_p^2}\left[1 - \left(\frac{V_p}{|V_g| + \psi_0}\right)^{1/2}\right] = \{(1.022 \times 10^{-3} \times 9)/(2 \times 2.3 \times 2.3)\}$$

$$\times (1 - 1.026) = -22.6\mu\text{mho}$$

c. Capacitance (C_g) between the gate and the source is given by Equation 14.33

$$C_g = LZ\varepsilon_s/2a = 10^{-3} \times 4 \times 10^{-5} \times 8.854 \times 10^{-14} \times 11.8/ (2 \times 10^{-5})$$

$$= 2.089544 \times 10^{-15} \text{ F}$$

EXAMPLE 14.7

Find the cutoff frequency and maximum operating frequency of a GaAs MESFET having gate metallisation resistance $R_g = 3\ \Omega$, input resistance $R_i = 2.6\ \Omega$, transconductance $g_m = 40$ m℧, drain resistance $R_d = 400\ \Omega$, source resistance $R_s = 2.5\ \Omega$ and gate–source capacitance $C_{gs} = 0.5$ pF.

Solution

Given $R_g = 3\ \Omega$, $R_i = 2.6\ \Omega$, $g_m = 40$ m℧, $R_d = 400\ \Omega$, $R_s = 2.5\ \Omega$ and $C_{gs} = 0.5$ pF.

The cutoff frequency is given by Equation 14.43

$$f_c = g_m/(2\pi C_{gs}) = 40 \times 10^{-3}/(2\pi \times 0.5 \times 10^{-12}) = 12.73 \text{ GHz}$$

The highest value of the operating frequency is given by Equation 14.49

$$f_{max} = \frac{f_c}{2}\left(\frac{R_d}{R_s + R_g + R_i}\right)^{1/2} = \{(12.73 \times 10^9)/2\}\{400/(2.5 + 3 + 2.6)\}^{1/2}$$

$$= 44.73\,\text{GHz}$$

EXAMPLE 14.8

Compute the (a) pinch-off voltage, (b) velocity ratio, (c) saturation current at $V_g = 0$ and (d) drain current I_d for a GaAs MESFET if the electron concentration $N_d = 6 \times 10^{17}$ cm^{-3}, electron mobility $\mu = 800$ cm^2/V·s, drain voltage $V_d = 6$ V, gate voltage $V_g = -3$ V, saturation drift velocity

$v_s = 2.5 \times 10^{-7}$ cm/s, channel height $a = 0.2\ \mu m$, channel length $L = 10\ \mu m$, channel width $Z = 30\ \mu m$ and the relative dielectric constant $\varepsilon_r = 13.1$.

Solution

Given $\quad N_d = 2 \times 10^{17}$ cm$^{-3} = 2 \times 10^{23}$ m^{-3}, $\quad \mu = 800$ cm^2/V·s = 0.08 m^2/V·s, $V_d = 5$ V, $V_g = -3$ V, $v_s = 2.5 \times 10^5$ m/s, $a = 0.2\ \mu m = 2 \times 10^{-7}$ m, $L = 10\ \mu m$ = 10^{-5} m, $Z = 30\ \mu m = 3 \times 10^{-5}$ m and $\varepsilon_r = 13.1$.

a. The pinch-off voltage V_p is given by Equation 14.36

$V_p = (qN_d a^2)/2\varepsilon_s$

$= \{1.6 \times 10^{-19} \times 2 \times 10^{23} \times (2 \times 10^{-7})^2\}/(2 \times 8.854 \times 10^{-12} \times 13.1) = 5.53$ V

b. The velocity ratio η is given by Equation 14.38d

$\eta = \{\mu|V_p|/(v_s L)\} = v/v_s$

$= (0.08 \times 5.53)/(2.5 \times 10^5 \times 10^{-5}) = 0.17696$

c. The saturation current at $V_g = 0$ is given by Equation 14.38a

$I_p = (q\ N_d\ \mu\ a\ Z\ V_p)/3\ L$

$= (1.6 \times 10^{-19} \times 6 \times 10^{23} \times 0.08 \times 2 \times 10^{-7} \times 3 \times 10^{-5} \times 5.53)/(3 \times 10^{-5})$

$= 8.5$ mA

d. The drain current I_d is given by Equation 14.37

$$I_d = I_p \frac{3(u^2 - \rho^2) - 2(u^3 - \rho^3)}{1 + \eta(u^2 - \rho^2)}\ A$$

In view of Equation 14.38c

$$\rho = (|V_g|/V_p)^{0.5} = \sqrt{(3/5.53)} = 0.736,\ \rho^2 = 0.543,\ \rho^3 = 0.4$$

In view of Equation 14.38b

$u = [\{V_d + |V_g|\}/V_p]^{0.5} = \sqrt{\{(5 + 3)/5.53\}} = 1.2,\ u^2 = 1.45,\ u^3 = 1.74$

$u^2 - \rho^2 = 1.45 - 0.543 = 0.907 \quad$ and $\quad u^3 - \rho^3 = 1.74 - 0.4 = 1.34$

$$I_d = 8.5 \times 10^{-3} \frac{3 \times 0.907 - 2 \times 1.34}{1 + 0.0885 \times 0.907} = 0.37 \text{mA}$$

EXAMPLE 14.9

Find the (a) surface potential (ψ_i) for strong inversion, (b) insulator capacitance (C_i) and (c) the threshold voltage for a p-channel MOSFET having doping concentration $N_a = 2 \times 10^{23}$ m^{-3}, relative dielectric constant $\varepsilon_r = 11.8$, relative dielectric constant of SiO$_2$ $\varepsilon_{ir} = 4$, insulator depth $d = 0.02\ \mu m$ and operating temperature $T = 300$ K.

Solution

Given $N_a = 2 \times 10^{23}$ m^{-3}, $\varepsilon_r = 13.1$, $\varepsilon_{ir} = 4$, $d = 0.02$ μm and $T = 300$ K.

a. The surface potential (ψ_i) for strong inversion is given by the relation

$$\psi_i = V_T \ln|N_a/n_i|, \text{ where } V_T (= 26 \text{ mV at } 300 \text{ K})$$
$$= 26 \times 10^{-3} \ln|(2 \times 10^{23}/1.5 \times 10^{10}| = 0.786 \text{ V}$$

b. The insulator capacitance per unit area is given by the relation

$$C_i = \varepsilon_{ir}/d = 4 \times 8.854 \times 10 - 12/2 \times 10 - 8 = 1.77 \text{ mF/m}^2$$

c. The threshold voltage is given by Equation 14.52b

$$V_{th} = 2\psi_i + (2/C_i)(\varepsilon_s q N_a \psi_i)^{1/2}$$
$$= 2 \times 0.786 + \{2/(1.77 \times 10^{-3})\}(8.854 \times 10^{-12}$$
$$\times 13.1 \times 1.6 \times 10^{-19} \times 2 \times 1023 \times 0.786)^{1/2} = 3.5 \text{ V}$$

EXAMPLE 14.10

Calculate the (a) insulator capacitance, (b) saturation drain current, (c) transconductance in saturation region and (d) maximum operating frequency in the saturation region for an n-channel MOSFET. The parameters of the device include gate voltage $V_g = 4$ V, threshold voltage $V_{th} = 0.2$ V, doping factor m = 1, electron mobility $\mu_n = 0.125$ m^2/V·s, electron velocity $v = 1.5 \times 10^5$ m/s, relative dielectric constant of SiO$_2$ $\varepsilon_{ir} = 4$, channel length $L = 4$ μm, channel depth $Z = 10$ μm and insulator thickness $d = 0.04$ μm.

Solution

Given $V_g = 4$ V, $V_{th} = 0.2$ V, $m = 1$, $\mu_n = 0.125$ m^2/V·s, $v = 1.5 \times 10^5$ m/s, $\varepsilon_{ir} = 4$, $L = 4$ μm, $Z = 10$ μm and $d = 0.04$ μm.

a. The insulator capacitance is given by the relation

$$C_i = \varepsilon_{ir}/d = 4 \times 8.854 \times 10^{-12}/4 \times 10^{-8} = 885.4 \ \mu F/m^2$$

b. The saturation drain current for non-idealised channel is given by Equation 14.57

$$I_{d \ sat} = Z C_i (V_g - V_{th}) v_s = 10^{-5} \times 885.4 \times 10^{-6}$$
$$\times (4-0.2) \times 1.5 \times 10^5 = 5.047 \text{ mA}$$

c. The transconductance in the saturation region is given by Equation 14.58

$$g_m = Z C_i v_s = 10^{-5} \times 885.4 \times 10^{-6} \times 1.5 \times 10^5 = 1.328 \text{ m}\mho$$

d. The maximum operating frequency in the saturation region is given by Equation 14.60b

$$f_{max} = v_s/2\pi L = 1.5 \times 10^5/(2\pi \times 4 \times 10^{-6}) = 5.97 \text{ GHz}$$

EXAMPLE 14.11

Calculate the drain current of an HEMT having gate width $W = 100$ μm, electron velocity $v(z) = 2 \times 10^5$ m/s and 2DEG density $n(z) = 5 \times 10^{15}$ m^{-2}.

Solution

Given $W = 100$ μm, $v(z) = 2 \times 10^5$ m/s and $n(z) = 5 \times 10^{15}$ m^{-2}
The drain current I_d is given by Equation 14.67

$$I_{ds} = q\, n(z)\, W\, v(z) = 1.6 \times 10^{-19} \times 5 \times 10^{15} \times 10^{-4} \times 2 \times 10^5 = 16 \text{ mA}$$

PROBLEMS

P-14.1 A microwave silicon transistor has saturation drift velocity $v_s = 4 \times 10^5$ cm/s, transit time cutoff frequency $f_T = 6$ GHz, reactance $X_C = 1$ ohm and $L = 20$ μm. Calculate the maximum electric field intensity E_m, maximum allowable applied voltage V_m, minimum value of distance L (L_{min}), collector–base capacitance C_o and maximum current of the device I_m.

P-14.2 A silicon transistor has $v_s = 5 \times 10^5$ cm/s, $f_T = 6$ GHz, reactance $X_C = 1$ ohm and $E_m = 2.5 \times 10^5$ V/cm. Calculate the maximum power carried by the transistor.

P-14.3 In the Ge–GaAs heterojunction transistor, the given parameters for Ge are marked with suffix 1 and for GaAs with suffix 2. These include electron affinity $\chi_1 = 3.99$, $\chi_2 = 4.02$ and energy gap $E_{g1} = 0.8$, $E_{g2} = 1.4$. Calculate the energy difference between Ge and GeAs for conduction and valence bands.

P-14.4 Calculate the pinch-off voltage (V_p) of a Si-JFET if its channel height $a = 0.1$ μm, relative permittivity $\varepsilon_r = 11.8$ and electron concentration $N_d = 5 \times 10^{17}$ cm^{-3}.

P-14.5 Calculate the (a) pinch-off voltage, (b) pinch-off current, (c) built-in voltage, (d) drain current, (e) saturation drain voltage and (f) cutoff frequency if the given parameters for a silicon JEFET at 300 K include electron density $N_d = 5 \times 10^{17}$ cm^{-3}, hole density $N_a = 3 \times 10^{19}$ cm^{-3}, $\varepsilon_r = 11.8$, channel length $L = 8 \times 10^{-4}$ cm, channel width $Z = 30 \times 10^{-4}$ cm, channel height $a = 0.1 \times 10^{-4}$ cm, drain voltage $V_d = 7$ V, gate voltage $V_g = -1$ V and electron mobility $\mu_n = 700$ cm^2/V·s.

P-14.6 Find the cutoff frequency and maximum operating frequency of a GaAs MESFET having gate metallisation resistance $R_g = 2.8\ \Omega$, input resistance $R_i = 2.5\ \Omega$, transconductance $g_m = 50$ m℧, drain resistance $R_d = 300\ \Omega$, source resistance $R_s = 2.2\ \Omega$ and gate–source capacitance $C_{gs} = 0.4$ pF.

P-14.7 Compute the (a) pinch-off voltage, (b) velocity ratio, (c) saturation current at $V_g = 0$ and (d) drain current I_d for a GaAs MESFET if the electron concentration $N_d = 5 \times 10^{17}$ cm^{-3}, electron mobility

$\mu = 800$ cm^2/V·s, drain voltage $V_d = 4$ V, gate voltage $V_g = -1.5$ V, saturation drift velocity $v_s = 2 \times 10^{-7}$ cm/s, channel height $a = 0.25$ μm, channel length $L = 8$ μm, channel width $Z = 25$ μm and the relative dielectric constant $\varepsilon_r = 13.1$.

P-14.8 Find the (a) surface potential (ψ_i) for strong inversion, (b) insulator capacitance (C_i) and (c) the threshold voltage for a p-channel MOSFET having doping concentration $N_a = 3 \times 10^{23}$ m^{-3}, relative dielectric constant $\varepsilon_r = 11.8$, relative dielectric constant of SiO$_2$ $\varepsilon_{ir} = 3.5$, insulator depth $d = 0.01$ μm and operating temperature $T = 300$ K.

P-14.9 Calculate (a) insulator capacitance, (b) saturation drain current, (c) transconductance in saturation region and (d) maximum operating frequency in the saturation region for an n-channel MOSFET. The parameters of the device include gate voltage $V_g = 5$ V, threshold voltage $V_{th} = 0.3$ V, doping factor $m = 1$, electron mobility $\mu_n = 0.115$ m^2/V·s, electron velocity $v = 1.7 \times 10^5$ m/s, relative dielectric constant of SiO$_2$ $\varepsilon_{ir} = 3.8$, channel length $L = 5$ μm, channel depth $Z = 8$ μm and insulator thickness $d = 0.03$ μm.

P-14.10 Calculate the drain current of an HEMT having gate width $W = 150$ μm, electron velocity $v(z) = 3 \times 10^5$ m/s and 2DEG density $n(z) = 4 \times 10^{15}$ m^{-2}.

Descriptive Questions

Q-14.1 Discuss the limitations of conventional transistors.

Q-14.2 List the types of transistors that are specifically used at microwave frequencies.

Q-14.3 Explain the working bipolar (n–p–n) junction transistors with the aid of its structure.

Q-14.4 Illustrate different transistor configurations and discuss their relative merits.

Q-14.5 Discuss different operational modes of an n–p–n transistor. Give its hybrid-pi equivalent model for CE configuration.

Q-14.6 Describe the limitations of BJTs.

Q-14.7 Illustrate the energy band diagrams of a heterojunction bipolar transistor when (i) n-Ge and p-GaAs are isolated and (ii) n-Ge and p-GaAs are jointed together.

Q-14.8 Illustrate the structure of an HBT transistor showing its different layers and terminal.

Q-14.9 Discuss the working of a junction field effect transistor (JFET).

Q-14.10 List the assessment parameters of JFET.

Q-14.11 Illustrate the physical structure of a MOSFET. Indicate its different components.

Q-14.12 Discuss the various modes of operation of a MOSFET. Add necessary figures.

Q-14.13 Illustrate MOSFET symbols for different channels and modes.

Q-14.14 Draw the diagram of GaAs MESFET. Name its components and terminals.

Q-14.15 Draw a basic HEMT structure and systematically explain its formation.

Q-14.16 How do NMOS and PMOS logics differ in their construction and operation?

Q-14.17 Discuss special features of the complementary metal-oxide-semiconductor logic.

Q-14.18 Discuss the power dissipation in CMOS logic.

Further Reading

1. Abe, M. et al., Recent advances in ultra-light speed HEMT technology. *IEEE J. Quantum Electronics*, QE-22(9), 1870–1879, September 1986.
2. Anderson, R. L., Experiment on Ge-GaAs heterojunction. *Solid State Electronics*, 5, 341, 1962.
3. Asai, S., Semiconductor memory trend. *Proc. IEEE*, 74(12), 1623–1635, December 1986.
4. Bahl, I., ed. *Microwave Solid-State Circuit Design*. John Wiley, New York, 1988.
5. Baker, R. J., *CMOS: Circuit Design, Layout, and Simulation*, 3rd ed. Wiley-IEEE Series on Microelectronic Systems, USA, 2010.
6. Chodorow, M. and Susskind, C., *Fundamentals of Microwave Electronics*. McGraw-Hill, New York, 1964.
7. Dingle, R. et al., Electron mobilities in modulation-doped semiconductor heterojunction super lattices. *Appl. Phys. Lett.*, 33, 665–667, 1978.
8. Early, J. M., Maximum rapidly switchable power density in junction triodes. *IRE Trans. Electron Devices*, ED-6, 322–335, 1959.
9. Eastman, L. F., *Gallium Arsenide Microwave Bulk and Transit-Time Devices*. Artech House, Dedham, MA, 1973.
10. *IEEE Transactions*. Special issues on microwave solid-state devices. MTT-21, No. 11, November 1973, MTT-24, No. 11, November 1976, MTT-27, No. 5, May 1979, MTT-28, No. 12, December 1980, MTT-30, No. 4, April 1982, MTT-30, No. 10, October 1982.
11. *IEEE Transactions on Electron Devices*. Special issues on MW solid-state devices. ED-27, No. 2, February 1980, ED-27, 6, June 1980, ED-28, 2, February 1981, ED-28, 8, August 1981.

12. *IEEE Proceedings*, Special issue on high power microwave tubes. 61, No. 3, March 1973.
13. *IEEE Proceedings*. Special issue on very fast solid-state technology. 70, No. 1, January 1982.
14. Johnson, E. O., Physical limitations on frequency and power parameters of transistor. *RCA Rev.*, 26(6), 163–177, June 1965.
15. Lee, T. H., *The Design of CMOS Radio-Frequency Integrated Circuits*. Cambridge University Press, New York, 1998.
16. Lehovec, K. and Zuleeg, R., Voltage-current characteristics of GaAs J-FET's in the hot electron range. *Solid-State Electronics*, 13, 1415–1426, 1970.
17. Liao, S. Y., *Microwave Solid-State Devices*. Prentice-Hall, Englewood Cliffs, NJ, 1985.
18. Liao, S. Y., *Semiconductor Electronic Devices*. Prentice-Hall, Englewood Cliffs, NJ, 1990.
19. Lio, S. Y., *Microwave Devices & Circuits*, 3rd ed. Prentice-Hall, New Delhi, 1995.
20. Manley, J. M. and Rowe, H. E., Some general properties of nonlinear elements— Pt. I, General energy relations. *Proc. IRE*, 44, 904–913, July 1956.
21. Mead, C. A., Schottky barrier gate-field effect transistor. *Proc. IEEE*, 54(2), 307–308, February 1966.
22. Mead, C. A. and Conway, L., *Introduction to VLSI Systems*. Addison-Wesley, Boston, 1980.
23. Nanavati, R. P., *Semiconductor Devices*. Intext Ed. Publishers, Scranton, PA, 1975.
24. Navon, D. H., *Semiconductor Micro Devices and Materials*. Holt, Rinehart and Winston, New York, 1986.
25. Pavlidis, D. and Wiess, M., The influence of device physical parameters on HEMT large-signal characteristics. *IEEE Trans.*, MTT-36(2), 239–249, February 1988.
26. Shockley, W., A unipolar field-effect transistor. *Proc. IRE*, 40(11), 1365–1376, November 1952.
27. Sidney, S., *Applications of Analog Integrated Circuits*. Prentice-Hall, New Delhi, 1990.
28. Shur, M., *Physics of Semiconductor Devices*. Prentice-Hall, New Delhi, 1990.
29. Streetman, B. G., *Solid-State Electronic Devices*, 3rd ed. Prentice-Hall, Englewood Cliffs, NJ, 1989.
30. Sze, S. M., *Physics of Semiconductor Devices*, 2nd ed. John Wiley, New York, 1981.
31. Sze, S. M., *Semiconductor Devices: Physics and Technology*. John Wiley, New York, 1985.
32. Togashi, K. et al., Reliability of low-noise microwave HEMTs made by MOCVD. *Microwave J.*, 123–132, April 1987.
33. Tsividis, Y., *Operation and Modeling of the Transistor*. Oxford University Press, UK, 1999.
34. Tyagi, M. S., *Introduction to Semiconductor Materials and Devices*. Wiley, New York, 1999.
35. Vanderziel, A. et al., Gate noise in field-effect transistor at moderately high frequencies. *Proc. IEEE*, 51(3), 461–467, March 1963.
36. Veendrick, H. J. M., *Nanometer CMOS ICs, from Basics to ASICs*. Springer, New York, 2008.
37. Weste, N. H. E. and Harris, D. M., *CMOS VLSI Design: A Circuits and Systems Perspective*, 4th ed. Pearson/Addison-Wesley, Boston, 2010.

15

Planar Transmission Lines

15.1 Introduction

A class of transmission lines referred to as planar transmission lines is used over a very wide frequency range starting from centimetric to the submillimetric zone to guide the electromagnetic waves. These have the same characterising parameters (viz. Z_0, γ, α, β, c, v_p and v_g) as applicable to other transmission structures. Their geometry allows the control of these parameters, particularly the characteristic impedance by defining dimensions in a single plane. This feature allows a complete transmission line circuit to be fabricated in one step by using thin film and photolithography techniques similar to those used for making printed circuits for low-frequency electronic circuitry.

Planar transmission line structures can be realised in a number of forms. In accordance with Figure 15.1 these may include striplines, microstrips, coupled microstrip, slot line, co-planar waveguide and fin lines. The topology of each of these is characterised by the modal distribution of their fields, as well as achievable levels of characteristic impedance, phase velocity and the number of modes supported by their structures. Their combinations can be used to produce circuit elements such as delay lines, crossovers, resonators, transitions, power splitters/combiners and filters. The strategies for their design, analysis and circuit implementation are well established. With the addition of normal metals, terminations and broadband absorbers can also be realised and used to tailor the circuit response required by the end applications.

To create high-performance passive circuit elements, an understanding of the behaviour and relationships between these structures is quite essential. The aspects for different structures can be summarised as below.

1. The field distribution in the microstrip lines is relatively well confined for a large line width over substrate height ratios and is well suited for realising elements with low characteristic impedance and low radiation loss.

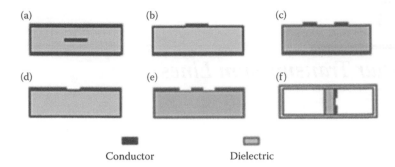

FIGURE 15.1
Cross-section views of the commonly used planar structures. (a) Stripline, (b) microstrip, (c) coupled microstrip, (d) slot line, (e) co-planar waveguide and (f) finline.

2. Slot line to a good approximation is the electromagnetic complement to a microstrip configuration. It has poor field confinement and can be susceptible to radiation losses. The complementary symmetry between slot line and microstrip topologies can be utilised for realising broadband antenna feed elements and 180° phase inversions.

3. If the lower metal ground plane region in the microstrip shrinks to the width of the upper line, parallel plate transmission line structure is formed that gives relatively high impedance.

4. The co-planar waveguide structure possessing three conductors supports two propagation modes. These modal field distributions can be characterised by their even and odd spatial symmetries. The existence of these two differing modes can be utilised from a design perspective to create a rich variety of coupling structures and hybrids. Sometimes, one of the two modes is intentionally suppressed to ensure single-mode propagation on the structure. This can be done, for example, with an 'air-bridge' connecting the two outer conductors that effectively sorts out one of the two modes. Topologically, this configuration is very similar to a microstrip with a slot in the ground plane. With appropriate care, this relationship can be used in creating microstrip structures with higher characteristic impedance levels if needed.

The simplest transmission line configuration is a coaxial line. Its simplicity lies in obtaining an exact solution for its impedance and propagation velocity in terms of the radii of its inner and outer conductors. In some of the situations, when a signal is routed through the printed circuit board, the coaxial lines provide an appropriate solution. The coaxial line equations help in matching these transmission lines to the surrounding circuitry and in

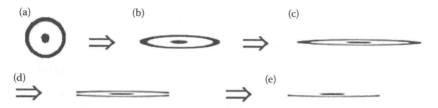

FIGURE 15.2
Transformation of a coaxial line into a stripline and a microstrip line. (a) Coaxial cable, (b) Squeezed coaxial cable, (c) coaxial cable further squeezed, (d) removal of edges from Figure 15.2c results in formation of stripline and (e) Removal of upper plane from Figure 15.2d results in microstrip line.

minimising the discontinuity. Very small coaxial lines are available in semi-rigid forms with quite small (down to 0.2 mm) outer diameters. These lines provide a fully shielded path for signals on printed circuit boards in low volume production, as rework and for prototyping. Figure 15.2 illustrates the transformation of a coaxial cable into two of the most commonly used planar lines, namely, (a) striplines and (b) microstrips.

In the subsequent sections, different types of planar transmission lines, some of which are illustrated in Figure 15.1, are studied. This study involves the evaluation of different characteristic parameters referred to above. The approximate values of some of these parameters can be obtained by solving the electrostatic problem for a given geometry. The analysis becomes simple if strip thickness is small enough or tends to zero. The exact values can, however, be obtained only by using conformal transformation method.

15.2 Striplines

A planar type of transmission line is well suited for microwave-integrated circuitry and photolithographic fabrication. Stripline is one of the forms of a planar transmission line. A conventional stripline is a balanced line wherein a flat conducting strip is symmetrically placed between two large ground planes. The space between these planes is filled with a homogeneous dielectric material.

Figure 15.3a illustrates the geometry of a stripline and Figure 15.3b shows the field distribution therein. As can be seen, almost all the field lines are concentrated near the central conductor. The field rapidly decays away from this strip. This property of stripline allows termination of the geometry in transverse direction without affecting its transmission characteristics.

Figure 15.4 shows the constructional details of a three-conductor line that can carry a wave of pure TEM mode. This line can also support

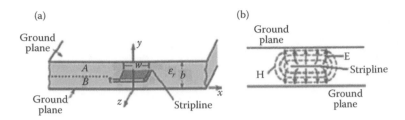

FIGURE 15.3
Stripline. (a) Geometry and (b) field distribution.

higher-order TE and TM modes, but all such modes are usually avoided in practice. Its two outer conductors act as ground planes maintained at zero potential. The central conductor, which is meant for TEM mode propagation, is the stripline. In general, the width of the strip (w) is much larger than its thickness (t). The order of thickness is kept between 1.4 and 2.8 mils. The space between two ground planes is filled by a dielectric material. No air pockets are allowed when segments A and B of the dielectric slabs are overlapped.

The stripline is basically a flattened coaxial line. It is also called a triplate or sandwich line. The presence of both the top and bottom shields provides good isolation from other signals on the printed circuit board. The coupled lines have equal even and odd mode phase velocities. These can be used to form high directivity couples for coupled line filters.

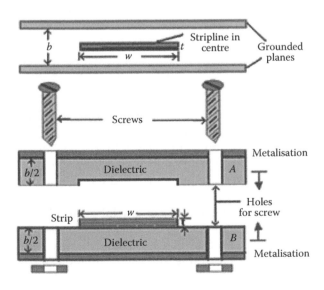

FIGURE 15.4
Constructional details of a stripline.

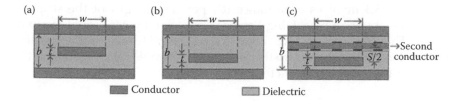

FIGURE 15.5
(a) Centered stripline, (b) off-centered stripline and (c) dual-orthogonal stripline.

15.2.1 Types of Striplines

Figure 15.5 shows three different versions of striplines. These include (a) centred stripline, (b) off-centred stripline and (c) dual-orthogonal stripline. A centred stripline is the ideal one; however, layout decisions may require an off-centred stripline. The dual-orthogonal stripline is useful in high-density routing situations and is accomplished by routing two off-centre layers at right angles to each other. In all these versions, b is the separation between outer conductors, t is the thickness and w is the width of the central conductor.

15.2.2 Assessment Parameters

Some useful relations for the lossless stripline are listed below:

- *Phase velocity* (v_p) for TEM mode:

$$v_p = 1/\sqrt{(\mu_0\varepsilon_0\varepsilon_r)} = v_0/\sqrt{\varepsilon_r} = c/\sqrt{\varepsilon_r} \qquad (15.1)$$

- Propagation constant (β_0) for free space:

$$\beta_0 \approx \omega\sqrt{(\mu_0\varepsilon_0)} \qquad (3.3c)$$

- Propagation constant (β) for a dielectric with a relative dielectric constant ε_r:

$$\beta = \omega/v_p = \omega\sqrt{(\mu_0\varepsilon_0\varepsilon_r)} = \beta_0\sqrt{\varepsilon_r} \qquad (15.2)$$

- Characteristic impedance:

$$Z_0 = \sqrt{(L/C)} = \{\sqrt{(LC)}\}/C = 1/(v_p \cdot C) \qquad (15.3)$$

Equation 15.3 involves capacitance (C) per unit length of the stripline. This can be obtained by solving Laplace's equation by conformal mapping method. This, however, involves complicated special functions. For practical computations of the characteristic impedance, a simple relation obtained by employing the curve-fitting method is given below.

$$Z_0 = \frac{30\pi}{\sqrt{\varepsilon_r}} \frac{b}{w_e + 0.441b} \qquad (15.4)$$

- *Effective width:* The effective width (w_e) of the centre conductor involved in Equation 15.4 is related to the actual width (w) given below:

$$w_e/b = w/b \quad \text{for } w/b > 0.35 \qquad (15.5a)$$

$$w_e/b = w/b - (0.35 - w/b)^2 \quad \text{for } w/b < 0.35 \qquad (15.5b)$$

The parameters w and b are shown in Figures 15.4 and 15.5.
- *Strip width:* If the characteristic impedance is known, the strip width can be obtained by the relation:

$$w/b = x \quad \text{for } \sqrt{\varepsilon_r}Z_0 > 120 \qquad (15.6a)$$

$$w/b = 0.85 - \sqrt{(0.6 - x)} \quad \text{for } \sqrt{\varepsilon_r}Z_0 < 120 \qquad (15.6b)$$

where

$$x = \frac{30\pi}{\sqrt{\varepsilon_r}Z_0} - 0.441 \qquad (15.7)$$

- *Attenuation due to the conductor loss (α_c):* It is given by the relation

$$\alpha_c = \frac{2.7 \times 10^{-3} R_s \varepsilon_r Z_0}{30\pi(b-t)} A \quad \text{for } \sqrt{\varepsilon_r}Z_0 > 120 \qquad (15.8a)$$

$$\alpha_c = \frac{0.16R_s}{Z_0 b} B \quad \text{for } \sqrt{\varepsilon_r}Z_0 < 120 \qquad (15.8b)$$

where the relation for surface resistance of the conductor (R_s) is

$$R_s = \sqrt{(\omega\mu/2\sigma)} \qquad (2.19b)$$

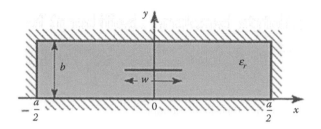

FIGURE 15.6
Modified geometry of an enclosed stripline.

The parameters A and B are given by

$$A = 1 + \frac{2w}{b-t} + \frac{1}{\pi}\frac{b+t}{b-t}\ln\left(\frac{2b-t}{t}\right)$$ (15.9a)

$$B = 1 + \frac{b}{(0.5w + 0.7t)}\left(0.5 + \frac{0.414t}{w} + \frac{1}{2\pi}\ln\frac{4\pi w}{t}\right)$$ (15.9b)

where w, b and t are shown in Figure 15.5.

As stated earlier, the property of rapid decay of the field allows termination of the geometry in transverse direction without affecting its transmission characteristics. Thus, the geometry of the stripline can be modified by truncating the plates after some distance by placing metal walls on the sides. This version shown in Figure 15.6 leads to an approximate solution to the problem.

15.3 Microstrips

If the upper ground plane of a stripline is removed, the resulting configuration is the microstrip line, often referred to only as microstrip. It is one of the most popular types of the lines that can be fabricated by photolithographic process. Figure 15.7 illustrates a microstrip with only one ground plane and that too at the bottom. Owing to the absence of the top ground plane, there is an easy access to the top surface that makes it convenient to mount discrete passive components and active microwave devices. Also, the signals are readily available for probing. Besides, minor adjustments can be made after fabrication of the circuit.

Owing to the advances in IC technology, microwave semiconductor devices are usually fabricated as semiconductor chips. In view of their small volumes (on the order of 0.008–0.08 mm³), microstrips (also called

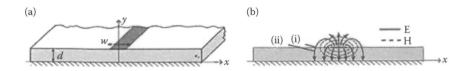

FIGURE 15.7
Microstrips. (a) Geometry and (b) field distribution.

open-strip lines) are commonly used for feeding into or extracting energy from, these chips. In most practical applications, the thickness (d) of the dielectric substrate is kept very small, that is, in electrical terms, $d \ll \lambda$. Owing to this thinness, the modes on the microstrip are quasi-transverse electric and magnetic (quasi-TEM) and microstrips cannot support a pure TEM wave. Thus, in the case of a microstrip, the theory of TEM-coupled lines applies only approximately.

Its disadvantage over the stripline is that some of the transmitted energy may be coupled into space or to the adjacent traces. In view of openness of the structure, there is likelihood of the radiation loss or interference from the nearby conductors. These losses are more significant at discontinuities, namely, short-circuit posts, corners and so on. Some remedial measures are therefore required to minimise these losses. To confine the field nearer to the strip, a substrate with higher permittivity is generally used. This results in the reduction of phase velocity and guide wavelength and hence circuit dimensions. In the case of microstrips, as the fields extend into the space (above the microstrip), the configuration becomes a mixed dielectric transmission structure that makes the analysis complicated.

Figure 15.7b shows that at the air–dielectric interface, as the electric field obliquely strikes, some reflections are bound to occur. These reflections need to be accounted in the analysis. This figure also shows that most of the electric field exists inside the substrate; thus, most of the energy remains confined inside. These two cases are marked as (i) and (ii) in the diagram.

Figure 15.8 shows an equivalent geometry of a quasi-TEM microstrip line, where the dielectric slab of thickness t and relative permittivity ε_r has been replaced with a homogeneous medium of effective relative permittivity ε_e.

15.3.1 Types of Microstrips

Figure 15.9 illustrates different versions of microstrips. These include (a) a common microstrip line, (b) an embedded microstrip line and (c) a covered microstrip line. The structure of the covered microstrip line is further shown in Figure 15.10. In an embedded microstrip, it is covered with solder mask or a thin layer of epoxy. Covered microstrip is a transitional structure somewhere between the microstrip and stripline.

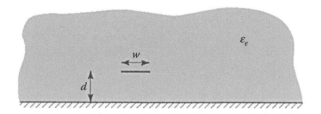

FIGURE 15.8
Equivalent quasi-TEM microstrip line.

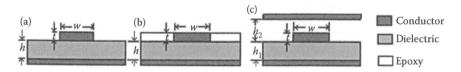

FIGURE 15.9
(a) Common microstrip, (b) embedded microstrip and (c) covered microstrip.

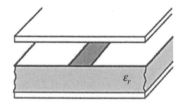

FIGURE 15.10
Covered microstrip line.

15.3.2 Assessment Parameters

Some useful relations for the microstrip are listed as follows:

Phase velocity (v_p)

$$v_p = c/\sqrt{\varepsilon_e} \quad \text{where } 1 < \varepsilon_e < \varepsilon_r \tag{15.10}$$

Propagation constant (β)

$$\beta = \beta_0 \sqrt{\varepsilon_e} \tag{15.11}$$

Effective dielectric constant (ε_e): The ε_e involved in Equations 15.10 and 15.11 is given by

$$\varepsilon_e = \frac{\varepsilon_r + 1}{2} + \frac{\varepsilon_r - 1}{2} \frac{1}{\sqrt{1 + 12d/w}} \tag{15.12}$$

Characteristic impedance (Z_0)

$$Z_0 = \frac{60}{\sqrt{\varepsilon_e}} \ln\left(\frac{8d}{w} + \frac{w}{4d}\right) \quad \text{for } w/d \leq 1 \quad \quad (15.13a)$$

$$Z_0 = \frac{120\pi}{\sqrt{\varepsilon_e}} \frac{1}{[(w/d) + 1.393 + 0.667 \ln\{(w/d) + 1.444\}]} \quad \text{for } w/d \geq 1 \quad (15.13b)$$

Strip width: If Z_0 and ε_r are known, the strip width can be obtained from the following relations:

$$w/d = \frac{8e^A}{e^{2A} - 2} \quad \text{for } w/d < 2 \quad \quad (15.14a)$$

$$w/d = \frac{2}{\pi}\left[B - 1 - \ln(2B - 1) + \frac{\varepsilon_r - 1}{2\varepsilon_r}\left\{\ln(B - 1) + 0.39 - \frac{0.61}{\varepsilon_r}\right\}\right] \quad \text{for } w/d > 2$$

$$(15.14b)$$

where

$$A = \frac{Z_0}{60}\sqrt{\frac{\varepsilon_r + 1}{2}} + \frac{\varepsilon_r - 1}{\varepsilon_r + 1}\left(0.23 + \frac{0.11}{\varepsilon_r}\right) \quad \quad (15.14c)$$

$$B = \frac{377\pi}{2Z_0\sqrt{\varepsilon_r}} \quad \quad (15.14d)$$

The attenuation due to dielectric loss (α_d)

$$\alpha_d = \frac{\beta_0\varepsilon_r(\varepsilon_e - 1)\tan\delta}{2\sqrt{\varepsilon_e}(\varepsilon_r - 1)} \quad \quad (15.15a)$$

In Equation 15.15a, the factor $\tan\delta$ is called the loss tangent. If ε_r becomes a complex quantity (i.e., $\varepsilon_r = \varepsilon_r' + \varepsilon_r''$), the loss tangent is given by

$$\tan\delta = \varepsilon_r''/\varepsilon_r' \quad \quad (15.15b)$$

The loss tangent can also be given as

$$\tan\delta = \sigma/\omega\varepsilon \quad \quad (15.15c)$$

The attenuation due to conductor loss (α_c)

$$\alpha_c = R_s/(Z_0 w) \quad \quad (15.16)$$

where R_s is the surface resistance of the conductor and was given by Equation 2.19b.

15.3.3 Approximate Electrostatic Solution

Consider the microstrip line shown in Figure 15.11. Let $\Phi(x, y)$ be the potential field that is a function of x and y. This field has to satisfy Laplace's equation $\nabla^2\Phi = 0$ within the space bounded by $-(a/2) \leq x \leq a/2$ and $0 \leq y \leq \infty$, which can be divided into two regions given as follows:

Region-I: $-a/2 \leq x \leq a/2$ and $0 \leq y \leq d$ and region-II: $-a/2 \leq x \leq a/2$ and $d \leq y \leq \infty$. Since these two regions are defined by air/dielectric interface, separate solutions $\Phi_I(x, y)$ and $\Phi_{II}(x, y)$ for $\nabla^2\Phi = 0$ are to be obtained. Using the method of separation of variables (used in Section 3.3.1) by taking $\phi(x, y) = X(x) Y(y)$, the solution obtained in Laplace's equation can be written as

$$X(x) = A^* \cos(kx) + B^* \sin(kx) \quad \text{for regions-I and II} \tag{15.17a}$$

$$Y(y) = C^* \cosh(ky) + D^* \sinh(ky) \quad \text{for region-I} \tag{15.17b}$$

$$Y(y) = E^* \exp(ky) + F^* \exp(-ky) \quad \text{for region-II} \tag{15.17c}$$

In Equation 15.17, A^*, B^*, C^*, D^*, E^* and F^* are the arbitrary constants and k is analogous to A and B used in Equations 3.12 and 3.13. In view of Equations 15.17a through 15.7c:

$$\phi_I(x, y) = X(x) Y(y) = \{(A^* \cos(kx) + B^* \sin(kx)\}$$
$$\times \{(C^* \cosh(ky) + D^* \sinh(ky)\} \tag{15.18a}$$

$$\phi_{II}(x, y) = X(x) Y(y) = \{(A^* \cos(kx) + B^* \sin(kx)\}$$
$$\{(E^* \exp(ky) + F^* \exp(-ky)\} \tag{15.18b}$$

The boundary conditions are separately applied on $X(x)$ and $Y(y)$ terms to satisfy $\phi_I(x, y) = 0$ or $\phi_{II}(x, y) = 0$. Thus, in view of the boundary condition,

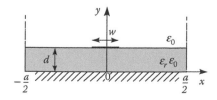

FIGURE 15.11
Geometry of a microstrip line with conducting walls.

$\phi_I(x, y) = 0$ at $y = 0$, $C^* = 0$. Also, as $\phi_{II}(x, y) = 0$ at $y = \infty$ field must vanish, E^* exp(ky) becomes infinity at $y = \infty$ and does not represent a physical field; thus, E^* can be taken as zero. Equations 15.18a and 15.18b can now be modified accordingly as

$$\phi_I(x, y) = \{(A^* \cos(kx) + B^* \sin(kx)\} D^* \sinh(ky) \qquad (15.19a)$$

$$\phi_{II}(x, y) = \{(A^* \cos(kx) + B^* \sin(kx)\} F^* \exp(-ky) \qquad (15.19b)$$

In view of Figure 15.11, $\phi(x, y) = 0$ at $x = a/2$ and $-a/2$ for all values of y. This potential field must have some non-zero value at $x = 0$ for all values of y except at $y = 0$ and thus must be of cosinusoidal nature. Such a field can be obtained only on substitution of $B^* = 0$. Thus, Equations 15.19a and 15.19b can be rewritten as

$$\phi_I(x, y) = A \cos(kx) \sinh(ky) \qquad (15.20a)$$

$$\phi_{II}(x, y) = B \cos(kx) \exp(-ky) \qquad (15.20b)$$

In Equations 15.20a and 15.20b, $A = A^* D^*$ and $B = A^* F^*$

As stated earlier, $\phi(x, y) = 0$ at $x = a/2$ and $-a/2$; this condition is met when $ka/2 = n\pi/2$ or $k = n\pi/a$ where n is an odd integer. On substitution of $k = n\pi/a$, Equation 15.20 can be written in the form of a series as

$$\phi_I(x,y) = \sum_{n-\text{odd}}^{\infty} A_n \cos(n\pi x/a)\sinh(n\pi y/a) \quad \text{for } 0 \le y \le d \qquad (15.21a)$$

$$\phi_{II}(x,y) = \sum_{n-\text{odd}}^{\infty} B_n \cos(n\pi x/a)e^{-(n\pi y/a)} \quad \text{for } d \le y \le \infty \qquad (15.21b)$$

In view of the continuity of Φ at the boundary between two regions, that is, at $y = d$

$$A_n \sinh(n\pi d/a) = B_n e^{-(n\pi d/a)}$$

or

$$B_n = A_n \sinh(n\pi d/a)e^{(n\pi d/a)} \qquad (15.22)$$

In view of Equation 15.22, Equation 15.21 can be rewritten as

$$\phi_I(x,y) = \sum_{n-\text{odd}}^{\infty} A_n \cos(n\pi x/a)\sinh(n\pi y/a) \quad \text{for } 0 \le y \le d \qquad (15.23a)$$

$$\phi_{II}(x,y) = \sum_{n-\text{odd}}^{\infty} A_n \cos(n\pi x/a)\sinh(n\pi d/a)e^{-n\pi(y-d)/a} \quad \text{for } d \le y \le \infty \quad (15.23b)$$

$$E_{yI} = \partial\phi/\partial y = \sum_{n-\text{odd}}^{\infty} A_n(n\pi/a)\cos(n\pi x/a)\cosh(n\pi y/a) \quad \text{for } 0 \le y \le d \quad (15.24a)$$

$$E_{yII} = \sum_{n-\text{odd}}^{\infty} -An(n\pi/a)\cos(n\pi x/a)\sinh(n\pi d/a)e^{-n\pi(y-d)/a} \quad \text{for } d \le y \le \infty$$

$$(15.24b)$$

The surface charge density at $y = d$

$$\rho_s = D_{yII} - D_{yI} = \varepsilon_0 E_{yII} - \varepsilon_0\varepsilon_r E_{yI} \quad (15.25a)$$

$$|\rho_s| = \varepsilon_0 \sum_{n-\text{odd}}^{\infty} An(n\pi/a)\cos(n\pi x/a)[\sinh(n\pi d/a) + \varepsilon_r \cosh(n\pi y/a)] \quad (15.25b)$$

Orthogonalising both sides of ρ_s (i.e., multiplying both sides of Equation 15.25b by $\cos(n\pi x/a)$ and integrating over the limits $-w/2$ to $+w/2$). Since $\rho_s(x) = 1$ for $|x| < a/2$ and $= 0$ for $|x| > a/2$

$$\int_{-w/2}^{w/2} \cos(n\pi x/a)\, dx = \varepsilon_0 A_n(n\pi/a) \int_{-a/2}^{a/2} \cos^2(n\pi x/a)$$
$$\times [\sinh(n\pi d/a) + \varepsilon_r \cosh(n\pi d/a]\, dx$$

$$(a/n\pi)\sin(n\pi x/a)\Big|_{-w/2}^{w/2} = \varepsilon_0 A_n(n\pi/a)(a/4)$$
$$\times [\sinh(n\pi d/a) + \varepsilon_r \cosh(n\pi d/a] \quad (15.26a)$$

$$A_n = \frac{4a\sin(n\pi w/2a)}{(n\pi)^2 \varepsilon_0[\sinh(n\pi d/a) + \varepsilon_r \cosh(n\pi d/a]} \quad (15.26b)$$

$$V = -\int_0^d E_y(x = 0, y)\, dy = \sum_{n=1}^{\infty} A_n \sinh\frac{n\pi b}{2a} \quad (15.27a)$$

$$Q = \int_{-W/2}^{W/2} \rho_s(x)\, dx = w \quad (15.27b)$$

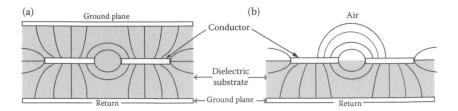

FIGURE 15.12
Electric field distribution in a (a) stripline and (b) microstrip.

$$C = \frac{Q}{V} = \frac{w}{\sum_{n=1}^{\infty} A_n \sinh(n\pi b/2a)} \tag{15.28}$$

$$Z_0 = \sqrt{\frac{L}{C}} = \frac{1}{v_p C} = \frac{\sqrt{\varepsilon_e}}{cC} \tag{15.29}$$

$$\varepsilon_e = C/C_0 \tag{15.30}$$

where C represents capacitance per unit length of the microstrip line with $\varepsilon_r \neq 1$ and C_0 represents capacitance per unit length of the microstrip line with $\varepsilon_r = 1$.

Figure 15.12 illustrates field lines in the stripline and microstrip. The figure does not illustrate the direction of the field as it will depend on (even or odd) mode. In the case of a stripline, these lines are confined between the two ground planes, that is, between the top and bottom conductors. In a microstrip, these lines are partly distributed in air. Thus, the conductor losses for the same dimensions of the microstrip and stripline are almost comparable, whereas, the dielectric loss in the microstrip is lower than the stripline as it sees a lower dissipation factor than the bulk laminate. Consequently, the microstrip has about 30% lower attenuation than the stripline and hence 30% higher interconnected bandwidth than that for the stripline.

Figure 15.13 shows the attenuation per unit length for an equal line width edge-coupled differential pair microstrip and stripline. The curves evidently show that there is an attenuation advantage for microstrip lines above 1 GHz.

15.3.4 Microstrip Design and Analysis

There are many CAD packages (such as 'Linecalc' or 'TX-LINE') for the microstrip design that calculate electrical parameters for the given line dimensions. Some CAD packages calculate these parameters in view of closed-form expressions. These simplified model expressions are based on

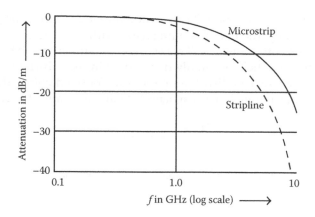

FIGURE 15.13
Relative attenuation in a stripline and microstrip.

some fitting factors and are in close agreement to some measured data. These models, however, do not work for the data outside some specified limits and for higher-order modes of propagation, that is, for non-TEM modes.

The fields in microstrips are too complicated and are commonly solved by using conformal mapping. In this method, the structure is mapped into a parallel plate guide by assuming that there is no fringing at the edges. Such a model (Figure 15.14) is often referred to as 'magnetic-wall model'. The microstrip fields are obtained by reverse mapping of these fields into the original microstrip domain.

The characteristic impedance '$Z_0(f)$' bears the following relation with the effective width (w_e or w_{eff}) and the effective dielectric constant (ε_e or ε_{eff}).

$$Z_0(f) = \frac{h\eta}{w_e(f)\sqrt{\varepsilon_e(f)}} \tag{15.31}$$

$$\varepsilon_e = \frac{\varepsilon_r + 1}{2} + \frac{\varepsilon_r + 1}{2}\left(1 + \frac{h}{w}\right)^{-0.555} \tag{15.32}$$

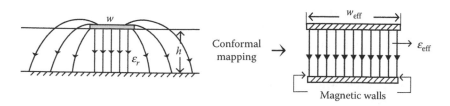

FIGURE 15.14
Magnetic-wall model obtained by using the conformal mapping method.

In view of the above relation, it can be noted that Z_0 changes (slightly) with frequency. Similarly, the ε_e varies with strip width 'W'. The effective permittivity also varies with frequency. This change remains confined between narrow lines (where the field is equally distributed in the substrate and air) and the wide lines (where field distribution remains confined within the substrate similar to that in a parallel plate capacitor). Mathematically, it is given by

$$\frac{\varepsilon_r + 1}{2} \le \varepsilon_e \le \varepsilon_r \tag{15.33}$$

The lowest-order transverse resonance cutoff frequency (f_{CT}) is given by the relation:

$$f_{CT} = \frac{C}{\sqrt{\varepsilon_r}(2w + 0.8h)} \tag{15.34}$$

It occurs when the width of the line (plus the fringing field component) approaches a half-wavelength in the dielectric by avoiding the wide lines. The TM mode in itself is not width dependent and is supported by the substrate with a lower cutoff frequency (f_{TM}) given by

$$f_{TM} = \frac{C \tan^{-1} \varepsilon_r}{\sqrt{2\pi h}\sqrt{\varepsilon_r - 1}} \tag{15.35}$$

Also, the maximum limit of substrate thickness (h_{max}) is given by

$$h_{max} = \frac{0.354 \lambda_0}{\sqrt{\varepsilon_r - 1}} \tag{15.36}$$

Thus, for higher and higher frequencies, the substrate has to be thinner and thinner till its further reduction is limited by tolerance and by losses due to the narrowness of conductors.

A microstrip can be used for over 100 GHz on MMICs (GaAs ICs). It has been demonstrated that a thin film microstrip with a deposited dielectric layer of a few microns can work up to several hundred gigahertz but such devices are found to be too lossy. For low loss operation beyond 100 GHz, rectangular waveguides, dielectric waveguides and some other forms of guiding structures may be employed.

15.3.5 Suspended Microstrips

Figure 15.5 illustrates two versions of suspended microstrips. In view of the involvement of air dielectric, these are of low loss. These also have a low parasitic in chip components, for example, wide-band dc block.

FIGURE 15.15
Two versions of suspended microstrips.

15.4 Coplanar Waveguides

The structure of a coplanar line consists of a signal conductor placed between two ground planes. A 3D view of a coplanar line is illustrated in Figure 15.16. The dominant mode of these lines is quasi-TEM and there is no low-frequency cutoff. These lines combine some of the advantageous features of both microstrips and slot lines. Microstrip geometry is convenient for series mounting of components across the gap in the strip conductor whereas slot lines are suitable for shunt mounting of components across the slots. In a coplanar line, both series and shunt-mounted components can be easily incorporated.

The transition from the coplanar line to the coaxial line and to a microstrip line is easily achieved. The dimensions of a coplanar line can be tapered, for uniform impedance, to connect to components of a wide to a minute dimension. It can have both balanced and unbalanced modes. Thus, a coplanar line finds extensive applications in devices (viz. balanced mixers, balanced modulators, etc.) where both balanced and unbalanced signals are present. The advantage of a circularly polarised magnetic field region of slot lines is available in the case of coplanar lines.

15.4.1 Types of Coplanar Waveguides

The coplanar line configurations may include (i) a coplanar waveguide and (ii) coplanar waveguide with ground. These are illustrated in Figure 15.17. Coplanar waveguides concentrate the field in the gap so that it will have the

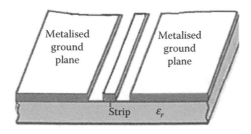

FIGURE 15.16
3D view of coplanar lines.

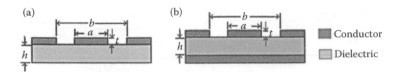

FIGURE 15.17
Different versions of coplanar lines. (a) Without ground and (b) with ground.

best ability to taper into a pin. It, however, requires special attention in case of transition from a microstrip line to a coplanar waveguide. For this situation, a coplanar waveguide with ground conductor that is coplanar with the signal conductor may provide an easier solution since a wide gap (microstrip) can more conveniently transition to the coplanar waveguide with ground configuration. Its impedance can be controlled by the signal line width and the ground gap. Thus, the impedance can be kept constant while the signal conductor's width is tapered down into a pin. This gives perfect matching to a component of pin width without changing the substrate thickness.

Its major attraction is that it does not require via holes through the substrate for the purpose of grounding. Besides, it has low dispersion and its chip components exhibit less parasitic capacitance. Tapers with constant Z_0 can be realised by varying the track width and gap combination. Its ground planes may provide some shielding between lines.

One of its major drawbacks include that its mode of propagation gets easily degenerated from quasi-TEM into a balanced coupled-slotline mode particularly at discontinuities. This, however, can be prevented by incorporating grounding straps between the ground planes either through bond wires or by using a second metallic layer.

15.5 Coplanar Strips

The coplanar strips consist of a pair of closely coupled parallel strips as shown in Figure 15.18. It is a planar equivalent to the twisted pair or a

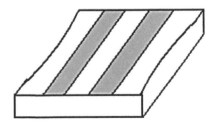

FIGURE 15.18
Coplanar strips.

lecher line. It is a balanced line and is ideally suited to balanced mixers and push–pull amplifiers. Currently, its use is restricted due to lack of design information.

15.6 Slot Line

A slot line consists of a pair of ground planes with a narrow slot between them. Figure 15.19 shows a slot line with two conductors in one plane that makes shunt mounting of the active or passive components quite easy. In slot lines, the signal propagates in TE mode. Thus, it is not an ideal general-purpose transmission medium. TE mode, however, makes it useful for circuits where push–pull operation is required such as in the case of balanced mixers and amplifiers. In the cross section of a slot line, there exists a location where the magnetic field is circularly polarised. This aspect makes it useful for the design of several components such as ferrite isolators.

A comparative study yields that in microstrip lines, high characteristic impedance is produced by a very narrow strip whereas in slot lines, high Z_0 requires a large slot width. In view of technological limitations on fabrication of very narrow strips, high Z_0 can be easily obtained in slot lines. As a coplanar waveguide has the quasi-TEM mode and slot line has a TE mode, their combination leads to the formation of useful hybrid junctions and transition circuits. The coplanar waveguide to the slot line transition is one of such most popular formations that has demonstrated balun operation over more than two octaves of bandwidth and has been used in miniaturised uniplanar mixers and amplifiers. However, the coaxial line to microstrip transition is relatively easier to fabricate than coaxial to slot line transition. Also, slot lines are more dispersive than microstrips, that is, the variations of Z_0 and λ_0/λ_g with frequency are larger.

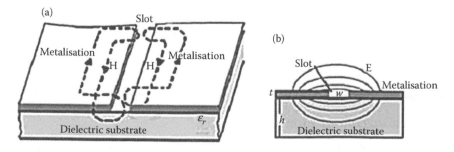

FIGURE 15.19
Slot line. (a) Geometry and H field; (b) E field distribution.

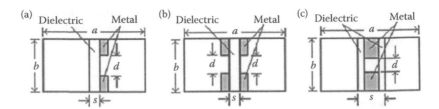

FIGURE 15.20
Integrated fin lines. (a) Unilateral, (b) bilateral and (c) insulated.

15.7 Fin Lines

The integrated fin lines are used for constructing low-cost millimetric wave circuits. It is basically an E-plane circuit in a rectangular waveguide in the form of capacitive loading. The metallic fins are incorporated onto the dielectric slabs using printed circuit technique. Its bandwidth is greater than an octave or above and attenuation is slightly greater than that of a microstrip line. Owing to high field concentration at edges, fin line structures are restricted to low- and medium-power applications. Figure 15.20 illustrates different versions of fin lines.

15.8 Micromachined Lines

Micromachined silicon components are developed by employing selective crystallographic etching methods. These are widely used as air-bag sensors, displays, disk drives and print heads. Such miniature components are classified as microelectromechanical systems (MEMS) or simply microsystems. This technology allows the realisation of moving parts for switching, tuning and steering. As these can be realised with air as the main dielectric, the structures have low loss. Thus, for microwave circuits, MEMS technology has special significance. Two of such components are shown in Figure 15.21.

15.9 Realisation of Lumped Elements

Microstrips and other planar lines can be used to realise lumped parameters, namely, resistors, inductors, capacitors and even transformers. These can also be used to form microwave components described in Chapter 7. The

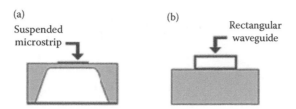

FIGURE 15.21
Micromachined lines. (a) Suspended microstrip and (b) rectangular waveguide.

steps involved in their fabrication process are discussed in Section 16.6. The realisation of some of these is briefly discussed below.

15.9.1 Resistors

Resistors on MICs use a deposited resistive layer. These are fabricated either with thin film (sputtering) technique or by thick film printing. Tantalum nitride, cermets and nickel chrome are the most commonly used materials. These are further discussed in Section 16.7.1.

15.9.2 Inductors

The availability of good inductors is one of the shortcomings of standard IC processes, particularly at higher frequencies. The synthesised active RF circuits always have higher noise, distortion and power consumption than the inductors made with turn of wires. MIC inductors can be realised through microstrips in the form of (i) straight narrow track (ribbon) inductors, (ii) as single-loop inductors or (iii) as multi-loop spiral inductors. The selection of the form depends on the required value of inductance.

15.9.2.1 Ribbon Inductors

Figure 15.22 shows a ribbon inductor and its equivalent circuit wherein L represents inductance, C represents parasitic capacitance, Z_0 represents characteristic impedance, f represents frequency and λ_g represents the guide wavelength. The values of L and C for length ($l < \lambda_g/4$) can be given as

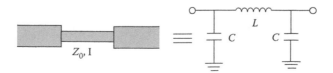

FIGURE 15.22
Microstrip ribbon inductor and its equivalent circuit.

$$L = \frac{Z_0}{2\pi f} \sin\left(\frac{2\pi l}{\lambda_g}\right) \tag{15.37a}$$

$$C = \frac{1}{2\pi f Z_0} \tan\left(\frac{\pi l}{\lambda_g}\right) \tag{15.37b}$$

A high value of inductance with parasitic capacitance can be obtained with a narrow track of high Z_0. The choice of track, however, is determined by the limits of fabrication, dc current-carrying capacity and by high resistance of a very narrow track.

15.9.2.2 Spiral Inductors

Inductance values exceeding 1 nH can be obtained by using spiral inductors with one or multi-turns. In the case of multi-turn inductors, the inductance significantly increases due to mutual inductance between the turns. Such an inductor also yields high Q. The centre of configuration is connected to either a bond wire or to a second metal layer.

The most widely used on-chip inductors are the planar spirals that may be of a different shape as shown in Figure 15.23. The selection of shape mainly depends on the convenience of accommodation. Octagonal or circular is moderately better than squares. Some more information about inductors is given in Section 16.7.2.

15.9.3 Capacitors

There are many forms of capacitors that can be realised by using IC technology. Figure 15.24 illustrates the following two such capacitors.

15.9.3.1 Inter-Digital Capacitor

It consists of a number of interleaved microstrip fingers coupled together. Its maximum value is limited by its physical size. The maximum usable operating frequency is governed by the distributed nature of fingers. These capacitors are ideal as tuning, coupling and matching elements. In case of the

(a)   (b)   (c)   (d)

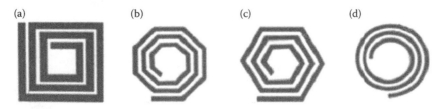

FIGURE 15.23
Shape of spiral inductors. (a) Square, (b) octagonal, (c) hexagonal and (d) circular.

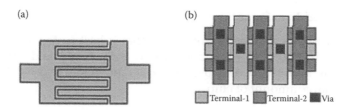

FIGURE 15.24
Microstrip capacitors. (a) Inter-digitated type and (b) woven type.

requirement of small and precise capacitance values, inter-digital capacitors meet the requirement. These, however, have inherent parasitic inductance that is surely its negative aspect.

15.9.3.2 Woven Capacitor

It can be used to achieve high capacitance density. It contains a number of vertical lines and another set of equal numbers of horizontal lines. The two sets are of different metals. The woven structure has much less inherent series inductance due to the current flow in different directions. This structure, therefore, results in higher self-resonant frequency. Its resistance contributed by vias is smaller whereas the capacitance density of this structure is less than that of the inter-digitated capacitor. Some more information about capacitors is included in Section 16.7.3.

15.9.4 Monolithic Transformers

Figure 15.25 illustrates configurations of monolithic transformers. These offer varying trade-offs in terms of self-inductance and series resistance of each port, mutual coupling coefficient, port-to-port and port-to-substrate capacitances, resonant frequencies, symmetry and dielectric area consumed. These can be configured as three, four or more terminal devices. Their desired characteristics are application dependent. These may be used for broadband or narrowband applications.

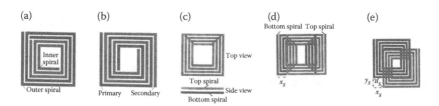

FIGURE 15.25
Monolithic transformers. (a) Monolithic, (b) tapped, (c) interleaved, (d) stacked with completely overlapped spirals and (e) stacked with offset top and bottom spirals.

15.9.4.1 Tapped Transformer

Figure 15.25a shows a tapped transformer that is best suited for three-port applications. In this, the self-inductance is maximised and port-to-port capacitance is minimised. It provides moderate coupling ($k = 0.3$–0.5) but consumes a large amount of chip area.

15.9.4.2 Interleaved Transformer

Figure 15.25b shows an interleaved transformer that is good for four-port applications. It permits high coupling ($k = 0.7$) but at the cost of reduced self-inductance. The coupling may be increased but at the cost of high series resistance by reducing the width of turns and the spacing between turns.

15.9.4.3 Stacked Transformer

The stacked transformer shown in Figure 15.25c is suitable for both three and four port. It has the best area efficiency, highest self-inductance and highest coupling ($k = 0.9$). It, however, has low self-resonant frequency due to high port-to-port capacitance. This capacitance can be reduced by displacing the centres of the stacked inductors in horizontal (Figure 15.25d) or vertical (Figure 15.25e) directions. Such displacement, however, results in the reduction of coupling.

15.10 Realisation of Microwave Components

15.10.1 Realisation of Basic Stripline Elements

Figure 15.26 shows some basic stripline shapes along with their equivalent circuit elements, which can be used in the formation of various microwave components. These include (a) an abrupt (low to high) stepped impedance, (b) an abrupt (high to low) stepped impedance, (c) a line with an abrupt end, (d) a hole or slit in a line, (e) a transverse half-slit across the line and (f) a gap in the line.

15.10.2 Realisation of Transmission Lines

Figure 15.27 shows some transmission line structures. These can also be used in the formation of filters and other components. The equivalent circuits are also illustrated for each case.

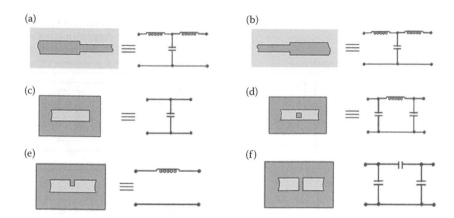

FIGURE 15.26
Stripline elements and their lumped-element equivalents. (a) An abrupt (low to high) stepped impedance, (b) an abrupt (high to low) stepped impedance, (c) a line with an abrupt end, (d) a hole or slit in a line, (e) a transverse half-slit across the line and (f) a gap in the line.

15.10.3 Realisation of Stubs

Stub lines were discussed in Section 2.29.5 along with their matching properties. Figure 15.28 shows the formation of various types of stubs through striplines and microstrips.

15.10.4 Realisation of Power Splitters

15.10.4.1 T-Junction Power Splitter

A simple T-junction power splitter (or divider) is shown in Figure 15.29 that can be implemented in planar format. T-junctions were the earliest power dividers. In view of its very poor isolation between its output ports, a large part of reflected power from port 2 appears at port 3 or vice versa. Theoretically, simultaneous matching of all ports of a passive, lossless three-port device is not possible. It is, however, possible in the case of four-port devices. Owing to this reason, four-port devices are used to implement three-port power dividers. Four-port devices can be designed for splitting power arriving at port 2 between ports 1 and 4 (that is match terminated) and no power goes to port 3 in the ideal case.

15.10.4.2 Wilkinson Power Splitter

The Wilkinson power splitter is shown in Figure 15.30. It consists of two quarter-wave matching sections, with characteristic impedance of 70.7 Ω (for a 50 Ω system) and a 100 Ω isolation resistor across its output offers a

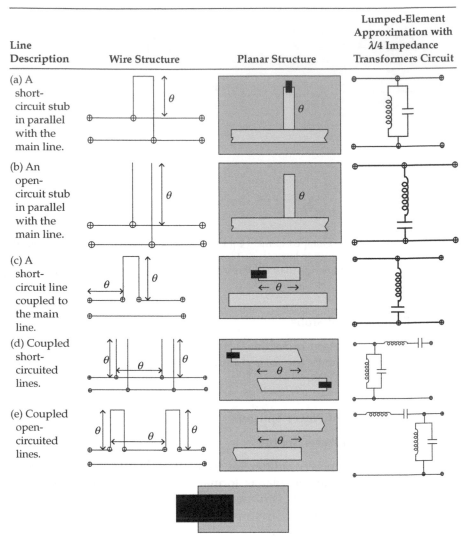

Line Description	Wire Structure	Planar Structure	Lumped-Element Approximation with λ/4 Impedance Transformers Circuit
(a) A short-circuit stub in parallel with the main line.			
(b) An open-circuit stub in parallel with the main line.			
(c) A short-circuit line coupled to the main line.			
(d) Coupled short-circuited lines.			
(e) Coupled open-circuited lines.			

In Figure 15.27a, c and d, a strap through the board for making connection with the ground plane underneath.

FIGURE 15.27
Different forms of lines implemented through planar lines. (a) Doubled stubs in parallel, (b) radial stub, (c) butterfly stub (paralleled radial stubs) and (d) clover-leaf stub (triple paralleled radial stubs).

convenient option. It can also be used as a power combiner. If two signals with equal phase and equal amplitude are fed to the two arms, there will be no voltage across the isolation resistor and the combining process will not result in any loss. In the case of non-identical inputs, a combining loss may occur.

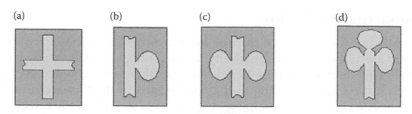

FIGURE 15.28
Various forms of stubs implemented through planar lines. (a) Doubled stubs in parallel, (b) radial stub, (c) butterfly stub (paralleled radial stubs) and (d) clover-leaf stub (triple paralleled radial stubs).

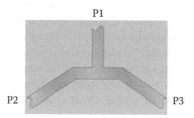

FIGURE 15.29
Simple T-junction power splitter.

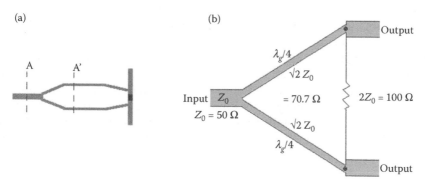

FIGURE 15.30
Wilkinson power splitter. (a) Two-way Wilkinson divider and (b) details between A-A'.

15.10.5 Realisation of Couplers

The directional couplers were discussed in Chapter 7 at length. A directional coupler may be used to separate signals according to their direction of travel for measurement and other purposes. Also, a coupler that requires a quadrature-(90°) equal power split is used in a balanced amplifier. There are many configurations that can be implemented through stripline and microstrips. Some of these are discussed below.

15.10.5.1 Proximity Couplers

A directional coupler can be realised in microstrip form wherein two conductors are placed in close proximity. A directional coupler of $\lambda/4$ length transfers maximum power in the reverse direction to the coupled port and there is a very small coupling to the port located in the forward direction or the isolated port. The remaining power travels straight to the direct port. This coupler often referred to as a backward wave coupler has exactly the 90° phase difference between the signals at coupled and direct ports over a wide frequency range. The amplitudes may, however, vary with frequency. Different forms of these couplers are described below.

Quarter-wave section couplers: Figure 15.31 shows a single-section $\lambda/4$ directional coupler. It is the most common form of a directional coupler wherein a pair of coupled transmission lines is involved. It can be realised by using a stripline or microstrip. In this, the power on the coupled line flows in the opposite direction to the power on the main line; thus, it is sometimes called a backward coupler. Depending on the requirement of the coupling factor and its accuracy, the two (main and coupled) lines can be printed on opposite sides of the dielectric rather than side by side. The coupling of the two lines across their width is larger than the coupling when they are side by side.

Short section couplers: The $\lambda/4$ coupled line design is good for coaxial and stripline implementations but does not work well in microstrip format as the microstrip is not a homogeneous medium and there are two different mediums above and below the transmission strip. Microstrips lead to a transmission mode other than the TEM mode. In these, the propagation velocities of even and odd modes are different leading to signal dispersion. In a microstrip implementation, a coupled line much shorter than $\lambda/4$, shown in Figure 15.32a, provides a solution. This too has a limitation as its coupling factor rises with frequency. This limitation can be circumvented by using a coupled line of higher impedance than the main line as shown in Figure 15.32b. This configuration is advantageous where the coupler is being fed to a detector for power monitoring. The higher impedance line results in a

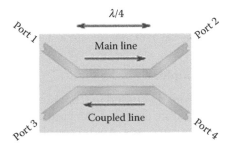

FIGURE 15.31
Single-section $\lambda/4$ directional coupler.

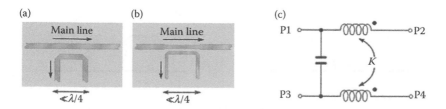

FIGURE 15.32
Short section couplers. (a) 50 Ω main line and coupled lines, (b) 50 Ω main line and 100 Ω coupled line and (c) equivalent circuit.

higher RF voltage for a given main line power, making the work of the detector diode easier. An equivalent circuit of the couplers in Figure 15.32a and b is shown in Figure 15.32c.

Multiple quarter-wave section couplers: A single $\lambda/4$ coupler provided bandwidths of less than an octave. For larger bandwidths, multiple $\lambda/4$ coupling sections are used. Such a configuration is shown in Figure 15.33.

Figure 15.34 illustrates a proximity coupler with port definitions and a cross section.

Even and odd modes: The working of the backward wave coupler can be explained by using the standing wave principle that involves the superposition of even and odd modes of propagation. In an even mode, the two conductors are at the equal potentials of the same polarity, while in an odd mode, the potential is equal but of opposite polarity. These potentials result in different field patterns as shown in Figure 15.35. These patterns result in different proportions of fields in air and dielectric, different velocities and

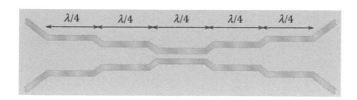

FIGURE 15.33
A five-section planar format coupler.

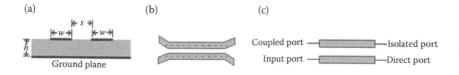

FIGURE 15.34
Microstrip couplers. (a) Cross-section with key parameters, (b) TEM proximity coupler (c) port definitions.

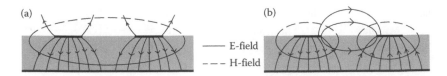

FIGURE 15.35
Approximate field patterns in a microstrip. (a) Even mode and (b) odd mode.

different characteristic impedances. In view of a homogeneous dielectric in coaxial and striplines, the fields are pure TEM and the two modes have identical phase velocities. Large isolation (e.g., 30 dB) can be more conveniently achieved if the phase velocities differ with a pure TEM transmission medium.

The characteristic impedances for (i) the even mode (Z_{0e}) and (ii) odd mode (Z_{0o}) must satisfy the following relation for the coupling factor (k in dB) up to approximately 10 dB:

$$k = 20\log\left[\frac{Z_{0e} - Z_{0o}}{Z_{0e} + Z_{0o}}\right] \tag{15.38}$$

Also,

$$Z_0^2 \approx Z_{0e}Z_{0o} \tag{15.39}$$

In view of the above relations,

$$Z_{0e} \approx Z_0\sqrt{\frac{1 + 10^{k/20}}{1 - 10^{k/20}}} \tag{15.40a}$$

$$Z_{0o} \approx Z_0\sqrt{\frac{1 - 10^{k/20}}{1 + 10^{k/20}}} \tag{15.40b}$$

The parallel coupled line structures are also used in filters, dc blocks, phase shifters and matching networks. In filters, Z_{0e} and Z_{0o} are the parameters that are synthesised from the filter theory in the same manner as inductances and capacitances are obtained in the lumped-element band-pass filters.

15.10.5.2 Lange Coupler

An equal (3 dB) power splitter cannot be realised from the parallel coupled microstrip structure even if the spacing between the two conductors is reduced to zero. The coupling remains limited to approximately 6 dB. To cross this barrier, the two lines are to be overlapped that require a multilayer structure.

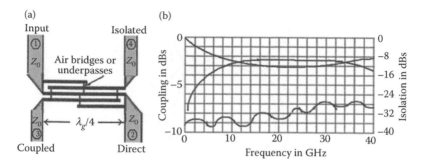

FIGURE 15.36
A 3 dB four fingers Lange coupler. (a) Layout and (b) frequency response.

This structure can be realised with ordinary microstrips by interleaving the two arms, each of which consists of a number of conducting fingers. These arms (with four, six or eight fingers) are arranged in an alternating manner. The fingers of each arm are maintained at the same potential by connecting them through bond wires, air bridges or underpasses. Figure 15.36 illustrates such a coupler proposed by Julius Lange along with its frequency response.

15.10.5.3 Branch-Line Coupler

Coupled line and Lange couplers require narrow tracks and/or gaps and thus, their realisation becomes a bit difficult. In the applications wherein a 90°, 3 dB split is required, a branch-line coupler provides a suitable alternative. This coupler can also be analysed in terms of odd and even modes, but more in terms of circuit terms than as fields. Figure 15.37a illustrates such a coupler wherein the ports can be interchanged to remove the step in width. A single-section branch-line coupler is narrowband. An enhanced bandwidth can be achieved by using a multi-section coupler. Figure 15.37b shows another version of a branch-line coupler called meandered branch-line coupler.

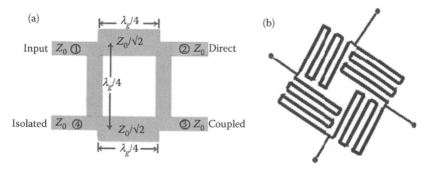

FIGURE 15.37
Branch-line couplers. (a) Microstrip type and (b) meandered type.

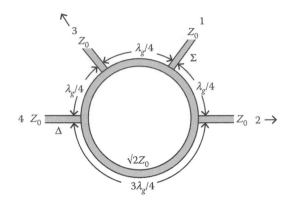

FIGURE 15.38
Microstrip rat-race coupler.

15.10.5.4 Rat-Race Coupler

A rat-race or hybrid ring coupler is a four-port 3-dB coupler consisting of a $3\lambda/2$ ring of transmission line with four lines at the intervals shown in Figure 15.38. Power fed at port 1 splits and travels both ways around the ring. At ports 2 and 3, signals arrive in phase and add, whereas, at port 4, it is out of phase and cancels. This figure shows its planar implementation. A coupler with a coupling factor different from 3 dB can be made by having each $\lambda/4$ section of the ring alternately of low and high impedance.

15.10.6 Resonators

Microstrips can also be used for the realisation of resonant devices. Three of such resonators are illustrated in Figure 15.39.

15.10.7 Filters

Microstrip filters are of low cost and can be easily integrated with active devices. These are, however, lossy with poor performance, low Q and low power-handling capacity. The structures of some of the microwave filters are given below.

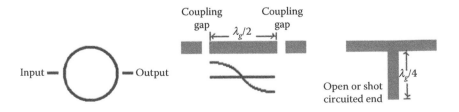

FIGURE 15.39
Microstrip resonators. (a) Ring resonator, (b) half-wave resonator and (c) quarter-wave resonator.

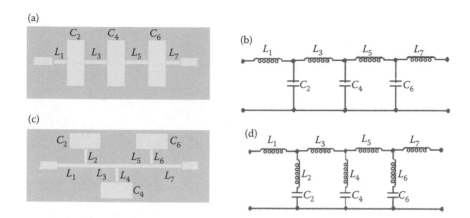

FIGURE 15.40
Stepped-impedance LPF. (a) LPF formed from alternate high and low impedance line sections. Also called Low Z/High Z, LPF. (b) Equivalent circuit of part (a). (c) LPF incorporating shunt resonators. (d) Equivalent circuit of part (c).

15.10.7.1 Low-Pass Filters

Figure 15.40 illustrates stepped-impedance low-pass filters (LPFs) along with their equivalent circuits.

Figure 15.41 also shows some structures of LPFs constructed from stubs.

15.10.7.2 Band-Pass Filters

Figure 15.42 shows eight different structures that exhibit the properties of band-pass filters (BPFs). All these configurations can be implemented by using microstrips or striplines.

15.10.8 Circuit Elements

Circuit elements that can be realised by using planar technology include transmission lines, bends, corners, steps, taper, tee junctions, cross-junctions and stubs. Some of these have already been discussed above.

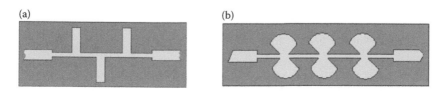

FIGURE 15.41
LPF constructed from stubs. (a) Constructed from standard stubs on alternating sides of main line λ/4 apart. (b) Constructed from butterfly stubs.

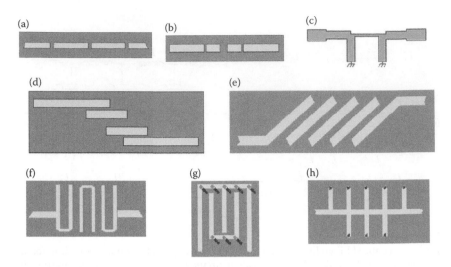

FIGURE 15.42
Different structures of BPFs. (a) Capacitive gap stripline filter, (b) end-coupled filter, (c) branch-line filter, (d) parallel-coupled filter, (e) stripline parallel-coupled lines filter, (f) stripline hairpin filter, (g) stripline inter-digital filter and (h) stripline stub filter composed of $\lambda/4$ short-circuit stubs.

EXAMPLE 15.1

The parameters involved in the construction of a planar transmission line (Figure 15.3) are $b = 8$ mils, $t = 2$ mils and $\varepsilon = 5.23$. Calculate its effective width and the characteristic impedance if (a) $w = 10$ mils and (b) $w = 2$ mils.

Solution

Given $b = 8$ mils $t = 2$ mils $\varepsilon_r = 5.23$ mils and (a) $w = 10$ mils (b) $w = 2$ mils.

Relations Required

The relation between effective width (w_e) and the actual width (w) is

$$w_e/b = w/b \quad \text{for } w/b > 0.35$$

and

$$w_e/b = w/b - (0.35 - w/b)^2 \quad \text{for } w/b < 0.35 \quad Z_0 = \frac{30\pi}{\sqrt{\varepsilon_r}} \frac{b}{w_e + 0.441b}$$

a. $w = 10$ mils, $w/b = 10/8 = 1.25 > 0.35$. Thus, $w_e = w = 10$ mils

$$Z_0 = \frac{30\pi}{\sqrt{5.23}} \frac{8}{10 + 0.441 \times 8} = 24.37 \ \Omega$$

b. $w = 2$ mils, $w/b = 2/8 = 0.25 < 0.35$ and $w_e/b = 0.25 - (0.35 - 0.25)^2$
$= 0.25 - 0.01 = 0.24$. Thus, $w_e = 0.24b = 0.24 \times 8 = 1.92$ mils

$$Z_0 = \frac{30\pi}{\sqrt{5.23}} \frac{8}{1.92 + 0.441 \times 8} = 60.51\ \Omega$$

EXAMPLE 15.2

A planar transmission line has characteristic impedance of 50 Ω. Calculate its width if $b = 8$ mils, $t = 2$ mils and (a) $\varepsilon_r = 5.23$, (b) $\varepsilon_r = 2.21$.

Solution

Given $b = 8$ mils, $t = 2$ mils (a) $\varepsilon_r = 5.23$ and (b) $\varepsilon_r = 6.25$

Relations Required

$w/b = x$ (for $\sqrt{\varepsilon_r}\, Z_0 > 120$) and $w/b = 0.85 - \sqrt{(0.6 - x)}$ (for $\sqrt{\varepsilon_r}\, Z_0 < 120$)

a. $\varepsilon_r = 5.23$

$$x = \frac{30\pi}{\sqrt{\varepsilon_r}\, Z_0} - 0.441 = \frac{30\pi}{\sqrt{5.23} \times 50} - 0.441 = 0.3832$$

$$\sqrt{\varepsilon_r}\, Z_0 = 50 \times \sqrt{5.23} = 114.346 < 120$$

$$w/b = 0.85 - \sqrt{(0.6 - 0.3832)} = 0.38438$$

$$w = 0.3844 \times 8 = 3.0752 \text{ mils}$$

b. $\varepsilon_r = 6.25$

$$x = \frac{30\pi}{\sqrt{6.25} \times 50} - 0.441 = 0.313$$

$$\sqrt{\varepsilon_r}\, Z_0 = 50 \times \sqrt{6.25} = 125 > 120$$

$$w/b = 0.313$$

$$w = 0.313 \times 8 = 2.504 \text{ mils}$$

EXAMPLE 15.3

A 10 GHz planar transmission line has characteristic impedance of 50 Ω. Calculate the attenuation due to conductor loss if its width is $b = 6$ mils, $t = 2$ mils, $w = 10$ mils, $\mu = \mu_0$ and $\sigma = 5.8 \times 10^7$ mhos/m and (a) $\varepsilon_r = 5.23$, (b) $\varepsilon_r = 2.21$.

Solution

Given $b = 8$ mils, $t = 2$ mils, $w = 10$ mils, $\mu = \mu_0$ and $\sigma = 5.8 \times 10^7$ mhos/m, $f = 10$ GHz, $Z_0 = 50\ \Omega$ and $\mu = \mu_0 = 4\pi \times 10^{-7}$
   (a) $\varepsilon_r = 5.23$ and (b) $\varepsilon_r = 6.25$

Relations Required

Surface resistance $R_s = \sqrt{(\omega\mu/2\sigma)}$

$\omega = 2\pi f = 2\pi \times 10^{10}$

$$(\omega\mu/2\sigma) = (2\pi \times 10^{10} \times 4\pi \times 10^{-7})/(2 \times 5.8 \times 10^7)$$

$$= (8\pi^2 \times 10^3)/(11.6 \times 10^7) = 0.6896\,\pi^2 \times 10^{-4}$$

$$R_s = \sqrt{(0.6896\,\pi^2 \times 10^{-4})} = 0.0261\ \Omega$$

$$A = 1 + \frac{2w}{b-t} + \frac{1}{\pi}\frac{b+t}{b-t}\ln\left(\frac{2b-t}{t}\right) = 1 + \frac{2 \times 10}{8-2} + \frac{1}{\pi}\frac{8+2}{8-2}\ln\frac{2 \times 8 - 2}{2} = 4.7813$$

$$B = 1 + \frac{b}{(0.5w + 0.7t)}\left(0.5 + \frac{0.414t}{w} + \frac{1}{2\pi}\ln\frac{4\pi w}{t}\right)$$

$$= 1 + \frac{8}{0.5 \times 10 + 0.7 \times 2}\left(0.5 + \frac{0.414 \times 2}{10} + \frac{1}{2\pi}\ln\frac{4\pi \times 10}{2}\right) = 1.3577$$

a. $\sqrt{\varepsilon_r}\ Z_0 = 50 \times \sqrt{5.23} = 114.346 < 120$

$$\alpha_c = \frac{0.16R_s}{Z_0 b}B = \frac{0.16 \times 0.0261}{50 \times 8} \times 1.3577 = 1.4175 \times 10^{-5}\ \text{Nep/m}$$

b. $\sqrt{\varepsilon_r}\ Z_0 = 50 \times \sqrt{6.25} = 125 > 120$

$$\alpha_c = \frac{2.7 \times 10^{-3}R_s\varepsilon_r Z_0}{30\pi(b-t)}A$$

$$= \frac{2.7 \times 10^{-3} \times 0.0261 \times 6.25 \times 50}{30\pi(8-2)} \times 4.7813$$

$$= 0.1862 \times 10^{-3}\ \text{Nep/m}$$

EXAMPLE 15.4

Calculate the (i) effective dielectric constant (ε_e), (ii) phase velocity (v_p), (iii) phase constant (β) and characteristic impedance (Z_0) of a 5-GHz microstrip if its depth $d = 5$ mils and (a) $w = 8$ mils, (b) $w = 3$ mils. $\mu = \mu_0$ and $\varepsilon_r = 5.23$.

Solution

Given $d = 5$ mils, $\mu = \mu_0 = 4\pi \times 10^{-7}$, $\varepsilon_0 = 8.854 \times 10^{-12}$, $\varepsilon_r = 5.23$, $f = 5$ GHz, $\omega = 2\pi \times 5 \times 10^9 = \pi \times 10^{10}$ and $c \approx 3 \times 10^8$

(a) $w = 8$ mils and (b) $w = 3$ mils

Propagation constant for free space

$$\beta_0 \approx \omega\sqrt{(\mu_0\varepsilon_0)} \approx \omega/c = \pi \times 10^{10}/3 \times 10^8 = 104.72$$

$$\varepsilon_e = \frac{\varepsilon_r + 1}{2} + \frac{\varepsilon_r - 1}{2} \frac{1}{\sqrt{1 + 12d/w}} = \frac{5.23 + 1}{2} + \frac{5.23 - 1}{2} \frac{1}{\sqrt{1 + 12 \times 5/w}}$$

$$= 3.115 + \frac{2.115}{\sqrt{1 + 60/w}}$$

	(a) $w = 8$ mils, $w/d = 8/5 = 1.6 > 1$	(b) $w = 3$ mils, $w/d = 3/5 = 0.6 < 1$
$\varepsilon_e = 3.115 + \dfrac{2.115}{\sqrt{1 + 60/w}}$	$\varepsilon_e = 3.115 + \dfrac{2.115}{\sqrt{1 + 60/8}}$	$\varepsilon_e = 3.115 + \dfrac{2.115}{\sqrt{1 + 60/3}}$
	$= 3.84$	$= 3.5765$
$v_p = c/\sqrt{\varepsilon_e}$	$v_p = 3 \times 10^8/\sqrt{3.84}$	$v_p = 3 \times 10^8/\sqrt{3.5765}$
where $1 < \varepsilon_e < \varepsilon_r$	$= 1.531 \times 10^8$ m/s	$= 1.586 \times 10^8$ m/s
$\beta = \beta_0 \sqrt{\varepsilon_e}$	$\beta = 104.72/\sqrt{3.84}$	$\beta = 104.72/\sqrt{3.5765}$
	$= 53.44$	$= 55.37$
$Z_0 = \dfrac{120\pi}{\sqrt{\varepsilon_e}} \dfrac{1}{[(w/d) + 1.393 + 0.667 \ln\{(w/d) + 1.444\}]}$		$Z_0 = \dfrac{60}{\sqrt{\varepsilon_e}} \ln\left(\dfrac{8d}{w} + \dfrac{w}{4d}\right)$
	$\dfrac{120\pi}{\sqrt{\varepsilon_e}} = 120\pi/\sqrt{3.84} = 192.382$	$\dfrac{60}{\sqrt{\varepsilon_e}} = 60/\sqrt{3.5765} = 31.726$
	$(w/d) + 1.393 + 0.667 \ln\{(w/d) + 1.444\}$	$8d/w = 8 \times 5/3 = 13.333$
	$= 1.6 + 1.393 + 0.667 \ln (1.6 + 1.444)$	$w/4d = 3/4 \times 5 = 0.15$
	$= 2.993 + 0.667 \ln (3.044)$	$(8d/w) + (w/4d) = 13.483$
	$= 2.993 + 0.667 \times 0.483445 = 3.315$	Ln $(13.483) = 1.1298$
	$1/[(w/d) + 1.393 + 0.667 \ln\{(w/d) + 1.444\}] = 0.316$	$Z_0 = 31.726 \times 1.1298$
	$Z_0 = 192.382 \times 0.316 = 58.025 \ \Omega$	$= 35.8436 \ \Omega$

EXAMPLE 15.5

Calculate the strip width of a microstrip for which $Z_0 = 50 \ \Omega$, $d = 5$ mils and relative dielectric constant (ε_r) is (a) 5.23 and (b) 2.25.

Solution

Given $d = 5$ mils, $\mu = \mu_0 = 4\pi \times 10^{-7}$, $\varepsilon_0 = 8.854 \times 10^{-12}$, $\varepsilon_r = 5.23$, $f = 5$ GHz, $\omega = 2\pi \times 5 \times 10^9 = \pi \times 10^{10}$ and $c \approx 3 \times 10^8$.

(a) $\varepsilon_r = 5.23$	(b) $\varepsilon_r = 2.25$ (polyethylene)
$A = \dfrac{Z_0}{60} \sqrt{\dfrac{\varepsilon_r + 1}{2}} + \dfrac{\varepsilon_r - 1}{\varepsilon_r + 1}\left(0.23 + \dfrac{0.11}{\varepsilon_r}\right)$	$A = \dfrac{Z_0}{60} \sqrt{\dfrac{\varepsilon_r + 1}{2}} + \dfrac{\varepsilon_r - 1}{\varepsilon_r + 1}\left(0.23 + \dfrac{0.11}{\varepsilon_r}\right)$
$= \dfrac{50}{60} \sqrt{\dfrac{5.23 + 1}{2}} + \dfrac{5.23 - 1}{5.23 + 1}\left(0.23 + \dfrac{0.11}{5.23}\right)$	$= \dfrac{50}{60} \sqrt{\dfrac{2.25 + 1}{2}} + \dfrac{2.25 - 1}{2.25 + 1}\left(0.23 + \dfrac{0.11}{2.25}\right)$
$= 1.64$	$= 1.17$

$w/d = \dfrac{8e^A}{e^{2A} - 2}$ for $w/d < 2$

$e^A = e^{1.64} = 5.155$

$8e^A = 8 \times 5.155 = 41.24$

$e^{2A} = e^{3.28} = 26.576$

$e^{2A} - 2 = 24.576$

$w/d = 41.24/24.576 = 1.68$

$w = 1.68 \times 5 = 8.39$

This relation is valid since $w/d = 1.68 < 2$

$B = \dfrac{377\pi}{2Z_0\sqrt{\varepsilon_r}} = \dfrac{377\pi}{2 \times 50\sqrt{5.23}} = 5.1789$

for $w/d > 2$

$2B - 1 = 2 \times 5.1789 - 1 = 9.3578$

$\text{Ln}(2B - 1) = 0.9711$

$B - 1 = 5.1789 - 1 = 4.1789$

$\text{Ln}(B - 1) = 0.621$

$(\varepsilon_r - 1)/\varepsilon_r = (5.23 - 1)/5.23 = 0.8088$

$0.61/5.23 = 0.1166$

$\left[B - 1 - \ln(2B - 1)\right] = 4.1789 - 0.9711$
$= 3.2078$

$\left[\dfrac{\varepsilon_r - 1}{2\varepsilon_r}\left\{\ln(B - 1) + 0.39 - \dfrac{0.61}{\varepsilon_r}\right\}\right]$

$= 0.8088(0.621 + 0.39 - 0.1166) = 0.7234$

$w/d = (2/\pi) \times (3.2078 + 0.7234) = 2.5$

$w = 2.5 \times 5 = 12.5$ mils

$w/d = \dfrac{8e^A}{e^{2A} - 2}$ for $w/d < 2$

$e^A = e^{1.17} = 3.222$

$8e^A = 8 \times 3.222 = 25.7758$

$e^{2A} = e^{2.34} = 10.38$

$e^{2A} - 2 = 10.38 - 2 = 8.38$

$w/d = 25.7758/8.38 = 3.076$

This relation is not valid for this case since $w/d > 2$

$B = \dfrac{377\pi}{2Z_0\sqrt{\varepsilon_r}} = \dfrac{377\pi}{2 \times 50\sqrt{2.25}} = 7.9$

$$w/d = \frac{2}{\pi}\left[B - 1 - \ln(2B - 1) + \frac{\varepsilon_r - 1}{2\varepsilon_r}\left\{\ln(B - 1) + 0.39 - \frac{0.61}{\varepsilon_r}\right\}\right]$$

for $w/d > 2$

$2B - 1 = 2 \times 7.9 - 1 = 14.8$

$\text{Ln}(2B - 1) = 1.17$

$B - 1 = 7.9 - 1 = 6.9$

$\text{Ln}(B - 1) = 0.8388$

$(\varepsilon_r - 1)/\varepsilon_r = (2.25 - 1)/2.25 = 0.5555$

$0.61/2.25 = 0.2711$

$\left[B - 1 - \ln(2B - 1)\right] = 6.9 - 1.17 = 5.73$

$\left[\dfrac{\varepsilon_r - 1}{2\varepsilon_r}\left\{\ln(B - 1) + 0.39 - \dfrac{0.61}{\varepsilon_r}\right\}\right]$

$= 0.5555(0.8388 + 0.39 - 0.2711) = 0.532$

$w/d = (2/\pi) \times (5.73 + 0.532) = 3.9865$

$w = 3.9865 \times 5 = 19.93$ mils

EXAMPLE 15.6

Calculate the attenuation due to (i) effective dielectric constant, (ii) dielectric loss (α_d) and (iii) conductor loss (α_c) for a 5-GHz, 50-Ω microstrip with depth $h = 5$ mils, $w = 8$ mils and $\varepsilon_r = 2.55$. For conductor $\mu = \mu_0$, $\sigma = 5.8 \times 10^7$ mhos/m

Solution

Given $h = 5$ mils, $w = 8$ mils, $\mu = \mu_0 = 4\pi \times 10^{-7}$, $\varepsilon_0 = 8.854 \times 10^{-12}$, $\sigma = 5.8 \times 10^7$ mhos/m, $\varepsilon_r = 2.55$, $f = 5$ GHz, $\omega = 2\pi \times 5 \times 10^9 = \pi \times 10^{10}$ and $c \approx 3 \times 10^8$

Propagation constant for free space

$$\beta_0 \approx \omega\sqrt{(\mu_0\varepsilon_0)} \approx \omega/c = \pi \times 10^{10}/3 \times 10^8 = 104.72$$

The surface resistance of the conductor:

$$R_s = \sqrt{(\omega\mu/2\sigma)} = \sqrt{[\{\pi \times 10^{10} \times 4\pi \times 10^{-7}\}/\{2 \times 5.8 \times 10^7\}]} = 0.01845 \ \Omega$$

$$\varepsilon_e = \frac{\varepsilon_r + 1}{2} + \frac{\varepsilon_r - 1}{2} \frac{1}{\sqrt{1 + 12h/w}} = \frac{2.55 + 1}{2} + \frac{2.55 - 1}{2} \frac{1}{\sqrt{1 + 12 \times 5/8}}$$

$$= 1.775 + 0.775 \times 0.343 = 2.040825$$

$$\tan \delta = \sigma/\omega\varepsilon_e = 5.8 \times 10^7/\{\pi \times 10^{10} \times 2.040825\} = 0.0009894$$

Attenuation due to dielectric loss

$$\alpha_d = \frac{\beta_0 \varepsilon_r (\varepsilon_e - 1)\tan\delta}{2\sqrt{\varepsilon_e}(\varepsilon_r - 1)} = \frac{104.72 \times 2.55(2.04 - 1) \times 0.0009894}{2\sqrt{2.0408} \times (2.55 - 1)} = 0.062$$

Attenuation due to conductor loss

$$\alpha_c = R_s/(Z_0 \, w) = 0.01845/(50 \times 8) = 0.0000464$$

PROBLEMS

P-15.1 The parameters involved in the construction of a planar transmission line (Figure 15.3) are $b = 6$ mils, $t = 1$ mil and $\varepsilon = 2.21$. Calculate its effective width and the characteristic impedance if (a) $w = 8$ mils and (b) $w = 4$ mils.

P-15.2 A planar transmission line has characteristic impedance of 75 Ω. Calculate its width if $b = 6$ mils, $t = 1$ mil and (a) $\varepsilon_r = 5.23$, (b) $\varepsilon_r = 2.21$.

P-15.3 An 8 GHz planar transmission line has characteristic impedance of 50 Ω. Calculate the attenuation due to conductor loss if its width is $b = 5$ mils, $t = 1$ mil, $w = 8$ mils, $\mu = \mu_0$ and $\sigma = 5.8 \times 10^7$ mhos/m and (a) $\varepsilon_r = 5.23$, (b) $\varepsilon_r = 2.21$.

P-15.4 Calculate the (i) effective dielectric constant (ε_e), (ii) phase velocity (v_p), (iii) phase constant (β) and characteristic impedance (Z_0) of a 10-GHz microstrip if its depth $d = 4$ mils and (a) $w = 6$ mils, (b) $w = 5$ mils. $\mu = \mu_0$ and $\varepsilon_r = 5.23$.

P-15.5 Calculate the strip width of a microstrip for which $Z_0 = 40 \ \Omega$, $d = 3$ mils and relative dielectric constant (ε_r) is (a) 5.23, (b) 2.25.

P-15.6 Calculate the attenuation due to (i) effective dielectric constant, (ii) dielectric loss (α_d) and (iii) conductor loss (α_c) for a 10-GHz, 75-Ω microstrip with depth $h = 6$ mils, $w = 10$ mils and $\varepsilon_r = 2.55$. For conductor $\mu = \mu_0$, $\sigma = 5.8 \times 10^7$ mhos/m.

Descriptive Questions

Q-15.1 Illustrate the cross-sectional views of the commonly used planar structures and describe their behaviour and relationships.

Q-15.2 Show the geometry and the electric and magnetic field distribution in a stripline. Also, illustrate its various forms.

Q-15.3 Illustrate different versions and types of microstrips.

Q-15.4 What are coplanar waveguides and what are their positive aspects?

Q-15.5 Draw the geometry of a slot line. Also, illustrate the H and E field distributions.

Q-15.6 How are the lumped elements realised by using microstrips and other planar lines?

Q-15.7 How are power splitters, couplers, filters and resonators formed by using planar lines? Give at least one example of each.

Further Reading

1. Bahl, I., Bhartia, P. and Gupta, K. C. *Microstrip Lines and Slotlines*. Artech House, Norwood, MA, 1996.
2. Bianco, B., Panini, L., Parodi, M. and Ridetlaj, S. Some considerations about the frequency dependence of the characteristic impedance of uniform microstrips. *IEEE Trans. Microwave Theory Tech.* MTT-26, 182–185, March 1978.
3. Brian, C. W. *Transmission Line Design Handbook*. Artech House, Norwood, MA, 1991.
4. Bryant, T. G. and Weiss, J. A. Parameters of microstrip transmission lines and of coupled pairs of microstrip lines. *IEEE Trans.*, MTT-6(12), 1021–1027, December 1968.
5. Chang, K., Bahl, I. and Nair, V. *RF and Microwave Circuits and Component Design for Wireless Systems*. John Wiley & Sons, New York, 2002.
6. Cohn, S. Characteristic impedance of the shielded-strip transmission line. *IRE Trans.*, MTT-2(7), 52, July 1954.
7. Cory, H. Dispersion characteristics of microstrip lines. *IEEE Trans. Microwave Theory Tech.*, MTT-29, 59–61, January 1981.
8. Denlinger, E. J. A frequency dependent solution for microstrip transmission lines. *IEEE Trans. Microwave Theory Tech.*, MTT-19, 30–39, January 1971.
9. Douville, R. J. P. and James, D. S. Experimental study of symmetric microstrip bends and their compensation. *IEEE Trans. Microwave Theory Tech.*, MTT-26, 175–182, March 1978.
10. Edward, T. C. *Foundation for Micro Strip Circuit Design*, 2nd ed. Wiley, Chichester, England, 1981, 1992.
11. Gardiol, F. *Micro Strip Circuits*. Wiley, New York, 1990.

12. Grieg, D. D. and Engelmann, H. F. Microstrip—A new transmission technique for the kilo-mega-cycle range. *Proc. IRE*, 40(12), 1644–1650, December 1952.
13. Gupta, K. C., Garg, R. and Bahl, I. J. *Micro Strip Lines and Slot Lines*. Artech House, Dedham, MA, 1979.
14. Hoffmann, R. K. *Handbook of Microwave Integrated Circuits*. Artech House, Norwood, MA, 1987.
15. Howe, Jr. H. *Strip Line Circuit Design*. Artech House, Dedham, MA, 1974.
16. Kaupp, H. R. Characteristics of microstrip transmission lines. *IEEE Trans. Electron. Comput.*, EC-16(2), 185–193, April 1967.
17. Lange, J. Interdigited stripline quadrature hybrid. *IEEE Trans.*, MTT-17(12), 1150–1151, December 1969.
18. Lee, T. H. *Planar Microwave Engineering*. Cambridge University Press, New York, pp. 173–174, 2004.
19. Lewin, L. Radiation from discontinuities in stripline. *Proc. IEE*, 107C, 163–170, February 1960.
20. Pucel, R. A., Masse, D. J. and Hartwig, C. P. Losses in microstrip. *IEEE Trans.*, MTT-16(6), 342–350, June 1968.
21. Pucel, R. A., Masse, D. J. and Hartwig, C. P. Correction to losses in microstrip. *IEEE Trans.*, MTT-16(12), 1064, December 1968.
22. Stinehelfer, H. E. An accurate calculation of uniform microstrip transmission lines. *IEEE Trans.*, MTT-16(7), 439–443, July 1968.
23. Vendeline, G. D. Limitations on stripline Q. *Microwave J.*, 63–69, May 1970.
24. Welch, J. D. and Pratt, H. G. Losses in microstrip transmission systems for integrated microwave circuits. *NEREM Rec.*, 8, 100–101, 1966.
25. Wheeler, H. A. Transmission-line properties of parallel wide strips by a conformal-mapping approximation. *IEEE Trans. Microwave Theory Tech.*, MTT-12, 280–289, May 1964.
26. Wheeler, H. A. Transmission-line properties of parallel strips separated by a dielectric sheet. *IEEE Trans.*, MTT-13, 172–185, March 1965.
27. Wheeler, H. A. Transmission-line properties of a strip on a dielectric sheet on a plane. *IEEE Trans.*, MTT-25, 631–647, August 1977.

16

Microwave Integrated Circuits

16.1 Introduction

Microwave integrated circuit (MIC) technology is an extension of integrated circuit (IC) technology to UHF and microwave range. Microwave assemblies form the major part of MIC technology. These assemblies greatly vary in complexity and may be of monolithic or hybrid nature. The MIC technology mainly encompasses (i) planar transmission lines and coplanar microstrip lines, (ii) photolithography and thin-film techniques and (iii) microwave semiconductor (SC) devices.

Electronic circuits are normally classified into three main categories, including discrete circuits (DCs), ICs and hybrid integrated circuits (HICs). In DCs, separately manufactured elements (R, L, C and SC devices) are connected through conducting wires. In ICs, a single-crystal SC chip contains both active and passive elements and their interconnections. HICs contain both the discrete elements and the ICs.

The ICs fabricated for use at microwave range are referred to as MICs. These can be of monolithic or hybrid nature, which are referred to as monolithic microwave integrated circuits (MMICs) and HICs. Monolithic integrated circuit is built on a single crystal by the process of epitaxial growth, masked impurity diffusion, oxidation growth and oxide etching. Both ICs and MMICs can be of monolithic or hybrid form. One of the basic differences between these two is of their packaging density. In case of conventional ICs, it is quite high, whereas in MMICs, it is likely to be quite low. The elements of an MMIC may be formed on an insulating substrate (e.g. glass or ceramic). It is referred to as film integrated circuit. If an MMIC contains a combination of two or more types of integrated circuit, such as monolithic or film, or one IC type with discrete elements, it is termed as a HIC.

16.2 Merits and Limitations of MICs

16.2.1 Merits

MICs have many distinct advantages, which make them suitable for wide applications. These are listed as follows:

- MICs greatly reduce the cross section of transmission line. As an example, the dimension of an X-band waveguide is 1″ × 0.5″ and that of an X-band stripline on alumina substrate for 50 Ω is 0.25″ × 0.25″. Microstrip line dimensions are also smaller than the dimensions of standard coaxial line.

- These also have the reduced guide wavelength as the use of high-ε substrate reduces guide wavelength (λ_g) by a factor equal to $\sqrt{\varepsilon_{eff}}$. For alumina circuits, a reduction in λ_g may be of the order of 2.5–3. The size of circuits using distributed elements may ultimately be reduced in the same ratio.

- Reduction in size is also achieved by using lumped elements in place of distributed components. As the dimensions of lumped elements are less than 10% of the wavelength, the lumped circuits are much smaller than the distributed circuits.

- MICs may also lead to the elimination of interconnections and to layout flexibility. In case of waveguides and coaxial lines, complicated circuits cannot be built in a single layout. In conventional subsystems consisting of a number of elements or circuits, coaxial connections, flanges or adopters are to be used. In MICs, even a large complicated circuit function can be produced by photo-etching of a single metallisation and a considerable volume occupied by connectors and so on is saved. Fabrication process allows the designer to locate input–output ports at the desired locations. Thus, the amount of hardware used for bends, corners and so on is saved.

- In MICs, it is possible to use uncapsulated devices that occupy much smaller space than the packaged ones. Packaged devices may require additional space for mounting.

- MIC fabrication leads to a very uniform product. Variations (if any) are caused by tolerance limits in SC device characteristics and substrate material. Methods to reduce or compensate for these variations may be applied. Microwave SC devices used in MICs have longer life as compared to that of the tubes. These require a fewer number of connections for achieving a given subsystem function. More durable connections between devices and circuits are obtained because of bonding techniques. The circuit construction of MICs is more rigid due to less possibility of relative motion of parts of the circuit.

- MICs lead to improvement in device circuit interface. In conventional microwave circuits, transverse dimensions of circuits are much larger than those when SC devices are used. This necessitates the use of mounting structures like post and so on, which introduces additional reactance and limits the circuit performance. In MICs, transverse dimensions of microwave structures are comparable with the size of device chips. As a result, the package and mounting reactances are eliminated. This elimination increases the circuit bandwidth and device impedance level. Ultra broadband PIN diode and switches have been realised through stripline and microstrip configurations.

- Lumped element resonators have wider bandwidth as compared to distributed element circuits since the equivalent L and C of distributed circuits are not constant with frequency. Thus, the use of lumped inductors allows a wider bandwidth in tunnel diode amplifiers and increases the range of tuning of varactor circuits.

- Combination of microstrip and slot line in the same circuit introduces another degree of flexibility in the circuit design. An increase in bandwidth of couplers (to the tune of 10 octaves) has been reported by using such a combination. Similarly, slot and coplanar lines have been combined usefully to design novel circuit configurations of balanced mixers.

- Fabrication of conventional microwave devices, components and circuits requires precise machining, which is time consuming and expensive. However, MICs can be made by using photo-etching, which are not only cheaper but also a quicker process. Besides, the cost of MIC fabrication is independent of circuit complexity to a large extent. Once the circuit is perfected and the mask is prepared, large-scale production does not require any additional tooling arrangements. MICs can use SC devices in chip form, which are much cheaper than the packaged devices.

- Computer-aided design (CAD) and computer-aided measurement techniques are well suited for MICs.

16.2.2 Limitations

- Once MICs are fabricated, the probability of adjustment of parameters and so on is quite low.

- After fabrication, inclusion of arrangements like tuning screws, variable shorts and so on become difficult. In case of any such inclusion, the desirability from the point of view of reliability and reproducibility is quite low.

- The quality factor of MIC microstrip resonators is quite low (≈ 100) as compared to that of conventional microwave resonators (≈ 1000).

Dielectric disk resonators, however, provide high Q and may be used in MICs.

- Power-handling capability of MICs (≈ 10 W) is much lower than that of waveguides.

- It is difficult to obtain high-frequency stability from solid-state oscillators in microstrip configurations.

- Precise characterisation of SC devices is needed for accurate circuit design. This poses considerable problems, especially when the devices to be used are in the form of chips or miniature packages. To ascertain desired circuit performance, computer-aided design techniques are to be employed for precise characterisation of SC devices.

16.3 Types of MICs

As very few microwave applications require densely packed arrays, thus, like LF ICs, MICs can be either of monolithic or of hybrid form. These are built on a single crystal and their packaging density in general is much less than that of conventional ICs. When elements are formed on an insulating substrate, these are called film ICs. The combination of two or more ICs, such as monolithic or film or an IC along with discrete elements, is referred to as hybrid IC. The basic features of these two forms are described in the subsequent subsections.

The basic building blocks for MICs are planar transmission lines discussed in Chapter 15. These are normally characterised by their characteristic impedance, phase velocity or effective dielectric constant, attenuation constant and power handling capability. These parameters are related to their dimensions and properties of the materials used.

16.3.1 Hybrid ICs

Hybrid technology is well suited to MICs in the 1–20 GHz range. These are commonly employed in satellite communication, phased array antennas and warfare systems. In hybrid circuits, active devices are attached to a glass, ceramic or substrate, which contains the passive circuitry. This passive circuitry may be composed of lumped elements or distributed elements or a combination thereof. These elements are fabricated on high-quality ceramic, glass or ferrite substrate. Passive circuit elements are deposited on the substrate and active devices are mounted on the substrate and connected to passive elements or circuits. Active devices may be either in the form of chip, chip carriers or small plastic packages. The distributed elements are formed by a single layer metallisation of a thin or thick film. The lumped elements are formed either by using multi-layer deposition and plating techniques or are attached to the substrate in chip form. Resistivity of MICs should be sufficiently greater than 1000 Ω-cm for good circuit performance.

ICs can further be categorised as *hybrid ICs* and *miniaturised hybrid ICs*. In the first type, the distributed circuit elements are formed by single-layer metallisation technique over a substrate and the other elements, namely, R, L, C and SC devices are externally connected with the substrate. In the second case, all passive components, namely, R, L and C are deposited on a common dielectric substrate and SC devices are attached to the substrate.

16.3.2 Monolithic Microwave Integrated Circuit

MMIC (sometimes pronounced as 'mimic') is a type of IC that operates over the frequency range from 300 MHz to 300 GHz. The functions of these devices include microwave mixing, power amplification, low noise amplification and high-frequency switching. As the inputs and outputs of MMIC devices are frequently matched to a characteristic impedance of 50 Ω, these can be used without any external matching network for their cascading. Additionally, most microwave test equipment is designed to operate in a 50 Ω environment. MMICs are dimensionally small (from around 1 to 10 mm²) and can be mass produced, which has allowed the proliferation of high-frequency devices such as cellular phones.

Besides their small size, MMICs are lightweight and low-cost devices. The reduction in cost is mainly due to large scale processing. As all components of MMICs are fabricated simultaneously without any soldered joints, they have high reliability, improved reproducibility and enhanced performance. These are suitable for space and military applications as they can successfully withstand shocks, temperature variations and severe vibrations. The available MMICs cover almost the entire range of microwave frequency.

Silicon (Si) is the traditional material for IC realisation, whereas gallium arsenide (GaAs), a III–V compound semiconductor was the base of the original MMIC fabrication. The fundamental advantages of GaAs over Si include (i) the device (transistor) speed and (ii) a semi-insulating substrate. Both these factors help in the design of HF circuit functions. During the course of time, the speed of Si-based technologies has increased as transistor feature sizes have reduced. Thus, the Si technology, which has relatively lower fabrication cost, can now also be used to fabricate MMICs. Silicon wafer diameters are larger (typically 8″ or 12″ compared with 4″ or 6″ for GaAs) and the wafer costs are lower; this led to the cheaper IC. Other III–V technologies, such as indium phosphide (InP) offer superior performance to GaAs in terms of higher gain, higher cutoff frequency and low noise. They are, however, more expensive due to smaller wafer sizes and increased material fragility. Silicon–germanium (SiGe) compound SC technology offers higher-speed transistors than conventional Si devices but with almost similar cost advantage. Gallium nitride (GaN) transistors can operate at much higher temperatures and work at much higher voltages than GaAs transistors. These transistors make ideal power amplifiers at microwave frequencies.

16.4 Materials Used

The materials required for MMICs include substrates, conductors, dielectrics and resistors.

16.4.1 Substrates

The selection of the substrate material depends on dissipation, function and type of circuit used. The properties to be possessed by these materials include (i) high dielectric constant usually greater than or equal to 9, (ii) low dissipation factor or loss tangent, (iii) constancy of ε and constancy of temperature, both over frequency range of interest, (iv) high purity, (v) constancy of thickness, (vi) high surface smoothness, (vii) high resistivity and dielectric strength and (viii) high thermal conductivity. The commonly used materials (along with their applications) include alumina (microstrip, suspended substrate), beryllia (compound substrate), ferrites/garnets (microstrip, coplanar compound substrate), gallium arsenide (HF microstrip, monolithic MICs), glass (lumped elements), rutile (microstrip) and sapphire (microstrip, lumped elements). Polytetra-fluoro-ethylene (PTFE) glass fibre, polyester glass, PTFE ceramic and hydrocarbon ceramic are the substrate materials used in microwave printed circuits. Quartz and glass/ceramic are employed in thin-film technology, whereas Al_2O_2, AlN and BeO are used in thin- and thick-film technology.

16.4.2 Conducting Materials

These are required to form conductor patterns on the ground planes. Conductor thickness should be 3–5 times the depth of penetration (δ) for small losses. Combination of more than one conducting material may be used to form a base for increasing adherence. A typical adhesive layer may have a surface resistivity ranging from 500 to 1000 Ω/m^2 without loss. Thick-film good conductors of about 10 μm are required. The required properties of conducting materials include (i) high conductivity, (ii) low temperature coefficient of resistance (TCR), (iii) good adhesion to the substrate, (iv) good etch-ability and solder-ability and (v) ease in deposition and electroplating. The conducting materials along with their properties (and method of deposition) include silver (Ag), copper (Cu), gold (Au), aluminium (Al), tungsten (W), molybdenum (Mo), chromium (Cr) and tantalum (Ta). The adherence to dielectric film or substrate of Ag, Cu and Au is poor and that of Al, W, Mo, Cr and Ta is good. The deposition techniques, including evaporation, screening, plating, sputtering, vapour phase and beam evaporation, differ from material to material. In reference to the technologies and processes, Cu is used in microwave printed circuits, Al, Au and Cu in thin-film technology and Au, Ag, Cu, PdAu, PdAg, PtAu, PtAg and PtPdAg in thick-film technology. Also, Au, Al and PdAgCu in co-fired ceramic (HTCC) and W and Mo in co-fired glass ceramic (LTCC) processes.

16.4.3 Dielectric Materials

These are used for blockers, capacitors and some coupled line structures. Their desirable properties include (i) reproducibility, (ii) capability to withstand high voltage, (iii) ability of processing without developing pinholes and (iv) low-RF dielectric loss. SiO_2 is the most commonly used dielectric material. The capacitors made of SiO_2 have high Q (usually $\gg 100$). SiO_2 film capacitors are in the range of 0.02–0.05 pF/mil^2. Thin-film SiO_2 is not very stable, and can be used only in non-critical applications, for example, bypass capacitors. In power MICs, capacitors may require breakdown voltages ≥ 200 V. Such capacitors can be realised with films of the order of 0.5–1.0 m thickness with low probability of pinholes or shorts. The other dielectric materials include SiO, Si_3N_4, Al_2O_3 and Ta_2O_5. Since these materials have different relative permittivities and dielectric strengths, the selection of material will depend on the application and method of deposition. These methods include evaporation, deposition, vapor phase sputtering and anodisation.

Dielectric materials used for different technologies and processes include (i) SiO_2, polyimide and benzo-cyclo-butene for thin film; (ii) glass–ceramics and recrystallising glasses for thick film; (iii) glass–ceramic tape for high-temperature co-fired ceramic (HTCC) and (iv) ceramic (Al_2O_3) tape for low-temperature co-fired ceramic (LTCC).

16.4.4 Resistive Materials

These are used for bias network, termination and attenuation. The thickness of the thick film is in the range of 1–500 μm, where the term 'thick film' refers to the process used and not to the physical thickness. Microwave thick-film metals are sometimes several micrometers thick, thicker than those of low-frequency ICs. Their required properties include (i) good stability, (ii) low TCR, (iii) adequate dissipation capacity and (iv) sheet resistivity in the range of 10–1000 Ω/m^2. Evaporated Ni and TaNi are most commonly used resistive materials. The exact TCR depends on film formation conditions. In general, the resistive materials include chromium (Cr), nichrome (NiCr), tantalum (Ta), Cr–SiO_2 and titanium (Ti). Data for their resistivity, TCR and stability are available in the literature. For most of these materials, evaporation is the common deposition method. Technologies and processes used for these resistive materials include (i) NiCr and TaN for thin film and (ii) RuO_2 and doped glass for thick film and HTCC.

16.5 Fabrication Techniques

Two most common MIC technologies are referred to as thin-film and thick-film technologies. These are used for the realisation of different types of MICs. These include the following.

16.5.1 Copper Clad Board

In this, copper is put on a large fibre glass or other boards using electro-deposition or rolling. Photo-resist is usually applied by laminating a pre-prepared film onto the substrate. It is then exposed to ultraviolet via a mask and developed. The copper is then etched away from the places it remained uncovered by photo-resist. Multi-layer boards are produced by laminating different layers together.

16.5.2 Thin-Film Fabrication

Thin-film MICs are made via a sputter-and-etch process. The MICs fabricated are robust to standard manufacturing process, exhibit repeatable performance and provide excellent device for millimetre-range applications. A thin-film technology variant called miniature hybrid MIC is based on thin film. In this, the multi-level passive circuits are batch fabricated on the substrate and only solid-state devices are externally attached to these circuits.

Besides sputtering, the evaporation technique is also employed for metal deposition. For increasing metal thickness, electroplating may also be employed. This technology gives the best pattern definition and highest performance in case materials used are appropriate. Some of such materials may include ceramic or other substrates (e.g., quartz) and integrated circuits using GaAs, InP, Si and so on. The required equipment is relatively expensive and the substrate must be in a vacuum. The advantages of this circuit technology are small size, lightweight, excellent heat dissipation and broadband performance. The time required for drooping of the chamber pressure and the limited substrate size are the main limitations of this technology, particularly when mass production is to be done.

The size of lumped elements can be greatly reduced by using thin-film technology. The operating limit of frequency can also be raised up to 20 GHz beyond which distributed parameters are preferred. In MMICs, lumped resistors may be used for resistive terminations of couplers, lumped capacitors for bypass applications and planar inductors for matching purposes. All these elements involve thin-film technology in their formation.

16.5.3 Thick-Film Fabrication

Thick-film MICs are fabricated using various inks pressed through patterned silk screen. In the process, metal and dielectric pastes are applied to ceramic base substrates using screen printing. The screen is a fine metal wire mesh with photographic emulsion applied and circuit pattern reproduced onto this emulsion layer. During printing, the paste is squeezed through the mesh where there are emulsion openings onto the substrate. The paste is then dried and fired at about 850°C. For the formation of a circuit multi-layer, layers can be printed successively. These MICs are inexpensive but are generally limited to the microwave range.

The thick-film printing can now be done by using LTCC technology wherein the ceramic (in the form of flexible sheet) is handled in its unfired state and via holes are made by punching or laser machining. The metal patterns are printed onto each individual sheet. The layers are stacked, gently pressed and dried. The co-firing step completes the process. LTCC technology is a thick-film variant equivalent to the hybrid miniature MIC. This process is similar to the thick-film process, except that it does not use a substrate. Dielectric layers are in the form of unfired ceramic tape instead of paste. This enables printing of reliable capacitors and resistors.

In view of its multi-layer process, LTCC technology offers several advantages over conventional thin-film, thick-film and HTCC technologies. These include a higher level of integration of components (e.g., resistors, inductors, capacitors, transmission lines and bias lines) and greater design flexibility in the realisation of different versions of planar lines discussed in Chapter 15.

16.5.3.1 Photoimageable Thick-Film Process

In the thick-film technology, the minimum conductor dimensions are limited to about 100 μm in view of the screen jagged edges. The photoimageable thick-film process (Figure 16.1) is an extension of conventional thick-film technology. In this, the standard thick-film paste is replaced by a photosensitive material. Such a paste is a combination of photosensitive vehicle and metal–glass powders. Both of these affect the electrical properties and resolution characteristics. This paste is first printed everywhere over the total substrate area. As the printing of sharp edge features is not required, the levelling properties of the photosensitive pastes are optimised to provide a uniform thick film with smooth and dense surface. The surface has to be free from pinholes and other printing defects. The exposure of this paste to ultraviolet light through a mask results in the desired conductor pattern.

FIGURE 16.1
Photoimageable thick-film process.

16.6 Fabrication Processes

MMICs involve the following processes in their fabrication.

16.6.1 Diffusion and Ion Implantation

These are two separate but complementary processes. Both are used to fabricate discrete and integrated devices. In the diffusion process, impurities are diffused into a pure material to alter its basic electrical characteristics. In ion implantation, high-energy ions are doped in substrate through the implantation process. In this process, precise control of doping is possible and there is improvement in reproducibility. Also, the process temperature in ion implantation is less than that in diffusion.

16.6.2 Oxidation and Film Deposition

By using the process of oxidation and thin-film deposition, discrete and integrated devices or circuits can be fabricated. The thin films used can be categorised into (i) thermal oxide, (ii) dielectric layers, (iii) polycrystalline silicon and (iv) metal films.

16.6.3 Epitaxial Growth Technique

In this single-crystal semiconductor, layers grow on a single-crystal semiconductor substrate. Through this technique, the doping profiles of devices can be suitably controlled to optimise the circuit performance. This technique can be subdivided into the following three categories.

16.6.3.1 Vapour-Phase Epitaxy

This is the most important technique for Silicon and GaAs devices.

16.6.3.2 Molecular-Beam Epitaxy

This process involves the reaction of one or more thermal beams of atoms or molecules with crystalline surface under ultra-high vacuum conditions. In this technique, both the chemical composition and doping profiles can be precisely controlled. Through this method, single-crystal multi-layer structures with dimensions of the order of atomic layers can be fabricated.

16.6.3.3 Liquid-Phase Epitaxy

It involves the growth of epitaxial layers on crystalline substrates by direct precipitation from the liquid phase. In view of its slow growth rate, it is suitable to grow thin epitaxial layers of sub-micron order. It is also suitable to grow multi-layered structures in which precise doping, composition controls are required.

16.6.4 Lithography

In this process, patterns of geometric shapes on the mask are transferred to a thin layer of radiation-sensitive material known as *resist*, for covering the surface of the SC wafer. The resist patterns are not permanent elements of the final device but only replicas of circuit features. Its various types include (i) electron beam lithography, (ii) ion beam lithography, (iii) optical lithography and (iv) x-ray lithography.

16.6.5 Etching and Photo-Resist

It provides selective removal of SiO_2 to form openings through which impurities can be diffused. In this process, a substrate (photosensitive emulsion) with a uniform film of Kodak photo-resist is coated. A mask with desired openings is placed over this photo-resist. It is then exposed to ultraviolet light. In view of this exposure, a polymerised photo-resist is developed. After removal of the mask, the unpolymerised portions are dissolved by using trichloroethylene. The uncovered SiO_2 can be removed by using the hydrofluoric acid. The thick-film process usually employs printing and silk-screening of silver or gold through metal mask in a glass frit. This is applied on the ceramic and fired at 850°C. The initial layers may be covered with gold after firing.

16.6.6 Deposition

The three common deposition methods used for making MMICs include (i) vacuum evaporation, (ii) electron beam evaporation and (iii) cathode (or dc) sputtering.

16.6.6.1 Vacuum Evaporation

In the vacuum evaporation technique, the impurity materials are located in a metallic boat through which a high current is passed. The substrate covered with a mask and the heated boat is located in a glass tube having high vacuum pressure. The heated impurities evaporate and its vapours deposit on a slightly heated substrate. This deposition results in the formation of a polycrystalline layer on the substrate.

16.6.6.2 Electron Beam Evaporation

In the electron beam evaporation method, a narrow beam of electrons is generated. This beam scans the substrate in the boat in order to vaporise the impurity.

16.6.6.3 Cathode (or dc) Sputtering

In this method, a crucible containing the impurity acts as the cathode and the substrate acts as the anode of a diode. Both the crucible and the substrate are located in a vacuum container with slight trace of argon gas. With sufficiently higher voltage applied between the two electrodes, a glow discharge of argon gas is formed. The positive argon ions with impurity atoms accelerate with enough energy towards the cathode and deposit these atoms on the substrate to which these conveniently adhere.

16.6.6.4 Processes Used in Different Technologies

Microwave printed circuit: Photolithography, collate sheets, bonding and etch.

Thin film: Sequentially vacuum deposit, spin coat and/or plate conductors, dielectric and resistor photolithography and etch.

Thick film: Sequentially print, dry and fire, conductor, dielectric and resistor pastes.

HTCC: Punch vias, print and dry conductors on tape, collate layers, laminate and co-fire.

LTCC: Same as HTCC.

16.7 Illustration of Fabrication by Photo-Resist Technique

The steps of the fabrication process by the photo-resist technique are illustrated in Figure 16.2.

16.7.1 Planar Resistors

A planar resistor consists of a thin film deposited on an insulating substrate. The materials used for such resistors may include aluminium, copper, gold, nichrome, titanium, tantalum and so on with resistivities varying between 30 and 1000 (Ω/m^2). Such resistors may be used for hybrid couplers, power combiners/dividers and bias-voltage circuits. While designing such a resistor, the sheet resistivity and thermal stability of the material, thermal resistance of the load and bandwidth are to be taken into consideration. Planar resistors can be classified as semiconductor films, deposited metal films and cermets. *Semiconductor films* can be fabricated by forming an isolated band of conducting epitaxial film on the substrate by mesa etching or by isolation implant of the surrounding conducting film. These can also be formed by implanting a high resistivity region within the semi-insulating substrate. For *metal film resistors,* a layer of desired pattern is formed by using the method

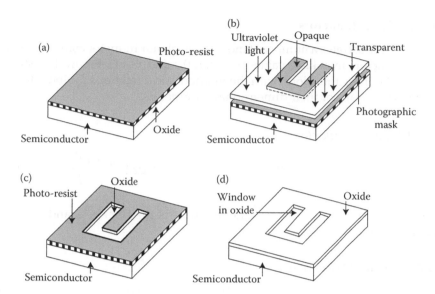

FIGURE 16.2
Illustration of fabrication steps. (a) (i) A semiconductor substrate is selected. (ii) It is covered by an oxide layer. (iii) The oxide layer is covered by a photo-resist layer. (b) (i) A precision photographic mask is placed on this oxidised photo-resist. (ii) It is then exposed to ultraviolet light. (c) The selected oxide region is removed through chemical etching, that is, by using hydrofluoric acid. (d) The photo-resist is finally dissolved with an organic solvent in the oxide, leaving the desired openings.

of evaporation and photolithography. For *cermet resistors,* a mixture of metal and dielectric is used to form the film. These resistors can have different configurations. Figure 16.3a shows the configuration of an implanted resistor, whereas Figure 16.3b illustrates a thin-film planar resistor with its dimensional parameters l (length of resistive film), w (width) and t (thickness).

The resistance of a planar resistor with resistivity ρ_s shown in Figure 16.3b can be given as

$$R = l\,\rho_s/w \cdot t \ \Omega \tag{16.1}$$

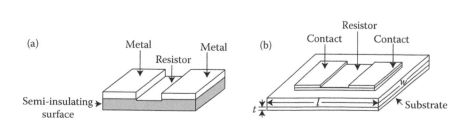

FIGURE 16.3
Planar resistors. (a) Implanted resistor and (b) thin-film resistors.

16.7.2 Planar Inductors

Figure 16.4 illustrates a number of monolithic planar inductor configurations. These inductors may have values between 0.5 and 10 nH. Figure 16.5 shows some of the inductors along with the mathematical relations for inductances (L in nH) in terms of their dimensional parameters.

16.7.2.1 Ribbon Inductor

$$L = 5.08 \times 10^{-3} l \left[\ln\left(\frac{l}{w + t}\right) + 1.19 + 0.022\left(\frac{w + t}{l}\right) \right] \qquad (16.2)$$

As shown in Figure 16.5a, l is the length, t is the thickness and w is the width of the ribbon.

16.7.2.2 Round-Wire Inductor

$$L = 5.08 \times 10^{-3} l [\ln(l/d) + 0.386] \qquad (16.3)$$

As shown in Figure 16.5b, d is the wire diameter and l is the wire length.

16.7.2.3 Single-Turn Flat Circular Loop Inductor

$$L = 5.08 \times 10^{-3} l \left[\ln\left(\frac{t}{w + t}\right) - 1.76 \right] \qquad (16.4)$$

As shown in Figure 16.5c, l is the length, t is the thickness and w is the width of the film.

16.7.2.4 Circular Spiral Inductor

$$L = 0.03125\, n^2 d_0 \qquad (16.5)$$

In view of Figure 16.5d, $d_0 = 5d_i = 2.5n(w + s)$, n is the number of turns, s is the separation between turns, d_i and d_0 are the inner and outer diameters of the loop and w is the film width.

(a)   (b)   (c)   (d)

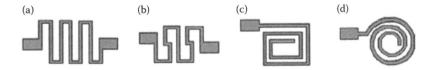

FIGURE 16.4
Planar inductor configurations. (a) Meander line, (b) S line, (c) square spiral and (d) circular spiral.

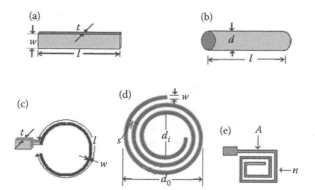

FIGURE 16.5
Inductors. (a) Ribbon, (b) round wire, (c) single wire circular loop, (d) circular spiral and (e) square spiral type.

16.7.2.5 Square Spiral Inductor

$$L = 8.5\, A^{0.5} n^{5/3} \qquad (16.6)$$

As shown in Figure 16.5e, A is the surface area in cm^2 and n is the number of turns.

The units of lengths, widths, thicknesses and diameters in all the above cases are in mils.

16.7.3 Planar Film Capacitor

The two most common planar capacitor configurations for MMICs are illustrated in Figure 16.6.

16.7.3.1 Metal-Oxide-Metal Capacitor

As shown in Figure 16.6a, it has two metal conducting layers and a middle dielectric sandwiched layer. Its capacitance (C in farads) is given by

$$C = \varepsilon_0 \varepsilon_r \frac{lw}{h} \qquad (16.7)$$

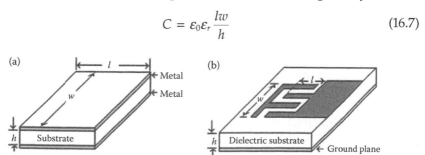

FIGURE 16.6
Planar film capacitors. (a) Metal-oxide-metal capacitor and (b) interdigitated capacitor.

where ε_0 is the permittivity of free space, ε_r is the relative dielectric constant of the dielectric material, w is the metal width, l is the metal length and h is the height of the dielectric material.

16.7.3.2 Interdigitated Capacitor

As illustrated in Figure 16.6b, it consists of a single-layer structure in which the vanes are fabricated as microstrip lines with capacitance values between 0.1 and 15 pF. The approximate value of capacitance (C in pF) can be obtained by

$$C = \frac{\varepsilon_r + 1}{w} l[(N - 3)A_1 + A_2] \tag{16.8}$$

where N is the number of fingers, l is the finger length, w is the finger-base width, A_1 (= 0.089 pF/cm) is the contribution of the interior finger for $h > w$ and A_2 (= 0.10 pF/cm) is the contribution of the two external fingers for $h > w$.

16.8 Fabrication of Devices

16.8.1 MOSFET

MOSFETs were discussed in Chapter 14. Its fabrication steps are shown in Figure 16.7.

16.8.2 Micro-Fabrication of a CMOS Inverter

The steps involved in a simplified process of micro-fabrication of a CMOS inverter on a p-type substrate in a semiconductor are shown in Figure 16.8.

EXAMPLE 16.1
Calculate the inductance of a ribbon inductor having $l = 10$ mils, $w = 5$ mils and $t = 0.2$ µm (1 mil = 2.54×10^{-3} cm or 1 cm = 393.7 mils, 1 m = 39,370 mils).

Solution
In view of Equation 16.2
$$t = 0.2 \text{ µm} = 0.2 \times 0.0393.7 = 0.007874 \text{ mils}$$

$$L = 5.08 \times 10^{-3} l \left[\ln\left(\frac{l}{w+t}\right) + 1.19 + 0.022\left(\frac{w+t}{l}\right) \right]$$

$$= 5.08 \times 10^{-3} \times 10 \times \left[\ln\left\{\frac{10}{(5+0.007874)}\right\} + 1.19 + 0.022\left\{\frac{(5+0.007874)}{10}\right\} \right]$$

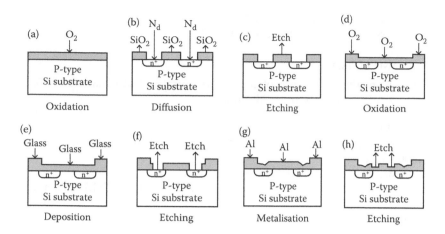

FIGURE 16.7

Fabrication of MOSFET. (a) (i) A p-type substrate (silicon) is selected. (ii) It is exposed to dry oxygen (O_2) to form a SiO_2 layer. This step is called *oxidation*. (b) (i) Two windows are opened by using the photo-resist technique. (ii) An n^+ layer is diffused through these windows. This step is called *diffusion*. (c) The centre oxide region is removed by using the photo-etching technique. This step is termed *etching*. (d) The entire surface is again exposed to dry oxygen to form a SiO_2 layer on the top surface. The step is again termed as *oxidation*. (e) Phosphorous glass is deposited over the entire surface to cover the oxide layer. The step is again termed as *deposition*. (f) Two windows are opened above two n^+ diffused regions by using photo-etching. The step is again called as *etching*. (g) Aluminium metallisation is carried out over the entire device surface. The step is referred to as *metallisation*. (h) (i) Unwanted metal is etched away. (ii) Metal contacts are attached to the diffused gate, drain and source regions. This step is again referred to as *etching*.

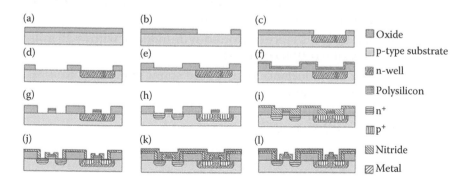

FIGURE 16.8

Micro-fabrication of a CMOS inverter. (a) (i) A p-type substrate is selected. (ii) Oxide is grown on the substrate. (b) A window for pMOSFET is etched in the oxide. (c) An n-well is diffused through the window. (d) A window for nMOSFET is etched in the oxide. (e) Gate oxide is grown over the entire surface. (f) Polysilicon is deposited over the entire surface. (g) Polysilicon and oxide are etched to form four windows. (h) Sources (n^+) and drains (p^+) are implanted through these four windows. (i) Nitride is grown over the entire surface. (j) Nitride is etched to form four small windows. (k) Metal is deposited over the entire surface. (l) Metal is etched from four locations.

$$= 5.08 \times 10^{-2} \, [0.6916 + 1.19 + 0.011]$$
$$= 5.08 \times 10^{-2} \times 1.8926 = 0.09614408 \text{ nH/mil}$$

EXAMPLE 16.2

Calculate the inductance of a circular spiral inductor having $n = 5$, $s = 50$ mils and $w = 30$ mils. Also find d_i.

Solution

In view of the given data, $d_0 = 5d_i = 2.5n(w + s) = 2.5 \times 5 \times (30 + 50) =$ 1000 mils

Since $5d_i = 1000$ mils, $d_i = 200$ mils

From Equation 16.5, $L = 0.03125n^2d_0 = 0.03125(5)^2 \times 1000 = 781.25$ nH/mil

EXAMPLE 16.3

Calculate the capacitance of a metal-oxide-metal capacitor with $l = 0.003$ cm, $w = 0.05$ cm, $h = 0.25$ cm and $\varepsilon_r = 5.23$.

Solution

In view of Equation 16.7, $C = \varepsilon_0 \varepsilon_r \{(lw)/h\}$

Thus,

$$C = 8.854 \times 10^{-14} \times 5.23 \times 0.003 \times 0.05/0.25 = 0.00027783852 \times 10^{-12}$$
$$= 0.00027783852 \text{ pF/cm}$$

PROBLEMS

P-16.1   Calculate the resistance of a planar resistor having sheet resistivity $\rho_s = 2.44 \times 10^{-8}$ Ω-m. The parameters shown in Figure 16.8 are $l = 20$ mm, $w = 8$ mm and $t = 0.25$ μm.

P-16.2   Calculate the inductance of a round-wire inductor with $l = 15$ mils and $d = 0.3$ μm (= 0.003937 mils).

P-16.3   Calculate the inductance of a square spiral inductor with $n = 6$ and $A = 3 \times 10^{-6}$ cm^2.

P-16.4   Calculate the capacitance of an interdigitated capacitor having $l = 0.005$, $w = 0.03$, $h = 0.05$ cm, $N = 8$ and $\varepsilon_r = 10.2$.

Descriptive Questions

Q-16.1   Discuss the merits and limitations of microwave integrated circuits.

Q-16.2   What do you understand by MMIC? How does it differ from a hybrid IC?

Q-16.3 Discuss the properties of materials required for substrates, conductors, dielectrics and resistors used in the fabrication of MMICs.

Q-16.4 Discuss the various techniques used to fabricate MMICs.

Q-16.5 List the processes involved in the fabrication of MMICs.

Q-16.6 Illustrate the steps involved in the fabrication process by the photo-resist technique.

Further Reading

1. Caulton, M. et al., Status of lumped elements in microwave integrated circuits: Present and future. *IEEE Trans.*, MTT-19(7), 588–599, July 1971.
2. Caulton, M. and Sobol, H., Microwave integrated circuit technology: A survey. *IEEE J. Solid-State Circuits*, SC-5(6), pp. 292–303, December 1970.
3. Niehenke, E.C., Pucel R. A. and Bahl, I. J., Microwave and millimeter-wave integrated circuits. *IEEE Trans.*, MTT- 50(3), 846-857, March 2002.
4. Frey, J., *Microwave Integrated Circuits*. Artech House, Dedham, MA, 1975.
5. Gupta, K. C. and Singh, A., *Microwave Integrated Circuits*. Wiley Eastern, New Delhi, 1974.
6. Keister, F. Z., An evaluation of materials and processes for integrated microwave circuits. *IEEE Trans., on Microwave Theory and Techniques*, MTT-16(7), 469–475, July 1968.
7. Lee, T. H., *The Design of CMOS Radio Frequency Integrated Circuits*. Cambridge University Press, New York, 2004.
8. Lio, S. Y., *Microwave Devices & Circuits*, 3rd ed. Prentice-Hall of India, New Delhi, 1995.
9. Parrillo, L. C., *VLSI Process Integration*. S. M. Sze (Ed.) VLSI Technology, John Wiley & Sons, New York, 1983.
10. Pucel, R. A., Design considerations for monolithic microwave circuits. *IEEE Trans. Microwave Theory Tech.*, 29(6), 513–534, June 1981.
11. Pulfrey, D. L. and Garry Tarr, N., *Introduction to Microelectronic Devices*. Prentice-Hall, Inc., Englewood Cliffs, NJ, 1989.
12. Sobol, H., Applications of integrated circuit technology to microwave frequencies. *Proc. IEEE*, 59(8), 1200–1211, August 1971.
13. Sobol, H., Technology and design of hybrid microwave integrated circuits. *Solid State Technology*, 13(2), 49–59, February 1970.
14. Sze, S. M., ed., *VLSI Technology*. John Wiley & Sons, New York, 1983.
15. Young, L., *Advances in Microwaves*. Academic Press, New York, 1974.

Appendix A1: Maxwell's and Other Equations

	Point Form		Integral Form	
Time-variant field	$\nabla \times H = J + \dfrac{\partial D}{\partial t}$	(A1.1a)	$\oint H \cdot dl = I + \iint \dfrac{\partial D}{\partial t} \cdot ds$	(A1.1b)
	$\nabla \times B = -\dfrac{\partial B}{\partial t}$	(A1.2a)	$\oint E \cdot dl = -\iint \dfrac{\partial B}{\partial t} \cdot ds$	(A1.2b)
Time-invariant field	$\nabla \times H = J$	(A1.3a)	$\oint H \cdot dl = I$	(A1.3b)
	$\nabla \times B = 0$	(A1.4a)	$\oint E \cdot dl = 0$	(A1.4b)
Region contains charges	$\nabla \cdot D = \rho$	(A1.5a)	$\iint D \cdot ds = \iiint \rho \, dv$	(A1.5b)
Region is charge free	$\nabla \cdot D = 0$	(A1.6a)	$\iint D \cdot ds = 0$	(A1.6b)
For all conditions	$\nabla \cdot B = 0$	(A1.7a)	$\iint B \cdot ds = 0$	(A1.7b)
Connecting equations	$D = \varepsilon E$ (A1.8a) $\quad B = \mu H$ (A1.8b) $\quad J = \sigma E$			(A1.8c)
Wave equations	$\nabla^2 E = \mu \, \varepsilon \, (\partial^2 E / \partial t^2)$ (A1.9a) $\quad \nabla^2 H = \mu \, \varepsilon \, (\partial^2 H / \partial t^2)$			(A1.9b)

Appendix A2: Solution to Equation 11.19

Equation 11.19 is reproduced and renumbered as below

$$\frac{dU_x}{dt} = \frac{d^2x}{dt^2} = -\frac{e}{m}E_x = -\frac{e}{m}\left(-\frac{V_o}{d}\right) = \frac{eV_o}{md} = k \qquad \text{(A2.1a)}$$

$$\frac{dU_y}{dt} = \frac{d^2y}{dt^2} = 0 \qquad \text{(A2.1b)}$$

and

$$\frac{dU_z}{dt} = \frac{d^2z}{dt^2} = 0 \qquad \text{(A2.1c)}$$

From Equation A2.1 we get

$$x = \frac{1}{2}kt^2 + A_1t + B_1 \qquad \text{(A2.2a)}$$

$$y = A_2t + B_2 \qquad \text{(A2.2b)}$$

and

$$z = A_3t + B_3 \qquad \text{(A2.2c)}$$

Initial condition at $t = 0$, $x = y = z = 0$ leads to $B_1 = B_2 = B_3 = 0$ and Equation A2.2 reduces to

$$x = \frac{1}{2}kt^2 + A_1t \qquad \text{(A2.3a)}$$

$$y = A_2t \qquad \text{(A2.3b)}$$

and

$$z = A_3t \qquad \text{(A2.3c)}$$

Since

$$\frac{dx}{dt} = kt + A_1 \qquad \text{(A2.4a)}$$

$$\frac{dy}{dt} = A_2 \tag{A2.4b}$$

and

$$\frac{dz}{dt} = A_3 \tag{A2.4c}$$

Application of another set of conditions at $t = 0$, $Ux = Ux_0$, $Uy = Uy_0$ and $Uz = 0$ gives

$$A_1 = Ux_0 \quad A_2 = Uy_0 \quad \text{and} \quad A_3 = 0.$$

Thus, Equation A2.3 reduces to

$$x = \frac{1}{2}kt^2 + Ux_0 t \tag{A2.5a}$$

$$y = Uy_0 t \tag{A2.5b}$$

and

$$z = 0 \tag{A2.5c}$$

Substituting the value of k from Equation A2.1a into Equation A2.5 we get

$$x = (eV_0/2md)\, t^2 + Ux_0 t \quad \text{or} \quad Ux_0 t = x - (eV_0/2md)\, t^2 \tag{A2.6a}$$

Also since

$$y = Uy_0 t \quad t = y/Uy_0 \tag{A2.6b}$$

Substitution of 't' from Equation A2.6b into Equation A2.6a gives

$$x = (eV_0/2md)\, (y/Uy_0)^2 + Ux_0(y/Uy_0) \tag{A2.6c}$$

Equation A2.6c is an *Equation of the parabola* in the x–y plane.
Now since

$$Ux = dx/dt = (eV_0/md)t + Ux_0, \quad Uy = dy/dt = Uy_0 \quad \text{and} \quad Uz = dz/dt = 0$$

$$U = \sqrt{(Ux^2 + Uy^2)} = \sqrt{\left[Uy_0^2 + \{Ux_0 + (eV_0/md)t\}^2\right]} \qquad \text{(A2.7a)}$$

This equation on substitution of Ux_0t from Equation A2.6a and with some manipulations reduces to

$$U = \sqrt{\left[Ux_0^2 + Uy_0^2 + (2eV_0/md)x\right]} \qquad \text{(A2.7b)}$$

Let KE_0 is the kinetic energy at $t = 0$
At $t = 0$, $x = 0$ thus from Equation A2.7b:

$$KE_0 = \frac{1}{2}mU^2 = \frac{1}{2}m(Ux_0^2 + Uy_0^2) \qquad \text{(A2.8)}$$

If KE_t is the kinetic energy at any arbitrary time 't' its value can be written as

$$KE_t = \frac{1}{2}m\{Ux_0^2 + Uy_0^2 + (2eV_0/md)x\} \qquad \text{(A2.9)}$$

The difference between kinetic energies (ΔKE) given by Equations A2.8 and A2.9 represents the energy acquired by the particle in time t. It equals

$$\Delta KE = \frac{1}{2}m(2eV_0/md)x = (eV_0/d)x \qquad \text{(A2.10)}$$

The potential energy of an electron at any displacement x is given by

$$-eV = -e(V_0/d)\,x = -\Delta KE \qquad \text{(A2.11)}$$

The minus sign in Equation A2.11 indicates that any increase in KE is compensated by the decrease in potential energy.

Now if the initial velocities are set to be zero at $t = 0$, that is ($U_{x0} = U_{Y0} = 0$)

$$U = \sqrt{(2eV/m)} \qquad \text{(A2.12)}$$

Appendix A3: Solution to Equation 11.25

Equation 11.25 is reproduced and renumbered as below:

$$-\frac{e}{m}E_r = \frac{dU_r}{dt} - U_\phi \frac{d\phi}{dt} \qquad (A3.1a)$$

$$-\frac{e}{m}E_\phi = U_r \frac{d\phi}{dt} + \frac{dU_\phi}{dt} \qquad (A3.1b)$$

$$-\frac{e}{m}E_z = \frac{dU_z}{dt} \qquad (A3.1c)$$

$$U = U_r a_r + U_\phi a_\phi + U_z a_z = \frac{dr}{dt} a_r + \frac{rd\phi}{dt} a_\phi + \frac{dz}{dt} a_z \qquad (A3.1d)$$

Since $d\phi/dt = \omega$, $U_\phi = r\omega$, Equation A3.1 gets modified to

$$-\frac{e}{m}E_r = \frac{d^2r}{dt^2} - r\omega^2 \qquad (A3.2a)$$

$$-\frac{e}{m}E_\phi = \omega \frac{dr}{dt} + \frac{d(\omega r)}{dt} = \frac{1}{r}\frac{d}{dt}(r^2\omega) \qquad (A3.2b)$$

$$-\frac{e}{m}E_z = \frac{d^2z}{dt} \qquad (A3.2c)$$

In this case the behaviour of a charged particle in the E-field is illustrated through a cylindrical diode shown in Figure 11.6. The two concentric cylinders with radii 'a' and 'b' are maintained at $V = 0$ and $V = V_0$ voltages, respectively. Since this case is similar to that of the coaxial cable which yields the following relation for potential for the given configuration:

$$V = V_0 \frac{\ln r/a}{\ln b/a} \qquad (A3.3a)$$

$$Er = -\frac{\partial V}{\partial r} = -V_0 \frac{1}{r\ln b/a} \qquad (A3.3b)$$

and

$$E_\phi = E_z = 0 \tag{A3.3c}$$

In view of Equation A3.3b, Equation A3.2 leads to

$$-\frac{eV_0}{mr\ln(b/a)} = k/r = \frac{d^2r}{dt^2} - r\omega^2 \tag{A3.4a}$$

$$d(r^2\omega)/dt = 0 \tag{A3.4b}$$

and

$$d^2z/dt^2 = 0 \tag{A3.4c}$$

Assume that an electron 'e' enters the E-field at time $t = 0$ at a physical location $r = a$, $\phi = 0$ and $z = 0$ with initial velocity $U = 0$ thus $z = 0$ for all t from Equation A3.2c. Also from Equation A3.2b let $r^2\omega = r \cdot r\omega = r \cdot U_\phi = A$. Since $U = 0$ at $t = 0$, $A = 0$ or $\omega = 0$ for all values of t.

Now Equation A3.2a can be written as

$$\frac{d^2r}{dt^2} = \frac{dU_r}{dt} = k/r \tag{A3.5a}$$

As $U_r = dr/dt$ thus $dt = dr/U_r$

$$dU_r = (k/r)dt = (k/r)(dr/U_r) \quad \text{or} \quad U_r dU_r = (k/r)dr \tag{A3.5b}$$

The integration of both sides of Equation A3.5b yields: $1/2U_r^2 = k\ln r + B$. In view of the conditions: $U_r = 0$ at $r = 0$, $B = -k\ln a$, where $\{k = (eV_0)/(m\ln(b/a)\}$ from Equation A3.4a

$$\frac{1}{2}U_r^2 = k\ln(r/a), \quad \text{or} \quad U_r^2 = 2k\ln(r/a) \quad \text{or} \quad U_r = \sqrt{\{2k\ln(r/a)\}} \tag{A3.6}$$

This finally yields $U = U_r \, a_r$, where,

$$U_r = \frac{dr}{dt} = \sqrt{\frac{2eV_0\ln(r/a)}{m\ln(b/a)}} \tag{A3.7a}$$

On substituting the values of e and m into Equation A3.7a,

$$U_r = \left[5.932 \times 10^5 \sqrt{\frac{\ln(r/a)}{\ln(b/a)}} \right] \sqrt{V_0}$$

(A3.7b)

At $r = b$, Equation A3.7b reduces to

$$U = 5.932 \times 10^5 \sqrt{V_0}$$

(A3.8)

Appendix A4: Solution to Equations 11.29 and 11.30

Equations 11.28 and 11.29 are reproduced and renumbered as below:

$$\frac{-e}{m}(U_y B_z - U_z B_y) = \frac{dU_x}{dt} \tag{A4.1a}$$

$$\frac{-e}{m}(U_z B_x - U_x B_z) = \frac{dU_y}{dt} \tag{A4.1b}$$

$$\frac{-e}{m}(U_x B_y - U_y B_x) = \frac{dU_z}{dt} \tag{A4.1c}$$

$$\frac{d^2x}{dt^2} = \frac{-e}{m}\left(B_z \frac{dy}{dt} - B_y \frac{dz}{dt}\right) \tag{A4.2a}$$

$$\frac{d^2y}{dt^2} = \frac{-e}{m}\left(B_x \frac{dz}{dt} - B_z \frac{dx}{dt}\right) \tag{A4.2b}$$

$$\frac{d^2z}{dt^2} = \frac{-e}{m}\left(B_y \frac{dx}{dt} - B_x \frac{dy}{dt}\right) \tag{A4.2c}$$

Assume that a uniform magnetic field $B = B_0 a_z$ is the only field present (i.e., $B_x = B_y = 0$). Further an electron is assumed to enter the field at $t = 0$ at $x = y = z = 0$ with an initial velocity $U = U_{y0}\, a_y$. As $U_z = 0$, $z = 0$ for all times. Also $U_x = 0$ at $t = 0$. In view of these assumptions Equation A4.2 reduces to

$$\frac{d^2x}{dt^2} = -\frac{e}{m}B_0 \frac{dy}{dt} \tag{A4.3a}$$

or

$$\frac{dy}{dt} = -\frac{m}{eB_0}\frac{d^2x}{dt^2} \tag{A4.3b}$$

$$\frac{d^2y}{dt^2} = \frac{e}{m}B_0 \frac{dx}{dt} \tag{A4.3c}$$

or

$$\frac{dx}{dt} = \frac{m}{eB_0} \frac{d^2y}{dt^2} \qquad (A4.3d)$$

and

$$\frac{d^2z}{dt^2} = 0 \qquad (A4.3e)$$

Substitution of Equation A4.3b into Equation A4.3c and Equation A4.3d into Equation A4.3a leads to

$$\frac{d^2y}{dt^2} = -\frac{m}{eB_0} \frac{d^3x}{dt^3} = \frac{eB_0}{m} \frac{dx}{dt} \qquad (A4.4a)$$

$$\frac{d^2x}{dt^2} = \frac{m}{eB_0} \frac{d^3y}{dt^3} = -\frac{eB_0}{m} \frac{dy}{dt} \qquad (A4.4b)$$

Equations A4.4a and A4.4b can be further manipulated to the forms:

$$\frac{d^2U_x}{dt^2} = -\left(\frac{eB_0}{m}\right)^2 U_x \quad \text{or} \quad \frac{d^2U_x}{dt^2} + \omega_0^2 U_x = 0 \qquad (A4.5a)$$

$$\frac{d^2U_y}{dt^2} = -\left(\frac{eB_0}{m}\right)^2 U_y \quad \text{or} \quad \frac{d^2U_x}{dt^2} + \omega_0^2 U_x = 0 \qquad (A4.5b)$$

where

$$\omega_0 = \frac{eB_0}{m} \qquad (A4.5c)$$

The solution to Equations A4.5a and A4.5b can be written as

$$U_x = A_1 \cos \omega_0 t + B_1 \sin \omega_0 t \qquad (A4.6a)$$

$$U_y = A_2 \cos \omega_0 t + B_2 \sin \omega_0 t \qquad (A4.6b)$$

Since at $t = 0$, $U_x = 0$, $A_1 = 0$ and Equation A4.6a reduces to

$$U_x = B_1 \sin \omega_0 t \qquad (A4.7a)$$

Also at $t = 0$, $U_y = U_{y0}$, $A_2 = U_{y0}$ and Equation A4.6b reduces to

$$U_y = U_{y0} \cos \omega_0 t + B_2 \sin \omega_0 t \qquad \text{(A4.7b)}$$

Now Equations A4.3a and A4.3c are rewritten as below:

$$\frac{d^2x}{dt^2} = -\frac{e}{m}B_0\frac{dy}{dt} \quad \text{or} \quad \frac{dU_x}{dt} = -\frac{e}{m}B_0 U_y = -\omega_0 U_y \qquad \text{(A4.8a)}$$

$$\frac{d^2y}{dt^2} = \frac{e}{m}B_0\frac{dx}{dt} \quad \text{or} \quad \frac{dU_y}{dt} = \frac{e}{m}B_0 U_x = \omega_0 U_x \qquad \text{(A4.8b)}$$

Substitution of Equations A4.7a and A4.7b into Equations A4.8a and A4.8b gives

$$\frac{dU_x}{dt} = -\omega_0 U_y \Rightarrow \omega_0 B_1 \cos \omega_0 t = -\omega_0(U_{y0} \cos \omega_0 t + B_2 \sin \omega_0 t) \quad \text{(A4.9a)}$$

$$\frac{dU_y}{dt} = \omega_0 U_x \Rightarrow -\omega_0 U_{y0} \sin \omega_0 t + \omega_0 B_2 \cos \omega_0 t = \omega_0 B_1 \sin \omega_0 t \quad \text{(A4.9b)}$$

Equations A4.9a and A4.9b are true for all times including $t = 0$. Thus from Equation A4.9 at $t = 0$, $B_1 = -U_{y0}$ and $B_2 = 0$, Equations A4.7a and A4.7b reduce to

$$U_x = -U_{y0} \sin \omega_0 t \qquad \text{(A4.10a)}$$

$$U_y = U_{y0} \cos \omega_0 t \qquad \text{(A4.10b)}$$

$$U = \sqrt{(U_x^2 + U_y^2)} = U_{y0} \qquad \text{(A4.10c)}$$

From Equation A4.10

$$U_x = dx/dt = -U_{y0} \sin \omega_0 t \Rightarrow x = (U_{y0}/\omega_0) \cos \omega_0 t + C_1 \qquad \text{(A4.11a)}$$

$$U_y = dy/dt = U_{y0} \cos \omega_0 t \Rightarrow y = (U_{y0}/\omega_0) \sin \omega_0 t + C_2 \qquad \text{(A4.11b)}$$

Since at $t = 0$, $x = 0$ and $y = 0$, $C_1 = -U_{y0}/\omega_0$ and $C_2 = 0$.
Thus,

$$x = (U_{y0}/\omega_0)(\cos \omega_0 t - 1) \qquad \text{(A4.12a)}$$

and

$$y = (U_{y0}/\omega_0) \sin \omega_0 t \tag{A4.12b}$$

Equations A4.12a and A4.12b are the parametric equations of a circle with radius 'a' where 'a' is given by

$$a = U_{y0}/\omega_0 = U/\omega_0 = m\, U/(eB_0) \tag{A4.13}$$

The center of the circle lies at $x = -a/2, y = a/2$.

The energy of the particle shall remain unchanged as expected in the case of constant (steady) magnetic field. The linear velocity of the particle bears a relation with the angular velocity which can be obtained from Equation A4.13

$$U = a\omega_0 \tag{A4.14a}$$

Or

$$U = aeB_0/m \tag{A4.14b}$$

And the radius of path is

$$a = mU/eB_0 \tag{A4.14c}$$

From Equations A4.14a and A4.14b the cyclotron angular frequency of the circular motion of electron is

$$\omega_0 = U/a = eB_0/m \tag{A4.14d}$$

The period (T) for one complete revolution can be given by

$$T = 2\pi/\omega_0 = 2\pi m/eB_0 \tag{A4.14e}$$

Appendix A5: Solution to Equation 11.42

Equation 11.42 is reproduced and renumbered as below:

$$\frac{d^2x}{dt^2} = -\frac{e}{m}\left[-\frac{V_0}{d}(1 + \alpha\cos\omega t) + B_0\frac{dy}{dt}\right] \tag{A5.1a}$$

$$\frac{d^2y}{dt^2} = -\frac{e}{m}\left(-B_0\frac{dx}{dt}\right) \tag{A5.1b}$$

and

$$\frac{d^2z}{dt^2} = 0 \tag{A5.1c}$$

Equation A5.1c gives that $z = 0$ for all times.
Let

$$\omega_0 = eB_0/m \quad\text{or}\quad B_0 = m\omega_0/e \tag{A5.2a}$$

and

$$eV_0/(md) = k \tag{A5.2b}$$

$$\frac{d^2x}{dt^2} = k(1 + \alpha\cos\omega t) - \omega_0\frac{dy}{dt} \tag{A5.3a}$$

$$\frac{d^2y}{dt^2} = \omega_0\frac{dx}{dt} \tag{A5.3b}$$

$$\frac{dU_x}{dt} = k(1 + \alpha\cos\omega t) - \omega_0 U_y \tag{A5.4a}$$

$$\frac{dU_y}{dt} = \omega_0 U_x \tag{A5.4b}$$

From Equations A5.4a and A5.4b

$$U_x = \frac{1}{\omega_0}\frac{dU_y}{dt} \tag{A5.5a}$$

$$U_y = \frac{k}{\omega_0}(1 + \alpha \cos \omega t) - \frac{1}{\omega_0}\frac{dU_x}{dt} \tag{A5.5b}$$

Differentiation of both sides of Equation A5.5a with time and use of Equation A5.4a gives

$$\frac{dU_x}{dt} = \frac{1}{\omega_0}\frac{d^2U_y}{dt^2} = k(1 + \alpha \cos \omega t) - \omega_0 U_y \tag{A5.6a}$$

Differentiation of both sides of Equation A5.5b with time and use of Equation A5.4b gives

$$\frac{dU_y}{dt} = -k\alpha\frac{\omega}{\omega_0}\sin \omega t - \frac{1}{\omega_0}\frac{d^2U_x}{dt^2} = \omega_0 U_x \tag{A5.6b}$$

From Equations A5.6a and A5.6b

$$\frac{d^2U_x}{dt^2} + k\alpha\omega \sin \omega t + \omega_0^2 U_x = 0 \tag{A5.7a}$$

$$\frac{d^2U_y}{dt^2} + \omega_0 U_y - k\omega_0(1 + \alpha \cos \omega t) = 0 \tag{A5.7b}$$

Solution of Equation A5.7a can be written as

$$U_x = A_1 \cos \omega_0 t + B_1 \sin \omega_0 t + C_1 \sin \omega t \tag{A5.7c}$$

Substitution of Equation A5.7c into Equation A5.7a gives

$$C_1 = \frac{\alpha k \omega}{\omega^2 - \omega_0^2} \tag{A5.8}$$

Thus,

$$U_x = A_1 \cos \omega_0 t + B_1 \sin \omega_0 t + \frac{\alpha k \omega}{\omega^2 - \omega_0^2}\sin \omega t \tag{A5.9a}$$

At $t = 0$, $U_x = 0$ thus

$$A_1 = 0 \quad \text{and} \quad U_x = B_1 \sin \omega_0 t + \frac{\alpha k \omega}{\omega^2 - \omega_0^2}\sin \omega t \tag{A5.9b}$$

In view of Equations A5.5b and A5.9b, U_y can be rewritten as

$$U_y = \frac{k}{\omega_0}(1 + \alpha\cos\omega t) - \frac{1}{\omega_0}\left\{B_1\omega_0\cos\omega_0 t + \frac{\alpha k\omega^2}{\omega^2 - \omega_0^2}\cos\omega t\right\}$$

$$= \frac{k}{\omega_0}\left(1 - \frac{\alpha\omega_0^2}{\omega^2 - \omega_0^2}\cos\omega t\right) - B_1\cos\omega_0 t \qquad (A5.10a)$$

At $t = 0$, $U_y = 0$ thus

$$B_1 = \frac{k}{\omega_0}\left(1 - \frac{\alpha\omega_0^2}{\omega^2 - \omega_0^2}\right) \qquad (A5.10b)$$

Substitution of value of B_1 from Equation A5.10b into Equations A5.9b and A5.10a gives

Thus,

$$U_x = \frac{k}{\omega_0}\left(1 - \frac{\alpha\omega_0^2}{\omega^2 - \omega_0^2}\right)\sin\omega_0 t + \frac{\alpha k\omega}{\omega^2 - \omega_0^2}\sin\omega t = \frac{dx}{dt} \qquad (A5.11a)$$

$$U_y = \frac{k}{\omega_0}\left\{1 - \left(1 - \frac{\alpha\omega_0^2}{\omega^2 - \omega_0^2}\cos\omega_0 t\right)\right\} - \frac{\alpha k\omega^2}{\omega^2 - \omega_0^2}\cos\omega t = \frac{dy}{dt} \qquad (A5.11b)$$

Finally, integration of Equations A5.11a and A5.11b leads to the expressions of x and y

$$x = \frac{k}{\omega_0^2}\left[\left(1 - \frac{\alpha\omega_0^2}{\omega^2 - \omega_0^2}\right)\cos\omega_0 t - \frac{\alpha\omega_0^2}{\omega^2 - \omega_0^2}\cos\omega t\right] \qquad (A5.12a)$$

$$y = \frac{k}{\omega_0^2}\left[\omega_0 t - \left(\left(1 - \frac{\alpha\omega_0^2}{\omega^2 - \omega_0^2}\right)\sin\omega_0 t\right) - \frac{\omega_0}{\omega}\frac{\alpha\omega_0^2}{\omega^2 - \omega_0^2}\sin\omega t\right] \qquad (A5.12b)$$

Index

A

AC, *see* Alternating current (AC)
Accutune magnetron, 480
Active reflector, 319–320
Aerial, *see* Antenna
A.F/D.C. substitution, 367–368
Agile
  excursion, 481
  frequency-agile magnetron, 409,
      480–481
  rate, 481
Alternating current (AC), 166
Ammonia maser, 568
Amplification; *see also* High field
      domain
  factor, 600
  process, 454–456
Amplifier; *see also* Parametric amplifiers
  BWAs, 407
  BWCFA, 409, 486–487
  CFA, 485
  degenerate parametric, 565
  forward wave crossed-field, 409
  FWAs, 407
  FWCFA, 409, 485–486
  multi-cavity klystron, 424–425
  mutual conductance in klystron, 422
  non-degenerative parametric
      amplifier, 562
  voltage gain of klystron, 423
Amplitron, *see* Backward wave crossed-
      field amplifier (BWCFA)
Antenna, 317; *see also* Microwave
      antenna; Reflector antenna
  bandwidth, 319
  characteristic parameters, 318
  efficiency, 318
  impedance equivalence, 317
  temperature, 319
  theorems, 317–318
Antenna arrays, 333
  broadside array, 335
  collinear array, 336

  coupling between elements, 334
  curved-shaped array, 334
  director, 337
  end-fire array, 336
  field patterns, 335
  parasitic arrays, 336–337
  radiating elements, 334
  rectangular planar array, 334
  reflector, 337
  2D planar arrays, 334
  for unidirectional radiation, 334
Antiferromagnetic materials, 164
  dipoles alignment, 165
Anti-transmit-receive (ATR), 271
Applegate diagram, 419
Approximate electrostatic solution,
      669–672
Arcing, 485
Array, 333–334
Atoms, 163
  electron spin, 166
  revolving and spinning electrons,
      164
ATR, *see* Anti-transmit-receive (ATR)
Attenuation, 51
  in circular waveguide, 120
  due to conductor loss, 664–665,
      668–669, 693–694, 697
  constant, 19, 96; *see also* Wave
      propagation
  constant for TE and TM waves, 103,
      119
  due to dielectric loss, 668, 696–697
  factor consideration, 213–214
  measurement of, 365–369
  in stripline and microstrip, 673
  variation of, 108
  variation with frequency, 52–53
  in walls of parallel plane guide,
      51–53
Attenuators, 259; *see also* Microwave
      components
  coaxial cutoff, 261
  fixed, 259

Attenuators, (*Continued*)
   flap-type variable, 260
   rotary vane, 261
   scattering matrices for, 303–304
   variable, 260–261
   waveguide-type variable, 260
Avalanche; *see also* BARITT diode;
        IMPATT diode; Read diodes;
        TRAPATT diode
   breakdown voltage, 542–543
   current, 549
   diode, 507
   multiplication, 542
   transit time devices, 6, 542
Axial electric field, 457

B

B, *see* Base (B); Common base (B)
Back-illuminated CCD (BI CCD), 646
Backward wave, 20
   coupler, 687
   M-type cross-field oscillator, 489
Backward wave amplifiers (BWAs), 407
Backward wave crossed-field amplifier
        (BWCFA), 409, 486–487
Backward wave crossed-field oscillator
        (BWCFO), 409
Backward wave oscillator (BWO),
        6, 453, 487, 489; *see also*
        Magnetron; Microwave tube
   applications and performance
        parameters, 490–491
   backward-wave *M*-type cross-field
        oscillator, 489
   bandwidth of, 488
   circular *M*-carcinotron oscillator, 489
   linear *M*-carcinotron oscillator, 489,
        490
   O-type, 488–489
Band-pass filters (BPFs), 691, 692
Bandwidth (BW), 385
   antenna, 319
   in backward-wave crossed-field
        amplifier, 486
   of BWO, 488
   in coupled-cavity structure, 444
   in different feed methods, 339
   of dispersive lens, 328
   enhanced, 689
   to increase, 242, 426
   input/output coupling impact, 449
   load voltage expression, 385
   lumped element resonators, 703
   of maser, 569
   to measure low level of signals, 367
   of MSAs, 340
   in multiple quarter-wave section
        couplers, 687
   of NRPA, 565, 566
   of parametric amplifier up
        converter, 564
   product limitation, 385
   of PUC, 566
   quarter-wave section, 88
   of rectangular waveguide, 233
   silicon BARITT amplifier, 556
   in TWT, 440
   in voltage tunable magnetron, 479
BARITT diode, 554
   advantages and limitations, 556
   applications, 556
   critical voltage, 556
   M-n-M diode, 555
   operation, 555
   power level comparison, 556
   structures, 555
Barraters, 354
Base (B), 579
Bathe-hole coupler, 250; *see also*
        Directional couplers
Bayonet Neill–Concelman connectors
        (BNC connectors), 343
BCCD, *see* Buried channel CCD (BCCD)
Beam
   coupling coefficient, 417
   current, 420–421, 433
   efficiency, 318
   voltage, 428
   width, 318
B-field, 394
BI CCD, *see* Back-illuminated CCD (BI
        CCD)
Bi-directional couplers, 252; *see also*
        Directional couplers
Bias-circuit oscillation mode, 534–535
Bifurcated waveguides, 234; *see also*
        Waveguides

Bipolar junction transistors (BJTs), 579, 581, 601; *see also* Field effect transistors (FETs); Heterojunction bipolar transistor (HBT)
  biasing, 584
  common base configuration, 583
  common collector configuration, 583, 584
  common emitter configuration, 583, 584
  current–frequency limitation, 589–590
  dimensional parameters, 582
  equivalent model, 585–587
  fabrication, 582, 591
  limitations of, 589
  operation modes, 584–585
  power–frequency limitation, 590
  power gain frequency limitation, 590
  transistor assessment parameters, 588
  transistor characteristics, 587–588
  transistor structures, 582
  voltage–frequency limitation, 589
Bipolar Static Random Access Memory (BSRAM), 636
BJTs, *see* Bipolar junction transistors (BJTs)
BNC connectors, *see* Bayonet Neill–Concelman connectors (BNC connectors)
Bolometer, 353
  barraters, 354
  mounts, 355
  thermistor, 354, 355
BPFs, *see* Band-pass filters (BPFs)
Branch guide couplers, 252–253; *see also* Directional couplers
Branch-line
  coupler, 689
  filter, 692
Breakdown voltage, 601
Brewster's angle, 39
Bridge method
  impedance measurement, 370
  power measurement, 371
Brillouin diagram, 451–454
Broadside arrays (BSAs), 335

BSAs, *see* Broadside arrays (BSAs)
BSRAM, *see* Bipolar Static Random Access Memory (BSRAM)
Buncher cavity, *see* Input cavity
Bunching, 405, 417, 418, 431, 449; *see also* Klystron
  Applegate diagram, 419
  beam current, 420–421, 433
  current modulation, 417, 433
  current-modulated electron beam, 433
  distance between buncher grid and bunching location, 418
  efficiency of klystron, 422
  electron bunching diagram, 432
  electronic admittance, 435
  equivalent circuit, 435–436
  impact on signal output, 420
  induced current, 421
  input and output voltages, 421
  mutual conductance of klystron amplifier, 422
  output equivalent circuit, 422
  output power, 421–422
  parameter, 420
  power output, 434
  power required for, 423
  quality factor, 424
  during retarding phase, 420
  round-trip transit angle, 433
  transit time, 420
  voltage gain of klystron amplifier, 423
Buried channel CCD (BCCD), 646
Bus bar, 89
Butterfly stub, 685
BW, *see* Bandwidth (BW)
BWAs, *see* Backward wave amplifiers (BWAs)
BWCFA, *see* Backward wave crossed-field amplifier (BWCFA)
BWCFO, *see* Backward wave crossed-field oscillator (BWCFO)
BWO, *see* Backward wave oscillator (BWO)

C

C, *see* Collector (C)
CAD, *see* Computer-aided design (CAD)

Cadmium telluride (CdTe), 541
  diode, 523
Calledean *H*-plane tee, 246
Capacitance, 600, 672
  applied voltage and capacitance
    variation, 522
  impact of, 383–385
  inter-electrode, 382–383
  junction, 508
  variation, 522
  voltage across non-linear, 557, 558
  –voltage relationship, 523
Capacitive gap stripline filter, 692
Capacitive loading, 468
Capacitive tuning, 469, 470–471
Capacitors, 680; *see also* Inductors;
    Lumped element realisation
  inter-digital, 680–681, 716
  metal-oxide-metal, 715–716
  planar film, 715
  single-transistor DRAM cell with
    storage, 637
  woven, 681
Cartesian coordinate system, 390
  electron movement in parallel plate
    diode, 390–391, 723–725
  motion in electric, magnetic and ac
    field, 398, 735–737
  motion in magnetic field in, 393–395,
    731–734
Cassegrain antenna, 323–324
Catcher cavity, *see* Output cavity
Cathode ray tube (CRT), 407
Cathode sputtering, 712
Cavity
  formation, 144–145
  resonance method, 368–369
  shapes, 140
  tuner, 413
  types, 140–141, 468
Cavity resonators, 139; *see also* Cavity;
    Dielectric resonator; Resonant
    circuit
  advantages and applications,
    155–156
  capacitance, 143
  capacitive tuning, 155
  characteristic impedance of coaxial
    cable, 142
circular cavity resonator, 148–150
coupled cavity and its equivalent,
    152
coupling mechanism, 153, 154
cylindrical, 145
descriptive questions, 160–161
examples, 158–160
exercises, 160
fields in cavity resonators, 145–146
fixed cavities, 156
geometry of, 146
group velocity, 147
inductance, 143
inductive tuning, 155
input impedance to coaxial line, 143
mode degeneracy, 148
mode designations, 150
physical length of cavity, 147
quality factor, 150–153
rectangular, 144, 146–148
reentrant cavities, 142–143
resonant frequency, 143–144
resonant length, 148
semi-circular cavity resonator, 150
tunable cavities, 156
tuning methods, 154–155
variation of amplitude of *E*, 151
variation of standing wave ratio, 153
volume tuning, 155
Cavity wave meter (CWM), 352
CB, *see* Conduction band (CB)
CC, *see* Common collector (CC)
CCD, *see* Charge-coupled device (CCD)
CCNR, *see* Current-controlled negative
    resistance (CCNR)
CdTe, *see* Cadmium telluride (CdTe)
CE, *see* Common emitter (CE)
CFA, *see* Crossed-field amplifiers (CFA)
Channel resistance, 598
Characteristic impedance, 56, 125, 188,
    663–664, 668, 688
Charge-coupled device (CCD), 642
  advantages, 648
  applications, 648
  charge transfer efficiency, 646–647
  depletion region, 643, 644
  dynamic range, 647, 648
  energy band diagram, 643
  frequency response, 647

linearity, 648
potential well, 643, 644
power dissipation, 647
quantum efficiency, 647
spectral ranges, 647
structure and working, 642–646
types of, 646
Charge storage diode, *see* Step-recovery
diode (SRD)
Charge transfer device (CTD), 642
Charge transfer efficiency, 646–647
Cheese antenna, 322–323
Circuit; *see also* Microwave integrated
circuit (MIC); Printed circuit
board (PCB)
bias-circuit oscillation mode,
534–535
for different electrical lengths and
terminations, 65
in different modes, 536
diode symbols, 506
efficiency, 475
elements, 691
equation, 457
equivalence of reflex klystron,
435–436
equivalence of waveguides, 103
multi-element circuit onto smith
chart, 210–211
open-circuit condition, 516
planar line, 684
symbols for n- and p-channels, 613,
614
of transmission line, 57
Circularly polarised wave, 25, 26; *see
also* Polarisation
Circular *M*-carcinotron oscillator, 489
Circular spiral inductor, 714
Circular waveguide (CWG), 74, 232,
109; *see also* Waveguides; Wave
propagation
attenuation in, 119, 120
characteristic parameters for TE and
TM waves, 117
cutoff frequencies, 119
example, 135
exercises, 136
field components, 111
field distribution, 118

field expressions, 114–116
flanges, 235
geometry of, 109
governing equations in differential
form, 110
input–output coupling, 243
Maxwell's equations, 109
mode designation, 117
mode excitation, 119
phase shifter, 268
solution of wave equation, 112–114
TE and TM modes in, 118
TE waves, 112
TEM modes in, 116
TM waves, 111
wave equations, 109
Circulators, 173; *see also* Non-reciprocal
ferrite devices
cascaded, 178
four-port, 175–178, 302
insertion loss, 174
isolation, 174
scattering matrices for, 301
three-port, 173–175, 302–303
CLA, *see* Collinear array (CLA)
Clover-leaf stub, 685
CMBR, *see* Cosmic microwave
background radiation (CMBR)
CMOS, *see* Complementary MOSFETs
(CMOS)
Coaxial
detector mount, 354
magnetron, 408, 477–478
Coaxial cables, 71; *see also* Wave
propagation
end views and field distributions, 73
example, 82
fixed attenuators, 259
parameters, 72
relative spacing in, 74
Coaxial line, 661–661
tuners, 266; *see also* Tuners
Collecting aperture, *see* Effective
aperture
Collector (C), 413, 579
Collinear array (CLA), 336
Common base (B), 579–580
Common collector (CC), 580
Common emitter (CE), 579

Complementary MOSFETs (CMOS),
        626, 629
    advantages, 633
    analog, 633
    applications, 633
    inverter micro-fabrication, 716–718
    operation, 631–632
    power dissipation, 632–633
    structures, 630–631
Complementary-symmetry metal-
        oxide-semiconductor (COS-
        MOS), 629, *see* Complementary
        MOSFETs (CMOS)
Computer-aided design (CAD), 703
Conducting lens, *see* WG-type lens
Conducting materials, 706–707
Conduction band (CB), 516, 524
Conductivity and current density, 526
Contact feeds, 338, 339
Convection current, 456
Conventional diodes, 506; *see also*
        Microwave diodes
    avalanche diode, 507
    circuit symbols, 506
    forward voltage drop, 507
    high-frequency limitations, 508
    junction capacitance, 508
    laser diodes, 508–509
    leakage current, 508
    light-emitting diodes, 508
    maximum forward current, 507
    peak inverse voltage, 507
    photodiodes, 508, 509
    PN junction diodes, 506
    zener diode, 507
Conventional transistor, 579; *see also*
        Microwave transistor
    limitations of, 580
Conventional tube limitations, 6, 381
    bandwidth product limitation, 385
    dielectric loss, 388
    electrostatic induction, 387–388
    inter-electrode capacitances, 382–383
    lead inductance, 383–385
    output-tuned circuit of pentode, 385
    skin effect, 387
    transit time effect, 386–387
Cookie-cutter tuner, 471
Coplanar strips, 676–677

Coplanar waveguides, 660
    drawback, 676
    types of, 675–676
Copper clad board, 708
Corner reflectors, 324–325
Cosmic microwave background
        radiation (CMBR), 6
COS-MOS, *see* Complementary-
        symmetry metal-oxide-
        semiconductor (COS-MOS)
Coupled-cavity structure, 444
Coupled microstrip, 660
Coupled opencircuited lines, 684
Coupled shortcircuited lines, 684
Coupler realisation, 685
    backward wave coupler, 687
    branch-line coupler, 689
    characteristic impedances, 688
    lange coupler, 688–689
    microstrip couplers, 687
    multiple quarter-wave section
        couplers, 687
    planar format coupler, 687
    proximity couplers, 686
    quarter-wave section couplers, 686
    rat-race coupler, 690
    short section couplers, 686–687
    single-section λ/4 directional
        coupler, 686
Coupling, 154
    coefficient, 253
    between elements, 334
Coupling factor, *see* Coupling—
        coefficient
Covered microstrip, 666, 667
Crossed-field amplifiers (CFA), 485
Crossed-field device, *see* Magnetron
Crown-of-thorns tuner, *see* Sprocket
        tuner
CRT, *see* Cathode ray tube (CRT)
Crystal, *see* Point-contact diode
    diode, 353
    mount, 354
CSWR, *see* Current standing wave ratio
        (CSWR)
CTD, *see* Charge transfer device (CTD)
Current
    density, 525
    gain parameter, 588

modulated electron beam, 433
modulation, 405–406, 417, 433
Current-controlled negative resistance (CCNR), 528
Current standing wave ratio (CSWR), 60, 189
Current–voltage (I–V), 587
  of GaAs MESFET, 604
  of HEMT amplifier, 624
  of MOSFETs, 613
  of n-channel MOSFET, 615
  of n–p–n transistor, 587
  of Si MESFET, 608
Curved-shaped array, 334
Cutoff
  field, 463
  frequency, 125
  mode, 618
  operation, 585
CWG, *see* Circular waveguide (CWG)
CWM, *see* Cavity wave meter (CWM)
Cycloidal path, 400
Cyclotron
  angular frequency, 395, 471, 473–474
  -frequency magnetron, 408
  resonance heating, 13
Cylindrical diode, 392
  cylindrical coordinate system, 391
  electron movement in, 392–393, 727–729
Cylindrical magnetron, 409
Cylindrical paraboloid antenna, 322

D

D, *see* Directivity (D)
DARS, *see* Digital Audio Radio Service (DARS)
DC, *see* Direct current (DC)
DCs, *see* Discrete circuits (DCs)
DD CCD, *see* Deep depletion CCD (DD CCD)
DE, *see* Driven element (DE)
Debunching, 438–439
Deep depletion CCD (DD CCD), 646
Degenerate parametric amplifiers, 565
Delay lenses, 331, 332
Dematron, 409
Depletion mode HEMT (D-HEMT), 623

Depletion-load NMOS logic, 627
Depletion-mode device (DMD), 620
Depletion region, 643, 644
Deposition, 711; *see also* Epitaxial growth technique
  cathode sputtering, 712
  electron beam evaporation, 711
  vacuum evaporation, 711
Depth of penetration, 26–27
  example, 75, 76
Device under test (DUT), 361
D-HEMT, *see* Depletion mode HEMT (D-HEMT)
Diamagnetic materials, 163, 164
Dielectric constant measurement, 369–370
Dielectric-loaded waveguides, 233–234; *see also* Waveguides
Dielectric loss, 129, 388
Dielectric materials, 707
Dielectric resonator, 156; *see also* Cavity resonators
  advantages, 158
  applications, 158
  resonant frequency, 157
  resonant modes, 157
Dielectric resonator antenna (DRA), 158
Dielectric resonator oscillator (DRO), 156
Dielectric sheet thickness measurement, 377
Dielectric waveguides, 119; *see also* Waveguides
  Maxwell's equations, 122–123
  rod waveguide, 123–124
  slab waveguide, 120–123
  types of, 120
  zigzag path of wave, 124
Differential negative resistance region, 527, 529–531
Digital Audio Radio Service (DARS), 11
Digital Video Broadcasting–Satellite services to Handhelds (DVB-SH), 11
Diode, 382; *see also* Conventional diodes
Diplexers, 271; *see also* Microwave components
  formation, 272

Dipole moments, 163
    alignments, 165
    electron spin, 166
    unmagnetised and magnetised
        material, 164
Direct current (DC), 166
Directional couplers, 249; *see also*
        Microwave components;
        Waveguide terminations
    backward, 250
    bathe-hole coupler, 250
    branch guide couplers, 252–253
    broadband applications, 255
    coupling coefficient, 253
    cross-guide, 251
    directivity, 255
    double-hole coupler, 251
    examples, 273–274
    exercises, 275–276
    figures of merit, 253–255
    formation of, 249
    forward, 250
    insertion loss, 254
    isolation, 254
    Moreno cross-guide coupler, 251
    multi-hole, 252, 298
    powers at different ports, 253
    scattering matrices for, 298–301
    Schwinger reversed-phase coupler,
        251–252
    short slot couplers, 252–253
    top wall couplers, 252–253
    unidirectional and bi-directional
        couplers, 252
Directional patterns equality, 317
Directive gain, 318
Directivity (D), 255
Director, 337
Discrete circuits (DCs), 701
Dish antenna, *see* Paraboloidal reflector
Dish reflector, *see* Paraboloidal reflector
Dispersive lens, 328
Dissipation factor, 22
Dither magnetron, 480
Dither-tuned magnetron, *see*
        Frequency—agile magnetron
DMD, *see* Depletion-mode device
        (DMD)
Dominant mode, 124; *see also* Modes

Doubled stubs, 685
Double-hole coupler, 251; *see also*
        Directional couplers
DRA, *see* Dielectric resonator antenna
        (DRA)
Drain
    conductance, 599
    current, 598
    resistance, 600
    saturation current, 589
    saturation voltage, 600
DRAM, *see* Dynamic Random Access
        Memories (DRAM)
Drift of current pulses, 549, 550
Drift velocity, 525
    curve for GaAs, 527
Driven element (DE), 336
Driving point, *see* Terminal impedance
DRO, *see* Dielectric resonator oscillator
        (DRO)
Duplexers, 270–271; *see also* Microwave
        components
DUT, *see* Device under test (DUT)
DVB-SH, *see* Digital Video
        Broadcasting–Satellite services
        to Handhelds (DVB-SH)
Dynamic Random Access Memories
        (DRAM), 634, 636–638

E

E, *see* Emitter (E)
E wave, *see* Transverse magnetic waves
        (TM waves)
ECM, *see* Electronic counter measure
        (ECM)
EEPROM, *see* Electrically Erasable
        Programmable Read-Only
        Memory (EEPROM)
EFA, *see* End-fire array (EFA)
Effective aperture, 318–319
Effective dielectric constant, 667,
        694–695
Effective length, 318
Effective width, 664, 692–693
E-HEMT, *see* Enhancement mode
        HEMT (E-HEMT)
EH tee, *see* Magic tee
EH tuner, 264–265; *see also* Tuners

EHT, *see* Extra high tension (EHT)
Electrically Erasable Programmable Read-Only Memory (EEPROM), 634, 635–636
Electric field
  distribution, 672
  on electrons, 403–404
  intensity measurement of, 373–374
Electron
  beam evaporation, 711
  bunching diagram, 432
  drift velocity, 526
  gun, 404, 413
  transfer between two valleys, 525
Electron motion, 471–472
  in *B* field, 462
  in combined *E* and *B* field, 462
  in cartesian coordinate system, 393–395, 731–734
  in cylindrical diode, 392–393, 727–729
  in *E* field, 461
  in electric, magnetic and ac field, 398, 735–737
  in magnetic field in, 393–395, 731–734
  in parallel plate diode, 390–391, 723–725
Electron-resonance magnetron, 409, 483–484
Electronic admittance, 435
  spiral, 437
Electronic circuit categories, 701
Electronic counter measure (ECM), 459
Electronic efficiency, 434, 475
Electrostatic coupling, 241
Electrostatic induction, 387–388
ELF, *see* Extremely low frequency (ELF)
Elliptically polarised wave, 25, 26; *see also* Polarisation
Elliptical waveguides, 232; *see also* Waveguides
Ellipticity, 24; *see also* Polarisation
Embedded microstrip, 666, 667
EMD, *see* Enhancement-mode device (EMD)
Emitter (E), 579
  current gain, 588
  efficiency, 588

field of *B*, ac and *E*, 398
  and *H* plane radiation patterns, 376
Empty state (ES), 517
End-coupled filter, 692
End-fire array (EFA), 336
Energy band diagram, 592
  built in voltages, 593
  electric flux, 594
  for n-Ge and p-GaAs, 592, 593
  tapering of bandgap, 594
Energy extraction, 406
Enhancement-mode device (EMD), 620
Enhancement mode HEMT (E-HEMT), 523
Enhancement modes, 617
  saturation mode, 619
  sub-modes, 618
  triode mode, 618–619
Epitaxial growth technique, 710; *see also* Deposition
  liquid-phase epitaxy, 710–711
  molecular-beam epitaxy, 710
  vapour-phase epitaxy, 710
E-plane tee, 245; *see also* Waveguide junctions
  scattering matrices for, 292–294
EPROM, *see* Erasable Programmable Read-Only Memory (EPROM)
Equality
  of directional patterns, 317
  of effective length, 318
Equivalent noise, *see* Antenna—temperature
Erasable Programmable Read-Only Memory (EPROM), 634, 635
ES, *see* Empty state (ES)
Extended interaction klystron, 412; *see also* Klystron
Extra high tension (EHT), 450
Extremely high frequency (EHF), 1, 2
  division of, 4
Extremely low frequency (ELF), 2

F

Faraday's rotation, 167–168; *see also* Circulators; Gyrators; Isolators
FB, *see* Forward bias (FB)

Feed systems, 323
Feeding element, *see* Driven element
    (DE)
FeRAM, *see* Ferroelectric RAM
    (FeRAM)
Ferrimagnetic materials, 164; *see also*
    Ferrites
  dipoles alignment, 165
Ferrites, 165; *see also* Ferrimagnetic
    materials; Non-reciprocal
    ferrite devices
  advantage of, 165
  attenuators, 178–179
  descriptive questions, 182–183
  device figures of merit, 180
  dipoles alignment, 165
  electron spin, 166
  examples, 181–182
  exercises, 182
  Faraday's rotation, 167–168
  isolator, 171
  magnetic flux density, 166
  permeability tensor, 166–167
  phase shifter, 178
  switches, 179
  YIG filters, 179
Ferroelectric RAM (FeRAM), 634, 640
Ferromagnetic substances, 163
  dipoles alignment, 165
  flux density values, 164
  magnetic moments, 164
FI CCD, *see* Front-illuminated CCD (FI
    CCD)
Field distribution, 391
  in microstrip lines, 659
Field effect CCD, *see* Charge-coupled
    device (CCD)
Field effect on electron motion, 388; *see*
    also Motion in electric field;
    Motion in magnetic field
  cartesian coordinates, 398–402
  in combined field, 398, 402
  cycloidal path, 400
  in cylindrical cavity magnetron, 402
  cylindrical Coordinates, 402–403
  E-, *B*- and ac field, 398
  in electric and magnetic field,
    397–398
  Lorentz force equation, 389

Field effect transistors (FETs), 6,
    581, 595; *see also* Bipolar
    junction transistors (BJTs);
    Junction field effect
    transistor (JFET)
  vs. BJTs, 601–602
  depletion mode FET, 597
  enhancement mode FET, 597
  low-noise FETs, 596
  meandered gate, 596
  modulation-doped FETs, 622
  power FETs, 596
  Schottky-barrier-gate cutoff
    frequency, 608
  switch FETs, 596
  transconductance, 604
Field patterns, 335
Filled state (FS), 517
Filters, 690–691
Fin lines, 660, 678
Flash memories, 638
  programming, 639
Flexible waveguides, 234; *see also*
    Waveguides
FM, *see* Frequency modulation (FM)
Focussing, 446–447
  magnet, 413
Forward bias (FB), 515, 584
Forward wave, 20
Forward wave amplifiers (FWAs), 407
Forward wave crossed-field amplifier
    (FWCFA), 409, 485–486
Four phase CCD, 646
Free space wave number, 19; *see also*
    Wave propagation
Frequency
  -agile coaxial magnetron, 480–481
  agile magnetron, 409
  measurement, 360–361
  pulling, 469
  pushing, 469
  -sensitive antennas, 333
Frequency meters, 350, 351
  calibration chart for, 351
  cavity wave meter, 352
Frequency modulation (FM), 3
Front-illuminated CCD (FI CCD), 646
Front-to-back ratio, 319
FS, *see* Filled state (FS)

FWAs, *see* Forward wave amplifiers
    (FWAs)
FWCFA, *see* Forward wave crossed-
    field amplifier (FWCFA)

G

Gain, 318
    coefficient, 19; *see also* Wave
        propagation
    measurement, 375–376
Gallium arsenide, 705
Gallium nitride transistors (GaN
    transistors), 705
GaN transistors, *see* Gallium nitride
    transistors (GaN transistors)
Gap factor, 476
Gate metallisation, 624
Global positioning satellite (GPS), 12
Global System for Mobile (GSM), 11
GP, *see* Ground plane (GP)
GPS, *see* Global positioning satellite
    (GPS)
Ground plane (GP), 337
Group velocity, 51, 125, 147
    variations, 127–128
GSM, *see* Global System for Mobile
    (GSM)
Guide wavelength, 125
Gunn diode, 537
    current fluctuations, 538
    mode, 534, 535
    peak power levels, 554
    relative aspects of IMPATT diode
        and, 552
Gunn effect, 537
Gyrators, 169; *see also* Non-reciprocal
    ferrite devices
    with input–output ports rotated by
        90°, 169–170
    90° with twist, 170
    scattering matrices for, 304
    with twist at waveguide input end,
        170

H

H wave, *see* Transverse electric waves
    (TE waves)

HBTs, *see* Hetero-junction bipolar
    transistors (HBTs)
Hetero-junction bipolar transistors
    (HBTs), 579, 591, 592; *see also*
    Bipolar junction transistors
    (BJTs)
    advantages, 594–595
    applications, 595
    energy band diagrams, 592–594
    materials, 591–592
HF, *see* High frequency (HF)
HICs, *see* Hybrid integrated circuits
    (HICs)
High electron mobility transistors
    (HEMTs), 579, 621
    advantages, 626
    applications, 626
    current–voltage characteristics,
        623–624
    exercise, 655
    fabrication, 624–625
    materials used, 625
    operation, 623
    structure, 622–623
High field domain, 531–532
    bias-circuit oscillation mode, 534–535
    delayed mode, 534
    growth factor, 536
    LSA mode, 534
    mode classification criterion,
        535–537
    properties of, 533
    stable amplification mode, 534
    transit-time mode, 534
High frequency (HF), 2; *see also* Maser
    capacitance per unit length, 72
    circuits, 185, 279
    devices, 705
    E field, 464
    flow of current, 27
    heterodyne systems, 350
    inductance per unit length, 72
    insulator, 88
    magnetic field application, 163
    operation requirements, 591
    signal transfer, 66
    voltage tunable magnetron, 479
High temperature co-fired ceramic
    (HTCC), 707, 712

Hole-and-slot magnetron, 461
Horn, 256
    antennas, 325–326
Hot carrier diode, *see* Schottky diode
Hot electron diode, *see* Schottky diode
H-plane tee, 246; *see also* Waveguide
        junctions
    scattering matrices for, 288–291
HTCC, *see* High temperature co-fired
        ceramic (HTCC)
Hull cutoff
    magnetic equation, 472
    voltage equation, 472–473
Hybrid integrated circuits (HICs), 701,
        704–705
Hybrid ring, 248–249; *see also*
        Waveguide junctions
    scattering matrices for, 301, 302
Hybrid tee, *see* Magic tee
Hydrogen maser, 569–570
Hyper-abrupt junction, 522

I

Idler circuit, 563
IEEE, *see* Institute of Electrical and
        Electronics Engineers (IEEE)
IL, *see* Insertion loss (IL)
IMPATT diode, 547; *see also* Read
        diodes
    advantages and limitations, 551
    applications, 552
    avalanche current, 549
    construction, 548
    drift of current pulses, 549, 550
    layers and field distribution in, 549
    n^+ layer, 548
    operation, 547–548
    p^+ layer, 548
    performance, 550
    relative aspects of Gunn diode and,
        552
    time dependence of growth and
        drift of holes, 551
    transit time, 548
    waveforms of voltage and currents,
        550
    working, 548–550
Impedance chart, 197; *see also* Smith chart

Impedance matching, 261; *see also*
        Microwave components
    capacitive irises, 262, 263
    inductive irises, 261–262
    magic tee, 247–248
    parallel resonant diaphragm, 263
    posts and screws, 263–264
    resonant windows, 262
Impedance measurement, 370–371
    bridge method, 370
    reference minima, 371
    slotted line method, 370–371
    terminated transmission line, 371
Indium phosphide (InP), 705
    diode, 523, 540–541
Induced current, 421
Inductance, 383
Inductive tuning, 469, 470
Inductors, 679; *see also* Capacitors;
        Lumped element realisation
    ribbon, 679–680
    spiral, 680
Industrial, Scientific and Medical
        bands (ISM bands), 11
InP, *see* Indium phosphide (InP)
Input cavity, 413
Input–output coupling, 243
Insertion attenuation, *see* Insertion loss
        (IL)
Insertion loss (IL), 254
    A.F/D.C. substitution, 367–368
    cavity resonance method, 368–369
    measurement, 365
    power ratio method, 366–367
    RF substitution method, 367
    scattering coefficient method, 369
    substitution using linear detection,
        368
    voltage contents of waves and
        impedances, 366
Institute of Electrical and Electronics
        Engineers (IEEE), 3
    band designation, 3
Insulated-gate field effect transistor
        (IGFET), 611
Integrated circuit (IC); *see also*
        Microwave integrated circuit
        (MIC)
    silicon for, 705

Interconnect metallisation, 624
Inter-digital capacitor, 716, 680
Inter-electrode capacitances, 382–383
Interleaved transformer, 682
International telecommunication union
(ITU), 2
International Thermonuclear
Experimental Reactor
(ITER), 13
Inverse mode operation, 585
Inverted magnetron, 409
coaxial, 478–479
Iris, 261
capacitive, 262, 263
inductive, 261–262
ISM bands, *see* Industrial, Scientific and
Medical bands (ISM bands)
Isolation, 254
Isolators, 170; *see also* Non-reciprocal
ferrite devices
Faraday's rotation, 171–172
ferrite, 171
resonance, 172–173, 174
scattering matrices for, 304–305
ITER, *see* International Thermonuclear
Experimental Reactor (ITER)
ITU, *see* International
telecommunication union
(ITU)
I–V, *see* Current–voltage (I–V)

J

JCCD, *see* Junction CCD (JCCD)
JFET, *see* Junction field effect transistor
(JFET)
Junction CCD (JCCD), 646
Junction field effect transistor (JFET),
579, 595
amplification factor, 600
breakdown voltage, 601
capacitance, 600
channel resistance, 598
comparison between FETs and BJTs,
601
cutoff frequency, 600
drain conductance, 599
drain current, 598
drain resistance, 600

drain saturation current, 589
exercise, 650–652
I–V characteristics of, 599
knee voltage, 601
mutual conductance at saturation,
600
pinch-off current, 598
pinch-off region, 600–601
pinch-off voltage, 597–598
principles of operation, 597
saturation drain voltage, 600
structure, 595–596
transconductance, 599
types of microwave FETs, 596

K

KE, *see* Kinetic energy (KE)
Kinetic energy (KE), 472
Klystron, 411; *see also* Bunching; Reflex
klystron; Microwave tube;
Two-cavity klystron
amplifier mutual conductance, 422
amplifier voltage gain, 423
efficiency, 422
extended interaction klystron, 412
multi-cavity klystron, 412
reflex, 346
Klystron power supply (KPS), 343, 344
carrier wave operation, 345
front panel of, 344–345
mount, 346–347
operating procedure, 345
operation with modulation, 346
reflex klystron, 346
Knee voltage, 601
KPS, *see* Klystron power supply (KPS)

L

LAN, *see* Local area network (LAN)
Lange coupler, 688–689
Laser diodes (LDs), 508–509
Latch configuration, 636
Law of sines, 38
LDs, *see* Laser diodes (LDs)
Lead inductances, 383
and inter-electrode capacitance
impact, 383–385

LEDs, *see* Light-emitting diodes (LEDs)
Lens antennas, 327; *see also* Microwave
    antenna
  classification of, 328
  continuous and discontinuous
      surfaces, 329
  delay lenses, 331, 332
  dispersive lens, 328
  with feeds, 330
  loaded lens, 333
  non-dispersive lens, 328
  to prevent reflected energy entry
      in source, 330
  WG-type lens, 331
  zoned lens, 328, 332
Less-than-lethal weaponry, 13; *see also*
    Microwave
LF, *see* Low frequency (LF)
Light-emitting diodes (LEDs), 508
Limited space-charge accumulation
    diode (LSA diode), 523
  peak power levels achieved, 554
Linear
  beam tubes, 407
  magnetron, 476–477, 408
  M-carcinotron oscillator, 489, 490
Linearly polarised wave, 25; *see also*
    Polarisation
Line of sight (LOS), 5
Lines of force in π-mode, 474
Liquid-phase epitaxy, 710–711
Lithography, 711
LO, *see* Local oscillator (LO)
Loaded lens, 333
Local area network (LAN), 11
Local oscillator (LO), 562
Loop coupling, 242
Lorentz force equation, 389
LOS, *see* Line of sight (LOS)
Loss tangent, *see* Power—factor
Low frequency (LF), 2
Low-frequency conventional diodes,
    506
  avalanche diode, 507
  PN junction diodes, 506
  rectifier diodes, 506
  signal diodes, 506
  Zener diode, 507
Low-pass filters (LPFs), 691

Low temperature co-fired ceramic
    (LTCC), 707, 712
Lower valley (LV), 524
Lowest-order transverse resonance
    cutoff frequency, 674
LPFs, *see* Low-pass filters (LPFs)
LSA diode, *see* Limited space-charge
    accumulation diode (LSA
    diode)
LTCC, *see* Low temperature co-fired
    ceramic (LTCC)
Lumped element realisation, 678; *see
    also* Planar transmission lines;
    Power splitter realisation
  capacitors, 680–681
  inductors, 679–680
  monolithic transformers, 681–682
  resistors, 679
LV, *see* Lower valley (LV)

M

Magic tee, 246; *see also* Waveguide
    junctions
  impedance matching, 247–248
  input–output at different ports, 247
  scattering matrices for, 295–298
Magnetic coupling, 242
Magnetic field, 164
  axial, 455, 479
  beam spread prevention, 448
  in circular waveguide, 116
  in cyclotron, 473
  DC, 168, 460, 465
  effect of, 461–462
  on electron, 482
  electron wobbling, 178
  in Faraday's rotation isolator, 171
  to focus electrons, 446
  in Gyrator, 170
  in high-frequency application, 163
  intensity, 72, 373
  in loop coupling, 242
  in metal cavity, 157
  mode designations, 124
  motion in, 393, 397, 398
  motion in Cartesian coordinate
      system, 393–395, 731–734
  restoring force in electron, 447

rotating, 168–169
in slot line, 667
solenoids, 488
static, 165–167
for TEM mode, 71
Magnetic materials, 164
Magnetic-wall model, 673
Magneto-motive-force (m.m.f.), 97
Magnetron, 408, 460; *see also* Klystron;
      Magnetron working
      mechanism; Microwave tube
agile excursion, 481
agile rate, 481
arcing, 485
baking-in procedure, 485
capacitive loading, 468
capacitive tuning, 469, 470–471
cavity types, 468
circuit efficiency, 475
coaxial magnetron, 408, 477–478
construction, 460
coupling methods, 469
as crossed-field device, 460
cyclotron angular frequency, 473–474
cyclotron-frequency magnetron, 408
cylindrical magnetron, 409
electron motion, 471–472
electron-resonance magnetron,
    483–484, 409
electronic efficiency, 475
equivalent circuit, 474–475
frequency-agile coaxial magnetron,
    480–481
frequency agile magnetron, 409
frequency pulling and pushing, 469
gap factor, 476
hole-and-slot magnetron, 461
hull cutoff equations, 472
inductive tuning, 469, 470
inverted coaxial magnetron, 478–479
inverted magnetron, 409
linear magnetron, 476–477, 408
lines of force in π-mode, 474
mode jumping, 467
negative-resistance magnetron,
    482–483
oscillation modes, 466–467
performance parameters and
    applications, 484

quality factor, 475
split anode magnetron, 408
strapping, 467–469
travelling-wave magnetrons, 408
voltage-tunable magnetron, 409,
    479–480
Magnetron working mechanism, 461
combined field effect, 462–463,
    464–466
cutoff field, 463
electric field effect, 461
electron motion in E, B and
    combined field, 461, 462
electron path, 464
high-frequency E field, 464
magnetic field effect, 461–462
resonant circuit, 463
RF field, 463
Majority carrier device, *see* Schottky
    diode
Manley–Rowe power relations,
    558–561
Masers, 566–567
ammonia maser, 568
applications, 570–574
hydrogen maser, 569–570
levels of, 567
operation, 567
Ruby maser, 568–569
Mask ROM, *see* Read-only memory
    (ROM)
Matter, 163
Maxwell's equations, 17, 721
Medium, 19
Medium frequency (MF), 2
Memory devices, 633
BSRAM, 636
classification, 633
DRAM, 636–638
EEPROM, 635
EPROM, 635
ferroelectric RAM, 640
flash memories, 638–639
ovonics unified memory, 642
phase change RAM, 641–642
PROM, 634, 635
resistance RAM, 640–641
ROM, 634–635
SRAM, 636

Memory varactor, *see* Step-recovery
    diode (SRD)
MEMS, *see* Microelectromechanical
    systems (MEMS)
MESFET, *see* Metal semiconductor field
    effect transistor (MESFET)
Metal-insulator-semiconductor FET
    (MISFET), 611
Metallic insulator, 88, 89
Metal-oxide-metal capacitor, 715–716
Metal-oxide-semiconductor (MOS),
    626, *see* Metal-oxide-
    semiconductor field effect
    transistor (MOSFET)
Metal-oxide-semiconductor field effect
    transistor (MOSFET), 579,
    610; *see also* Complementary
    MOSFETs (CMOS); Insulated-
    gate field effect transistor
    (IGFET); Junction field effect
    transistor (JFET); Metal-
    insulator-semiconductor
    FET (MISFET); Metal
    semiconductor field effect
    transistor (MESFET);
    N-channel MOSFETs (NMOS);
    P-channel MOSFETs (PMOS)
  advantages and limitations, 620
  applications, 620
  vs. BJTs, 621
  body effect, 620
  channel conductance, 616
  channel length modulation
      parameter, 619
  circuit symbols, 613
  cutoff mode, *617*
  depletion mode HEMT, 623
  depletion-mode MOSFETs, 619
  drain current, 613, 615–616
  enhancement mode, 617
  equivalent circuit, 617
  I–V characteristics of n-channel, 614,
      615
  materials, 620
  maximum operating frequency, 617
  n-channel operation, 612
  non-idealised n-channel, 616
  n- or p-type, 610
  operation modes, 612–613

saturation current, 616
structure, 610–611
symbols, 614
transconductance, 616
Metal-oxide-semiconductor
    transistor, *see* Metal-oxide-
    semiconductor field effect
    transistor (MOSFET)
Metal semiconductor field effect
    transistor (MESFET), 579, 602;
    see also Field effect transistors
    (FETs); Junction field effect
    transistor (JFET)
  applications, 610
  characteristics, 606
  cutoff frequency, 608
  drain current, 607
  electron mobility, 622
  GaAs MESFET, 602, 603–604, 606,
      652–653
  gate–source capacitance, 609
  I–V characteristics, 607
  maximum oscillation frequency, 609
  operation principles, 603
  pinch-off voltage, 604
  Schottky-barrier diode's advantages
      and limitations, 610
  small-signal equivalent circuit,
      605–606
  structure, 602–603
MF, *see* Medium frequency (MF)
MIC, *see* Microwave integrated circuit
    (MIC)
Microelectromechanical systems
    (MEMS), 678
Micromachined lines, 678, 679
Microstrip, 660, 665; *see also* Striplines
  approximate electrostatic solution,
      669–672
  attenuation, 668–669, 672, 673,
      693–694, 696–697
  capacitance, 672
  characteristic impedance, 668
  couplers, 687
  covered microstrip, 666, 667
  design and analysis, 672–674
  disadvantage, 666
  effective dielectric constant, 667,
      694–695

electric field distribution, 672
embedded microstrip, 666, 667
field distribution, 659, 674
field patterns, 688
line geometry, 669
lowest-order transverse resonance
    cutoff frequency, 674
magnetic-wall model, 673
phase velocity, 667
propagation constant, 667
quasi-TEM microstrip line, 667
strip width, 668, 695–696
substrate thickness limit, 674
suspended, 674–675
Microstrip antenna (MSA), 337; *see also*
    Microwave antenna
applications, 340
configurations of, 338
contact feeds, 338, 339
demerits of, 340
merits of, 340
non-contacting feeds, 338, 339
radar cross section, 339
return loss, 339
Microsystems, *see*
    Microelectromechanical
    systems (MEMS)
Microwave, 1
advantages, 1, 3, 5
descriptive questions, 14
detectors, 352–353
devices and components, 7
dish, *see* Paraboloidal reflector
ferrite devices, 163; *see also* Ferrites;
    Non-reciprocal ferrite devices
frequency band, 2–3, 4
generation, 5–6
health hazards, 13–14
history, 1
IEEE band designation, 3
limitations, 2
new U.S. military microwave bands, 4
ovens, 12–13
printed circuit, 712
processing, 6
thick-film metals, 707
transistors, 6
transmission, 6, 8
tubes, 6

Microwave antenna, 9, 317; *see also*
    Antenna; Antenna arrays;
    Lens antenna; Microstrip
    antenna (MSA); Reflector
    antenna
exercise, 340–341
frequency-sensitive antennas, 333
horn antennas, 325–326
slot antennas, 326–327
Microwave applications, 9, 10, 12
communication, 9, 11
home appliances, 12–13
industrial heating, 13
navigation, 12
plasma generation, 13
radars, 11
radio astronomy, 11
spectroscopy, 13
weaponry system, 13
Microwave components, 231; *see*
    also Attenuators; Coupler
    realisation; Diplexers;
    Directional couplers;
    Duplexers; Impedance
    matching; Microwave
    antenna; Microwave filters;
    Microwave measurement;
    Phase shifters; Power
    splitter realisation; Tuners;
    Waveguide input–output
    methods; Waveguide
    junctions; Waveguide
    terminations; Waveguides
band-pass filters, 691, 692
butterfly stub, 685
circuit elements, 691
clover-leaf stub, 685
descriptive questions, 276–277
doubled stubs, 685
examples, 273–274
exercises, 275–276
filters, 690–691
low-pass filters, 691
mode suppressors, 271
opencircuited lines, 684
opencircuit stub, 684
radar system containing different,
    232
radial stub, 685

Microwave components, *(Continued)*
   realisation, 682
   resonators, 690
   shortcircuit line, 684
   shortcircuit stub, 684
   stripline element realisation, 682
   stub realisation, 683
   system components, 5
   transmission line realisation, 682
Microwave diodes, 6, 503, 509; *see also*
      Avalanche—transit time
      devices; Conventional diodes;
      Masers; Parametric devices;
      Point-contact diode; Schottky
      diode; Tunnel diode; Varactor
   exercise, 574–576
   mechanism, 505
   negative resistance significance,
      505–506
   semiconducting material properties,
      503–504
   step-recovery diode, 511–512
Microwave FETs types, 596
Microwave filters, 269; *see also*
      Microwave components
   band-pass filter, 269
   high-pass filter, 270
   low-pass filter, 269
   parallel resonant filter, 270
Microwave integrated circuit (MIC),
      701; *see also* Monolithic
      microwave integrated circuit
      (MMIC)
   copper clad board, 708
   exercises, 718–719
   hybrid ICs, 704–705
   limitations, 703–704, 718
   merits of, 702–703, 718
   photoimageable thick-film process,
      709
   planar resistor, 718
   thick-film fabrication, 708–709
   thin-film fabrication, 708
Microwave measurements, 8–9, 343; *see*
      also Impedance measurement;
      Klystron power supply (KPS);
      Voltage standing wave ratio
      (VSWR)
   device quality, 355–356

   dielectric constant measurement,
      369–370
   dielectric sheet thickness
      measurement, 377
   E and H plane radiation patterns,
      376
   electric field intensity measurement
      of, 373–374
   equipment components, 343
   exercise, 379
   frequency measurement, 360–361
   gain measurement, 375–376
   IL measurement, 365
   metallic sheet thickness
      measurement, 376–377
   moisture measurement, 377–378
   norms, 358–359
   phase shift measurement, 362–364
   power measurement, 371–373
   precautions, 356–358
   quality factor measurement, 364
   reflection coefficient measurement,
      374–375
   scattering parameters measurement,
      365
   transmission coefficient
      measurement, 375
   travelling wave detection, 348
   wavelength measurement, 359–360
   wire diameter measurement, 377
Microwave radiation, 6, 11; *see also*
      Maser
   CMBR, 6
   in corner-reflector antenna, 324–325
   direction of maximum radiation,
      335, 336
   for directive radiation, 326
   E and H Plane radiation patterns,
      376
   effects, 14
   intensity, 318
   losses, 130, 142, 156
   in microwave oven, 12–13
   pattern, 318, 323, 324, 333
   produced by, 337
   radiation-sensitive material, 711
   in radio astronomy, 11
   rearward, 328
   resistance, 318

in spectroscopy, 13
from sun, 6
for unidirectional, 334
Microwave transistor, 581; *see also*
      Bipolar junction transistors
      (BJTs); Charge-coupled
      device (CCD); Heterojunction
      bipolar transistor (HBT); High
      electron mobility transistors
      (HEMTs); Junction Field
      Effect Transistor (JFET);
      Memory devices; Metal-
      oxide-semiconductor field
      effect transistor (MOSFET);
      Metal semiconductor field
      effect transistor (MESFET);
      Transistors
  examples, 648–655
  exercises, 655–657
Microwave tube, 6, 411; *see also*
      Microwave tube; Conventional
      tube limitations; Field effect
      on electron motion; Klystron;
      Magnetron; Travelling
      wave tube (TWT); Velocity
      modulation
  basics, 381
  classification, 406
  dematron, 409
  examples, 491–498
  exercise, 409–410, 498–500
  linear beam tubes, 407
  wave crossed-field amplifier, 409,
    486–487
Mid-valley (MV), 524
m.m.f, *see* Magneto-motive-force
  (m.m.f.)
MMIC, *see* Monolithic microwave
  integrated circuit (MMIC)
M-n-M diode, 555
Modes, 124; *see also* Microwave
      components
  designations, 124
  disturbance, 126
  dominant, 124
  filters, 272–273
  jumping, 467
  propagation, 126–127
  spurious, 128

suppressors, 271
  transformation, 126
Moisture measurement, 377
  in liquids, 378
Molecular-beam epitaxy, 710
Monolithic microwave integrated
    circuit (MMIC), 701, 705–706;
    see also Deposition; Epitaxial
    growth technique; Planar
    film capacitor; Photo-resist
    technique
  CMOS inverter micro-fabrication,
    716–718
  conducting materials, 706–707
  dielectric materials, 707
  diffusion and ion implantation, 710
  etching and photo-resist, 711
  gallium arsenide for, 705
  lithography, 711
  oxidation and film deposition, 710
  processes used, 712
  resistive materials, 707
  substrate material for, 706
Monolithic transformers, 681; *see also*
    Lumped element realisation
  interleaved, 682
  stacked, 682
  tapped, 682
Moreno cross-guide coupler, 251; *see*
    also Directional couplers
MOS, *see* Metal-oxide-semiconductor
  (MOS)
MOSFET, *see* Metal-oxide-
  semiconductor field effect
  transistor (MOSFET)
Motion in electric field, 390
  cartesian coordinate system,
    390–391
  cylindrical coordinate system, 391
  in cylindrical diode, 392
  electron movement, 391, 392, 393
  field distribution, 391
Motion in magnetic field, 393
  B-field, 394
  cartesian coordinate system, 393
  cylindrical coordinate system,
    395–396
  motion movement, 394
  particle energy, 395

Movable shorts, 350
MSA, *see* Microstrip antenna (MSA)
Multi-cavity klystron, 412; *see also* Klystron
    amplifiers, 424–425
Multi-hole directional coupler, 252; *see also* Directional couplers
Multimode propagation, 126–127; *see also* Modes
Multiple quarter-wave section couplers, 687
Mutual conductance at saturation, 600
MV, *see* Mid-valley (MV)

N

NATO, *see* North Atlantic Treaty Organization (NATO)
N-channel
    depletion mode, 613
    enhancement mode, 612–613
N-channel MOSFETs (NMOS), 626, 627
    example, 654
    I–V characteristics, 628
    logic, 627
    operation, 628–629
    structures, 627–628
Negative differential conductivity effect, 528
Negative resistance (NR), 518, 528
    magnetron, 482–483
    parametric amplifier, 562, 564–565
    regions, 526
n+ Layer, 548
NMOS, *see* N-channel MOSFETs (NMOS)
Noise figure, 565
Non-contacting feeds, 338, 339
Non-degenerative parametric amplifier, 562
Non-dispersive lens, 328
Non-reciprocal ferrite devices, 168; *see also* Circulators; Ferrites; Gyrators; Isolators
Non-volatile read-write memories (NVRWM), 634
Normal mode operation, 585
North Atlantic Treaty Organization (NATO), 3

NR, *see* Negative resistance (NR)
NVRWM, *see* Non-volatile read-write memories (NVRWM)

O

OC, *see* Open circuit (OC)
OEIC, *see* Optoelectronic-integrated circuit (OEIC)
Ohmic contact formation, 624
One-sided abrupt junction, 522
One-sided linearly graded junction, 522
Open circuit (OC), 197
    stub, 684
Open-strip lines, *see* Microstrip
Opening gate windows, 624
Optoelectronic-integrated circuit (OEIC), 340
Orange-peel paraboloid antenna, 322
Oscillation
    modes, 466–467
    prevention, 459
Oscillator, 426; *see also* Degenerate parametric amplifiers
    backward wave crossed-field oscillator, 409
    O-type BWOs, 488–489
Output cavity, 413
Output power, 421–422
    gain, 458
Output-tuned circuit, 385
Ovonics unified memory, 642

P

Parabolic dish, *see* Paraboloidal reflector
Parabolic reflector, *see* Paraboloidal reflector
Paraboloidal reflector, 320, 321
Parallel plane guide, 40; *see also* Wave propagation
    attenuation in walls of, 51–53
    configuration of two parallel planes, 41
    example, 78
    Maxwell and wave equations, 41–42
    unguided waves, 40
Parallel plate magnetron with *E*-, *B*- and ac field, 398

Parallel-coupled filter, 692
Paramagnetic materials, 163
   magnetic moments, 164
   relative values of flux densities, 164
Parametric amplifiers, 562; *see also*
      Oscillator
   bandwidth, 565
   degenerate, 565
   equivalent circuit of, 563
   idler circuit, 563
   negative resistance, 564–565
   noise figure, 565
   non-degenerative parametric
      amplifier, 562
   PDC and PUC, 563, 564
Parametric devices, 556; *see also*
      Parametric amplifiers
   classification of, 561
   down converter, 562
   Manley–Rowe power relations,
      558–561
   NR parametric amplifier, 562
   power gain, 521
   relative merits, 566
   solid-state varactor diode, 557
   sum-frequency parametric
      amplifier, 562
   voltage across non-linear
      capacitance, 557, 558
Parametric down converter (PDC), 563,
   564
Parametric up converter (PUC),
   563–564
Parasitic arrays, 336–337
Passive reflector, 319
PCB, *see* Printed circuit board (PCB)
p-Channel depletion mode, 613
p-Channel enhancement mode, 613
P-channel MOSFETs (PMOS), 626, 627
   exercise, 653–654
   logic, 629
PDC, *see* Parametric down converter
   (PDC)
PDN, *see* Pull-down network' (PDN)
PDs, *see* Photodiodes (PDs)
Pentode, 382
Periodic permanent magnet (PPM), 448
Phase change RAM (PCRAM), 641–642,
   634

Phase shifters, 267; *see also* Microwave
      components
   circular waveguide, 268–269
   dielectric vane 267, 268
   linear, 268
   line stretcher, 267
Phase shift measurement, 362–364
Phase velocity, 21, 125, 663, 667
Photodiodes (PDs), 508, 509
Photoimageable thick-film process, 709
Photo-resist technique; *see also*
      Monolithic microwave
      integrated circuit (MMIC)
   circular spiral inductor, 714
   planar resistor, 712–713, 718
   ribbon inductor, 714
   round-wire inductor, 714, 718
   single turn flat circular loop
      inductor, 714
   square spiral inductor, 715, 718
Pillbox antenna, 322
Pinch-off
   current, 598
   region, 600–601
   voltage, 597–598, 604
PIN diode, *see* P-type—Intrinsic—N-
      type diode (PIN diode)
Planar film capacitor, 715; *see also*
      Photo-resist technique
   exercise, 718
   interdigitated capacitor, 716
   metal-oxide-metal capacitor, 715–716
Planar format coupler, 687
Planar resistor, 712–713, 718
Planar transmission lines, 659; *see also*
      Lumped element realisation;
      Microstrip; Microwave
      components; Striplines
   coaxial line, 661–661
   coplanar strips, 676–677
   coplanar waveguides, 660, 675–676
   coupled microstrip, 660
   examples, 692–698
   fin lines, 660, 678
   micromachined lines, 678, 679
   slot line, 660, 677
Plane wave, 20
   front, 320
p^+ Layer, 548

PMOS, *see* P-channel MOSFETs
        (PMOS)
PN junction diodes, 506
Point-contact diode, 509; *see also*
        Schottky diode
    construction, 509–510
    operation, 510–511
Polar chart, 189; *see also* Smith chart
    complex reflection coefficient, 191
    domain of reflection coefficient, 189
    equivalent normalised impedance,
        191
    normalised resistance and
        reactance, 190
Polarisation, 24, 319; *see also* Reflection;
        Wave propagation
    circularly polarised wave, 25, 26
    elliptically polarised wave, 25, 26
    ellipticity, 24
    forms of, 36
    linearly polarised wave, 25
    orientation, 24
    parallelly polarised wave, 36
    perpendicularly polarised wave, 36
    with reference to earth surface, 26
    rotation, 168
    twisters, 324
    of uniform plane wave, 25
Polytetra-fluoro-ethylene (PTFE), 706
Potential well, 643
Power
    bolometer bridge method of
        measurement, 371
    calorimetric method of
        measurement, 372
    factor, 23
    gain, 318, 521
    output, 434
    ratio method, 366–367
Power splitter realisation, 683; *see also*
        Lumped element realisation
    T-junction power splitter, 683, 685
    Wilkinson power splitter, 683, 684,
        685
Poynting theorem, 27–29
Poynting vector, 28
    components of, 29
PPM, *see* Periodic permanent magnet
        (PPM)

Printed circuit board (PCB), 8, 337
Probe coupling, 241–242
Processor, 6
Programmable read-only memory
        (PROM), 634, 635
PROM, *see* Programmable read-only
        memory (PROM)
Propagation constant, 18, 663, 667; *see
        also* Wave propagation
Proximity couplers, 686
PTFE, *see* Polytetra-fluoro-ethylene
        (PTFE)
P-type—Intrinsic—N-type diode (PIN
        diode), 512, 513
    applications, 515
    construction, 512–513
    doping profile and field distribution,
        513
    equivalent circuit, 514, 515
    operation, 514
PUC, *see* Parametric up converter (PUC)
Pull-down network (PDN), 627

Q

Quality factor, 108–109, 475
    measurement, 364
Quantum efficiency, 647
Quarter-wave section couplers, 686
Quasi-TEM, *see* Quasi-transverse
        electric and magnetic
        (Quasi-TEM)
Quasi-TEM microstrip line, 667
Quasi-transverse electric and magnetic
        (Quasi-TEM), 666

R

Radar, 11; *see also* Microwave
    waveguide system for, 231, 232
Radar cross section (RCS), 339
Radial stub, 685
Radiating elements, 334
Radio astronomy, 11; *see also* Microwave
Radio frequencies (RFs), 167, 512
    field, 463
    signals, 5
    substitution method, 367
Rat-race coupler, 690

Rat race tee, *see* Hybrid ring
Ray, 20
RB, *see* Reverse bias (RB)
RCS, *see* Radar cross section (RCS)
Read diodes, 543; *see also* IMPATT diode
  distribution and doping profile in,
    544
  electric field reduction, 546
  operation, 543
  output power, 546
  quality factor, 547
  resonant frequency in cavity, 546
  voltage and currents in, 545
Read-only memory (ROM), 633,
    634–635
Read-write memories (RWM), 633
Rectangular planar array, 334
Rectangular waveguide, 90; *see also*
    Waveguides
  attenuation, 96, 107–108
  boundary conditions, 93
  characteristic impedance variation,
    102
  characteristic parameters for TE and
    TM waves, 103
  circuit equivalence of, 103–106
  cutoff frequencies, 97, 103
  cutoff wave number, 95
  equivalent circuits of TE and TM
    waves, 105
  example, 130–134
  excitation of modes, 100–101
  exercises, 135–136
  field distribution for modes, 98–100
  fixed attenuators, 259
  flanges, 235
  geometry of, 91
  Maxwell's equations, 91–92
  phase shift constant, 97
  possible and impossible modes, 97–98
  power transmission in, 106–107
  probe coupling, 241
  quality factor, 108–109
  TE wave, 96, 98, 99, 103
  TM wave, 93–95, 100, 103
  wave behaviour with frequency
    variation, 96–97
  wave equations, 92
  wave impedances, 101

Rectifier diodes, 506
Reentrant cavities, 142–143; *see also*
    Cavity resonators
Reference minima, 371
Reflection, 29–31; *see also* Polarisation;
    Wave propagation
  cases of reflection, 32–33
  example, 77
  normal incidence cases, 33–35
  oblique incidence, 36–40
Reflection coefficient, 30, 188; *see also*
    Smith chart
  complex, 191
  domain of, 189
  measurement, 374–375
Reflector, 320, 337
Reflector antenna, 319; *see also* Lens
    antennas; Microwave antenna
  active reflector, 319–320
  cassegrain antenna, 323–324
  cheese antenna, 322–323
  corner reflectors, 324–325
  cylindrical paraboloid, 322
  feed systems, 323
  orange-peel paraboloid, 322
  paraboloidal reflector, 320, 321
  passive reflector, 319
  pillbox, 322
  polarisation twisters, 324
  reflector, 320
  spherical reflector, 323
  transreflector, 324
  truncated paraboloid, 321–322
  wave front, 320
Reflex klystron, 346, 411–412, 426, 429;
    see also Bunching; Klystron
  applications, 439–440
  constructional features, 427
  debunching, 438–439
  electric field, 429
  electronic admittance spiral, 437
  electronic efficiency of, 434
  functional diagram, 428
  operation modes, 436–438
  power sources required, 428
  transit time, 437
  tuning, 439, 440
  velocity modulation, 428–431
Refraction coefficient, 30

Resistance RAM (RRAM), 634, 640
    V–I characteristics, 641
Resistive materials, 707
Resonant cavity, *see* Cavity resonators
Resonant circuit, 139, 463, 543; *see*
        also Cavity resonators;
        Transmission lines
    alternating voltage effect, 405
    cavity as, 463
    inductance of, 470
    lumped component, 142
    plasma in, 553
    quality factor, 150
    RLC, 140
    unloaded quality factor, 475
Resonant frequency in cavity, 546
Resonators, 690
Return loss, 339
Reverse bias (RB), 515, 584
    gate-to-channel junction, 608
RFs, *see* Radio frequencies (RFs)
Ribbon inductors, 679–680, 714
Ridged waveguides, 233; *see also*
        Waveguides
Ridley–watkins–helsum theory, 528
Ring-bar structure, 445
Ring-loop structure, 444–445
ROM, *see* Read-only memory (ROM)
Round-trip transit angle, 433
Round-wire inductor, 714, 718
RRAM, *see* Resistance RAM (RRAM)
Ruby maser, 568–569
RWM, *see* Read-write memories (RWM)

S

Satellite-Digital Multimedia
        Broadcasting (S-DMB), 11
Saturation drain voltage, 600
Saturation mode operation, 585
SC, *see* Short circuit (SC)
Scattering coefficient method, 369
Scattering matrices, 288; *see also*
        Scattering parameters
        (*s* parameters)
    attenuator, 303–304
    circulators, 301
    directional coupler, 298–301
    EH tee, 295–298

E-plane tee, 292–294
gyrator, 304
H-plane tee, 288–291
hybrid ring, 301, 302
isolator, 304–305
order and nature, 281
phase shift property, 282
properties, 281
symmetry property, 281
unity property, 281
zero property, 281–282
Scattering parameters (*s* parameters),
        279; *see also* Scattering
        matrices; Scattering transfer
        parameters (*T* parameters);
        2-Port network
    cascaded networks, 286
    coefficients, 309
    descriptive questions, 314–315
    examples, 312–314
    exercises, 314
    large-signal, 287
    lossless network, 287
    lossy network, 287
    measurement, 365
    mixed-mode, 288
    network nature, 287
    for networks with different ports,
        282
    N-port network, 280, 286
    1-port network, 282
    pulsed, 288
    small-signal, 287
    and Smith chart, 307–311
    3-port network, 285, 286
    types of, 287
Scattering transfer parameters
        (*T* parameters), 311; *see*
        also Scattering parameters
        (*s* parameters)
    advantage, 311
    matrix, 311
    to S-parameters, 312
    S-parameters to, 312
SCCD, *see* Surface channel CCD (SCCD)
Schottky diode, 518, 519, 610; *see also*
        Point-contact diode
    advantages, 520
    applications, 521

characteristics, 520
construction, 519
fabrication, 519
with guard ring, 520
limitations, 520
NPN transistor with, 520
Schwinger reversed-phase coupler,
251–252; *see also* Directional
couplers
S-DMB, *see* Satellite-Digital Multimedia
Broadcasting (S-DMB)
Selective dry etching, 624
Self-impedance, 319
Semiconducting material properties,
503–504
Semiconductor (SC); *see also* Microwave
integrated circuit (MIC)
memories, *see* Memory devices
regions, 579
Semiconductor devices, 503; *see also*
Microwave diodes
energy band diagram, 504
material properties, 503
Series tee, *see* E-plane tee
SHF, *see* Super-high frequency (SHF)
Short circuit (SC), 197
line, 684
stub, 684
Short section couplers, 686–687
Short slot couplers, 252–253; *see also*
Directional couplers
Shunt tee, *see* H-plane tee
Si-chip storage capacity, 637, 638
Signal diodes, 506
Signal growth, 449
Signal-to-noise ratio (SNR), 319
Silicon, 705
Silicon–germanium SC technology, 705
Simply crossed-field amplifier, *see*
Forward wave crossed-field
amplifier (FWCFA)
Single
-layer stacked-capacitor cell, 638
-section $\lambda/4$ directional coupler,
686
stub matching, 217, 219
-transistor DRAM cell with storage
capacitor, 637
turn flat circular loop inductor, 714

Skin effect, 129, 387
SLF, *see* Super low frequency (SLF)
Slide screw tuner, 265; *see also* Tuners
Slot
antennas, 326–327
coupling, 243
line, 660, 677
Slotted line method, 370–371
Slotted transmission line, 349
Slow wave structures, 442
Smith chart, 185; *see also* Impedance
chart; Polar chart; Reflection
coefficient; Stub matching;
Transmission lines
for admittance mapping, 195–197,
201
advantages of, 203–204
attenuation factor, 213–214
characteristic impedance, 188
characteristics, 185
circuit parameters mapped on, 211
descriptive questions, 229
d_{max} and d_{min} evaluation, 206–207
equivalence of movement on
transmission line, 202–203
examples, 222–227
exercises, 228
Γ (d) evaluation, 204–206
impedance/admittance inversion,
202
for impedance mapping, 191–195,
201
information imparted by, 197
for lossless transmission lines,
204
multi-element circuit mapping,
210–211
normalised impedances, 197–198,
199, 212
reflection and transmission
coefficients, 198, 199
rotation by 180°, 198
two half-wave peripheral scales,
200–201
voltage standing wave ratio, 200
VSWR and Γ circles, 200
VSWR evaluation, 207–208
wavelength scales on, 201
$Y(d)$ evaluation, 208–210

Snap-off' diode, *see* Step-recovery
    diode (SRD)
Snell's Law, *see* Law of sines
SNR, *see* Signal-to-noise ratio (SNR)
Solid
    angle, 318
    -state varactor diode, 557
s parameters, *see* Scattering parameters
    (*s* parameters)
Spatial frequency, *see* Free space wave
    number
Spherical reflector, 323
Spherical wave front, 320
Spiral inductors, 680
Split anode magnetron, 408; *see*
    also Negative resistance
    (NR)—magnetron
Sprocket tuner, 470
Spurious modes, 128; *see also* Modes
Square spiral inductor, 715, 718
SRAM, *see* Static Random Access
    Memory (SRAM)
SRD, *see* Step-recovery diode (SRD)
Stable amplification mode, 534
Stacked transformer, 682
Staggered tuning, 426
Standing wave, 21
    ratio, 60
Static Random Access Memory
    (SRAM), 636
Step-recovery diode (SRD), 505, 511
    applications, 512
    construction, 511–512
    operation, 512
Strapping, 467–469
Striking wave, 30
Striplines, 660, 661; *see also* Microstrip
    attenuation, 664–665, 672, 673
    characteristic impedance, 663–664
    constructional details of, 662
    effective width, 664, 692–693
    electric field distribution, 672
    element realisation, 682
    geometry of, 665
    hairpin filter, 692
    inter-digital filter, 692
    and lumped-element equivalents,
        683
    parallel-coupled lines filter, 692

phase velocity, 663
propagation constant, 663
strip width, 664, 668, 695–696
stub filter, 692
types of, 663
Stub matching, 214; *see also* Smith chart
    conjugate matching problems, 222
    double stub matching
        implementation, 220–221
    entry on Smith chart, 215
    example, 227–228
    impedance rotation impact, 217, 218
    main-line signal phase changing,
        215–216
    open-circuited stubs, 214
    phase delays, 216
    requirements for proper matching,
        216–217
    single stub matching, 217, 219–220
    stub type and length selection,
        217–218
    susceptance variation, 215, 216
Stub realisation, 683
Substrate material for MIMIC, 706
Substrate thickness limit, 674
Sub-threshold, *see* Cutoff—mode
Sum-frequency parametric amplifier, 562
Super low frequency (SLF), 2
Super paramagnetic materials, 164
Super-high frequency (SHF), 1, 2
    division of, 4
Surface barrier diode, *see* Schottky
    diode
Surface channel CCD (SCCD), 646
Surface impedance, 27
Suspended microstrip, 674–675
Synchronous tuning, 425

T

Tapped transformer, 682
TCR, *see* Temperature coefficient of
    resistance (TCR)
Television (TV), 2
Temperature coefficient of resistance
    (TCR), 706
Tera-hertz (THz), 74
Tera-hertz frequency (THF), 2
Terminal impedance, 319

Terminated transmission line, 371
Terminating impedance, 56
Tetrode, 382
TE waves, *see* Transverse electric waves
        (TE waves)
Thermistor, 354, 355
THF, *see* Tera-hertz frequency (THF)
Thick film, 712
        fabrication, 708–709
Thickness measurement, 376–377
Thin film, 712
        fabrication, 708
Three phase CCD, 646
Three-cavity klystron, 424; *see also*
        Klystron
THz, *see* Tera-hertz (THz)
Time-invariant field, 721
Time-variant field, 721
T-junction power splitter, 683, 685
TM waves, *see* Transverse magnetic
        waves (TM waves)
Top wall couplers, 252–253; *see also*
        Directional couplers
T parameters, *see* Scattering transfer
        parameters (*T* parameters)
TR, *see* Transmit-receive (TR)
Transconductance, 599
Transferred electron devices (TEDs),
        6, 505, 523, 591; *see also* Gunn
        diode; High field domain;
        Valley band structure
    advantages, 541
    applications, 541
    cadmium telluride diodes, 541
    indium phosphide diodes, 540–541
    LSA diode, 538–541
Transistors, 579; *see also* Microwave
        transistor
    p–n–p and n–p–n, 580, 583
Transit time, 386, 420, 437, 548
        effect, 386–387
Transmission coefficient, 30
Transmission lines, 53, 185; *see*
        also Resonant circuit;
        Smith chart; Transverse
        electromagnetic waves (TEM
        waves); Waveguides; Wave
        propagation
    applications of transmission lines, 66

    characteristic impedance, 56
    characteristic parameters of
        uniform, 185–189
    circuit representation of, 57
    complex impedance, 60–61
    condition for maximum power
        transfer, 186
    configuration of, 53
    descriptive questions, 86
    double stub matching, 70
    end views and field distributions, 73
    equations governing, 54–56
    equation solutions, 56–58
    example, 80
    exercises, 84–85
    impact of frequency change, 89–90
    as harmonic suppressors, 68–69
    impedance transformation, 67
    impedance transformer, 67
    input impedance of, 57
    limitations of, 73
    lossless RF and UHF lines, 58
    method of VSWR measurement, 361
    quarter-wave auxiliary line, 88
    realisation, 682
    reflection phenomena, 61–62
    single-stub matching, 69–70
    standing wave ratio, 60
    for stub matching, 69–70
    terminating impedance, 56
    as tuned circuit, 66
    tuned line, 66–67
    two-wire, 185, 186
    types, 54, 70–71
    voltage transformer, 67–68
    waveguide and, 87–89
    wave voltage and current phasors,
        187
Transmit-receive (TR), 271
Transport factor, 588
Transreflector, 324
Transverse electric waves (TE waves),
        42; *see also* Transverse waves
    characteristic parameters, 103, 117
    characteristics of, 45–46
    circular waveguide, 112
    example, 79
    field configuration for TE mode, 44
    mode, 43

Transverse electric waves, (*Continued*)
   rectangular waveguide, 96
   TE and TM wave comparison, 103
   values of attenuation constant, 52
   wave equation, 42
   zig-zag paths and field components
      of, 49
Transverse electromagnetic waves
      (TEM waves), 20, 47; *see also*
      Transmission lines; Wave
      propagation
   attenuation constant values, 51
   equations for, 47
   field distribution for, 47
Transverse magnetic waves (TM
      waves), 42; *see also* Transverse
      waves
   attenuation constant, 52
   characteristic parameters, 103, 117
   characteristics of, 45–46
   circular waveguide, 111
   cutoff wave number, 95
   example, 79
   field configurations, 45
   modes, 44
   rectangular waveguide, 93–95
   TE and TM wave comparison, 103
   wave equation, 44
   zig-zag paths and field components
      of, 49
Transverse waves, 42; *see also*
      Transverse electric waves (TE
      waves); Transverse magnetic
      waves (TM waves); Wave
      propagation
TRAPATT diode, *see* TRApped
      Plasma Avalanche Trigger
      Transit-time diode
      (TRAPATT diode)
TRApped Plasma Avalanche Trigger
      Transit-time diode (TRAPATT
      diode), 552
   advantages and limitations, 553
   applications, 554
   comparison, 554
   cycle, 553
   operation, 552–553
   peak power levels achieved, 554
   performance, 553

Travelling wave, 20; *see also* Bolometer;
      Frequency meters
   crystal diode, 353
   detection, 348
   in lossy media, 22
   magnetrons, 408
   microwave detectors, 352–353
   movable shorts, 350
   slotted section, 348–349
   slotted transmission line, 349
   tunable crystal detector mounts,
      353, 354
   tunable probes, 349–350
Travelling wave tube (TWT), 6, 440; *see*
      also Microwave tube
   amplification process, 454–456
   applications, 459–460
   axial electric field, 457
   Brillouin diagram, 451–454
   characteristics, 458–459
   circuit equation, 457
   construction, 445–446
   convection current, 456
   coupled-cavity structure, 444
   electric field parallel to electron
      beam, 443
   focussing, 446–447
   helical structure, 441–443
   helix geometry, 442
   input–output, 446t, 448
   mathematical analysis, 450
   operating principle, 448–449
   output power gain, 458
   power supply, 449–450
   prevention of oscillations, 459
   ring-bar structure, 445
   ring-loop structure, 444–445
   signal growth and bunching, 449
   simplified model, 441
   slow wave structures, 442
   wave modes, 457–458
   wave propagation, 446
Trench cell structure, 638
Trifurcated waveguides, 234; *see also*
      Waveguides
Triode, 382
   circuit, 383
Truncated paraboloid antenna,
      321–322

Tunable crystal detector mounts, 353, 354
Tunable probes, 349–350
Tunable/dither magnetron, 480
Tuners, 264; *see also* Microwave
    components
  coaxial line tuners, 266
  EH tuner, 264–265
  slide screw tuner, 265
  tuning stub, 265–266
  waveguide slug tuner, 266–267
Tunnel diode, 505, 515
  advantages, 518
  applications, 518
  construction, 515–516
  current flow through, 517
  forward bias condition, 516, 517
  open-circuit condition, 516
  operation, 516
  V–I characteristics, 517–518
TV, *see* Television (TV)
Twice minimum method of VSWR
    measurement, 361
Two-cavity klystron, 411, 412; *see also*
    Bunching; Klystron
  as amplifier, 413
  applications, 426
  assumptions in mathematical
    analysis, 415
  beam coupling coefficient, 417
  cavity tuner, 413
  collector, 413
  constructional features, 413
  electron gun, 413
  electronic efficiency, 434
  with feedback arrangement, 414
  focussing magnet
  functional view of, 414
  input cavity, 413
  multi-cavity klystron amplifiers,
    424–425
  operation, 414–415
  oscillator, 426
  output cavity, 413
  performance, 426
  staggered tuning, 426
  synchronous tuning, 425
  velocity modulation, 415–417
2D Planar arrays, 334
Two phase CCD, 646

2-Port network, 282–285; *see also*
    Scattering parameters
    (*s* parameters)
  complex linear gain, 305
  electrical properties, 305
  input return loss, 306
  insertion loss, 306
  normalised incident and normalised
    reflected voltages, 308
  output return loss, 306
  reflection coefficient, 310
  return loss, 306
  reverse gain and reverse isolation,
    306
  scalar linear gain, 305
  scalar logarithmic gain, 305–306
  and Smith chart, 307–311
  s-parameter coefficients, 309
  voltage reflection coefficient, 307
  voltage standing wave ratio, 307
Two-valley semiconductors, 526, 527
TWT, *see* Travelling wave tube (TWT)

U

UHF, *see* Ultra-high frequency (UHF)
ULF, *see* Ultralow frequency (ULF)
Ultra-high frequency (UHF), 1, 2
  connector, 344
  division of, 4
Ultralow frequency (ULF), 2
Unguided waves, 40
Unidirectional couplers, 252; *see also*
    Directional couplers
Unidirectional radiation, 334
Uniform plane wave, 20
  example, 75–76
  polarisation of, 25
Up converter, *see* Sum-frequency
    parametric amplifier
Upper valley (UV), 524
Usable frequency range, 125
UV, *see* Upper valley (UV)

V

Vacuum
  evaporation, 711
  tubes, 579

Valence band (VB), 516, 524
Valley band structure, 524
    conductivity, 526, 527
    current density, 525, 526, 527
    data for two-valley semiconductors,
        526, 527
    differential negative resistance
        region, 527, 529–531
    electron drift velocity, 525, 526, 527
    electron transfer between two
        valleys, 525
    negative differential conductivity
        effect, 528
    negative resistance of sample, 528
    Ridley–watkins–helsum theory,
        528
    VCNR and CCNR modes, 528
Vapour-phase epitaxy, 710
Varactor, 521
    applications, 523
    capacitance–voltage relationship, 523
    characteristics, 522
    equivalent circuit, 523
    operation, 521–522
    reverse biased p–n junction, 522
    symbol of, 522
Varicap diode; *see also* Varactor
VB, *see* Valence band (VB)
VCNR, *see* Voltage-controlled negative
        resistance (VCNR)
Velocity modulation, 403, 406, 415–417,
        428–431; *see also* Velocity-
        modulated tubes
    electric field on electrons, 403–404
Velocity-modulated tubes, 404
    bunching, 405
    current modulation, 405–406
    electron gun, 404
    energy extraction, 406
Very high frequency (VHF), 2
Very low frequency (VLF), 2
VHF, *see* Very high frequency (VHF)
VLF, *see* Very low frequency (VLF)
Voltage standing wave ratio (VSWR),
        60, 189, 343
    meter, 347–348
    reflectometre method of
        measurement, 362

transmission line method of
        measurement, 361
    twice minimum method of
        measurement, 361
Voltage tunable magnetron, 479–480
Voltage-controlled negative resistance
        (VCNR), 528
Voltage-tunable magnetron (VTM),
        409
VSWR, *see* Voltage standing wave ratio
        (VSWR)
VTM, *see* Voltage-tunable magnetron
        (VTM)

W

Wave; *see also* Wave propagation
    equations, 18, 721
    forms, 20
    front, 128, 320
    modes, 457–458
    velocity, 21
Wave impedances, 48; *see also* Wave
        propagation
    classification, 48
    components of Poynting vector, 49
    group velocity, 51
    TE and TM waves, 49
    variation with frequency, 51
Wave propagation, 17, 446; *see also*
        Coaxial cables; Parallel plane
        guide; Polarisation; Reflection;
        Transmission lines; Transverse
        electromagnetic waves (TEM
        waves); Transverse waves;
        Wave impedances
    attenuation constant, 19
    basic equations and parameters, 17–19
    circular waveguide, 74
    conductors and dielectrics, 22–24
    depth of penetration, 26–27
    descriptive questions, 85–86
    direction cosines, 31
    E-field on plane of incidence, 33
    examples, 74–83
    exercises, 84–85
    forms of wave, 20
    gain coefficient, 19

limitations of guiding structures, 73
Maxwell's equations, 17
nature of media, 19
phase velocity, 21, 31
Poynting theorem, 27–29
propagation constant, 18
refraction, 29–31
resonance phenomena in line
    sections, 62–63
resonant section quality factor, 63
sinusoidal time variation, 18
surface impedance, 27
UHF lines as circuit elements,
    63–65
wave equations, 18
wave in loss less media, 19–21
wave in lossy media, 21–22
wave velocity, 21
wavelength, 31
Wave terminology, 124; *see also*
    Waveguides
characteristic impedance, 125
cutoff frequency, 125
dominant mode, 124
group velocity, 125, 127
guide wavelength, 125
mode designations, 124
mode disturbance, 126
modes, 124
multimode propagation, 126
phase velocity, 125
spurious modes, 128
transformation of modes, 126
usable frequency range, 125
wavefront movement, 128
Waveguide input–output methods,
    240; *see also* Microwave
    components; Waveguides
input–output coupling, 243
loop coupling, 242
probe coupling, 241–242
slot/aperture coupling, 243
Waveguide junctions, 245; *see also*
    Microwave components;
    Waveguides
E-plane tee, 245
H-plane tee, 246
hybrid ring, 248–249

impedance matching, 247–248
magic tee, 246
Waveguides, 87, 232; *see also* Circular
    waveguide; Dielectric
    waveguides; Microwave
    components; Rectangular
    waveguide; Transmission line;
    Wave terminology
bends, 237–238
bifurcated and trifurcated, 234
bus bar, 89
coaxial to waveguide adapter,
    243–244
copper losses, 129
corners, 238–239
cutoff frequency of, 90
descriptive questions, 136–137
dielectric loaded, 233–234
dielectric loss, 129
dimensional change impact,
    234–235
example, 130–135
exercises, 135–136
flanges, 235–236
flexible, 234
impact of frequency change,
    89–90
insulation breakdown, 129–130
interrelation between transmission
    line and, 87–89
joints, 235–237
limitations of, 130
match detector mount, 354
merits of, 129
metallic insulator, 88, 89
power-handling capability, 130
quarter-wave auxiliary line, 88
radiation losses, 130
ridged, 233
rotating joints, 236, 237
shapes, 232
skin effect, 129
slug tuner, 266–267; *see also* Tuners
stands, 237
system, 231, 232
transitions, 240
tunable detector mount, 354
twists, 239–240

Waveguide terminations, 256; *see*
         also Directional couplers;
         Microwave components
    different forms, 256
    horns, 256
    low power loads, 258
    matched load, 257
    moving load, 258
    water loads, 258, 259
    waveguide and coaxial matched
         terminations, 258
Wavelength measurement, 359–360
Weak-inversion mode, *see*
         Cutoff—mode
WG-type lens, 331
Wilkinson power splitter, 683, 684, 685
WIMAX, *see* Worldwide
         interoperability for microwave
         access (WIMAX)

Wire diameter measurement, 377
Worldwide interoperability
         for microwave access
         (WIMAX), 11
Woven capacitor, 681

Y

YIGs, *see* Yttrium iron garnets
         (YIGs)
Yttrium iron garnets (YIGs), 179
    idealised loop coupled to YIG
         resonator, 180

Z

Zener diode, 507
Zoned lens, 328, 332

Printed and bound by CPI Group (UK) Ltd, Croydon, CR0 4YY

18/10/2024

01776257-0019